IMPORTANT NAMES, DATES AND ADDRESSES

This book belongs to
L. E. KANDALEC
8320 PARKVIEW Ave
Munster, IN 46321
phone 219 838 7451

examination date: ____________ hours: ________

examination location: ____________________

tape your cancelled check here

phone number of your registration board: ____________________

address of your registration board: ____________________

names of contacts at your registration board: ____________________

tape your proof of mailing here

date you sent your application: ____________

registered/certified mail receipt number: ____________

date confirmation was received: ____________

names of examination proctors: ____________________

tape dime here

booklet number: ____________ (A.M.) ____________ (P.M.)

tape dime here

problems you disagreed with on the examination

problem no.	reason

MECHANICAL ENGINEERING REFERENCE MANUAL

Eighth Edition

Michael R. Lindeburg, P.E.

PROFESSIONAL PUBLICATIONS, INC.
Belmont, CA 94002

In the ENGINEERING REVIEW MANUAL SERIES

Engineer-In-Training Review Manual
    Engineering Fundamentals Quick Reference Cards
    Mini-Exams for the E-I-T Exam
    1001 Solved Engineering Fundamentals Problems
    E-I-T Review: A Study Guide
Civil Engineering Reference Manual
    Civil Engineering Quick Reference Cards
    Civil Engineering Sample Examination
    Civil Engineering Review Course on Cassettes
    Seismic Design for the Civil P.E. Exam
    Timber Design for the Civil P.E. Exam
Structural Engineering Practice Problem Manual
Mechanical Engineering Reference Manual
    Mechanical Engineering Quick Reference Cards
    Mechanical Engineering Sample Examination
    101 Solved Mechanical Engineering Problems
    Mechanical Engineering Review Course on Cassettes
    Consolidated Gas Dynamics Tables
Electrical Engineering Reference Manual
Chemical Engineering Reference Manual
    Chemical Engineering Practice Exam Set
Land Surveyor Reference Manual
Metallurgical Engineering Practice Problem Manual
Petroleum Engineering Practice Problem Manual
Expanded Interest Tables
Engineering Law, Design Liability, and Professional Ethics
Engineering Unit Conversions

In the ENGINEERING CAREER ADVANCEMENT SERIES

How to Become a Professional Engineer
The Expert Witness Handbook—A Guide for Engineers
Getting Started as a Consulting Engineer
Intellectual Property Protection—A Guide for Engineers
E-I-T/P.E. Course Coordinator's Handbook
Becoming a Professional Engineer

Distributed by: Professional Publications, Inc.
             1250 Fifth Avenue
             Department 77
             Belmont, CA 94002
             (415) 593-9119

MECHANICAL ENGINEERING REFERENCE MANUAL
Eighth Edition

Printed in the United States of America

ISBN: 0-912045-17-5

Professional Publications, Inc.
1250 Fifth Avenue, Belmont, CA 94002

Current printing of this edition (last number): 5 4 3 2 1

TABLE OF CONTENTS

viii TABLE OF CONTENTS

Concentrate Your Studies . . .

Studying for the exam requires a significant amount of time and effort on your part. Make every minute count towards your success with these additional study materials in the Engineering Review Manual Series:

Solutions Manual for the Mechanical Engineering Reference Manual

134 pages, $8\frac{1}{2} \times 11$, paper   0-932276-57-1

Don't forget that there is a companion **Solutions Manual** that provides step-by-step solutions to the practice problems given at the end of each chapter in this reference manual. This important study aid will provide immediate feedback on your progress. Without the **Solutions Manual**, you may never know if your methods are correct.

Mechanical Engineering Quick Reference Cards

56 pages, $8\frac{1}{2} \times 11$, paper   0-932276-45-8

Because speed is important during the exam, you will welcome the advantage provided by the **Mechanical Engineering Quick Reference Cards**. This handy resource gives you quick access to equations, methods, and data needed during the exam. The cards are divided into specific exam subjects following the **Mechanical Engineering Reference Manual** in organization and nomenclature.

Mechanical Engineering Sample Exam

24 pages, $8\frac{1}{2} \times 11$, paper   0-932276-72-5

Increase your speed and confidence in solving the types of questions that may appear on the exam by taking the **Mechanical Engineering Sample Exam**. This facsimile examination includes 20 typical problems (with complete, worked-out solutions) divided into two 10-problem sets to acquaint you with the exam format and teach you the correct and most efficient methods for solving Mechanical Engineering exam problems.

Mechanical Engineering Review Course on Cassettes

The discipline of a formalized class is provided through the **Mechanical Engineering Review Course on Cassettes** taken directly from Michael Lindeburg's outstanding lecture series. All of the most important exam subjects are reviewed, using a well-tested format combining lecture and problem-solving. Each $2\frac{1}{2}$ hour class reviews a crucial exam topic, based on the **Mechanical Engineering Reference Manual**. This strong review structure provides you with the focus and organization you need for successful exam preparation. The tapes are accompanied by a complete set of class notes, containing all the hand-outs and chalkboard work presented during class.

The cassette course has been divided into two sections, so that you can select the review program best suited to your needs.

- **Mechanics and Design Series** (13 tapes)

Approximate lecture time: 18 hours   0-932276-86-5

Introduction to the exam	Machine design
Engineering economy	Dynamics
Mechanics of materials	HVAC
Failure theories	

- **Fluids and Thermodynamics Series** (16 tapes)

Approximate lecture time: 22 hours   0-932276-87-3

Introduction to the exam	Power cycles
Fluid statics and dynamics	Gas dynamics
Hydraulic machines	Combustion
Fans and ductwork	Heat transfer
Thermodynamics	

Professional Publications, Inc.
1250 Fifth Avenue
Department 77
Belmont, CA 94002
(415) 593-9119

PREFACE
to the Eighth Edition

Regardless of whether you are studying on your own or taking a review course, this book will enable you to prepare effectively for the NCEES Professional Engineering examination (also known as the Principles and Practice examination) in Mechanical Engineering. The book's format, content, examples, and practice problems have been extensively tested and continually refined by thousands of engineers studying at home and in hundreds of classroom review courses across the country.

While the breadth of coverage has not changed significantly, more than a thousand changes distinguish this eighth edition of the *Mechanical Engineering Reference Manual* from the seventh edition of the *Mechanical Engineering Review Manual*. In addition to a change in title, tables have been expanded and updated; explanations have been clarified; figures have been rearranged to show more detail; and several new problems have been added.

This eighth edition is consistent with and explains the new examination format introduced in the Fall 1990 exam. In addition, all problem categories from the 1982 NCEES task analysis are covered.

Even though 50% of the P.E. exam is multiple choice, the nature of the exam—its level of difficulty and subjects covered—is unchanged. It is still important to solve as many representative practice problems as you can in order to get a feel for the level of difficulty of P.E. exam problems. For that reason, most chapters of this book end with numerous practice problems in three increasing levels of difficulty.

I have attempted to write the *Mechanical Engineering Reference Manual* so that it will be your most significant resource when you prepare for and take your P.E. exam. Nevertheless, if you are like most people who become familiar with this book, you will continue to refer to it long after the exam is over.

Good luck!

Michael R. Lindeburg, P.E.
Belmont, CA

INTRODUCTION

Purpose of Registration

As an engineer, you may have to obtain your professional engineering license through procedures that have been established by the state in which you work. These procedures are designed to protect the public by preventing unqualified individuals from legally practicing as engineers.

There are many reasons for wanting to become a professional engineer. Among them are the following:

• You may wish to become an independent consultant. By law, consulting engineers must be registered.

• Your company may require a professional engineering license as a requisite for employment or advancement.

• Your state may require registration as a professional engineer if you want to use the title *engineer*.

The Registration Procedure

The registration procedure is similar in most states. You probably will take two eight-hour written examinations. The first examination is the *Engineer-In-Training* examination, also known as the *Intern Engineer* exam and the *Fundamentals of Engineering* exam. The initials E-I-T, I.E., and F.E. also are used. The second examination is the *Professional Engineering* (P.E.) exam, which differs from the E-I-T exam in format and content.

If you have advanced degrees, other licenses, or significant experience in engineering, you may be allowed to skip the E-I-T examination. However, actual details of registration, experience requirements, minimum education levels, fees, and examination schedules vary from state to state. You should contact your state's Board of Registration for Professional Engineers.

Reciprocity Among States

All states use the NCEES P.E. examination.[1] If you take and pass the P.E. examination in one state, your license probably will be honored by other states that have used the same NCEES examination. It will not be necessary to retake the P.E. examination.

The simultaneous administration of identical examinations in multiple states has led to the term *Uniform Examination*. However, each state is free to choose its own minimum passing score or to add special questions to the NCEES examination. Therefore, this Uniform Examination does not automatically ensure reciprocity among states.

Of course, you may apply for and receive a professional engineering license from another state. However, a license from one state will not permit you to practice engineering in another state. You must have a professional engineering license from each state in which you work.

Applying for the Examination

Each state charges different fees, requires different qualifications, and uses different forms. Therefore, it will be necessary for you to request an application and an information packet from the state in which you plan to take the exam. It generally is sufficient to phone for this information. Telephone numbers for all of the U.S. state boards of registration are given in the accompanying listing.

Phone Numbers of State Boards of Registration

Alabama	(205) 261-5568
Alaska	(907) 465-2540

[1] The National Council of Examiners for Engineering and Surveying (NCEES) in Clemson, South Carolina produces, distributes, and grades the national examinations. It does not distribute applications to take the P.E. examination.

Arizona	(602) 255-4053
Arkansas	(501) 371-2517
California	(916) 920-7466
Colorado	(303) 866-2396
Connecticut	(203) 566-3386
Delaware	(302) 656-7311
District of Columbia	(202) 727-7454
Florida	(904) 488-9912
Georgia	(404) 656-3926
Guam	(671) 646-1079
Hawaii	(808) 548-4100
Idaho	(208) 334-3860
Illinois	(217) 782-8556
Indiana	(317) 232-2980
Iowa	(515) 281-5602
Kansas	(913) 296-3053
Kentucky	(502) 564-2680
Louisiana	(504) 568-8450
Maine	(207) 289-3236
Maryland	(301) 333-6322
Massachusetts	(617) 727-3055
Michigan	(517) 335-1669
Minnesota	(612) 296-2388
Mississippi	(601) 359-6160
Missouri	(314) 751-2334
Montana	(406) 444-4285
Nebraska	(402) 471-2021
Nevada	(702) 789-0231
New Hampshire	(603) 271-2219
New Jersey	(201) 648-2660
New Mexico	(505) 827-7316
New York	(518) 474-3846
North Carolina	(919) 781-9499
North Dakota	(701) 258-0786
Ohio	(614) 466-8948
Oklahoma	(405) 521-2874
Oregon	(503) 378-4180
Pennsylvania	(717) 783-7049
Puerto Rico	(809) 722-2121
Rhode Island	(401) 277-2565
South Carolina	(803) 734-9166
South Dakota	(605) 394-2510
Tennessee	(615) 741-3221
Texas	(512) 440-7723
Utah	(801) 530-6628
Vermont	(802) 828-2363
Virgin Islands	(809) 774-3130
Virginia	(804) 367-8512
Washington	(206) 753-6966
West Virginia	(304) 348-3554
Wisconsin	(608) 266-1397
Wyoming	(307) 777-6155

Examination Dates

The NCEES examinations are administered on the same weekend in all states. Each state determines independently whether to offer the examination on Thursday, Friday, or Saturday of the examination period. The upcoming examination dates are given in the accompanying listing.

National P.E. Examination Dates

year	Spring dates	Fall dates
1990	April 19–21	October 25–27
1991	April 11–13	October 24–26
1992	April 9–11	October 29–31
1993	April 15–17	October 28–30
1994	April 14–16	October 27–29
1995	April 6–8	October 26–28
1996	April 18–20	October 24–26
1997	April 17–19	Oct. 30–31, Nov. 1
1998	April 23–25	October 29–31
1999	April 22–24	October 28–30

Examination Format

The NCEES Professional Engineering examination in Mechanical Engineering consists of two four-hour sessions separated by a one-hour lunch period. Both the morning and the afternoon sessions contain ten problems. Most states do not have required problems.

Each examinee is given an exam booklet that contains problems for civil, mechanical, electrical, and chemical engineers. Some states, such as California, will allow you to work problems only from the mechanical part of the booklet. Other states will allow you to work problems from the entire booklet. Read the examination instructions on this point carefully.

In 1982, NCEES completed a task analysis of engineering activities. This study was intended to determine all the important activities in which engineers engage, and then to suggest changes in the NCEES examinations. The following subjects and numbers of problems resulted from the task analysis. These subjects are rather broad, and the NCEES examinations are not obligated to follow the task analysis guidelines.

Results of M.E. Task Analysis

examination subject	number of problems
mechanical design	8
management	1
energy systems	6
control systems	1
thermal and fluid processes	3
engineering economics	1

Since the examination structure is not rigid, it is not possible to give the exact number of problems that will appear in each subject area. Only engineering economics can be considered a permanent part of the examination. There is no guarantee that any other single subject will appear.

The examination is open book. With rare exceptions, all forms of solution aids are allowed in the examination, including nomographs, specialty slide rules, and pre-programmed and programmable calculators. Since their use says little about the depth of your knowledge, such aids should be used only to check your work. For example, using a psychrometric chart is an acceptable method of solving an HVAC problem; however, very few points will be earned if a pre-programmed calculator is used to solve a design problem, since your method is not shown.

Most states do not limit the number and types of books you can bring into the exam.[2] Loose-leaf papers (including Post-it™ notes) and writing tablets are usually forbidden, although you may be able to bring in loose reference pages in a three-ring binder. References used in the afternoon session need not be the same as for the morning session.

Any battery-powered, silent calculator may be used. Most states have no restrictions on programmable or pre-programmed calculators. To ensure exam security, however, printers and calculators with alphanumeric memories and other word processing functions are forbidden.

You will not be permitted to share books, calculators, or any other items with other examinees during the examination.

You will receive the results of your examination by mail. Allow at least four months for notification. Your score may or may not be revealed to you, depending on your state's procedure.

Objectively-Scored Problems

Objectively-scored problems appear in multiple-choice, true/false, and data-selection formats and constitute 50% of the examination.

The single engineering economics problem on the exam is one of the problems that has been converted to objective scoring. In addition, NCEES has specifically targeted the following subjects for objective scoring:

- energy use and generation
- power plants
- HVAC

- machine design
- applied mechanics
- thermodynamics
- fluids
- heat transfer

Grading by Criterion

NCEES has named its grading method the *Criterion-Referenced Method*. The *criteria* are those specific elements that you must include in your solution. These criteria are determined in advance, prior to the administration of the examination. As an example of a criterion-graded problem, consider a heat-transfer problem involving a hot duct passing through a cold room. Five of the criteria might require you to:

- recognize that convection and radiation are both present and operate in parallel
- recognize the heat balance between supply and loss
- use ΔT, not LMTD (characteristic of HVAC)
- obtain numerical values of film coefficients that deviate no more than $\pm 10\%$ from an answer key
- set up all equations correctly, though it would not be necessary to complete the problem

Getting full credit on a problem requires solving it completely and correctly, with no mathematical errors in the solution. For each mathematical error, a point or two would be lost from the total possible score of ten.

Each state still is free to specify its own cutoff score and passing requirements. NCEES's recommended passing score of 48 points does not have to be used (although most states do so).

Preparing for the Exam

You should develop an examination strategy early in the preparation process. This strategy will depend on your background. One of the following two general strategies is typically used:

- A broad approach has been successful for examinees who have recently completed academic studies. Their strategy has been to review the fundamentals in a broad range of undergraduate mechanical engineering subjects. The examination includes enough fundamental problems to give merit to this strategy.
- Engineers who have been away from classroom work for a long time have found it better to concentrate on the subjects in which they have had extensive professional experience. By studying the list of examination

---

[2] Check with your state to see if review books can be brought into the examination. Most states do not have any restrictions. Some states ban only collections of solved problems, such as Schaum's Outline Series. A few states prohibit all review books.

subjects, they have been able to choose those that will give them a high probability of finding enough problems that they can solve.

Do not make the mistake of studying only a few subjects in hopes of finding enough problems to work. The more subjects you are familiar with, the better will be your chances of passing the examination. More important than strategy are fast recall and stamina. You must be able to recall quickly solution procedures, formulas, and important data; and this sharpness must be maintained for eight hours.

In order to develop this recall and stamina, you should work the sample problems at the end of each chapter and compare your answers to the solutions in the accompanying *Solutions Manual*. This will enable you to become familar with problem types and solution methods. You will not have time in the exam to derive solution methods; you must know them instinctively.

It is imperative that you develop and adhere to a review outline and schedule. If you are not taking a classroom review course where the order of preparation is determined by the lectures, you should use the accompanying *Outline of Subjects for Self-Study* (located at the end of this chapter) to schedule your preparation.

It is unnecessary to take a large quantity of books to the examination. This book, a dictionary, and three to five other references of your choice should be sufficient. The examination is very fast-paced. You will not have time to use books with which you are not thoroughly familiar.

To minimize time spent searching for often-used formulas and data, you should prepare a one-page summary of all the important formulas and information in each subject area.[3] You can then use these summaries during the examination instead of searching for the correct page in your book.

Items to Get for the Examination

- Obtain ten sheets of each of the following types of graph paper: 10 squares to the inch grid, semi-log (3 cycles × 10 squares to the inch grid), and full-log (3 cycles × 3 cycles).

- Obtain a long, flexible, clear plastic ruler marked in tenths of an inch or in millimeters.

- Obtain the following tables, documents, and data:

  charts for solving transient heat flow problems (for simple solids other than spheres)

- charts of arrangement factors (F_a) for radiation heat transfer problems

- charts of correction factors (F_c) for multiple-pass heat exchanger problems

- periodic chart of the elements

- large Mollier diagrams in both English and SI units

- detailed steam tables in both English and SI units

- tables of isentropic flow and normal shock factors[4]

- detailed air table in both English and SI units

- static regain chart for R=.90 for air duct design

- tables of HVAC data including:

  k or U values for various wall constructions

  outside conditions versus location (winter temperature, winter degree days, summer temperature, summer degree days, average temperature swing, and wind velocity)

  infiltration coefficients

  slab-edge method conductance coefficients

  shading coefficients

  solar heat gain factors

  energy reflection of indoor shading devices

  equivalent temperature differences

- enthalpies of formation for hydrocarbon and combustion-related components

● Obtain a set of at least ten psychrometric charts, including several low-pressure and low- and high-temperature charts.

● From Professional Publications, Inc., the following special study aids will prove valuable:

  · EXPANDED INTEREST TABLES for Economic Analysis Problems

  · ENGINEERING LAW, DESIGN LIABILITY, AND PROFESSIONAL ETHICS

  · CONSOLIDATED GAS DYNAMICS TABLES

  · MECHANICAL ENGINEERING QUICK REFERENCE CARDS

[3] *Mechanical Engineering Quick Reference Cards*, published by Professional Publications, Inc., is an excellent prepared summary of important equations, methods, and data needed during the exam.

[4] *Consolidated Gas Dynamics Tables*, published by Professional Publications, Inc., provides tables of data for isentropic flow, normal shock waves, Fanno flow, and Rayleigh flow for specific heat (k) values of 1.1, 1.2, 1.3, 1.4, and 1.67.

· MECHANICAL ENGINEERING SAMPLE
  EXAMINATION

· 101 SOLVED MECHANICAL ENGINEERING
  PROBLEMS

· MECHANICAL ENGINEERING REVIEW
  COURSE ON CASSETTES

What to Do Before the Exam

The engineers who have taken the P.E. exam in previous
years have made the suggestions listed below. These
suggestions will make your examination experience as
comfortable and successful as possible.

• Keep a copy of your examination application. Send
  the original application by certified mail and request
  a receipt of delivery. Tape your delivery receipt in
  the space indicated on the first page of this book.

• Visit the exam site the day before your examination.
  This is especially important if you are not familiar
  with the area. Find the examination room, the parking area, and the rest rooms.

• Plan on arriving 30–60 minutes before the examination starts. This will assure you a convenient parking
  place and adequate time for site, room, and seating
  changes.

• If you live a considerable distance from the examination site, consider getting a hotel room in which to
  spend the night before.

• Take off the day before the examination to relax.
  Don't cram the last night. Rather, get a good night's
  sleep.

• Be prepared to find that the examination room is
  not ready at the designated time. Take an interesting
  novel or magazine to read in the interim and at lunch.

• If you make arrangements for babysitters or transportation, allow for a delayed completion.

• Prepare your examination kit the day before. Here is
  a checklist of items to take with you to the examination.

  [ ] copy of your application

  [ ] proof of delivery receipt

  [ ] letter admitting you to the exam

  [ ] photographic identification

  [ ] this book

  [ ] STANDARD HANDBOOK FOR
       MECHANICAL ENGINEERS (Baumeister)

[ ] other books of your choice, including:

  [ ] machine design book

  [ ] HVAC book

  [ ] air and gas properties book

  [ ] steam properties book

  [ ] scientific dictionary

  [ ] standard English dictionary

[ ] review course notes in a binder

[ ] calculator and a spare

[ ] spare calculator batteries or battery pack

[ ] battery charger and 20 foot extension cord

[ ] chair cushions (a large, thick bath mat works
     well)

[ ] earplugs

[ ] desk expander—if you are taking the exam
     in theater chairs with tiny, fold-up writing
     surfaces, you should take a long, wide board
     to place across the armrests

[ ] a cardboard box cut to fit your references

[ ] twist-to-advance pencils with extra leads

[ ] number 2 pencils

[ ] large eraser

[ ] snacks

[ ] beverage in a thermos

[ ] light lunch

[ ] collection of graph paper

[ ] transparent and masking tapes

[ ] sunglasses

[ ] extra prescription glasses, if you wear them

[ ] aspirin

[ ] travel pack of Kleenex and other personal
     needs

[ ] $2.00 in change, $10.00 in cash

[ ] a light, comfortable sweater

[ ] comfortable shoes or slippers for the exam
     room

[ ] raincoat, boots, gloves, hat, and umbrella

[ ] local street maps

[ ] note to the parking patrol for your windshield

[ ] pad of three-hole punched scratch paper

[ ] straightedge, ruler, compass, protractor, and French curves

[ ] battery-powered desk lamp

[ ] watch

[ ] extra car keys

What to Do During the Exam

Previous examinees have reported that the following strategies and techniques have helped them considerably.

• Read through all of the problems before starting your first solution. In order to save you from rereading and reevaluating each problem later in the day, you should classify each problem at the beginning of the four hour session. The following categories are suggested:

   · problems you can do easily

   · problems you can do with effort

   · problems for which you can get partial credit

   · problems you cannot do

• Do all of the problems in order of increasing difficulty. All problems on the examination are worth ten points. There is nothing to be gained by attempting the difficult or long problems if easier or shorter problems are available.

• Follow these guidelines when solving a problem:

   · Do not rewrite the problem statement.

   · Do not unnecessarily redraw any figures.

   · Use pencil only.

   · Be neat. (Print all text. Use a straightedge or template where possible.)

   · Draw a box around each answer.

   · Label each answer with a symbol.

   · Give the units.

   · List your sources whenever you use obscure solution methods or data.

   · Write on one side of the page only.

   · Use one page per problem, no matter how short the solution is.

   · Go through all calculations a second time and check for mathematical errors, or solve the problem by an alternate method.

• Remember the details of any problem that you think is impossible to solve with the information given. Your ability to point out an error may later give you the margin needed to pass.

A Personal Note to Instructors Using This Book

This book started as a series of handouts for a P.E. review course. It originally was intended as a reference for all of the long formulas, illustrations, and tables of data that I did not have time to put on the chalkboard. Starting with the fourth edition, however, the chapters were rewritten to more closely parallel the organization and content of my lectures.

If you are unfamiliar with the P.E. examination, you can use the chapters in this book as guides to preparing your lectures. You should emphasize the subjects in each chapter and avoid subjects omitted. You can feel confident that subjects omitted from this book are relatively rare on the exam.

The end-of-chapter timed problems are representative of the actual examination. The types of problems, format, emphasis, and degree of difficulty are equivalent to what your students will experience in the actual P.E. examination.

There are between 20 and 30 practice problems for each major examination subject. Every problem is assigned in my review courses. This requires between five and ten hours of preparation on the part of the students each week.

If you were to assign ten hours of practice problems and a student could put in only eight hours of preparation, then that student will have worked to capacity. *Capacity assignment* is the goal in my courses. After the actual P.E. examination, your students will honestly say they could not have prepared any more than they did.

Assignments in my classes are not individually graded. The students have the solutions to all practice problems in the solutions manual. However, each student must turn in the completed set of problems for credit each week. Any special problems or questions written on the assignment are answered at that time.

I have found that a 14-week format works well for a P.E. review course. Each week contains one three-hour lecture with a short intermediate break. The course format is shown in the accompanying table.

Mechanical Engineering
P.E. Review Course Outline

meeting	subject covered
1	introduction to the examination
2	engineering economics

3	heat transfer
4	fluids and hydraulic machines
5	fans and ductwork
6	thermodynamics
7	power cycles
8	HVAC
9	combustion
10	compressible fluid flow
11	mechanics of materials #1
12	mechanics of materials #2
13	machine design
14	kinematics of machinery

I have tried to order the subjects in a logical, progressive manner. For example, HVAC is dependent on some thermodynamic principles, and machine design requires mechanics of materials.

Some examination subjects are not covered at all by my lectures. These omissions are intentional; they are not the result of scheduling limitations. For example, I have found that very few people try to learn modeling of engineering systems and the related subjects of feedback and control systems. Your students' time can be better spent covering other subjects.

Other chapters cover support subjects (e.g., materials science, noise control, nuclear engineering, and management) which rarely contribute to the examination.

All of the skipped chapters are assigned as floating assignments to be made up on the students' free time.

I do not use a practice examination as an in-class exercise. Since the review course usually ends only a few days prior to the real examination, I cannot see making students sit for four hours in the late evening. A take-home practice examination is assigned.[5]

If a practice examination is to be used as an indication of preparedness, your students should not even look at it prior to taking it. Looking at the examination or otherwise using it to direct their study will produce unwarranted specialization in subjects contained in the practice examination.

There are many ways to organize a P.E. course depending on the amount of time available for the instructor, students, and course. However, all good organizations have the same result: the students complain about the work load during the course ... and they breeze through the examination after the course.

---

[5] The *Mechanical Engineering Sample Exam*, available from Professional Publications, Inc., contains 20 representative problems (and complete, worked-out solutions) designed to help you develop the skills and strategies critical to success in taking the real exam.

Outline of Subjects for Self-Study
(Subjects do not have to be studied in the order listed.)

subject	chapter	recommended number of weeks	date to be started	date to be completed	complete
mathematics	1	2	_____	_____	☐
engineering economic analysis	2	1	_____	_____	☐
fluid statics and dynamics	3	1	_____	_____	☐
hydraulic machines	4	1	_____	_____	☐
fans and ductwork	5	1	_____	_____	☐
thermodynamics	6	2	_____	_____	☐
power and refrigeration cycles	7	2	_____	_____	☐
compressible fluid flow	8	1	_____	_____	☐
combustion	9	1	_____	_____	☐
heat transfer	10	2	_____	_____	☐
HVAC	11	2	_____	_____	☐
statics	12	1	_____	_____	☐
materials science	13	1	_____	_____	☐
mechanics of materials	14	1	_____	_____	☐
machine design	15	2	_____	_____	☐
dynamics	16	1	_____	_____	☐
noise control	17	1	_____	_____	☐
nuclear engineering	18	1	_____	_____	☐
modeling of engineering systems	19	1	_____	_____	☐
management	20	0	_____	_____	☐
miscellaneous topics	21	0	_____	_____	☐
systems of units	22	0	_____	_____	☐
postscripts	23	0	_____	_____	☐

ACKNOWLEDGMENTS

I want to express my gratitude for the contributions made by the following individuals. Lisa Rominger, supervisor of Professional Publications' production department, timed and managed all aspects of the revision. Mary Christensson keyboarded all of the new material, Yves Martin revised illustrations and did the page composition, and Shelley Axmaker proofread both text and illustrations.

Behind the scenes are many loyal review course coordinators and thousands of engineers who have used previous editions of the *Mechanical Engineering Review Manual*. Their suggestions for improvement helped me immensely and will surely be appreciated by subsequent generations of examinees.

Last, but not least, the support of my family has been consistent throughout the years. My wife, Elizabeth, and daughters, Jennifer and Katherine, never complained about the weekends and long hours past midnight spent researching, writing, and proofreading. The time we missed cannot ever be replaced; that has been a real sacrifice.

Thanks to you for your support!

Michael R. Lindeburg, P.E.
Belmont, CA
February, 1990

---

pages 3-20 and 3-36: Figure 3.13 and Appendix B, "Moody Friction Factor Chart" and "Properties of Air at Atmospheric Pressure," reprinted from *Fluid Mechanics and Hydraulics*, 2nd ed., Ranald V. Giles, McGraw-Hill Book Company, 1962

page 3-28: Figure 3.25, "Flow Coefficients for Orifice Plates," reprinted from *Elementary Fluid Mechanics*, 2nd ed., John K. Vennard, John Wiley & Sons, Inc., 1947

page 3-36: Appendix A, "Properties of Water at Atmospheric Pressure," reprinted with permission from *ASCE Manual of Engineering Practice #25*, American Society of Civil Engineers, 1942

pages 3-38, 3-45, and 4-4: Appendix E, Appendix I, and Figure 4.9, "Properties of Liquids," "Specific Gravity of Hydrocarbons," and "Vapor Pressure of Hydrocarbons" reprinted with permission from *Cameron Hydraulic Data*, 15th ed., C.R. Westaway and A.W. Loomis, Ingersoll-Rand, 1977

page 4-7: Figure 4.11, "Hydrocarbon NPSHR Correction Factor," reprinted with permission from *Hydraulic Handbook*, 10th ed., Colt Industries, 1977

pages 4-18, 4-19, and 4-21: Appendixes B, C, and E, "Upper Limits of Specific Speeds" and "Pump Performance Correction Factor Chart," reprinted by permission of the Hydraulic Institute from *Hydraulic Institute Standards*, 14th ed., 1983

page 5-8: Table 5.3, "Minor Loss Coefficients," reprinted with permission from *ASHRAE Guide and Data Book*, American Society of Heating, Refrigerating, and Air Conditioning Engineers, 1969

page 5-11: Figure 5.6, "Low-Velocity Static Regain in Air Ducts," reprinted with permission from *Handbook of Air Conditioning System Design*, McGraw-Hill Book Company, 1965

page 6-16: Figures 6.11 and 6.12, "Compressibility Factors for Low Pressures" and "Compressibility Factors for High Pressures," reprinted with permission from ASME TRANSACTIONS, Vol. 76, Edward F. Obert and L.C. Nelson, American Society of Mechanical Engineers, 1954

pages 6-19 and 11-27: Figures 6.14 and Appendix B, "Psychrometric Chart," reproduced by permission of Carrier Corporation, copyright 1947 by Carrier Corporation

page 6-28: "Table of Specific Heat Gas Constants," reproduced from *Engineering Thermodynamics*, C.O. MacKay, W.N. Barnard, and F.O. Ellenwood, John Wiley and Sons, Inc., 1957

pages 6-29 through 6-33: Appendixes A, B, C, and D, properties of saturated and superheated steam, reprinted with permission from *Thermodynamic Properties of Steam*, Joseph H. Keenan and Frederick G. Keyes, John Wiley & Sons, Inc., 1937

pages 6-34, 9-8, 9-12, and 9-22: Appendix E, Tables 9.4 and 9.12, and Appendix A, "Mollier Diagram," "Selected U.S. Coals," "Typical Amounts of Excess Air," and "Heats of Combustion" reprinted with permission from *STEAM: Its Generation and Use*, 39th ed., Babcock and Wilcox Company, 1978

page 6-35: Appendix F, "Air Table," reprinted with permission from *Gas Tables: Thermodynamic Properties of Air*, Joseph H. Keenan and Joseph Kaye, John Wiley & Sons, Inc., 1945 pages 6-39 and 6-40: Appendixes I and J, "Saturated Freon-12" and "Superheated Freon-12," reprinted with permission from *Thermodynamic Properties of Freon-12*, Bulletin T-12, E. I. Du Pont Nemours & Company, Inc., 1956

pages 7-34 and 7-35: Appendix B, "Physical Properties of Pipe Materials," reprinted from *Design Manual: Mechanical Engineering*, (NAVFAC DM-3), Naval Facilities Engineering Command, Department of the Navy, 1972

page 9-9: Tables 9.6 and 9.7, "Properties of Fuel Oils," and "Fuel Oil Grade vs. Firing Rate," reprinted with permission from *1989 ASHRAE Handbook—Fundamentals*, American Society of Heating, Refrigerating, and Air Conditioning Engineers

page 9-9: Table 9.8, "Burner Type and Atomizing Viscosity," reprinted with permission from *ASHRAE Guide and Data Book: Equipment*, American Society of Heating, Refrigerating, and Air Conditioning Engineers, 1967

page 9-11: Table 9.11, "Consolidated Combustion Data," reprinted from POWER, July 1951, with permission. Copyright, McGraw-Hill, Inc., 1984

pages 9-12 and 9-22: Table 9.12 and Appendix A, "Typical Amounts of Excess Air" and "Heats of Combustion," reprinted with permission from *STEAM: Its Generation and Use*, 39th ed., Babcock and Wilcox Company, 1978

page 10-6: Figure 10.3, "Transient Temperature Charts for Spheres," reprinted with permission from *Heat Transfer*, 3rd ed., L.M.K. Boelter, V.H. Cherry, and H.A. Johnson, University of California (Berkeley) Press, 1942

page 10-16: Figure 10.13, "Multiple Tube Correction Factor," reprinted with permission from *Standards of the Tubular Exchanger Manufacturers Association*, 6th ed., copyright 1978 by Tubular Exchanger Manufacturers Association

page 10-18: Figure 10.15, "Configuration Factors for Parallel Squares, Rectangles, and Disks," reprinted with permission from MECHANICAL ENGINEERING, Vol. 52, 1930, H.C. Hottel, American Society of Mechanical Engineers

pages 10-23 through 10-26: Appendixes C, D, E, and F, "Properties of Saturated Water," "Properties of Air at One Atmosphere," "Properties of Steam at One Atmosphere," and "Emissivities of Various Surfaces," reprinted from *Principles of Heat Transfer*, 2nd ed., Frank Kreith, International Textbook Company, 1968

pages 11-2, 11-3, and 11-4: Tables 11.2, 11.3, and 11.4, "Typical Thermal Conductances of Air Spaces," "Wind Correction Factors," and "Typical Infiltration Coefficients," reprinted with permission from *Heating, Ventilating, and Air Conditioning Guide*, American Society of Heating, Refrigerating, and Air Conditioning Engineers, 1956

page 11-8: Table 11.8, "Approximate Values of TLV and LEL," reprinted with permission from *Industrial Ventilation*, 14th ed., American Conference of Governmental Industrial Hygienists, 1976

pages 11-18, 11-19, and 11-20: Tables 11.10, 11.11, 11.12, and 11.13, "Typical Total Equivalent Temperature Differences," "Typical Shading Coefficients," "Typical Solar Gain Factors," and "Heat Gain from Occupants," abridged with permission from *1989 ASHRAE Handbook—Fundamentals*, American Society of Heating, Refrigerating, and Air Conditioning Engineers

pages 11-25 and 11-26: Appendix A, "Typical Heat Loss Factors," reprinted by permission of Pitman Publishing Ltd., London, from *Modern Air Conditioning, Heating, and Ventilating*, 3rd ed., Willis H. Carrier, Realto E. Cherne, and Walter A. Grant, 1965

page 11-28: Appendix C, "Sample Outdoor Design Data," reprinted with permission from *Electric Heater Catalog*, SD 149 (12/76), Square D Company, 1976

pages 14-29, 14-30, and 14-31: Appendix B, "Mechanical Properties of Representative Metals," M.F. Spotts, *Design of Machine Elements*, 2nd ed., copyright 1953, renewed 1981, p. 478. Adapted by permission of Prentice-Hall, Inc., Englewood Cliffs, N.J.

page 14-33: Appendix D, "Stress Concentration Factors," reprinted from MACHINE DESIGN, Vol. 23, 1951, copyright 1951 by Penton/IPC, Inc., Cleveland, Ohio

page 15-22: Table 15.8, "Values of C*," reprinted by permission from *Dynamic Loads on Gear Teeth*, Earle Buckingham, American Society of Mechanical Engineers (Research Publications), 1931

page 15-34: Figures 15.45 and 15.46, "Coefficient of Friction Variable" and "Film Thickness Variable and Eccentricity Ratio," reprinted from ASLE TRANSACTIONS, Vol. 1, No. 1, A.A. Raimondi and John Boyd, by permission of the American Society of Lubrication Engineers, publishers of ASLE TRANSACTIONS and LUBRICATION ENGINEERING

page 17-9: Appendix A, "NRC and α Coefficients," reprinted from *Performance Data for Acoustical Materials Bulletin*, Acoustical and Board Products Association, 1975

page 17-10: Appendix B, "Transmission Loss of Common Materials," reprinted from *Industrial Noise Control Manual*, HEW Publication NIOSH 75-183, U.S. Government Printing Office, 1975

page 18-5: Table 18.4, "Typical Non 1/v Factors," from *Introduction to Nuclear Engineering*, John R. Lamarsh, © 1983, Addison-Wesley Publishing Co., Inc., Reading, Massachusetts. Reprinted with permission of the publisher.

page 18-7: Table 18.6 and 18.7, "Mass Attenuation Coefficients" and "Energy Absorption Coefficients," reprinted from *Reactor Physics Constants*, 2nd ed., (ANL-5800), L.T. Templin, Editor, Argonne National Laboratory, 1963

pages 18-8 and 18-9: Tables 18.8 and 18.9, "Values of B_m for a Plane Monodirectional Source," and "Values of B_p for Isotropic Point Sources," from *Fundamental Aspects of Reactor Shielding*, Herbert Goldstein, © 1959, Addison-Wesley Publishing Co., Inc., Reading, Massachusetts. Reprinted with permission of the publisher.

1 MATHEMATICS AND RELATED SUBJECTS

1 INTRODUCTION

Most engineering problems on the mechanical engineering examination require only algebra and simple trigonometry. Due to the limited time, difficult and time-consuming mathematical techniques seldom are required.

The probability of a purely mathematical problem appearing on the mechanical engineering exam is extremely small. You will not be asked for the roots of a cubic equation or the determinant of a matrix. However, you may have to use these techniques as part of a solution to a much more complex problem. This chapter is designed as a reference for formulas and techniques required to solve most of the mechanical engineering problems.

2 SYMBOLS USED IN THIS BOOK

Many symbols, letters, and Greek characters are used to represent variables in the formulas used throughout this book. These symbols and characters are defined in the nomenclature section of each chapter. However, some of the symbols which are used as operators in this book are listed here.

**Table 1.1
Symbols Used in This Book**

Symbol	Name	Use	Example
$\sum$	sigma	series addition	$\sum_{i=1}^{3} x_i = x_1 + x_2 + x_3$
$\prod$	pi	series multiplication	$\prod_{i=1}^{3} x_i = x_1 x_2 x_3$
Δ	delta	change in quantity	$\Delta h = h_2 - h_1$
$-$	over bar	average value	$\overline{x}$
$\cdot$	over dot	per unit time	$\dot{Q} =$ quantity flowing per second
!	factorial		$x! = x(x-1)(x-2)\cdots(2)(1)$
| |	absolute value		$\|-3\| = +3$
$\approx$	approximately equal to		$x \approx 1.5$
$\propto$	proportional to		$x \propto y$
∞	infinity		$x \to \infty$
log	base 10 logarithm		$\log(5.74)$
ln	natural logarithm		$\ln(5.74)$
EE	scientific notation		$EE - 4$
exp	exponential power		$\exp(x) = e^x$

3 THE GREEK ALPHABET

Table 1.2
The Greek Alphabet

A	α	alpha	N	ν	nu
B	β	beta	Ξ	ξ	xi
Γ	γ	gamma	O	o	omicron
Δ	δ	delta	Π	π	pi
E	ϵ	epsilon	P	ρ	rho
Z	ς	zeta	Σ	σ	sigma
H	η	eta	T	τ	tau
Θ	θ	theta	Υ	υ	upsilon
I	ι	iota	Φ	ϕ	phi
K	κ	kappa	X	χ	chi
Λ	λ	lambda	Ψ	ψ	psi
M	μ	mu	Ω	ω	omega

4 MENSURATION

Nomenclature

A   total surface area
d   distance
h   height
p   perimeter
r   radius
s   side (edge) length, arc length
V   volume
θ vertex angle, in radians
ϕ central angle, in radians

Circle

$$p = 2\pi r \tag{1.1}$$

$$A = \pi r^2 = \frac{p^2}{4\pi} \tag{1.2}$$

Circular Segment

$$A = \frac{1}{2}r^2\left(\phi - \sin\phi\right) \tag{1.3}$$

$$\phi = \frac{s}{r} = 2\left(\arccos\frac{r-d}{r}\right) \tag{1.4}$$

Triangle

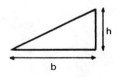

$$A = \frac{1}{2}bh \tag{1.5}$$

Parabola

$$A = \frac{2bh}{3} \tag{1.6}$$

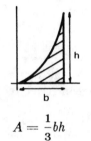

$$A = \frac{1}{3}bh \tag{1.7}$$

Circular Sector

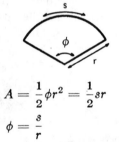

$$A = \frac{1}{2}\phi r^2 = \frac{1}{2}sr \tag{1.8}$$

$$\phi = \frac{s}{r} \tag{1.9}$$

Ellipse

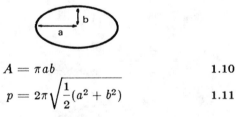

$$A = \pi ab \tag{1.10}$$

$$p = 2\pi\sqrt{\frac{1}{2}(a^2 + b^2)} \tag{1.11}$$

Trapezoid

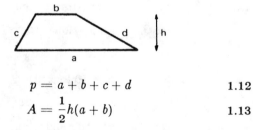

$$p = a + b + c + d \tag{1.12}$$

$$A = \frac{1}{2}h(a + b) \tag{1.13}$$

The trapezoid is isosceles if $c = d$.

Parallelogram

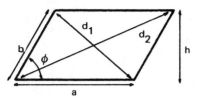

$$p = 2(a + b) \qquad 1.14$$

$$d_1 = \sqrt{a^2 + b^2 - 2ab(\cos\phi)} \qquad 1.15$$

$$d_2 = \sqrt{a^2 + b^2 + 2ab(\cos\phi)} \qquad 1.16$$

$$d_1^2 + d_2^2 = 2(a^2 + b^2) \qquad 1.17$$

$$A = ah = ab(\sin\phi) \qquad 1.18$$

If $a = b$, the parallelogram is a rhombus.

Regular Polygon

(n equal sides)

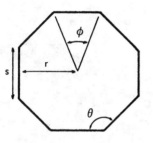

$$\phi = \frac{2\pi}{n} \qquad 1.19$$

$$\theta = \frac{\pi(n - 2)}{n} \qquad 1.20$$

$$p = ns \qquad 1.21$$

$$s = 2r\left(\tan\left(\frac{\phi}{2}\right)\right) \qquad 1.22$$

$$A = \frac{1}{2}nsr \qquad 1.23$$

Table 1.3
Polygons

Number of Sides	Name of Polygon
3	triangle
4	rectangle
5	pentagon
6	hexagon
7	heptagon
8	octagon
9	nonagon
10	decagon

Sphere

$$V = \frac{4\pi r^3}{3} \qquad 1.24$$

$$A = 4\pi r^2 \qquad 1.25$$

Right Circular Cone

$$V = \frac{\pi r^2 h}{3} \qquad 1.26$$

$$A = \pi r\sqrt{r^2 + h^2} \qquad 1.27$$

(does not include base area)

Right Circular Cylinder

$$V = \pi h r^2 \qquad 1.28$$

$$A = 2\pi r h \qquad 1.29$$

(does not include end area)

Paraboloid of Revolution

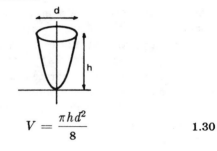

$$V = \frac{\pi h d^2}{8} \qquad 1.30$$

Regular Polyhedron

The radius of an inscribed sphere is

$$r = \frac{3V}{A_{\text{total}}} \qquad 1.31$$

Table 1.4
Polyhedrons

Number of Faces	Form of Faces	Total Surface Area	Volume
4	equilateral triangle	$1.7321\ s^2$	$0.1179\ s^3$
6	square	$6.0000\ s^2$	$1.0000\ s^3$
8	equilateral triangle	$3.4641\ s^2$	$0.4714\ s^3$
12	regular pentagon	$20.6457\ s^2$	$7.6631\ s^3$
20	equilateral triangle	$8.6603\ s^2$	$2.1817\ s^3$

Example 1.1

What is the hydraulic radius of a 6" pipe filled to a depth of 2"?

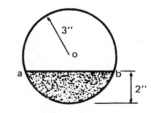

The hydraulic radius is defined as

$$r_h = \frac{\text{area in flow}}{\text{length of wetted perimeter}} = \frac{A}{s}$$

Points o, a, and b may be used to find the central angle of the circular segment.

$$\frac{1}{2}(\text{angle } aob) = \arccos\left(\frac{1}{3}\right) = 70.53°$$
$$\phi = 141.06° = 2.46 \text{ radians}$$

Then,

$$A = \frac{1}{2}(3)^2(2.46 - .63) = 8.235 \text{ in}^2$$
$$s = (3)(2.46) = 7.38 \text{ in}$$
$$r_h = \frac{8.235}{7.38} = 1.12 \text{ in}$$

5 SIGNIFICANT DIGITS

The significant digits in a number include the left-most, non-zero digits to the right-most digit written. Final answers from computations should be rounded off to the number of decimal places justified by the data. The answer can be no more accurate than the least accurate number in the data. Of course, rounding should be done on final calculation results only. It should not be done on interim results.

number as written	number of significant digits	implied range
341	3	340.5 to 341.5
34.1	3	34.05 to 34.15
.00341	3	.003405 to .003415
3410.	4	3409.5 to 3410.5
341 EE7	3	340.5 EE7 to 341.5 EE7
3.41 EE−2	3	3.405 EE−2 to 3.415 EE−2

6 ALGEBRA

Algebra provides the rules which allow complex mathematical relationships to be expanded or condensed. Algebraic laws may be applied to complex numbers, variables, and numbers. The general rules for changing the form of a mathematical relationship are given here:

Commutative law for addition:
$$a + b = b + a \qquad\qquad 1.32$$

Commutative law for multiplication:
$$ab = ba \qquad\qquad 1.33$$

Associative law for addition:
$$a + (b + c) = (a + b) + c \qquad\qquad 1.34$$

Associative law for multiplication:
$$a(bc) = (ab)c \qquad\qquad 1.35$$

Distributive law:
$$a(b + c) = ab + ac \qquad\qquad 1.36$$

A. POLYNOMIAL EQUATIONS

1 Standard Forms

$$(a + b)(a - b) = a^2 - b^2 \qquad\qquad 1.37$$
$$(a \pm b)^2 = a^2 \pm 2ab + b^2 \qquad\qquad 1.38$$
$$(a \pm b)^3 = a^3 \pm 3a^2 b + 3ab^2 \pm b^3 \qquad\qquad 1.39$$
$$(a^3 \pm b^3) = (a \pm b)(a^2 \mp ab + b^2) \qquad\qquad 1.40$$
$$(a^n + b^n) = (a + b)(a^{n-1} - a^{n-2}b + \cdots + b^{n-1})(\text{for } n \text{ odd}) \qquad 1.41$$
$$(a^n - b^n) = (a - b)(a^{n-1} + a^{n-2}b + \cdots + b^{n-1}) \qquad 1.42$$

2 Quadratic Equations

Given a quadratic equation $ax^2 + bx + c = 0$, the roots x_1^* and x_2^* may be found from

$$x_1^*, x_2^* = \frac{-b \pm \sqrt{b^2 - 4ac}}{2a} \qquad\qquad 1.43$$
$$x_1^* + x_2^* = -\frac{b}{a} \qquad\qquad 1.44$$
$$x_1^* x_2^* = \frac{c}{a} \qquad\qquad 1.45$$

3 Cubic Equations

Cubic and higher order equations occur infrequently in most engineering problems. However, they usually are difficult to factor when they do occur. Trial and error solutions are usually unsatisfactory except for finding the general region in which a root occurs. Graphical means can be used to obtain only a fair approximation to the root.

Numerical analysis techniques must be used if extreme accuracy is needed. The more efficient numerical analysis techniques are too complicated to present here. However, the bisection method illustrated in example 1.2 usually can provide the required accuracy with only a few simple iterations.

The bisection method starts out with two values of the independent variable, L_0 and R_0, which straddle a root. Since the function has a value of zero at a root, $f(L_0)$ and $f(R_0)$ will have opposite signs. The following algorithm describes the remainder of the bisection method:

Let n be the iteration number. Then, for $n = 0, 1, 2, \ldots$ perform the following steps until sufficient accuracy is attained.

Set $m = \frac{1}{2}(L_n + R_n)$

Calculate $f(m)$

If $f(L_n)f(m) \leqslant 0$, set $L_{n+1} = L_n$ and $R_{n+1} = m$

Otherwise, set $L_{n+1} = m$ and $R_{n+1} = R_n$

$f(x)$ has at least one root in the interval (L_{n+1}, R_{n+1})

The estimated value of that root, x^*, is

$$x^* \approx \frac{1}{2}(L_{n+1} + R_{n+1}) \qquad \qquad \textbf{1.46}$$

The maximum error is $1/2(R_{n+1} - L_{n+1})$. The iterations continue until the maximum error is reasonable for the accuracy of the problem.

Example 1.2

Use the bisection method to find the roots of

$$f(x) = x^3 - 2x - 7$$

The first step is to find L_0 and R_0, which are the values of x which straddle a root and have opposite signs. A table can be made and values of $f(x)$ calculated for random values of x.

x	-2	-1	0	$+1$	$+2$	$+3$
$f(x)$	-11	-6	-7	-8	-3	$+14$

Since $f(x)$ changes sign between $x = 2$ and $x = 3$,

$$L_0 = 2 \text{ and } R_0 = 3$$

Iteration 0:

$$m = \frac{1}{2}(2 + 3) = 2.5$$
$$f(2.5) = (2.5)^3 - 2(2.5) - 7 = 3.625$$

Since $f(2.5)$ is positive, a root must exist in the interval $(2, 2.5)$. Therefore,

$$L_1 = 2 \text{ and } R_1 = 2.5$$

At this point, the best estimate of the root is

$$x^* \approx \frac{1}{2}(2 + 2.5) = 2.25$$

The maximum error is $\frac{1}{2}(2.5 - 2) = .25$.

Iteration 1:

$$m = \frac{1}{2}(2 + 2.5) = 2.25$$
$$f(2.25) = -.1094$$

Since $f(m)$ is negative, a root must exist in the interval $(2.25, 2.5)$. Therefore,

$$L_2 = 2.25 \text{ and } R_2 = 2.5$$

The best estimate of the root is

$$x^* \approx \frac{1}{2}(2.25 + 2.5) = 2.375$$

The maximum error is $\frac{1}{2}(2.5 - 2.25) = .125$.

This procedure continues until the maximum error is acceptable. Of course, this method does not automatically find any other roots that may exist on the real number line.

4 Finding Roots to General Expressions

There is no specific technique that will work with all general expressions for which roots are needed. If graphical means are not used, some combination of factoring and algebraic simplification must be used. However, multiplying each side of an equation by a power of a variable may introduce extraneous roots. Such an extraneous root will not satisfy the original equation, even though it was derived correctly according to the rules of algebra.

Although it is always a good idea to check your work, this step is particularly necessary whenever you have squared an expression or multiplied it by a variable.

Example 1.3

Find the value of x which will satisfy the following expression:
$$\sqrt{x - 2} = \sqrt{x} + 2$$

First, square both sides.

$$x - 2 = x + 4\sqrt{x} + 4$$

Next, subtract x from both sides and combine constants.

$$4\sqrt{x} = -6$$

Solving for x yields $x^* = 9/4$. However $9/4$ does not satisfy the original expression since it is an extraneous root.

B. SIMULTANEOUS LINEAR EQUATIONS

Given n independent equations and n unknowns, the n values which simultaneously solve all n equations can be found by the methods illustrated in example 1.4.

1 By Substitution (shown by example)

Example 1.4

Solve

$$2x + 3y = 12 \, (a)$$
$$3x + 4y = 8 \, (b)$$

step 1: From equation (a), solve for $x = 6 - 1.5y$

step 2: Substitute $(6 - 1.5y)$ into equation (b) wherever x appears. $3(6 - 1.5y) + 4y = 8$ or $y^* = 20$

step 3: Solve for x^* from either equation:

$$x^* = 6 - 1.5(20) = -24$$

step 4: Check that $(-24, 20)$ solves both original equations.

2 By Reduction (same example)

step 1: Multiply each equation by a number chosen to make the coefficient of one of the variables the same in each equation.

$3 \times$ equation (a): $6x + 9y = 36 \, (c)$

$2 \times$ equation (b): $6x + 8y = 16 \, (d)$

step 2: Subtract one equation from the other. Solve for one of the variables.

$$(c) - (d): \, y^* = 20$$

step 3: Solve for the remaining variable.

step 4: Check that the calculated values of (x^*, y^*) solve both original equations.

3 By Cramer's Rule

This method is best for 3 or more simultaneous equations. (The calculation of determinants is covered later in this chapter.)

To find x^* and y^* which satisfy

$$a_1 x + b_1 y = c_1$$
$$a_2 x + b_2 y = c_2$$

calculate the determinants

$$D_1 = \begin{vmatrix} a_1 & b_1 \\ a_2 & b_2 \end{vmatrix} \qquad 1.47$$

$$D_2 = \begin{vmatrix} c_1 & b_1 \\ c_2 & b_2 \end{vmatrix} \qquad 1.48$$

$$D_3 = \begin{vmatrix} a_1 & c_1 \\ a_2 & c_2 \end{vmatrix} \qquad 1.49$$

Then, if $D_1 \neq 0$, the unique numbers satisfying the two simultaneous equations are:

$$x^* = \frac{D_2}{D_1} \qquad 1.50$$

$$y^* = \frac{D_3}{D_1} \qquad 1.51$$

If D_1 (the determinant of the coefficients matrix) is zero, the system of simultaneous equations may still have a solution. However, Cramer's rule cannot be used to find that solution. If the system is homogeneous (i.e., has the general form $\mathbf{Ax} = \mathbf{0}$), then a non-zero solution exists if and only if D_1 is zero.

Example 1.5

Solve the following system of simultaneous equations:

$$2x + 3y - 4z = 1$$
$$3x - y - 2z = 4$$
$$4x - 7y - 6z = -7$$

Calculate the determinants:

$$D_1 = \begin{vmatrix} 2 & 3 & -4 \\ 3 & -1 & -2 \\ 4 & -7 & -6 \end{vmatrix} = 82$$

$$D_2 = \begin{vmatrix} 1 & 3 & -4 \\ 4 & -1 & -2 \\ -7 & -7 & -6 \end{vmatrix} = 246$$

$$D_3 = \begin{vmatrix} 2 & 1 & -4 \\ 3 & 4 & -2 \\ 4 & -7 & -6 \end{vmatrix} = 82$$

$$D_4 = \begin{vmatrix} 2 & 3 & 1 \\ 3 & -1 & 4 \\ 4 & -7 & -7 \end{vmatrix} = 164$$

Then,

$$x^* = \frac{D_2}{D_1} = 3$$

$$y^* = \frac{D_3}{D_1} = 1$$

$$z^* = \frac{D_4}{D_1} = 2$$

C. SIMULTANEOUS QUADRATIC EQUATIONS

Although simultaneous non-linear equations are best solved graphically, a specialized method exists for simultaneous quadratic equations. This method is known as *Eliminating the Constant Term.*

step 1: Isolate the constant terms of both equations on the right-hand side of the equalities.

step 2:   Multiply both sides of one equation by a number chosen to make the constant terms of both equations the same.

step 3:   Subtract one equation from the other to obtain a difference equation.

step 4:   Factor the difference equation into terms.

step 5:   Solve for one of the variables from one of the factor terms.

step 6:   Substitute the formula for the variable into one of the original equations and complete the solution.

step 7:   Check the solution.

Example 1.6

Solve for the simultaneous values of x and y:

step 1:

$$2x^2 - 3xy + y^2 = 15$$
$$x^2 - 2xy + y^2 = 9$$

steps 2 & 3:

$$\begin{aligned} 6x^2 - 9xy + 3y^2 &= 45 \\ -(5x^2 - 10xy + 5y^2) &= 45 \\ \hline x^2 + xy - 2y^2 &= 0 \end{aligned}$$

steps 4 & 5: $x^2 + xy - 2y^2$ factors into $(x + 2y)(x - y)$ from which we obtain $x = -2y$.

step 6:   Substituting $x = -2y$ into $(2x^2 - 3xy + y^2 = 15)$ gives $y^* = \pm 1$, from which $x^* = \pm 2$ can be derived by further substitution.

D. EXPONENTIATION

(x is any variable or constant)

$x^m x^n = x^{(n+m)}$	1.52
$\dfrac{x^m}{x^n} = x^{(m-n)}$	1.53
$(x^n)^m = x^{(mn)}$	1.54
$a^{m/n} = \sqrt[n]{a^m}$	1.55
$\left(\dfrac{a}{b}\right)^n = \dfrac{a^n}{b^n}$	1.56
$\sqrt[n]{x} = (x)^{1/n}$	1.57
$x^{-n} = \dfrac{1}{x^n}$	1.58
$x^0 = 1$	1.59

E. LOGARITHMS

Logarithms are exponents. That is, the exponent x in the expression $b^x = n$ is the logarithm of n to the base b. Therefore, $(\log_b n) = x$ is equivalent to $(b^x = n)$.

The base for common logs is 10. Usually, *log* will be written when common logs are desired, although $\log_{10}$ appears occasionally. The base for *natural (Napierian) logs* is 2.718..., a number that is given the symbol e. When natural logs are desired, usually *ln* will be written, although $\log_e$ is also used.

Most logarithms will contain an integer part (the *characteristic*) and a fractional part (the *mantissa*). The logarithm of any number less than one is negative. If the number is greater than one, its logarithm is positive. Although the logarithm may be negative, the mantissa is always positive.

For common logarithms of numbers greater than one, the characteristics will be positive and equal to one less than the number of digits in front of the decimal. If the number is less than one, the characteristic will be negative and equal to one more than the number of zeros immediately following the decimal point.

Example 1.7

What is $\log_{10}(.05)$?

Since the number is less than one and there is one leading zero, the characteristic is -2. From the logarithm tables, the mantissa of 5.0 is .699. Two ways of combining the mantissa and characteristic are possible:

Method 1: $\bar{2}.699$

Method 2: $8.699 - 10$

If the logarithm is to be used in a calculation, it must be converted to operational form: $-2 + .699 = -1.301$. Notice that -1.301 is not the same as $\bar{1}.301$.

F. LOGARITHM IDENTITIES

$x^a = \text{antilog}[a \log(x)]$	1.60
$\log(x^a) = a \log(x)$	1.61
$\log(xy) = \log(x) + \log(y)$	1.62
$\log\left(\dfrac{x}{y}\right) = \log(x) - \log(y)$	1.63
$ln(x) = \dfrac{\log_{10} x}{\log_{10} e}$	
$\approx 2.3(\log_{10} x)$	1.64
$\log_b(b) = 1$	1.65
$\log(1) = 0$	1.66
$\log_b(b^n) = n$	1.67

Example 1.8

The surviving fraction, x, of a radioactive isotope is given by

$$x = e^{-.005t}$$

For what value of t will the surviving fraction be 7%?

$$.07 = e^{-.005t}$$

Taking the natural log of both sides,

$$ln(.07) = ln(e^{-.005t})$$
$$-2.66 = -.005t$$
$$t = 532$$

G. PARTIAL FRACTIONS

Given some rational fraction $H(x) = P(x)/Q(x)$ where $P(x)$ and $Q(x)$ are polynomials, the polynomials and constants A_i and $Y_i(x)$ are needed such that

$$H(x) = \sum_i \frac{A_i}{Y_i(x)} \qquad 1.68$$

case 1: $Q(x)$ factors into n different linear terms. That is,

$$Q(x) = (x - a_1)(x - a_2)\cdots(x - a_n) \quad 1.69$$

Then,

$$H(x) = \sum_{i=1}^{n} \frac{A_i}{x - a_i} \qquad 1.70$$

case 2: $Q(x)$ factors into n identical linear terms. That is,

$$Q(x) = (x - a)(x - a)\cdots(x - a) \quad 1.71$$

Then,

$$H(x) = \sum_{i=1}^{n} \frac{A_i}{(x - a)^i} \qquad 1.72$$

case 3: $Q(x)$ factors into n different quadratic terms, $(x^2 + p_i x + q_i)$. Then,

$$H(x) = \sum_{i=1}^{n} \frac{A_i x + B_i}{x^2 + p_i x + q_i} \qquad 1.73$$

case 4: $Q(x)$ factors into n identical quadratic terms, $(x^2 + px + q)$. Then,

$$H(x) = \sum_{i=1}^{n} \frac{A_i x + B_i}{(x^2 + px + q)^i} \qquad 1.74$$

case 5: $Q(x)$ factors into any combination of the above. The solution is illustrated by example 1.9.

Example 1.9

Resolve

$$H(x) = \frac{x^2 + 2x + 3}{x^4 + x^3 + 2x^2}$$

into partial fractions.

Here, $Q(x) = x^4 + x^3 + 2x^2$ which factors into $x^2(x^2 + x + 2)$. This is a combination of cases 2 and 3. We set

$$H(x) = \frac{A_1}{x} + \frac{A_2}{x^2} + \frac{A_3 + A_4 x}{x^2 + x + 2}$$

Cross multiplying to obtain a common denominator yields

$$\frac{(A_1 + A_4)x^3 + (A_1 + A_2 + A_3)x^2 + (2A_1 + A_2)x + 2A_2}{x^4 + x^3 + 2x^2}$$

Since the original numerator is known, the following simultaneous equations result:

$$A_1 + A_4 = 0$$
$$A_1 + A_2 + A_3 = 1$$
$$2A_1 + A_2 = 2$$
$$2A_2 = 3$$

The solutions are: $A_1^* = .25$; $A_2^* = 1.5$; $A_3^* = -.75$; $A_4^* = -.25$. So,

$$H(x) = \frac{1}{4x} + \frac{3}{2x^2} - \frac{x + 3}{4(x^2 + x + 2)}$$

H. LINEAR AND MATRIX ALGEBRA

A matrix is a rectangular collection of variables or scalars contained within a set of square or round brackets. In the discussion that follows, matrix **A** will be assumed to have m rows and n columns. There are several classifications of matrices:

If $n = m$, the matrix is *square*.

A *diagonal* matrix is a square matrix with all zero values except for the a_{ij} values, for all $i = j$.

An *identity* matrix is a diagonal matrix with all non-zero entries equal to '1'. (This usually is designated as '**I**'.)

A *scalar* matrix is a square diagonal matrix with all non-zero entries equal to some constant.

A *triangular* matrix has zeros in all positions above or below the diagonal. This is not the same as an *echelon* matrix since the diagonal entries are non-zero.

Matrices are used to simplify the presentation and solution of sets of linear equations (hence the name 'linear algebra'). For example, the system of equations in example 1.4 can be written in matrix form as:

$$\begin{pmatrix} 2 & 3 \\ 3 & 4 \end{pmatrix} \begin{pmatrix} x \\ y \end{pmatrix} = \begin{pmatrix} 12 \\ 8 \end{pmatrix}$$

The above expression implies that there is a set of algebraic operations that can be performed with matrices. The important algebraic operations are listed here, along with their extensions to linear algebra.

(a) *Equality of Matrices*: For two matrices to be equal, they must have the same number of rows and columns. Corresponding entries must all be the same.

(b) *Inequality of Matrices*: There are no 'less-than' or 'greater than' relationships in linear algebra.

(c) *Addition and Subtraction of Matrices*: Addition (or subtraction) of two matrices can be accomplished by adding (or subtracting) the corresponding entries of two matrices which have the same shape.

(d) *Multiplication of Matrices*: Multiplication can be done only if the left-hand matrix has the same number of columns as the right-hand matrix has rows. Multiplication is accomplished by multiplying the elements in each right-hand matrix column, adding the products, and then placing the sum at the intersection point of the involved row and column. This is illustrated by example 1.10.

(e) *Division of Matrices*: Division can be accomplished only by multiplying by the inverse of the denominator matrix.

Example 1.10

$$\begin{pmatrix} 1 & 4 & 3 \\ 5 & 2 & 6 \end{pmatrix} \begin{pmatrix} 7 & 12 \\ 11 & 8 \\ 9 & 10 \end{pmatrix} = C$$

$$[(1)(7) + (4)(11) + (3)(9)] = 78$$
$$[(1)(12) + (4)(8) + (3)(10)] = 74$$
$$[(5)(7) + (2)(11) + (6)(9)] = 111$$
$$[(5)(12) + (2)(8) + (6)(10)] = 136$$

$$C = \begin{pmatrix} 78 & 74 \\ 111 & 136 \end{pmatrix}$$

Other operations which can be performed on a matrix are described and illustrated below.

1. The *transpose* is an $(n \times m)$ matrix formed from the original $(m \times n)$ matrix by taking the ith row and making it the ith column. The diagonal is unchanged in this operation. The transpose of a matrix A is indicated as A^t.

Example 1.11

What is the transpose of

$$A = \begin{pmatrix} 1 & 6 & 9 \\ 2 & 3 & 4 \\ 7 & 1 & 5 \end{pmatrix}?$$

$$A^t = \begin{pmatrix} 1 & 2 & 7 \\ 6 & 3 & 1 \\ 9 & 4 & 5 \end{pmatrix}$$

2. The *determinant*, D, is a scalar calculated from a square matrix. The determinant of a matrix is indicated by enclosing the matrix in vertical lines.

For a (2×2) matrix,

$$A = \begin{pmatrix} a & b \\ c & d \end{pmatrix}$$

$$D = \begin{vmatrix} a & b \\ c & d \end{vmatrix} = ad - bc \qquad 1.75$$

For a (3×3) matrix,

$$A = \begin{pmatrix} a & b & c \\ d & e & f \\ g & h & i \end{pmatrix}$$

$$D = a \begin{vmatrix} e & f \\ h & i \end{vmatrix} - d \begin{vmatrix} b & c \\ h & i \end{vmatrix} + g \begin{vmatrix} b & c \\ e & f \end{vmatrix} \qquad 1.76$$

There are several rules governing the calculation of determinants:

- If A has a row or column of zeros, the determinant is zero.

- If A has two identical rows or columns, the determinant is zero.

- If A is triangular, the determinant is equal to the product of the diagonal entries.

- If B is obtained from A by multiplying a row or column by a scalar k, then $D_B = k(D_A)$.

- If B is obtained from A by switching two rows or columns, then $D_B = -D_A$.

- If B is obtained from A by adding a multiple of a row or column to another, then $D_B = D_A$.

Example 1.12

What is the determinant of

$$\begin{pmatrix} 2 & 3 & -4 \\ 3 & -1 & -2 \\ 4 & -7 & -6 \end{pmatrix}?$$

$$\mathbf{D} = 2\begin{vmatrix} -1 & -2 \\ -7 & -6 \end{vmatrix} - 3\begin{vmatrix} 3 & -4 \\ -7 & -6 \end{vmatrix} + 4\begin{vmatrix} 3 & -4 \\ -1 & -2 \end{vmatrix}$$
$$= 2(6-14) - 3(-18-28) + 4(-6-4)$$
$$= 82$$

3. The *cofactor* of an entry in a matrix is the determinant of the matrix formed by omitting the entry's row and column in the original matrix. The sign of the cofactor is determined from the following positional matrices:

For a (2 × 2) matrix,

$$\begin{pmatrix} + & - \\ - & + \end{pmatrix}$$

For a (3 × 3) matrix,

$$\begin{pmatrix} + & - & + \\ - & + & - \\ + & - & + \end{pmatrix}$$

Example 1.13

What is the cofactor of the (−3) in the following matrix?

$$\begin{pmatrix} 2 & 9 & 1 \\ -3 & 4 & 0 \\ 7 & 5 & 9 \end{pmatrix}$$

The resulting matrix is

$$\begin{pmatrix} 9 & 1 \\ 5 & 9 \end{pmatrix}$$

with determinant 76. The cofactor is −76.

4. The *classical adjoint* is a matrix formed from the transposed cofactor matrix with the conventional sign arrangement. The resulting matrix is represented as $\mathbf{A}_{adj}$.

Example 1.14

What is the classical adjoint of

$$\begin{pmatrix} 2 & 3 & -4 \\ 0 & -4 & 2 \\ 1 & -1 & 5 \end{pmatrix}?$$

The matrix of cofactors (considering the sign convention) is

$$\begin{pmatrix} -18 & 2 & 4 \\ -11 & 14 & 5 \\ -10 & -4 & -8 \end{pmatrix}$$

The transposed cofactor matrix is

$$\mathbf{A}_{adj} = \begin{pmatrix} -18 & -11 & -10 \\ 2 & 14 & -4 \\ 4 & 5 & -8 \end{pmatrix}$$

5. The *inverse*, $\mathbf{A}^{-1}$, of $\mathbf{A}$ is a matrix such that $(\mathbf{A})(\mathbf{A}^{-1}) = \mathbf{I}$. ($\mathbf{I}$ is a square matrix with ones along the left-to-right diagonal and zeros elsewhere.)

For a (2 × 2) matrix

$$\begin{pmatrix} a & b \\ c & d \end{pmatrix}$$

the inverse is

$$\frac{1}{\mathbf{D}}\begin{pmatrix} d & -b \\ -c & a \end{pmatrix} \qquad 1.77$$

For larger matrices, the inverse is best calculated by dividing every entry in the classical adjoint by the determinant of the original matrix.

Example 1.15

What is the inverse of

$$\begin{pmatrix} 4 & 5 \\ 2 & 3 \end{pmatrix}?$$

The determinant is 2. The inverse is

$$\frac{1}{2}\begin{pmatrix} 3 & -5 \\ -2 & 4 \end{pmatrix} = \begin{pmatrix} \frac{3}{2} & -\frac{5}{2} \\ -1 & 2 \end{pmatrix}$$

7 TRIGONOMETRY

A. DEGREES AND RADIANS

360 degrees = one complete circle = 2π radians

90 degrees = right angle = $\frac{1}{2}\pi$ radians

one radian = 57.3 degrees

one degree = .0175 radians

multiply degrees by $\left(\frac{\pi}{180}\right)$ to obtain radians

multiply radians by $\left(\frac{180}{\pi}\right)$ to obtain degrees

B. RIGHT TRIANGLES

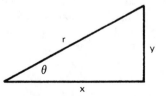

Figure 1.1 A Right Triangle

1 Pythagorean Theorem

$$x^2 + y^2 = r^2 \qquad 1.78$$

2 Trigonometric Functions

$$\sin \theta = \frac{y}{r} \qquad 1.79$$

$$\cos \theta = \frac{x}{r} \qquad 1.80$$

$$\tan \theta = \frac{y}{x} \qquad 1.81$$

$$\cot \theta = \frac{x}{y} \qquad 1.82$$

$$\csc \theta = \frac{r}{y} \qquad 1.83$$

$$\sec \theta = \frac{r}{x} \qquad 1.84$$

3 Relationship of the Trigonometric Functions to the Unit Circle

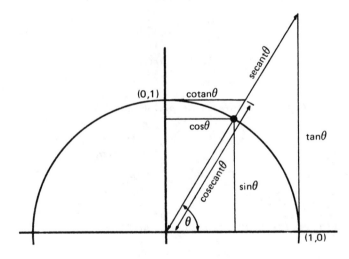

Figure 1.2 The Unit Circle

4 Signs of the Trigonometric Functions

quadrants		quadrant	I	II	III	IV
II	I	sin	+	+	−	−
		cos	+	−	−	+
III	IV	tan	+	−	+	−

5 Functions of the Related Angles

$f(\theta)$	$-\theta$	$90 - \theta$	$90 + \theta$	$180 - \theta$	$180 + \theta$
sin	$-\sin \theta$	$\cos \theta$	$\cos \theta$	$\sin \theta$	$-\sin \theta$
cos	$\cos \theta$	$\sin \theta$	$-\sin \theta$	$-\cos \theta$	$-\cos \theta$
tan	$-\tan \theta$	$\cot \theta$	$-\cot \theta$	$-\tan \theta$	$\tan \theta$

6 Trigonometric Identities

$$\sin^2 \theta + \cos^2 \theta = 1 \qquad 1.85$$

$$1 + \tan^2 \theta = \sec^2 \theta \qquad 1.86$$

$$1 + \cot^2 \theta = \csc^2 \theta \qquad 1.87$$

$$\sin 2\theta = 2 (\sin \theta)(\cos \theta) \qquad 1.88$$

$$\cos 2\theta = \cos^2 \theta - \sin^2 \theta = 1 - 2\sin^2 \theta \qquad 1.89$$

$$\sin \theta = 2 \left[\sin\left(\frac{\theta}{2}\right)\cos\left(\frac{\theta}{2}\right)\right] \qquad 1.90$$

$$\sin \left(\frac{\theta}{2}\right) = \pm\sqrt{\frac{1}{2}(1 - \cos\theta)} \qquad 1.91$$

7 Two-Angle Formulas

$$\sin(\theta + \phi) = [\sin \theta][\cos \phi] + [\cos \theta][\sin \phi] \qquad 1.92$$

$$\sin(\theta - \phi) = [\sin \theta][\cos \phi] - [\cos \theta][\sin \phi] \qquad 1.93$$

$$\cos(\theta + \phi) = [\cos \theta][\cos \phi] - [\sin \theta][\sin \phi] \qquad 1.94$$

$$\cos(\theta - \phi) = [\cos \theta][\cos \phi] + [\sin \theta][\sin \phi] \qquad 1.95$$

C. GENERAL TRIANGLES

Law of Sines: $\dfrac{\sin A}{a} = \dfrac{\sin B}{b} = \dfrac{\sin C}{c}$ 1.96

Law of Cosines: $a^2 = b^2 + c^2 - 2bc(\cos A)$ 1.97

$$\text{Area} = \frac{1}{2}ab(\sin C) \qquad 1.98$$

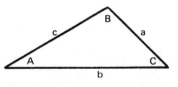

Figure 1.3 A General Triangle

D. HYPERBOLIC FUNCTIONS

Hyperbolic functions are specific equations containing the terms e^x and e^{-x}. These combinations of e^x and e^{-x} appear regularly in certain types of problems. In order to simplify the mathematical equations in which they appear, these hyperbolic functions are given special names and symbols.

$$\sinh x = \frac{e^x - e^{-x}}{2} \qquad 1.99$$

$$\cosh x = \frac{e^x + e^{-x}}{2} \qquad 1.100$$

$$\tanh x = \frac{e^x - e^{-x}}{e^x + e^{-x}} = \frac{\sinh x}{\cosh x} \qquad 1.101$$

$$\coth x = \frac{e^x + e^{-x}}{e^x - e^{-x}} = \frac{\cosh x}{\sinh x} \qquad 1.102$$

$$\operatorname{sech} x = \frac{2}{e^x + e^{-x}} = \frac{1}{\cosh x} \qquad 1.103$$

$$\operatorname{csch} x = \frac{2}{e^x - e^{-x}} = \frac{1}{\sinh x} \qquad 1.104$$

The hyperbolic identities are somewhat different from the standard trigonometric identities. Several of the most common identities are presented below.

$$\cosh^2 x - \sinh^2 x = 1 \qquad 1.105$$

$$1 - \tanh^2 x = \operatorname{sech}^2 x \qquad 1.106$$

$$1 - \coth^2 x = \operatorname{csch}^2 x \qquad 1.107$$

$$\cosh x + \sinh x = e^x \qquad 1.108$$

$$\cosh x - \sinh x = e^{-x} \qquad 1.109$$

$$\sinh(x + y) = [\sinh x][\cosh y] + [\cosh x][\sinh y] \qquad 1.110$$

$$\cosh(x + y) = [\cosh x][\cosh y] + [\sinh x][\sinh y] \qquad 1.111$$

8 STRAIGHT LINE ANALYTIC GEOMETRY

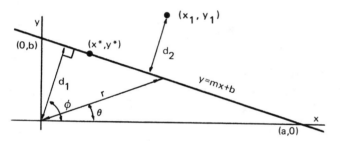

Figure 1.4 A Straight Line

A. EQUATIONS OF A STRAIGHT LINE

General Form: $\quad Ax + By + C = 0 \qquad 1.112$

Slope Form: $\quad y = mx + b \qquad 1.113$

Point-Slope Form: $\quad (y - y^*) = m(x - x^*) \qquad 1.114$

(x^*, y^*) is any point on the line

Intercept Form: $\quad \dfrac{x}{a} + \dfrac{y}{b} = 1 \qquad 1.115$

Two-Point Form: $\quad \dfrac{y - y_1^*}{x - x_1^*} = \dfrac{y_2^* - y_1^*}{x_2^* - x_1^*} \qquad 1.116$

Normal Form: $\quad x(\cos\phi) + y(\sin\phi) - d_1 = 0 \qquad 1.117$

Polar Form: $\quad r = \dfrac{d_1}{\cos(\phi - \theta)} \qquad 1.118$

B. POINTS, LINES, AND DISTANCES

The distance d_2 between a point and a line is:

$$d_2 = \frac{|Ax_1 + By_1 + C|}{\sqrt{A^2 + B^2}} \qquad 1.119$$

The distance between two points is:

$$d = \sqrt{(x_2 - x_1)^2 + (y_2 - y_1)^2} \qquad 1.120$$

Parallel lines:

$$\frac{A_1}{A_2} = \frac{B_1}{B_2} \qquad 1.121$$

$$m_1 = m_2 \qquad 1.122$$

Perpendicular lines:

$$A_1 A_2 = -B_1 B_2 \qquad 1.123$$

$$m_1 = \frac{-1}{m_2} \qquad 1.124$$

Point of intersection of two lines:

$$x_1 = \frac{B_2 C_1 - B_1 C_2}{A_2 B_1 - A_1 B_2} \qquad 1.125$$

$$y_1 = \frac{A_1 C_2 - A_2 C_1}{A_2 B_1 - A_1 B_2} \qquad 1.126$$

Smaller angle between two intersecting lines:

$$\tan\phi = \frac{A_1 B_2 - A_2 B_1}{A_1 A_2 + B_1 B_2} = \frac{m_2 - m_1}{1 + m_1 m_2} \qquad 1.127$$

$$\phi = |\arctan(m_1) - \arctan(m_2)|. \qquad 1.128$$

Example 1.16

What is the angle between the lines?

$$y_1 = -.577x + 2$$

$$y_2 = +.577x - 5$$

method 1:

$$\arctan\left[\frac{m_2 - m_1}{1 + m_1 m_2}\right] =$$

$$\arctan\left[\frac{.577 - (-.577)}{1 + (.577)(-.577)}\right] = 60°$$

method 2: Write both equations in general form:

$$-.577x - y_1 + 2 = 0$$

$$.577x - y_2 - 5 = 0$$

$$\arctan\left[\frac{A_1 B_2 - A_2 B_1}{A_1 A_2 + B_1 B_2}\right] =$$

$$\arctan\left[\frac{(-.577)(-1) - (.577)(-1)}{(-.577)(.577) + (-1)(-1)}\right] = 60°$$

method 3:

$$\phi = |\arctan(-.577) - \arctan(.577)|$$
$$= |-30° - 30°| = 60°$$

C. LINEAR AND CURVILINEAR REGRESSION

If it is necessary to draw a straight line through n data points $(x_1, y_1), (x_2, y_2), \ldots, (x_n, y_n)$, the following method based on the theory of least squares can be used:

step 1: Calculate the following quantities.

$$\sum x_i \quad \sum x_i^2 \quad (\sum x_i)^2 \quad \bar{x} = \left(\frac{\sum x_i}{n}\right) \quad \sum x_i y_i$$
$$\sum y_i \quad \sum y_i^2 \quad (\sum y_i)^2 \quad \bar{y} = \left(\frac{\sum y_i}{n}\right)$$

step 2: Calculate the slope of the line $y = mx + b$.

$$m = \frac{n \sum(x_i y_i) - (\sum x_i)(\sum y_i)}{n \sum x_i^2 - (\sum x_i)^2} \qquad 1.129$$

step 3: Calculate the y intercept.

$$b = \bar{y} - m\bar{x} \qquad 1.130$$

step 4: To determine the goodness of fit, calculate the correlation coefficient.

$$r = \frac{n \sum(x_i y_i) - (\sum x_i)(\sum y_i)}{\sqrt{[n \sum x_i^2 - (\sum x_i)^2][n \sum y_i^2 - (\sum y_i)^2]}} \qquad 1.131$$

If m is positive, r will be positive. If m is negative, r will be negative. As a general rule, if the absolute value of r exceeds .85, the fit is good. Otherwise, the fit is poor. r equals 1.0 if the fit is a perfect straight line.

Example 1.17

An experiment is performed in which the dependent variable (y) is measured against the independent variable (x). The results are as follows:

x	y
1.2	.602
4.7	5.107
8.3	6.984
20.9	10.031

What is the least squares straight line equation which represents this data?

step 1:

$$\sum x_i = 35.1$$
$$\sum y_i = 22.72$$
$$\sum x_i^2 = 529.23$$
$$\sum y_i^2 = 175.84$$

$$\left(\sum x_i\right)^2 = 1232.01$$
$$\left(\sum y_i\right)^2 = 516.38$$
$$\bar{x} = 8.775$$
$$\bar{y} = 5.681$$
$$\sum x_i y_i = 292.34$$
$$n = 4$$

step 2:

$$m = \frac{(4)(292.34) - (35.1)(22.72)}{(4)(529.23) - (35.1)^2} = .42$$

step 3:

$$b = 5.681 - (.42)(8.775) = 2.0$$

step 4: From equation 1.131, $r = .91$.

A low value of r does not eliminate the possibility of a non-linear relationship existing between x and y. It is possible that the data describes a parabolic, logarithmic, or other non-linear relationship. (Usually this will be apparent if the data are graphed.) It may be necessary to convert one or both variables to new variables by taking squares, square roots, cubes, or logs, to name a few of the possibilities.

The apparent shape of the line through the data will give a clue to the type of variable transformation that is required. The following curves may be used as guides to some of the simpler variable transformations.

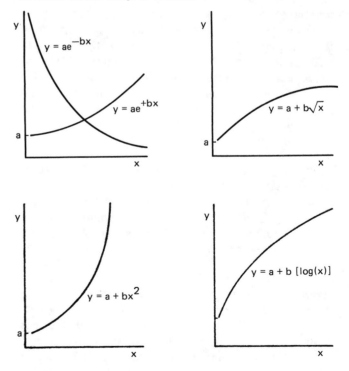

Figure 1.5 Non-Linear Data Plots

Example 1.18

Repeat example **1.17** assuming that the relationship between the variables is non-linear.

The first step is to graph the data. Since the graph has the appearance of the fourth case, it can be assumed that the relationship between the variables has the form of $y = a + b[\log(x)]$. Therefore, the variable change $z = \log(x)$ is made, resulting in the following set of data:

z	y
.0792	.602
.672	5.107
.919	6.984
1.32	10.031

If the regression analysis is performed on this set of data, the resulting equation and correlation coefficient are:

$$y = -.036 + 7.65z$$

$$r = .999$$

This is a very good fit. The relationship between the variable x and y is approximately

$$y = -.036 + 7.65[\log(x)]$$

Figure 1.6 illustrates several common problems encountered in trying to fit and evaluate curves from experimental data. Figure 1.6(a) shows a graph of clustered data with several extreme points. There will be moderate correlation due to the weighting of the extreme points, although there is little actual correlation at low values of the variables. The extreme data should be excluded or the range should be extended by obtaining more data.

Figure 1.6(b) shows that good correlation exists in general, but extreme points are missed, and the overall correlation is moderate. If the results within the small linear range can be used, the extreme points should be excluded. Otherwise, additional data points are needed, and curvilinear relationships should be investigated.

Figure 1.6(c) illustrates the problem of drawing conclusions of cause and effect. There may be a predictable relationship between variables, but that does not imply a cause and effect relationship. In the case shown, both variables are functions of a third variable, the city population. But, there is no direct relationship between the plotted variables.

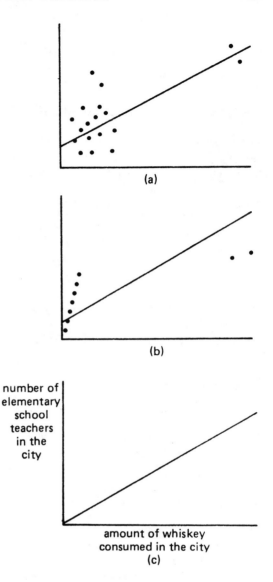

Figure 1.6 Common Regression Difficulties

D. VECTOR OPERATIONS

A vector is a directed straight line of a given magnitude. Two directed straight lines with the same magnitudes and directions are said to be *equivalent*. Thus, the actual end points of a vector often are irrelevant as long as the direction and magnitude are known.

A vector defined by its end points and direction is designated as

$$\mathbf{V} = \overrightarrow{p_1}\overrightarrow{p_2}$$

Usually, p_1 will be the origin, in which case $\mathbf{V}$ will be designated by its end-point, $p_2 = (x, y)$. Such a zero-based vector is equivalent to all other vectors of the same magnitude and direction. Any vector $p_1 p_2$ can be transformed into a zero-based vector by subtracting (x_1, y_1) from all points along the vector line.

A vector also can be specified in the terms of the unit vectors $(\mathbf{i}, \mathbf{j}, \mathbf{k})$. Thus,

$$\mathbf{V} = (x, y) = (x\mathbf{i} + y\mathbf{j})$$

Or,

$$\mathbf{V} = (x, y, z) = (x\mathbf{i} + y\mathbf{j} + z\mathbf{k})$$

Important operations on vectors based at the origin are:

$$c\mathbf{V} = (cx, cy) \quad \text{(vector multiplication by a scalar)}$$

$$\mathbf{V}_1 + \mathbf{V}_2 = (x_1 + x_2, y_1 + y_2) \qquad \textbf{1.132}$$

$$|\mathbf{V}| = \sqrt{x^2 + y^2} \quad \text{(vector magnitude)} \qquad \textbf{1.133}$$

α = angle between vector $\mathbf{V}$ and x axis

$$= \arccos\left(\frac{x}{|\mathbf{V}|}\right) = \arcsin\left(\frac{y}{|\mathbf{V}|}\right) \qquad \textbf{1.134}$$

$$m = \text{slope of vector} = \frac{y}{x} \qquad \textbf{1.135}$$

$$\theta = \left\{\begin{array}{c}\text{angle between}\\ \text{two vectors}\end{array}\right\}$$

$$= \arccos\frac{(x_1 x_2 + y_1 y_2)}{|\mathbf{V}_1||\mathbf{V}_2|} \qquad \textbf{1.136}$$

$\mathbf{V}_1 \cdot \mathbf{V}_2$ = dot product

$$= |\mathbf{V}_1||\mathbf{V}_2|\cos\theta = x_1 x_2 + y_1 y_2 \qquad \textbf{1.137}$$

When equation **1.137** is solved for $\cos\theta$, it is known as the *Cauchy-Schwartz theorem*.

$$\cos\theta = \frac{x_1 x_2 + y_1 y_2}{|\mathbf{V}_1||\mathbf{V}_2|} \qquad \textbf{1.138}$$

$$\mathbf{V}_1 \times \mathbf{V}_2 = \text{cross product} = \begin{vmatrix} \mathbf{i} & x_1 & x_2 \\ \mathbf{j} & y_1 & y_2 \\ \mathbf{k} & z_1 & z_2 \end{vmatrix} \qquad \textbf{1.139}$$

$$\mathbf{V}_1 \times \mathbf{V}_2 = -\mathbf{V}_2 \times \mathbf{V}_1 \qquad \textbf{1.140}$$

$$|\mathbf{V}_1 \times \mathbf{V}_2| = |\mathbf{V}_1||\mathbf{V}_2|\sin\theta \qquad \textbf{1.141}$$

Example 1.19

What is the angle between the vectors $\mathbf{V}_1 = (-\sqrt{3}, 1)$ and $\mathbf{V}_2 = (2\sqrt{3}, 2)$?

$$\cos\theta = \frac{\mathbf{V}_1 \cdot \mathbf{V}_2}{|\mathbf{V}_1||\mathbf{V}_2|} = \frac{(-\sqrt{3})(2\sqrt{3}) + (1)(2)}{\sqrt{3+1}\sqrt{12+4}} = -\frac{1}{2}$$

$$\theta = 120°$$

(Graph and compare this result to example **1.16**, in which the lines were not directed.)

Example 1.20

Find a unit vector orthogonal to $\mathbf{V}_1 = \mathbf{i} - \mathbf{j} + 2\mathbf{k}$ and $\mathbf{V}_2 = 3\mathbf{j} - \mathbf{k}$.

The cross product is orthogonal to $\mathbf{V}_1$ and $\mathbf{V}_2$, although its length may not be equal to one.

$$\mathbf{V}_1 \times \mathbf{V}_2 = \begin{vmatrix} \mathbf{i} & 1 & 0 \\ \mathbf{j} & -1 & 3 \\ \mathbf{k} & 2 & -1 \end{vmatrix} = -5\mathbf{i} + \mathbf{j} + 3\mathbf{k}$$

Since the length of $|\mathbf{V}_1 \times \mathbf{V}_2|$ is $\sqrt{35}$, it is necessary to divide $|\mathbf{V}_1 \times \mathbf{V}_2|$ by this amount to obtain a unit vector. Thus,

$$\mathbf{V}_3 = \frac{-5\mathbf{i} + \mathbf{j} + 3\mathbf{k}}{\sqrt{35}}$$

The orthogonality can be proved from

$$\mathbf{V}_1 \cdot \mathbf{V}_3 = 0 \text{ and } \mathbf{V}_2 \cdot \mathbf{V}_3 = 0$$

That $\mathbf{V}_3$ is a unit vector can be proved from

$$\mathbf{V}_3 \cdot \mathbf{V}_3 = +1$$

E. DIRECTION NUMBERS, DIRECTION ANGLES, AND DIRECTION COSINES

Given a directed line from (x_1, y_1, z_1) to (x_2, y_2, z_2), the direction numbers are:

$$L = x_2 - x_1 \qquad \textbf{1.142}$$
$$M = y_2 - y_1 \qquad \textbf{1.143}$$
$$N = z_2 - z_1 \qquad \textbf{1.144}$$

The distance between two points is:

$$d = \sqrt{L^2 + M^2 + N^2} \qquad \textbf{1.145}$$

The direction cosines are:

$$\cos\alpha = \frac{L}{d} \qquad \textbf{1.146}$$

$$\cos\beta = \frac{M}{d} \qquad \textbf{1.147}$$

$$\cos\gamma = \frac{N}{d} \qquad \textbf{1.148}$$

Note that

$$\cos^2\alpha + \cos^2\beta + \cos^2\gamma = 1 \qquad \textbf{1.149}$$

The direction angles are the angles between the axes and the lines. They are found from the inverse functions of the direction cosines. That is,

$$\alpha = \arccos\left(\frac{L}{d}\right) \qquad \textbf{1.150}$$

$$\beta = \arccos\left(\frac{M}{d}\right) \qquad \textbf{1.151}$$

$$\gamma = \arccos\left(\frac{N}{d}\right) \qquad \textbf{1.152}$$

Once the direction cosines have been found, they can be used to write the equation of the straight line in terms of the unit vectors. The line $\mathbf{R}$ would be defined as

$$\mathbf{R} = \mathbf{i}\cos\alpha + \mathbf{j}\cos\beta + \mathbf{k}\cos\gamma \qquad 1.153$$

Similarly, the line may be written in terms of its direction numbers,

$$\mathbf{R} = L\mathbf{i} + M\mathbf{j} + N\mathbf{k} \qquad 1.154$$

Given two directed lines, $\mathbf{R}_1$ and $\mathbf{R}_2$, the angle between $\mathbf{R}_1$ and $\mathbf{R}_2$ is defined as the angle between the two arrow heads.

$$\cos\phi = \cos\alpha_1\cos\alpha_2 + \cos\beta_1\cos\beta_2 + \cos\gamma_1\cos\gamma_2$$
$$= \frac{L_1 L_2 + M_1 M_2 + N_1 N_2}{d_1 d_2} \qquad 1.155$$

If $\mathbf{R}_1$ and $\mathbf{R}_2$ are parallel and in the same direction, then

$$\alpha_1 = \alpha_2$$
$$\beta_1 = \beta_2$$
$$\gamma_1 = \gamma_2$$

If $\mathbf{R}_1$ and $\mathbf{R}_2$ are parallel but in opposite directions, then

$$\alpha_1 + \alpha_2 = 180 \,(\text{etc.})$$

If $\mathbf{R}_1$ and $\mathbf{R}_2$ are normal to each other, then

$$\phi = 90° \text{ and } \cos\phi = 0$$

Example 1.21

A line passes through the points (4,7,9) and (0,1,6). Write the equation of the line in terms of its direction cosines and direction numbers.

$$L = 4 - 0 = 4$$
$$M = 7 - 1 = 6$$
$$N = 9 - 6 = 3$$

Example 1.22

Now, the line may be written in terms of its direction numbers.

$$\mathbf{R} = 4\mathbf{i} + 6\mathbf{j} + 3\mathbf{k}$$

The distance between the two points is

$$d = \sqrt{(4)^2 + (6)^2 + (3)^2} = 7.81$$

The line now may be written in terms of its direction cosines.

$$\mathbf{R} = \frac{4\mathbf{i} + 6\mathbf{j} + 3\mathbf{k}}{7.81}$$
$$= .512\mathbf{i} + .768\mathbf{j} + .384\mathbf{k}$$

F. CURVILINEAR INTERPOLATION

A situation which occurs frequently is one in which a function value must be interpolated from other data along the curve. Straight-line interpolation typically is used because of its simplicity and speed. However, straight-line interpolation ignores all but two of the points on the curve and is, therefore, unable to include any effects of curvature.

A more powerful technique is the *Lagrangian Interpolating Polynomial*. It is assumed that $(n + 1)$ values of $f(x)$ are known (for $x_0, x_1, x_2, \ldots, x_n$) and that $f(x)$ is a continuous, real-valued function on the interval (x_0, x_n). The value of $f(x)$ at x^* can be estimated from the following equations:

$$f(x^*) = \sum_{k=0}^{n} f(x_k) L_k(x^*) \qquad 1.156$$

where the Lagrangian Interpolating Polynomial is

$$L_k(x^*) = \prod_{\substack{i=0 \\ i \neq k}}^{n} \frac{x^* - x_i}{x_k - x_i} \qquad 1.157$$

A real-valued function has the following values:

$$f(1) = 1.5709 \quad f(4) = 1.5727 \quad f(6) = 1.5751$$

What is $f(3.5)$?

$$k = 0: \quad L_0(3.5) = \left(\frac{3.5 - 1}{1 - 1}\right)\left(\frac{3.5 - 4}{1 - 4}\right)\left(\frac{3.5 - 6}{1 - 6}\right) = .08333$$

$$k = 1: \quad L_1(3.5) = \left(\frac{3.5 - 1}{4 - 1}\right)\left(\frac{3.5 - 4}{4 - 4}\right)\left(\frac{3.5 - 6}{4 - 6}\right) = 1.04167$$

$$k = 2: \quad L_2(3.5) = \left(\frac{3.5 - 1}{6 - 1}\right)\left(\frac{3.5 - 4}{6 - 4}\right)\left(\frac{3.5 - 6}{6 - 6}\right) = -.12500$$

$$f(x^*) = (1.5709)(.08333) + (1.5727)(1.04167)$$
$$+ (1.5751)(-.12500)$$
$$= 1.57225$$

9 TENSORS

A scalar has magnitude only. A vector has magnitude and a definite direction. A *tensor* has magnitude in a specific direction, but the direction is not unique. An example of a tensor is stress. From the combined stress equation, stress at a point in a solid depends on the direction of the plane passing through that point. Tensors frequently are associated with *anisotropic materials* which have different properties in different directions. Other examples are dielectric constant and magnetic susceptibility.

A vector in a three-dimensional space is defined completely by three quantities, F_x, F_y, and F_z. A tensor in three-dimensional space requires nine quantities for complete definition. These nine values are given in matrix form. The tensor definition for stress at a point is

$$\begin{pmatrix} \sigma_{xx} \sigma_{xy} \sigma_{xz} \\ \sigma_{yx} \sigma_{yy} \sigma_{yz} \\ \sigma_{zx} \sigma_{zy} \sigma_{zz} \end{pmatrix} \qquad 1.158$$

10 PLANES

A plane **P** is uniquely determined by one of three combinations of parameters:

1. three non-collinear points in space

2. two non-parallel vectors (V_1 and V_2) and their intersection point p_0

3. a point p_0 and a normal vector **N**

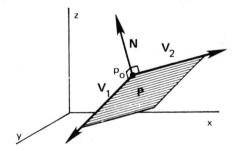

Figure 1.7 A Plane in 3-Space

The plane consists of all points such that the coordinates can be written as a linear combination of V_1 and V_2. That is, points in the plane can be written as

$$(x, y, z) = sV_1 + tV_2 \qquad 1.159$$

where s and t are constants and

$$V_1 = a_1 i + b_1 j + c_1 k \qquad 1.160$$
$$V_2 = a_2 i + b_2 j + c_2 k \qquad 1.161$$

If the intersection point $p_0 = (x_0, y_0, z_0)$ is known, then points in the plane can be represented by the parametric equations given below. Notice the similarity to the slope form of an equation for a straight line.

$$x = sa_1 + ta_2 + x_0 \qquad 1.162$$
$$y = sb_1 + tb_2 + y_0 \qquad 1.163$$
$$z = sc_1 + tc_2 + z_0 \qquad 1.164$$

The plane also is defined by its rectangular equations:

$$A(x - x_0) + B(y - y_0) + C(z - z_0) = 0 \qquad 1.165$$

or

$$Ax + By + Cz + D = 0 \qquad 1.166$$

where

$$D = -(Ax_0 + By_0 + Cz_0) \qquad 1.167$$

Constants A, B, and C are found from the cross product giving the normal vector **N**.

$$N = V_1 \times V_2 = Ai + Bj + Ck \qquad 1.168$$

Example 1.23

A plane is defined by a point $(2, 1, -4)$ and two vectors:

$$V_1 = (2i - 3j + k) \qquad V_2 = (2j - 4k)$$

Find the parametric and rectangular plane equations.

The parametric equations (for any values of s and t) are:

$$x = 2 + 2s$$
$$y = 1 - 3s + 2t$$
$$z = -4 + s - 4t$$

The normal vector is found by evaluating the determinant

$$N = \begin{vmatrix} i & 2 & 0 \\ j & -3 & 2 \\ k & 1 & -4 \end{vmatrix}$$

$$= i(12 - 2) - 2(-4j - 2k) = 10i + 8j + 4k$$

One form of the rectangular equation is

$$10(x - 2) + 8(y - 1) + 4(z + 4) = 0$$

Another form can be derived from equations 1.166 and 1.167

$$D = -[(10)(2) + (8)(1) + (4)(-4)] = -12$$
$$\mathbf{P} = 10x + 8y + 4z - 12 = 0$$

Three noncollinear points can be used to describe a plane with the following procedure:

step 1: Form vectors $\mathbf{V}_1$ and $\mathbf{V}_2$ from two pairs of the points.

step 2: Find the normal vector $\mathbf{N} = \mathbf{V}_1 \times \mathbf{V}_2$.

step 3: Write the rectangular form of the plane using A, B, and C from the normal vector and any one of the three points.

If the rectangular form of the plane is known, it can be used to write parametric equations. In this case, two of the three variables (x, y, z) replace the parameters s and t.

Example 1.24

Find the rectangular and parametric equations of a plane containing the following points: $(2, 1, -4)$; $(4, -2, -3)$; $(2, 3, -8)$.

Use the first two points to find $\mathbf{V}_1$:

$$\mathbf{V}_1 = (4 - 2)\mathbf{i} + (-2 - 1)\mathbf{j} + (-3 - (-4))\mathbf{k}$$
$$= 2\mathbf{i} - 3\mathbf{j} + \mathbf{k}$$

Similarly,

$$\mathbf{V}_2 = (2 - 2)\mathbf{i} + (3 - 1)\mathbf{j} + (-8 - (-4))\mathbf{k}$$
$$= 2\mathbf{j} - 4\mathbf{k}$$

From the previous example,

$$\mathbf{N} = 10\mathbf{i} + 8\mathbf{j} + 4\mathbf{k}$$
$$\mathbf{P} = 10x + 8y + 4z - 12 = 0$$

Dividing the rectangular form by 4 gives

$$2.5x + 2y + z - 3 = 0$$

or

$$z = 3 - 2y - 2.5x$$

Using x and y as the parameters, the parametric equations are

$$x = x$$
$$y = y$$
$$z = 3 - 2y - 2.5x$$

The angle between two planes is the same as the angle between their normal vectors, as calculated from the following equation:

$$\cos\phi = \frac{|\mathbf{N}_1 \cdot \mathbf{N}_2|}{|\mathbf{N}_1||\mathbf{N}_2|}$$
$$= \frac{|A_1 A_2 + B_1 B_2 + C_1 C_2|}{\sqrt{A_1^2 + B_1^2 + C_1^2}\sqrt{A_2^2 + B_2^2 + C_2^2}} \qquad 1.169$$

A vector equation of the line formed by the intersection of two planes is given by the cross product $(\mathbf{N}_1 \times \mathbf{N}_2)$. The distance from a point (x', y', z') to a plane is given by

$$d = \frac{Ax' + By' + Cz' + D}{\sqrt{A^2 + B^2 + C^2}} \qquad 1.170$$

11 CONIC SECTIONS

A. CIRCLE

The center-radius form of a circle with radius r and center at (h, k) is

$$(x - h)^2 + (y - k)^2 = r^2 \qquad 1.171$$

The x-intercept is found by letting $y = 0$ and solving for x. The y-intercept is found similarly.

The general form is

$$x^2 + y^2 + Dx + Ey + F = 0 \qquad 1.172$$

This can be converted to the center-radius form.

$$\left(x + \frac{D}{2}\right)^2 + \left(y + \frac{E}{2}\right)^2 = \frac{1}{4}(D^2 + E^2 - 4F) \qquad 1.173$$

If the right-hand side is greater than zero, the equation is that of a circle with center at $(-\frac{1}{2}D, -\frac{1}{2}E)$ and radius given by the square root of the right-hand side. If the right-hand side is zero, the equation is that of a point. If the right-hand side is negative, the plot is imaginary.

B. PARABOLA

A parabola is formed by a locus of points equidistant from point F and the *directrix*.

$$(y - k)^2 = 4p(x - h) \qquad 1.174$$

Equation 1.174 represents a parabola with *vertex* at (h, k), focus at $(p + h, k)$, and directrix equation $x = h - p$. The parabola points to the left if $p > 0$ and points to the right if $p < 0$.

$$(x - h)^2 = 4p(y - k) \qquad 1.175$$

Equation 1.175 represents a parabola with vertex at (h, k), focus at $(h, p + k)$, and directrix equation $y = k - p$. The parabola points down if $p > 0$ and points up if $p < 0$.

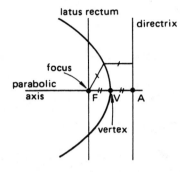

Figure 1.8 A Parabola

An alternate form of the vertically-oriented parabola is

$$Ax^2 + Bx + C = 0 \qquad 1.176$$

This parabola has a vertex at

$$\left(\frac{-B}{2A}, C - \frac{B^2}{4A}\right)$$

and points down if $A > 0$ and points up if $A < 0$.

C. ELLIPSE

An ellipse is formed from a locus of points such that the sum of distances from the two foci is constant. The distance between the two foci is $2c$. The sum of those distances is

$$F_1 P + P F_2 = 2a \qquad 1.177$$

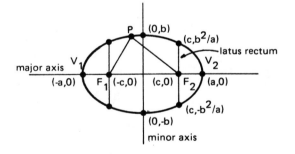

Figure 1.9 An Ellipse

The eccentricity of an ellipse is less than 1, and is equal to

$$e = \frac{\sqrt{a^2 - b^2}}{a} \qquad 1.178$$

For an ellipse centered at the origin,

$$\left(\frac{x}{a}\right)^2 + \left(\frac{y}{b}\right)^2 = 1 \qquad 1.179$$

$$b^2 = a^2 - c^2 \qquad 1.180$$

If $a > b$, the ellipse is wider than it is tall. If $a < b$, it is taller than it is wide.

For an ellipse centered at (h, k),

$$\frac{(x - h)^2}{a^2} + \frac{(y - k)^2}{b^2} = 1 \qquad 1.181$$

The general form of an ellipse is

$$Ax^2 + Cy^2 + Dx + Ey + F = 0 \qquad 1.182$$

If $A \neq C$ and both have the same sign, the general form can be written as

$$A\left(x + \frac{D}{2A}\right)^2 + C\left(y + \frac{E}{2C}\right)^2 = M \qquad 1.183$$

$$M = \frac{D^2}{4A} + \frac{E^2}{4C} - F \qquad 1.184$$

If $M = 0$, the graph is a single point at

$$\left(\frac{-D}{2A}, \frac{-E}{2C}\right)$$

If $M < 0$, the graph is the null set.

If $M > 0$, then the ellipse is centered at

$$\left(-\frac{D}{2A}, -\frac{E}{2C}\right)$$

and the equation can be rewritten

$$\frac{\left(x + \frac{D}{2A}\right)^2}{\frac{M}{A}} + \frac{\left(y + \frac{E}{2C}\right)^2}{\frac{M}{C}} = 1 \qquad 1.185$$

D. HYPERBOLA

A hyperbola is a locus of points such that $F_1 P - P F_2 = 2a$. The distance between the foci is $2c$, and $a < c$.

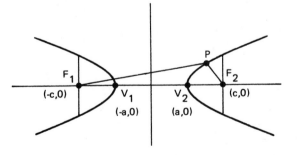

Figure 1.10 A Hyperbola

For a hyperbola centered at the origin with foci on the x-axis,

$$\left(\frac{x}{a}\right)^2 - \left(\frac{y}{b}\right)^2 = 1 \quad \text{with} \quad b^2 = c^2 - a^2 \qquad 1.186$$

If the foci are on the y-axis,

$$\left(\frac{y}{a}\right)^2 - \left(\frac{x}{b}\right)^2 = 1 \qquad \textbf{1.187}$$

The coordinates and length of the *latus recta* are the same as for the ellipse. The hyperbola is asymptotic to the lines

$$y = \pm\left(\frac{b}{a}\right)x \qquad \textbf{1.188}$$

The asymptotes need not be perpendicular, but if they are, the hyperbola is known as a *rectangular hyperbola*. If the asymptotes are the x and y axes, the equation of the hyperbola is

$$xy = \pm a^2 \qquad \textbf{1.189}$$

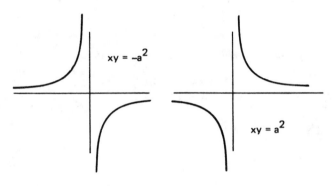

Figure 1.11 Rectangular Hyperbolas

In general, for a hyperbola with transverse axis parallel to the x-axis and center at (h, k),

$$\frac{(x-h)^2}{a^2} - \frac{(y-k)^2}{b^2} = 1 \qquad \textbf{1.190}$$

The general form of the hyperbolic equation is

$$Ax^2 + Cy^2 + Dx + Ey + F = 0 \qquad \textbf{1.191}$$

If $AC < 0$, the equation can be rewritten as

$$A\left(x + \frac{D}{2A}\right)^2 + C\left(y + \frac{E}{2C}\right)^2 = M \qquad \textbf{1.192}$$

where

$$M = \frac{D^2}{4A} + \frac{E^2}{4C} - F \qquad \textbf{1.193}$$

If $M = 0$, the graph is two intersecting lines.

If $M \neq 0$, the graph is a hyperbola with center at

$$\left(-\frac{D}{2A}, \quad -\frac{E}{2C}\right)$$

The transverse axis is horizontal if (M/A) is positive. It is vertical if (M/C) is positive.

12 SPHERES

The equation of a sphere whose center is at the point (h, k, l) and whose radius is r is

$$(x - h)^2 + (y - k)^2 + (z - l)^2 = r^2 \qquad \textbf{1.194}$$

If the sphere is centered at the origin, its equation is

$$x^2 + y^2 + z^2 = r^2 \qquad \textbf{1.195}$$

13 PERMUTATIONS AND COMBINATIONS

Suppose you have n objects, and you wish to work with a subset of r of them. An order-conscious arrangement of n objects taken r at a time is known as *permutation*. The permutation is said to be order-conscious because the arrangement of two objects (say A and B) as AB is different from the arrangement BA. There are a number of ways of taking n objects r at a time. The total number of possible permutations is

$$P(n, r) = \frac{n!}{(n - r)!} \qquad \textbf{1.196}$$

Example 1.25

A shelf has room for only three vases. If four different vases are available, how many ways can the shelf be arranged?

$$P(4, 3) = \frac{4!}{(4 - 3)!} = \frac{(4)(3)(2)(1)}{(1)} = 24$$

The special cases of n objects taken n at a time are illustrated by the following examples.

Example 1.26

How many ways can seven resistors be connected end-to-end into a single unit?

$$P(7, 7) = \frac{7!}{(7 - 7)!} = \frac{7!}{0!} = 7! = 5040$$

Example 1.27

Five people are to sit at a round table with five chairs. How many ways can these five people be arranged so that they all have different companions?

This is known as a *ring permutation*. Since the starting point of the arrangement around the circle does not affect the number of permutations, the answer is

$$(5 - 1)! = 4! = 24$$

An arrangement of n objects taken r at a time is known as a *combination* if the arrangement is not order-conscious. The total number of possible combinations is

$$C(n, r) = \frac{n!}{(n-r)!r!} \qquad 1.197$$

Example 1.28

How many possible ways can six people fit into a four-seat boat?

$$C(6, 4) = \frac{6!}{(6-4)!4!} = \frac{(6)(5)(4)(3)(2)(1)}{(2)(1)(4)(3)(2)(1)} = 15$$

14 PROBABILITY AND STATISTICS

A. PROBABILITY RULES

The following rules are applied to sample spaces **A** and **B**:

$$\mathbf{A} = [A_1, A_2, A_3, \ldots, A_n] \text{ and } \mathbf{B} = [B_1, B_2, B_3, \ldots, B_n]$$

where the A_i and B_i are independent.

Rule 1:

$$p\{\emptyset\} = \text{probability of an impossible event} = 0 \quad 1.198$$

Example 1.29

An urn contains five white balls, two red balls, and three green balls. What is the probability of drawing a blue ball from the urn?

$$p\{\text{blue ball}\} = p\{\emptyset\} = 0$$

Rule 2:

$$p\{A_1 \text{ or } A_2 \text{ or } \ldots \text{ or } A_n\}$$
$$= p\{A_1\} + p\{A_2\} + \cdots + p\{A_n\} \qquad 1.199$$

Example 1.30

Returning to the urn described in example 1.29, what is the probability of getting either a white ball or a red ball in one draw from the urn?

$$p\{\text{red or white}\} = p\{\text{red}\} + p\{\text{white}\} = .2 + .5 = .7$$

Rule 3:

$$p\{A_i \text{ and } B_i \text{ and} \ldots Z_i\} = \qquad 1.200$$
$$p\{A_i\}p\{B_i\}\ldots p\{Z_i\}$$

Example 1.31

Given two identical urns (as described in example 1.29), what is the probability of getting a red ball from the

first urn and a green ball from the second urn, given one draw from each urn?

$$p\{\text{red and green}\} = p\{\text{red}\}\, p\{\text{green}\} = (.2)(.3) = .06$$

Rule 4:

$$p\{\text{not A}\} = \text{probability of event A not occurring}$$
$$= 1 - p\{A\} \qquad 1.201$$

Example 1.32

Given the urn of example 1.29, what is the probability of not getting a red ball from the urn in one draw?

$$p\{\text{not red}\} = 1 - p\{\text{red}\} = 1 - .2 = .8$$

Rule 5:

$$p\{A_i \text{ or } B_i\} = p\{A_i\} + p\{B_i\} - p\{A_i\}p\{B_i\} \quad 1.202$$

Example 1.33

Given one urn as described in example 1.29 and a second urn containing eight red balls and two black balls, what is the probability of drawing either a white ball from the first urn or a red ball from the second urn, given one draw from each?

$$p\{\text{white or red}\} = p\{\text{white}\} + p\{\text{red}\}$$
$$- p\{\text{white}\}p\{\text{red}\}$$
$$= .5 + .8 - (.5)(.8) = .9$$

Rule 6:

$$p\{A|B\} = \text{probability that A will occur given that}$$
B has already occurred, where the two events are dependent.

$$= \frac{p\{A \text{ and } B\}}{p\{B\}} \qquad 1.203$$

The above equation is known as *Bayes Theorem*.

B. PROBABILITY DENSITY FUNCTIONS

Probability density functions are mathematical functions giving the probabilities of numerical events. A *numerical event* is any occurrence that can be described by an integer or real number. For example, obtaining heads in a coin toss is not a numerical event. However, a concrete sample having a compressive strength less than 5000 psi is a numerical event.

Discrete density functions give the probability that the event x will occur. That is,

$$f\{x\} = \text{probability of a process having a value of } x$$

Important discrete functions are the binomial and Poisson distributions.

1 Binomial

n is the number of trials

x is the number of successes

p is the probability of a success in a single trial

q is the probability of failure, $1 - p$

$\binom{n}{x}$ is the binomial coefficient $= \frac{n!}{(n-x)!x!}$

$x! = x(x-1)(x-2)\ldots(2)(1)$

Then, the probability of obtaining x successes in n trials is

$$f\{x\} = \binom{n}{x}p^x q^{(n-x)} \qquad 1.204$$

The mean of the binomial distribution is np. The variance of the distribution is npq.

Example 1.34

In a large quantity of items, 5% are defective. If seven items are sampled, what is the probability that exactly three will be defective?

$$f\{3\} = \binom{7}{3}(.05)^3(.95)^4 = .0036$$

2 Poisson

Suppose an event occurs, on the average, λ times per period. The probability that the event will occur x times per period is

$$f\{x\} = \frac{e^{-\lambda}\lambda^x}{x!} \qquad 1.205$$

λ is both the distribution mean and the variance. λ must be a number greater than zero.

Example 1.35

The number of customers arriving in some period is distributed as Poisson with a mean of eight. What is the probability that six customers will arrive in any given period?

$$f\{6\} = \frac{e^{-8}8^6}{6!} = .122$$

Continuous probability density functions are used to find the cumulative distribution functions, $F\{x\}$. Cumulative distribution functions give the probability of event x or less occurring.

$x = $ any value, not necessarily an integer

$$f\{x\} = \frac{dF\{x\}}{dx} \qquad 1.206$$

$F\{x\} = $ probability of x or less occurring

3 Exponential

$$f\{x\} = u(e^{-ux}) \qquad 1.207$$
$$F\{x\} = 1 - e^{-ux} \qquad 1.208$$

The mean of the exponential distribution is $\frac{1}{u}$. The variance is $\left(\frac{1}{u}\right)^2$.

Example 1.36

The reliability of a unit is exponentially distributed with mean time to failure (MTBF) of 1000 hours. What is the probability that the unit will be operational at $t = 1200$ hours?

The reliability of an item is $(1 - $ probability of failing before time $t)$. Therefore,

$$R\{t\} = 1 - F\{t\} = 1 - (1 - e^{-ux}) = e^{-ux}$$
$$u = \frac{1}{\text{MTBF}} = \frac{1}{1000} = .001$$
$$R\{1200\} = e^{-(.001)(1200)} = .3$$

4 Normal

Although $f\{x\}$ can be expressed mathematically for the normal distribution, tables are used to evaluate $F\{x\}$ since $f\{x\}$ cannot be easily integrated. Since the x axis of the normal distribution will seldom correspond to actual sample variables, the sample values are converted into standard values. Given the mean, u, and the standard deviation, σ, the standard normal variable is

$$z = \frac{\text{sample value} - u}{\sigma} \qquad 1.209$$

Then, the probability of a sample exceeding the given sample value is equal to the area in the tail past point z.

Example 1.37

Given a population that is normally distributed with mean of 66 and standard deviation of five, what percent of the population exceeds 72?

$$z = \frac{72 - 66}{5} = 1.2$$

Then, from table 1.5,

$$p\{\text{exceeding } 72\} = .5 - .3849 = .1151 \text{ or } 11.5\%$$

C. STATISTICAL ANALYSIS OF EXPERIMENTAL DATA

Experiments can take on many forms. An experiment might consist of measuring the weight of one cubic foot of concrete. Or, an experiment might consist of measuring the speed of a car on a roadway. Generally, such experiments are performed more than once to increase the precision and accuracy of the results.

Of course, the intrinsic variability of the process being measured will cause the observations to vary, and we would not expect the experiment to yield the same result each time it was performed. Eventually, a collection of experimental outcomes (observations) will be available for analysis.

One fundamental technique for organizing random observations is the *frequency distribution*. The frequency distribution is a systematic method for ordering the observations from small to large, according to some convenient numerical characteristic.

Example 1.38

The number of cars that travel through an intersection between 12 noon and 1 p.m. is measured for 30 consecutive working days. The results of the 30 observatons are:

79,66,72,70,68,66,68,76,73,71,74,70,71,69,67,

74,70,68,69,64,75,70,68,69,64,69,62,63,63,61

What is the frequency distribution using an interval of 2 cars per hour?

cars per hour	frequency of occurrence
60-61	1
62-63	3
64-65	2
66-67	3
68-69	8
70-71	6
72-73	2
74-75	3
76-77	1
78-79	1

In example 1.38, two cars per hour is known as the *step interval*. The step interval should be chosen so that the data is presented in a meaningful manner. If there are too many intervals, many of them will have zero frequencies. If there are too few intervals, the frequency distribution will have little value. Generally, 10 to 15 intervals are used.

Once the frequency distribution is complete, it can be represented graphically as a histogram. The procedure in drawing a histogram is to mark off the interval limits on a number line and then draw bars with lengths that are proportional to the frequencies in the intervals. If it is necessary to show the continuous nature of the data, a frequency polygon can be drawn.

Example 1.39

Draw the frequency histogram and frequency polygon for the data given in example 1.38.

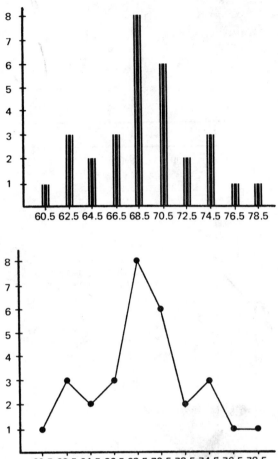

If it is necessary to know the number or percentage of observations that occur up to and including some value, the cumulative frequency table can be formed. This procedure is illustrated in the following example.

Example 1.40

Form the cumulative frequency distribution and graph for the data given in example 1.38.

cars per hour	frequency	cumulative frequency	cumulative percent
60-61	1	1	3
62-63	3	4	13
64-65	2	6	20
66-67	3	9	30
68-69	8	17	57
70-71	6	23	77
72-73	2	25	83
74-75	3	28	93
76-77	1	29	97
78-79	1	30	100

Table 1.5
Areas Under The Standard Normal Curve
(0 to z)

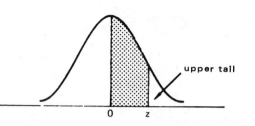

z	0	1	2	3	4	5	6	7	8	9
0.0	.0000	.0040	.0080	.0120	.0160	.0199	.0239	.0279	.0319	.0359
0.1	.0398	.0438	.0478	.0517	.0557	.0596	.0636	.0675	.0714	.0754
0.2	.0793	.0832	.0871	.0910	.0948	.0987	.1026	.1064	.1103	.1141
0.3	.1179	.1217	.1255	.1293	.1331	.1368	.1406	.1443	.1480	.1517
0.4	.1554	.1591	.1628	.1664	.1700	.1736	.1772	.1808	.1844	.1879
0.5	.1915	.1950	.1985	.2019	.2054	.2088	.2123	.2157	.2190	.2224
0.6	.2258	.2291	.2324	.2357	.2389	.2422	.2454	.2486	.2518	.2549
0.7	.2580	.2612	.2642	.2673	.2704	.2734	.2764	.2794	.2823	.2852
0.8	.2881	.2910	.2939	.2967	.2996	.3023	.3051	.3078	.3106	.3133
0.9	.3159	.3186	.3212	.3238	.3264	.3289	.3315	.3340	.3365	.3389
1.0	.3413	.3438	.3461	.3485	.3508	.3531	.3554	.3577	.3599	.3621
1.1	.3643	.3665	.3686	.3708	.3729	.3749	.3770	.3790	.3810	.3830
1.2	.3849	.3869	.3888	.3907	.3925	.3944	.3962	.3980	.3997	.4015
1.3	.4032	.4049	.4066	.4082	.4099	.4115	.4131	.4147	.4162	.4177
1.4	.4192	.4207	.4222	.4236	.4251	.4265	.4279	.4292	.4306	.4319
1.5	.4332	.4345	.4357	.4370	.4382	.4394	.4406	.4418	.4429	.4441
1.6	.4452	.4463	.4474	.4484	.4495	.4505	.4515	.4525	.4535	.4545
1.7	.4554	.4564	.4573	.4582	.4591	.4599	.4608	.4616	.4625	.4633
1.8	.4641	.4649	.4656	.4664	.4671	.4678	.4686	.4693	.4699	.4706
1.9	.4713	.4719	.4726	.4732	.4738	.4744	.4750	.4756	.4761	.4767
2.0	.4772	.4778	.4783	.4788	.4793	.4798	.4803	.4808	.4812	.4817
2.1	.4821	.4826	.4830	.4834	.4838	.4842	.4846	.4850	.4854	.4857
2.2	.4861	.4864	.4868	.4871	.4875	.4878	.4881	.4884	.4887	.4890
2.3	.4893	.4896	.4898	.4901	.4904	.4906	.4909	.4911	.4913	.4916
2.4	.4918	.4920	.4922	.4925	.4927	.4929	.4931	.4932	.4934	.4936
2.5	.4938	.4940	.4941	.4943	.4945	.4946	.4948	.4949	.4951	.4952
2.6	.4953	.4955	.4956	.4957	.4959	.4960	.4961	.4962	.4963	.4964
2.7	.4965	.4966	.4967	.4968	.4969	.4970	.4971	.4972	.4973	.4974
2.8	.4974	.4975	.4976	.4977	.4977	.4978	.4979	.4979	.4980	.4981
2.9	.4981	.4982	.4982	.4983	.4984	.4984	.4985	.4985	.4986	.4986
3.0	.4987	.4987	.4987	.4988	.4988	.4989	.4989	.4989	.4990	.4990
3.1	.4990	.4991	.4991	.4991	.4992	.4992	.4992	.4992	.4993	.4993
3.2	.4993	.4993	.4994	.4994	.4994	.4994	.4994	.4995	.4995	.4995
3.3	.4995	.4995	.4995	.4996	.4996	.4996	.4996	.4996	.4996	.4997
3.4	.4997	.4997	.4997	.4997	.4997	.4997	.4997	.4997	.4997	.4998
3.5	.4998	.4998	.4998	.4998	.4998	.4998	.4998	.4998	.4998	.4998
3.6	.4998	.4998	.4999	.4999	.4999	.4999	.4999	.4999	.4999	.4999
3.7	.4999	.4999	.4999	.4999	.4999	.4999	.4999	.4999	.4999	.4999
3.8	.4999	.4999	.4999	.4999	.4999	.4999	.4999	.4999	.4999	.4999
3.9	.5000	.5000	.5000	.5000	.5000	.5000	.5000	.5000	.5000	.5000

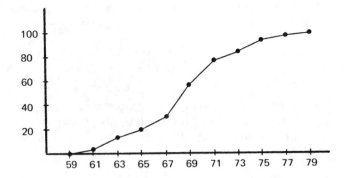

It is often unnecessary to present the experimental data in its entirety, either in tabular or graphical form. In such cases, the data and distribution can be represented by various parameters. One type of parameter is a measure of *central tendency*. Mode, median, and mean are measures of central tendency. The other type of parameter is a measure of dispersion. Standard deviation and variance are measures of dispersion.

The *mode* is the observed value which occurs most frequently. The mode may vary greatly between series of observations. Therefore, its main use is as a quick measure of the central value since no computation is required to find it. Beyond this, the usefulness of the mode is limited.

The *median* is the point in the distribution which divides the total observations into two parts containing equal numbers of observations. It is not influenced by the extremity of scores on either side of the distribution. The median is found by counting up (from either end of the frequency distribution) until half of the observations have been accounted for. The procedure is more difficult if the median falls within an interval, as illustrated in example 1.41.

Similar in concept to the median are *percentile ranks*, *quartiles*, and *deciles*. The median could also have been called the *50th percentile* observation. Similarly, the 80th percentile would be the number of cars per hour for which the cumulative frequency was 80%. The quartile and decile points on the distribution divide the observations or distribution into segments of 25% and 10%, respectively.

The *arithmetic mean* is the arithmetic average of the observations. The *mean* may be found without ordering the data (which was necessary to find the mode and median). The mean can be found from the following formula:

$$\bar{x} = \left(\frac{1}{n}\right)(x_1 + x_2 + \cdots + x_n) = \frac{\sum x_i}{n} \qquad 1.210$$

The *geometric mean* is used occasionally when it is necessary to average ratios. The geometric mean is calculated as

$$\text{geometric mean} = \sqrt[n]{x_1 x_2 x_3 \ldots x_n} \qquad 1.211$$

The *harmonic mean* is defined as

$$\text{harmonic mean} = \frac{n}{\frac{1}{x_1} + \frac{1}{x_2} + \cdots + \frac{1}{x_n}} \qquad 1.212$$

The *root-mean-squared (rms) value* of a series of observations is defined as

$$x_{rms} = \sqrt{\frac{\sum x_i^2}{n}} \qquad 1.213$$

Example 1.41

Find the mode, median, and arithmetic mean of the distribution represented by the data given in example 1.38.

The mode is the interval 68-69, since this interval has the highest frequency. If 68.5 is taken as the interval center, then 68.5 would be the mode.

Since there are 30 observations, the median is the value which separates the observations into two groups of 15. From example 1.40, the median occurs someplace within the 68-69 interval. Up through interval 66-67, there are nine observations, so six more are needed to make 15. Interval 68-69 has eight observations, so the median is found to be $\left(\frac{6}{8}\right)$ or $\left(\frac{3}{4}\right)$ of the way through the interval. Since the real limits of the interval are 67.5 and 69.5, the median is located at

$$67.5 + \frac{3}{4}(69.5 - 67.5) = 69$$

The mean can be found from the raw data or from the grouped data using the interval center as the assumed observation value. Using the raw data,

$$\bar{x} = \frac{\sum x}{n} = \frac{2069}{30} = 68.97$$

The simplest statistical parameter which describes the variation in observed data is the *range*. The range is found by subtracting the smallest value from the largest. Since the range is influenced by extreme (low probability) observations, its use as a measure of variability is limited.

The *standard deviation* is a better estimate of variability because it considers every observation. The standard deviation can be found from:

$$\sigma = \sqrt{\frac{\sum (x_i - \bar{x})^2}{n}} = \sqrt{\frac{\sum x_i^2}{n} - (\bar{x})^2} \qquad 1.214$$

The above formula assumes that n is a large number, such as above 50. Theoretically, n is the size of the entire population. If a small sample (less than 50) is used to calculate the standard deviation of the distribution, the formulas are changed. The *sample standard deviation* is

$$s = \sqrt{\frac{\sum (x_i - \bar{x})^2}{n-1}} = \sqrt{\frac{\sum x_i^2 - \frac{(\sum x_i)^2}{n}}{n-1}} \qquad 1.215$$

The difference is small when n is large, but care must be taken in reading the problem. If the *standard deviation of the sample* is requested, calculate σ. If an estimate of the *population standard deviation* or *sample standard deviation* is requested, calculate s. (Note that the standard deviation of the sample is not the same as the sample standard deviation.)

The *relative dispersion* is defined as a measure of dispersion divided by a measure of central tendency. The *coefficient of variation* is a relative dispersion calculated from the standard deviation and the mean. That is,

$$\text{coefficient of variation} = \frac{s}{\bar{x}} \qquad 1.216$$

Skewness is a measure of a frequency distribution's lack of symmetry. It is calculated as

$$\text{skewness} = \frac{\bar{x} - mode}{s} \qquad 1.217$$
$$\approx \frac{3(\bar{x} - \text{median})}{s} \qquad 1.218$$

Example 1.42

Calculate the range, standard deviation of the sample, and population variance from the data given in example 1.38.

$$\sum x = 2069 \quad \left(\sum x\right)^2 = 4280761 \quad \sum x^2 = 143225$$

$$n = 30 \quad \bar{x} = 68.97$$

$$\sigma = \sqrt{\frac{143225}{30} - (68.97)^2} = 4.16$$

$$s = \sqrt{\frac{143225 - \frac{(4280761)}{30}}{29}} = 4.29$$

$$s^2 = 18.4 \quad \text{(sample variance)}$$

$$\sigma^2 = 17.3 \quad \text{(population variance)}$$

$$R = 79 - 61 = 18$$

Referring again to example 1.38, suppose that the hourly through-put for 15 similar intersections is measured over a 30 day period. At the end of the 30 day period, there will be 15 ranges, 15 medians, 15 means, 15 standard deviations, and so on. These parameters themselves constitute distributions.

The mean of the sample means is an excellent estimator of the average hourly through-put of an intersection, μ.

$$\mu = \left(\frac{1}{15}\right) \sum \bar{x}$$

The standard deviation of the sample means is known as the *standard error of the mean* to distinguish it from the standard deviation of the raw data. The standard error is written as $\sigma_{\bar{x}}$.

The standard error is not a good estimator of the population standard deviation, σ'.

In general, if k sets of n observations each are used to estimate the population mean (μ) and the population standard deviation (σ'), then

$$\mu \approx \left(\frac{1}{k}\right) \sum \bar{x} \qquad 1.219$$

$$\sigma' \approx \sqrt{k} \sigma_{\bar{x}} \qquad 1.220$$

15 BASIC HYPOTHESIS TESTING

Suppose a distribution is $\approx N(\mu, \sigma'^2)$.[1] If samples of size n are taken k times, the values of the sample means $\bar{x}$ will form a distribution themselves. These means also will be distributed normally with the form

$$\sim N\left(\mu, \frac{\sigma'^2}{n}\right)$$

That is, the mean of the sample means will be identical to the original population, but the variance and standard deviation will be much smaller. This is known as the *Central Limit Theorem*.

Thus, the probability that $\bar{x}$ exceeds some value, say x^*, is

$$P\left\{z > \frac{x^* - \mu}{\frac{\sigma'}{\sqrt{n}}}\right\} \qquad 1.221$$

This can be solved as an *exceedance problem* (see example 1.37), or a hypothesis test can be performed. A *hypothesis test* has the following characteristics:

- a sample is taken in an experiment
- a parameter (usually $\bar{x}$) is measured
- it is desired to know if the sample could have come from a population $\sim N(\mu, \sigma'^2)$

---

[1] This is the standard method of saying the distribution is normally distributed with mean μ and variance σ'^2.

There are many types of hypothesis tests, depending on the type of population (i.e., whether or not normal), the parameter being tested (i.e., central tendency or dispersion), and the size of the sample.

If the sample size is not much greater than 30, if the native population is assumed to be normal, and if μ and σ' are known, the following procedure can be used.

step 1: Assume random sampling from a normal population.

step 2: Choose the desired confidence level, C. Usually, a 95% confidence level result is said to be *significant*. 99% test results are said to be *highly significant*.

step 3: Decide on a 1-tail or 2-tail test. If the question is worded as "Has the population mean changed?" or "Are the populations the same?", then a *2-tail test* is needed. If the question is "Has the mean increased?" or "...decreased?", then a *1-tail test* is needed.

step 4: From the normal table, find the value z' for a table entry equal to

$$\frac{1-C}{\#\text{tails in the test}} \qquad 1.222$$

step 5: Calculate

$$z = \left| \frac{\overline{x} - \mu}{\frac{\sigma'}{\sqrt{n}}} \right| \qquad 1.223$$

If $z \geqslant z'$, then the distributions are not the same.

Example 1.43

When operating properly, a chemical plant has a product output which is normally distributed with mean 880 tons/day and standard deviation of 21 tons. The output is measured on 50 consecutive days, and the mean output is 871 tons/day. Is the plant operating correctly?

step 1: Assume random sampling from the normal distribution

step 2: Choose $C = .95$ for significant results.

step 3: Wanting to know if the plant is operating correctly is the same as asking, "Has anything changed?" There is no mention of *direction* (i.e., the question was not, "Has the output decreased?"). Therefore, choose a 2-tail test.

step 4: $1 - C = .05$. The .05 outside lower limit in **table 1.5 is $z' = 1.96$. This corresponds to an area under the curve of 0.5 − 0.025 = 0.475.**

step 5:

$$z = \left| \frac{871 - 880}{\frac{21}{\sqrt{50}}} \right| = 3.03$$

Since $3.03 > 1.96$, the distributions are not the same. There is a 95% chance that the plant is not operating correctly.

16 DIFFERENTIAL CALCULUS

A. TERMINOLOGY

Given y, a function of x, the first derivative with respect to x may be written as

$$\mathbf{D}y, \; y', \; \text{or} \; \left(\frac{dy}{dx} \right)$$

The first derivative corresponds to the slope of the line described by the function y. The second derivative may be written as

$$\mathbf{D}^2 y, \; \left(\frac{d^2 y}{dx^2} \right), \; \text{or} \; y''$$

B. BASIC OPERATIONS

In the formulas that follow, f and g are functions of x. $\mathbf{D}$ is the derivative operator. a is a constant.

$$\mathbf{D}(a) = 0 \qquad 1.224$$
$$\mathbf{D}(af) = a\mathbf{D}(f) \qquad 1.225$$
$$\mathbf{D}(f + g) = \mathbf{D}(f) + \mathbf{D}(g) \qquad 1.226$$
$$\mathbf{D}(f - g) = \mathbf{D}(f) - \mathbf{D}(g) \qquad 1.227$$
$$\mathbf{D}(f \cdot g) = f\mathbf{D}(g) + g\mathbf{D}(f) \qquad 1.228$$
$$\mathbf{D}\left(\frac{f}{g} \right) = \frac{g\mathbf{D}(f) - f\mathbf{D}(g)}{g^2} \qquad 1.229$$
$$\mathbf{D}(x^n) = nx^{n-1} \qquad 1.230$$
$$\mathbf{D}(f^n) = nf^{n-1}\mathbf{D}(f) \qquad 1.231$$
$$\mathbf{D}(f(g)) = \frac{df(g)}{dg}\mathbf{D}(g) \qquad 1.232$$
$$\mathbf{D}(lnx) = \frac{1}{x} \qquad 1.233$$
$$\mathbf{D}(e^{ax}) = ae^{ax} \qquad 1.234$$

Example 1.44

A function is given as $f(x) = x^3 - 2x$. What is the slope of the line at $x = 3$?

$$y' = 3x^2 - 2$$
$$y'(3) = 27 - 2 = 25$$

C. TRANSCENDENTAL FUNCTIONS

$$\mathbf{D}(\sin x) = \cos x \qquad\qquad 1.235$$

$$\mathbf{D}(\cos x) = -\sin x \qquad\qquad 1.236$$

$$\mathbf{D}(\tan x) = \sec^2 x \qquad\qquad 1.237$$

$$\mathbf{D}(\cot x) = -\csc^2 x \qquad\qquad 1.238$$

$$\mathbf{D}(\sec x) = (\sec x)(\tan x) \qquad\qquad 1.239$$

$$\mathbf{D}(\csc x) = (-\csc x)(\cot x) \qquad\qquad 1.240$$

$$\mathbf{D}(\arcsin x) = \frac{1}{\sqrt{1-x^2}} \qquad\qquad 1.241$$

$$\mathbf{D}(\arctan x) = \frac{1}{(1+x^2)} \qquad\qquad 1.242$$

$$\mathbf{D}(\text{arcsec } x) = \frac{1}{x\sqrt{x^2-1}} \qquad\qquad 1.243$$

$$\mathbf{D}(\arccos x) = -\mathbf{D}(\arcsin x) \qquad\qquad 1.244$$

$$\mathbf{D}(\text{arccot } x) = -\mathbf{D}(\arctan x) \qquad\qquad 1.245$$

$$\mathbf{D}(\text{arccsc } x) = -\mathbf{D}(\text{arcsec } x) \qquad\qquad 1.246$$

D. VARIATIONS ON DIFFERENTIATION

1 Partial Differentiation

If the function has two or more independent variables, a partial derivative is found by considering all extraneous variables as constants. The geometric interpretation of the partial derivative $(\partial z/\partial x)$ is the slope of a line tangent to the 3-dimensional surface in a plane of constant y and parallel to the x axis. Similarly, the interpretation of $(\partial z/\partial y)$ is the slope of a line tangent to the surface in a plane of constant x and parallel to the y axis.

Example 1.45

A surface has the equation $x^2 + y^2 + z^2 = 9$. What is the slope of a line tangent to (1,2,2) and parallel to the x axis?

$$z = \sqrt{9 - x^2 - y^2}$$
$$\frac{\partial z}{\partial x} = \frac{-x}{\sqrt{9 - x^2 - y^2}}$$

At the point (1,2,2),

$$\frac{\partial z}{\partial x} = -\frac{1}{2}$$

2 Implicit Differentiation

If a relationship between n variables cannot be manipulated to yield an explicit function of $(n-1)$ independent variables, the relationship implicitly defines the nth remaining variable. The derivative of the implicit variable taken with respect to any other variable is found by a process known as implicit differentiation.

If $f(x, y) = 0$ is a function, the implicit derivative is

$$\frac{dy}{dx} = -\frac{\partial f}{\partial x} \bigg/ \frac{\partial f}{\partial y} \qquad\qquad 1.247$$

If $f(x, y, z) = 0$ is a function, the implicit derivatives are

$$\frac{\partial z}{\partial x} = -\frac{\partial f}{\partial x} \bigg/ \frac{\partial f}{\partial z} \qquad\qquad 1.248$$

$$\frac{\partial z}{\partial y} = -\frac{\partial f}{\partial y} \bigg/ \frac{\partial f}{\partial z} \qquad\qquad 1.249$$

Example 1.46

If $f = x^2 + xy + y^3$, what is $\frac{dy}{dx}$?

Since this function cannot be written as an explicit function of x, implicit differentiation is required.

$$\frac{\partial f}{\partial x} = 2x + y$$
$$\frac{\partial f}{\partial y} = x + 3y^2$$
$$\frac{dy}{dx} = \frac{-(2x + y)}{(x + 3y^2)}$$

Example 1.47

Solve example 1.45 using implicit differentiation.

$$f = x^2 + y^2 + z^2 - 9$$
$$\frac{\partial f}{\partial x} = 2x$$
$$\frac{\partial f}{\partial z} = 2z$$
$$\frac{\partial z}{\partial x} = \frac{-2x}{2z} = -\frac{x}{z}$$

and at (1,2,2),

$$\frac{\partial z}{\partial x} = -\frac{1}{2}$$

3 The Gradient Vector

The slope of a function is defined as the change in one variable with respect to a distance in another direction. Usually, this direction is parallel to an axis. However, the maximum slope at a point on a 3-dimensional object may not be in a direction parallel to one of the coordinate axes.

The gradient vector function $\nabla f(x, y, z)$ (pronounced "del f") gives the maximum rate of change of the function f(x,y,z). The gradient vector function is defined as

$$\nabla f(x, y, z) = \frac{\partial f(x, y, z)}{\partial x}\mathbf{i} + \frac{\partial f(x, y, z)}{\partial y}\mathbf{j}$$
$$+ \frac{\partial f(x, y, z)}{\partial z}\mathbf{k} \qquad\qquad 1.250$$

Example 1.48

Find the maximum slope of $f(x, y) = 2x^2 - y^2 + 3x - y$ at the point $(1, -2)$. What is the equation of the maximum-slope tangent?

This is a 2-dimensional problem.

$$\frac{\partial f(x, y)}{\partial x} = 4x + 3$$

$$\frac{\partial f(x, y)}{\partial y} = -2y - 1$$

$$\nabla f(x, y) = (4x + 3)\mathbf{i} + (-2y - 1)\mathbf{j}$$

The equation of the maximum-slope tangent is

$$\nabla f(1, -2) = 7\mathbf{i} + 3\mathbf{j}$$

The magnitude of the slope is

$$\sqrt{(7)^2 + (3)^2} = \sqrt{58}$$

4 The Directional Derivative

The rate of change of a function in the direction of some given vector $\mathbf{U}$ can be found from the directional derivative function, $\nabla_u f(x, y, z)$. This directional derivative function depends on the gradient vector and the direction cosines of the vector $\mathbf{U}$.

$$\nabla_u f(x, y, z) = \frac{\partial f(x, y, z)}{\partial x} \cos \alpha + \frac{\partial f(x, y, z)}{\partial y} \cos \beta$$
$$+ \frac{\partial f(x, y, z)}{\partial z} \cos \gamma$$

$$1.251$$

Example 1.49

What is the rate of change of $f(x, y) = 3x^2 + xy - 2y^2$ at the point $(1, -2)$ in the direction $4\mathbf{i} + 3\mathbf{j}$?

$$\cos \alpha = \frac{4}{\sqrt{(4)^2 + (3)^2}} = \frac{4}{5}$$

$$\cos \beta = \frac{3}{5}$$

$$\frac{\partial f(x, y)}{\partial x} = 6x + y$$

$$\frac{\partial f(x, y)}{\partial y} = x - 4y$$

$$\nabla_u f(x, y) = \left(\frac{4}{5}\right)(6x + y) + \left(\frac{3}{5}\right)(x - 4y)$$

$$\nabla_u f(1, -2) = \left(\frac{4}{5}\right)[(6)(1) - 2] + \left(\frac{3}{5}\right)[1 - (4)(-2)]$$

$$= 8.6$$

5 Tangent Plane Function

Partial derivatives can be used to find the tangent plane to a 3-dimensional surface at some point p_o. If the surface is defined by the function $f(x, y, z) = 0$, the equation of the tangent plane is

$$T(x_o, y_o, z_o) = (x - x_o)\frac{\partial f(x, y, z)}{\partial x}\Big|_{p_o}$$

$$+ (y - y_o)\frac{\partial f(x, y, z)}{\partial y}\Big|_{p_o}$$

$$1.252$$

$$+ (z - z_o)\frac{\partial f(x, y, z)}{\partial z}\Big|_{p_o}$$

Example 1.50

What is the equation of the plane tangent to $f(x, y, z) = 4x^2 + y^2 - 16z$ at the point $(2,4,2)$?

$$\frac{\partial f(x, y, z)}{\partial x}\Big|_{p_o} = 8x\Big|_{(2,4,2)} = (8)(2) = 16$$

$$\frac{\partial f(x, y, z)}{\partial y}\Big|_{p_o} = 2y\Big|_{(2,4,2)} = (2)(4) = 8$$

$$\frac{\partial f(x, y, z)}{\partial z}\Big|_{p_o} = -16\Big|_{(2,4,2)} = -16$$

Therefore,

$$\mathbf{T}(2, 4, 2) = 16(x - 2) + 8(y - 4) - 16(z - 2)$$
$$= 2x + y - 2z - 4$$

6 Normal Line Function

Partial derivatives can be used to find the equation of a straight line normal to a 3-dimensional surface at some point p_o. If the surface is defined by the function $f(x, y, z) = 0$, the equation of the normal line is

$$\mathbf{N} = A\mathbf{i} + B\mathbf{j} + C\mathbf{k} \qquad 1.253$$

where

$$A = \frac{\partial f(x, y, z)}{\partial x}\Big|_{p_o} \qquad 1.254$$

$$B = \frac{\partial f(x, y, z)}{\partial y}\Big|_{p_o} \qquad 1.255$$

$$C = \frac{\partial f(x, y, z)}{\partial z}\Big|_{p_o} \qquad 1.256$$

7 Extrema and Optimization

Derivatives can be used to locate local *maxima, minima,* and *points of inflection.* No distinction is made between local and global extrema. The end points of the interval always should be checked against the local extrema located by the method below. The following rules define the extreme points.

$$f'(x) = 0 \text{ at any extrema}$$
$$f''(x) = 0 \text{ at an inflection point}$$
$$f''(x) \text{ is negative at a maximum}$$
$$f''(x) \text{ is positive at a minimum}$$

There is always an inflection point between a maximum and a minimum.

Example 1.51

Find the global extreme points of the function $f(x) = x^3 + x^2 - x + 1$ on the interval $[-2, +2]$.

$$f'(x) = 3x^2 + 2x - 1$$

$$f'(x) = 0 \text{ at } x = \frac{1}{3} \text{ and } x = -1$$

$$f''(x) = 6x + 2$$
$$f(-1) = 2$$
$$f''(-1) = -4$$

So, $x = -1$ is a maximum.

$$f\left(\frac{1}{3}\right) = \frac{22}{27}$$
$$f''\left(\frac{1}{3}\right) = +4$$

So, $x = -\frac{1}{3}$ is a minimum.

Checking the end points,
$$f(-2) = -1$$
$$f(+2) = +11$$

Therefore, the absolute extreme points are the end points.

17 INTEGRAL CALCULUS

A. FUNDAMENTAL THEOREM

The *Fundamental Theorem of Calculus* is

$$\int_{x_1}^{x_2} f'(x) = f(x_2) - f(x_1) \qquad 1.257$$

B. INTEGRATION BY PARTS

If f and g are functions, then

$$\int f \, dg = fg - \int g \, df \qquad 1.258$$

Example 1.52

Evaluate the following integral: $\int xe^x \, dx$

Use integration by parts.

Let $f = x$. Then, $df = dx$

Let $dg = e^x \, dx$. Then, $g = \int e^x \, dx = e^x$.

Therefore,

$$\int xe^x \, dx = xe^x - \int e^x \, dx + c$$
$$= xe^x - e^x + c$$

C. INDEFINITE INTEGRALS   ("··· + C" omitted)

$$\int dx = x \qquad 1.259$$

$$\int au \, dx = a \int u \, dx \qquad 1.260$$

$$\int (u + v)dx = \int u \, dx + \int v \, dx \qquad 1.261$$

$$\int x^m \, dx = \frac{x^{(m+1)}}{m+1} \qquad m \neq -1 \qquad 1.262$$

$$\int \frac{dx}{x} = ln|x| \qquad 1.263$$

$$\int e^{ax} \, dx = \frac{1}{a}e^{ax} \qquad 1.264$$

$$\int xe^{ax} \, dx = \frac{1}{a^2}e^{ax}(ax - 1) \qquad 1.265$$

$$\int \cosh x \, dx = \sinh x \qquad 1.266$$

$$\int \sinh x \, dx = \cosh x \qquad 1.267$$

$$\int \sin x \, dx = -\cos x \qquad 1.268$$

$$\int \cos x \, dx = \sin x \qquad 1.269$$

$$\int \tan x \, dx = ln(\sec x) \qquad 1.270$$

$$\int \cot x \, dx = ln(\sin x) \qquad 1.271$$

$$\int \sec x \, dx = ln(\sec x + \tan x) \qquad 1.272$$

$$\int \csc x \, dx = ln(\csc x - \cot x) \qquad 1.273$$

$$\int \frac{dx}{1+x^2} = \arctan x \qquad 1.274$$

$$\int \frac{dx}{\sqrt{1-x^2}} = \arcsin x \qquad 1.275$$

$$\int \frac{dx}{x\sqrt{x^2-1}} = \text{arcsec } x \qquad 1.276$$

D. USES OF INTEGRALS

1 Finding Areas

The area bounded by $x = a$, $x = b$, $f_1(x)$ above, and $f_2(x)$ below is given by

$$A = \int_a^b [f_1(x) - f_2(x)]dx \qquad \text{1.277}$$

2 Surfaces of Revolution

The surface area obtained by rotating $f(x)$ about the x axis is

$$A_s = 2\pi \int_a^b f(x)\sqrt{1 + [f'(x)]^2}\,dx \qquad \text{1.278}$$

3 Rotation of a Function

The volume of a function rotated about the x axis is

$$V = \pi \int_a^b (f(x))^2\,dx \qquad \text{1.279}$$

The volume of a function rotated about the y axis is

$$V = 2\pi \int_a^b xf(x)dx \qquad \text{1.280}$$

4 Length of a Curve

The length of a curve given by f(x) is

$$L = \int_a^b \sqrt{1 + (f'(x))^2}\,dx \qquad \text{1.281}$$

Example 1.53

For the shaded area shown, find (a) the area, and (b) the volume enclosed by the curve rotated about the x axis.

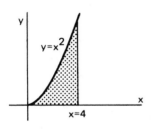

(a)
$$f_2(x) = 0 \qquad f_1(x) = x^2$$
$$A = \int_0^4 x^2\,dx = \left[\frac{x^3}{3}\right]_0^4 = 21.33$$

(b)
$$V = \pi \int_0^4 (x^2)^2\,dx = \pi\left[\frac{x^5}{5}\right]_0^4 = 204.8\pi$$

18 DIFFERENTIAL EQUATIONS

A differential equation is a mathematical expression containing a dependent variable and one or more of that variable's derivatives. First order differential equations contain only the first derivative of the dependent variable. Second order equations contain the second derivative.

The differential equation is said to be *linear* if all terms containing the dependent variable are multiplied only by real scalars. The equation is said to be *homogeneous* if there are no terms which do not contain the dependent variable or one of its derivatives.

Most differential equations are difficult to solve. However, there are several forms which are fairly simple. These are presented here.

A. FIRST ORDER LINEAR

The first order linear differential equation has the general form given by equation 1.282. $p(t)$ and $g(t)$ may be constants or any function of t.

$$x' + p(t)x = g(t) \qquad \text{1.282}$$

The solution depends on an *integrating factor* defined as

$$u = \exp[\int p(t)\,dt] \qquad \text{1.283}$$

The solution to the first order linear differential equation is

$$x = \frac{1}{u}[\int ug(t)\,dt + c] \qquad \text{1.284}$$

Example 1.54

Find a solution to the differential equation

$$x' - x = 2te^{2t} \qquad x(0) = 1$$

This meets the definition of a first order linear equation with

$$p(t) = -1 \text{ and } g(t) = 2te^{2t}$$

The integrating constant is

$$u = \exp[\int -1\,dt] = e^{-t}$$

Then, x is

$$x = \left(\frac{1}{e^{-t}}\right)[\int e^{-t}2te^{2t}dt + c]$$
$$= e^t[\int 2te^t dt + c]$$
$$= e^t[2te^t - 2e^t + c]$$

But, $x(0) = 1$, so

$$c = +3, \text{ and}$$
$$x = e^t[2e^t(t-1) + 3]$$

B. SECOND ORDER HOMOGENEOUS WITH CONSTANT COEFFICIENTS

This type of differential equation has the following general form:

$$c_1 x'' + c_2 x' + c_3 x = 0 \qquad 1.285$$

The solution can be found by first solving the characteristic quadratic equation for its roots k_1^* and k_2^*. This characteristic equation is derived directly from the differential equation:

$$c_1 k^2 + c_2 k + c_3 \qquad 1.286$$

The form of the solution depends on the values of k_1^* and k_2^*. If $k_1^* \neq k_2^*$ and both are real, then

$$x = a_1\left(e^{k_1^* t}\right) + a_2\left(e^{k_2^* t}\right) \qquad 1.287$$

If $k_1^* = k_2^*$, then

$$x = a_1\left(e^{k_1^* t}\right) + a_2 t\left(e^{k_2^* t}\right) \qquad 1.288$$

If $k^* = (r \pm iu)$, then

$$x = a_1(e^{rt})\cos(ut) + a_2(e^{rt})\sin(ut) \qquad 1.289$$

In all three cases, a_1 and a_2 must be found from the given initial conditions.

Example 1.55

Solve the following differential equation for x.

$$x'' + 6x' + 9x = 0 \quad x(0) = 0, \ x'(0) = 1$$

The characteristic equation is

$$k^2 + 6k + 9 = 0$$

This has roots of $k_1^* = k_2^* = -3$; therefore, the solution has the form

$$x(t) = a_1 e^{-3t} + a_2 t e^{-3t}$$

But, $x(0) = 0$,

$$0 = a_1(e^0) + a_2(0)(e^0)$$
$$0 = a_1(1) + 0$$
$$0 = a_1$$

Also, $x'(0) = 1$. The derivative of x(t) is

$$x'(t) = -3a_2 t e^{-3t} + a_2 e^{-3t}$$
$$1 = -3a_2(0)(e^0) + a_2 e^0$$
$$1 = 0 + a_2$$
$$1 = a_2$$

The final solution is

$$x = t e^{-3t}$$

19 LAPLACE TRANSFORMS

Traditional methods of solving non-homogeneous differential equations are very difficult. The Laplace transformation can be used to reduce the solution of many complex differential equations to simple algebra.

Every mathematical function can be converted into a Laplace function by use of the following transformation definition.

$$\mathcal{L}[f(t)] = \int_0^\infty e^{-st} f(t)dt \qquad 1.290$$

The variable s is equivalent to the derivative operator. However, it may be thought of as a simple variable.

Example 1.56

Let f(t) be the unit step. That is, $f(t) = 0$ for $t < 0$ and $f(t) = 1$ for $t \geqslant 0$.

Then, the Laplace transform of $f(t) = 1$ is

$$\mathcal{L}[f(t)] = \int_0^\infty e^{-st}(1)dt = -\frac{e^{-st}}{s}\Big]_0^\infty$$
$$= 0 - \left(-\frac{1}{s}\right) = \frac{1}{s}$$

Example 1.57

What is the Laplace transformation of $f(t) = e^{at}$?

$$\mathcal{L}[e^{at}] = \int_0^\infty e^{-st} e^{at} dt$$
$$= \int_0^\infty e^{-(s-a)t} dt$$
$$= -\frac{e^{-(s-a)t}}{(s-a)}\Big]_0^\infty$$
$$= \frac{1}{s-a}$$

Generally it is unnecessary to actually obtain a function's Laplace transform by use of equation 1.290. Tables of these transforms are readily available. A small collection of the most frequently required transforms is given at the end of this chapter.

The Laplace transform method can be used with any linear differential equation with constant coefficients. Assuming the dependent variable is x, the basic procedure is as follows:

step 1: Put the differential equation in standard form.

step 2: Use superposition and take the Laplace transform of each term.

step 3: Use the following relationships to expand terms.

$$\mathcal{L}(x'') = s^2 \mathcal{L}(x) - sx_0 - x_0' \qquad 1.291$$
$$\mathcal{L}(x') = s\mathcal{L}(x) - x_0 \qquad 1.292$$

step 4: Solve for $\mathcal{L}(x)$. Simplify the resulting expression using partial fractions.

step 5: Find x by applying the inverse transform.

This method reduces the solutions of differential equations to simple algebra. However, a complete set of transforms is required.

Working with Laplace transforms is simplified by the following two theorems:

Linearity Theorem: If c is constant, then

$$\mathcal{L}[cf(t)] = c\mathcal{L}[f(t)] \qquad 1.293$$

Superposition Theorem: If $f(t)$ and $g(t)$ are different functions, then

$$\mathcal{L}[f(t) \pm g(t)] = \mathcal{L}[f(t)] \pm \mathcal{L}[g(t)] \qquad 1.294$$

Example 1.58

Suppose the following differential equation results from the analysis of a mechanical system:

$$x'' + 2x' + 2x = \cos(t)$$
$$x_0 = 1, \ x_0' = 0$$

x is the dependent variable. Start by taking the Laplace transform of both sides:

$$\mathcal{L}(x'') + 2\mathcal{L}(x') + 2\mathcal{L}(x) = \mathcal{L}(\cos(t))$$
$$s^2\mathcal{L}(x) - sx_0 - x_0' + 2s\mathcal{L}(x) - 2x_0 + 2\mathcal{L}(x) = \mathcal{L}\cos(t)$$

But, $x_0 = 1$ and $x_0' = 0$. Also, the Laplace transform of $\cos(t)$ can be found from the appendix of this chapter.

$$s^2\mathcal{L}(x) - s + 2s\mathcal{L}(x) - 2 + 2\mathcal{L}(x) = \frac{s}{s^2+1}$$
$$\mathcal{L}(x)[s^2+2s+2] - s - 2 = \frac{s}{s^2+1}$$
$$\mathcal{L}(x) = \frac{s^3 + 2s^2 + 2s + 2}{(s^2+1)(s^2+2s+2)}$$

This is now expanded by partial fractions:

$$\frac{s^3 + 2s^2 + 2s + 2}{(s^2+1)(s^2+2s+2)}$$
$$= \frac{A_1 s + B_1}{s^2+1} + \frac{A_2 s + B_2}{s^2+2s+2}$$
$$= [s^3(A_1 + A_2) + s^2(2A_1 + B_1 + B_2)$$
$$+ s(2A_1 + 2B_1 + A_2) + 2B_1 + B_2]$$
$$\div [(s^2+1)(s^2+2s+2)]$$

The following simultaneous equations result:

$$\begin{aligned} A_1 + A_2 &= 1 \\ 2A_1 + B_1 + B_2 &= 2 \\ 2A_1 + A_2 + 2B_1 &= 2 \\ 2B_1 + B_2 &= 2 \end{aligned}$$

These equations have the solutions

$$A_1^* = \frac{1}{5} \quad A_2^* = \frac{4}{5}$$
$$B_1^* = \frac{2}{5} \quad B_2^* = \frac{6}{5}$$

Therefore, x can be found by taking the following inverse transform:

$$x = \mathcal{L}^{-1}\left[\frac{\frac{s}{5}+\frac{2}{5}}{s^2+1} + \frac{\frac{4s}{5}+\frac{6}{5}}{s^2+2s+2}\right]$$

The solution is

$$x = \frac{1}{5}\cos(t) + \frac{2}{5}\sin(t) + \frac{4}{5}e^{-t}\cos(t) + \frac{2}{5}e^{-t}\sin(t)$$

20 APPLICATIONS OF DIFFERENTIAL EQUATIONS

A. FLUID MIXTURE PROBLEMS

The typical fluid mixing problem involves a tank containing some liquid. There may be an initial solute in the liquid, or the liquid may be pure. Liquid and solute are added at known rates. A drain usually removes some of the liquid which is assumed to be thoroughly mixed. The problem is to find the weight or concentration of solute in the tank at some time t. The following symbols are used.

$C(t)$ concentration of solute in tank at time t
$I(t)$ liquid inflow rate from all sources at time t
k a constant
$\Phi(t)$ liquid outflow rate due to all drains at time t
$S_1(t)$ solute inflow rate at time t (this may have to be calculated from the incoming concentration and $I(t)$)
$S_2(t)$ solute outflow rate at time t

V_o original volume in tank at time = 0

$V(t)$ volume in tank at time = t (equal to $V_o + \int I(t)dt - \int \Phi(t)dt$)

W_o initial weight of solute in tank at $t = 0$

$W(t)$ weight of solute in tank at time = t

case 1: Constant Volume

$$W'(t) = S_1(t) - S_2(t)$$
$$= S_1(t) - \frac{\Phi(t)W(t)}{V_o} \qquad 1.295$$

The differential equation is

$$W'(t) + \frac{\Phi W(t)}{V_o} = S_1(t) \qquad 1.296$$

This is a first order linear equation because Φ, V_o, and S_1 are constants.

case 2: Changing Volume

$$W'(t) = S_1(t) - S_2(t) \qquad 1.297$$
$$= S_1(t) - \frac{\Phi(t)W(t)}{V(t)} \qquad 1.298$$

The differential equation is

$$W'(t) + \frac{\Phi(t)W(t)}{V(t)} = S_1(t) \qquad 1.299$$

Example 1.59

A tank contains 100 gallons of pure water at the beginning of an experiment. 1 gpm of pure water flows into the tank, as does 1 gpm of water containing $\frac{1}{4}$ pound of salt per gallon. A perfectly mixed solution drains from the tank at the rate of 2 gpm. How much salt is in the tank 8 minutes after the experiment has begun?

Choose $W(t)$ as the variable giving the weight of salt in the tank at time t. $\frac{1}{4}$ pound of salt enters the tank per minute. What goes out depends on the concentration in the tank. Specifically, the leaving salt is

salt leaving = $(2\,gpm)(\#$ lbs salt per gallon$)$
$$= (2\,gpm)\left(\frac{\#\text{ lbs salt total}}{100}\right) = .02W(t)$$

The difference between the inflow and the outflow is given by equation 1.295.

$$W'(t) = \frac{1}{4} - .02W(t)$$

This is a first order linear differential equation. It can be solved using the integrating factor (equation 1.283), simple constant coefficient methods, or Laplace transforms. The solution is

$$W(t) = 12.5 - 12.5e^{-.02t}$$

At $t = 8$, $W(t) = 1.85$ lbs.

B. DECAY PROBLEMS

A given quantity is known to decrease at a rate proportional to the amount present. The original amount is known, and the amount at some time t is desired.

k a negative proportionality constant

Q_o original amount present

$Q(t)$ amount present at time t

t time

$t_{1/2}$ half-life

The differential equation is

$$Q'(t) = kQ(t) \qquad 1.300$$

The solution is

$$Q(t) = Q_o e^{kt} \qquad 1.301$$

If Q^* is known for some time t^*, k can be found from

$$k = \left(\frac{1}{t^*}\right) ln\left(\frac{Q^*}{Q_0}\right) \qquad 1.302$$

k also can be found from the half-life:

$$k = \frac{-.693}{t_{1/2}} \qquad 1.303$$

C. SURFACE TEMPERATURE

k a constant

t time

T absolute temperature of the surface

T_o ambient temperature

Assuming that the surface temperature changes at a rate proportional to the difference in surface and ambient temperatures, the differential equation is

$$\frac{dT}{dt} = k(T - T_o) \qquad 1.304$$

Equation 1.304 is known as *Newton's Law of Cooling*.

D. SURFACE EVAPORATION

A exposed surface area

k proportionality constant

r radius

s side length

t time

V object volume

The equation is

$$\frac{dV}{dt} = -kA \qquad 1.305$$

For a spherical drop, this reduces to

$$\frac{dr}{dt} = -k \qquad 1.306$$

For a cube, this reduces to

$$\frac{ds}{dt} = -2k \qquad 1.307$$

21 FOURIER ANALYSIS

Any periodic waveform can be written as the sum of an infinite number of sinusoidal terms. In practice, it is possible to obtain a close approximation to the original waveform with a limited number of sinusoidal terms since most series converge rapidly.

Fourier's theorem is given in equation 1.308. The object of Fourier analysis is to determine the coefficients a_n and b_n.

$$f(t) = a_o + a_1 \cos \omega_o t + a_2 \cos 2\omega_o t + \cdots$$
$$+ b_1 \sin \omega_o t + b_2 \sin 2\omega_o t + \cdots \qquad 1.308$$

ω_o is known as the *fundamental frequency* of the waveform. It depends on the actual waveform period.

$$\omega_o = \frac{2\pi}{T} \qquad 1.309$$

To simplify the analysis, the time domain can be normalized to the radian scale. The normalized scale is obtained by dividing all frequencies by ω_o. Then, the Fourier series becomes

$$f(t) = a_o + a_1 \cos t + a_2 \cos 2t + \cdots$$
$$+ b_1 \sin t + b_2 \sin 2t + \cdots \qquad 1.310$$

The coefficients a_n and b_n can be found from the following relationships:

$$a_o = \frac{1}{2\pi} \int_o^{2\pi} f(t)\, dt \qquad 1.311$$

$$a_n = \frac{1}{\pi} \int_o^{2\pi} f(t) \cos nt\, dt \qquad 1.312$$

$$b_n = \frac{1}{\pi} \int_o^{2\pi} f(t) \sin nt\, dt \qquad 1.313$$

Notice that a_o is the average value of the function. Usually, this average value can be determined by observation without having to go through the integration process. The equation for a_n cannot be used to find a_o.

Example 1.60

Find the Fourier series for

$$f(t) = \begin{cases} 1 & 0 < t < \pi \\ 0 & \pi < t < 2\pi \end{cases}$$

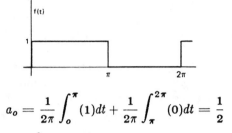

$$a_o = \frac{1}{2\pi} \int_o^{\pi} (1)dt + \frac{1}{2\pi} \int_{\pi}^{2\pi} (0)dt = \frac{1}{2}$$

This value of $\frac{1}{2}$ corresponds to the average value of $f(t)$. It could have been found by observation.

$$a_1 = \frac{1}{\pi} \int_o^{\pi} (1) \cos t\, dt + \frac{1}{\pi} \int_{\pi}^{2\pi} (0) \cos t\, dt$$
$$= \frac{1}{\pi} [\sin t]_o^{\pi} + 0 = 0$$

In general,

$$a_n = \frac{1}{\pi} \left[\frac{\sin nt}{n} \right]_o^{\pi} = 0$$

$$b_1 = \frac{1}{\pi} \int_o^{\pi} (1) \sin t\, dt + \frac{1}{\pi} \int_{\pi}^{2\pi} (0) \sin t\, dt$$
$$= \frac{1}{\pi} [-\cos t]_o^{\pi} = \frac{2}{\pi}$$

In general,

$$b_n = \frac{1}{\pi} \left[\frac{-\cos nt}{n} \right]_o^{\pi} = \begin{cases} 0 \text{ for n even} \\ \frac{2}{\pi n} \text{ for n odd} \end{cases}$$

The series is

$$f(t) = \frac{1}{2} + \frac{2}{\pi} \left[\sin t + \frac{1}{3} \sin 3t + \frac{1}{5} \sin 5t + \cdots \right]$$

The sum of the first few terms is illustrated.

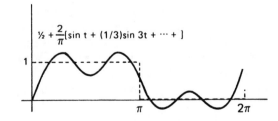

It may be possible to eliminate some of the a_n or b_n coefficients if the function $f(t)$ is symmetrical. There are four types of *symmetry*.

A function is said to have *even symmetry* if $f(t) = f(-t)$. The cosine is an example of this type of waveform. Even symmetry can be detected from the graph of the function. The function to the left of $t = 0$ is a reflection of the function to the right of $t = 0$. With even symmetry, all b_n terms are zero.

A function is said to have *odd symmetry* if $f(t) = -f(-t)$. The sine is an example of this type of waveform. With odd symmetry, all a_n terms are zero (but not necessarily a_o).

A function is said to have *rotational symmetry* or *half-wave symmetry* if $f(t) = -f(t + \pi)$. Functions of this type are identical on alternate $\frac{1}{2}$-cycles, except for a sign reversal. All a_n and b_n are zero for even values of n.

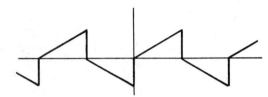

Figure 1.12 Rotational Symmetry

These types of symmetry are not mutually exclusive. For example, it is possible for a function with rotational symmetry to have either odd or even symmetry also. Such a case is known as *quarter-wave symmetry*.

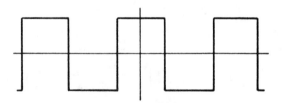

Figure 1.13 Quarter-Wave Symmetry

22 CRITICAL PATH TECHNIQUES

A. INTRODUCTION

Critical path techniques are used to represent graphically the multiple relationships between stages in complicated projects. The graphical networks show the dependencies or *precedence relationships* between the various activities and can be used to control and monitor the progress, cost, and resources of projects. Critical path techniques will also identify the most critical activities in projects.

B. DEFINITIONS

Activity: Any subdivision of a project whose execution requires time and other resources.

Critical path: A path connecting all activities which have minimum or zero slack times. The critical path is the longest path through the network.

Duration: The time required to perform an activity. All durations are *normal durations* unless otherwise referred to as *crash durations*.

Event: The beginning or completion of an activity.

Event time: Actual time at which an event occurs.

Float: Same as slack time.

Slack time: The maximum time that an activity can be delayed without causing the project to fall behind schedule. Slack time is always minimum or zero along the critical path.

Critical path techniques use directed graphs to represent a project. These graphs are made up of arcs (arrows) and nodes (junctions). The placement of the arcs and nodes completely specifies the precedences of the project. Durations and precedences usually are given in a *precedence table* or matrix.

One specific technique is known as the *Critical Path Method, CPM*. This deterministic method is applicable when all activity durations are known in advance. CPM usually is represented as an *activity-on-node model*. Arcs are used to specify precedence, and the nodes actually represent the activities. Events are not present on the graph, other than as the heads and tails of the arcs. Two dummy nodes taking zero time can be used to specify the start and finish of the project.

Example 1.61

Given the project listed in the precedence table, construct the precedence matrix and draw an activity-on-node-network.

Activity	Time (days)	Predecessors
A, start	0	–
B	7	A
C	6	A
D	3	B
E	9	B,C
F	1	D,E
G	4	C
H, finish	0	F,G

The precedence matrix is given below.

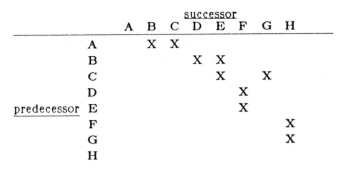

	A	B	C	D	E	F	G	H
A		X	X					
B				X	X			
C					X		X	
D						X		
E						X		
F								X
G								X
H								

(successor across top; predecessor down side)

The activity-on-node-network is shown.

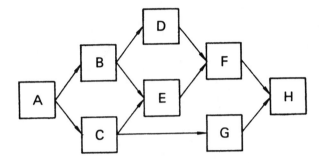

C. SOLVING A CPM PROBLEM

The solution of a critical path problem results in a knowledge of the earliest and latest times that an activity can be started and finished. It also identifies the critical path and generates the slack time for each activity.

To facilitate the solution method, each node should be replaced by a square which has been quartered. The compartments have the meanings indicated by the following key.

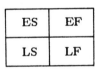

ES	EF
LS	LF

Key
ES: Earliest Start
EF: Earliest Finish
LS: Latest Start
LF: Latest Finish

The following procedure will find the earliest and latest starts and finishes of each node.

1. Place the project start time or date in the ES and EF positions of the start activity. The start time is zero for relative calculations.

2. Consider any unmarked activity whose predecessors have all been marked in the EF and ES positions. (Go to step 4 if there is none.) Mark in its ES position the largest number marked in the EF position of those predecessors.

3. Add the activity time to the ES time and write this in the EF box. Return to step 2.

4. Place the value of the latest finish date in the LS and LF boxes of the finish node.

5. Consider any unmarked activity whose successors have all been marked in the LS and LF positions. The LF is the smallest LS of the successors. Go to step 7 if there are no unmarked activities.

6. The LS for the new node is LF minus its activity time. Return to step 5.

7. The slack for each node is (LS-ES) or (LF-EF).

8. The critical path encompasses nodes for which the slack equals (LS-ES) from the start node. There may be more than one critical path.

Example 1.62

Complete the network for the previous example and find the critical path. Assume the desired completion date is in 19 days.

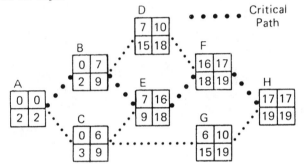

D. PROBABILISTIC CRITICAL PATH MODELS

Probabilistic networks differ from deterministic networks only in the way in which the activity durations are found. Whereas durations are known explicitly for a deterministic network, the time for a probabilistic activity is distributed as a random variable.

This variable nature complicates the problem greatly since the actual distribution of times often is unknown. For this reason, such a problem usually is solved as a deterministic model using the mean of the duration distribution as the activity duration.

The most common probabilistic critical path model is *PERT*, which stands for *Program Evaluation and Review Technique*. In PERT, all duration variables are assumed to come from a beta distribution, with mean and standard deviation given as:

$$t_{\text{mean}} = \left(\frac{1}{6}\right)\left(t_{\text{minimum}} + 4t_{\text{most likely}} + t_{\text{maximum}}\right) \qquad 1.314$$

$$\sigma = \frac{1}{6}\left(t_{\text{maximum}} - t_{\text{minimum}}\right) \qquad 1.315$$

The project completion time for large projects is assumed to be distributed normally with mean (μ) equal to the critical path length and overall variance (σ^2) equal to the sum of the variances along the critical path.

If necessary, the probability that a project duration will exceed some length (D) can be found from the normal table and the following relationship:

$$P\{\text{duration} > D\} = p\{x > z\} \qquad \textbf{1.316}$$

where z is the standard normal variable equal to

$$z = \frac{D - \mu}{\sigma} \qquad \textbf{1.317}$$

Appendix A
Laplace Transforms

$f(t)$	$\mathscr{L}[f(t)]$
Unit impulse at $t=0$	1
Unit impulse at $t=c$	e^{-cs}
Unit step at $t=0$	$(1/s)$
Unit step at $t=c$	$\dfrac{e^{-cs}}{s}$
t	$\dfrac{1}{s^2}$
$\dfrac{t^{n-1}}{(n-1)!}$	$\dfrac{1}{s^n}$
$\sin At$	$\dfrac{A}{s^2+A^2}$
$At-\sin At$	$\dfrac{A^3}{s^2(s^2+A^2)}$
$\sinh(At)$	$\dfrac{A}{s^2-A^2}$
$t\sin At$	$\dfrac{2As}{(s^2+A^2)^2}$
$\cos At$	$\dfrac{s}{s^2+A^2}$
$1-\cos At$	$\dfrac{A^2}{s(s^2+A^2)}$
$\cosh(At)$	$\dfrac{s}{s^2-A^2}$
$t\cos At$	$\dfrac{s^2-A^2}{(s^2+A^2)^2}$
t^n (n is a positive integer)	$\dfrac{n!}{s^{(n+1)}}$
e^{At}	$\dfrac{1}{s-A}$
$e^{At}\sin Bt$	$\dfrac{B}{(s-A)^2+B^2}$
$e^{At}\cos Bt$	$\dfrac{s-A}{(s-A)^2+B^2}$
$e^{At}t^n$ (n is positive integer)	$\dfrac{n!}{(s-A)^{n+1}}$
$1-e^{-At}$	$\dfrac{A}{s(s+A)}$
$e^{-At}+At-1$	$\dfrac{A^2}{s^2(s+A)}$
$\dfrac{e^{-At}-e^{-Bt}}{B-A}$	$\dfrac{1}{(s+A)(s+B)}$
$\dfrac{(C-A)e^{-At}-(C-B)e^{-Bt}}{B-A}$	$\dfrac{s+C}{(s+A)(s+B)}$
$\dfrac{1}{AB}+\dfrac{Be^{-At}-Ae^{-Bt}}{AB(A-B)}$	$\dfrac{1}{s(s+A)(s+B)}$

Appendix B
Conversion Factors

To Convert	Into	Multiply by
Acres	hectares	0.4047
Acres	square feet	43,560.0
Acres	square miles	1.562 EE−3
Ampere hours	coulombs	3,600.0
Angstrom units	inches	3.937 EE−9
Angstrom units	microns	1 EE−4
Astronomical units	kilometers	1.495 EE8
Atmospheres	cms of mercury	76.0
BTU's	horsepower-hrs	3.931 EE−4
BTU's	kilowatt-hrs	2.928 EE−4
BTU/hr	watts	0.2931
Bushels	cubic inches	2,150.4
Calories, gram (mean)	BTU (mean)	3.9685 EE−3
Centares	square meters	1.0
Centimeters	kilometers	1 EE−5
Centimeters	meters	1 EE−2
Centimeters	millimeters	10.0
Centimeters	feet	3.281 EE−2
Centimeters	inches	0.3937
Chains	inches	792.0
Coulombs	faradays	1.036 EE−5
Cubic centimeters	cubic inches	0.06102
Cubic centimeters	pints (U.S. liq.)	2.113 EE−3
Cubic feet	cubic meters	0.02832
Cubic feet/min.	pounds water/min.	62.43
Cubic feet/sec.	gallons/min.	448.831
Cubits	inches	18.0
Days	seconds	86,400.0
Degrees (angle)	radians	1.745 EE−2
Degrees/sec.	revolutions/min.	0.1667
Dynes	grams	1.020 EE−3
Dynes	joules/meter (newtons)	1 EE−5
Ells	inches	45.0
Ergs	BTU's	9.480 EE−11
Ergs	foot-pounds	7.3670 EE−8
Ergs	kilowatt-hours	2.778 EE−14
Faradays/sec.	amperes (absolute)	96,500
Fathoms	feet	6.0
Feet	centimeters	30.48
Feet	meters	0.3048
Feet	miles (nautical)	1.645 EE−4
Feet	miles (statute)	1.894 EE−4
Feet/min.	centimeters/sec.	0.5080
Feet/sec.	knots	0.5921
Feet/sec.	miles/hour	0.6818
Foot-pounds	BTU's	1.286 EE−3
Foot-pounds	kilowatt-hours	3.766 EE−7
Furlongs	miles (U.S.)	0.125
Furlongs	feet	660.0
Gallons	liters	3.785
Gallons of water	pounds of water	8.3453
Gallons/min.	cubic feet/hour	8.0208
Grams	ounces (avoirdupois)	3.527 EE−2
Grams	ounces (troy)	3.215 EE−2
Grams	pounds	2.205 EE−3
Hectares	acres	2.471
Hectares	square feet	1.076 EE5
Horsepower	BTU/min.	42.42
Horsepower	kilowatts	0.7457
Horsepower	watts	745.7
Hours	days	4.167 EE−2
Hours	weeks	5.952 EE−3
Inches	centimeters	2.540
Inches	miles	1.578 EE−5
Joules	BTU's	9.480 EE−4
Joules	ergs	1 EE7
Kilograms	pounds	2.205
Kilometers	feet	3,281.0
Kilometers	meters	1,000.0
Kilometers	miles	0.6214
Kilometers/hr.	knots	0.5396
Kilowatts	horsepower	1.341
Kilowatt-hours	BTU'S	3,413.0
Knots	feet/hour	6,080.0
Knots	nautical miles/hr.	1.0
Knots	statute miles/hr.	1.151
Light years	miles	5.9 EE12
Links (surveyor's)	inches	7.92
Liters	cubic centimeters	1,000.0
Liters	cubic inches	61.02
Liters	gallons (U.S. liq.)	0.2642
Liters	milliliters	1,000.0
Liters	pints (U.S. liq.)	2.113
Meters	centimeters	100.0
Meters	feet	3.281
Meters	kilometers	1 EE−3
Meters	miles (nautical)	5.396 EE−4
Meters	miles (statute)	6.214 EE−4
Meters	millimeters	1,000.0
Microns	meters	1 EE−6
Miles (nautical)	feet	6,080.27
Miles (statute)	feet	5,280.0
Miles (nautical)	kilometers	1.853
Miles (statute)	kilometers	1.609
Miles (nautical)	miles (statute)	1.1516
Miles (statute)	miles (nautical)	0.8684
Miles/hour	feet/min.	88.0
Milligrams/liter	parts/million	1.0
Milliliters	liters	1 EE−3
Millimeters	inches	3.937 EE−2
Newtons	dynes	1 EE5
Ohms (international)	ohms (absolute)	1.0005
Ounces	grams	28.349527
Ounces	pounds	6.25 EE−2
Ounces (troy)	ounces (avoirdupois)	1.09714
Parsecs	miles	19 EE12
Parsecs	kilometers	3.084 EE13
Pints (liq.)	cubic centimeters	473.2
Pints (liq.)	cubic inches	28.87
Pints (liq.)	gallons	0.125
Pints (liq.)	quarts (liq.)	0.5
Pounds	kilograms	0.4536
Pounds	ounces	16.0
Pounds	ounces (troy)	14.5833
Pounds	pounds (troy)	1.21528
Quarts (dry)	cubic inches	67.20

To Convert	Into	Multiply by
Quarts (liq.)	cubic inches	57.75
Quarts (liq.)	gallons	0.25
Quarts (liq.)	liters	0.9463
Radians	degrees	57.30
Radians	minutes	3,438.0
Revolutions	degrees	360.0
Revolutions/min.	degrees/sec.	6.0
Rods	meters	5.029
Rods	feet	16.5
Rods (surveyor's measure)	yards	5.5
Seconds	minutes	1.667 EE−2
Slugs	pounds	32.17
Tons (long)	kilograms	1,016.0
Tons (short)	kilograms	907.1848
Tons (long)	pounds	2,240.0
Tons (short)	pounds	2,000.0
Tons (long)	tons (short)	1.120
Tons (short)	tons (long)	0.89287
Volt (absolute)	statvolts	3.336 EE−3
Watts	BTU/hour	3.4129
Watts	horsepower	1.341 EE−3
Yards	meters	0.9144
Yards	miles (nautical)	4.934 EE−4
Yards	miles (statute)	5.682 EE−4

Appendix C
Computational Values of Fundamental Constants

Constant	SI	English
charge on electron	-1.602 EE-19 C	
charge on proton	$+1.602$ EE-19 C	
atomic mass unit	1.66 EE-27 kg	
electron rest mass	9.11 EE-31 kg	
proton rest mass	1.673 EE-27 kg	
neutron rest mass	1.675 EE-27 kg	
earth weight		1.32 EE25 lb
earth mass	6.00 EE24 kg	4.11 EE23 slug
mean earth radius	6.37 EE3 km	2.09 EE7 ft
mean earth density	5.52 EE3 kg/m³	3.45 lbm/ft³
earth escape velocity	1.12 EE4 m/s	3.67 EE4 ft/sec
distance from sun	1.49 EE11 m	4.89 EE11 ft
Boltzmann constant	1.381 EE-23 J/°K	5.65 EE-24 $\frac{\text{ft}-\text{lbf}}{°\text{R}}$
permeability of a vacuum	1.257 EE-6 H/m	
permittivity of a vacuum	8.854 EE-12 F/m	
Planck constant	6.626 EE-34 J·s	
Avogadro's number	6.022 EE23 molecules/gmole	2.73 EE26 $\frac{\text{molecules}}{\text{pmole}}$
Faraday's constant	9.648 EE4 C/gmole	
Stefan-Boltzmann constant	5.670 EE-8 W/m²$-$K⁴	1.71 EE-9 $\frac{\text{BTU}}{\text{ft}^2-\text{hr}-°\text{R}^4}$
gravitational constant (G)	6.672 EE-11 m³/s²$-$kg	3.44 EE-8 $\frac{\text{ft}^4}{\text{lbf}-\text{sec}^4}$
universal gas constant	8.314 J/°K$-$gmole	1545 $\frac{\text{ft}-\text{lbf}}{°\text{R}-\text{pmole}}$
speed of light	3.00 EE8 m/s	9.84 EE8 ft/sec
speed of sound, air, STP	3.31 EE2 m/s	1.09 EE3 ft/sec
speed of sound, air, 70°F, one atmosphere	3.44 EE2 m/s	1.13 EE3 ft/sec
standard atmosphere	1.013 EE5 N/m²	14.7 psia
standard temperature	0°C	32°F
molar ideal gas volume (STP)	22.4138 EE-3 m³/gmole	359 ft³/pmole
standard water density	1 EE3 kg/m³	62.4 lbm/ft³
air density, STP	1.29 kg/m³	8.05 EE-2 lbm/ft³
air density, 70°F, 1 atm	1.20 kg/m³	7.49 EE-2 lbm/ft³
mercury density	1.360 EE4 kg/m³	8.49 EE2 lbm/ft³
gravity on moon	1.67 m/s²	5.47 ft/sec²
gravity on earth	9.81 m/s²	32.17 ft/sec²

PRACTICE PROBLEMS: MATHEMATICS

Warm-ups

1. Calculate the following sum:

$$\sum_{j=1}^{5} [(j+1)^2 - 1]$$

2. A function is given as $y = 3x^{.93} + 4.2$. What is the percent error if the value of y at $x = 2.7$ is found by using straight-line interpolation between $x = 2$ and $x = 3$?

3. The diameter of a sphere and the base of a cone are equal. What percentage of that diameter must the cone's height be so that both volumes are equal?

4. A 5-pound block sits on a 20° incline without slipping. Draw the freebody with respect to axes parallel and perpendicular to the surface of the incline.

5. What is the determinant of the following matrix?

$$\begin{vmatrix} 7 & 2 & 6 \\ 4 & 0 & 3 \\ 9 & 0 & 5 \end{vmatrix}$$

6. Convert 250°F to Celsius and Rankine.

7. Convert the Stefan-Boltzmann constant from English to SI units.

8. Given the following data points, find y by straight-line interpolation for $x = 2.75$.

(x,	y)
(1,	4)
(2,	6)
(3,	2)
(4,	−14)

9. Find the point-slope form of the straight line passing through (1.7, 3.4) and (8.3, 9.5).

10. If every 0.1 second a quantity increases by 0.1% of its current value, calculate the doubling time.

Concentrates

1. Solve for the values of x, y, and z which simultaneously satisfy the following equations:

$$x + y = -4$$
$$x + z - 1 = 0$$
$$2z - y + 3x = 4$$

2. Find the best equation for a line passing through the points given. Find the correlation coefficient.

(x,	y)
(400,	370)
(800,	780)
(1250,	1210)
(1600,	1560)
(2000,	1980)
(2500,	2450)
(4000,	3950)

3. Find the best equation for a line passing through the points given.

(s,	t)
(20,	43)
(18,	141)
(16,	385)
(14,	1099)

4. Solve the differential equation for y:

$$y' - y = 2xe^{2x} \qquad y(0) = 1$$

5. Solve the differential equation for y:

$$y'' - 4y' - 12y = 0$$

6. A tank contains 100 gallons of brine made by dissolving 60 pounds of salt in water. Salt water containing 1 pound of salt per gallon runs in at the rate of 2 gallons per minute. A well-stirred mixture runs out at the rate of 3 gallons per minute. Find the amount of salt in the tank at the end of 1 hour.

7. What are the values of a, b, and c in the expression below such that $n(\infty) = 100$, $n(0) = 10$, and $dn(0)/dt = .5$?

$$n(t) = \frac{a}{1 + be^{ct}}$$

8. Find all minima, maxima, and inflection points for

$$y = x^3 - 9x^2 - 3$$

9. How many tons of 13,000 BTU/lbm coal must be burned to produce as much energy from the nuclear conversion (complete) of 1 gram of its mass?

10. An automatic screw machine turns out 200 washers each hour. The size of the hole in the washer is normally distributed with a mean of .502 inches and standard deviation of .005 inches. Washers are defective if the hole diameter is less than .497 or more than .507 inches diameter.

(a) What is the probability that a washer chosen at random will be defective? (b) What is the probability that 2 out of 3 randomly sampled washers will be defective? (c) How many washers per 8-hour day will be defective?

11. California law requires a statistical analysis of the average speed driven by motorists on a road prior to the use of radar speed control. The following speeds were observed in a random sample of 40 cars:

44, 48, 26, 25, 20, 43, 40, 42, 29, 39, 23, 26, 24, 47, 45, 28, 29, 41, 38, 36, 27, 44, 42, 43, 29, 37, 34, 31, 33, 30, 42, 43, 28, 41, 29, 36, 35, 30, 32, 31 (all in mph)

a) Tabulate the frequency distribution of the above data

b) Draw the frequency histogram

c) Draw the frequency polygon

d) Tabulate the cumulative frequency distribution

e) Draw the cumulative frequency graph

f) What is the upper quartile speed?

g) What are the mode, median, and mean speeds?

h) What is the standard deviation of the sample data?

i) What is the sample standard deviation?

j) What is the sample variance?

12. Activities constituting a project are listed below. The project starts at time zero.

Activity	Predecessors	Successors	Duration
start	—	A	0
A	start	B,C,D	7
B	A	G	6
C	A	E,F	5
D	A	G	2
E	C	H	13
F	C	H,I	4
G	D,B	I	18
H	E,F	finish	7
I	F,G	finish	5
finish	H,I	—	0

a) Draw the CPM network

b) Indicate the critical path

c) What is the earliest finish?

d) What is the latest finish?

e) What is the slack along the critical path?

f) What is the float along the critical path?

13. Activities constituting a short project are listed here.

Activity	Predecessors	Successors	t_{min}	t_{likely}	t_{max}
start	—	A	0	0	0
A	start	B,D	1	2	5
B	A	C	7	9	20
C	B	D	5	12	18
D	A,C	finish	2	4	7
finish	D	—	0	0	0

If the project starts at $t = 15$, what is the probability that the project will be completed by $t = 42$ or sooner?

14. The number of cars entering a toll plaza on a bridge during the hour following midnight is distributed as Poisson with a mean of 20. What is the probability that 17 cars will pass through the toll plaza during that hour on any given night? What is the probability that 3 or fewer cars will pass through the toll plaza at that hour on any given night?

15. The time taken by a toll taker to collect the toll from vehicles crossing a bridge is an exponential distribution with mean of 23 seconds when a line of vehicles waiting to enter the toll booth exists. What is the probability that a random vehicle will be processed in 25 seconds or more (i.e., will take longer than 25 seconds)?

16. The number of vehicles lining up behind a flashing railroad crossing has been observed for five trains of different lengths, as given below. What is the mathematical formula which relates the two variables?

no. cars in train	no. vehicles
2	14.8
5	18.0
8	20.4
12	23.0
27	29.9

17. The oscillation exhibited by a certain 1-story building in free motion is given by the following differential equation:

$$x'' + 2x' + 2x = 0 \qquad x(0) = 0; \ x'(0) = 1$$

(a) What is x as a function of time? (b) What is the building's natural frequency of vibration? (c) What is the amplitude of oscillation? (d) What is x as a function of time if a lateral wind load is applied with form of $\sin(t)$?

Timed (1 hour allowed for each)

1. 100 bearings were tested to failure. The average life was 1520 hours and the standard deviation was 120 hours. The manufacturer advertises a 1600-hour life. Should the company change its claim? Evaluate using confidence limits of 95% and 99%.

2. An 8-pound weight hangs from a suspended spring, stretching it to 5.9″. The weight is also attached to a dashpot with a damping coefficient of .50 lbf-sec/ft. A forcing function of $4(\cos(2t))$ lbf is applied to the weight. What is the weight's response assuming the system is at rest initially?

3. A survey field crew measures one leg of a traverse four times. The following results are obtained.

repetition	measurement	direction
1	1249.529	forward
2	1249.494	backward
3	1249.384	forward
4	1249.348	backward

The crew chief is under orders to obtain readings with confidence limits of 90%.

(a) Which readings are acceptable?

(b) Which readings are not acceptable?

(c) Explain how to determine which readings are not acceptable.

(d) What is the most probable value of the distance?

(e) What is the error in the most probable value (at 90% confidence)?

(f) If the distance is one side of a square traverse hose sides are all equal, what is the most probable closure error?

(g) What is the probable error of part (f) expressed as a fraction?

(h) Is this error of closure within the second order of accuracy?

(i) Define accuracy and distinguish it from precision.

(j) Give an example of systematic error.

4. A 90-pound bag of a chemical is accidentally dropped in an aerating lagoon. The chemical is water soluble and non-reacting. The lagoon is 120 feet in diameter and filled to a depth of 10 feet. The aerators circulate and distribute the chemical evenly throughout the lagoon.

Water enters the lagoon at the rate of 30 gallons per minute. Fully mixed water is pumped into a reservoir at the rate of 30 gpm.

The established safe concentration of this chemical is 1 ppb (part per billion). How long (in days) will it take for the concentration of the discharge water to reach this level?

RESERVED FOR FUTURE USE

2 ENGINEERING ECONOMIC ANALYSIS

Nomenclature

A	annual amount or annuity	$
B	present worth of all benefits	$
BV_j	book value at the end of the jth year	$
C	cost, or present worth of all costs	$
d	declining balance depreciation rate	decimal
D_j	depreciation in year j	$
$D.R.$	present worth of after-tax depreciation recovery	$
e	natural logarithm base (2.718)	–
EAA	equivalent annual amount	$
EUAC	equivalent uniform annual cost	$
f	federal income tax rate	decimal
F	future amount, or future worth	$
G	uniform gradient amount	$
i	effective rate per period (usually per year)	decimal
k	number of compounding periods per year	–
n	number of compounding periods, or life of asset	–
P	present worth, or present value	$
P_t	present worth after taxes	$
ROR	rate of return	decimal
ROI	return on investments	$
r	nominal rate per year (rate per annum)	decimal
s	state income tax rate	decimal
S_n	expected salvage value in year n	$
t	composite tax rate, or time	decimal
z	a factor equal to $\dfrac{1+i}{1-d}$	decimal
ϕ	effective rate per period	decimal

1 EQUIVALENCE

Industrial decision-makers using engineering economics are concerned with the timing of a project's cash flows as well as with the total profitability of that project. In this situation, a method is required to compare projects involving receipts and disbursements occurring at different times.

By way of illustration, consider $100 placed in a bank account that pays 5% effective annual interest at the end of each year. After the first year, the account will have grown to $105. After the second year, the account will have grown to $110.25.

Assume that you will have no need for money during the next two years and that any money received would immediately go into your 5% bank account. Then, which of the following options would be more desirable?

> **option a:** $100 now
> **option b:** $105 to be delivered in one year
> **option c:** $110.25 to be delivered in two years

In light of the previous illustration, none of the options is superior under the assumptions given. If the first option is chosen, you will immediately place $100 into a 5% account, and in two years the account will have grown to $110.25. In fact, the account will contain $110.25 at the end of two years regardless of the option chosen. Therefore, these alternatives are said to be *equivalent*.

2 CASH FLOW DIAGRAMS

Although they are not always necessary in simple problems (and they are often unwieldy in very complex problems), *cash flow diagrams* may be drawn to help visualize and simplify problems having diverse receipts and disbursements.

The conventions below are used to standardize cash flow diagrams.

- The horizontal (time) axis is marked off in equal increments, one per period, up to the duration or horizon of the project.

- All disbursements and receipts (cash flows) are assumed to take place at the end of the year in which they occur. This is known as the *year-end convention*. The exception to the year-end convention is any initial cost (purchase cost) which occurs at $t = 0$.

- Two or more transfers in the same year are placed end-to-end, and these may be combined.

- Expenses incurred before $t = 0$ are called *sunk costs*. Sunk costs are not relevant to the problem.

- Receipts are represented by arrows directed upward. Disbursements are represented by arrows directed downward. The arrow length is proportional to the magnitude of the cash flow.

Example 2.1

A mechanical device will cost $20,000 when purchased. Maintenance will cost $1000 each year. The device will generate revenues of $5000 each year for five years after which the salvage value is expected to be $7000. Draw and simplify the cash flow diagram.

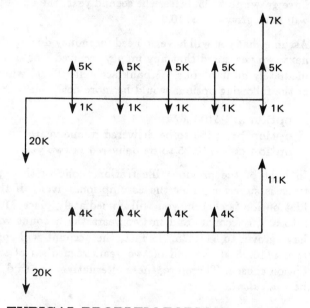

3 TYPICAL PROBLEM FORMAT

With the exception of some investment and rate of return problems, the typical problem involving engineering economics will have the following characteristics:

- An interest rate will be given.

- Two or more alternatives will be competing for funding.

- Each alternative will have its own cash flows.

- It is necessary to select the best alternative.

Example 2.2

Investment A costs $10,000 today and pays back $11,500 two years from now. Investment B costs $8000 today and pays back $4500 each year for two

years. If an interest rate of 5% is used, which alternative is superior?

The solution to this example is not difficult, but it will be postponed until methods of calculating equivalence have been covered.

4 CALCULATING EQUIVALENCE

It was previously illustrated that $100 now is equivalent at 5% interest to $105 in one year. The equivalence of any present amount, P, at $t = 0$ to any future amount, F, at $t = n$ is called the *future worth* and can be calculated from equation 2.1.

$$F = P(1+i)^n \qquad 2.1$$

The factor $(1+i)^n$ is known as the *compound amount factor* and has been tabulated at the end of this chapter for various combinations of i and n. Rather than actually writing the formula for the compound amount factor, the convention is to use the standard functional notation $(F/P, i\%, n)$. Thus,

$$F = P(F/P, i\%, n) \qquad 2.2$$

Similarly, the equivalence of any future amount to any present amount is called the *present worth* and can be calculated from

$$P = F(1+i)^{-n} = F(P/F, i\%, n) \qquad 2.3$$

The factor $(1+i)^{-n}$ is known as the *present worth factor*, with functional notation $(P/F, i\%, n)$. Tabulated values are also given for this factor at the end of this chapter.

Example 2.3

How much should you put into a 10% savings account in order to have $10,000 in five years?

This problem could also be stated: What is the equivalent present worth of $10,000 five years from now if money is worth 10%?

$$P = F(1+i)^{-n} = (10,000)(1+.10)^{-5}$$
$$= 6209$$

The factor .6209 would usually be obtained from the tables.

A cash flow that repeats regularly each year is known as an *annual amount*. When annual costs are incurred due to the functioning of a piece of equipment, they are often known as *operating and maintenance* (O&M) *costs*. The annual costs associated with operating a business in general are known as *general, selling, and administrative* (GS&A) *expenses*. Although the equivalent value for each of the n annual amounts could be calculated and then summed, it is much easier to use one of the *uniform series factors*, as illustrated in example 2.4.

Example 2.4

Maintenance costs for a machine are $250 each year. What is the present worth of these maintenance costs over a 12-year period if the interest rate is 8% ?

Notice that

$$(P/A, 8\%, 12) = (P/F, 8\%, 1) + (P/F, 8\%, 2)$$
$$+ \cdots + (P/F, 8\%, 12)$$

Then,

$$P = A(P/A, i\%, n) = (-250)(7.5361)$$
$$= -1884$$

A common complication involves a uniformly increasing cash flow. Such an increasing cash flow should be handled with the *uniform gradient factor*, $(P/G, i\%, n)$. The uniform gradient factor finds the present worth of a uniformly increasing cash flow that starts in year 2 (not year 1) as shown in example 2.5.

Example 2.5

Maintenance on an old machine is $100 this year but is expected to increase by $25 each year thereafter. What is the present worth of five years of maintenance? Use an interest rate of 10%.

In this problem, the cash flow must be broken down into parts. Notice that the five-year gradient factor is used even though there are only four non-zero gradient cash flows.

$$P = A(P/A, 10\%, 5) + G(P/G, 10\%, 5)$$
$$= (-100)(3.7908) - (25)(6.8618)$$
$$= -551$$

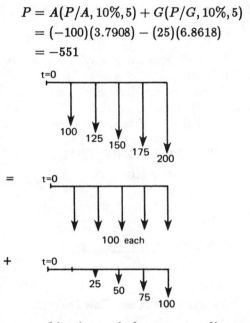

Various combinations of the compounding and discounting factors are possible. For instance, the annual cash flow that would be equivalent to a uniform gradient may be found from

$$A = G(P/G, i\%, n)(A/P, i\%, n) \qquad 2.4$$

Formulas for all of the compounding and discounting factors are contained in table 2.1. Normally, it will not

Table 2.1
Discount Factors for Discrete Compounding

factor name	converts	symbol	formula
single payment compound amount	P to F	$(F/P, i\%, n)$	$(1+i)^n$
present worth	F to P	$(P/F, i\%, n)$	$(1+i)^{-n}$
uniform series sinking fund	F to A	$(A/F, i\%, n)$	$\dfrac{i}{(1+i)^n - 1}$
capital recovery	P to A	$(A/P, i\%, n)$	$\dfrac{i(1+i)^n}{(1+i)^n - 1}$
compound amount	A to F	$(F/A, i\%, n)$	$\dfrac{(1+i)^n - 1}{i}$
equal series present worth	A to P	$(P/A, i\%, n)$	$\dfrac{(1+i)^n - 1}{i(1+1)^n}$
uniform gradient	G to P	$(P/G, i\%, n)$	$\dfrac{(1+i)^n - 1}{i^2(1+i)^n} - \dfrac{n}{i(1+i)^n}$

be necessary to calculate factors from the formulas. The tables at the end of this chapter are adequate for solving most problems.

5 THE MEANING OF PRESENT WORTH AND i

It is clear that $100 invested in a 5% bank account will allow you to remove $105 one year from now. If this investment is made, you will clearly receive a *return on investment* (ROI) of $5. The cash flow diagram and the present worth of the two transactions are

$$P = -100 + (105)(P/F, 5\%, 1)$$
$$= -100 + (105)(.9524)$$
$$= 0$$

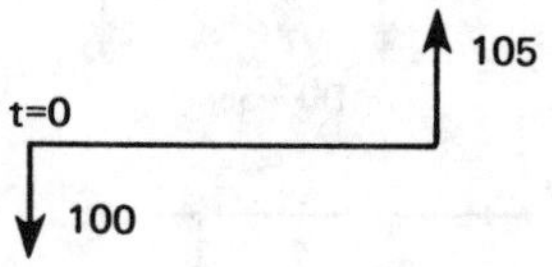

Figure 2.1 Cash Flow Diagram

Notice that the present worth is zero even though you did receive a 5% return on your investment.

However, if you are offered $120 for the use of $100 over a one-year period, the cash flow diagram and present worth (at 5%) would be

$$P = -100 + (120)(P/F, 5\%, 1)$$
$$= -100 + (120)(.9524)$$
$$= 14.29$$

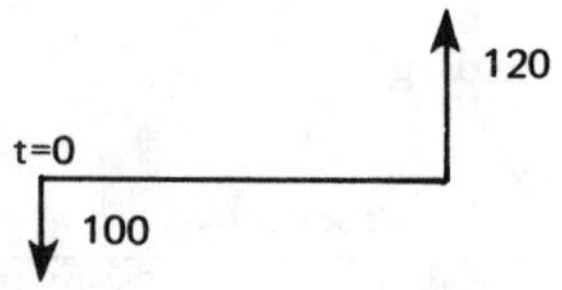

Figure 2.2 Cash Flow Diagram

Therefore, it appears that the present worth of an alternative is equal to the equivalent value at $t = 0$ of the increase in return above that which you would be able to earn in an investment offering $i\%$ per period. In the above case, $14.29 is the present worth of ($20−$5), the difference in the two ROIs.

Alternatively, the actual earned interest rate, called *rate of return*, ROR, can be defined as the rate that makes the present worth of the alternative zero.

The *present worth* is also the amount that you would have to be given to dissuade you from making an investment, since placing the initial investment amount along

with the present worth into a bank account earning $i\%$ will yield the same eventual ROI. Relating this to the previous paragraphs, you could be dissuaded against investing $100 in an alternative that would return $120 in one year by a $t = 0$ payment of $14.29. Clearly, ($100 + $14.29) invested at $t = 0$ will also yield $120 in one year at 5%.

The selection of the interest rate is difficult in engineering economics problems. Usually it is taken as the average rate of return that an individual or business organization has realized in past investments. Fortunately, an interest rate is usually given. A company may not know what effective interest rate to use in an economic analysis. In such a case, the company can establish a minimum acceptable return on its investment. This *minimum attractive rate of return* (MARR) should be used as the effective interest rate i in economic analyses.

It should be obvious that alternatives with negative present worths are undesirable, and that alternatives with positive present worths are desirable because they increase the average earning power of invested capital.

6 CHOICE BETWEEN ALTERNATIVES

A variety of methods exist for selecting a superior alternative from among a group of proposals. Each method has its own merits and applications.

A. PRESENT WORTH METHOD

The *present worth method* has already been implied. When two or more alternatives are capable of performing the same functions, the superior alternative will have the largest present worth. This method is suitable for ranking the desirability of alternatives. The present worth method is restricted to evaluating alternatives that are mutually exclusive and that have the same lives.

Returning to example 2.2, the present worth of each alternative should be found in order to determine which alternative is superior.

Example 2.2, continued

$$P(A) = -10,000 + (11,500)(P/F, 5\%, 2)$$
$$= 431$$
$$P(B) = -8000 + (4500)(P/A, 5\%, 2)$$
$$= 367$$

Alternative A is superior and should be chosen.

B. CAPITALIZED COST METHOD

The present worth of a project with an infinite life is known as the *capitalized cost* or *life cycle cost*. Capitalized cost is the amount of money at $t = 0$ needed to perpetually support the project on the earned interest only. Capitalized cost is a positive number when expenses exceed income.

$$\frac{\text{capitalized}}{\text{cost}} = \frac{\text{initial}}{\text{cost}} + \frac{\text{annual costs}}{i} \qquad 2.5$$

Capitalized cost is the present worth of an infinitely-lived project. Normally, it would be difficult to work with an infinite stream of cash flows since most economics tables don't list factors for periods in excess of 100 years. However, the (A/P) discounting factor approaches the interest rate as n becomes large. Since the (P/A) and (A/P) factors are reciprocals of each other, we would expect to divide an infinite series of equal cash flows by the interest rate in order to calculate the present worth of the infinite series. This is the basis of equation 2.5.

Equation 2.5 can be used when the annual costs are equal in every year. The annual cost in that equation is assumed to be the same each year. If the operating and maintenance costs occur irregularly instead of annually, or if the costs vary from year to year, it will be necessary to somehow determine a cash flow of *equal annual amounts* (EAA) which is equivalent to the stream of original costs.

The equal annual amount may be calculated in the usual manner by first finding the present worth of all the actual costs, and then multiplying the present worth by the interest rate (the (A/P) factor for an infinite series). However, it is not necessary to convert the present worth to an equal annual amount, since equation 2.6 will convert the equal annual amount back to the present worth.

$$\frac{\text{capitalized}}{\text{cost}} = \frac{\text{initial}}{\text{cost}} + \frac{\text{EAA}}{i} \qquad 2.6$$

In comparing two alternatives, each of which is infinitely lived, the superior alternative will have the lowest capitalized cost.

C. ANNUAL COST METHOD

Alternatives that accomplish the same purpose but that have unequal lives must be compared by the *annual cost method*. The annual cost method assumes that each alternative will be replaced by an identical twin at the end of its useful life (infinite renewal). This method, which may also be used to rank alternatives according to their desirability, is also called the *annual return method* and *capital recovery method*.

Restrictions are that the alternatives must be mutually exclusive and infinitely renewed up to the dura-

tion of the longest-lived alternative. The calculated annual cost is known as the *equivalent uniform annual cost* (EUAC). Cost is a positive number when expenses exceed income.

Example 2.6

Which of the following alternatives is superior over a 30-year period if the interest rate is 7%?

	A	B
type	brick	wood
life	30 years	10 years
cost	$1800	$450
maintenance	$5/year	$20/year

$$\text{EUAC}(A) = (1800)(A/P, 7\%, 30) + 5 = 150$$
$$\text{EUAC}(B) = (450)(A/P, 7\%, 10) + 20 = 84$$

Alternative B is superior since its annual cost of operation is the lowest. It is assumed that three wood facilities, each with a life of 10 years and a cost of $450, will be built to span the 30-year period.

D. BENEFIT-COST RATIO METHOD

The *benefit-cost ratio method* is often used in municipal project evaluations where benefits and costs accrue to different segments of the community. With this method, the present worth of all benefits (regardless of the beneficiary) is divided by the present worth of all costs. The project is considered acceptable if the ratio exceeds *one*.

When the benefit-cost ratio method is used, disbursements by the initiators or sponsors are *costs*. Disbursements by the users of the project are known as *disbenefits*. It is often difficult to determine whether a cash flow is a cost or a disbenefit (whether to place it in the numerator or denominator of the benefit-cost ratio calculation).

Regardless of where the cash flow is placed, an acceptable project will always have a benefit-cost ratio greater than one, although the actual numerical result will depend on the placement. For this reason, the benefit-cost ratio method should not be used to rank competing projects.

The benefit-cost ratio method may be used to rank alternative proposals only if an *incremental analysis* is used. First, determine that the ratio is greater than one for each alternative. Then, calculate the ratio of benefits to costs for each possible pair of alternatives by using equation 2.7. If the ratio exceeds one, alternative 2 is superior to alternative 1. Otherwise, alternative 1 is superior.

$$\text{benefit-to-cost ratio} = \frac{B_2 - B_1}{C_2 - C_1} \qquad 2.7$$

E. RATE OF RETURN METHOD

Perhaps no method of analysis is less understood than the *rate of return method*, (ROR). As was stated previously, the ROR is the interest rate that would yield identical profits if all money were invested at that rate. The present worth of any such investment is zero.

The ROR is defined as the interest rate that will discount all cash flows to a total present worth equal to the initial required investment. This definition is used to determine the ROR of an alternative. The advantage of the ROR method is that no knowledge of an interest rate is required.

To find the ROR of an alternative, proceed as follows:

step 1:   Set up the problem as if to calculate the present worth.

step 2:   Arbitrarily select a reasonable value for i. Calculate the present worth.

step 3:   Choose another value of i (not too close to the original value) and again solve for the present worth.

step 4:   Interpolate or extrapolate the value of i that gives a zero present worth.

step 5:   For increased accuracy, repeat steps (2) and (3) with two more values that straddle the value found in step (4).

A common, although incorrect, method of calculating the ROR involves dividing the annual receipts or returns by the initial investment. This technique ignores such items as salvage, depreciation, taxes, and the time value of money. This technique also fails when the annual returns vary.

Once a rate of return is known for an investment alternative, it is typically compared to the *minimum attractive rate of return* (MARR) specified by a company. However, ROR should not be used to rank alternatives. When two alternatives have RORs exceeding the MARR, it is not sufficient to select the alternative with the higher ROR.

An *incremental analysis*, also known as a *rate of return on added investment study*, should be performed if ROR is to be used to select between investments. In an incremental analysis, the cash flows for the investment with the lower initial cost are subtracted from the cash flows for the higher-priced alternative on a year-by-year basis. This produces, in effect, a third alternative representing the cost and benefits of the added investment.

The added expense of the higher-priced investment is not warranted unless the ROR of this third alternative exceeds the MARR as well.

Example 2.7

What is the return on invested capital if $1000 is invested now with $500 being returned in year 4 and $1000 being returned in year 8?

First, set up the problem as a present worth calculation.

$$P = -1000 + (500)(P/F, i\%, 4) + (1000)(P/F, i\%, 8)$$

Arbitrarily select $i = 5\%$. The present worth is then found to be $88.15. Next take a higher value of i to reduce the present worth. If $i = 10\%$, the present worth is $-$192. The ROR is found from simple interpolation to be approximately 6.6%. (A more exact solution is 6.37%.)

7 TREATMENT OF SALVAGE VALUE IN REPLACEMENT STUDIES

An investigation into the retirement of an existing process or piece of equipment is known as a *replacement study*. Replacement studies are similar in most respects to other alternative comparison problems: an interest rate is given, two alternatives exist, and one of the previously mentioned methods of comparing alternatives is used to choose the superior alternative.

In replacement studies, the existing process or piece of equipment is known as the *defender*. The new process or piece of equipment being considered for purchase is known as the *challenger*.

Because most defenders still have some market value when they are retired, the problem of what to do with the salvage arises. It seems logical to use the salvage value of the defender to reduce the initial purchase cost of the challenger. This is consistent with what would actually happen if the defender were to be retired.

By convention, however, the salvage value is subtracted from the defender's present value. This does not seem logical, but it is done to keep all costs and benefits related to the defender with the defender. In this case, the salvage value is treated as an opportunity cost that would be incurred if the defender is not retired.

If the defender and the challenger have the same lives and a present worth study is used to choose the superior alternative, the placement of the salvage value will have no effect on the net difference between present worths for the challenger and defender. Although the values of the two present worths will be different depending on the placement, the difference in present worths will be the same.

If the defender and the challenger have different lives, an annual cost comparison must be made. Since the salvage value would be spread over a different number of years depending on its placement, it is important to abide by the conventions listed in this section.

There are a number of ways to handle salvage value. The best way is to think of the EUAC of the defender as the cost of keeping the defender from now until next year. In addition to the usual operating and maintenance costs, that cost would include an opportunity interest cost incurred by not selling the defender and also a drop in the salvage value if the defender is kept for one additional year. Specifically,

$$
\begin{aligned}
\text{EUAC(defender)} = {}& \text{maintenance costs} \\
& + i \,(\text{current salvage value}) \\
& + (\text{current salvage} - \text{next} \\
& \quad \text{year's salvage}) \qquad 2.8
\end{aligned}
$$

It is important in retirement studies not to double count the salvage value. That is, it would be incorrect to add the salvage value to the defender and at the same time subtract it from the challenger.

8 BASIC INCOME TAX CONSIDERATIONS

Assume that an organization pays $f\%$ of its profits to the federal government as income taxes. If the organization also pays a state income tax of $s\%$, and if state taxes paid are recognized by the federal government as expenses, then the composite tax rate is

$$
t = s + f - sf \qquad 2.9
$$

The basic principles used to incorporate taxation into economic analyses are listed below.

- Initial purchase cost is unaffected by income taxes.

- Salvage value is unaffected by income taxes.

- Deductible expenses, such as operating costs, maintenance costs, and interest payments, are reduced by $t\%$ (i.e., multiplied by the quantity $(1-t)$).

- Revenues are reduced by $t\%$ (i.e., multiplied by the quantity $(1-t)$).

- Depreciation is multiplied by t and added to the appropriate year's cash flow, increasing that year's present worth.

Income taxes and depreciation have no bearing on municipal or governmental projects since municipalities, states, and the U.S. government pay no taxes.

Example 2.8

A corporation that pays 53% of its revenue in income taxes invests \$10,000 in a project that will result in \$3000 annual revenue for eight years. If the annual expenses are \$700, salvage after eight years is \$500, and 9% interest is used, what is the after-tax present worth? Disregard depreciation.

$$
\begin{aligned}
P_t = {}& -10{,}000 + (3000)(P/A, 9\%, 8)(1 - .53) \\
& - (700)(P/A, 9\%, 8)(1 - .53) \\
& + (500)(P/F, 9\%, 8) \\
= {}& -3766
\end{aligned}
$$

It is interesting that the alternative evaluated in example 2.8 is undesirable if income taxes are considered but is desirable if income taxes are omitted.

9 DEPRECIATION

Although depreciation calculations may be considered independently, it is important to recognize that depreciation has no effect on engineering economic calculations unless income taxes are also considered.

Generally, tax regulations do not allow the cost of equipment[1] to be treated as a deductible expense in the year of purchase. Rather, portions of the cost may be allocated to each year of the item's economic life (which may be different from the actual useful life). Each year, the book value (which is initially equal to the purchase price) is reduced by the depreciation in that year. Theoretically, the book value of an item will equal the market value at any time within the economic life of that item.

Since tax regulations allow the depreciation in any year to be handled as if it were an actual operating expense, and since operating expenses are deductible from the income base prior to taxation, the after-tax profits will be increased. If D is the depreciation, the net result to the after-tax cash flow will be the addition of tD.

The present worth of all depreciation over the economic life of the item is called the *depreciation recovery*. Although originally established to do so, depreciation recovery can never fully replace an item at the end of its life.

---

[1] The IRS tax regulations allow depreciation on almost all forms of *property* except land. The following types of property are distinguished: *real* (e.g., buildings used for business, etc.), *residential* (e.g., buildings used as rental property), and *personal* (e.g., equipment used for business). Personal property does *not* include items for personal use, despite its name. *Tangible* personal property is distinguished from *intangible* property (e.g., goodwill, copyrights, patents, trademarks, franchises, and agreements not to compete).

Depreciation is often confused with amortization and depletion. While depreciation spreads the cost of a fixed asset over a number of years, *amortization* spreads the cost of an intangible asset (e.g., a patent) over some basis such as time or expected units of production.

Depletion is another artificial deductible operating expense designed to compensate mining organizations for decreasing mineral reserves. Since original and remaining quantities of minerals are seldom known accurately, the *depletion allowance* is calculated as a fixed percentage of the organization's gross income. These percentages are usually in the 10% to 20% range and apply to such mineral deposits as oil, natural gas, coal, uranium, and most metal ores.

There are four common methods of calculating depreciation. The book value of an asset depreciated with the *straight line* (SL) *method* (also known as the *fixed percentage method*) decreases linearly from the initial purchase at $t = 0$ to the estimated salvage at $t = n$. The depreciated amount is the same each year. The quantity $(C - S_n)$ in equation 2.10 is known as the *depreciation base*.

$$D_j = \frac{C - S_n}{n} \qquad 2.10$$

Double declining balance[2] (DDB) depreciation is independent of salvage value. Furthermore, the book value never stops decreasing, although the depreciation decreases in magnitude. Usually, any remaining book value is written off in the last year of the asset's estimated life. Unlike any of the other depreciation methods, DDB depends on accumulated depreciation.

$$D_j = \frac{(2)(C - \sum_{i=1}^{j-1} D_i)}{n} \qquad 2.11$$

In *sum-of-the-years' digits* (SOYD) depreciation, the digits from 1 to n, inclusive, are summed. The total, T, can also be calculated from

$$T = \frac{1}{2} n(n + 1) \qquad 2.12$$

The depreciation can be found from

$$D_j = \frac{(C - S_n)(n - j + 1)}{T} \qquad 2.13$$

---

[2] Double declining balance depreciation is a particular form of *declining balance depreciation*, as defined by the IRS tax regulations. Declining balance depreciation also includes 125% declining balance and 150% declining balance depreciations which can be calculated by substituting 1.25 and 1.50, respectively for the 2 in equation 2.10.

The *sinking fund method* is seldom used in industry because the initial depreciation is low. The formula for sinking fund depreciation (which increases each year) is

$$D_j = (C - S_n)(A/F, i\%, n)(F/P, i\%, j\text{-}1) \qquad 2.14$$

The previous discussion gives the impression that any form of depreciation may be chosen regardless of the nature and circumstances of the purchase. In reality, the IRS tax regulations place restrictions on the higher-rate ("accelerated") methods such as DDB and SOYD. Furthermore, the *Economic Recovery Act of 1981* substantially changed the laws relating to personal and corporate income taxes.

Property placed into service in 1981 or after must use an *accelerated cost recovery system* (ACRS or MACRS). Other methods (straight-line, declining balance, etc.) cannot be used except in special cases.

Property placed into service in 1980 or before must continue to be depreciated according to the method originally chosen (e.g., straight-line, declining balance, or sum-of-the-years' digits). ACRS and MACRS cannot be used.

Under ACRS and MACRS, the cost recovery amount in the jth year of an asset's cost recovery period is calculated by multiplying the initial cost by a factor.

$$D_j = (\text{initial cost})(\text{factor}) \qquad 2.15$$

The initial cost used is not reduced by the asset's salvage value for either the regular or alternate ACRS and MACRS calculations. The factor used depends on the asset's cost recovery period. Such factors are subject to continuing legislation changes. Current tax publications should be consulted before using an accelerated method.

Example 2.9

An asset is purchased for $9000. Its estimated economic life is 10 years, after which it will be sold for $1000. Find the depreciation in the first three years using SL, DDB, and SOYD.

SL: $\quad D = \dfrac{9000 - 1000}{10}$

$\qquad\qquad = 800$ each year

DDB: $\quad D_1 = \dfrac{(2)(9000)}{10}$

$\qquad\qquad = 1800$ in year 1

$$D_2 = \frac{(2)(9000 - 1800)}{10}$$

$$= 1440 \text{ in year } 2$$

$$D_3 = \frac{(2)(9000 - 3240)}{10}$$

$$= 1152 \text{ in year } 3$$

SOYD: $\quad T = \frac{1}{2}(10)(11) = 55$

$$D_1 = \left(\frac{10}{55}\right)(9000 - 1000)$$

$$= 1455 \text{ in year } 1$$

$$D_2 = \left(\frac{9}{55}\right)(8000)$$

$$= 1309 \text{ in year } 2$$

$$D_3 = \left(\frac{8}{55}\right)(8000)$$

$$= 1164 \text{ in year } 3$$

Example 2.10

For the asset described in example 2.9, calculate the book value during the first three years if SOYD depreciation is used.

The book value at the beginning of year 1 is $9000. Then,

$$BV_1 = 9000 - 1455 = 7545$$
$$BV_2 = 7545 - 1309 = 6236$$
$$BV_3 = 6236 - 1164 = 5072$$

Example 2.11

For the asset described in example 2.9, calculate the after-tax depreciation recovery with SL and SOYD depreciation methods. Use 6% interest with 48% income taxes.

SL: $\quad D.R. = (.48)(800)(P/A, 6\%, 10)$
$$= 2826$$

SOYD: The depreciation series can be thought of as a constant 1454 term with a negative 145 gradient.

$$D.R. = (.48)(1454)(P/A, 6\%, 10)$$
$$- (.48)(145)(P/G, 6\%, 10)$$
$$= 3076$$

Finding book values, depreciation, and depreciation recovery is particularly difficult with DDB depreciation, since all previous years' quantities seem to be required. It appears that the depreciation in the sixth year cannot

be calculated unless the values of depreciation for the first five years are calculated first. Questions asking for depreciation or book value in the middle or at the end of an asset's economic life may be solved from the following equations:

$$d = \frac{2}{n} \qquad\qquad 2.16$$

$$z = \frac{1+i}{1-d} \qquad\qquad 2.17$$

$$(P/EG) = \frac{z^n - 1}{z^n(z - 1)} \qquad\qquad 2.18$$

Then, assuming that the remaining book value after n periods is written off in one lump sum, the present worth of the depreciation recovery is

$$D.R. = t\left[\frac{(d)(C)}{1-d}(P/EG) + 1 - d^n(C)(P/F, i\%, n)\right]$$
$$2.19$$

$$D_j = (d)(C)(1-d)^{j-1} \qquad\qquad 2.20$$

$$BV_j = C(1-d)^j \qquad\qquad 2.21$$

Example 2.12

What is the after-tax present worth of the asset described in example 2.8 if SL, SOYD, and DDB depreciation methods are used?

The after-tax present worth, neglecting depreciation, was previously found to be -3766.

Using SL, the depreciation recovery is

$$D.R. = (.53)\left(\frac{10,000 - 500}{8}\right)(P/A, 9\%, 8)$$

$$= 3483$$

Using SOYD, the depreciation recovery is calculated as follows:

$$T = \frac{1}{2}(8)(9) = 36$$

Depreciation base $= (10,000 - 500) = 9500$

$$D_1 = \frac{8}{36}(9500) = 2111$$

$$G = \text{gradient} = \frac{1}{36}(9500)$$

$$= 264$$

$$D.R. = (.53)\left[2111(P/A, 9\%, 8) \right.$$
$$\left. - 264(P/G, 9\%, 8)\right]$$
$$= 3829$$

Using DDB, the depreciation recovery is calculated as follows:

$$d = \frac{2}{8} = .25$$

$$z = \frac{1.09}{.75} = 1.453$$

$$(P/EG) = \frac{(1.453)^8 - 1}{(1.453)^8(.453)} = 2.096$$

$$D.R. = (.53)\left[\frac{(.25)(10,000)}{.75}(2.096)\right.$$
$$\left. + (.75)^8(10,000)(P/F, 9\%, 8)\right]$$
$$= 3969$$

The after-tax present worths, including depreciation recovery, are:

SL: $P_t = -3766 + 3483 = -283$
SOYD: $P_t = -3766 + 3829 = 63$
DDB: $P_t = -3766 + 3969 = 203$

10 ADVANCED INCOME TAX CONSIDERATIONS

There are a number of specialized techniques that are needed infrequently. These techniques are related more to the accounting profession than to the engineering profession.

A. TAX CREDIT

A *tax credit* (also known as an *investment tax credit* or an *investment credit*) is a one-time credit against income taxes. The investment tax credit is calculated as a fraction of the initial purchase price of certain types of equipment purchased for industrial, commercial, and manufacturing use.

$$\text{credit} = (\text{initial cost})(\text{fraction}) \qquad 2.22$$

Since the investment tax credit reduces the buyer's tax liability, the credit should only be used in after-tax analyses.

B. GAIN (OR LOSS) ON THE SALE OF A DEPRECIATED ASSET

If an asset is sold for more (or less) than its current book value, the difference between selling price and book value is taxable income (deductible expense). The gain is taxed at capital gains rates.

11 RATE AND PERIOD CHANGES

All of the foregoing calculations were based on compounding once a year at an *effective interest rate, i.* However, some problems specify compounding more frequently than annually. In such cases, a *nominal interest rate, r*, will be given. The nominal rate does not include the effect of compounding and is not the same as the effective rate, *i*. A nominal rate may be used to calculate the effective rate by using equation 2.23 or 2.24.

$$i = \left(1 + \frac{r}{k}\right)^k - 1 \qquad 2.23$$

$$= (1 + \phi)^k - 1 \qquad 2.24$$

A problem may also specify an effective rate per period, ϕ, (e.g., per month). However, that will be a simple problem since compounding for n periods at an effective rate per period is not affected by the definition or length of the period.

The following rules may be used to determine which interest rate is given in a problem:

- Unless specifically qualified in the problem, the interest rate given is an annual rate.

- If the compounding is annually, the rate given is the effective rate. If compounding is other than annually, the rate given is the nominal rate.

- If the type of compounding is not specified, assume annual compounding.

In the case of continuous compounding, the appropriate discount factors may be calculated from the formulas in table 2.2.

Table 2.2
Discount Factors for Continuous Compounding

(F/P)	e^{rn}
(P/F)	e^{-rn}
(A/F)	$\dfrac{e^r - 1}{e^{rn} - 1}$
(F/A)	$\dfrac{e^{rn} - 1}{e^r - 1}$
(A/P)	$\dfrac{e^r - 1}{1 - e^{-rn}}$
(P/A)	$\dfrac{1 - e^{-rn}}{e^r - 1}$

Example 2.13

A savings and loan offers $5\frac{1}{4}\%$ interest compounded daily. What is the annual effective rate?

method 1:

$$r = .0525, \quad k = 365$$

$$i = \left(1 + \frac{.0525}{365}\right)^{365} - 1 = .0539$$

method 2: Assume daily compounding is the same as continuous compounding.

$$i = (F/P) - 1$$
$$= e^{.0525} - 1 = .0539$$

12 PROBABILISTIC PROBLEMS

Thus far, all of the cash flows included in the examples have been known exactly. If the cash flows are not known exactly but are given by some implicit or explicit probability distribution, the problem is *probabilistic*.

Probabilistic problems typically possess the following characteristics:

- There is a chance of extreme loss that must be minimized.

- There are multiple alternatives that must be chosen from. Each alternative gives a different degree of protection against the loss or failure.

- The outcome is independent of the alternative chosen. Thus, as illustrated in example 2.15 the size of the dam that is chosen for construction will not alter the rainfall in successive years. However, it will alter the effects on the downstream watershed areas.

Probabilistic problems are typically solved using annual costs and expected values. An *expected value* is similar to an average value since it is calculated as the mean of the given probability distribution. If cost 1 has a probability of occurrence of p_1, cost 2 has a probability of occurrence of p_2, and so on, the expected value is

$$E(\text{cost}) = p_1(\text{cost } 1) + p_2(\text{cost } 2) + \cdots \qquad 2.25$$

Example 2.14

Flood damage in any year is given according to the table below. What is the present worth of flood damage for a 10-year period? Use 6%.

damage	probability
0	.75
$10,000	.20
$20,000	.04
$30,000	.01

The expected value of flood damage is

$$E(\text{damage}) = (0)(.75) + (10,000)(.20)$$
$$+ (20,000)(.04) + (30,000)(.01)$$
$$= 3100$$
$$\text{present worth} = (3100)(P/A, 6\%, 10)$$
$$= 22,816$$

Probabilities in probabilistic problems may be given in the problem (as in the example above) or they may have to be obtained from some named probability distribution. In either case, the probabilities are known explicitly and such problems are known as *explicit probability problems*.

Example 2.15

A dam is being considered on a river which periodically overflows and causes $600,000 damage. The damage is essentially the same each time the river causes flooding. The project horizon is 40 years. A 10% interest rate is being used.

Three different designs are available, each with different costs and storage capacities.

design alternative	cost	maximum capacity
A	500,000	1 unit
B	625,000	1.5 units
C	900,000	2.0 units

The U.S. Weather Service has provided a statistical analysis of annual rainfall in the area draining into the river.

units annual rainfall	probability
0	.10
.1–.5	.60
.6–1.0	.15
1.1–1.5	.10
1.6–2.0	.04
2.1 or more	.01

Which design alternative would you choose assuming the dam is essentially empty at the start of each rainfall season?

The sum of the construction cost and the expected damage needs to be minimized. If alternative A is chosen, it will have a capacity of 1 unit. Its capacity will be exceeded (causing \$600,000 damage) when the annual rainfall exceeds 1 unit. Therefore, the annual cost of A is

$$EUAC(A) = (500,000)(A/P, 10\%, 40)$$
$$+ (600,000)(.10 + .04 + .01)$$
$$= 141,150$$

Similarly,

$$EUAC(B) = (625,000)(A/P, 10\%, 40)$$
$$+ (600,000)(.04 + .01)$$
$$= 93,940$$

$$EUAC(C) = (900,000)(A/P, 10\%, 40)$$
$$+ (600,000)(.01)$$
$$= 98,070$$

Alternative B should be chosen.

In other problems, a probability distribution will not be given even though some parameter (such as the life of an alternative) is not known with certainty. Such problems are known as *implicit probability problems* since they require a reasonable assumption about the probability distribution.

Implicit probability problems typically involve items whose *expected time to failure* are known. The key to such problems is in recognizing that an expected time to failure is not the same as a fixed life.

Reasonable assumptions can be made about the form of probability distributions in implicit probability problems.

One such reasonable assumption is that of a *rectangular distribution*. A rectangular distribution is one that is assumed to give an equal probability of failure in each year. Such an assumption is illustrated in example 2.16.

Example 2.16

A bridge is needed for 20 years. Failure of the bridge at any time will require a 50% reinvestment. Assume that each alternative has an annual probability of failure that is inversely proportional to its expected time to failure. Evaluate the two design alternatives below using 6% interest.

design alternative	initial cost	expected time to failure	annual costs	salvage at $t = 20$
A	15,000	9 years	1200	0
B	22,000	43 years	1000	0

For alternative A, the probability of failure in any year is 1/9. Similarly, the annual failure probability for alternative B is 1/43.

$$EUAC(A) = (15,000)(A/P, 6\%, 20)$$
$$+ (15,000)(.5)\left(\frac{1}{9}\right) + 1200$$
$$= 3341$$

$$EUAC(B) = (22,000)(A/P, 6\%, 20)$$
$$+ (22,000)(.5)\left(\frac{1}{43}\right) + 1000$$
$$= 3174$$

Alternative B should be chosen.

13 ESTIMATING ECONOMIC LIFE

As assets grow older, their operating and maintenance costs typically increase each year. Eventually, the cost to keep an asset in operation becomes prohibitive, and the asset is retired or replaced. However, it is not always obvious when an asset should be retired or replaced.

As the asset's maintenance is increasing each year, the amortized cost of its initial purchase is decreasing. It is the sum of these two costs that should be evaluated to determine the point at which the asset should be retired or replaced. Since an asset's initial purchase price is likely to be high, the amortized cost will be the controlling factor in those years when the maintenance costs are low. Therefore, the EUAC of the asset will decrease in the initial part of its life.

However, as the asset grows older, the change in its amortized cost decreases while maintenance increases. Eventually the sum of the two costs reaches a minimum and then starts to increase. The age of the asset at the minimum cost point is known as the *economic life* of the asset. The economic life is, generally, less than the mission and technological lifetimes of the asset.

The determination of an asset's economic life is illustrated by example 2.17.

Example 2.17

A bus in a municipal transit system has the characteristics listed below. When should the city replace its buses if money can be borrowed at 8% ?

initial cost: $120,000

year	maintenance cost	salvage value
1	35,000	60,000
2	38,000	55,000
3	43,000	45,000
4	50,000	25,000
5	65,000	15,000

If the bus is kept for one year and then sold, the annual cost will be

$$EUAC(1) = (120,000)(A/P, 8\%, 1)$$
$$+ (35,000)(A/F, 8\%, 1)$$
$$- (60,000)(A/F, 8\%, 1)$$
$$= 104,600$$

If the bus is kept for two years and then sold, the annual cost will be

$$EUAC(2) = [120,000 + (35,000)(P/F, 8\%, 1)](A/P, 8\%, 2)$$
$$+ (38,000 - 55,000)(A/F, 8\%, 2)$$
$$= 77,300$$

If the bus is kept for three years and then sold, the annual cost will be

$$EUAC(3) = [120,000 + (35,000)(P/F, 8\%, 1)$$
$$+ (38,000)(P/F, 8\%, 2)](A/P, 8\%, 3)$$
$$+ (43,000 - 45,000)(A/F, 8\%, 3)$$
$$= 71,200$$

This process is continued until EUAC begins to increase. In this example, EUAC(4) is 71,700. Therefore, the bus should be retired after three years.

14 BASIC COST ACCOUNTING

Cost accounting is the system that determines the cost of manufactured products. Cost accounting is called *job cost accounting* if costs are accumulated by part number or contract. It is called *process cost accounting* if costs are accumulated by departments or manufacturing processes.

Three types of costs (direct material, direct labor, and all indirect costs) make up the total manufacturing cost of a product.

Direct material costs are the costs of all materials that go into the product, priced at the original purchase cost.

Indirect material and labor costs are generally limited to costs incurred in the factory, excluding costs incurred in the office area. Examples of indirect materials are cleaning fluids, assembly lubricants, and temporary routing tags. Examples of indirect labor are stock-picking, inspection, expediting, and supervision labor.

Here are some important points concerning basic cost accounting:

- The sum of direct material and direct labor costs is known as the *prime cost*.
- Indirect costs may be called *indirect manufacturing expenses* (IME).
- Indirect costs may also include the overhead sector of the company (e.g., secretaries, engineers, and corporate administration). In this case, the indirect cost is usually called *burden* or *overhead*. Burden may also include the EUAC of non-regular costs which must be spread evenly over several years.
- The cost of a product is usually known in advance from previous manufacturing runs or by estimation. Any deviation from this known cost is called a *variance*. Variance may be broken down into *labor variance* and *material variance*.
- Indirect cost per item is not easily measured. The method of allocating indirect costs to a product is as follows:

step 1: Estimate the total expected indirect (and over-head) costs for the upcoming year.

step 2: Decide on some convenient vehicle for allocating the overhead to production. Usually, this vehicle is either the number of units expected to be produced or the number of direct hours expected to be worked in the upcoming year.

step 3: Estimate the quantity or size of the overhead vehicle.

step 4: Divide expected overhead costs by the expected overhead vehicle to obtain the unit overhead.

step 5: Regardless of the true size of the overhead vehicle during the upcoming year, one unit of overhead cost is allocated per product.

• Although estimates of production for the next year are always somewhat inaccurate, the cost of the product is assumed to be independent of forecasting errors. Any difference between true cost and calculated cost goes into a variance account.

• *Burden (overhead) variance* will be caused by errors in forecasting both the actual overhead for the upcoming year and the vehicle size. In the former case, the variance is called *burden budget variance*; in the latter, it is called *burden capacity variance*.

Example 2.18

A small company expects to produce 8000 items in the upcoming year. The current material cost is $4.54 each. 16 minutes of direct labor are required per unit. Workers are paid $7.50 per hour. 2133 direct labor hours are forecast for the product. Miscellaneous overhead costs are estimated at $45,000.

Find the expected direct material cost, the direct labor cost, the prime cost, the burden as a function of production and direct labor, and the total cost.

The direct material cost was given as $4.54.

The direct labor cost is $(16/60)\,(\$7.50) = \$2.00$.

The prime cost is $\$4.54 + \$2.00 = \$6.54$.

If the burden vehicle is production, the burden rate is $\$45,000/8000 = \$5.63$ per item, making the total cost $\$4.54 + \$2.00 + \$5.63 = \$12.17$.

If the burden vehicle is direct labor hours, the burden rate is $(45,000/2133) = \$21.10$ per hour, making the total cost $\$4.54 + \$2.00 + (16/60)(\$21.10) = \$12.17$.

Example 2.19

The actual performance of the company in example 2.18 is given by the following figures:

actual production: 7560

actual overhead costs: $47,000

What are the burden budget variance and the burden capacity variance?

The burden capacity variance is

$$\$45,000 - (7560)(\$5.63) = \$2437$$

The burden budget variance is

$$\$47,000 - \$45,000 = \$2000$$

The overall burden variance is

$$\$47,000 - (7560)(\$5.63) = \$4437$$

15 BREAK-EVEN ANALYSIS

Break-even analysis is a method of determining when costs exactly equal revenue. If the manufactured quantity is less than the break-even quantity, a loss is incurred. If the manufactured quantity is greater than the break-even quantity, a profit is incurred.

Consider the following special variables:

f a fixed cost that does not vary with production

a an incremental cost that is the cost to produce one additional item. It may also be called the *marginal cost* or *differential cost*.

Q the quantity sold

p the incremental revenue

R the total revenue

C the total cost

Assuming no change in the inventory, the *break-even point* can be found from $C = R$, where

$$C = f + aQ \qquad 2.26$$
$$R = pQ \qquad 2.27$$

An alternate form of the break-even problem is to find the number of units per period for which two alternatives have the same total costs. Fixed costs are to be spread over a period longer than one year. One of the alternatives will have a lower cost if production is less than the break-even point. The other will have a lower cost for production greater than the break-even point.

The *cost per unit* problem is a variation of the break-even problem. In the typical cost per unit problem, data will be available to determine the direct labor and material costs per unit, but some method is needed to additionally allocate part of the annual overhead (burden) and initial facility purchase/construction costs.

Annual overhead is allocated to the unit cost simply by dividing the overhead by the number of units produced each year. The initial purchase/construction cost is multiplied by the appropriate (A/P) factor before similarly dividing by the production rate. The total unit cost is the sum of the direct labor, direct material, pro rata share of overhead, and pro rata share of the equivalent annual facility investment costs.

Example 2.20

Two plans are available for a company to obtain automobiles for its salesmen. How many miles must the cars be driven each year for the two plans to have the same costs? Use an interest rate of 10%. Assume the year-end convention applies to the insurance.

Plan A Lease the cars and pay $.15 per mile.

Plan B Purchase the cars for $5000. Each car has an economic life of three years, after which it can be sold for $1,200. Gas and oil cost $.04 per mile. Insurance is $500 per year. Assume the year-end convention applies to the insurance.

Let x be the number of miles driven per year. Then, the EUAC for both alternatives is

$$\text{EUAC}(A) = .15x$$
$$\text{EUAC}(B) = .04x + 500 + (5000)(A/P, 10\%, 3)$$
$$- (1200)(A/F, 10\%, 3)$$
$$= .04x + 2148$$

Setting EUAC(A) and EUAC(B) equal and solving for x yields 19,527 miles per year as the break-even point.

16 HANDLING INFLATION

It is important to perform economic studies in terms of *constant value dollars*. One method of converting all cash flows to constant value dollars is to divide the flows by some annual *economic indicator* or price index. Such indicators would normally be given to you as part of a problem.

If indicators are not available, this method can still be used by assuming that inflation is relatively constant at a decimal rate e per year. Then, all cash flows can be converted to $t = 0$ dollars by dividing by $(1+e)^n$ where n is the year of the cash flow.

Example 2.21

What is the uninflated present worth of $2000 in two years if the average inflation rate is 6% and i is 10% ?

$$P = \frac{\$2000}{(1.10)^2(1.06)^2} = \$1471.07$$

An alternative is to replace i with a value corrected for inflation. This corrected value, i', is

$$i' = i + e + ie \qquad\qquad 2.28$$

This method has the advantage of simplifying the calculations. However, pre-calculated factors may not be available for the non-integer values of i'. Therefore, table 2.1 will have to be used to calculate the factors.

Example 2.22

Repeat example 2.21 using i'.

$$i' = .10 + .06 + (.10)(.06)$$
$$= .166$$
$$P = \frac{\$2000}{(1.166)^2} = \$1471.07$$

17 LEARNING CURVES

The more products that are made, the more efficient the operation becomes due to experience gained. Therefore, direct labor costs decrease. Usually, a *learning curve* is specified by the decrease in cost each time the quantity produced doubles. If there is a 20% decrease per doubling, the curve is said to be an 80% learning curve.

Consider the following special variables:

T_1	time or cost for the first item
T_n	time or cost for the nth item
n	total number of items produced
b	learning curve constant

Table 2.3
Learning Curve Constants

learning curve	b
80%	.322
85%	.234
90%	.152
95%	.074

Then, the time to produce the nth item is given by

$$T_n = T_1(n)^{-b} \qquad 2.29$$

The total time to produce units from quantity n_1 to n_2 inclusive is

$$\int_{n_1}^{n_2} T_n \, dn \approx \frac{T_1}{1-b}$$

$$\times \left[\left(n_2 + \frac{1}{2} \right)^{1-b} - \left(n_1 - \frac{1}{2} \right)^{1-b} \right] \qquad 2.30$$

The average time per unit over the production from n_1 to n_2 is the above total time from equation 2.30 divided by the quantity produced, $(n_2 - n_1 + 1)$.

It is important to remember that learning curve reductions apply only to direct labor costs. They are not applied to indirect labor or direct material costs.

Example 2.23

A 70% learning curve is used with an item whose first production time was 1.47 hours. How long will it take to produce the 11th item? How long will it take to produce the 11th through 27th items?

First, find b.

$$\frac{T_2}{T_1} = .7 = (2)^{-b}$$

$$b = .515$$

Then,

$$T_{11} = (1.47)(11)^{-.515} = .428 \text{ hours}$$

The time to produce the 11th item through 27th item is approximately

$$T = \frac{1.47}{1 - .515} \left[(27.5)^{1-.515} - (10.5)^{1-.515} \right]$$

$$= 5.643 \text{ hours}$$

18 ECONOMIC ORDER QUANTITY

The *economic order quantity* (EOQ) is the order quantity that minimizes the inventory costs per unit time. Although there are many different EOQ models, the simplest is based on the following assumptions:

- Reordering is instantaneous. The time between order placement and receipt is zero.

- Shortages are not allowed.

- Demand for the inventory item is deterministic (i.e., is not a random variable).

- Demand is constant with respect to time.

- An order is placed when the on-hand quantity is zero.

The following special variables are used:

a the constant depletion rate $\left(\frac{\text{items}}{\text{unit time}} \right)$

h the inventory storage cost $\left(\frac{\$}{\text{item-unit time}} \right)$

H the total inventory storage cost between orders ($)

K the fixed cost of placing an order ($)

Q_0 the order quantity

If the original quantity on hand is Q_0, the stock will be depleted at

$$t^* = \frac{Q_0}{a} \qquad 2.31$$

The total inventory storage cost between t_0 and t^* is

$$H = \frac{1}{2} h \frac{Q_0^2}{a} \qquad 2.32$$

The total inventory and ordering cost per unit time is

$$C_t = \frac{aK}{Q_0} + \frac{1}{2} h Q_0 \qquad 2.33$$

C_t can be minimized with respect to Q_0. The EOQ and time between orders are:

$$Q_0^* = \sqrt{2 \frac{aK}{h}} \qquad 2.34$$

$$t^* = \frac{Q_0^*}{a} \qquad 2.35$$

19 CONSUMER LOANS

Many consumer loans cannot be handled by the equivalence formulas presented up to this point. Many different arrangements can be made between lender and borrower. Four of the most common consumer loan arrangements are presented below. Refer to a real estate or investment analysis book for more complex loans.

A. SIMPLE INTEREST

Interest due does not compound with a *simple interest* loan. The interest due is merely proportional to the length of time the principal is outstanding. Because of this, simple interest loans are seldom made for long periods (e.g., longer than one year).

Example 2.24

A \$12,000 simple interest loan is taken out at 16% per year. The loan matures in one year with no intermediate payments. How much will be due at the end of the year?

$$\text{amount due} = (1 + .16)(\$12,000)$$
$$= \$13,920$$

For loans less than one year, it is commonly assumed that a year consists of 12 months of 30 days each.

Example 2.25

\$4000 is borrowed for 75 days at 16% per annum simple interest. How much will be due at the end of 75 days?

$$\text{amount due} = \$4000 + (.16)\left(\frac{75}{360}\right)(4000)$$
$$= \$4133$$

B. LOANS WITH CONSTANT AMOUNT PAID TOWARDS PRINCIPAL

With this loan type, the payment is not the same each period. The amount paid towards the principal is constant, but the interest varies from period to period. The following special symbols are used:

BAL_j balance after the jth payment
LV total value loaned (cost minus down payment)
j payment or period number
N total number of payments to pay off the loan
PI_j jth interest payment
PP_j jth principal payment

PT_j jth total payment
ϕ effective rate per period (r/k)

The equations which govern this type of loan are:

$$BAL_j = LV - (j)(PP) \qquad 2.36$$
$$PI_j = \phi(BAL_{j-1}) \qquad 2.37$$
$$PT_j = PP + PI_j \qquad 2.38$$

C. DIRECT REDUCTION LOANS

This is the typical "interest paid on unpaid balance" loan. The amount of the periodic payment is constant, but the amounts paid towards the principal and interest both vary.

The same symbols are used with this type of loan as are listed above.

$$N = -\frac{ln\left[\frac{-\phi(LV)}{PT} + 1\right]}{ln(1 + \phi)} \qquad 2.39$$

$$BAL_{j-1} = PT\left[\frac{1 - (1+\phi)^{j-1-N}}{\phi}\right] \qquad 2.40$$

$$PI_j = \phi(BAL_{j-1}) \qquad 2.41$$

$$PP_j = PT - PI_j \qquad 2.42$$

$$BAL_j = BAL_{j-1} - PP_j \qquad 2.43$$

Example 2.26

A \$45,000 loan is financed at 9.25% per annum. The monthly payment is \$385. What are the amounts paid toward interest and principal in the 14th period? What is the remaining principal balance after the 14th payment has been made?

The effective rate per month is

$$\phi = \frac{r}{k} = \frac{.0925}{12}$$
$$= .007708$$

$$N = \frac{-ln\left[\frac{-(.007708)(45,000)}{385} + 1\right]}{ln(1 + .007708)} = 301$$

$$BAL_{13} = (385)\left[\frac{1 - (1 + .007708)^{14-1-301}}{.007708}\right]$$
$$= \$44,476.39$$

$$PI_{14} = (.007708)(\$44,476.39) = \$342.82$$

$$PP_{14} = \$385 - \$342.82 = \$42.18$$

$$BAL_{14} = \$44,476.39 - \$42.18 = \$44,434.21$$

Equation 2.39 calculates the number of payments necessary to pay off a loan. This equation can be solved with effort for the total periodic payment (PT) or the initial value of the loan (LV). It is easier, however, to use the $(A/P, \phi, n)$ factor to find the payment and loan value.

$$PT = (LV)(A/P, \phi\%, n) \qquad 2.44$$

If the loan is repaid in yearly installments, then i is the effective annual rate. If the loan is paid off monthly, then i should be replaced by the effective rate per month (ϕ from equation 2.24). For monthly payments, n is the number of months in the payback period.

D. DIRECT REDUCTION LOAN WITH BALLOON PAYMENT

This type of loan has a constant periodic payment, but the duration of the loan is insufficient to completely pay back the principal. Therefore, all remaining unpaid principal must be paid back in a lump sum when the loan matures. This large payment is known as a *balloon payment*.

Equations 2.39 through 2.43 can also be used with this type of loan. The remaining balance after the last payment is the balloon payment. This balloon payment must be repaid along with the last regular payment calculated.

20 SENSITIVITY ANALYSIS

Data analysis and forecasts in economic studies represent judgment on costs that will occur in the future. There are always uncertainties about these costs. However, these uncertainties are insufficient reason not to make the best possible estimates of the costs. Nevertheless, a decision between alternatives often can be made more confidently if it is known whether or not the conclusion is sensitive to moderate changes in data forecasts. Sensitivity analysis provides this extra dimension to an economic analysis.

The sensitivity of a decision is determined by inserting a range of estimates for critical cash flows. If radical changes can be made to a cash flow without changing the decision, the decision is said to be insensitive to uncertainties regarding that cash flow. However, if a small change in the estimate of a cash flow will alter the decision, that decision is said to be very sensitive to changes in the estimate.

An established semantic tradition distinguishes between risk analysis and uncertainty analysis. Risk analysis addresses variables that have a known or estimated probability distribution. In this regard, statistics and probability theory can be used to determine the probability of a cash flow varying between given limits. On the other hand, uncertainty analysis is concerned with situations in which there is not enough information to determine the probability or frequency distribution for the variables involved.

As a first step, sensitivity analysis should be applied one at a time to the dominant cost factors. Dominant cost factors are those which have the most significant impact on the present value of the alternative. If warranted, additional investigation can be used to determine the sensitivity to several cash flows varying simultaneously. Significant judgment is needed, however, to successfully determine the proper combinations of cash flows to vary.

It is common to plot the dependency of the present value on the cash flow being varied on a two-dimensional graph. Simple linear interpolation is used (within reason) to determine the critical value of the cash flow being varied.

PRACTICE PROBLEMS: ENGINEERING ECONOMICS

Warm-ups

1. How much will be accumulated at 6% if $1000 is invested for 10 years?

2. What is the present worth of $2000 at 6% which becomes available in four years?

3. How much will it take to accumulate $2000 in 20 years at 6%?

4. What year-end annual amount over seven years at 6% is equivalent to $500 invested now?

5. $50 is invested at the end of each year for 10 years. What will be the accumulated amount at the end of 10 years at 6%?

6. How much should be deposited at 6% at the start of each year for 10 years in order to empty the fund by drawing out $200 at the end of each year for 10 years?

7. How much should be deposited at 6% at the start of each year for five years to accumulate $2000 on the date of the last deposit?

8. How much will be accumulated at 6% in 10 years if three payments of $100 are deposited every other year for four years, with the first payment occuring at $t = 0$?

9. $500 is compounded monthly at 6% annual rate. How much will be accumulated in five years?

10. What is the rate of return on an $80 investment that pays back $120 in seven years?

Concentrates

1. A new machine will cost $17,000 and will have a value of $14,000 in five years. Special tooling will cost $5000 and it will have a resale value of $2500 after five years. Maintenance will be $200 per year. What will the average cost of ownership be during the next five years if interest is 6%?

2. An old highway bridge can be strengthened at a cost of $9000, or it can be replaced for $40,000. The present salvage value of the old bridge is $13,000. It is estimated that the reinforced bridge will last for 20 years with an annual cost of $500 and will have a salvage value of $10,000 at the end of 20 years. The estimated salvage of the new bridge after 25 years is $15,000. The maintenance for the new bridge will be $100 annually. Which is the best alternative at 8% interest?

3. A firm expects to receive $32,000 each year for 15 years from the sale of a product. It will require an initial investment of $150,000. Expenses will run $7530 per year. Salvage is zero and straight-line depreciation is used. The tax rate is 48%. What is the after-tax rate of return?

4. A public works project has initial costs of $1,000,000, benefits of $1,500,000, and disbenefits of $300,000. (a) What is the benefit/cost ratio? (b) What is the excess of benefits over costs?

5. A speculator in land pays $14,000 for property that he expects to hold for 10 years. $1000 is spent in renovation and a monthly rent of $75 is collected from the tenants. Taxes are $150 per year and maintenance costs are $250. What must be the sale price in 10 years to realize a 10% rate of return? Use the year-end convention.

6. What is the effective interest rate for a payment plan of 30 equal payments of $89.30 per month when a lump sum of $2000 would have been an outright purchase?

7. An apartment complex is purchased for $500,000. What is the depreciation in each of the first three years if the salvage value is $100,000 in 25 years? Use (a) straight line, (b) sum-of-the-years' digits, and (c) double declining balance.

8. Equipment is purchased for $12,000 and is expected to be sold after 10 years for $2000. The estimated maintenance is $1000 for the first year, but is expected to increase $200 each year thereafter. Using 10%, find the present worth and the annual cost.

9. One of five grades of pipe with average lives (in years) and costs (in dollars) of (9, 1500), (14, 1600), (30, 1750), (52, 1900), and (86, 2100) is to be chosen for a 20-year project. A failure of the pipe at any time during the project will result in a cost equal to 35% of the original cost. Annual costs are 4% of the initial cost, and the pipes are not recoverable. At 6%, which pipe is superior? Note: The lives are average values, not absolute replacement times.

10. A grain combine with a 20-year life can remove seven pounds of rock from its harvest per hour. Any rocks left in its output will cause $25,000 damage in subsequent processes. Several investments are available to increase its removal capacity. At 10%, what should be done?

rock rate (lb/hr)	probability of exceeding rock rate	required investment to meet rock rate
7	.15	no cost
8	.10	$15,000
9	.07	$20,000
10	.03	$30,000

Timed (1 hour allowed for each)

1. A structure that costs $10,000 has the operating costs and salvage values given.

> year 1: maintenance $2000, salvage $8000
> year 2: maintenance $3000, salvage $7000
> year 3: maintenance $4000, salvage $6000
> year 4: maintenance $5000, salvage $5000
> year 5: maintenance $6000, salvage $4000

(a) What is the economic life of the structure? (b) Assuming that the structure has been owned and operated for four years, what is the cost of owning the structure for exactly one more year? Use 20% as the interest rate.

2. A man purchases a car for $5000 for personal use, intending to drive 15,000 miles per year. It costs him $200 per year for insurance and $150 per year for maintenance. He gets 15 mpg and gasoline costs $0.60 per gallon. The resale value after five years is $1000. Because of unexpected business driving (5000 miles per year extra), his insurance is increased to $300 per year and maintenance to $200. Salvage is reduced to $500. Use 10% interest to answer the following questions. (a) The man's company offers $0.10 per mile reimbursement. Is that adequate? (b) How many miles must be driven per year at $0.10 per mile to justify the company buying a car for the man's use? The cost would be $5000, but insurance, maintenance, and salvage would be $250, $200, and $800, respectively.

3. Two alternatives are to be evaluated. The cash flows for these two alternatives are presented below. Use a 10% interest rate to determine which alternative is superior.

	A	B
First cost	$80,000	$35,000
Life	20 years	10 years
Salvage value	$7,000	0
Annual costs		
years 1-5	$1,000	$3,000
years 6-10	$1,500	$4,000
years 11-20	$2,000	—
Additional cost at		
$t = 10$	$5,000	—

4. Two plans are available for the use of automobiles over a three-year period. Find the number of miles the cars would have to be driven each year to make the plans have the same cost. Use 10% as an interest rate.

Option A:	Lease cars at $.15 per mile
Option B:	Purchase cars for $5,000
	Life: 3 years
	Salvage: $1,200
	Gas and Oil: $.04 per mile
	Other costs: $500 per year

5. Two methods are being considered to meet air pollution control requirements. Method 1 is based on an installation with a 10-year life. Method 2 is based on an interim installation with a 5-year life, to be replaced by another installation with a 5-year life. Use an interest rate of 7%.

	A	B 1st	B 2nd
Installation cost	$13,000	$6,000	$7,000
Equipment cost	$10,000	$2,000	$2,200
Operating cost			
(per hour)	$10.50	$8.00	$8.00
Salvage value	$5,000	$2,000	$2,000
Capacity	50 tons/yr	20 tons/yr	20 tons/yr
Life	10 years	5 years	5 years

(a) Find the uniform annual cost per ton for both methods. Assume 24 hours per day, 365 days per year. (b) Find the ranges (in tons/year) for which each method has the minimum annual cost.

6. A city transportation agency has asked for your assistance in determining the proper fare for its bus system. The following information was compiled for your study.

> Current cost per bus: $60,000
>
> Expected life: 20 years
>
> Salvage value after 20 years: $10,000
>
> Use: 37,440 miles per year, and 80,000 passengers per year
>
> Costs: $1.00/mile per bus the first year, increasing by $.10 each year thereafter
>
> Interest rate: 7%

(a) Determine the break-even fare to assess for each passenger trip if the fare is to remain constant for 20 years. Express your answer in dollars per passenger trip.

(b) If the agency institutes a fare of $.35 the first year, what is the uniform gradient which should be applied to the fares in following years if the agency is to break even after 20 years? Express your answer in dollars per passenger trip.

(c) If the fare is $.35 the first year, and the uniform gradient is set at $.05 per passenger trip, what is the value of the required government subsidy? Express your answer in dollars per passenger trip.

STANDARD CASH FLOW FACTORS

MULTIPLY	BY	TO OBTAIN

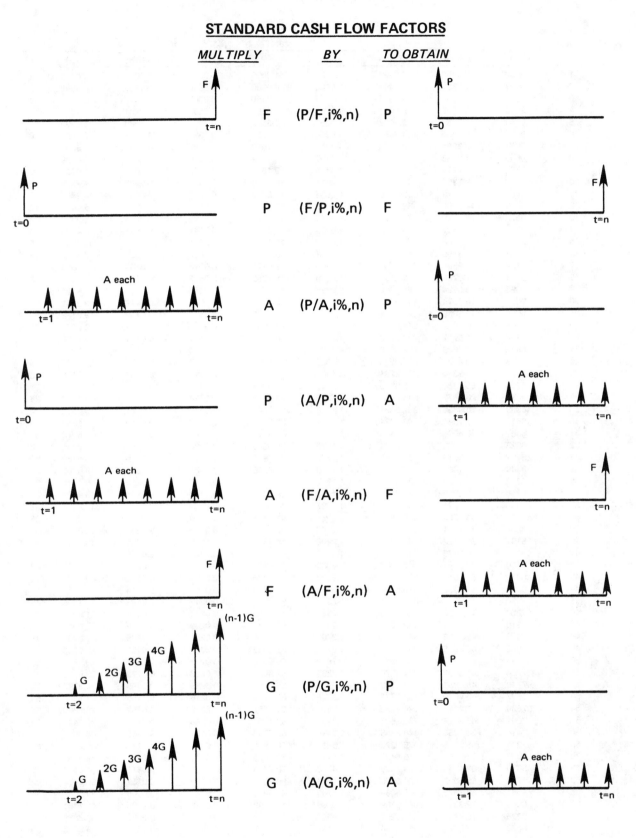

F	(P/F,i%,n)	P
P	(F/P,i%,n)	F
A	(P/A,i%,n)	P
P	(A/P,i%,n)	A
A	(F/A,i%,n)	F
F	(A/F,i%,n)	A
G	(P/G,i%,n)	P
G	(A/G,i%,n)	A

ENGINEERING ECONOMIC ANALYSIS

I = 0.50 %

N	(P/F)	(P/A)	(P/G)	(F/P)	(F/A)	(A/P)	(A/F)	(A/G)	N
1	.9950	0.9950	− 0.0000	1.0050	1.0000	1.0050	1.0000	− 0.0000	1
2	.9901	1.9851	0.9901	1.0100	2.0050	0.5038	0.4988	0.4988	2
3	.9851	2.9702	2.9604	1.0151	3.0150	0.3367	0.3317	0.9967	3
4	.9802	3.9505	5.9011	1.0202	4.0301	0.2531	0.2481	1.4938	4
5	.9754	4.9259	9.8026	1.0253	5.0503	0.2030	0.1980	1.9900	5
6	.9705	5.8964	14.6552	1.0304	6.0755	0.1696	0.1646	2.4855	6
7	.9657	6.8621	20.4493	1.0355	7.1059	0.1457	0.1407	2.9801	7
8	.9609	7.8230	27.1755	1.0407	8.1414	0.1278	0.1228	3.4738	8
9	.9561	8.7791	34.8244	1.0459	9.1821	0.1139	0.1089	3.9668	9
10	.9513	9.7304	43.3865	1.0511	10.2280	0.1028	0.0978	4.4589	10
11	.9466	10.6770	52.8526	1.0564	11.2792	0.0937	0.0887	4.9501	11
12	.9419	11.6189	63.2136	1.0617	12.3356	0.0861	0.0811	5.4406	12
13	.9372	12.5562	74.4602	1.0670	13.3972	0.0796	0.0746	5.9302	13
14	.9326	13.4887	86.5835	1.0723	14.4642	0.0741	0.0691	6.4190	14
15	.9279	14.4166	99.5743	1.0777	15.5365	0.0694	0.0644	6.9069	15
16	.9233	15.3399	113.4238	1.0831	16.6142	0.0652	0.0602	7.3940	16
17	.9187	16.2586	128.1231	1.0885	17.6973	0.0615	0.0565	7.8803	17
18	.9141	17.1728	143.6634	1.0939	18.7858	0.0582	0.0532	8.3658	18
19	.9096	18.0824	160.0360	1.0994	19.8797	0.0553	0.0503	8.8504	19
20	.9051	18.9874	177.2322	1.1049	20.9791	0.0527	0.0477	9.3342	20
21	.9006	19.8880	195.2434	1.1104	22.0840	0.0503	0.0453	9.8172	21
22	.8961	20.7841	214.0611	1.1160	23.1944	0.0481	0.0431	10.2993	22
23	.8916	21.6757	233.6768	1.1216	24.3104	0.0461	0.0411	10.7806	23
24	.8872	22.5629	254.0820	1.1272	25.4320	0.0443	0.0393	11.2611	24
25	.8828	23.4456	275.2686	1.1328	26.5591	0.0427	0.0377	11.7407	25
26	.8784	24.3240	297.2281	1.1385	27.6919	0.0411	0.0361	12.2195	26
27	.8740	25.1980	319.9523	1.1442	28.8304	0.0397	0.0347	12.6975	27
28	.8697	26.0677	343.4332	1.1499	29.9745	0.0384	0.0334	13.1747	28
29	.8653	26.9330	367.6625	1.1556	31.1244	0.0371	0.0321	13.6510	29
30	.8610	27.7941	392.6324	1.1614	32.2800	0.0360	0.0310	14.1265	30
31	.8567	28.6508	418.3348	1.1672	33.4414	0.0349	0.0299	14.6012	31
32	.8525	29.5033	444.7618	1.1730	34.6086	0.0339	0.0289	15.0750	32
33	.8482	30.3515	471.9055	1.1789	35.7817	0.0329	0.0279	15.5480	33
34	.8440	31.1955	499.7583	1.1848	36.9606	0.0321	0.0271	16.0202	34
35	.8398	32.0354	528.3123	1.1907	38.1454	0.0312	0.0262	16.4915	35
36	.8356	32.8710	557.5598	1.1967	39.3361	0.0304	0.0254	16.9621	36
37	.8315	33.7025	587.4934	1.2027	40.5328	0.0297	0.0247	17.4317	37
38	.8274	34.5299	618.1054	1.2087	41.7354	0.0290	0.0240	17.9006	38
39	.8232	35.3531	649.3883	1.2147	42.9441	0.0283	0.0233	18.3686	39
40	.8191	36.1722	681.3347	1.2208	44.1588	0.0276	0.0226	18.8359	40
41	.8151	36.9873	713.9372	1.2269	45.3796	0.0270	0.0220	19.3022	41
42	.8110	37.7983	747.1886	1.2330	46.6065	0.0265	0.0215	19.7678	42
43	.8070	38.6053	781.0815	1.2392	47.8396	0.0259	0.0209	20.2325	43
44	.8030	39.4082	815.6087	1.2454	49.0788	0.0254	0.0204	20.6964	44
45	.7990	40.2072	850.7631	1.2516	50.3242	0.0249	0.0199	21.1595	45
46	.7950	41.0022	886.5376	1.2579	51.5758	0.0244	0.0194	21.6217	46
47	.7910	41.7932	922.9252	1.2642	52.8337	0.0239	0.0189	22.0831	47
48	.7871	42.5803	959.9188	1.2705	54.0978	0.0235	0.0185	22.5437	48
49	.7832	43.3635	997.5116	1.2768	55.3683	0.0231	0.0181	23.0035	49
50	.7793	44.1428	1035.6966	1.2832	56.6452	0.0227	0.0177	23.4624	50
51	.7754	44.9182	1074.4670	1.2896	57.9284	0.0223	0.0173	23.9205	51
52	.7716	45.6897	1113.8162	1.2961	59.2180	0.0219	0.0169	24.3778	52
53	.7677	46.4575	1153.7372	1.3026	60.5141	0.0215	0.0165	24.8343	53
54	.7639	47.2214	1194.2236	1.3091	61.8167	0.0212	0.0162	25.2899	54
55	.7601	47.9814	1235.2686	1.3156	63.1258	0.0208	0.0158	25.7447	55
60	.7414	51.7256	1448.6458	1.3489	69.7700	0.0193	0.0143	28.0064	60
65	.7231	55.3775	1675.0272	1.3829	76.5821	0.0181	0.0131	30.2475	65
70	.7053	58.9394	1913.6427	1.4178	83.5661	0.0170	0.0120	32.4680	70
75	.6879	62.4136	2163.7525	1.4536	90.7265	0.0160	0.0110	34.6679	75
80	.6710	65.8023	2424.6455	1.4903	98.0677	0.0152	0.0102	36.8474	80
85	.6545	69.1075	2695.6389	1.5280	105.5943	0.0145	0.0095	39.0065	85
90	.6383	72.3313	2976.0769	1.5666	113.3109	0.0138	0.0088	41.1451	90
95	.6226	75.4757	3265.3298	1.6061	121.2224	0.0132	0.0082	43.2633	95
100	.6073	78.5426	3562.7934	1.6467	129.3337	0.0127	0.0077	45.3613	100

I = 0.75 %

N	(P/F)	(P/A)	(P/G)	(F/P)	(F/A)	(A/P)	(A/F)	(A/G)	N
1	.9926	0.9926	-0.0000	1.0075	1.0000	1.0075	1.0000	-0.0000	1
2	.9852	1.9777	0.9852	1.0151	2.0075	0.5056	0.4981	0.4981	2
3	.9778	2.9556	2.9408	1.0227	3.0226	0.3383	0.3308	0.9950	3
4	.9706	3.9261	5.8525	1.0303	4.0452	0.2547	0.2472	1.4907	4
5	.9633	4.8894	9.7058	1.0381	5.0756	0.2045	0.1970	1.9851	5
6	.9562	5.8456	14.4866	1.0459	6.1136	0.1711	0.1636	2.4782	6
7	.9490	6.7946	20.1808	1.0537	7.1595	0.1472	0.1397	2.9701	7
8	.9420	7.7366	26.7747	1.0616	8.2132	0.1293	0.1218	3.4608	8
9	.9350	8.6716	34.2544	1.0696	9.2748	0.1153	0.1078	3.9502	9
10	.9280	9.5996	42.6064	1.0776	10.3443	0.1042	0.0967	4.4384	10
11	.9211	10.5207	51.8174	1.0857	11.4219	0.0951	0.0876	4.9253	11
12	.9142	11.4349	61.8740	1.0938	12.5076	0.0875	0.0800	5.4110	12
13	.9074	12.3423	72.7632	1.1020	13.6014	0.0810	0.0735	5.8954	13
14	.9007	13.2430	84.4720	1.1103	14.7034	0.0755	0.0680	6.3786	14
15	.8940	14.1370	96.9876	1.1186	15.8137	0.0707	0.0632	6.8606	15
16	.8873	15.0243	110.2973	1.1270	16.9323	0.0666	0.0591	7.3413	16
17	.8807	15.9050	124.3887	1.1354	18.0593	0.0629	0.0554	7.8207	17
18	.8742	16.7792	139.2494	1.1440	19.1947	0.0596	0.0521	8.2989	18
19	.8676	17.6468	154.8671	1.1525	20.3387	0.0567	0.0492	8.7759	19
20	.8612	18.5080	171.2297	1.1612	21.4912	0.0540	0.0465	9.2516	20
21	.8548	19.3628	188.3253	1.1699	22.6524	0.0516	0.0441	9.7261	21
22	.8484	20.2112	206.1420	1.1787	23.8223	0.0495	0.0420	10.1994	22
23	.8421	21.0533	224.6682	1.1875	25.0010	0.0475	0.0400	10.6714	23
24	.8358	21.8891	243.8923	1.1964	26.1885	0.0457	0.0382	11.1422	24
25	.8296	22.7188	263.8029	1.2054	27.3849	0.0440	0.0365	11.6117	25
26	.8234	23.5422	284.3888	1.2144	28.5903	0.0425	0.0350	12.0800	26
27	.8173	24.3595	305.6387	1.2235	29.8047	0.0411	0.0336	12.5470	27
28	.8112	25.1707	327.5416	1.2327	31.0282	0.0397	0.0322	13.0128	28
29	.8052	25.9759	350.0867	1.2420	32.2609	0.0385	0.0310	13.4774	29
30	.7992	26.7751	373.2631	1.2513	33.5029	0.0373	0.0298	13.9407	30
31	.7932	27.5683	397.0602	1.2607	34.7542	0.0363	0.0288	14.4028	31
32	.7873	28.3557	421.4675	1.2701	36.0148	0.0353	0.0278	14.8636	32
33	.7815	29.1371	446.4746	1.2796	37.2849	0.0343	0.0268	15.3232	33
34	.7757	29.9128	472.0712	1.2892	38.5646	0.0334	0.0259	15.7816	34
35	.7699	30.6827	498.2471	1.2989	39.8538	0.0326	0.0251	16.2387	35
36	.7641	31.4468	524.9924	1.3086	41.1527	0.0318	0.0243	16.6946	36
37	.7585	32.2053	552.2969	1.3185	42.4614	0.0311	0.0236	17.1493	37
38	.7528	32.9581	580.1511	1.3283	43.7798	0.0303	0.0228	17.6027	38
39	.7472	33.7053	608.5451	1.3383	45.1082	0.0297	0.0222	18.0549	39
40	.7416	34.4469	637.4693	1.3483	46.4465	0.0290	0.0215	18.5058	40
41	.7361	35.1831	666.9144	1.3585	47.7948	0.0284	0.0209	18.9556	41
42	.7306	35.9137	696.8709	1.3686	49.1533	0.0278	0.0203	19.4040	42
43	.7252	36.6389	727.3297	1.3789	50.5219	0.0273	0.0198	19.8513	43
44	.7198	37.3587	758.2815	1.3893	51.9009	0.0268	0.0193	20.2973	44
45	.7145	38.0732	789.7173	1.3997	53.2901	0.0263	0.0188	20.7421	45
46	.7091	38.7823	821.6283	1.4102	54.6898	0.0258	0.0183	21.1856	46
47	.7039	39.4862	854.0056	1.4207	56.1000	0.0253	0.0178	21.6280	47
48	.6986	40.1848	886.8404	1.4314	57.5207	0.0249	0.0174	22.0691	48
49	.6934	40.8782	920.1243	1.4421	58.9521	0.0245	0.0170	22.5089	49
50	.6883	41.5664	953.8486	1.4530	60.3943	0.0241	0.0166	22.9476	50
51	.6831	42.2496	988.0050	1.4639	61.8472	0.0237	0.0162	23.3850	51
52	.6780	42.9276	1022.5852	1.4748	63.3111	0.0233	0.0158	23.8211	52
53	.6730	43.6006	1057.5810	1.4859	64.7859	0.0229	0.0154	24.2561	53
54	.6680	44.2686	1092.9842	1.4970	66.2718	0.0226	0.0151	24.6898	54
55	.6630	44.9316	1128.7869	1.5083	67.7688	0.0223	0.0148	25.1223	55
60	.6387	48.1734	1313.5189	1.5657	75.4241	0.0208	0.0133	27.2665	60
65	.6153	51.2963	1507.0910	1.6253	83.3709	0.0195	0.0120	29.3801	65
70	.5927	54.3046	1708.6065	1.6872	91.6201	0.0184	0.0109	31.4634	70
75	.5710	57.2027	1917.2225	1.7514	100.1833	0.0175	0.0100	33.5163	75
80	.5500	59.9944	2132.1472	1.8180	109.0725	0.0167	0.0092	35.5391	80
85	.5299	62.6838	2352.6375	1.8873	118.3001	0.0160	0.0085	37.5318	85
90	.5104	65.2746	2577.9961	1.9591	127.8790	0.0153	0.0078	39.4946	90
95	.4917	67.7704	2807.5694	2.0337	137.8225	0.0148	0.0073	41.4277	95
100	.4737	70.1746	3040.7453	2.1111	148.1445	0.0143	0.0068	43.3311	100

ENGINEERING ECONOMIC ANALYSIS

I = 1.00 %

N	(P/F)	(P/A)	(P/G)	(F/P)	(F/A)	(A/P)	(A/F)	(A/G)	N
1	.9901	0.9901	-0.0000	1.0100	1.0000	1.0100	1.0000	-0.0000	1
2	.9803	1.9704	0.9803	1.0201	2.0100	0.5075	0.4975	0.4975	2
3	.9706	2.9410	2.9215	1.0303	3.0301	0.3400	0.3300	0.9934	3
4	.9610	3.9020	5.8044	1.0406	4.0604	0.2563	0.2463	1.4876	4
5	.9515	4.8534	9.6103	1.0510	5.1010	0.2060	0.1960	1.9801	5
6	.9420	5.7955	14.3205	1.0615	6.1520	0.1725	0.1625	2.4710	6
7	.9327	6.7282	19.9168	1.0721	7.2135	0.1486	0.1386	2.9602	7
8	.9235	7.6517	26.3812	1.0829	8.2857	0.1307	0.1207	3.4478	8
9	.9143	8.5660	33.6959	1.0937	9.3685	0.1167	0.1067	3.9337	9
10	.9053	9.4713	41.8435	1.1046	10.4622	0.1056	0.0956	4.4179	10
11	.8963	10.3676	50.8067	1.1157	11.5668	0.0965	0.0865	4.9005	11
12	.8874	11.2551	60.5687	1.1268	12.6825	0.0888	0.0788	5.3815	12
13	.8787	12.1337	71.1126	1.1381	13.8093	0.0824	0.0724	5.8607	13
14	.8700	13.0037	82.4221	1.1495	14.9474	0.0769	0.0669	6.3384	14
15	.8613	13.8651	94.4810	1.1610	16.0969	0.0721	0.0621	6.8143	15
16	.8528	14.7179	107.2734	1.1726	17.2579	0.0679	0.0579	7.2886	16
17	.8444	15.5623	120.7834	1.1843	18.4304	0.0643	0.0543	7.7613	17
18	.8360	16.3983	134.9957	1.1961	19.6147	0.0610	0.0510	8.2323	18
19	.8277	17.2260	149.8950	1.2081	20.8109	0.0581	0.0481	8.7017	19
20	.8195	18.0456	165.4664	1.2202	22.0190	0.0554	0.0454	9.1694	20
21	.8114	18.8570	181.6950	1.2324	23.2392	0.0530	0.0430	9.6354	21
22	.8034	19.6604	198.5663	1.2447	24.4716	0.0509	0.0409	10.0998	22
23	.7954	20.4558	216.0660	1.2572	25.7163	0.0489	0.0389	10.5626	23
24	.7876	21.2434	234.1800	1.2697	26.9735	0.0471	0.0371	11.0237	24
25	.7798	22.0232	252.8945	1.2824	28.2432	0.0454	0.0354	11.4831	25
26	.7720	22.7952	272.1957	1.2953	29.5256	0.0439	0.0339	11.9409	26
27	.7644	23.5596	292.0702	1.3082	30.8209	0.0424	0.0324	12.3971	27
28	.7568	24.3164	312.5047	1.3213	32.1291	0.0411	0.0311	12.8516	28
29	.7493	25.0658	333.4863	1.3345	33.4504	0.0399	0.0299	13.3044	29
30	.7419	25.8077	355.0021	1.3478	34.7849	0.0387	0.0287	13.7557	30
31	.7346	26.5423	377.0394	1.3613	36.1327	0.0377	0.0277	14.2052	31
32	.7273	27.2696	399.5858	1.3749	37.4941	0.0367	0.0267	14.6532	32
33	.7201	27.9897	422.6291	1.3887	38.8690	0.0357	0.0257	15.0995	33
34	.7130	28.7027	446.1572	1.4026	40.2577	0.0348	0.0248	15.5441	34
35	.7059	29.4086	470.1583	1.4166	41.6603	0.0340	0.0240	15.9871	35
36	.6989	30.1075	494.6207	1.4308	43.0769	0.0332	0.0232	16.4285	36
37	.6920	30.7995	519.5329	1.4451	44.5076	0.0325	0.0225	16.8682	37
38	.6852	31.4847	544.8835	1.4595	45.9527	0.0318	0.0218	17.3063	38
39	.6784	32.1630	570.6616	1.4741	47.4123	0.0311	0.0211	17.7428	39
40	.6717	32.8347	596.8561	1.4889	48.8864	0.0305	0.0205	18.1776	40
41	.6650	33.4997	623.4562	1.5038	50.3752	0.0299	0.0199	18.6108	41
42	.6584	34.1581	650.4514	1.5188	51.8790	0.0293	0.0193	19.0424	42
43	.6519	34.8100	677.8312	1.5340	53.3978	0.0287	0.0187	19.4723	43
44	.6454	35.4555	705.5853	1.5493	54.9318	0.0282	0.0182	19.9006	44
45	.6391	36.0945	733.7037	1.5648	56.4811	0.0277	0.0177	20.3273	45
46	.6327	36.7272	762.1765	1.5805	58.0459	0.0272	0.0172	20.7524	46
47	.6265	37.3537	790.9938	1.5963	59.6263	0.0268	0.0168	21.1758	47
48	.6203	37.9740	820.1460	1.6122	61.2226	0.0263	0.0163	21.5976	48
49	.6141	38.5881	849.6237	1.6283	62.8348	0.0259	0.0159	22.0178	49
50	.6080	39.1961	879.4176	1.6446	64.4632	0.0255	0.0155	22.4363	50
51	.6020	39.7981	909.5186	1.6611	66.1078	0.0251	0.0151	22.8533	51
52	.5961	40.3942	939.9175	1.6777	67.7689	0.0248	0.0148	23.2686	52
53	.5902	40.9844	970.6057	1.6945	69.4466	0.0244	0.0144	23.6823	53
54	.5843	41.5687	1001.5743	1.7114	71.1410	0.0241	0.0141	24.0945	54
55	.5785	42.1472	1032.8148	1.7285	72.8525	0.0237	0.0137	24.5049	55
60	.5504	44.9550	1192.8061	1.8167	81.6697	0.0222	0.0122	26.5333	60
65	.5237	47.6266	1358.3903	1.9094	90.9366	0.0210	0.0110	28.5217	65
70	.4983	50.1685	1528.6474	2.0068	100.6763	0.0199	0.0099	30.4703	70
75	.4741	52.5871	1702.7340	2.1091	110.9128	0.0190	0.0090	32.3793	75
80	.4511	54.8882	1879.8771	2.2167	121.6715	0.0182	0.0082	34.2492	80
85	.4292	57.0777	2059.3701	2.3298	132.9790	0.0175	0.0075	36.0801	85
90	.4084	59.1609	2240.5675	2.4486	144.8633	0.0169	0.0069	37.8724	90
95	.3886	61.1430	2422.8811	2.5735	157.3538	0.0164	0.0064	39.6265	95
100	.3697	63.0289	2605.7758	2.7048	170.4814	0.0159	0.0059	41.3426	100

I = 1.50 %

N	(P/F)	(P/A)	(P/G)	(F/P)	(F/A)	(A/P)	(A/F)	(A/G)	N
1	.9852	0.9852	-0.0000	1.0150	1.0000	1.0150	1.0000	-0.0000	1
2	.9707	1.9559	0.9707	1.0302	2.0150	0.5113	0.4963	0.4963	2
3	.9563	2.9122	2.8833	1.0457	3.0452	0.3434	0.3284	0.9901	3
4	.9422	3.8544	5.7098	1.0614	4.0909	0.2594	0.2444	1.4814	4
5	.9283	4.7826	9.4229	1.0773	5.1523	0.2091	0.1941	1.9702	5
6	.9145	5.6972	13.9956	1.0934	6.2296	0.1755	0.1605	2.4566	6
7	.9010	6.5982	19.4018	1.1098	7.3230	0.1516	0.1366	2.9405	7
8	.8877	7.4859	25.6157	1.1265	8.4328	0.1336	0.1186	3.4219	8
9	.8746	8.3605	32.6125	1.1434	9.5593	0.1196	0.1046	3.9008	9
10	.8617	9.2222	40.3675	1.1605	10.7027	0.1084	0.0934	4.3772	10
11	.8489	10.0711	48.8568	1.1779	11.8633	0.0993	0.0843	4.8512	11
12	.8364	10.9075	58.0571	1.1956	13.0412	0.0917	0.0767	5.3227	12
13	.8240	11.7315	67.9454	1.2136	14.2368	0.0852	0.0702	5.7917	13
14	.8118	12.5434	78.4994	1.2318	15.4504	0.0797	0.0647	6.2582	14
15	.7999	13.3432	89.6974	1.2502	16.6821	0.0749	0.0599	6.7223	15
16	.7880	14.1313	101.5178	1.2690	17.9324	0.0708	0.0558	7.1839	16
17	.7764	14.9076	113.9400	1.2880	19.2014	0.0671	0.0521	7.6431	17
18	.7649	15.6726	126.9435	1.3073	20.4894	0.0638	0.0488	8.0997	18
19	.7536	16.4262	140.5084	1.3270	21.7967	0.0609	0.0459	8.5539	19
20	.7425	17.1686	154.6154	1.3469	23.1237	0.0582	0.0432	9.0057	20
21	.7315	17.9001	169.2453	1.3671	24.4705	0.0559	0.0409	9.4550	21
22	.7207	18.6208	184.3798	1.3876	25.8376	0.0537	0.0387	9.9018	22
23	.7100	19.3309	200.0006	1.4084	27.2251	0.0517	0.0367	10.3462	23
24	.6995	20.0304	216.0901	1.4295	28.6335	0.0499	0.0349	10.7881	24
25	.6892	20.7196	232.6310	1.4509	30.0630	0.0483	0.0333	11.2276	25
26	.6790	21.3986	249.6065	1.4727	31.5140	0.0467	0.0317	11.6646	26
27	.6690	22.0676	267.0002	1.4948	32.9867	0.0453	0.0303	12.0992	27
28	.6591	22.7267	284.7958	1.5172	34.4815	0.0440	0.0290	12.5313	28
29	.6494	23.3761	302.9779	1.5400	35.9987	0.0428	0.0278	12.9610	29
30	.6398	24.0158	321.5310	1.5631	37.5387	0.0416	0.0266	13.3883	30
31	.6303	24.6461	340.4402	1.5865	39.1018	0.0406	0.0256	13.8131	31
32	.6210	25.2671	359.6910	1.6103	40.6883	0.0396	0.0246	14.2355	32
33	.6118	25.8790	379.2691	1.6345	42.2986	0.0386	0.0236	14.6555	33
34	.6028	26.4817	399.1607	1.6590	43.9331	0.0378	0.0228	15.0731	34
35	.5939	27.0756	419.3521	1.6839	45.5921	0.0369	0.0219	15.4882	35
36	.5851	27.6607	439.8303	1.7091	47.2760	0.0362	0.0212	15.9009	36
37	.5764	28.2371	460.5822	1.7348	48.9851	0.0354	0.0204	16.3112	37
38	.5679	28.8051	481.5954	1.7608	50.7199	0.0347	0.0197	16.7191	38
39	.5595	29.3646	502.8576	1.7872	52.4807	0.0341	0.0191	17.1246	39
40	.5513	29.9158	524.3568	1.8140	54.2679	0.0334	0.0184	17.5277	40
41	.5431	30.4590	546.0814	1.8412	56.0819	0.0328	0.0178	17.9284	41
42	.5351	30.9941	568.0201	1.8688	57.9231	0.0323	0.0173	18.3267	42
43	.5272	31.5212	590.1617	1.8969	59.7920	0.0317	0.0167	18.7227	43
44	.5194	32.0406	612.4955	1.9253	61.6889	0.0312	0.0162	19.1162	44
45	.5117	32.5523	635.0110	1.9542	63.6142	0.0307	0.0157	19.5074	45
46	.5042	33.0565	657.6979	1.9835	65.5684	0.0303	0.0153	19.8962	46
47	.4967	33.5532	680.5462	2.0133	67.5519	0.0298	0.0148	20.2826	47
48	.4894	34.0426	703.5462	2.0435	69.5652	0.0294	0.0144	20.6667	48
49	.4821	34.5247	726.6884	2.0741	71.6087	0.0290	0.0140	21.0484	49
50	.4750	34.9997	749.9636	2.1052	73.6828	0.0286	0.0136	21.4277	50
51	.4680	35.4677	773.3629	2.1368	75.7881	0.0282	0.0132	21.8047	51
52	.4611	35.9287	796.8774	2.1689	77.9249	0.0278	0.0128	22.1794	52
53	.4543	36.3830	820.4986	2.2014	80.0938	0.0275	0.0125	22.5517	53
54	.4475	36.8305	844.2184	2.2344	82.2952	0.0272	0.0122	22.9217	54
55	.4409	37.2715	868.0285	2.2679	84.5296	0.0268	0.0118	23.2894	55
60	.4093	39.3803	988.1674	2.4432	96.2147	0.0254	0.0104	25.0930	60
65	.3799	41.3378	1109.4752	2.6320	108.8028	0.0242	0.0092	26.8393	65
70	.3527	43.1549	1231.1658	2.8355	122.3638	0.0232	0.0082	28.5290	70
75	.3274	44.8416	1352.5600	3.0546	136.9728	0.0223	0.0073	30.1631	75
80	.3039	46.4073	1473.0741	3.2907	152.7109	0.0215	0.0065	31.7423	80
85	.2821	47.8607	1592.2095	3.5450	169.6652	0.0209	0.0059	33.2676	85
90	.2619	49.2099	1709.5439	3.8189	187.9299	0.0203	0.0053	34.7399	90
95	.2431	50.4622	1824.7224	4.1141	207.6061	0.0198	0.0048	36.1602	95
100	.2256	51.6247	1937.4506	4.4320	228.8030	0.0194	0.0044	37.5295	100

ENGINEERING ECONOMIC ANALYSIS

I = 2.00 %

N	(P/F)	(P/A)	(P/G)	(F/P)	(F/A)	(A/P)	(A/F)	(A/G)	N
1	.9804	0.9804	-0.0000	1.0200	1.0000	1.0200	1.0000	-0.0000	1
2	.9612	1.9416	0.9612	1.0404	2.0200	0.5150	0.4950	0.4950	2
3	.9423	2.8839	2.8458	1.0612	3.0604	0.3468	0.3268	0.9868	3
4	.9238	3.8077	5.6173	1.0824	4.1216	0.2626	0.2426	1.4752	4
5	.9057	4.7135	9.2403	1.1041	5.2040	0.2122	0.1922	1.9604	5
6	.8880	5.6014	13.6801	1.1262	6.3081	0.1785	0.1585	2.4423	6
7	.8706	6.4720	18.9035	1.1487	7.4343	0.1545	0.1345	2.9208	7
8	.8535	7.3255	24.8779	1.1717	8.5830	0.1365	0.1165	3.3961	8
9	.8368	8.1622	31.5720	1.1951	9.7546	0.1225	0.1025	3.8681	9
10	.8203	8.9826	38.9551	1.2190	10.9497	0.1113	0.0913	4.3367	10
11	.8043	9.7868	46.9977	1.2434	12.1687	0.1022	0.0822	4.8021	11
12	.7885	10.5753	55.6712	1.2682	13.4121	0.0946	0.0746	5.2642	12
13	.7730	11.3484	64.9475	1.2936	14.6803	0.0881	0.0681	5.7231	13
14	.7579	12.1062	74.7999	1.3195	15.9739	0.0826	0.0626	6.1786	14
15	.7430	12.8493	85.2021	1.3459	17.2934	0.0778	0.0578	6.6309	15
16	.7284	13.5777	96.1288	1.3728	18.6393	0.0737	0.0537	7.0799	16
17	.7142	14.2919	107.5554	1.4002	20.0121	0.0700	0.0500	7.5256	17
18	.7002	14.9920	119.4581	1.4282	21.4123	0.0667	0.0467	7.9681	18
19	.6864	15.6785	131.8139	1.4568	22.8406	0.0638	0.0438	8.4073	19
20	.6730	16.3514	144.6003	1.4859	24.2974	0.0612	0.0412	8.8433	20
21	.6598	17.0112	157.7959	1.5157	25.7833	0.0588	0.0388	9.2760	21
22	.6468	17.6580	171.3795	1.5460	27.2990	0.0566	0.0366	9.7055	22
23	.6342	18.2922	185.3309	1.5769	28.8450	0.0547	0.0347	10.1317	23
24	.6217	18.9139	199.6305	1.6084	30.4219	0.0529	0.0329	10.5547	24
25	.6095	19.5235	214.2592	1.6406	32.0303	0.0512	0.0312	10.9745	25
26	.5976	20.1210	229.1987	1.6734	33.6709	0.0497	0.0297	11.3910	26
27	.5859	20.7069	244.4311	1.7069	35.3443	0.0483	0.0283	11.8043	27
28	.5744	21.2813	259.9392	1.7410	37.0512	0.0470	0.0270	12.2145	28
29	.5631	21.8444	275.7064	1.7758	38.7922	0.0458	0.0258	12.6214	29
30	.5521	22.3965	291.7164	1.8114	40.5681	0.0446	0.0246	13.0251	30
31	.5412	22.9377	307.9538	1.8476	42.3794	0.0436	0.0236	13.4257	31
32	.5306	23.4683	324.4035	1.8845	44.2270	0.0426	0.0226	13.8230	32
33	.5202	23.9886	341.0508	1.9222	46.1116	0.0417	0.0217	14.2172	33
34	.5100	24.4986	357.8817	1.9607	48.0338	0.0408	0.0208	14.6083	34
35	.5000	24.9986	374.8826	1.9999	49.9945	0.0400	0.0200	14.9961	35
36	.4902	25.4888	392.0405	2.0399	51.9944	0.0392	0.0192	15.3809	36
37	.4806	25.9695	409.3424	2.0807	54.0343	0.0385	0.0185	15.7625	37
38	.4712	26.4406	426.7764	2.1223	56.1149	0.0378	0.0178	16.1409	38
39	.4619	26.9026	444.3304	2.1647	58.2372	0.0372	0.0172	16.5163	39
40	.4529	27.3555	461.9931	2.2080	60.4020	0.0366	0.0166	16.8885	40
41	.4440	27.7995	479.7535	2.2522	62.6100	0.0360	0.0160	17.2576	41
42	.4353	28.2348	497.6010	2.2972	64.8622	0.0354	0.0154	17.6237	42
43	.4268	28.6616	515.5253	2.3432	67.1595	0.0349	0.0149	17.9866	43
44	.4184	29.0800	533.5165	2.3901	69.5027	0.0344	0.0144	18.3465	44
45	.4102	29.4902	551.5652	2.4379	71.8927	0.0339	0.0139	18.7034	45
46	.4022	29.8923	569.6621	2.4866	74.3306	0.0335	0.0135	19.0571	46
47	.3943	30.2866	587.7985	2.5363	76.8172	0.0330	0.0130	19.4079	47
48	.3865	30.6731	605.9657	2.5871	79.3535	0.0326	0.0126	19.7556	48
49	.3790	31.0521	624.1557	2.6388	81.9406	0.0322	0.0122	20.1003	49
50	.3715	31.4236	642.3606	2.6916	84.5794	0.0318	0.0118	20.4420	50
51	.3642	31.7878	660.5727	2.7454	87.2710	0.0315	0.0115	20.7807	51
52	.3571	32.1449	678.7849	2.8003	90.0164	0.0311	0.0111	21.1164	52
53	.3501	32.4950	696.9900	2.8563	92.8167	0.0308	0.0108	21.4491	53
54	.3432	32.8383	715.1815	2.9135	95.6731	0.0305	0.0105	21.7789	54
55	.3365	33.1748	733.3527	2.9717	98.5865	0.0301	0.0101	22.1057	55
60	.3048	34.7609	823.6975	3.2810	114.0515	0.0288	0.0088	23.6961	60
65	.2761	36.1975	912.7085	3.6225	131.1262	0.0276	0.0076	25.2147	65
70	.2500	37.4986	999.8343	3.9996	149.9779	0.0267	0.0067	26.6632	70
75	.2265	38.6771	1084.6393	4.4158	170.7918	0.0259	0.0059	28.0434	75
80	.2051	39.7445	1166.7868	4.8754	193.7720	0.0252	0.0052	29.3572	80
85	.1858	40.7113	1246.0241	5.3829	219.1439	0.0246	0.0046	30.6064	85
90	.1683	41.5869	1322.1701	5.9431	247.1567	0.0240	0.0040	31.7929	90
95	.1524	42.3800	1395.1033	6.5617	278.0850	0.0236	0.0036	32.9189	95
100	.1380	43.0984	1464.7527	7.2446	312.2323	0.0232	0.0032	33.9863	100

I = 3.00 %

N	(P/F)	(P/A)	(P/G)	(F/P)	(F/A)	(A/P)	(A/F)	(A/G)	N
1	.9709	0.9709	-0.0000	1.0300	1.0000	1.0300	1.0000	-0.0000	1
2	.9426	1.9135	0.9426	1.0609	2.0300	0.5226	0.4926	0.4926	2
3	.9151	2.8286	2.7729	1.0927	3.0909	0.3535	0.3235	0.9803	3
4	.8885	3.7171	5.4383	1.1255	4.1836	0.2690	0.2390	1.4631	4
5	.8626	4.5797	8.8888	1.1593	5.3091	0.2184	0.1884	1.9409	5
6	.8375	5.4172	13.0762	1.1941	6.4684	0.1846	0.1546	2.4138	6
7	.8131	6.2303	17.9547	1.2299	7.6625	0.1605	0.1305	2.8819	7
8	.7894	7.0197	23.4806	1.2668	8.8923	0.1425	0.1125	3.3450	8
9	.7664	7.7861	29.6119	1.3048	10.1591	0.1284	0.0984	3.8032	9
10	.7441	8.5302	36.3088	1.3439	11.4639	0.1172	0.0872	4.2565	10
11	.7224	9.2526	43.5330	1.3842	12.8078	0.1081	0.0781	4.7049	11
12	.7014	9.9540	51.2482	1.4258	14.1920	0.1005	0.0705	5.1485	12
13	.6810	10.6350	59.4196	1.4685	15.6178	0.0940	0.0640	5.5872	13
14	.6611	11.2961	68.0141	1.5126	17.0863	0.0885	0.0585	6.0210	14
15	.6419	11.9379	77.0002	1.5580	18.5989	0.0838	0.0538	6.4500	15
16	.6232	12.5611	86.3477	1.6047	20.1569	0.0796	0.0496	6.8742	16
17	.6050	13.1661	96.0280	1.6528	21.7616	0.0760	0.0460	7.2936	17
18	.5874	13.7535	106.0137	1.7024	23.4144	0.0727	0.0427	7.7081	18
19	.5703	14.3238	116.2788	1.7535	25.1169	0.0698	0.0398	8.1179	19
20	.5537	14.8775	126.7987	1.8061	26.8704	0.0672	0.0372	8.5229	20
21	.5375	15.4150	137.5496	1.8603	28.6765	0.0649	0.0349	8.9231	21
22	.5219	15.9369	148.5094	1.9161	30.5368	0.0627	0.0327	9.3186	22
23	.5067	16.4436	159.6566	1.9736	32.4529	0.0608	0.0308	9.7093	23
24	.4919	16.9355	170.9711	2.0328	34.4265	0.0590	0.0290	10.0954	24
25	.4776	17.4131	182.4336	2.0938	36.4593	0.0574	0.0274	10.4768	25
26	.4637	17.8768	194.0260	2.1566	38.5530	0.0559	0.0259	10.8535	26
27	.4502	18.3270	205.7309	2.2213	40.7096	0.0546	0.0246	11.2255	27
28	.4371	18.7641	217.5320	2.2879	42.9309	0.0533	0.0233	11.5930	28
29	.4243	19.1885	229.4137	2.3566	45.2189	0.0521	0.0221	11.9558	29
30	.4120	19.6004	241.3613	2.4273	47.5754	0.0510	0.0210	12.3141	30
31	.4000	20.0004	253.3609	2.5001	50.0027	0.0500	0.0200	12.6678	31
32	.3883	20.3888	265.3993	2.5751	52.5028	0.0490	0.0190	13.0169	32
33	.3770	20.7658	277.4642	2.6523	55.0778	0.0482	0.0182	13.3616	33
34	.3660	21.1318	289.5437	2.7319	57.7302	0.0473	0.0173	13.7018	34
35	.3554	21.4872	301.6267	2.8139	60.4621	0.0465	0.0165	14.0375	35
36	.3450	21.8323	313.7028	2.8983	63.2759	0.0458	0.0158	14.3688	36
37	.3350	22.1672	325.7622	2.9852	66.1742	0.0451	0.0151	14.6957	37
38	.3252	22.4925	337.7956	3.0748	69.1594	0.0445	0.0145	15.0182	38
39	.3158	22.8082	349.7942	3.1670	72.2342	0.0438	0.0138	15.3363	39
40	.3066	23.1148	361.7499	3.2620	75.4013	0.0433	0.0133	15.6502	40
41	.2976	23.4124	373.6551	3.3599	78.6633	0.0427	0.0127	15.9597	41
42	.2890	23.7014	385.5024	3.4607	82.0232	0.0422	0.0122	16.2650	42
43	.2805	23.9819	397.2852	3.5645	85.4839	0.0417	0.0117	16.5660	43
44	.2724	24.2543	408.9972	3.6715	89.0484	0.0412	0.0112	16.8629	44
45	.2644	24.5187	420.6325	3.7816	92.7199	0.0408	0.0108	17.1556	45
46	.2567	24.7754	432.1856	3.8950	96.5015	0.0404	0.0104	17.4441	46
47	.2493	25.0247	443.6515	4.0119	100.3965	0.0400	0.0100	17.7285	47
48	.2420	25.2667	455.0255	4.1323	104.4084	0.0396	0.0096	18.0089	48
49	.2350	25.5017	466.3031	4.2562	108.5406	0.0392	0.0092	18.2852	49
50	.2281	25.7298	477.4803	4.3839	112.7969	0.0389	0.0089	18.5575	50
51	.2215	25.9512	488.5535	4.5154	117.1808	0.0385	0.0085	18.8258	51
52	.2150	26.1662	499.5191	4.6509	121.6962	0.0382	0.0082	19.0902	52
53	.2088	26.3750	510.3742	4.7904	126.3471	0.0379	0.0079	19.3507	53
54	.2027	26.5777	521.1157	4.9341	131.1375	0.0376	0.0076	19.6073	54
55	.1968	26.7744	531.7411	5.0821	136.0716	0.0373	0.0073	19.8600	55
60	.1697	27.6756	583.0526	5.8916	163.0534	0.0361	0.0061	21.0674	60
65	.1464	28.4529	631.2010	6.8300	194.3328	0.0351	0.0051	22.1841	65
70	.1263	29.1234	676.0869	7.9178	230.5941	0.0343	0.0043	23.2145	70
75	.1089	29.7018	717.6978	9.1789	272.6309	0.0337	0.0037	24.1634	75
80	.0940	30.2008	756.0865	10.6409	321.3630	0.0331	0.0031	25.0353	80
85	.0811	30.6312	791.3529	12.3357	377.8570	0.0326	0.0026	25.8349	85
90	.0699	31.0024	823.6302	14.3005	443.3489	0.0323	0.0023	26.5667	90
95	.0603	31.3227	853.0742	16.5782	519.2720	0.0319	0.0019	27.2351	95
100	.0520	31.5989	879.8540	19.2186	607.2877	0.0316	0.0016	27.8444	100

ENGINEERING ECONOMIC ANALYSIS

I = 4.00 %

N	(P/F)	(P/A)	(P/G)	(F/P)	(F/A)	(A/P)	(A/F)	(A/G)	N
1	.9615	0.9615	-0.0000	1.0400	1.0000	1.0400	1.0000	-0.0000	1
2	.9246	1.8861	0.9246	1.0816	2.0400	0.5302	0.4902	0.4902	2
3	.8890	2.7751	2.7025	1.1249	3.1216	0.3603	0.3203	0.9739	3
4	.8548	3.6299	5.2670	1.1699	4.2465	0.2755	0.2355	1.4510	4
5	.8219	4.4518	8.5547	1.2167	5.4163	0.2246	0.1846	1.9216	5
6	.7903	5.2421	12.5062	1.2653	6.6330	0.1908	0.1508	2.3857	6
7	.7599	6.0021	17.0657	1.3159	7.8983	0.1666	0.1266	2.8433	7
8	.7307	6.7327	22.1806	1.3686	9.2142	0.1485	0.1085	3.2944	8
9	.7026	7.4353	27.8013	1.4233	10.5828	0.1345	0.0945	3.7391	9
10	.6756	8.1109	33.8814	1.4802	12.0061	0.1233	0.0833	4.1773	10
11	.6496	8.7605	40.3772	1.5395	13.4864	0.1141	0.0741	4.6090	11
12	.6246	9.3851	47.2477	1.6010	15.0258	0.1066	0.0666	5.0343	12
13	.6006	9.9856	54.4546	1.6651	16.6268	0.1001	0.0601	5.4533	13
14	.5775	10.5631	61.9618	1.7317	18.2919	0.0947	0.0547	5.8659	14
15	.5553	11.1184	69.7355	1.8009	20.0236	0.0899	0.0499	6.2721	15
16	.5339	11.6523	77.7441	1.8730	21.8245	0.0858	0.0458	6.6720	16
17	.5134	12.1657	85.9581	1.9479	23.6975	0.0822	0.0422	7.0656	17
18	.4936	12.6593	94.3498	2.0258	25.6454	0.0790	0.0390	7.4530	18
19	.4746	13.1339	102.8933	2.1068	27.6712	0.0761	0.0361	7.8342	19
20	.4564	13.5903	111.5647	2.1911	29.7781	0.0736	0.0336	8.2091	20
21	.4388	14.0292	120.3414	2.2788	31.9692	0.0713	0.0313	8.5779	21
22	.4220	14.4511	129.2024	2.3699	34.2480	0.0692	0.0292	8.9407	22
23	.4057	14.8568	138.1284	2.4647	36.6179	0.0673	0.0273	9.2973	23
24	.3901	15.2470	147.1012	2.5633	39.0826	0.0656	0.0256	9.6479	24
25	.3751	15.6221	156.1040	2.6658	41.6459	0.0640	0.0240	9.9925	25
26	.3607	15.9828	165.1212	2.7725	44.3117	0.0626	0.0226	10.3312	26
27	.3468	16.3296	174.1385	2.8834	47.0842	0.0612	0.0212	10.6640	27
28	.3335	16.6631	183.1424	2.9987	49.9676	0.0600	0.0200	10.9909	28
29	.3207	16.9837	192.1206	3.1187	52.9663	0.0589	0.0189	11.3120	29
30	.3083	17.2920	201.0618	3.2434	56.0849	0.0578	0.0178	11.6274	30
31	.2965	17.5885	209.9556	3.3731	59.3283	0.0569	0.0169	11.9371	31
32	.2851	17.8736	218.7924	3.5081	62.7015	0.0559	0.0159	12.2411	32
33	.2741	18.1476	227.5634	3.6484	66.2095	0.0551	0.0151	12.5396	33
34	.2636	18.4112	236.2607	3.7943	69.8579	0.0543	0.0143	12.8324	34
35	.2534	18.6646	244.8768	3.9461	73.6522	0.0536	0.0136	13.1198	35
36	.2437	18.9083	253.4052	4.1039	77.5983	0.0529	0.0129	13.4018	36
37	.2343	19.1426	261.8399	4.2681	81.7022	0.0522	0.0122	13.6784	37
38	.2253	19.3679	270.1754	4.4388	85.9703	0.0516	0.0116	13.9497	38
39	.2166	19.5845	278.4070	4.6164	90.4091	0.0511	0.0111	14.2157	39
40	.2083	19.7928	286.5303	4.8010	95.0255	0.0505	0.0105	14.4765	40
41	.2003	19.9931	294.5414	4.9931	99.8265	0.0500	0.0100	14.7322	41
42	.1926	20.1856	302.4370	5.1928	104.8196	0.0495	0.0095	14.9828	42
43	.1852	20.3708	310.2141	5.4005	110.0124	0.0491	0.0091	15.2284	43
44	.1780	20.5488	317.8700	5.6165	115.4129	0.0487	0.0087	15.4690	44
45	.1712	20.7200	325.4028	5.8412	121.0294	0.0483	0.0083	15.7047	45
46	.1646	20.8847	332.8104	6.0748	126.8706	0.0479	0.0079	15.9356	46
47	.1583	21.0429	340.0914	6.3178	132.9454	0.0475	0.0075	16.1618	47
48	.1522	21.1951	347.2446	6.5705	139.2632	0.0472	0.0072	16.3832	48
49	.1463	21.3415	354.2689	6.8333	145.8337	0.0469	0.0069	16.6000	49
50	.1407	21.4822	361.1638	7.1067	152.6671	0.0466	0.0066	16.8122	50
51	.1353	21.6175	367.9289	7.3910	159.7738	0.0463	0.0063	17.0200	51
52	.1301	21.7476	374.5638	7.6866	167.1647	0.0460	0.0060	17.2232	52
53	.1251	21.8727	381.0686	7.9941	174.8513	0.0457	0.0057	17.4221	53
54	.1203	21.9930	387.4436	8.3138	182.8454	0.0455	0.0055	17.6167	54
55	.1157	22.1086	393.6890	8.6464	191.1592	0.0452	0.0052	17.8070	55
60	.0951	22.6235	422.9966	10.5196	237.9907	0.0442	0.0042	18.6972	60
65	.0781	23.0467	449.2014	12.7987	294.9684	0.0434	0.0034	19.4909	65
70	.0642	23.3945	472.4789	15.5716	364.2905	0.0427	0.0027	20.1961	70
75	.0528	23.6804	493.0408	18.9453	448.6314	0.0422	0.0022	20.8206	75
80	.0434	23.9154	511.1161	23.0498	551.2450	0.0418	0.0018	21.3718	80
85	.0357	24.1085	526.9384	28.0436	676.0901	0.0415	0.0015	21.8569	85
90	.0293	24.2673	540.7369	34.1193	827.9833	0.0412	0.0012	22.2826	90
95	.0241	24.3978	552.7307	41.5114	1012.7846	0.0410	0.0010	22.6550	95
100	.0198	24.5050	563.1249	50.5049	1237.6237	0.0408	0.0008	22.9800	100

I = 5.00 %

N	(P/F)	(P/A)	(P/G)	(F/P)	(F/A)	(A/P)	(A/F)	(A/G)	N
1	.9524	0.9524	-0.0000	1.0500	1.0000	1.0500	1.0000	-0.0000	1
2	.9070	1.8594	0.9070	1.1025	2.0500	0.5378	0.4878	0.4878	2
3	.8638	2.7232	2.6347	1.1576	3.1525	0.3672	0.3172	0.9675	3
4	.8227	3.5460	5.1028	1.2155	4.3101	0.2820	0.2320	1.4391	4
5	.7835	4.3295	8.2369	1.2763	5.5256	0.2310	0.1810	1.9025	5
6	.7462	5.0757	11.9680	1.3401	6.8019	0.1970	0.1470	2.3579	6
7	.7107	5.7864	16.2321	1.4071	8.1420	0.1728	0.1228	2.8052	7
8	.6768	6.4632	20.9700	1.4775	9.5491	0.1547	0.1047	3.2445	8
9	.6446	7.1078	26.1268	1.5513	11.0266	0.1407	0.0907	3.6758	9
10	.6139	7.7217	31.6520	1.6289	12.5779	0.1295	0.0795	4.0991	10
11	.5847	8.3064	37.4988	1.7103	14.2068	0.1204	0.0704	4.5144	11
12	.5568	8.8633	43.6241	1.7959	15.9171	0.1128	0.0628	4.9219	12
13	.5303	9.3936	49.9879	1.8856	17.7130	0.1065	0.0565	5.3215	13
14	.5051	9.8986	56.5538	1.9799	19.5986	0.1010	0.0510	5.7133	14
15	.4810	10.3797	63.2880	2.0789	21.5786	0.0963	0.0463	6.0973	15
16	.4581	10.8378	70.1597	2.1829	23.6575	0.0923	0.0423	6.4736	16
17	.4363	11.2741	77.1405	2.2920	25.8404	0.0887	0.0387	6.8423	17
18	.4155	11.6896	84.2043	2.4066	28.1324	0.0855	0.0355	7.2034	18
19	.3957	12.0853	91.3275	2.5270	30.5390	0.0827	0.0327	7.5569	19
20	.3769	12.4622	98.4884	2.6533	33.0660	0.0802	0.0302	7.9030	20
21	.3589	12.8212	105.6673	2.7860	35.7193	0.0780	0.0280	8.2416	21
22	.3418	13.1630	112.8461	2.9253	38.5052	0.0760	0.0260	8.5730	22
23	.3256	13.4886	120.0087	3.0715	41.4305	0.0741	0.0241	8.8971	23
24	.3101	13.7986	127.1402	3.2251	44.5020	0.0725	0.0225	9.2140	24
25	.2953	14.0939	134.2275	3.3864	47.7271	0.0710	0.0210	9.5238	25
26	.2812	14.3752	141.2585	3.5557	51.1135	0.0696	0.0196	9.8266	26
27	.2678	14.6430	148.2226	3.7335	54.6691	0.0683	0.0183	10.1224	27
28	.2551	14.8981	155.1101	3.9201	58.4026	0.0671	0.0171	10.4114	28
29	.2429	15.1411	161.9126	4.1161	62.3227	0.0660	0.0160	10.6936	29
30	.2314	15.3725	168.6226	4.3219	66.4388	0.0651	0.0151	10.9691	30
31	.2204	15.5928	175.2333	4.5380	70.7608	0.0641	0.0141	11.2381	31
32	.2099	15.8027	181.7392	4.7649	75.2988	0.0633	0.0133	11.5005	32
33	.1999	16.0025	188.1351	5.0032	80.0638	0.0625	0.0125	11.7566	33
34	.1904	16.1929	194.4168	5.2533	85.0670	0.0618	0.0118	12.0063	34
35	.1813	16.3742	200.5807	5.5160	90.3203	0.0611	0.0111	12.2498	35
36	.1727	16.5469	206.6237	5.7918	95.8363	0.0604	0.0104	12.4872	36
37	.1644	16.7113	212.5434	6.0814	101.6281	0.0598	0.0098	12.7186	37
38	.1566	16.8679	218.3378	6.3855	107.7095	0.0593	0.0093	12.9440	38
39	.1491	17.0170	224.0054	6.7048	114.0950	0.0588	0.0088	13.1636	39
40	.1420	17.1591	229.5452	7.0400	120.7998	0.0583	0.0083	13.3775	40
41	.1353	17.2944	234.9564	7.3920	127.8398	0.0578	0.0078	13.5857	41
42	.1288	17.4232	240.2389	7.7616	135.2318	0.0574	0.0074	13.7884	42
43	.1227	17.5459	245.3925	8.1497	142.9933	0.0570	0.0070	13.9857	43
44	.1169	17.6628	250.4175	8.5572	151.1430	0.0566	0.0066	14.1777	44
45	.1113	17.7741	255.3145	8.9850	159.7002	0.0563	0.0063	14.3644	45
46	.1060	17.8801	260.0844	9.4343	168.6852	0.0559	0.0059	14.5461	46
47	.1009	17.9810	264.7281	9.9060	178.1194	0.0556	0.0056	14.7226	47
48	.0961	18.0772	269.2467	10.4013	188.0254	0.0553	0.0053	14.8943	48
49	.0916	18.1687	273.6418	10.9213	198.4267	0.0550	0.0050	15.0611	49
50	.0872	18.2559	277.9148	11.4674	209.3480	0.0548	0.0048	15.2233	50
51	.0831	18.3390	282.0673	12.0408	220.8154	0.0545	0.0045	15.3808	51
52	.0791	18.4181	286.1013	12.6428	232.8562	0.0543	0.0043	15.5337	52
53	.0753	18.4934	290.0184	13.2749	245.4990	0.0541	0.0041	15.6823	53
54	.0717	18.5651	293.8208	13.9387	258.7739	0.0539	0.0039	15.8265	54
55	.0683	18.6335	297.5104	14.6356	272.7126	0.0537	0.0037	15.9664	55
60	.0535	18.9293	314.3432	18.6792	353.5837	0.0528	0.0028	16.6062	60
65	.0419	19.1611	328.6910	23.8399	456.7980	0.0522	0.0022	17.1541	65
70	.0329	19.3427	340.8409	30.4264	588.5285	0.0517	0.0017	17.6212	70
75	.0258	19.4850	351.0721	38.8327	756.6537	0.0513	0.0013	18.0176	75
80	.0202	19.5965	359.6460	49.5614	971.2288	0.0510	0.0010	18.3526	80
85	.0158	19.6838	366.8007	63.2544	1245.0871	0.0508	0.0008	18.6346	85
90	.0124	19.7523	372.7488	80.7304	1594.6073	0.0506	0.0006	18.8712	90
95	.0097	19.8059	377.6774	103.0347	2040.6935	0.0505	0.0005	19.0689	95
100	.0076	19.8479	381.7492	131.5013	2610.0252	0.0504	0.0004	19.2337	100

ENGINEERING ECONOMIC ANALYSIS

I = 6.00 %

N	(P/F)	(P/A)	(P/G)	(F/P)	(F/A)	(A/P)	(A/F)	(A/G)	N
1	.9434	0.9434	-0.0000	1.0600	1.0000	1.0600	1.0000	-0.0000	1
2	.8900	1.8334	0.8900	1.1236	2.0600	0.5454	0.4854	0.4854	2
3	.8396	2.6730	2.5692	1.1910	3.1836	0.3741	0.3141	0.9612	3
4	.7921	3.4651	4.9455	1.2625	4.3746	0.2886	0.2286	1.4272	4
5	.7473	4.2124	7.9345	1.3382	5.6371	0.2374	0.1774	1.8836	5
6	.7050	4.9173	11.4594	1.4185	6.9753	0.2034	0.1434	2.3304	6
7	.6651	5.5824	15.4497	1.5036	8.3938	0.1791	0.1191	2.7676	7
8	.6274	6.2098	19.8416	1.5938	9.8975	0.1610	0.1010	3.1952	8
9	.5919	6.8017	24.5768	1.6895	11.4913	0.1470	0.0870	3.6133	9
10	.5584	7.3601	29.6023	1.7908	13.1808	0.1359	0.0759	4.0220	10
11	.5268	7.8869	34.8702	1.8983	14.9716	0.1268	0.0668	4.4213	11
12	.4970	8.3838	40.3369	2.0122	16.8699	0.1193	0.0593	4.8113	12
13	.4688	8.8527	45.9629	2.1329	18.8821	0.1130	0.0530	5.1920	13
14	.4423	9.2950	51.7128	2.2609	21.0151	0.1076	0.0476	5.5635	14
15	.4173	9.7122	57.5546	2.3966	23.2760	0.1030	0.0430	5.9260	15
16	.3936	10.1059	63.4592	2.5404	25.6725	0.0990	0.0390	6.2794	16
17	.3714	10.4773	69.4011	2.6928	28.2129	0.0954	0.0354	6.6240	17
18	.3503	10.8276	75.3569	2.8543	30.9057	0.0924	0.0324	6.9597	18
19	.3305	11.1581	81.3062	3.0256	33.7600	0.0896	0.0296	7.2867	19
20	.3118	11.4699	87.2304	3.2071	36.7856	0.0872	0.0272	7.6051	20
21	.2942	11.7641	93.1136	3.3996	39.9927	0.0850	0.0250	7.9151	21
22	.2775	12.0416	98.9412	3.6035	43.3923	0.0830	0.0230	8.2166	22
23	.2618	12.3034	104.7007	3.8197	46.9958	0.0813	0.0213	8.5099	23
24	.2470	12.5504	110.3812	4.0489	50.8156	0.0797	0.0197	8.7951	24
25	.2330	12.7834	115.9732	4.2919	54.8645	0.0782	0.0182	9.0722	25
26	.2198	13.0032	121.4684	4.5494	59.1564	0.0769	0.0169	9.3414	26
27	.2074	13.2105	126.8600	4.8223	63.7058	0.0757	0.0157	9.6029	27
28	.1956	13.4062	132.1420	5.1117	68.5281	0.0746	0.0146	9.8568	28
29	.1846	13.5907	137.3096	5.4184	73.6398	0.0736	0.0136	10.1032	29
30	.1741	13.7648	142.3588	5.7435	79.0582	0.0726	0.0126	10.3422	30
31	.1643	13.9291	147.2864	6.0881	84.8017	0.0718	0.0118	10.5740	31
32	.1550	14.0840	152.0901	6.4534	90.8898	0.0710	0.0110	10.7988	32
33	.1462	14.2302	156.7681	6.8406	97.3432	0.0703	0.0103	11.0166	33
34	.1379	14.3681	161.3192	7.2510	104.1838	0.0696	0.0096	11.2276	34
35	.1301	14.4982	165.7427	7.6861	111.4348	0.0690	0.0090	11.4319	35
36	.1227	14.6210	170.0387	8.1473	119.1209	0.0684	0.0084	11.6298	36
37	.1158	14.7368	174.2072	8.6361	127.2681	0.0679	0.0079	11.8213	37
38	.1092	14.8460	178.2490	9.1543	135.9042	0.0674	0.0074	12.0065	38
39	.1031	14.9491	182.1652	9.7035	145.0585	0.0669	0.0069	12.1857	39
40	.0972	15.0463	185.9568	10.2857	154.7620	0.0665	0.0065	12.3590	40
41	.0917	15.1380	189.6256	10.9029	165.0477	0.0661	0.0061	12.5264	41
42	.0865	15.2245	193.1732	11.5570	175.9505	0.0657	0.0057	12.6883	42
43	.0816	15.3062	196.6017	12.2505	187.5076	0.0653	0.0053	12.8446	43
44	.0770	15.3832	199.9130	12.9855	199.7580	0.0650	0.0050	12.9956	44
45	.0727	15.4558	203.1096	13.7646	212.7435	0.0647	0.0047	13.1413	45
46	.0685	15.5244	206.1938	14.5905	226.5081	0.0644	0.0044	13.2819	46
47	.0647	15.5890	209.1681	15.4659	241.0986	0.0641	0.0041	13.4177	47
48	.0610	15.6500	212.0351	16.3939	256.5645	0.0639	0.0039	13.5485	48
49	.0575	15.7076	214.7972	17.3775	272.9584	0.0637	0.0037	13.6748	49
50	.0543	15.7619	217.4574	18.4202	290.3359	0.0634	0.0034	13.7964	50
51	.0512	15.8131	220.0181	19.5254	308.7561	0.0632	0.0032	13.9137	51
52	.0483	15.8614	222.4823	20.6969	328.2814	0.0630	0.0030	14.0267	52
53	.0456	15.9070	224.8525	21.9387	348.9783	0.0629	0.0029	14.1355	53
54	.0430	15.9500	227.1316	23.2550	370.9170	0.0627	0.0027	14.2402	54
55	.0406	15.9905	229.3222	24.6503	394.1720	0.0625	0.0025	14.3411	55
60	.0303	16.1614	239.0428	32.9877	533.1282	0.0619	0.0019	14.7909	60
65	.0227	16.2891	246.9450	44.1450	719.0829	0.0614	0.0014	15.1601	65
70	.0169	16.3845	253.3271	59.0759	967.9322	0.0610	0.0010	15.4613	70
75	.0126	16.4558	258.4527	79.0569	1300.9487	0.0608	0.0008	15.7058	75
80	.0095	16.5091	262.5493	105.7960	1746.5999	0.0606	0.0006	15.9033	80
85	.0071	16.5489	265.8096	141.5789	2342.9817	0.0604	0.0004	16.0620	85
90	.0053	16.5787	268.3946	189.4645	3141.0752	0.0603	0.0003	16.1891	90
95	.0039	16.6009	270.4375	253.5463	4209.1042	0.0602	0.0002	16.2905	95
100	.0029	16.6175	272.0471	339.3021	5638.3681	0.0602	0.0002	16.3711	100

I = 7.00 %

N	(P/F)	(P/A)	(P/G)	(F/P)	(F/A)	(A/P)	(A/F)	(A/G)	N
1	.9346	0.9346	-0.0000	1.0700	1.0000	1.0700	1.0000	-0.0000	1
2	.8734	1.8080	0.8734	1.1449	2.0700	0.5531	0.4831	0.4831	2
3	.8163	2.6243	2.5060	1.2250	3.2149	0.3811	0.3111	0.9549	3
4	.7629	3.3872	4.7947	1.3108	4.4399	0.2952	0.2252	1.4155	4
5	.7130	4.1002	7.6467	1.4026	5.7507	0.2439	0.1739	1.8650	5
6	.6663	4.7665	10.9784	1.5007	7.1533	0.2098	0.1398	2.3032	6
7	.6227	5.3893	14.7149	1.6058	8.6540	0.1856	0.1156	2.7304	7
8	.5820	5.9713	18.7889	1.7182	10.2598	0.1675	0.0975	3.1465	8
9	.5439	6.5152	23.1404	1.8385	11.9780	0.1535	0.0835	3.5517	9
10	.5083	7.0236	27.7156	1.9672	13.8164	0.1424	0.0724	3.9461	10
11	.4751	7.4987	32.4665	2.1049	15.7836	0.1334	0.0634	4.3296	11
12	.4440	7.9427	37.3506	2.2522	17.8885	0.1259	0.0559	4.7025	12
13	.4150	8.3577	42.3302	2.4098	20.1406	0.1197	0.0497	5.0648	13
14	.3878	8.7455	47.3718	2.5785	22.5505	0.1143	0.0443	5.4167	14
15	.3624	9.1079	52.4461	2.7590	25.1290	0.1098	0.0398	5.7583	15
16	.3387	9.4466	57.5271	2.9522	27.8881	0.1059	0.0359	6.0897	16
17	.3166	9.7632	62.5923	3.1588	30.8402	0.1024	0.0324	6.4110	17
18	.2959	10.0591	67.6219	3.3799	33.9990	0.0994	0.0294	6.7225	18
19	.2765	10.3356	72.5991	3.6165	37.3790	0.0968	0.0268	7.0242	19
20	.2584	10.5940	77.5091	3.8697	40.9955	0.0944	0.0244	7.3163	20
21	.2415	10.8355	82.3393	4.1406	44.8652	0.0923	0.0223	7.5990	21
22	.2257	11.0612	87.0793	4.4304	49.0057	0.0904	0.0204	7.8725	22
23	.2109	11.2722	91.7201	4.7405	53.4361	0.0887	0.0187	8.1369	23
24	.1971	11.4693	96.2545	5.0724	58.1767	0.0872	0.0172	8.3923	24
25	.1842	11.6536	100.6765	5.4274	63.2490	0.0858	0.0158	8.6391	25
26	.1722	11.8258	104.9814	5.8074	68.6765	0.0846	0.0146	8.8773	26
27	.1609	11.9867	109.1656	6.2139	74.4838	0.0834	0.0134	9.1072	27
28	.1504	12.1371	113.2264	6.6488	80.6977	0.0824	0.0124	9.3289	28
29	.1406	12.2777	117.1622	7.1143	87.3465	0.0814	0.0114	9.5427	29
30	.1314	12.4090	120.9718	7.6123	94.4608	0.0806	0.0106	9.7487	30
31	.1228	12.5318	124.6550	8.1451	102.0730	0.0798	0.0098	9.9471	31
32	.1147	12.6466	128.2120	8.7153	110.2182	0.0791	0.0091	10.1381	32
33	.1072	12.7538	131.6435	9.3253	118.9334	0.0784	0.0084	10.3219	33
34	.1002	12.8540	134.9507	9.9781	128.2588	0.0778	0.0078	10.4987	34
35	.0937	12.9477	138.1353	10.6766	138.2369	0.0772	0.0072	10.6687	35
36	.0875	13.0352	141.1990	11.4239	148.9135	0.0767	0.0067	10.8321	36
37	.0818	13.1170	144.1441	12.2236	160.3374	0.0762	0.0062	10.9891	37
38	.0765	13.1935	146.9730	13.0793	172.5610	0.0758	0.0058	11.1398	38
39	.0715	13.2649	149.6883	13.9948	185.6403	0.0754	0.0054	11.2845	39
40	.0668	13.3317	152.2928	14.9745	199.6351	0.0750	0.0050	11.4233	40
41	.0624	13.3941	154.7892	16.0227	214.6096	0.0747	0.0047	11.5565	41
42	.0583	13.4524	157.1807	17.1443	230.6322	0.0743	0.0043	11.6842	42
43	.0545	13.5070	159.4702	18.3444	247.7765	0.0740	0.0040	11.8065	43
44	.0509	13.5579	161.6609	19.6285	266.1209	0.0738	0.0038	11.9237	44
45	.0476	13.6055	163.7559	21.0025	285.7493	0.0735	0.0035	12.0360	45
46	.0445	13.6500	165.7584	22.4726	306.7518	0.0733	0.0033	12.1435	46
47	.0416	13.6916	167.6714	24.0457	329.2244	0.0730	0.0030	12.2463	47
48	.0389	13.7305	169.4981	25.7289	353.2701	0.0728	0.0028	12.3447	48
49	.0363	13.7668	171.2417	27.5299	378.9990	0.0726	0.0026	12.4387	49
50	.0339	13.8007	172.9051	29.4570	406.5289	0.0725	0.0025	12.5287	50
51	.0317	13.8325	174.4915	31.5190	435.9860	0.0723	0.0023	12.6146	51
52	.0297	13.8621	176.0037	33.7253	467.5050	0.0721	0.0021	12.6967	52
53	.0277	13.8898	177.4447	36.0861	501.2303	0.0720	0.0020	12.7751	53
54	.0259	13.9157	178.8173	38.6122	537.3164	0.0719	0.0019	12.8500	54
55	.0242	13.9399	180.1243	41.3150	575.9286	0.0717	0.0017	12.9215	55
60	.0173	14.0392	185.7677	57.9464	813.5204	0.0712	0.0012	13.2321	60
65	.0123	14.1099	190.1452	81.2729	1146.7552	0.0709	0.0009	13.4760	65
70	.0088	14.1604	193.5185	113.9894	1614.1342	0.0706	0.0006	13.6662	70
75	.0063	14.1964	196.1035	159.8760	2269.6574	0.0704	0.0004	13.8136	75
80	.0045	14.2220	198.0748	224.2344	3189.0627	0.0703	0.0003	13.9273	80
85	.0032	14.2403	199.5717	314.5003	4478.5761	0.0702	0.0002	14.0146	85
90	.0023	14.2533	200.7042	441.1030	6287.1854	0.0702	0.0002	14.0812	90
95	.0016	14.2626	201.5581	618.6697	8823.8535	0.0701	0.0001	14.1319	95
100	.0012	14.2693	202.2001	867.7163	12381.6618	0.0701	0.0001	14.1703	100

ENGINEERING ECONOMIC ANALYSIS

I = 8.00 %

N	(P/F)	(P/A)	(P/G)	(F/P)	(F/A)	(A/P)	(A/F)	(A/G)	N
1	.9259	0.9259	-0.0000	1.0800	1.0000	1.0800	1.0000	-0.0000	1
2	.8573	1.7833	0.8573	1.1664	2.0800	0.5608	0.4808	0.4808	2
3	.7938	2.5771	2.4450	1.2597	3.2464	0.3880	0.3080	0.9487	3
4	.7350	3.3121	4.6501	1.3605	4.5061	0.3019	0.2219	1.4040	4
5	.6806	3.9927	7.3724	1.4693	5.8666	0.2505	0.1705	1.8465	5
6	.6302	4.6229	10.5233	1.5869	7.3359	0.2163	0.1363	2.2763	6
7	.5835	5.2064	14.0242	1.7138	8.9228	0.1921	0.1121	2.6937	7
8	.5403	5.7466	17.8061	1.8509	10.6366	0.1740	0.0940	3.0985	8
9	.5002	6.2469	21.8081	1.9990	12.4876	0.1601	0.0801	3.4910	9
10	.4632	6.7101	25.9768	2.1589	14.4866	0.1490	0.0690	3.8713	10
11	.4289	7.1390	30.2657	2.3316	16.6455	0.1401	0.0601	4.2395	11
12	.3971	7.5361	34.6339	2.5182	18.9771	0.1327	0.0527	4.5957	12
13	.3677	7.9038	39.0463	2.7196	21.4953	0.1265	0.0465	4.9402	13
14	.3405	8.2442	43.4723	2.9372	24.2149	0.1213	0.0413	5.2731	14
15	.3152	8.5595	47.8857	3.1722	27.1521	0.1168	0.0368	5.5945	15
16	.2919	8.8514	52.2640	3.4259	30.3243	0.1130	0.0330	5.9046	16
17	.2703	9.1216	56.5883	3.7000	33.7502	0.1096	0.0296	6.2037	17
18	.2502	9.3719	60.8426	3.9960	37.4502	0.1067	0.0267	6.4920	18
19	.2317	9.6036	65.0134	4.3157	41.4463	0.1041	0.0241	6.7697	19
20	.2145	9.8181	69.0898	4.6610	45.7620	0.1019	0.0219	7.0369	20
21	.1987	10.0168	73.0629	5.0338	50.4229	0.0998	0.0198	7.2940	21
22	.1839	10.2007	76.9257	5.4365	55.4568	0.0980	0.0180	7.5412	22
23	.1703	10.3711	80.6726	5.8715	60.8933	0.0964	0.0164	7.7786	23
24	.1577	10.5288	84.2997	6.3412	66.7648	0.0950	0.0150	8.0066	24
25	.1460	10.6748	87.8041	6.8485	73.1059	0.0937	0.0137	8.2254	25
26	.1352	10.8100	91.1842	7.3964	79.9544	0.0925	0.0125	8.4352	26
27	.1252	10.9352	94.4390	7.9881	87.3508	0.0914	0.0114	8.6363	27
28	.1159	11.0511	97.5687	8.6271	95.3388	0.0905	0.0105	8.8289	28
29	.1073	11.1584	100.5738	9.3173	103.9659	0.0896	0.0096	9.0133	29
30	.0994	11.2578	103.4558	10.0627	113.2832	0.0888	0.0088	9.1897	30
31	.0920	11.3498	106.2163	10.8677	123.3459	0.0881	0.0081	9.3584	31
32	.0852	11.4350	108.8575	11.7371	134.2135	0.0875	0.0075	9.5197	32
33	.0789	11.5139	111.3819	12.6760	145.9506	0.0869	0.0069	9.6737	33
34	.0730	11.5869	113.7924	13.6901	158.6267	0.0863	0.0063	9.8208	34
35	.0676	11.6546	116.0920	14.7853	172.3168	0.0858	0.0058	9.9611	35
36	.0626	11.7172	118.2839	15.9682	187.1021	0.0853	0.0053	10.0949	36
37	.0580	11.7752	120.3713	17.2456	203.0703	0.0849	0.0049	10.2225	37
38	.0537	11.8289	122.3579	18.6253	220.3159	0.0845	0.0045	10.3440	38
39	.0497	11.8786	124.2470	20.1153	238.9412	0.0842	0.0042	10.4597	39
40	.0460	11.9246	126.0422	21.7245	259.0565	0.0839	0.0039	10.5699	40
41	.0426	11.9672	127.7470	23.4625	280.7810	0.0836	0.0036	10.6747	41
42	.0395	12.0067	129.3651	25.3395	304.2435	0.0833	0.0033	10.7744	42
43	.0365	12.0432	130.8998	27.3666	329.5830	0.0830	0.0030	10.8692	43
44	.0338	12.0771	132.3547	29.5560	356.9496	0.0828	0.0028	10.9592	44
45	.0313	12.1084	133.7331	31.9204	386.5056	0.0826	0.0026	11.0447	45
46	.0290	12.1374	135.0384	34.4741	418.4261	0.0824	0.0024	11.1258	46
47	.0269	12.1643	136.2739	37.2320	452.9002	0.0822	0.0022	11.2028	47
48	.0249	12.1891	137.4428	40.2106	490.1322	0.0820	0.0020	11.2758	48
49	.0230	12.2122	138.5480	43.4274	530.3427	0.0819	0.0019	11.3451	49
50	.0213	12.2335	139.5928	46.9016	573.7702	0.0817	0.0017	11.4107	50
51	.0197	12.2532	140.5799	50.6537	620.6718	0.0816	0.0016	11.4729	51
52	.0183	12.2715	141.5121	54.7060	671.3255	0.0815	0.0015	11.5318	52
53	.0169	12.2884	142.3923	59.0825	726.0316	0.0814	0.0014	11.5875	53
54	.0157	12.3041	143.2229	63.8091	785.1141	0.0813	0.0013	11.6403	54
55	.0145	12.3186	144.0065	68.9139	848.9232	0.0812	0.0012	11.6902	55
60	.0099	12.3766	147.3000	101.2571	1253.2133	0.0808	0.0008	11.9015	60
65	.0067	12.4160	149.7387	148.7798	1847.2481	0.0805	0.0005	12.0602	65
70	.0046	12.4428	151.5326	218.6064	2720.0801	0.0804	0.0004	12.1783	70
75	.0031	12.4611	152.8448	321.2045	4002.5566	0.0802	0.0002	12.2658	75
80	.0021	12.4735	153.8001	471.9548	5886.9354	0.0802	0.0002	12.3301	80
85	.0014	12.4820	154.4925	693.4565	8655.7061	0.0801	0.0001	12.3772	85
90	.0010	12.4877	154.9925	1018.9151	12723.9386	0.0801	0.0001	12.4116	90
95	.0007	12.4917	155.3524	1497.1205	18701.5069	0.0801	0.0001	12.4365	95
100	.0005	12.4943	155.6107	2199.7613	27484.5157	0.0800	0.0000	12.4545	100

I = 9.00 %

N	(P/F)	(P/A)	(P/G)	(F/P)	(F/A)	(A/P)	(A/F)	(A/G)	N
1	.9174	0.9174	-0.0000	1.0900	1.0000	1.0900	1.0000	-0.0000	1
2	.8417	1.7591	0.8417	1.1881	2.0900	0.5685	0.4785	0.4785	2
3	.7722	2.5313	2.3860	1.2950	3.2781	0.3951	0.3051	0.9426	3
4	.7084	3.2397	4.5113	1.4116	4.5731	0.3087	0.2187	1.3925	4
5	.6499	3.8897	7.1110	1.5386	5.9847	0.2571	0.1671	1.8282	5
6	.5963	4.4859	10.0924	1.6771	7.5233	0.2229	0.1329	2.2498	6
7	.5470	5.0330	13.3746	1.8280	9.2004	0.1987	0.1087	2.6574	7
8	.5019	5.5348	16.8877	1.9926	11.0285	0.1807	0.0907	3.0512	8
9	.4604	5.9952	20.5711	2.1719	13.0210	0.1668	0.0768	3.4312	9
10	.4224	6.4177	24.3728	2.3674	15.1929	0.1558	0.0658	3.7978	10
11	.3875	6.8052	28.2481	2.5804	17.5603	0.1469	0.0569	4.1510	11
12	.3555	7.1607	32.1590	2.8127	20.1407	0.1397	0.0497	4.4910	12
13	.3262	7.4869	36.0731	3.0658	22.9534	0.1336	0.0436	4.8182	13
14	.2992	7.7862	39.9633	3.3417	26.0192	0.1284	0.0384	5.1326	14
15	.2745	8.0607	43.8069	3.6425	29.3609	0.1241	0.0341	5.4346	15
16	.2519	8.3126	47.5849	3.9703	33.0034	0.1203	0.0303	5.7245	16
17	.2311	8.5436	51.2821	4.3276	36.9737	0.1170	0.0270	6.0024	17
18	.2120	8.7556	54.8860	4.7171	41.3013	0.1142	0.0242	6.2687	18
19	.1945	8.9501	58.3868	5.1417	46.0185	0.1117	0.0217	6.5236	19
20	.1784	9.1285	61.7770	5.6044	51.1601	0.1095	0.0195	6.7674	20
21	.1637	9.2922	65.0509	6.1088	56.7645	0.1076	0.0176	7.0006	21
22	.1502	9.4424	68.2048	6.6586	62.8733	0.1059	0.0159	7.2232	22
23	.1378	9.5802	71.2359	7.2579	69.5319	0.1044	0.0144	7.4357	23
24	.1264	9.7066	74.1433	7.9111	76.7898	0.1030	0.0130	7.6384	24
25	.1160	9.8226	76.9265	8.6231	84.7009	0.1018	0.0118	7.8316	25
26	.1064	9.9290	79.5863	9.3992	93.3240	0.1007	0.0107	8.0156	26
27	.0976	10.0266	82.1241	10.2451	102.7231	0.0997	0.0097	8.1906	27
28	.0895	10.1161	84.5419	11.1671	112.9682	0.0989	0.0089	8.3571	28
29	.0822	10.1983	86.8422	12.1722	124.1354	0.0981	0.0081	8.5154	29
30	.0754	10.2737	89.0280	13.2677	136.3075	0.0973	0.0073	8.6657	30
31	.0691	10.3428	91.1024	14.4618	149.5752	0.0967	0.0067	8.8083	31
32	.0634	10.4062	93.0690	15.7633	164.0370	0.0961	0.0061	8.9436	32
33	.0582	10.4644	94.9314	17.1820	179.8003	0.0956	0.0056	9.0718	33
34	.0534	10.5178	96.6935	18.7284	196.9823	0.0951	0.0051	9.1933	34
35	.0490	10.5668	98.3590	20.4140	215.7108	0.0946	0.0046	9.3083	35
36	.0449	10.6118	99.9319	22.2512	236.1247	0.0942	0.0042	9.4171	36
37	.0412	10.6530	101.4162	24.2538	258.3759	0.0939	0.0039	9.5200	37
38	.0378	10.6908	102.8158	26.4367	282.6298	0.0935	0.0035	9.6172	38
39	.0347	10.7255	104.1345	28.8160	309.0665	0.0932	0.0032	9.7090	39
40	.0318	10.7574	105.3762	31.4094	337.8824	0.0930	0.0030	9.7957	40
41	.0292	10.7866	106.5445	34.2363	369.2919	0.0927	0.0027	9.8775	41
42	.0268	10.8134	107.6432	37.3175	403.5281	0.0925	0.0025	9.9546	42
43	.0246	10.8380	108.6758	40.6761	440.8457	0.0923	0.0023	10.0273	43
44	.0226	10.8605	109.6456	44.3370	481.5218	0.0921	0.0021	10.0958	44
45	.0207	10.8812	110.5561	48.3273	525.8587	0.0919	0.0019	10.1603	45
46	.0190	10.9002	111.4103	52.6767	574.1860	0.0917	0.0017	10.2210	46
47	.0174	10.9176	112.2115	57.4176	626.8628	0.0916	0.0016	10.2780	47
48	.0160	10.9336	112.9625	62.5852	684.2804	0.0915	0.0015	10.3317	48
49	.0147	10.9482	113.6661	68.2179	746.8656	0.0913	0.0013	10.3821	49
50	.0134	10.9617	114.3251	74.3575	815.0836	0.0912	0.0012	10.4295	50
51	.0123	10.9740	114.9420	81.0497	889.4411	0.0911	0.0011	10.4740	51
52	.0113	10.9853	115.5193	88.3442	970.4908	0.0910	0.0010	10.5158	52
53	.0104	10.9957	116.0593	96.2951	1058.8349	0.0909	0.0009	10.5549	53
54	.0095	11.0053	116.5642	104.9617	1155.1301	0.0909	0.0009	10.5917	54
55	.0087	11.0140	117.0362	114.4083	1260.0918	0.0908	0.0008	10.6261	55
60	.0057	11.0480	118.9683	176.0313	1944.7921	0.0905	0.0005	10.7683	60
65	.0037	11.0701	120.3344	270.8460	2998.2885	0.0903	0.0003	10.8702	65
70	.0024	11.0844	121.2942	416.7301	4619.2232	0.0902	0.0002	10.9427	70
75	.0016	11.0938	121.9646	641.1909	7113.2321	0.0901	0.0001	10.9940	75
80	.0010	11.0998	122.4306	986.5517	10950.5741	0.0901	0.0001	11.0299	80
85	.0007	11.1038	122.7533	1517.9320	16854.8003	0.0901	0.0001	11.0551	85
90	.0004	11.1064	122.9758	2335.5266	25939.1842	0.0900	0.0000	11.0726	90
95	.0003	11.1080	123.1287	3593.4971	39916.6350	0.0900	0.0000	11.0847	95
100	.0002	11.1091	123.2335	5529.0408	61422.6755	0.0900	0.0000	11.0930	100

ENGINEERING ECONOMIC ANALYSIS

I = 10.00 %

N	(P/F)	(P/A)	(P/G)	(F/P)	(F/A)	(A/P)	(A/F)	(A/G)	N
1	.9091	0.9091	−0.0000	1.1000	1.0000	1.1000	1.0000	−0.0000	1
2	.8264	1.7355	0.8264	1.2100	2.1000	0.5762	0.4762	0.4762	2
3	.7513	2.4869	2.3291	1.3310	3.3100	0.4021	0.3021	0.9366	3
4	.6830	3.1699	4.3781	1.4641	4.6410	0.3155	0.2155	1.3812	4
5	.6209	3.7908	6.8618	1.6105	6.1051	0.2638	0.1638	1.8101	5
6	.5645	4.3553	9.6842	1.7716	7.7156	0.2296	0.1296	2.2236	6
7	.5132	4.8684	12.7631	1.9487	9.4872	0.2054	0.1054	2.6216	7
8	.4665	5.3349	16.0287	2.1436	11.4359	0.1874	0.0874	3.0045	8
9	.4241	5.7590	19.4215	2.3579	13.5795	0.1736	0.0736	3.3724	9
10	.3855	6.1446	22.8913	2.5937	15.9374	0.1627	0.0627	3.7255	10
11	.3505	6.4951	26.3963	2.8531	18.5312	0.1540	0.0540	4.0641	11
12	.3186	6.8137	29.9012	3.1384	21.3843	0.1468	0.0468	4.3884	12
13	.2897	7.1034	33.3772	3.4523	24.5227	0.1408	0.0408	4.6988	13
14	.2633	7.3667	36.8005	3.7975	27.9750	0.1357	0.0357	4.9955	14
15	.2394	7.6061	40.1520	4.1772	31.7725	0.1315	0.0315	5.2789	15
16	.2176	7.8237	43.4164	4.5950	35.9497	0.1278	0.0278	5.5493	16
17	.1978	8.0216	46.5819	5.0545	40.5447	0.1247	0.0247	5.8071	17
18	.1799	8.2014	49.6395	5.5599	45.5992	0.1219	0.0219	6.0526	18
19	.1635	8.3649	52.5827	6.1159	51.1591	0.1195	0.0195	6.2861	19
20	.1486	8.5136	55.4069	6.7275	57.2750	0.1175	0.0175	6.5081	20
21	.1351	8.6487	58.1095	7.4002	64.0025	0.1156	0.0156	6.7189	21
22	.1228	8.7715	60.6893	8.1403	71.4027	0.1140	0.0140	6.9189	22
23	.1117	8.8832	63.1462	8.9543	79.5430	0.1126	0.0126	7.1085	23
24	.1015	8.9847	65.4813	9.8497	88.4973	0.1113	0.0113	7.2881	24
25	.0923	9.0770	67.6964	10.8347	98.3471	0.1102	0.0102	7.4580	25
26	.0839	9.1609	69.7940	11.9182	109.1818	0.1092	0.0092	7.6186	26
27	.0763	9.2372	71.7773	13.1100	121.0999	0.1083	0.0083	7.7704	27
28	.0693	9.3066	73.6495	14.4210	134.2099	0.1075	0.0075	7.9137	28
29	.0630	9.3696	75.4146	15.8631	148.6309	0.1067	0.0067	8.0489	29
30	.0573	9.4269	77.0766	17.4494	164.4940	0.1061	0.0061	8.1762	30
31	.0521	9.4790	78.6395	19.1943	181.9434	0.1055	0.0055	8.2962	31
32	.0474	9.5264	80.1078	21.1138	201.1378	0.1050	0.0050	8.4091	32
33	.0431	9.5694	81.4856	23.2252	222.2515	0.1045	0.0045	8.5152	33
34	.0391	9.6086	82.7773	25.5477	245.4767	0.1041	0.0041	8.6149	34
35	.0356	9.6442	83.9872	28.1024	271.0244	0.1037	0.0037	8.7086	35
36	.0323	9.6765	85.1194	30.9127	299.1268	0.1033	0.0033	8.7965	36
37	.0294	9.7059	86.1781	34.0039	330.0395	0.1030	0.0030	8.8789	37
38	.0267	9.7327	87.1673	37.4043	364.0434	0.1027	0.0027	8.9562	38
39	.0243	9.7570	88.0908	41.1448	401.4478	0.1025	0.0025	9.0285	39
40	.0221	9.7791	88.9525	45.2593	442.5926	0.1023	0.0023	9.0962	40
41	.0201	9.7991	89.7560	49.7852	487.8518	0.1020	0.0020	9.1596	41
42	.0183	9.8174	90.5047	54.7637	537.6370	0.1019	0.0019	9.2188	42
43	.0166	9.8340	91.2019	60.2401	592.4007	0.1017	0.0017	9.2741	43
44	.0151	9.8491	91.8508	66.2641	652.6408	0.1015	0.0015	9.3258	44
45	.0137	9.8628	92.4544	72.8905	718.9048	0.1014	0.0014	9.3740	45
46	.0125	9.8753	93.0157	80.1795	791.7953	0.1013	0.0013	9.4190	46
47	.0113	9.8866	93.5372	88.1975	871.9749	0.1011	0.0011	9.4610	47
48	.0103	9.8969	94.0217	97.0172	960.1723	0.1010	0.0010	9.5001	48
49	.0094	9.9063	94.4715	106.7190	1057.1896	0.1009	0.0009	9.5365	49
50	.0085	9.9148	94.8889	117.3909	1163.9085	0.1009	0.0009	9.5704	50
51	.0077	9.9226	95.2761	129.1299	1281.2994	0.1008	0.0008	9.6020	51
52	.0070	9.9296	95.6351	142.0429	1410.4293	0.1007	0.0007	9.6313	52
53	.0064	9.9360	95.9679	156.2472	1552.4723	0.1006	0.0006	9.6586	53
54	.0058	9.9418	96.2763	171.8719	1708.7195	0.1006	0.0006	9.6840	54
55	.0053	9.9471	96.5619	189.0591	1880.5914	0.1005	0.0005	9.7075	55
60	.0033	9.9672	97.7010	304.4816	3034.8164	0.1003	0.0003	9.8023	60
65	.0020	9.9796	98.4705	490.3707	4893.7073	0.1002	0.0002	9.8672	65
70	.0013	9.9873	98.9870	789.7470	7887.4696	0.1001	0.0001	9.9113	70
75	.0008	9.9921	99.3317	1271.8954	12708.9537	0.1001	0.0001	9.9410	75
80	.0005	9.9951	99.5606	2048.4002	20474.0021	0.1000	0.0000	9.9609	80
85	.0003	9.9970	99.7120	3298.9690	32979.6903	0.1000	0.0000	9.9742	85
90	.0002	9.9981	99.8118	5313.0226	53120.2261	0.1000	0.0000	9.9831	90
95	.0001	9.9988	99.8773	8556.6760	85556.7605	0.1000	0.0000	9.9889	95
100	.0001	9.9993	99.9202	13780.6123	137796.1234	0.1000	0.0000	9.9927	100

I = 12.00 %

N	(P/F)	(P/A)	(P/G)	(F/P)	(F/A)	(A/P)	(A/F)	(A/G)	N
1	.8929	0.8929	-0.0000	1.1200	1.0000	1.1200	1.0000	-0.0000	1
2	.7972	1.6901	0.7972	1.2544	2.1200	0.5917	0.4717	0.4717	2
3	.7118	2.4018	2.2208	1.4049	3.3744	0.4163	0.2963	0.9246	3
4	.6355	3.0373	4.1273	1.5735	4.7793	0.3292	0.2092	1.3589	4
5	.5674	3.6048	6.3970	1.7623	6.3528	0.2774	0.1574	1.7746	5
6	.5066	4.1114	8.9302	1.9738	8.1152	0.2432	0.1232	2.1720	6
7	.4523	4.5638	11.6443	2.2107	10.0890	0.2191	0.0991	2.5515	7
8	.4039	4.9676	14.4714	2.4760	12.2997	0.2013	0.0813	2.9131	8
9	.3606	5.3282	17.3563	2.7731	14.7757	0.1877	0.0677	3.2574	9
10	.3220	5.6502	20.2541	3.1058	17.5487	0.1770	0.0570	3.5847	10
11	.2875	5.9377	23.1288	3.4785	20.6546	0.1684	0.0484	3.8953	11
12	.2567	6.1944	25.9523	3.8960	24.1331	0.1614	0.0414	4.1897	12
13	.2292	6.4235	28.7024	4.3635	28.0291	0.1557	0.0357	4.4683	13
14	.2046	6.6282	31.3624	4.8871	32.3926	0.1509	0.0309	4.7317	14
15	.1827	6.8109	33.9202	5.4736	37.2797	0.1468	0.0268	4.9803	15
16	.1631	6.9740	36.3670	6.1304	42.7533	0.1434	0.0234	5.2147	16
17	.1456	7.1196	38.6973	6.8660	48.8837	0.1405	0.0205	5.4353	17
18	.1300	7.2497	40.9080	7.6900	55.7497	0.1379	0.0179	5.6427	18
19	.1161	7.3658	42.9979	8.6128	63.4397	0.1358	0.0158	5.8375	19
20	.1037	7.4694	44.9676	9.6463	72.0524	0.1339	0.0139	6.0202	20
21	.0926	7.5620	46.8188	10.8038	81.6987	0.1322	0.0122	6.1913	21
22	.0826	7.6446	48.5543	12.1003	92.5026	0.1308	0.0108	6.3514	22
23	.0738	7.7184	50.1776	13.5523	104.6029	0.1296	0.0096	6.5010	23
24	.0659	7.7843	51.6929	15.1786	118.1552	0.1285	0.0085	6.6406	24
25	.0588	7.8431	53.1046	17.0001	133.3339	0.1275	0.0075	6.7708	25
26	.0525	7.8957	54.4177	19.0401	150.3339	0.1267	0.0067	6.8921	26
27	.0469	7.9426	55.6369	21.3249	169.3740	0.1259	0.0059	7.0049	27
28	.0419	7.9844	56.7674	23.8839	190.6989	0.1252	0.0052	7.1098	28
29	.0374	8.0218	57.8141	26.7499	214.5828	0.1247	0.0047	7.2071	29
30	.0334	8.0552	58.7821	29.9599	241.3327	0.1241	0.0041	7.2974	30
31	.0298	8.0850	59.6761	33.5551	271.2926	0.1237	0.0037	7.3811	31
32	.0266	8.1116	60.5010	37.5817	304.8477	0.1233	0.0033	7.4586	32
33	.0238	8.1354	61.2612	42.0915	342.4294	0.1229	0.0029	7.5302	33
34	.0212	8.1566	61.9612	47.1425	384.5210	0.1226	0.0026	7.5965	34
35	.0189	8.1755	62.6052	52.7996	431.6635	0.1223	0.0023	7.6577	35
36	.0169	8.1924	63.1970	59.1356	484.4631	0.1221	0.0021	7.7141	36
37	.0151	8.2075	63.7406	66.2318	543.5987	0.1218	0.0018	7.7661	37
38	.0135	8.2210	64.2394	74.1797	609.8305	0.1216	0.0016	7.8141	38
39	.0120	8.2330	64.6967	83.0812	684.0102	0.1215	0.0015	7.8582	39
40	.0107	8.2438	65.1159	93.0510	767.0914	0.1213	0.0013	7.8988	40
41	.0096	8.2534	65.4997	104.2171	860.1424	0.1212	0.0012	7.9361	41
42	.0086	8.2619	65.8509	116.7231	964.3595	0.1210	0.0010	7.9704	42
43	.0076	8.2696	66.1722	130.7299	1081.0826	0.1209	0.0009	8.0019	43
44	.0068	8.2764	66.4659	146.4175	1211.8125	0.1208	0.0008	8.0308	44
45	.0061	8.2825	66.7342	163.9876	1358.2300	0.1207	0.0007	8.0572	45
46	.0054	8.2880	66.9792	183.6661	1522.2176	0.1207	0.0007	8.0815	46
47	.0049	8.2928	67.2028	205.7061	1705.8838	0.1206	0.0006	8.1037	47
48	.0043	8.2972	67.4068	230.3908	1911.5898	0.1205	0.0005	8.1241	48
49	.0039	8.3010	67.5929	258.0377	2141.9806	0.1205	0.0005	8.1427	49
50	.0035	8.3045	67.7624	289.0022	2400.0182	0.1204	0.0004	8.1597	50
51	.0031	8.3076	67.9169	323.6825	2689.0204	0.1204	0.0004	8.1753	51
52	.0028	8.3103	68.0576	362.5243	3012.7029	0.1203	0.0003	8.1895	52
53	.0025	8.3128	68.1856	406.0273	3375.2272	0.1203	0.0003	8.2025	53
54	.0022	8.3150	68.3022	454.7505	3781.2545	0.1203	0.0003	8.2143	54
55	.0020	8.3170	68.4082	509.3206	4236.0050	0.1202	0.0002	8.2251	55
60	.0011	8.3240	68.8100	897.5969	7471.6411	0.1201	0.0001	8.2664	60
65	.0006	8.3281	69.0581	1581.8725	13173.9374	0.1201	0.0001	8.2922	65
70	.0004	8.3303	69.2103	2787.7998	23223.3319	0.1200	0.0000	8.3082	70
75	.0002	8.3316	69.3031	4913.0558	40933.7987	0.1200	0.0000	8.3181	75
80	.0001	8.3324	69.3594	8658.4831	72145.6925	0.1200	0.0000	8.3241	80
85	.0001	8.3328	69.3935	15259.2057	127151.7140	0.1200	0.0000	8.3278	85
90	.0000	8.3330	69.4140	26891.9342	224091.1185	0.1200	0.0000	8.3300	90
95	.0000	8.3332	69.4263	47392.7766	394931.4719	0.1200	0.0000	8.3313	95
100	.0000	8.3332	69.4336	83522.2657	696010.5477	0.1200	0.0000	8.3321	100

ENGINEERING ECONOMIC ANALYSIS

I = 15.00 %

N	(P/F)	(P/A)	(P/G)	(F/P)	(F/A)	(A/P)	(A/F)	(A/G)	N
1	.8696	0.8696	-0.0000	1.1500	1.0000	1.1500	1.0000	-0.0000	1
2	.7561	1.6257	0.7561	1.3225	2.1500	0.6151	0.4651	0.4651	2
3	.6575	2.2832	2.0712	1.5209	3.4725	0.4380	0.2880	0.9071	3
4	.5718	2.8550	3.7864	1.7490	4.9934	0.3503	0.2003	1.3263	4
5	.4972	3.3522	5.7751	2.0114	6.7424	0.2983	0.1483	1.7228	5
6	.4323	3.7845	7.9368	2.3131	8.7537	0.2642	0.1142	2.0972	6
7	.3759	4.1604	10.1924	2.6600	11.0668	0.2404	0.0904	2.4498	7
8	.3269	4.4873	12.4807	3.0590	13.7268	0.2229	0.0729	2.7813	8
9	.2843	4.7716	14.7548	3.5179	16.7858	0.2096	0.0596	3.0922	9
10	.2472	5.0188	16.9795	4.0456	20.3037	0.1993	0.0493	3.3832	10
11	.2149	5.2337	19.1289	4.6524	24.3493	0.1911	0.0411	3.6549	11
12	.1869	5.4206	21.1849	5.3503	29.0017	0.1845	0.0345	3.9082	12
13	.1625	5.5831	23.1352	6.1528	34.3519	0.1791	0.0291	4.1438	13
14	.1413	5.7245	24.9725	7.0757	40.5047	0.1747	0.0247	4.3624	14
15	.1229	5.8474	26.6930	8.1371	47.5804	0.1710	0.0210	4.5650	15
16	.1069	5.9542	28.2960	9.3576	55.7175	0.1679	0.0179	4.7522	16
17	.0929	6.0472	29.7828	10.7613	65.0751	0.1654	0.0154	4.9251	17
18	.0808	6.1280	31.1565	12.3755	75.8364	0.1632	0.0132	5.0843	18
19	.0703	6.1982	32.4213	14.2318	88.2118	0.1613	0.0113	5.2307	19
20	.0611	6.2593	33.5822	16.3665	102.4436	0.1598	0.0098	5.3651	20
21	.0531	6.3125	34.6448	18.8215	118.8101	0.1584	0.0084	5.4883	21
22	.0462	6.3587	35.6150	21.6447	137.6316	0.1573	0.0073	5.6010	22
23	.0402	6.3988	36.4988	24.8915	159.2764	0.1563	0.0063	5.7040	23
24	.0349	6.4338	37.3023	28.6252	184.1678	0.1554	0.0054	5.7979	24
25	.0304	6.4641	38.0314	32.9190	212.7930	0.1547	0.0047	5.8834	25
26	.0264	6.4906	38.6918	37.8568	245.7120	0.1541	0.0041	5.9612	26
27	.0230	6.5135	39.2890	43.5353	283.5688	0.1535	0.0035	6.0319	27
28	.0200	6.5335	39.8283	50.0656	327.1041	0.1531	0.0031	6.0960	28
29	.0174	6.5509	40.3146	57.5755	377.1697	0.1527	0.0027	6.1541	29
30	.0151	6.5660	40.7526	66.2118	434.7451	0.1523	0.0023	6.2066	30
31	.0131	6.5791	41.1466	76.1435	500.9569	0.1520	0.0020	6.2541	31
32	.0114	6.5905	41.5006	87.5651	577.1005	0.1517	0.0017	6.2970	32
33	.0099	6.6005	41.8184	100.6998	664.6655	0.1515	0.0015	6.3357	33
34	.0086	6.6091	42.1033	115.8048	765.3654	0.1513	0.0013	6.3705	34
35	.0075	6.6166	42.3586	133.1755	881.1702	0.1511	0.0011	6.4019	35
36	.0065	6.6231	42.5872	153.1519	1014.3457	0.1510	0.0010	6.4301	36
37	.0057	6.6288	42.7916	176.1246	1167.4975	0.1509	0.0009	6.4554	37
38	.0049	6.6338	42.9743	202.5433	1343.6222	0.1507	0.0007	6.4781	38
39	.0043	6.6380	43.1374	232.9248	1546.1655	0.1506	0.0006	6.4985	39
40	.0037	6.6418	43.2830	267.8635	1779.0903	0.1506	0.0006	6.5168	40
41	.0032	6.6450	43.4128	308.0431	2046.9539	0.1505	0.0005	6.5331	41
42	.0028	6.6478	43.5286	354.2495	2354.9969	0.1504	0.0004	6.5478	42
43	.0025	6.6503	43.6317	407.3870	2709.2465	0.1504	0.0004	6.5609	43
44	.0021	6.6524	43.7235	468.4950	3116.6334	0.1503	0.0003	6.5725	44
45	.0019	6.6543	43.8051	538.7693	3585.1285	0.1503	0.0003	6.5830	45
46	.0016	6.6559	43.8778	619.5847	4123.8977	0.1502	0.0002	6.5923	46
47	.0014	6.6573	43.9423	712.5224	4743.4824	0.1502	0.0002	6.6006	47
48	.0012	6.6585	43.9997	819.4007	5456.6047	0.1502	0.0002	6.6080	48
49	.0011	6.6596	44.0506	942.3108	6275.4055	0.1502	0.0002	6.6146	49
50	.0009	6.6605	44.0958	1083.6574	7217.7163	0.1501	0.0001	6.6205	50
51	.0008	6.6613	44.1360	1246.2061	8301.3737	0.1501	0.0001	6.6257	51
52	.0007	6.6620	44.1715	1433.1370	9547.5798	0.1501	0.0001	6.6304	52
53	.0006	6.6626	44.2031	1648.1075	10980.7167	0.1501	0.0001	6.6345	53
54	.0005	6.6631	44.2311	1895.3236	12628.8243	0.1501	0.0001	6.6382	54
55	.0005	6.6636	44.2558	2179.6222	14524.1479	0.1501	0.0001	6.6414	55
60	.0002	6.6651	44.3431	4383.9987	29219.9916	0.1500	0.0000	6.6530	60
65	.0001	6.6659	44.3903	8817.7874	58778.5826	0.1500	0.0000	6.6593	65
70	.0001	6.6663	44.4156	17735.7200	118231.4669	0.1500	0.0000	6.6627	70
75	.0000	6.6665	44.4292	35672.8680	237812.4532	0.1500	0.0000	6.6646	75
80	.0000	6.6666	44.4364	71750.8794	478332.5293	0.1500	0.0000	6.6656	80
85	.0000	6.6666	44.4402	144316.6470	962104.3133	0.1500	0.0000	6.6661	85
90	.0000	6.6666	44.4422	290272.3252	1935142.1680	0.1500	0.0000	6.6664	90
95	.0000	6.6667	44.4433	583841.3276	3892268.8509	0.1500	0.0000	6.6665	95
100	.0000	6.6667	44.4438	1174313.4507	7828749.6713	0.1500	0.0000	6.6666	100

I = 20.00 %

N	(P/F)	(P/A)	(P/G)	(F/P)	(F/A)	(A/P)	(A/F)	(A/G)	N
1	.8333	0.8333	-0.0000	1.2000	1.0000	1.2000	1.0000	-0.0000	1
2	.6944	1.5278	0.6944	1.4400	2.2000	0.6545	0.4545	0.4545	2
3	.5787	2.1065	1.8519	1.7280	3.6400	0.4747	0.2747	0.8791	3
4	.4823	2.5887	3.2986	2.0736	5.3680	0.3863	0.1863	1.2742	4
5	.4019	2.9906	4.9061	2.4883	7.4416	0.3344	0.1344	1.6405	5
6	.3349	3.3255	6.5806	2.9860	9.9299	0.3007	0.1007	1.9788	6
7	.2791	3.6046	8.2551	3.5832	12.9159	0.2774	0.0774	2.2902	7
8	.2326	3.8372	9.8831	4.2998	16.4991	0.2606	0.0606	2.5756	8
9	.1938	4.0310	11.4335	5.1598	20.7989	0.2481	0.0481	2.8364	9
10	.1615	4.1925	12.8871	6.1917	25.9587	0.2385	0.0385	3.0739	10
11	.1346	4.3271	14.2330	7.4301	32.1504	0.2311	0.0311	3.2893	11
12	.1122	4.4392	15.4667	8.9161	39.5805	0.2253	0.0253	3.4841	12
13	.0935	4.5327	16.5883	10.6993	48.4966	0.2206	0.0206	3.6597	13
14	.0779	4.6106	17.6008	12.8392	59.1959	0.2169	0.0169	3.8175	14
15	.0649	4.6755	18.5095	15.4070	72.0351	0.2139	0.0139	3.9588	15
16	.0541	4.7296	19.3208	18.4884	87.4421	0.2114	0.0114	4.0851	16
17	.0451	4.7746	20.0419	22.1861	105.9306	0.2094	0.0094	4.1976	17
18	.0376	4.8122	20.6805	26.6233	128.1167	0.2078	0.0078	4.2975	18
19	.0313	4.8435	21.2439	31.9480	154.7400	0.2065	0.0065	4.3861	19
20	.0261	4.8696	21.7395	38.3376	186.6880	0.2054	0.0054	4.4643	20
21	.0217	4.8913	22.1742	46.0051	225.0256	0.2044	0.0044	4.5334	21
22	.0181	4.9094	22.5546	55.2061	271.0307	0.2037	0.0037	4.5941	22
23	.0151	4.9245	22.8867	66.2474	326.2369	0.2031	0.0031	4.6475	23
24	.0126	4.9371	23.1760	79.4968	392.4842	0.2025	0.0025	4.6943	24
25	.0105	4.9476	23.4276	95.3962	471.9811	0.2021	0.0021	4.7352	25
26	.0087	4.9563	23.6460	114.4755	567.3773	0.2018	0.0018	4.7709	26
27	.0073	4.9636	23.8353	137.3706	681.8528	0.2015	0.0015	4.8020	27
28	.0061	4.9697	23.9991	164.8447	819.2233	0.2012	0.0012	4.8291	28
29	.0051	4.9747	24.1406	197.8136	984.0680	0.2010	0.0010	4.8527	29
30	.0042	4.9789	24.2628	237.3763	1181.8816	0.2008	0.0008	4.8731	30
31	.0035	4.9824	24.3681	284.8516	1419.2579	0.2007	0.0007	4.8908	31
32	.0029	4.9854	24.4588	341.8219	1704.1095	0.2006	0.0006	4.9061	32
33	.0024	4.9878	24.5368	410.1863	2045.9314	0.2005	0.0005	4.9194	33
34	.0020	4.9898	24.6038	492.2235	2456.1176	0.2004	0.0004	4.9308	34
35	.0017	4.9915	24.6614	590.6682	2948.3411	0.2003	0.0003	4.9406	35
36	.0014	4.9929	24.7108	708.8019	3539.0094	0.2003	0.0003	4.9491	36
37	.0012	4.9941	24.7531	850.5622	4247.8112	0.2002	0.0002	4.9564	37
38	.0010	4.9951	24.7894	1020.6747	5098.3735	0.2002	0.0002	4.9627	38
39	.0008	4.9959	24.8204	1224.8096	6119.0482	0.2002	0.0002	4.9681	39
40	.0007	4.9966	24.8469	1469.7716	7343.8578	0.2001	0.0001	4.9728	40
41	.0006	4.9972	24.8696	1763.7259	8813.6294	0.2001	0.0001	4.9767	41
42	.0005	4.9976	24.8890	2116.4711	10577.3553	0.2001	0.0001	4.9801	42
43	.0004	4.9980	24.9055	2539.7653	12693.8263	0.2001	0.0001	4.9831	43
44	.0003	4.9984	24.9196	3047.7183	15233.5916	0.2001	0.0001	4.9856	44
45	.0003	4.9986	24.9316	3657.2620	18281.3099	0.2001	0.0001	4.9877	45
46	.0002	4.9989	24.9419	4388.7144	21938.5719	0.2000	0.0000	4.9895	46
47	.0002	4.9991	24.9506	5266.4573	26327.2863	0.2000	0.0000	4.9911	47
48	.0002	4.9992	24.9581	6319.7487	31593.7436	0.2000	0.0000	4.9924	48
49	.0001	4.9993	24.9644	7583.6985	37913.4923	0.2000	0.0000	4.9935	49
50	.0001	4.9995	24.9698	9100.4382	45497.1908	0.2000	0.0000	4.9945	50
51	.0001	4.9995	24.9744	10920.5258	54597.6289	0.2000	0.0000	4.9953	51
52	.0001	4.9996	24.9783	13104.6309	65518.1547	0.2000	0.0000	4.9960	52
53	.0001	4.9997	24.9816	15725.5571	78622.7856	0.2000	0.0000	4.9966	53
54	.0001	4.9997	24.9844	18870.6685	94348.3427	0.2000	0.0000	4.9971	54
55	.0000	4.9998	24.9868	22644.8023	113219.0113	0.2000	0.0000	4.9976	55
60	.0000	4.9999	24.9942	56347.5144	281732.5718	0.2000	0.0000	4.9989	60
65	.0000	5.0000	24.9975	140210.6469	701048.2346	0.2000	0.0000	4.9995	65
70	.0000	5.0000	24.9989	348888.9569	1744439.7847	0.2000	0.0000	4.9998	70
75	.0000	5.0000	24.9995	868147.3693	4340731.8466	0.2000	0.0000	4.9999	75

ENGINEERING ECONOMIC ANALYSIS

I = 25.00 %

N	(P/F)	(P/A)	(P/G)	(F/P)	(F/A)	(A/P)	(A/F)	(A/G)	N
1	.8000	0.8000	0.0	1.2500	1.0000	1.2500	1.0000	0.0	1
2	.6400	1.4400	0.6400	1.5625	2.2500	0.6944	0.4444	0.4444	2
3	.5120	1.9520	1.6640	1.9531	3.8125	0.5123	0.2623	0.8525	3
4	.4096	2.3616	2.8928	2.4414	5.7656	0.4234	0.1734	1.2249	4
5	.3277	2.6893	4.2035	3.0518	8.2070	0.3718	0.1218	1.5631	5
6	.2621	2.9514	5.5142	3.8147	11.2588	0.3388	0.0888	1.8683	6
7	.2097	3.1611	6.7725	4.7684	15.0735	0.3163	0.0663	2.1424	7
8	.1678	3.3289	7.9469	5.9605	19.8419	0.3004	0.0504	2.3872	8
9	.1342	3.4631	9.0207	7.4506	25.8023	0.2888	0.0388	2.6048	9
10	.1074	3.5705	9.9870	9.3132	33.2529	0.2801	0.0301	2.7971	10
11	.0859	3.6564	10.8460	11.6415	42.5661	0.2735	0.0235	2.9663	11
12	.0687	3.7251	11.6020	14.5519	54.2077	0.2684	0.0184	3.1145	12
13	.0550	3.7801	12.2617	18.1899	68.7596	0.2645	0.0145	3.2437	13
14	.0440	3.8241	12.8334	22.7374	86.9495	0.2615	0.0115	3.3559	14
15	.0352	3.8593	13.3260	28.4217	109.6868	0.2591	0.0091	3.4530	15
16	.0281	3.8874	13.7482	35.5271	138.1085	0.2572	0.0072	3.5366	16
17	.0225	3.9099	14.1085	44.4089	173.6357	0.2558	0.0058	3.6084	17
18	.0180	3.9279	14.4147	55.5112	218.0446	0.2546	0.0046	3.6698	18
19	.0144	3.9424	14.6741	69.3889	273.5558	0.2537	0.0037	3.7222	19
20	.0115	3.9539	14.8932	86.7362	342.9447	0.2529	0.0029	3.7667	20
21	.0092	3.9631	15.0777	108.4202	429.6809	0.2523	0.0023	3.8045	21
22	.0074	3.9705	15.2326	135.5253	538.1011	0.2519	0.0019	3.8365	22
23	.0059	3.9764	15.3625	169.4066	673.6264	0.2515	0.0015	3.8634	23
24	.0047	3.9811	15.4711	211.7582	843.0329	0.2512	0.0012	3.8861	24
25	.0038	3.9849	15.5618	264.6978	1054.7912	0.2509	0.0009	3.9052	25
26	.0030	3.9879	15.6373	330.8722	1319.4890	0.2508	0.0008	3.9212	26
27	.0024	3.9903	15.7002	413.5903	1650.3612	0.2506	0.0006	3.9346	27
28	.0019	3.9923	15.7524	516.9879	2063.9515	0.2505	0.0005	3.9457	28
29	.0015	3.9938	15.7957	646.2349	2580.9394	0.2504	0.0004	3.9551	29
30	.0012	3.9950	15.8316	807.7936	3227.1743	0.2503	0.0003	3.9628	30
31	.0010	3.9960	15.8614	1009.7420	4034.9678	0.2502	0.0002	3.9693	31
32	.0008	3.9968	15.8859	1262.1774	5044.7098	0.2502	0.0002	3.9746	32
33	.0006	3.9975	15.9062	1577.7218	6306.8872	0.2502	0.0002	3.9791	33
34	.0005	3.9980	15.9229	1972.1523	7884.6091	0.2501	0.0001	3.9828	34
35	.0004	3.9984	15.9367	2465.1903	9856.7613	0.2501	0.0001	3.9858	35
36	.0003	3.9987	15.9481	3081.4879	12321.9516	0.2501	0.0001	3.9883	36
37	.0003	3.9990	15.9574	3851.8599	15403.4396	0.2501	0.0001	3.9904	37
38	.0002	3.9992	15.9651	4814.8249	19255.2994	0.2501	0.0001	3.9921	38
39	.0002	3.9993	15.9714	6018.5311	24070.1243	0.2500	0.0000	3.9935	39
40	.0001	3.9995	15.9766	7523.1638	30088.6554	0.2500	0.0000	3.9947	40
41	.0001	3.9996	15.9809	9403.9548	37611.8192	0.2500	0.0000	3.9956	41
42	.0001	3.9997	15.9843	11754.9435	47015.7740	0.2500	0.0000	3.9964	42
43	.0001	3.9997	15.9872	14693.6794	58770.7175	0.2500	0.0000	3.9971	43
44	.0001	3.9998	15.9895	18367.0992	73464.3969	0.2500	0.0000	3.9976	44
45	.0000	3.9998	15.9915	22958.8740	91831.4962	0.2500	0.0000	3.9980	45
46	.0000	3.9999	15.9930	28698.5925	114790.3702	0.2500	0.0000	3.9984	46
47	.0000	3.9999	15.9943	35873.2407	143488.9627	0.2500	0.0000	3.9987	47
48	.0000	3.9999	15.9954	44841.5509	179362.2034	0.2500	0.0000	3.9989	48
49	.0000	3.9999	15.9962	56051.9386	224203.7543	0.2500	0.0000	3.9991	49
50	.0000	3.9999	15.9969	70064.9232	280255.6929	0.2500	0.0000	3.9993	50
51	.0000	4.0000	15.9975	87581.1540	350320.6161	0.2500	0.0000	3.9994	51
52	.0000	4.0000	15.9980	109476.4425	437901.7701	0.2500	0.0000	3.9995	52
53	.0000	4.0000	15.9983	136845.5532	547378.2126	0.2500	0.0000	3.9996	53
54	.0000	4.0000	15.9986	171056.9414	684223.7658	0.2500	0.0000	3.9997	54
55	.0000	4.0000	15.9989	213821.1768	855280.7072	0.2500	0.0000	3.9997	55
60	.0000	4.0000	15.9996	652530.4468	2610117.7872	0.2500	0.0000	3.9999	60

EXPANDED INTEREST TABLES

I = 30.00 %

n	(P/F)	(P/A)	(P/G)	(F/P)	(F/A)	(A/P)	(A/F)	(A/G)	n
1	0.7692	0.7692	0.0000	1.3000	1.0000	1.3000	1.0000	0.000	1
2	0.5917	1.3609	0.5917	1.6900	2.3000	0.7348	0.4348	0.434	2
3	0.4552	1.8161	1.5020	2.1970	3.9900	0.5506	0.2506	0.827	3
4	0.3501	2.1662	2.5524	2.8561	6.1870	0.4616	0.1616	1.178	4
5	0.2693	2.4356	3.6297	3.7129	9.0431	0.4106	0.1106	1.490	5
6	0.2072	2.6427	4.6656	4.8268	12.7560	0.3784	0.0784	1.765	6
7	0.1594	2.8021	5.6218	6.2749	17.5828	0.3569	0.0569	2.006	7
8	0.1226	2.9247	6.4800	8.1573	23.8577	0.3419	0.0419	2.215	8
9	0.0943	3.0190	7.2343	10.6045	32.0150	0.3312	0.0312	2.396	9
10	0.0725	3.0915	7.8872	13.7858	42.6195	0.3235	0.0235	2.551	10
11	0.0558	3.1473	8.4452	17.9216	56.4053	0.3177	0.0177	2.683	11
12	0.0429	3.1903	8.9173	23.2981	74.3270	0.3135	0.0135	2.795	12
13	0.0330	3.2233	9.3135	30.2875	97.6250	0.3102	0.0102	2.889	13
14	0.0254	3.2487	9.6437	39.3738	127.9125	0.3078	0.0078	2.968	14
15	0.0195	3.2682	9.9172	51.1859	167.2863	0.3060	0.0060	3.034	15
16	0.0150	3.2832	10.1426	66.5417	218.4722	0.3046	0.0046	3.089	16
17	0.0116	3.2948	10.3276	86.5042	285.0139	0.3035	0.0035	3.134	17
18	0.0089	3.3037	10.4788	112.4554	371.5180	0.3027	0.0027	3.171	18
19	0.0068	3.3105	10.6019	146.1920	483.9734	0.3021	0.0021	3.202	19
20	0.0053	3.3158	10.7019	190.0496	630.1655	0.3016	0.0016	3.227	20
21	0.0040	3.3198	10.7828	247.0645	820.2151	0.3012	0.0012	3.248	21
22	0.0031	3.3230	10.8482	321.1839	1067.2796	0.3009	0.0009	3.264	22
23	0.0024	3.3254	10.9009	417.5391	1388.4635	0.3007	0.0007	3.278	23
24	0.0018	3.3272	10.9433	542.8008	1806.0026	0.3006	0.0006	3.289	24
25	0.0014	3.3286	10.9773	705.6410	2348.8033	0.3004	0.0004	3.297	25
26	0.0011	3.3297	11.0045	917.3333	3054.4443	0.3003	0.0003	3.305	26
27	0.0008	3.3305	11.0263	1192.5333	3971.7776	0.3003	0.0003	3.310	27
28	0.0006	3.3312	11.0437	1550.2933	5164.3109	0.3002	0.0002	3.315	28
29	0.0005	3.3317	11.0576	2015.3813	6714.6042	0.3001	0.0001	3.318	29
30	0.0004	3.3321	11.0687	2619.9956	8729.9855	0.3001	0.0001	3.321	30
31	0.0003	3.3324	11.0775	3405.9943	11349.9811	0.3001	0.0001	3.324	31
32	0.0002	3.3326	11.0845	4427.7926	14755.9755	0.3001	0.0001	3.326	32
33	0.0002	3.3328	11.0901	5756.1304	19183.7681	0.3001	0.0001	3.327	33
34	0.0001	3.3329	11.0945	7482.9696	24939.8985	0.3000	0.0000	3.328	34
35	0.0001	3.3330	11.0980	9727.8604	32422.8681	0.3000	0.0000	3.329	35
36	0.0001	3.3331	11.1007	12646.2186	42150.7285	0.3000	0.0000	3.330	36
37	0.0001	3.3331	11.1029	16440.0841	54796.9471	0.3000	0.0000	3.331	37
38	0.0000	3.3332	11.1047	21372.1094	71237.0312	0.3000	0.0000	3.331	38
39	0.0000	3.3332	11.1060	27783.7422	92609.1405	0.3000	0.0000	3.331	39
40	0.0000	3.3332	11.1071	36118.8648	120392.8827	0.3000	0.0000	3.332	40
41	0.0000	3.3333	11.1080	46954.5243	156511.7475	0.3000	0.0000	3.332	41
42	0.0000	3.3333	11.1086	61040.8815	203466.2718	0.3000	0.0000	3.332	42
43	0.0000	3.3333	11.1092	79353.1460	264507.1533	0.3000	0.0000	3.332	43
44	0.0000	3.3333	11.1096	103159.0898	343860.2993	0.3000	0.0000	3.332	44
45	0.0000	3.3333	11.1099	134106.8167	447019.3890	0.3000	0.0000	3.333	45
46	0.0000	3.3333	11.1102	174338.8617	581126.2058	0.3000	0.0000	3.333	46
47	0.0000	3.3333	11.1104	226640.5202	755465.0675	0.3000	0.0000	3.333	47
48	0.0000	3.3333	11.1105	294632.6763	982105.5877	0.3000	0.0000	3.333	48
49	0.0000	3.3333	11.1107	383022.4792	1276738.2640	0.3000	0.0000	3.333	49
50	0.0000	3.3333	11.1108	497929.2230	1659760.7433	0.3000	0.0000	3.333	50

ENGINEERING ECONOMIC ANALYSIS

EXPANDED INTEREST TABLES

I = 40.00 %

n	(P/F)	(P/A)	(P/G)	(F/P)	(F/A)	(A/P)	(A/F)	(A/G)	n
1	0.7143	0.7143	0.0000	1.4000	1.0000	1.4000	1.0000	0.000	1
2	0.5102	1.2245	0.5102	1.9600	2.4000	0.8167	0.4167	0.416	2
3	0.3644	1.5889	1.2391	2.7440	4.3600	0.6294	0.2294	0.779	3
4	0.2603	1.8492	2.0200	3.8416	7.1040	0.5408	0.1408	1.092	4
5	0.1859	2.0352	2.7637	5.3782	10.9456	0.4914	0.0914	1.358	5
6	0.1328	2.1680	3.4278	7.5295	16.3238	0.4613	0.0613	1.581	6
7	0.0949	2.2628	3.9970	10.5414	23.8534	0.4419	0.0419	1.766	7
8	0.0678	2.3306	4.4713	14.7579	34.3947	0.4291	0.0291	1.918	8
9	0.0484	2.3790	4.8585	20.6610	49.1526	0.4203	0.0203	2.042	9
10	0.0346	2.4136	5.1696	28.9255	69.8137	0.4143	0.0143	2.141	10
11	0.0247	2.4383	5.4166	40.4957	98.7391	0.4101	0.0101	2.221	11
12	0.0176	2.4559	5.6106	56.6939	139.2348	0.4072	0.0072	2.284	12
13	0.0126	2.4685	5.7618	79.3715	195.9287	0.4051	0.0051	2.334	13
14	0.0090	2.4775	5.8788	111.1201	275.3002	0.4036	0.0036	2.372	14
15	0.0064	2.4839	5.9688	155.5681	386.4202	0.4026	0.0026	2.403	15
16	0.0046	2.4885	6.0376	217.7953	541.9883	0.4018	0.0018	2.426	16
17	0.0033	2.4918	6.0901	304.9135	759.7837	0.4013	0.0013	2.444	17
18	0.0023	2.4941	6.1299	426.8789	1064.6971	0.4009	0.0009	2.457	18
19	0.0017	2.4958	6.1601	597.6304	1491.5760	0.4007	0.0007	2.468	19
20	0.0012	2.4970	6.1828	836.6826	2089.2064	0.4005	0.0005	2.476	20
21	0.0009	2.4979	6.1998	1171.3556	2925.8889	0.4003	0.0003	2.482	21
22	0.0006	2.4985	6.2127	1639.8978	4097.2445	0.4002	0.0002	2.486	22
23	0.0004	2.4989	6.2222	2295.8569	5737.1423	0.4002	0.0002	2.490	23
24	0.0003	2.4992	6.2294	3214.1997	8032.9993	0.4001	0.0001	2.492	24
25	0.0002	2.4994	6.2347	4499.8796	11247.1990	0.4001	0.0001	2.494	25
26	0.0002	2.4996	6.2387	6299.8314	15747.0785	0.4001	0.0001	2.495	26
27	0.0001	2.4997	6.2416	8819.7640	22046.9099	0.4000	0.0000	2.496	27
28	0.0001	2.4998	6.2438	12347.6696	30866.6739	0.4000	0.0000	2.497	28
29	0.0001	2.4999	6.2454	17286.7374	43214.3435	0.4000	0.0000	2.498	29
30	0.0000	2.4999	6.2466	24201.4324	60501.0809	0.4000	0.0000	2.498	30
31	0.0000	2.4999	6.2475	33882.0053	84702.5132	0.4000	0.0000	2.499	31
32	0.0000	2.4999	6.2482	47434.8074	118584.5185	0.4000	0.0000	2.499	32
33	0.0000	2.5000	6.2487	66408.7304	166019.3260	0.4000	0.0000	2.499	33
34	0.0000	2.5000	6.2490	92972.2225	232428.0563	0.4000	0.0000	2.499	34
35	0.0000	2.5000	6.2493	130161.1116	325400.2789	0.4000	0.0000	2.499	35
36	0.0000	2.5000	6.2495	182225.5562	455561.3904	0.4000	0.0000	2.499	36
37	0.0000	2.5000	6.2496	255115.7786	637786.9466	0.4000	0.0000	2.499	37
38	0.0000	2.5000	6.2497	357162.0901	892902.7252	0.4000	0.0000	2.499	38
39	0.0000	2.5000	6.2498	500026.9261	1250064.8153	0.4000	0.0000	2.499	39
40	0.0000	2.5000	6.2498	700037.6966	1750091.7415	0.4000	0.0000	2.499	40
41	0.0000	2.5000	6.2499	980052.7752	2450129.4381	0.4000	0.0000	2.500	41
42	0.0000	2.5000	6.2499	1372073.8853	3430182.2133	0.4000	0.0000	2.500	42
43	0.0000	2.5000	6.2499	1920903.4394	4802256.0986	0.4000	0.0000	2.500	43
44	0.0000	2.5000	6.2500	2689264.8152	6723159.5381	0.4000	0.0000	2.500	44
45	0.0000	2.5000	6.2500	3764970.7413	9412424.3533	0.4000	0.0000	2.500	45

3

FLUID STATICS AND DYNAMICS

Nomenclature

a	acceleration	ft/sec^2
bhp	brake horsepower	hp
A	area	ft^2
c	speed of sound in fluid	ft/sec
C	compressibility, Hazen-Williams constant, or coefficient	ft^2/lbf, –, –
d	depth, diameter	ft
D	diameter, drag	ft, lbf
ehp	electrical horsepower	hp
E	bulk modulus, energy	lbf/ft^2, ft-lbf
f	Darcy friction factor	–
fhp	friction horsepower	hp
F	force	lbf
F_{va}	velocity of approach factor	–
g	local gravitational acceleration	ft/sec^2
g_c	gravitational constant (32.2)	lbm-ft/lbf-sec^2
G	mass flow rate per unit area	lbm/sec-ft^2
h	fluid height, head, depth	ft
H	total head	ft
I	moment of inertia	ft^4
k	ratio of specific heats	–
K	minor loss coefficient	–
L	length of pipe, lift	ft, lbf
m	mass	lbm
ṁ	mass flow rate	lbm/sec
n	rotational speed	rpm
n_s	specific speed	–
N_{Fr}	Froude number	
N_{Re}	Reynolds number	–
N_W	Weber number	–
p	pressure	lbf/ft^2
P	power	ft-lbf/sec
Q	flow rate	gpm
r	radius	ft
r_h	hydraulic radius	ft
R	specific gas **constant**	ft-lbf/lbm-°R

s	length	ft
S.G.	specific gravity	–
t	time	sec
T	absolute temperature	°R
v	velocity	ft/sec
V	volume	ft^3
V̇	volumetric flow rate	ft^3/sec
w	weight	lbf
whp	water horsepower	hp
x	x-coordinate	ft
y	distance, y-coordinate	ft
z	height above datum	ft

Symbols

β	contact angle, beta ratio	°
γ	specific weight	lbf/ft^3
ϵ	specific roughness	ft
η	efficiency	
θ	angle	°
μ	absolute viscosity	lbf-sec/ft^2
ν	kinematic viscosity	ft^2/sec
ρ	density	lbm/ft^3
τ	shear stress	lbf/ft^2
T	surface tension	lbf/ft
υ	specific volume	ft^3/lbm
ϕ	angle, deflection angle	°
ω	rotational speed	rad/sec

Subscripts

a	atmospheric
A	added
b	blade
c	centroid, contraction
d	discharge
D	drag
e	equivalent, entrance
E	English, extracted
f	friction, flow
i	inside, inlet

j	jet
k	kinetic
L	lift
m	manometer fluid, metacentric, model, motor
M	metric
n	nozzle
o	outside, outlet
p	static pressure, pump
r	ratio
R	resultant
s	stagnation
STP	standard temperature and pressure
t	total, tank, true
v	velocity
vp	vapor pressure

PART 1: Fluid Properties

Fluids are generally divided into two categories: ideal and real. *Ideal fluids* are those which have zero viscosity and shearing forces, are incompressible, and have uniform velocity distributions when flowing.

Real fluids are divided into Newtonian and non-Newtonian fluids. *Newtonian fluids* are typified by gases, thin liquids, and most fluids having simple chemical formulas. *Non-Newtonian fluids* are typified by gels, emulsions, and suspensions. Both Newtonian and non-Newtonian fluids exhibit finite viscosities and non-uniform velocity distributions. However, Newtonian fluids exhibit viscosities which are independent of the rate of change of shear stress, while non-Newtonian fluids exhibit viscosities dependent on the rate of change of shear stress.

Most fluid problems assume Newtonian fluid characteristics.

1 FLUID DENSITY

Most fluid flow calculations are based on an inconsistent system of units which measures density in pounds-mass per cubic foot. That convention is followed in this chapter when the symbol ρ is employed.

$$\rho = \text{fluid density in lbm/ft}^3 \qquad 3.1$$

Hydrostatic pressure and energy conservation equations given in this chapter require a standard local gravity of 32.2 ft/sec^2. However, a short discussion of situations with non-standard gravity is given at the ends of parts 2 and 4 in this chapter.

The density of a fluid in liquid form is usually given, known in advance, or easily obtained from a table (see end of this chapter). The density of a gas can be found from the following formula, which has been derived from the ideal gas law:

$$\rho = \frac{p}{RT} \qquad 3.2$$

2 SPECIFIC VOLUME

Specific volume is the volume occupied by a pound of fluid. It is the reciprocal of the density.

$$v = \frac{1}{\rho} \qquad 3.3$$

3 SPECIFIC GRAVITY

Specific gravity is the ratio of a fluid's density to some specified reference density. For liquids, the reference density is the density of pure water. There is some confusion about this reference since the density of water varies with temperature, and various reference temperatures have been used (e.g., 39°F, 60°F, 70°F, etc.).

Strictly speaking, specific gravity of a liquid cannot be given without specifying the reference temperature at which the water's density was evaluated. However, the reference temperature is often omitted since water's density is fairly constant over the normal ambient temperature range. Using three significant digits, this reference density is 62.4 lbm/ft^3.

$$S.G._{\text{liquid}} = \frac{\rho}{62.4} \qquad 3.4$$

Specific gravities of petroleum products and aqueous acid solutions can be found from hydrometer readings. There are two basic hydrometer scales. The Baumé scale has been used widely in the past. Now, however, the API (American Petroleum Institute) scale is recommended for use with all liquids.

For liquids lighter than water, specific gravity may be found from the Baumé hydrometer reading:

$$S.G. = \frac{140.0}{130.0 + °\text{Baumé}} \qquad 3.5$$

For liquids heavier than water, specific gravity may be found from the Baumé hydrometer reading:

$$S.G. = \frac{145.0}{145.0 - {}^{\circ}\text{Baumé}} \qquad 3.6$$

The modern API scale may be used with all liquids:

$$S.G. = \frac{141.5}{131.5 + {}^{\circ}\text{API}} \qquad 3.7$$

Specific gravities also can be given for gases. The reference density is the density of air at specified conditions of pressure and temperature. The density of air evaluated at STP is approximately .075 lbm/ft^3. Therefore,

$$S.G._{\text{.gas}} = \frac{\rho_{\text{STP}}}{.075} \qquad 3.8$$

If the gas and air densities both are evaluated at the same temperature and pressure, the specific gravity is the inverse ratio of specific gas constants.

$$S.G._{\text{.gas}} = \frac{R_{\text{air}}}{R_{\text{gas}}} = \frac{53.3}{R_{\text{gas}}} \qquad 3.9$$

Example 3.1

Determine the specific gravity of carbon dioxide (150°F, 20 psia) using STP air as a reference.

The specific gas constant for carbon dioxide is approximately 35.1 ft-lbf/lbm-°R. Using equation 3.2, the density is

$$\rho = \frac{(20)(144)}{(35.1)(150+460)} = .135 \text{ lbm/ft}^3$$

From equation 3.8,

$$S.G. = \frac{.135}{.075} = 1.8$$

4 VISCOSITY

Viscosity of a fluid is a measure of its resistance to flow. Consider two plates separated by a viscous fluid with thickness equal to y. The bottom plate is fixed. The top plate is kept in motion at a constant velocity v by a constant force F.

Experiments with Newtonian fluids have shown that the force required to maintain the velocity is proportional to the velocity and inversely proportional to the separation of the plates. That is,

$$\frac{F}{A} \propto \frac{dv}{dy} \qquad 3.10$$

The constant of proportionality is known as the *absolute*[1] *viscosity.* Recognizing that the quantity (F/A) is the *fluid shear stress* allows the following equation to be written.

$$\tau = \mu \frac{dv}{dy} \qquad 3.11$$

Another quantity using the name *viscosity* is the combination of units given by equation 3.12. This combination of units, known as the kinematic viscosity, appears sufficiently often in fluids problems to warrant its own symbol and name.

$$\nu = \frac{\mu g_c}{\rho} \qquad 3.12$$

There are a number of different units used to measure viscosity. Table 3.1 lists the most commonly used units in the English and SI systems.

Table 3.1
Typical Viscosity Units

	Absolute	Kinematic
English	lbf-sec/ft^2 (slug/ft-sec)	ft^2/sec
Conventional Metric	dyne-sec/cm^2 (poise)	cm^2/sec (stoke)
SI	Pascal-second (N-s/m^2)	m^2/s

Conversions between the two types of viscosities and between the English and various metric systems can be accomplished with table 3.2.

Example 3.2

Water at 60°F has a specific gravity of .999 and a kinematic viscosity of 1.12 centistokes. What is the absolute viscosity in lbf-sec/ft^2?

$$\nu_M = \frac{1.12}{100} = .0112 \text{ stokes}$$
$$\mu_M = (.0112)(.999) = .01119 \text{ poise}$$
$$\mu_E = \frac{.01119}{478.8} = 2.34 \text{ EE} - 5 \text{ lbf-sec/ft}^2$$

Viscosity also can be measured by a viscometer. A viscometer essentially is a container which allows the fluid to leak out through a small hole. The more viscous the fluid, the more time will be required to leak out a given quantity. Viscosity measured in this indirect manner has the units of seconds. Seconds Saybolt Universal (SSU) and Seconds Saybolt Furol (SSF) are two systems of indirect viscosity measurement.

[1] Another name for *absolute* viscosity is *dynamic* viscosity. The term *absolute* is preferred.

Table 3.2
Viscosity Conversions

To Obtain	Multiply	By	and Divide by
ft^2/sec	$lbf\text{-}sec/ft^2$	32.2	density
ft^2/sec	stokes	1.076 EE−3	1
$lbf\text{-}sec/ft^2$	ft^2/sec	density	32.2
$lbf\text{-}sec/ft^2$	poise	1	478.8
m^2/s	centistokes	1 EE−6	1
m^2/s	stokes	1 EE−4	1
m^2/s	ft^2/sec	9.29 EE−2	1
pascal-sec	centipoise	1 EE−3	1
pascal-sec	lbm/ft-sec	1.488	1
pascal-sec	$lbf\text{-}sec/ft^2$	47.88	1
pascal-sec	poise	.1	1
pascal-sec	slug/ft-sec	47.88	1
poise	$lbf\text{-}sec/ft^2$	478.8	1
poise	stokes	specific gravity	1
reyns	$lbf\text{-}sec/ft^2$	1	144
stokes	ft^2/sec	929	1
stokes	poise	1	specific gravity

In liquids, molecular cohesion is the dominating cause of viscosity. As the temperature of a liquid increases, these cohesive forces decrease, resulting in an absolute viscosity decrease.

In gases, the dominating cause of viscosity is random collisions between gas molecules. This molecular agitation increases with increases in temperature. Therefore, viscosity in gases increases with temperature.

The absolute viscosity of both gases and liquids is independent of changes in pressure. Of course, kinematic viscosity greatly depends on both temperature and pressure since these variables affect density.

5 VAPOR PRESSURE

Molecular activity in a liquid tends to free some surface molecules. This tendency toward vaporization is dependent on temperature. The partial pressure exerted at the surface by these free molecules is known as the *vapor pressure*. Boiling occurs when the vapor pressure is increased (by increasing the fluid temperature) to the local ambient pressure. Thus, a liquid's boiling point depends on both the temperature and the external pressure. Liquids with low vapor pressures are used in accurate barometers.

Vapor pressure is a function of temperature only. Typical values are given in table 3.3.

Table 3.3
Typical Vapor Pressures at 68°F

Fluid	Vapor Pressure
Ethyl alcohol	122.4 psf
Turpentine	1.115
Water	48.9
Ether	1231.
Mercury	.00362

6 SURFACE TENSION

The skin which seems to form on the free surface of a fluid is due to the intermolecular cohesive and adhesive forces known as *surface tension*. Surface tension is the amount of work required to form a new unit of surface area. The units, therefore, are $ft\text{-}lbf/ft^2$ or just lbf/ft.

Surface tension can be measured as the tension between two points on the surface separted by a foot. It decreases with temperature increases and depends on the gas contacting the free surface. Surface tension values usually are quoted for air contact.

Table 3.4
Typical Surface Tensions
(68°F, air contact)

Fluid	T
ethyl alcohol	.001527 lbf/ft
turpentine	.001857
water	.004985
mercury	.03562
n-octane	.00144
acetone	.00192
benzene	.00192
carbon tetrachloride	.00180

The relationship between surface tension and the pressure in a bubble surrounded by gas is given by equation 3.13. r is the radius of the bubble.

$$T = \frac{1}{4} r \left(p_{\text{inside}} - p_{\text{outside}} \right) \qquad 3.13$$

The surface tension in a full spherical droplet or in a bubble in a liquid is given by equation 3.14.

$$T = \frac{1}{2} r \left(p_{\text{inside}} - p_{\text{outside}} \right) \qquad 3.14$$

7 CAPILLARITY

Surface tension is the cause of capillarity which occurs whenever a liquid comes into contact with a vertical solid surface. In water, adhesive forces dominate. They cause water to attach itself readily to a vertical surface, to climb the wall. In a thin-bore tube, water will rise above the general level as it tries to wet the interior surface.

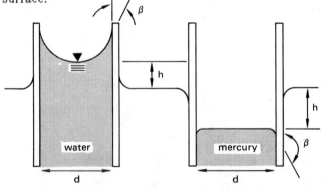

Figure 3.1 Capillarity in Thin-Wall Tubes

On the other hand, cohesive forces dominate in mercury, since mercury molecules have a great affinity for each other. The curved surface called the *meniscus* formed inside a thin-bore tube inserted into a container of mercury will be below the general level.

Whether adhesive or cohesive forces dominate can be determined by the *angle of contact*, β, as shown in figure 3.1. For a contact angle less than 90°, adhesive forces dominate. Typical values of β are given in table 3.5.

If the tube diameter is less than .1″, the surface tension inside a circular capillary tube can be approximated by equation 3.15. The meniscus is assumed spherical with radius r. Equation 3.15 also can be used to estimate the capillary rise in a capillary tube, as illustrated in figure 3.1.

$$T = \frac{h\rho d}{4\cos\beta} \qquad 3.15$$

$$r = \frac{d}{2\cos\beta} \qquad 3.16$$

Table 3.5 can be used to determine T and β for various combinations of contacting liquids.

8 COMPRESSIBILITY

Usually, fluids are considered to be incompressible. Actually, fluids are somewhat compressible. Compressibility is the percentage change in a unit volume per unit change in pressure.

$$C = \frac{\frac{\Delta V}{V}}{\Delta p} \qquad 3.17$$

9 BULK MODULUS

The bulk modulus of a liquid is the reciprocal of the compressibility.

$$E = \frac{1}{C} \qquad 3.18$$

The bulk modulus of an ideal gas is given by equation 3.19, where k is the ratio of specific heats. k is equal to 1.4 for air.

$$E = kp \qquad 3.19$$

Table 3.5
Capillary Constants

Combination	Surface Tension, T	Contact Angle, β
Mercury-vacuum-glass	3.29 EE−2 lbf/ft	140°
Mercury-air-glass	2.02 EE−2	140°
Mercury-water-glass	2.60 EE−2	140°
Water-air-glass	5.00 EE−3	0°

10 SPEED OF SOUND

The speed of sound in a pure liquid or gas is given by equations 3.20 and 3.21.

$$c_{\text{liquid}} = \sqrt{\frac{Eg_c}{\rho}} = \sqrt{\frac{g_c}{C\rho}} \qquad 3.20$$

$$c_{\text{gas}} = \sqrt{kg_c RT} = \sqrt{\frac{kpg_c}{\rho}} \qquad 3.21$$

The temperature term in equation 3.21 must be in degrees absolute.

Example 3.3

What is the velocity of sound in 150°F water?

From the tables at the end of this chapter, the density of water at 150°F is 61.2 lbm/ft^3. Similarly, the bulk modulus is 328 EE3 psi. From equation 3.20,

$$c = \sqrt{\frac{(328\ \text{EE3})(144)(32.2)}{61.2}} = 4985\ \text{ft/sec}$$

Example 3.4

What is the velocity of sound in 150°F air at atmospheric pressure?

The specific gas constant for air is 53.3 ft-lbf/lbm-°R. Using equation 3.21,

$$c = \sqrt{(1.4)(32.2)(53.3)(150 + 460)}$$
$$= 1210.7\ \text{ft/sec}$$

$$\rho_{\text{WATER}} = .0361\ \text{lb/in}^3$$

$$\rho_{\text{mercury}} = .491\ \text{lb/in}^3$$

PART 2: Fluid Statics

1 MEASURING PRESSURES

The value of pressure, regardless of the device used to measure it, is dependent on the reference point chosen. Two such reference points exist: zero absolute pressure and standard atmospheric pressure.

If standard atmospheric pressure (approximately 14.7 psia) is chosen as the reference, pressures are known as *gage* pressures. Positive gage pressures always are pressures above atmospheric pressure. Vacuum (negative gage pressure) is the pressure below atmospheric. Maximum vacuum, according to this convention, is −14.7 psig. The term *gage* is somewhat misleading, as a mechanical gauge may not be used to measure gage pressures.

If zero absolute pressure is chosen as the reference, the pressures are known as *absolute* pressures. The barometer is a common device for measuring the absolute pressure of the atmosphere. It is constructed by filling a long, hollow tube, open at one end, with mercury, and inverting it such that the open end is below the level of a mercury-filled container. If the vapor pressure is neglected, the mercury will be supported only by the atmospheric pressure transmitted through the container fluid at the lower, open end. The equation balancing the weight of the fluid against the atmospheric force is:

$$p_a = .491(h)(144) \qquad 3.22$$

h is the height of the mercury column in inches, and .491 is the density of mercury in pounds per cubic inch.

Any fluid can be used to measure atmospheric pressure, although vapor pressure may be significant. For any fluid used in a barometer,

$$p_a = [(.0361)(S.G.)(h) + p_v](144) \qquad 3.23$$

.0361 is the density of water in pounds per cubic inch. p_v should be given in psi in equation 3.23.

Example 3.5

A vacuum pump is used to drain a flooded mine shaft of 68°F water. The pump is incapable of lifting the water beyond 400 inches. What is the atmospheric pressure?

From table 3.3, the vapor pressure of 68°F water is

$$\frac{48.9}{144} = .34 \text{ psi}$$

Then, the atmospheric pressure is

$$p_a = [(.0361)(1)(400) + .34](144) = 2128.3 \text{ psf}$$

This is 14.78 psia.

2 MANOMETERS

Manometers are used frequently to measure pressure differentials. Figure 3.2 shows a simple U-tube manometer whose ends are connected to two pressure vessels. Often, one end will be open to the atmosphere, which then determines that end's pressure.

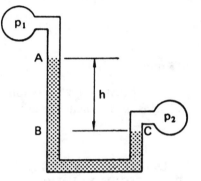

Figure 3.2 A Simple Manometer

Since the pressure at point B is the same as at point C, the pressure differential produces the fluid column of height h.

$$\Delta p = p_2 - p_1 = \rho_m h \qquad 3.24$$

Equation 3.24 assumes that the manometer is small and that only low density gases fill the tubes above the measuring fluid. If a high density fluid (such as water) is present above the measuring fluid, or if the gas columns h_1 or h_2 are very long, corrections must be made:

$$\Delta p = \rho_m h + \rho_1 h_1 - \rho_2 h_2 \qquad 3.25$$

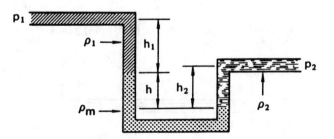

Figure 3.3 A Manometer Requiring Corrections

Corrections for capillarity are seldom needed since manometer tubes generally are large in diameter.

Example 3.6

What is the pressure at the bottom of the water tank?

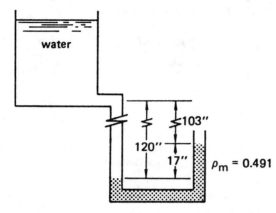

water

$\leq 103''$

$120''$

$17''$ $\rho_m = 0.491$

Using equation 3.25, the pressure differential is

$$\Delta p = p_{\text{tank bottom}} - p_a = (.491)(17) - (.0361)(120)$$
$$= 8.347 - 4.332$$
$$= 4.015 \text{ psig}$$

The third term in equation 3.25 was omitted because the density of air is much smaller than that of water or mercury.

3 HYDROSTATIC PRESSURE DUE TO INCOMPRESSIBLE FLUIDS

Hydrostatic pressure is the pressure which a fluid exerts on an object or container walls. It always acts through the center of pressure and is normal to the exposed surface, regardless of the object's orientation or shape. It varies linearly with depth and is a function of depth and density only.

A. FORCE ON A HORIZONTAL PLANE SURFACE

In the case of a horizontal surface, such as the bottom of a container, the pressure is uniform, and the center of pressure corresponds to the centroid of the plane surface. The gage pressure is

$$p = \rho h \qquad 3.26$$

The total vertical force on the horizontal plane is

$$R = pA \qquad 3.27$$

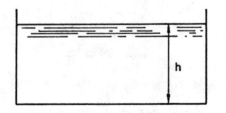

h

Figure 3.4 Horizontal Plane Surface

B. VERTICAL AND INCLINED RECTANGULAR PLANE SURFACES

If a rectangular plate is vertical or inclined within a fluid body, the linear variation in pressure with depth is maintained. The pressures at the top and bottom of the plate are

$$p_1 = \rho h_1 \sin \theta \qquad 3.28$$

$$p_2 = \rho h_2 \sin \theta \qquad 3.29$$

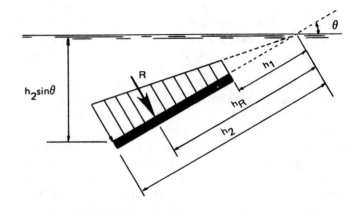

Figure 3.5 Inclined Rectangular Plane

The average pressure occurs at the average depth $(\frac{1}{2})(h_1 + h_2) \sin \theta$. The average pressure over the entire vertical or inclined surface is

$$\bar{p} = \frac{1}{2}\rho(h_1 + h_2)\sin \theta \qquad 3.30$$

The total resultant force on the inclined plane is

$$R = \bar{p}A \qquad 3.31$$

The center of pressure is not located at the average depth but is located at the centroid of the triangular or trapezoidal pressure distribution. That depth is

$$h_R = \frac{2}{3}\left[h_1 + h_2 - \frac{h_1 h_2}{(h_1 + h_2)}\right] \qquad 3.32$$

If the object is inclined, h_R must be measured parallel to the object's surface (e.g., an inclined length).

Example 3.7

The tank shown is filled with water. What is the force on a one-foot length of the inclined portion of the wall? Where is the resultant located on the inclined section?

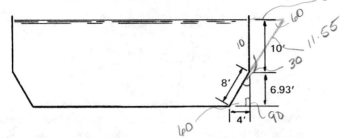

The sin θ terms in equations 3.28, 3.29, and 3.30 convert the inclined distances to vertical distances. Therefore, the sin θ terms may be omitted if the vertical distances are known. The average pressure in the inclined section is

$$\bar{p} = \frac{1}{2}(62.4)(10 + 16.93) = 840.2 \text{ psf}$$

The total force is

$$R = (840.2)(8)(1) = 6721.6 \text{ lbf}$$

To determine h_R, θ must be known in order to calculate h_1 and h_2.

$$\theta = \arctan\left(\frac{6.93}{4}\right) = 60°$$
$$h_1 = \frac{10}{\sin 60°} = 11.55 \text{ ft}$$
$$h_2 = \frac{16.93}{\sin 60°} = 19.55 \text{ ft}$$

h_R can be calculated from equation 3.32 by substituting the inclined distances.

$$h_R = \frac{2}{3}\left[11.55 + 19.55 - \frac{(11.55)(19.55)}{11.55+19.55}\right]$$
$$= 15.89 \text{ (inclined)}$$

C. GENERAL PLANE SURFACE

For any non-rectangular plane surface, the average pressure depends on the location of the surface's centroid, h_c.

$$\bar{p} = \rho h_c \sin \theta \qquad 3.33$$
$$R = \bar{p}A \qquad 3.34$$

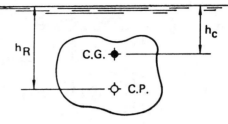

Figure 3.6 General Plane Surface

The resultant is normal to the surface, acting at depth h_R.

$$h_R = h_c + \frac{I_c}{Ah_c} \qquad 3.35$$

I_c is the moment of inertia about an axis parallel to the surface through the area's centroid. As with the previous case, h_c and h_r must be measured parallel to the area's surface. That is, if the plane is inclined, h_c and h_R also must be the inclined distances.

Example 3.8

What is the force on a one-foot diameter circular sighthole whose top edge is located 4' below the water surface? Where does the resultant act?

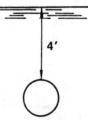

$$h_c = 4.5 \text{ ft}$$
$$A = \frac{1}{4}\pi(1)^2 = .7854 \text{ ft}^2$$
$$I_c = \frac{1}{4}\pi r^4 = .049 \text{ ft}^4$$
$$\bar{p} = (62.4)(4.5) = 280.8 \text{ psf}$$
$$R = (280.8)(.7854) = 220.5 \text{ lbf}$$
$$h_R = 4.5 + \frac{.049}{(.7854)(4.5)} = 4.514 \text{ ft}$$

D. CURVED SURFACES

The horizontal component of the resultant force acting on a curved surface can be found by the same method used for a vertical plane surface. The vertical component of force on an area usually will equal the weight of the liquid above it. In figure 3.7, the vertical force on length AB is the weight of area ABCD, with a line of action passing through the centroid of the area ABCD.

The resultant magnitude and direction may be found from conventional component composition.

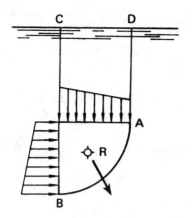

Figure 3.7 Forces On a Curved Surface

Example 3.9

What is the total force on a one-foot section of the wall described in example 3.7?

The centroid of section ABCD (with point B serving as the reference) is

$$\overline{x} = \frac{\sum A_i \overline{x}_i}{\sum A_i}$$

$$= \frac{(4)(10)(2) + \left(\frac{1}{2}\right)(4)(6.93)\left(\frac{1}{3}\right)(4)}{40 + 13.86}$$

$$= 1.83$$

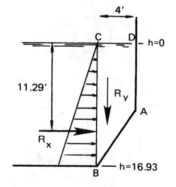

The average depth is $\frac{1}{2}(0 + 16.93) = 8.465$

Using equations 3.30 and 3.31, the average pressure and horizontal component of the resultant are

$$\overline{p} = 62.4(8.465) = 528.2 \text{ psf}$$
$$R_x = (16.93)(1)(528.2) = 8942.4 \text{ lbf}$$

From equation 3.32, the horizontal component acts $\left(\frac{2}{3}\right)(16.93) = 11.29$ ft from the top.

The volume of a one foot section of area ABCD is

$$(1)\left[(4)(10) + \frac{1}{2}(4)(6.93)\right] = 53.86 \text{ ft}^3$$

Therefore, the vertical component is

$$R_y = (62.4)(53.86) = 3360.9 \text{ lbf}$$

The resultant of R_x and R_y is

$$R = \sqrt{(8942.4)^2 + (3360.9)^2} = 9553.1 \text{ lbf}$$
$$\phi = \arctan\left(\frac{3360.9}{8942.4}\right) = 20.6°$$

In general, it is not correct to calculate the vertical component of force on a submerged surface as being the weight of the fluid above it, as was done in example 3.9. This procedure is valid only when there is no change in the cross section of the tank area.

The *hydrostatic paradox* is illustrated by figure 3.8. The pressure anywhere on the bottom of either container is the same. This pressure is dependent on only the maximum height of the fluid, not the volume.

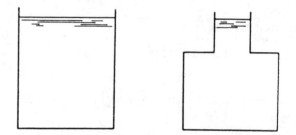

Figure 3.8 Hydrostatic Paradox

4 HYDROSTATIC PRESSURE DUE TO COMPRESSIBLE FLUIDS

Equation 3.26 is a special case of a more general equation known as the *Fundamental Equation of Fluid Statics*, presented as equation 3.36. As defined in the nomenclature, h is a variable representing height, and it is assumed that h_2 is greater than h_1. The minus sign in equation 3.36 indicates that pressure decreases when height increases.

$$\int_1^2 \frac{dp}{\rho} = -(h_2 - h_1) \qquad 3.36$$

If the fluid is a compressible layer of perfect gas, and if compression is assumed to be isothermal, equation 3.36 becomes

$$h_2 - h_1 = RT \ln\left(\frac{p_1}{p_2}\right) \qquad 3.37$$

The pressure at height h_2 in a layer of gas which has been isothermally compressed is

$$p_2 = p_1\left[e^{\frac{h_1 - h_2}{RT}}\right] \qquad 3.38$$

The following relationships assume a polytropic compression of the gas layer. These three relationships can be used for adiabatic compression by substituting k for n.

$$h_2 - h_1 = \frac{n}{n-1} RT_1 \left[1 - \left(\frac{p_2}{p_1} \right)^{\frac{n-1}{n}} \right] \qquad 3.39$$

$$p_2 = p_1 \left[1 - \left(\frac{n-1}{n} \right) \left(\frac{h_2 - h_1}{RT_1} \right) \right]^{\frac{n}{n-1}} \qquad 3.40$$

$$T_2 = T_1 \left[1 - \left(\frac{n-1}{n} \right) \left(\frac{h_2 - h_1}{RT_1} \right) \right] \qquad 3.41$$

Example 3.10

The pressure at sea level is 14.7 psia. Assume 70°F isothermal compression, and calculate the pressure at 5000 feet altitude.

$R = 53.3$ ft-lbf/lbm- °R for air. $T = (70 + 460) = 530$ °R.

From equation 3.38,

$$p_{5000} = 14.7 \left[e^{\frac{0-5000}{(53.3)(530)}} \right]$$
$$= 12.32 \text{ psia}$$

5 BUOYANCY

The buoyancy theorem, also known as *Archimedes' principle*, states that the upward force on an immersed object is equal to the weight of the displaced fluid. A buoyant force due to displaced air also is relevant in the case of partially-submerged objects. For lighter-than-air crafts, the buoyant force results entirely from displaced air.

$$F_{\text{buoyant}} = \left(\begin{array}{c} \text{displaced} \\ \text{volume} \end{array} \right) \left(\begin{array}{c} \text{density of} \\ \text{displaced fluid} \end{array} \right) \qquad 3.42$$

In the case of floating or submerged objects not moving vertically, the buoyant force and weight are equal. If the forces are not in equilibrium, the object will rise or fall until some equilibrium is reached. The object will sink until it is supported by the bottom or until the density of the supporting fluid increases sufficiently. It will rise until the weight of the displaced fluid is reduced, either by a decrease in the fluid density or by breaking the surface.

Example 3.11

An empty polyethylene telemetry balloon with payload has a mass of 500 pounds. It is charged with helium when the atmospheric conditions are 60°F and 14.8 psia.

What volume of helium is required for liftoff from a sea level platform? The specific gas constant of helium is 386.3 ft-lbf/lbm-°R.

The gas densities are

$$\rho_{\text{air}} = \frac{p}{RT} = \frac{(14.8)(144)}{(53.3)(60+460)} = .07689 \text{ lbm/ft}^3$$

$$\rho_{\text{helium}} = \frac{(14.8)(144)}{(386.3)(60+460)} = .01061 \text{ lbm/ft}^3$$

The total mass of the balloon, payload, and helium is

$$m = 500 + (.01061) \left(\begin{array}{c} \text{helium} \\ \text{volume} \end{array} \right)$$

The buoyant force is the weight of the displaced air.

$$F = (.07689) \left(\begin{array}{c} \text{helium} \\ \text{volume} \end{array} \right)$$

Equating F and m results in a helium volume of 7544 ft^3.

Example 3.12

An evacuated 6-foot diameter sphere floats half-submerged in sea water. How much concrete (in lbm) is required as an external anchor to just submerge the sphere completely?

Assume the densities are 64.0 lbm/ft^3 for sea water and 150 lbm/ft^3 for concrete.

The weight of the sphere can be calculated from the buoyant force required to support it when half-submerged. Both the displaced sea water and the displaced air contribute to the buoyant force.

$$V_{\text{sphere}} = \frac{4\pi(3)^3}{3} = 113.1 \text{ ft}^3$$

$$w_{\text{sphere}} = \left(\frac{1}{2} \right)(113.1)(64) + \left(\frac{1}{2} \right)(113.1)(.075)$$
$$= 3623.4 \text{ lbf}$$

The buoyant force equation for a fully-submerged sphere and anchor can be solved for the concrete volume.

$$w_{\text{sphere}} + w_{\text{concrete}} = \left(V_{\text{sphere}} + V_{\text{concrete}} \right)(64.0)$$
$$3623.4 + 150 V_{\text{concrete}} = (113.1 + V_{\text{concrete}})(64.0)$$
$$V_{\text{concrete}} = 42.0 \text{ ft}^3$$
$$m_{\text{concrete}} = (42)(150) = 6305 \text{ lbm}$$

6 STABILITY OF FLOATING OBJECTS

The buoyant force on a floating object acts upward through the center of gravity of the displaced *volume*, known as the *center of buoyancy*. The weight acts downward through the center of gravity of the *object*. For totally submerged objects such as balloons and submarines, the center of buoyancy must be above the center of gravity for stability. For partially submerged vessels, the metacenter must be above the center of gravity. Stability exists because a righting moment is created if the vessel heels over, since the center of buoyancy moves outboard of the center of gravity.

Refer to figure 3.9. If the floating body heels through an angle ϕ, the location of the center of gravity does not change. However, the center of buoyancy will shift to the center of gravity of the new submerged section 123. The center of buoyancy and gravity are no longer in line. The couple thus formed tends to resist further overturning.

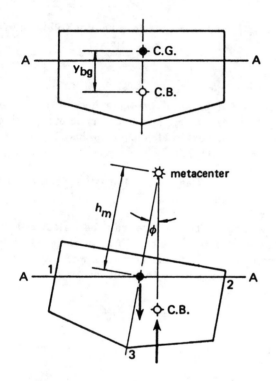

Figure 3.9 Locating the Metacenter

This righting couple exists when the extension of the buoyant force F intersects line O-O above the center of gravity at M, the *metacenter*. If M lies below the center of gravity, an overturning couple will exist. The distance between the center of gravity and the metacenter is called the *metacentric height*, and it is reasonably constant for heel angles less than 10 degrees.

The metacentric height, h_m, can be found from equation 3.43 where I is the moment of inertia of the submerged portion about the line A-A, and V is the displaced volume.

$$h_m = \frac{I}{V} \pm y_{bg} \qquad 3.43$$

7 FLUID MASSES UNDER ACCELERATION

The pressures obtained thus far have assumed that the fluid is subjected only to gravitational acceleration. As soon as the fluid is subjected to any other acceleration, additional forces which change hydrostatic pressures are imposed.

If the fluid is subjected to constant accelerations in the vertical and/or horizontal directions, the fluid behavior will be given by equations 3.44 and 3.45.

$$\phi = \arctan\left(\frac{a_x}{a_y + g}\right) \qquad 3.44$$

$$p_h = \rho h \left(1 + \frac{a_y}{g}\right) \qquad 3.45$$

a_y is negative if the acceleration is downward. Notice that a plane of equal pressure also is inclined if the fluid mass experiences a horizontal acceleration.

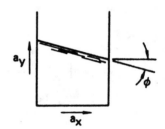

Figure 3.10 Constant Linear Acceleration

If the fluid mass is rotated about a vertical axis, a parabolic fluid surface will result. The distance h is measured from the lowermost part of the fluid during rotation. h is not measured from the original level of the stationary fluid. h is the height of the fluid at a distance r from the center of rotation.

$$\phi = \arctan\left(\frac{\omega^2 r}{g}\right) \qquad 3.46$$

$$h = \frac{(\omega r)^2}{2g} = \frac{v^2}{2g} \qquad 3.47$$

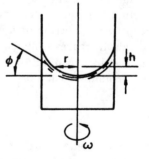

Figure 3.11 Constant Rotational Acceleration

8 MODIFICATION FOR OTHER GRAVITIES

All of the equations in Part 2 of this chapter can be used in a standard gravitational field of 32.2 ft/sec². In such a standard field, the terms *lbm* and *lbf* can be cancelled freely. However, a modification to the formulas is needed if the local gravity deviates from standard.

The modification that must be made is that of converting the mass density in lbm/ft³ to a *specific weight* in lbf/ft³. This can be done through the use of equation 3.48.

$$\gamma = \frac{\rho g_{local}}{g_c} \qquad 3.48$$

g_{local} is the actual gravitational acceleration in ft/sec². g_c is the dimensional conversion factor presented in chapter 22. g_c is approximately equal to 32.2 lbm-ft/lbf-sec².

The various hydrostatic pressure formulas can be used in non-standard gravitational fields by substituting γ for ρ.

Example 3.13

What is the maximum height that a vacuum pump can lift 60°F water if the atmospheric pressure is 14.6 psia and the local gravity is 28 ft/sec²?

At 60°F, the vapor pressure of water is .2563 psia. The effective pressure which can be used to lift water is $(14.6 - .2563) = 14.3437$ psi.

$$p = \left(\frac{g}{g_c}\right)\rho h$$

$$h = \frac{(14.3437)(144)\left(\frac{32.2}{28.0}\right)}{62.4} = 38.07 \text{ ft}$$

PART 3: Fluid Flow Parameters

1 INTRODUCTION

Pressure commonly is measured in pounds per square inch (psi) or pounds per square foot (psf). However, pressure may be changed into a new variable called *head* by dividing by the density of the fluid. This operation does more than just scale down the pressure by a factor equal to the reciprocal of the density. Since density itself possesses dimensional units, the units of head are not the same as the units of pressure.

$$(h \text{ in ft }) = \frac{(p \text{ in lbf/ft}^2)}{(\rho \text{ in lbm/ft}^3)} \qquad 3.49$$

Equation 3.49 is, of course, the same as equation 3.26. As long as the fluid density and local gravitational acceleration remain constant, there is complete interchangeability between the variables of pressure and head.

When Bernoulli's equation is introduced in this chapter, head also will be used as a measure of energy. Actually, head is used as a measure of *specific* energy. This is commonly justified by equation 3.50.

$$(h \text{ in ft}) = \frac{(E \text{ in ft-lbf})}{(m \text{ in lbm})} \qquad 3.50$$

A certain amount of care in the use of equations 3.49 and 3.50 is required since lbf is being cancelled completely by lbm. The actual operation being performed is given by equation 3.51.

$$(h \text{ in ft}) = \frac{\left(g_c \text{ in } \frac{\text{lbm-ft}}{\text{lbf-sec}^2}\right)(p \text{ in lbf/ft}^2)}{\left(g \text{ in } \frac{\text{ft}}{\text{sec}^2}\right)\left(\rho \text{ in } \frac{\text{lbm}}{\text{ft}^3}\right)} \qquad 3.51$$

As g_c always equals 32.2, it can be seen from equation 3.51 that equations 3.49 and 3.50 will give the correct numerical value for head as long as the local gravitational acceleration is 32.2 ft/sec^2.

2 FLUID ENERGY

A fluid can possess energy in three forms[2] — as pressure, kinetic, or potential energies. The energy (work) that must be put into a fluid to raise its pressure is known as the *pressure energy* or *static energy*. (This form of energy also is known as *flow work* or *flow

[2]Another important energy form is *thermal* energy. Thermal energy terms (internal energy and enthalpy) are not included in this analysis since it is assumed that the temperature of the fluid remains constant.

energy*). In keeping with the common convention to put fluid energy terms into units of feet, the *pressure head* or *static head* is defined by equation 3.52.

$$h_p = \frac{p}{\rho} \qquad 3.52$$

Energy also is required to accelerate fluid to velocity v. The specific kinetic energy with units of feet is known as the *velocity head* or *dynamic head*. Although the units in equation 3.53 actually do yield feet, you should remember that specific energy is in foot-pounds per pound mass of fluid.

$$h_v = \frac{v^2}{2g_c} \qquad 3.53$$

Potential energy also is given with units of feet. This results in a very simple expression for *potential head* or *gravitational head*. z in equation 3.54 is the height of the fluid above some arbitrary reference point. Cancelling the weight (lbf) of the fluid by its mass (lbm) is acceptable under the constraints previously given.

$$h_z = \frac{wz}{m} \qquad 3.54$$

3 BERNOULLI EQUATION

The Bernoulli equation is an energy conservation equation. It states that the total energy of a fluid flowing without losses in a pipe cannot change. The total energy possessed by a fluid is the sum of its pressure, kinetic, and potential energies.

$$\frac{p_1}{\rho} + \frac{v_1^2}{2g_c} + z_1 = \frac{p_2}{\rho} + \frac{v_2^2}{2g_c} + z_2 \qquad 3.55$$

Equation 3.55 is valid for laminar and turbulent flow. It can be used for gases as well as liquids if the gases are incompressible.[3] It is assumed that the flow between points 1 and 2 is frictionless and adiabatic.

The sum of the three head terms is known as the total head, H. The total pressure can be calculated from the total head.

$$H = \frac{p}{\rho} + \frac{v^2}{2g_c} + z \qquad 3.56$$

$$p_t = \rho H \qquad 3.57$$

[3]A gas can be considered to be incompressible as long as its pressure does not change more than 10% between points 1 and 2, and its velocity is less than Mach .3 everywhere.

The total energy of the fluid stream has two definitions. The *total specific energy* is the same as total head, H. The total energy of all fluid flowing can be calculated from equation 3.58.

$$E_t = mH \qquad 3.58$$

A graph of the total specific energy versus distance along a pipe is known as a *total energy line*. In a frictionless pipe without pumps or turbines, the total specific energy will remain constant. Total specific energy will decrease if fluid friction is present.

Example 3.14

A pipe takes water from the reservoir as shown and discharges it freely 100 feet below. The flow is frictionless. (a) What is the total specific energy at point B? (b) What is the velocity at point C?

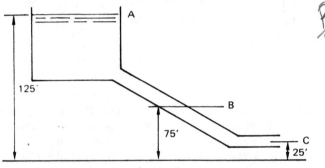

energy reference line

(a) At point A, the velocity and the gage pressure both are zero, so the total specific energy with respect to the reference line is

$$H_A = 0 + 0 + 125 = 125 \text{ ft}$$

At point B, the fluid is moving and possesses kinetic energy. The fluid is also under hydrostatic pressure. However, the flow is frictionless, and the total specific energy is constant (see equation 3.56). The velocity and static heads have increased at the expense of the potential head. Therefore,

$$H_B = H_A = 125 \text{ ft}$$

(b) At point C, the pressure head is again zero since the discharge is at atmospheric pressure. The potential head with respect to the energy reference line is 25 ft. From equation 3.56,

$$125 = 0 + \frac{v^2}{2g_c} + 25$$
$$v^2 = (2)(32.2)(100)$$
$$v = 80.2 \text{ ft/sec}$$

Example 3.15

Water is pumped at a rate of 3 cfs through the piping system illustrated. If the pump has a discharge pressure of 150 psig, to what elevation can the tank be raised? Assume the head loss due to friction is 10 feet.

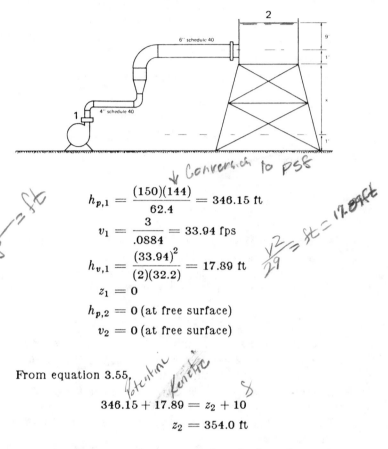

$$h_{p,1} = \frac{(150)(144)}{62.4} = 346.15 \text{ ft}$$
$$v_1 = \frac{3}{.0884} = 33.94 \text{ fps}$$
$$h_{v,1} = \frac{(33.94)^2}{(2)(32.2)} = 17.89 \text{ ft}$$
$$z_1 = 0$$
$$h_{p,2} = 0 \text{ (at free surface)}$$
$$v_2 = 0 \text{ (at free surface)}$$

From equation 3.55,

$$346.15 + 17.89 = z_2 + 10$$
$$z_2 = 354.0 \text{ ft}$$

The tank bottom can be raised to $(354.0 - 10 + 1) = 345$ feet above the ground.

4 IMPACT ENERGY

Impact energy (also known as stagnation or total[4] energy) is the sum of the kinetic and pressure energies. The impact head is

$$h_s = \frac{p}{\rho} + \frac{v^2}{2g_c} = h_p + h_v \qquad 3.59$$

Impact head represents the effective head in a fluid which has been brought to rest (stagnated) in an adiabatic and reversible manner. Equation 3.59 can be used with a gas as long as the velocity is low—less than 400 ft/sec.

[4]There is confusion about *total head* as defined by equations 3.56 and 3.59. The effective pressure in a fluid which has been brought to rest adiabatically does not depend on the potential energy term, z. The application will determine which definition of total head is intended.

Impact head can be measured directly by using a pitot tube. This is illustrated in figure 3.12. Equation 3.59 can be used with a pitot tube.

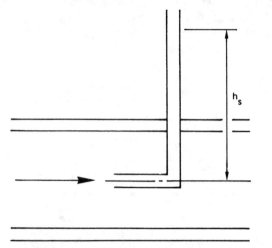

Figure 3.12 A Pitot Tube

A mercury manometer must be used if the stagnation properties of a gas or high-pressure liquid are being measured. Measurement of stagnation properties is covered in greater detail in part 4 of this chapter.

Example 3.16

The static pressure of air ($\rho = .075$ lbm/ft^3) flowing in a pipe is measured by a precision gage to be 10.0 psig. A pitot tube manometer indicates 20.6 inches of mercury. What is the velocity of the air in the pipe?

The pitot tube measures stagnation pressure. From equation 3.24,

$$p_s = (20.6)(0.491) = 10.1146 \text{ psig}$$

Since stagnation pressure is the sum of static and velocity pressures, the velocity pressure is

$$p_v = p_s - p_p = 10.1146 - 10.0 = 0.1146 \text{ psig}$$

The velocity head is

$$h_v = \frac{p_v}{\rho} = \frac{(0.1146)(144)}{0.075} = 220.0 \text{ ft}$$

From equation 3.53,

$$v = \sqrt{(2)(32.2)(220)} = 119.0 \text{ ft/sec}$$

5 HYDRAULIC GRADE LINE

The hydraulic grade line is a graphical representation of the sum of the static and potential heads versus position along the pipeline

$$\frac{\text{hydraulic}}{\text{grade}} = z + h_p \qquad 3.60$$

Since the pressure head can increase at the expense of the velocity head, the hydraulic grade line can increase if an increase in flow area is encountered.

6 REYNOLDS NUMBER

The Reynolds number is a dimensionless ratio of the inertial flow forces to the viscous forces within the fluid. Two expressions for Reynolds number are used, one requiring absolute viscosity, the other kinematic viscosity:

$$N_{Re} = \frac{D_e v \rho}{\mu g_c} \qquad 3.61$$

$$= \frac{D_e v}{\nu} \qquad 3.62$$

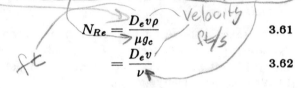

The Reynolds number also can be calculated from the mass flow rate per unit area, G. G must have the units of lbm/sec-ft^2.

$$N_{Re} = \frac{D_e G}{\mu g_c} \qquad 3.63$$

The Reynolds number is an important indicator in many types of problems. In addition to being used quantitatively in many equations, the Reynolds number also is used to determine whether fluid flow is laminar or turbulent.

A Reynolds number of 2000 or less indicates *laminar flow*. Fluid particles in laminar flow move in straight paths parallel to the flow direction. Viscous effects are dominant, resulting in a parabolic velocity distribution with a maximum velocity along the fluid flow centerline.

The fluid is said to be *turbulent* if the Reynolds number is greater than 2000.[5] Turbulent flow is characterized by random movement of fluid particles. The velocity distribution is essentially uniform with turbulent flow.

7 EQUIVALENT DIAMETER

The equivalent diameter, D_e, used in equations 3.61 and 3.62, is equal to the inside diameter of a circular pipe. The equivalent diameters of other cross sections in flow are given by table 3.6.

[5] The beginning of the turbulent region is difficult to predict. There actually is a transition region between Reynolds numbers 2000 to 4000. In most fluid problems, however, flow is well within the turbulent region.

Table 3.6
Equivalent Diameters

Conduit Cross Section	D_e
Flowing Full	
annulus	$D_o - D_i$
square	L
rectangle	$\dfrac{2L_1 L_2}{L_1 + L_2}$
Flowing Partially Full	
half-filled circle	D
rectangle (h deep, L wide)	$\dfrac{4hL}{L + 2h}$
wide, shallow stream (h deep)	$4h$
triangle (h deep, L broad, s side)	$\dfrac{hL}{s}$
trapezoid (h deep, a wide at top, b wide at bottom, s side)	$\dfrac{2h(a + b)}{b + 2s}$

Example 3.17

Determine the equivalent diameter of the open trapezoidal channel shown.

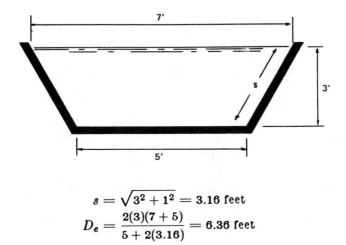

$$s = \sqrt{3^2 + 1^2} = 3.16 \text{ feet}$$

$$D_e = \frac{2(3)(7 + 5)}{5 + 2(3.16)} = 6.36 \text{ feet}$$

8 HYDRAULIC RADIUS

The equivalent diameter also can be found from the hydraulic radius, which is defined as the area in flow divided by the wetted perimeter. The wetted perimeter does not include free fluid surface.

$$D_e = 4r_h \qquad\qquad 3.64$$

$$r_h = \frac{\text{area in flow}}{\text{wetted perimeter}} \qquad 3.65$$

Consider a circular pipe flowing full. The area in flow is πr^2. The wetted perimeter is the entire circumference, $2\pi r$. The hydraulic radius is

$$\frac{\pi r^2}{2\pi r} = \frac{1}{2}r$$

Therefore, the hydraulic radius and the pipe radius are not the same. (The hydraulic radius of a pipe flowing half full is also $\frac{1}{2}r$, as the flow area and the wetted perimeter both are halved.)

The hydraulic radius of a pipe flowing less than full can be found from table 3.7.

Example 3.18

What is the hydraulic radius of the trapezoidal channel described in example 3.17?

From equation 3.65,

$$r_H = \frac{(5)(3) + (3)(1)}{3.16 + 5 + 3.16} = 1.59 \text{ feet}$$

Using the results of the previous example and equation 3.64

$$r_H = \frac{6.36}{4} = 1.59 \text{ feet}$$

FLUID STATICS AND DYNAMICS

Table 3.7
Hydraulic Radius
of Partially Filled Circular Pipes

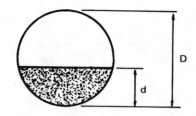

$\dfrac{d}{D}$	$\dfrac{\text{hyd. rad.}}{D}$	$\dfrac{d}{D}$	$\dfrac{\text{hyd. rad.}}{D}$
0.05	0.0326	0.55	0.2649
0.10	0.0635	0.60	0.2776
0.15	0.0929	0.65	0.2881
0.20	0.1206	0.70	0.2962
0.25	0.1466	0.75	0.3017
0.30	0.1709	0.80	0.3042
0.35	0.1935	0.85	0.3033
0.40	0.2142	0.90	0.2980
0.45	0.2331	0.95	0.2864
0.50	0.2500	1.00	0.2500

PART 4: Fluid Dynamics

1 FLUID CONSERVATION LAWS

Many fluid flow problems can be solved by using the principles of conservation of mass and energy.

When applied to fluid flow, the principle of mass conservation is known as the *continuity equation*:

$$\rho_1 A_1 v_1 = \rho_2 A_2 v_2 \qquad 3.66$$
$$\dot{m}_1 = \dot{m}_2 \qquad 3.67$$

If the fluid is incompressible, $\rho_1 = \rho_2$, so

$$A_1 v_1 = A_2 v_2 \qquad 3.68$$
$$\dot{V}_1 = \dot{V}_2 \qquad 3.69$$

The energy conservation principle is based on the Bernoulli equation. However, terms for friction loss and hydraulic machines must be included.

$$\left(\frac{p_1}{\rho} + \frac{v_1^2}{2g_c} + z_1\right) + h_A = \left(\frac{p_2}{\rho} + \frac{v_2^2}{2g_c} + z_2\right) + h_E + h_f$$
$$3.70$$

2 HEAD LOSS DUE TO FRICTION

The most common expression for calculating head loss due to friction (h_f) is the *Darcy formula*:

$$h_f = \frac{fLv^2}{2Dg_c} \qquad 3.71$$

The *Moody friction factor chart* (figure 3.13) probably is the most convenient method of determining the friction factor, f.

The basic parameter required to use the Moody friction factor chart is the Reynolds number. If the Reynolds number is less than 2000, the friction factor is given by equation 3.72.

$$f = \frac{64}{N_{Re}} \qquad 3.72$$

For turbulent flow ($N_{Re} > 2000$), the friction factor depends on the relative roughness of the pipe. This roughness is expressed by the ratio $\frac{\epsilon}{D}$, where ϵ is the specific surface roughness and D is the inside diameter. Values of ϵ for various types of pipe are found in table 3.8.

Another method for finding the friction head loss is the *Hazen-Williams formula*. The Hazen-Williams formula gives good results for liquids that have kinematic viscosities around 1.2 EE−5 ft²/sec (corresponding to 60°F water). At extremely high and low temperatures, the Hazen-Williams formula can be as much as 20% in er-

ror for water. The Hazen-Williams formula should be used only for turbulent flow.

The Hazen-Williams head loss is

$$h_f = \frac{(3.022)(v)^{1.85}L}{(C)^{1.85}(D)^{1.165}} \qquad 3.73$$

Or, in terms of other units,

$$h_f = (10.44)(L)\frac{(gpm)^{1.85}}{(C)^{1.85}(d_{\text{inches}})^{4.8655}} \qquad 3.74$$

Use of these formulas requires a knowledge of the Hazen-Williams coefficient, C, which is assumed to be independent of the Reynolds number. Table 3.8 gives values of C for various types of pipe.

Values of f and h_f are appropriate for clean, new pipe. As some pipes age, it is not uncommon for scale build-up to decrease the equivalent flow diameter. This diameter decrease produces a dramatic increase in the friction loss.

$$\frac{h_{f,\text{scaled}}}{h_{f,\text{new}}} = \left(\frac{D_{\text{new}}}{D_{\text{scaled}}}\right)^5 \qquad 3.75$$

Because of this scale effect, an uprating factor of 10-30% is commonly applied to f or h_f in anticipation of future service conditions.

Example 3.19

50°F water is pumped through 4″ schedule 40 welded steel pipe ($\epsilon = .0002$) at the rate of 300 gpm. What is the friction head loss calculated by the Darcy formula for 1000 feet of pipe?

First, it is necessary to collect data on the pipe and water. The fluid viscosity and pipe dimensions can be found from tables at the end of the chapter.

kinematic viscosity = 1.41 EE − 5 ft²/sec
inside diameter = .3355 ft
flow area = .0884 ft²

The flow quantity is

$$(300)(.002228) = .6684 \text{ cfs}$$

The velocity is

$$v = \frac{\dot{V}}{A} = \frac{.6684}{.0884} = 7.56 \text{ fps}$$

The Reynolds number is

$$N_{Re} = \frac{(.3355)(7.56)}{1.41 \text{ EE} - 5} = 1.8 \text{ EE5}$$

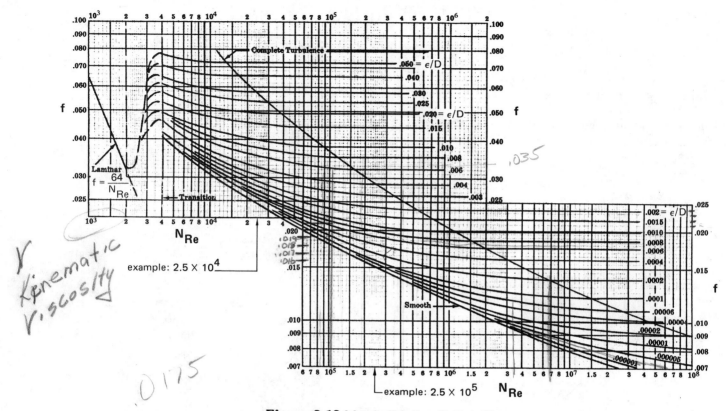

Figure 3.13 Moody Friction Factor Chart

Table 3.8
Specific Roughness and Hazen-Williams Constants for Various Pipe Materials

Type of pipe or surface	Roughness ϵ(ft) Range	Design	Hazen-Williams C Constants Range	Clean	Design
STEEL					
welded and seamless	.0001–.0003	.0002	150–80	140	100
interior riveted, no projecting rivets				139	100
projecting girth rivets				130	100
projecting girth and horizontal rivets				115	100
vitrified, spiral-riveted, flow with lap				110	100
vitrified, spiral-riveted, flow against lap				100	90
corrugated				60	60
MINERAL					
concrete	.001–.01	.004	152–85	120	100
cement-asbestos			160–140	150	140
vitrified clays					110
brick sewer					100
IRON					
cast, plain	.0004–.002	.0008	150–80	130	100
cast, tar (asphalt) coated	.0002–.0006	.0004	145–50	130	100
cast, cement lined	.000008	.000008		150	140
cast, bituminous lined	.000008	.000008	160–130	148	140
cast, centrifugally spun	.00001	.00001			
galvanized, plain	.0002–.0008	.0005			
wrought, plain	.0001–.0003	.0002	150–80	130	100
MISCELLANEOUS					
fiber				150	140
copper and brass	.000005	.000005	150–120	140	130
wood stave	.0006–.003	.002	145–110	120	110
transite	.000008	.000008			
lead, tin, glass		.000005	150–120	140	130
plastic (ABS, PVC, etc.)		.000005	150–120	140	130

The relative roughness is

$$\frac{\epsilon}{D} = \frac{.0002}{.3355} = .0006$$

From the Moody friction factor chart, $f = .0195$.
From equation 3.71,

$$h_f = \frac{(.0195)(1000)(7.56)^2}{(2)(.3355)(32.2)} = 51.6 \text{ ft}$$

Example 3.20

Repeat example 3.19 using the Hazen-Williams formula.
Assume $C = 100$.

Using equation 3.73,

$$h_f = \frac{(3.022)(7.56)^{1.85}(1000)}{(100)^{1.85}(.3355)^{1.165}} = 90.8 \text{ ft}$$

Using equation 3.74,

$$h_f = (10.44)(1000)\frac{(300)^{1.85}}{(100)^{1.85}(4.026)^{4.8655}} = 90.9 \text{ ft}$$

3 MINOR LOSSES

In addition to the head loss caused by friction between the fluid and the pipe wall, losses also are caused by obstructions in the line, changes in direction, and changes in flow area. These losses are named *minor losses* because they are much smaller in magnitude than the h_f term. Two methods are used to determine these losses: the method of equivalent lengths and the method of loss coefficients.

The method of *equivalent lengths* uses a table to convert each valve and fitting into an equivalent length of straight pipe. This length is added to the actual pipeline length and substituted into the Darcy equation for L_e.

$$h_f = \frac{fL_e v^2}{2Dg_c} \qquad 3.76$$

Example 3.21

Using table 3.9, determine the equivalent length of the piping network shown.

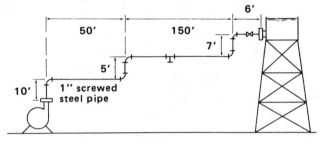

The line consists of:

1 gate valve	.84
5 90° standard elbows	5.2 × 5
1 tee run	3.2
straight pipe	228
$L_e =$	258 feet

The alternative is to use a loss coefficient, K. This loss coefficient, when multiplied by the velocity head, will give the head loss in feet. This method must be used to find exit and entrance losses.

$$h_f = K\frac{v^2}{2g_c} \qquad 3.77$$

Table 3.9
Typical Equivalent Lengths of Schedule 40 Straight Pipe
For Steel Fittings and Valves
(For any fluid in turbulent flow)

Equivalent Length, ft

Fitting Type	Pipe Size* 1″	2″	4″	6″	(flanged pipe) 8″
Standard Radius 90° Elbow	5.2	8.5	13.0	8.9	12.0
Long Radius 90° Elbow	2.7	3.6	4.6	5.7	7.0
Regular 45° Elbow	1.3	2.7	5.5	5.6	7.7
Tee, flow through line (run)	3.2	7.7	17.0	3.8	4.7
Tee, flow through stem	6.6	12.0	21.0	18.0	24.0
180° Return Bend	5.2	8.5	13.0	8.9	12.0
Globe Valve, open	29.0	54.0	110.0	190.0	260.0
Gate Valve, open	.84	1.5	2.5	3.2	3.2
Angle Valve, open	17.0	18.0	18.0	63.0	90.0
Swing Check Valve	11.0	19.0	38.0	63.0	90.0
Coupling or Union	.29	.45	.65	—	—

*Screwed pipe and fittings unless flanged indicated.

Values of K are widely tabulated, but they also can be calculated from the following formulas.

Valves and Fittings: Refer to the manufacturer's data, or calculate from the equivalent length.

$$K = \frac{fL_e}{D} \qquad 3.78$$

Sudden Enlargements:

$$K = \left[1 - \left(\frac{D_1}{D_2}\right)^2\right]^2 \qquad 3.79$$

Sudden Contractions:

$$K = \frac{1}{2}\left[1 - \left(\frac{D_1}{D_2}\right)^2\right] \qquad 3.80$$

Pipe Exit: (projecting exit, sharp-edged, and rounded)

$$K = 1.0$$

Pipe Entrance:

Reentrant: $K = .78$
Sharp edged: $K = .5$
Rounded:

$\frac{r}{D}$	K
.02	.28
.04	.24
.06	.15
.10	.09
.15	.04

Tapered Diameter Changes:

$$\beta = \frac{\text{small diameter}}{\text{large diameter}}$$

ϕ = wall-to-horizontal angle

	Gradual, $\phi < 22°$	Sudden, $\phi > 22°$	
Enlargement	$2.6(\sin\phi)(1-\beta^2)^2$	$(1-\beta^2)^2$	3.81
Contraction	$.8(\sin\phi)(1-\beta^2)$	$\frac{1}{2}(1-\beta^2)\sqrt{\sin\phi}$	3.82

4 HEAD ADDITIONS/EXTRACTIONS

A pump adds head (energy) to the fluid stream. A turbine extracts head from the fluid stream. The amount of head added or extracted can be found by evaluating Bernoulli's equation (equation 3.55) on both sides of the device.

$$h_A = (H_2 - H_1) \quad \text{(pumps)} \qquad 3.83$$
$$h_E = (H_1 - H_2) \quad \text{(turbines)} \qquad 3.84$$

The head increase from a pump is given by equation 3.85.

$$h_A = \frac{(550)(\text{pump input horsepower})\eta_{\text{pump}}}{\dot{m}} \qquad 3.85$$

Bernoulli's equation also can be used to calculate the power available to a turbine in a fluid stream by multiplying the total energy by the mass flow rate. This is called the *water horsepower.*

$$P = \dot{m}H = \dot{m}\left(\frac{p}{\rho} + \frac{v^2}{2g_c} + z\right) \qquad 3.86$$
$$\dot{m} = \rho A v \qquad 3.87$$
$$whp = \frac{P}{550} \qquad 3.88$$

Pumps and turbines are covered in greater detail in chapter 4.

5 DISCHARGE FROM TANKS

Flow from a tank discharging liquid to the atmosphere through an opening in the tank wall (figure 3.14) is affected by both the area and the shape of the opening. At the orifice, the total head of the fluid is converted into kinetic energy according to equation 3.89.[6]

$$v_o = C_v\sqrt{2gh} \qquad 3.89$$

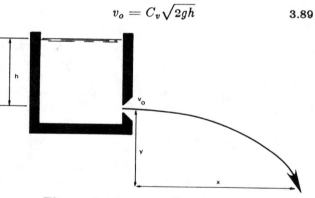

Figure 3.14 Discharge From a Tank

C_v is the *coefficient of velocity* which can be calculated from the *coefficients of discharge* and *contraction.* Typical values of C_v, C_d, and C_c are given in table 3.10.

$$C_v = \frac{C_d}{C_c} \qquad 3.90$$

The discharge from the orifice is

$$Q_o = (C_cA_o)v_o = C_cA_oC_v\sqrt{2gh}$$
$$= C_dA_o\sqrt{2gh} \qquad 3.91$$

[6]Although the term g_c appears in the equation for velocity head (equation 3.53), here it is the local gravity, g, which appears in equation 3.89. An analysis of equation 3.197 will show you why this is so.

Table 3.10
Orifice Coefficients for Water
(fully turbulent*)

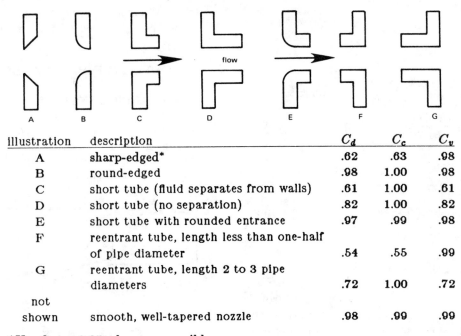

illustration	description	C_d	C_c	C_v
A	sharp-edged*	.62	.63	.98
B	round-edged	.98	1.00	.98
C	short tube (fluid separates from walls)	.61	1.00	.61
D	short tube (no separation)	.82	1.00	.82
E	short tube with rounded entrance	.97	.99	.98
F	reentrant tube, length less than one-half of pipe diameter	.54	.55	.99
G	reentrant tube, length 2 to 3 pipe diameters	.72	1.00	.72
not shown	smooth, well-tapered nozzle	.98	.99	.99

*Use figure 3.25 whenever possible

The head loss due to turbulence at the orifice is

$$h_f = \left(\frac{1}{C_v^2} - 1 \right) \frac{v_o^2}{2g_c} \qquad 3.92$$

The discharge stream coordinates (see figure 3.14) are

$$x = v_o t = v_o \sqrt{\frac{2y}{g}} = 2C_v \sqrt{hy} \qquad 3.93$$

$$y = \frac{gt^2}{2} = \frac{g}{2}\left(\frac{x}{v_o} \right)^2 \qquad 3.94$$

The fluid velocity at a point downstream of the orifice is

$$v_x = v_o \qquad 3.95$$

$$v_y = gt \qquad 3.96$$

If the liquid in a tank is not being replenished constantly, the static head forcing discharge through the orifice will decrease. For a tank with a constant cross-sectional area, the time required to lower the fluid level from level h_1 to h_2 is calculated from equation 3.97.

$$t = \frac{2A_t(\sqrt{h_1} - \sqrt{h_2})}{C_d A_o \sqrt{2g}} \qquad 3.97$$

If the tank has a varying cross section, the following basic relationship holds.

$$Q\, dt = -A_t\, dh \qquad 3.98$$

An expression for the tank area, A_t, as a function of h, must be determined. Then, the time to empty the tank from height h_1 to lower height h_2 is

$$t = \int_{h_1}^{h_2} \frac{A_t\, dh}{C_d A_o \sqrt{2gh}} \qquad 3.99$$

For a tank being fed at a rate, $\dot{V}_{in}$, which is less than the discharge through the orifice, the time to empty expression is

$$t = \int_{h_1}^{h_2} \frac{A_t\, dh}{(C_d A_o \sqrt{2gh}) - \dot{V}_{in}} \qquad 3.100$$

When a tank is being fed at a rate greater than the discharge, equation 3.100 will become positive, indicating a rising head. t then will be the time it takes to raise the fluid level from h_1 to h_2.

The preceding discussion has assumed that the tank has been open or vented to the atmosphere. If the fluid is discharging from a pressurized tank, the total head will be increased by the gage pressure converted to head of fluid by means of equation 3.52.

Example 3.22

A 15$'$ diameter tank discharges 150°F water through a sharp edged 1$''$ diameter orifice. If the original water depth is 12' and the tank is continually pressurized to 50 psig, find the time to empty the tank.

At 150°F, $\rho = 61.20 \ \text{lbm/ft}^3$

For the orifice,

$$A_o = .00545 \ \text{ft}^2, \ C_d = .62$$
$$h_1 = 12 + \frac{(50)(144)}{61.2} = 129.65 \ \text{ft}$$
$$h_2 = \frac{(50)(144)}{61.20} = 117.65 \ \text{ft}$$

From equation 3.97,

$$t = \frac{2\,[\pi(7.5)^2](\sqrt{129.65} - \sqrt{117.65})}{(.62)(.00545)\sqrt{(2)(32.2)}} = 7035 \ \text{seconds}$$

6 CULVERTS AND SIPHONS

A culvert is a water path used to drain runoff from an obstructing geographical feature. Most culvert designs are empirical. However, if the entrance and exit of the culvert both are submerged, the discharge will be independent of the barrel slope. In that case, equation 3.101 can be used to evaluate the discharge. h is the difference in surface levels of the headwater and tailwater.

$$\dot{V} = C_d A \sqrt{2gh} \qquad\qquad 3.101$$

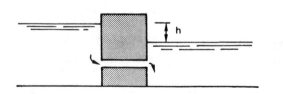

Figure 3.15 A Simple Pipe Culvert

If the culvert length is greater than 50 feet or if the entrance is not smooth, the available energy will be divided between friction and velocity heads. The effective head to be used in equation 3.101 is:

$$h' = h - h_{\text{entrance}} - h_f \qquad\qquad 3.102$$

The entrance head loss is calculated using loss coefficients:

$$h_{\text{entrance}} = K_e \left(\frac{v^2}{2g_c} \right) \qquad\qquad 3.103$$

Typical values of K_e are:

.08 for a smooth and tapered entrance
.10 for a flush concrete groove or bell design
.15 for a projecting concrete groove or bell design
.50 for a flush square-edged entrance
.90 for a projecting square-edged entrance

The friction loss, h_f, can be found in the usual manner, either from the Darcy equation and Moody friction factor chart or from the Hazen-Williams equation. A trial and error solution may be necessary since v is not known but is needed to find the friction factor.

7 MULTIPLE PIPE SYSTEMS

A. SERIES PIPE SYSTEMS

A series pipe system has one or more diameters along its run. If Q or v is known in any part of the system, the friction loss can be found easily as the sum of the friction losses in the sections.

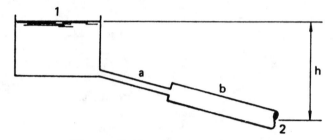

Figure 3.16 A Series Pipe System

If both v and Q are unknown, a trial and error solution method is required. The following procedure can be used with the Darcy friction factor.

step 1: Using the Moody diagram with ϵ_a, ϵ_b, D_a, and D_b, find f_a and f_b for fully turbulent flow.

step 2: Write all of the velocities in terms of one unknown velocity.

$$v_a = v_a \qquad\qquad 3.104$$
$$v_b = \left(\frac{A_a}{A_b} \right) v_a \qquad\qquad 3.105$$

step 3: Write the friction loss in terms of the unknown velocity.

$$h_{f'} \text{ total} = \frac{f_a L_a v_a^2}{2 D_a g_c} + \frac{f_b L_b}{2 D_b g_c} \left(\frac{A_a}{A_b} \right)^2 v_a^2 \qquad 3.106$$
$$= \frac{v_a^2}{2g_c} \left[\frac{f_a L_a}{D_a} + \frac{f_b L_b}{D_b} \left(\frac{A_a}{A_b} \right)^2 \right] \qquad 3.107$$

step 4: Solve for the unknown velocity using Bernoulli's equation between points 1 and 2. Include the pipe friction but, for convenience, ignore minor losses.

$$h = \frac{v_b^2}{2g_c} + h_f \qquad\qquad 3.108$$
$$= \frac{v_a^2}{2g_c} \left[\left(\frac{A_a}{A_b} \right)^2 \left(1 + \frac{f_b L_b}{D} \right) + \frac{f_a L_a}{D_a} \right] \qquad 3.109$$

step 5:   Using the values of v_a and v_b found from step
4, check the values of f_a and f_b. Repeat steps
3 and 4 using the new values of f_a and f_b if
necessary.

B. PARALLEL PIPE SYSTEMS

A common method of increasing the capacity of an
existing line is to install a second line parallel to the
first. If that is done, the flow will divide in such a
manner as to make the friction loss the same in both
branches.

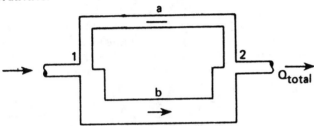

Figure 3.17 A Parallel Pipe System

If the parallel system has only two branches, a simultaneous solution approach can be taken.

$$h_{f,a} = h_{f,b} = \frac{f_a L_a v_a^2}{2 D_a g_c} = \frac{f_b L_b v_b^2}{2 D_b g_c} \qquad 3.110$$

$$Q_a + Q_b = Q_c \qquad 3.111$$

$$\frac{1}{4}\pi\left(D_a^2 v_a + D_b^2 v_b\right) = Q_c \qquad 3.112$$

However, if the parallel system has three or more
branches, it is easier to use the following procedure.

step 1:   Write $h_f = \frac{fLv^2}{2Dg_c}$ for each branch.

step 2:   Solve for v for each branch.

$$v = \sqrt{\frac{2Dg}{fL}h_f} \qquad 3.113$$

step 3:   Solve for Q for each branch. (There will be a
different value of K' for each branch.)

$$Q = Av = A\sqrt{\frac{2Dg}{fL}h_f} = K'\sqrt{h_f} \qquad 3.114$$

$$K' = A\sqrt{\frac{2Dg}{fL}}$$

step 4:
$$Q_{total} = Q_1 + Q_2 + Q_3 \qquad 3.115$$
$$= (K_1' + K_2' + K_3')\sqrt{h_f} \qquad 3.116$$

Since Q_{total}, K_1', K_2', and K_3' are known, it is possible
to solve for the friction loss.

step 5:   Check the values of f and repeat as necessary.

8 FLOW MEASURING DEVICES

The total energy in a fluid flow is the sum of pressure
head, velocity head, and gravitational head.

$$H = \frac{p}{\rho} + \frac{v^2}{2g_c} + z \qquad 3.117$$

Change in gravitational head within a flow-measuring
instrument is negligible. Therefore, if two of the three
remaining variables (H, p, or v) are known, the third
can be found from subtraction. The flow measuring
devices discussed in this section are capable of measuring total head (H) or pressure head (p).

A. VELOCITY MEASUREMENT

Velocity of a fluid stream is determined by measuring
the difference between the static and the stagnation
pressures, then solving for the velocity head.

A *piezometer tap* can be used to measure the pressure
head directly in feet of fluid.

$$h_p = \frac{p}{\rho} \qquad 3.118$$

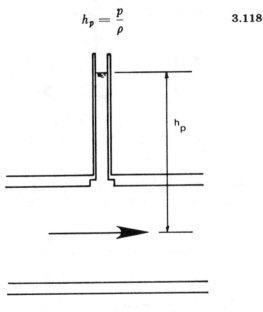

Figure 3.18 Piezometer Tap

For liquids with pressures higher than the capability of
the direct reading tap, a manometer can be used with a
piezometer tap or with a *static probe* as shown in figure
3.19.

For either configuration of figure 3.19, the static pressure is

$$p = \rho_m \Delta h_m - \rho y \qquad 3.119$$
$$h_p = \frac{\rho_m \Delta h_m}{\rho} - y \qquad 3.120$$

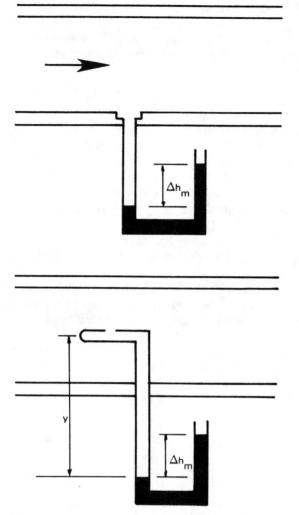

Figure 3.19 Use of Manometers to Measure
Static Pressure

Stagnation pressure, also known as *total pressure* or *impact pressure*, can be measured directly in feet of fluid by using a pitot tube as shown in figure 3.12.

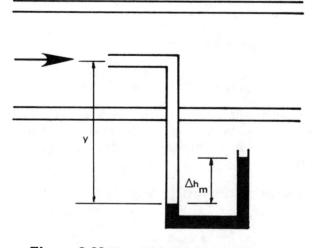

Figure 3.20 Use of Manometer to Measure
Total Pressure

Using the results of the measurements above, the velocity head can be calculated from equation 3.123.

For high-pressure fluids, a manometer must be used to measure stagnation pressure:

$$p_s = \rho h_s = \rho_m \Delta h_m - \rho y \qquad 3.121$$

$$h_s = \frac{p}{\rho} + \frac{v^2}{2g_c} = \frac{\rho_m \Delta h_m}{\rho} - y \qquad 3.122$$

$$v = \sqrt{2g(h_s - h_p)} = \sqrt{\frac{2g(p_s - p)}{\rho}} \qquad 3.123$$

If the piezometer tap of figure 3.18 and the pitot tube of figure 3.12 are placed at the same point, the velocity head in feet of fluid can be read directly.

$$\frac{v^2}{2g_c} = \Delta h \qquad 3.124$$

$$v = \sqrt{2g\Delta h} \qquad 3.125$$

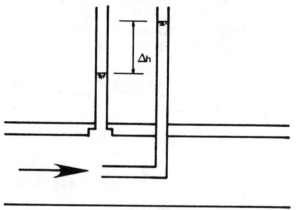

Figure 3.21 Comparative Velocity Head Measurement

The instrumentation arrangement of figures 3.19 and 3.20 can be combined into a single instrument to provide a measurement of velocity head as shown in figure 3.22.

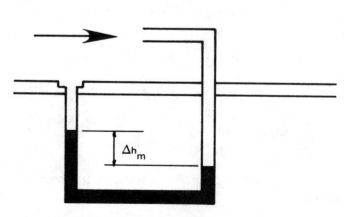

Figure 3.22 Velocity Head Measurement

$$\frac{v^2}{2g_c} = \frac{\Delta h_m(\rho_m - \rho)}{\rho} \qquad 3.126$$

$$v = \sqrt{\frac{2g(\rho_m - \rho)\Delta h_m}{\rho}} \qquad 3.127$$

Example 3.23

50°F water is flowing through a pipe. A pitot-static gage registers a 3″ deflection of mercury. What is the velocity within the pipe? (The density of mercury is 848.6 pcf.)

Using equation 3.127,

$$v = \sqrt{\frac{2(32.2)(848.6 - 62.4)\left(\frac{3}{12}\right)}{62.4}} = 14.24 \text{ fps}$$

B. FLOW MEASUREMENT

Using the techniques described in the preceding section, the flow rate in a line can be determined by measuring the pressure drop across a restriction. Once the geometry of the restriction is known, the Bernoulli equation, along with empirically determined correction coefficients, can be applied to obtain an expression directly relating flow rate with pressure drop.

If potential head is neglected, Bernoulli's equation becomes

$$\frac{p_1}{\rho} + \frac{v_1^2}{2g_c} = \frac{p_2}{\rho} + \frac{v_2^2}{2g_c} \qquad 3.128$$

But, v_1 and v_2 are related. From equation 3.68,

$$v_1 = v_2\left(\frac{A_2}{A_1}\right) \qquad 3.129$$

Combining equations 3.128 and 3.129 yields the standard flow measurement equation.

$$v_2 = \frac{\sqrt{2g\left(\frac{p_1 - p_2}{\rho}\right)}}{\sqrt{1 - \left(\frac{A_2}{A_1}\right)^2}} \qquad 3.130$$

The flow quantity can be found from

$$\dot{V} = v_2 A_2 \qquad 3.131$$

The reciprocal of the denominator of equation 3.130 is known as the *velocity of approach factor*, F_{va}. The *beta ratio* can be incorporated into the formula for F_{va}.

$$F_{va} = \frac{1}{\sqrt{1 - \beta^4}} \qquad 3.132$$

$$\beta = \frac{D_2}{D_1} \qquad 3.133$$

The simplest fluid flow measuring device is the *orifice plate*. This consists of a thin plate or diaphragm with a central hole through which the fluid flows.

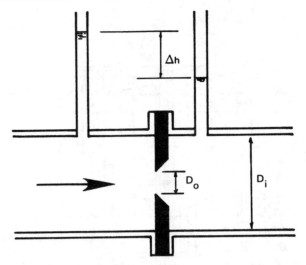

Figure 3.23 Comparative Reading Orifice Plate

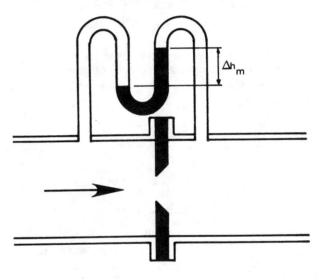

Figure 3.24 Direct Reading Orifice Plate

3.134 and 3.135 are the governing orifice plate equations for liquid flow.

$$\dot{V} = F_{va}C_d A_o\sqrt{\frac{2g(p_1 - p_2)}{\rho}} \qquad 3.134$$

$$= F_{va}C_d A_o\sqrt{\frac{2g(\rho_m - \rho)\Delta h_m}{\rho}} \qquad 3.135$$

The definition of the velocity of approach factor is modified slightly for the orifice plate.

$$F_{va} = \frac{1}{\sqrt{1 - \left(\frac{C_c A_o}{A_i}\right)^2}} \qquad 3.136$$

The *flow coefficient* depends on the velocity of approach factor and the discharge coefficient. It also can be obtained from figure 3.25.

$$C_f = F_{va}C_d \qquad \text{3.137}$$

$$C_d = C_v C_c \qquad \text{3.138}$$

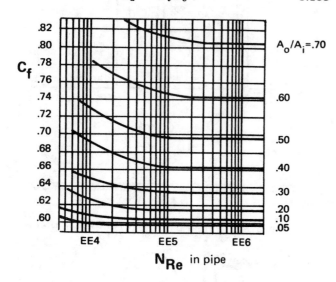

C_f

$A_o/A_i = .70$

N_{Re} in pipe

Figure 3.25 Flow Coefficients for I.S.A. (German Standard) Orifice Plates

The flow coefficients can be used to rewrite equations 3.134 and 3.135.

$$\dot{V} = C_f A_o \sqrt{\frac{2g(p_1 - p_2)}{\rho}} \qquad \text{3.139}$$

$$= C_f A_o \sqrt{\frac{2g(\rho_m - \rho)\Delta h_m}{\rho}} \qquad \text{3.140}$$

Operating on the same principles as the orifice plate, the *venturi meter* induces a smaller pressure drop. It is, however, mechanically more complex, as shown by figure 3.26.

The governing equations are similar to those for orifice plates. C_c usually is 1.0 for venturi meters.

$$v_2 = F_{va} \sqrt{\frac{2g(p_1 - p_2)}{\rho}} \qquad \text{3.141}$$

$$\dot{V} = F_{va} C_d A_2 \sqrt{\frac{2g(p_1 - p_2)}{\rho}}$$

$$= C_f A_2 \sqrt{\frac{2g(p_1 - p_2)}{\rho}} \qquad \text{3.142}$$

$$F_{va} = \frac{1}{\sqrt{1 - \left(\frac{A_2}{A_1}\right)^2}} \qquad \text{3.143}$$

$$C_d = C_v C_c \qquad \text{3.144}$$

$$C_f = F_{va} C_d \qquad \text{3.145}$$

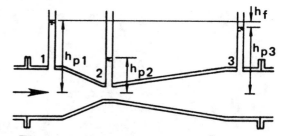

Figure 3.26 Venturi Meter with Wall Taps

Table 3.11
C_d for Venturi Meters

$$2 < (A_1/A_2) < 3$$

C_d	N_{Re}
.94	6,000
.95	10,000
.96	20,000
.97	50,000
.98	200,000
.99	2,000,000

Example 3.24

150°F water is flowing in an 8″ schedule 40 steel pipe at 2.23 cfs. If a 5.75″ sharp-edged orifice plate is bolted across the line, what manometer deflection in inches of mercury would be expected? (Mercury has a density of 848.6 pcf.)

	5.75″ orifice	8″ schedule 40
flow area	.180 ft^2	.3474 ft^2
diameter	.479 ft	.6651 ft

The initial approach velocity is

$$v_i = \frac{Q}{A} = \frac{2.23 \text{ cfs}}{.3474} = 6.419 \text{ ft/sec}$$

For 150°F water, $\nu = .476 \text{ EE} - 5 \text{ ft}^2/\text{sec}$ and $\rho = 61.2$ lbm/ft^3.

$$N_{Re} = \frac{vD}{\nu} = \frac{(6.419)(.6651)}{.476 \text{ EE} - 5} = 8.97 \text{ EE5}$$

$$\frac{A_o}{A_i} = \frac{.180}{.3474} = .518$$

From figure 3.25, $C_f \approx .70$

From equation 3.140,

$$\Delta h_m = \frac{\rho \left(\frac{\dot{V}}{C_f A_o}\right)^2}{2g(\rho_m - \rho)} = \frac{(61.2)\left[\frac{2.23}{(.70)(.180)}\right]^2 \left(12\frac{\text{in}}{\text{ft}}\right)}{(2)(32.2)(848.6 - 61.2)}$$

$$= 4.54 \text{ in}$$

9 THE IMPULSE/MOMENTUM PRINCIPLE

A force is required to cause a direction or velocity change in a flowing fluid. Conventions necessary to determine such a force are:

1. $\Delta v = v_2 - v_1$

2. A positive Δv indicates an increase in velocity. A negative Δv indicates a decrease in velocity.

3. F and x are positive to the right. F and y are positive upward.

4. F is the force on the fluid. The force on the walls or support has the same magnitude but opposite direction.

5. The fluid is assumed to flow horizontally from left to right and is assumed to possess no y-component of velocity.

The *momentum* possessed by a moving fluid is defined as the product of mass (in slugs) and velocity (in ft/sec). The g_c term in equation 3.146 is needed to convert pounds-mass into slugs.

$$\text{momentum} = \frac{mv}{g_c} \qquad 3.146$$

Impulse is defined as the product of a force and the length of time the force is applied.

$$\text{impulse} = F\Delta t \qquad 3.147$$

The *impulse-momentum principle* states that the impulse applied to a moving body is equal to the change in momentum. This is expressed by equation 3.148.

$$F\Delta t = \frac{m\Delta v}{g_c} \qquad 3.148$$

Solving for F and combining m and Δt yields equation 3.149.

$$F = \frac{m\Delta v}{g_c \Delta t} = \frac{\dot{m}\Delta v}{g_c} \qquad 3.149$$

Since F is a vector, it can be broken into its components

$$F_x = \frac{\dot{m}\Delta v_x}{g_c} \qquad 3.150$$

$$F_y = \frac{\dot{m}\Delta v_y}{g_c} \qquad 3.151$$

If the fluid flow is directed through an angle ϕ,

$$\Delta v_x = v(\cos \phi - 1) \qquad 3.152$$

$$\Delta v_y = v \sin \phi \qquad 3.153$$

There are several fluid applications of the impulse-momentum principle.

A. JET PROPULSION

$$\dot{m}_2 = \dot{m}_1 + \dot{m}_{\text{fuel}} = \dot{V}_1\rho_1 + \dot{V}_{\text{fuel}}\rho_{\text{fuel}} \qquad 3.154$$

$$F_x = \frac{\dot{V}_2\rho_2 v_{2x} - \dot{V}_1\rho_1 v_{1x}}{g_c} \qquad 3.155$$

$$F_y = \frac{\dot{V}_2\rho_2 v_{2y} - \dot{V}_1\rho_1 v_{1y}}{g_c} \qquad 3.156$$

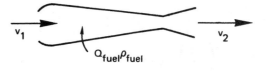

Figure 3.27 Jet Propulsion

B. OPEN JET ON VERTICAL FLAT PLATE

$$\Delta v_y = 0 \qquad 3.157$$

$$\Delta v_x = -v \qquad 3.158$$

$$F_x = \frac{-\dot{m}v}{g_c} = \frac{-\dot{V}\rho v}{g_c} \qquad 3.159$$

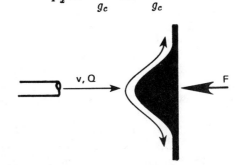

Figure 3.28 Open Jet on Vertical Plate

C. OPEN JET ON HORIZONTAL FLAT PLATE

As the jet travels upwards, its velocity decreases since gravity is working against it. By the time the liquid has reached the plate, the velocity has become

$$v_y = \sqrt{v_0^2 - 2gh} \qquad 3.160$$

$$\Delta v_x = 0 \qquad 3.161$$

$$\Delta v_y = -\sqrt{v_0^2 - 2gh} \qquad 3.162$$

$$F = \left(\frac{-\dot{m}}{g_c}\right)\sqrt{v_o^2 - 2gh}$$
$$= \left(\frac{\dot{V}\rho}{g_c}\right)\sqrt{v_o^2 - 2gh} \qquad 3.163$$

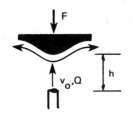

Figure 3.29 Open Jet on Horizontal Plate

D. OPEN JET ON SINGLE STATIONARY BLADE

v_2 may not be the same as v_1 if friction is present. If no information is given, assume that $v_2 = v_1$.

$$\Delta v_x = v_2 \cos \phi - v_1 \qquad 3.164$$

$$\Delta v_y = v_2 \sin \phi \qquad 3.165$$

$$F_x = \left(\frac{\dot{m}}{g_c}\right)(v_2 \cos \phi - v_1)$$

$$= \left(\frac{\dot{V}\rho}{g_c}\right)(v_2 \cos \phi - v_1) \qquad 3.166$$

$$F_y = \left(\frac{\dot{m}}{g_c}\right)(v_2 \sin \phi) = \left(\frac{\dot{V}\rho}{g_c}\right)(v_2 \sin \phi) \qquad 3.167$$

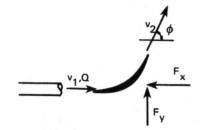

Figure 3.30 Open Jet on Stationary Blade

E. OPEN JET ON SINGLE MOVING BLADE

v_b is the blade velocity. For simplicity, friction is ignored. The discharge overtaking the moving blade is $\dot{V}'$.

$$\dot{V}' = \left(\frac{v - v_b}{v}\right)\dot{V} \qquad 3.168$$

$$\Delta v_x = (v - v_b)(\cos \phi - 1) \qquad 3.169$$

$$\Delta v_y = (v - v_b)(\sin \phi) \qquad 3.170$$

$$F_x = \frac{\dot{m}'\Delta v_x}{g_c} = \left(\frac{\dot{V}'\rho}{g_c}\right)\Delta v_x \qquad 3.171$$

$$F_y = \frac{\dot{m}'\Delta v_y}{g_c} = \left(\frac{\dot{V}'\rho}{g_c}\right)\Delta v_y \qquad 3.172$$

The power transferred to the blade is given by equation 3.173. Power is maximized when $\phi = 180°$ and $v_b = \frac{1}{2}v$.

$$P = F_x v_b \qquad 3.173$$

Equations 3.169 and 3.173 can be used with a *multiple-bladed* wheel by using the full $\dot{V}$ instead of $\dot{V}'$.

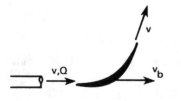

Figure 3.31 Open Jet on Moving Blade

F. CONFINED STREAMS IN PIPE BENDS

Since the fluid is confined, the forces caused by static pressure must be included along with the force from momentum changes. Using gage pressures and neglecting the fluid weight,

$$F_x = p_2 A_2 \cos \phi - p_1 A_1$$
$$+ \left(\frac{\dot{V}\rho}{g_c}\right)(v_2 \cos \phi - v_1) \qquad 3.174$$

$$F_y = \left[p_2 A_2 + \frac{\dot{V}\rho v_2}{g}\right]\sin \phi \qquad 3.175$$

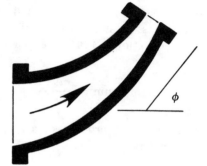

Figure 3.32 A Pipe Bend

G. WATER HAMMER

Water hammer is an increase in pressure in a pipe caused by a sudden velocity decrease. The sudden velocity decrease usually will be caused by a valve's closing.

Assuming the pipe material is inelastic, the time required for the water hammer shock wave to travel from a valve to the end of a pipe and back is given by

$$t = \frac{2L}{c} \qquad 3.176$$

The fluid pressure increase resulting from this shock wave is

$$\Delta p = \frac{\rho c \Delta v}{g_c} \qquad 3.177$$

Example 3.25

60°F water at 40 psig flowing at 8 ft/sec enters a 12" × 8" reducing elbow as shown and is turned 30°. (a) What is the resultant force on the water? (b) What other forces should be considered in the design of supports for the fitting?

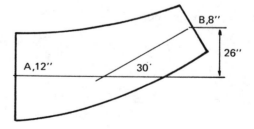

(a) The total head at point A is

$$\frac{(40)(144)}{(62.4)} + \frac{(8)^2}{(2)(32.2)} + 0 = 93.3 \text{ ft}$$

At point B, the velocity is

$$(8)\left(\frac{12}{8}\right)^2 = 18 \text{ ft/sec}$$

The pressure at B can be found from the Bernoulli equation.

$$93.3 = \frac{p_B(144)}{62.4} + \frac{(18)^2}{(2)(32.2)} + \frac{26}{12}$$

So, $p_B = 37.3$ psig

$$V = vA = (8)\left(\frac{1}{4}\right)\pi\left(\frac{12}{12}\right)^2 = 6.28 \text{ cfs}$$

From equation 3.174,

$$\begin{aligned}
F_x = &-(40)(144)\left(\frac{1}{4}\right)\pi\left(\frac{12}{12}\right)^2 \\
&+ (37.3)(144)\left(\frac{1}{4}\right)\pi\left(\frac{8}{12}\right)^2 \cos 30° \\
&+ \left(\frac{(6.28)(62.4)}{32.2}\right)[(18)(\cos 30°) - 8] \\
= &-2808 \text{ lbf}
\end{aligned}$$

From equation 3.175,

$$\begin{aligned}
F_y = &\left[(37.3)(144)\left(\frac{1}{4}\right)\pi\left(\frac{8}{12}\right)^2 \right. \\
&\left. + \left(\frac{(6.28)(62.4)(18)}{32.2}\right)\right] \sin 30° \\
= &\ 1047 \text{ lbf}
\end{aligned}$$

The resultant force on the water is

$$R = \sqrt{(-2808)^2 + (1047)^2} = 2997 \text{ lbf}$$

(b) The support also should be designed to carry the weight of the water in the pipe and the bend, and the weight of the pipe and the bend itself.

Example 3.26

40°F water is flowing at 10 ft/sec through a 4″ schedule 40 welded steel pipe. A valve suddenly is closed. What increase in fluid pressure will occur?

Assume that the closing valve completely stops the flow. Therefore, Δv is 10 ft/sec.

At 40°F, $E = 294$ EE3 psi, and $\rho = 62.43$ lbm/ft³. From equation 3.20, the speed of sound in the water is

$$c = \sqrt{\frac{(294 \text{ EE3})(144)(32.2)}{62.43}} = 4673 \text{ fps}$$

From equation 3.177,

$$\Delta p = \frac{(62.43)(4673)(10)}{32.2} = 90,600 \text{ psf}$$

H. OPEN JET ON INCLINED PLATE

An open jet will be diverted both up and down a stationary, inclined flat plate. The velocity in each diverted flow will be v, the same as in the approaching jet. The fractions f_1 and f_2 of the jet which are diverted up and down can be found from equations 3.178 and 3.179.

$$f_1 = \frac{1 + \cos \phi}{2} \qquad\qquad 3.178$$

$$f_2 = \frac{1 - \cos \phi}{2} \qquad\qquad 3.179$$

$$f_1 - f_2 = \cos \phi \qquad\qquad 3.180$$

$$f_1 + f_2 = 1 \qquad\qquad 3.181$$

As the flow along the plate is assumed to be frictionless, there will be no force component parallel to the plate.

The force perpendicular to the plate is

$$F = \left(\frac{\dot{V}\rho}{g_c}\right)v \sin \phi \qquad\qquad 3.182$$

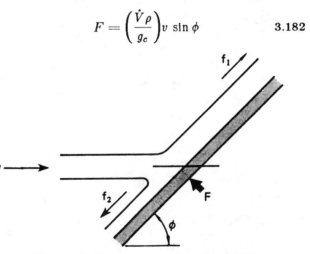

Figure 3.33 Open Jet on Inclined Plate

10 LIFT AND DRAG

Lift and drag both are forces exerted on an object as it passes through a fluid. For example, lift on the wing of an airplane forces the plane upward, and drag tries to slow it down. Lift and drag are the components of the resultant force on an object, as shown in figure 3.34.

The amounts of lift and drag on an object depend on the shape of the object. *Coefficients of lift* and *drag* are used to measure the effectiveness of the object in producing lift and drag. Lift and drag may be calculated from equations 3.183 and 3.184.

$$L = \frac{C_L A \rho v^2}{2 g_c} \qquad \text{3.183}$$

$$D = \frac{C_D A \rho v^2}{2 g_c} \qquad \text{3.184}$$

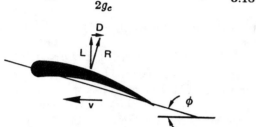

Figure 3.34 Lift and Drag on an Airfoil

A, in equation 3.183, is the object's area projected onto a plane parallel to the direction of motion. In equation 3.184, A is the area projected onto a plane normal to the direction of motion.

Values of C_L for various airfoil sections have been correlated with N_{Re}. No simple relationship can be given for airfoils in general. However, the theoretical relationship for a thin flat plate inclined at an angle ϕ is

$$C_L = 2\pi \sin \phi \qquad \text{3.185}$$

The drag coefficient for a sphere moving with N_{Re} less than .4 is predicted by *Stokes' law*, equation 3.186. The same equation may be used for circular disks. Values of C_D for other shapes are given in table 3.12.

$$C_D = \frac{24}{N_{Re}} \qquad \text{3.186}$$

Table 3.12
Approximate Drag Coefficients

(Do not interpolate between $N_{Re} = EE5$ and $N_{Re} = EE6$)

Reynolds Number, N_{Re}

Body Shape, and (Characteristic Dimension)	EE0	EE1	EE2	EE3	EE4	EE5	EE6	fully turbulent EE6-EE7
sphere (diameter)	(a)	4	1.0	.45	.40	.55	.25	.2
flat disk (diameter)	(a)	4	1.5	1.9(b)	1.1	1.1	1.1	1.1
flat plate, normal to flow, (short side)								
length/breadth= 1				(b)	1.16	1.16	1.16	1.16
4				(b)	1.17	1.17	1.17	1.17
8				(b)	1.23	1.23	1.23	1.23
12.5				(b)	1.34	1.34	1.34	1.34
20				(b)	1.50	1.50	1.50	1.50
25				(b)	1.57	1.57	1.57	1.57
50				(b)	1.76	1.76	1.76	1.76
∞				(b)	2.0	2.0	2.0	2.0
circular cylinder, axis normal to flow (diameter)								
length/diameter= 1				.6	.6	.6		.35
5				.7	.9	.9		
20				.9	.9	.9		
∞	10	2.5	1.3	.9	1.1	1.4	.37	.33
circular cylinder, axis parallel to flow (diameter)								
length/diameter= 1				.91	.91	.91		
2				.85	.85	.85		
4				.87	.87	.87		
7				.99	.99	.99		

Note a: Use Stokes' law, equation 3.186

Note b: Becomes fully turbulent at $N_{Re} = 3$ EE3

11 SIMILARITY

Similarity between a model (subscript *m*) and a full-sized object (subscript *t*) implies that the model can be used to predict the performance of the full-sized object. Such a model is said to be *mechanically similar* to the full-sized object.

Complete mechanical similarity requires geometric and dynamic similarity.[7] *Geometric similarity* means that the model is true to scale in length, area, and volume. *Dynamic similarity* means that the ratios of all types of forces are equal. These forces result from inertia, gravity, viscosity, elasticity (fluid compressibility), surface tension, and pressure.

The *model scale* or *length ratio* is

$$L_r = \frac{\text{size of model}}{\text{full size}} \qquad 3.187$$

The area and volume ratios are based on the length ratio.

$$\frac{A_m}{A_t} = (L_r)^2 \qquad 3.188$$

$$\frac{V_m}{V_t} = (L_r)^3 \qquad 3.189$$

The number of possible ratios of forces is large. Fortunately, some force ratios may be ignored because the forces are negligible or self-canceling. Three important cases where the analysis can be simplified are dominant viscous and inertial forces, dominant inertial and gravitational forces, and dominant surface tension.

A. VISCOUS AND INERTIAL FORCES DOMINATE

Consider the testing of a completely submerged object such as a submarine. Surface tension effects are negligible. The fluid is assumed incompressible for low velocities. Gravity does not change the path of the fluid particles significantly during the time the submarine is near.

Only inertial, viscous, and pressure forces are significant. Being the only significant ones, these three forces are in equilibrium. Since they are in equilibrium, knowing any two will define the third completely. Since it is not an independent force, pressure is omitted from the similarity analysis.

The ratio of the inertial forces to the viscous forces is the Reynolds number. Setting the model and full-size Reynolds numbers equal will ensure similarity. That is,

$$(N_{Re})_m = (N_{Re})_t \qquad 3.190$$

[7]Complete mechanical similarity also requires kinematic and thermal similarity, which are not discussed in this book.

This approach works for problems involving fans, pumps, turbines, drainage through holes in tanks, closed-pipe flow with no free surfaces (in the turbulent region with the same relative roughness), and for completely submerged objects such as torpedoes, airfoils, and submarines. It is assumed that the drag coefficients are the same for the model and for the full-size object.

Example 3.27

A 1/30th size model is tested in a wind tunnel at 120 mph. The wind tunnel conditions are 50 psia and 100°F. What would be the equivalent speed of a prototype traveling at 14.0 psia, 40°F still air?

Start by setting the Reynolds number of the model and its prototype.

$$\frac{v_m L_m}{\nu_m} = \frac{v_p L_p}{\nu_p}$$

$$v_p = v_m \left(\frac{L_m}{L_p}\right)\left(\frac{\nu_p}{\nu_m}\right) = (120)\left(\frac{1}{30}\right)\left(\frac{\nu_p}{\nu_m}\right)$$

Air viscosity terms must be evaluated at the respective temperatures and pressures. As tables of viscosities are not readily available, the viscosities must be calculated.

Absolute viscosity essentially is independent of pressure. In Appendix B, the absolute viscosity of air is

$$\mu_p @ 40° = 3.62 \, EE - 7$$
$$\mu_m @ 100°F = 3.96 \, EE - 7$$

The density of air at the two conditions is

$$\rho_p = \frac{(14.0)(144)}{(53.3)(500)} = 0.0756 \, \text{lbm/ft}^3$$

$$\rho_m = \frac{(50)(144)}{(53.3)(560)} = 0.2412$$

The kinematic viscosity can be calculated from the absolute viscosity. (The g_c terms are omitted as they ultimately cancel out.)

$$\nu = \frac{\mu g_c}{\rho}$$

$$\nu_p = \frac{3.62 \, EE - 7}{0.0756} = 4.79 \, EE - 6$$

$$\nu_m = \frac{3.96 \, EE - 7}{0.2412} = 1.64 \, EE - 6$$

Then, the prototype velocity is

$$v_p = (120)\left(\frac{1}{30}\right)\left(\frac{4.79}{1.64}\right) = 11.7 \, \text{mph}$$

B. INERTIAL AND GRAVITATIONAL FORCES DOMINATE

Elasticity and surface tension can be neglected in the analysis of large surface vessels. This leaves pressure, inertia, viscosity, and gravity. Pressure, again, is omitted as being dependent.

There are only two possible combinations of the remaining three forces. The ratio of inertial and viscous forces is recognized again as the Reynolds number. The ratio of the inertial forces to the gravitational forces is known as the *Froude number*.

$$N_{Fr} = \frac{v^2}{Lg} \qquad 3.191$$

Similarity is ensured when equations 3.192 and 3.193 are satisfied.

$$(N_{Re})_m = (N_{Re})_t \qquad 3.192$$
$$(N_{Fr})_m = (N_{Fr})_t \qquad 3.193$$

As an alternative, equations 3.191 and 3.62 can be solved simultaneously. This results in the following requirement for complete similarity.

$$\frac{\nu_m}{\nu_t} = \left(\frac{L_m}{L_t}\right)^{3/2} = (L_r)^{3/2} \qquad 3.194$$

Sometimes it is not possible to satisfy equation 3.193 or 3.194. This occurs when a model viscosity that is not available is called for. If only equation 3.192 is satisfied, the model is said to be *partially similar*.

This analysis is valid for surface ships, seaplane hulls, and open channels with varying surface levels such as weirs and spillways.

C. SURFACE TENSION DOMINATES

Problems involving waves, droplets, bubbles, and air entrainment can be solved by setting the *Weber numbers* equal.

$$N_W = \frac{v^2 L \rho}{T} \qquad 3.195$$
$$(N_W)_m = (N_W)_t \qquad 3.196$$

12 EFFECTS OF NON-STANDARD GRAVITY

Most of the equations in part 4 are based on Bernoulli's equation. This equation can be modified to allow for non-standard gravities. Assuming an incompressible fluid, Bernoulli's equation becomes

$$\frac{p_1}{\rho} + \frac{v_1^2}{2g_c} + \frac{gz_1}{g_c} = \frac{p_2}{\rho} + \frac{v_2^2}{2g_c} + \frac{gz_2}{g_c} \qquad 3.197$$

13 CHOICE OF PIPING MATERIALS

Steel and copper are commonly used in piping systems. Each material is available in several configurations. For example, steel can be uncoated or galvanized. Copper tubing can be hard or soft. Table 3.13 can be used to select an appropriate pipe material.

Steel pipe is specified by its nominal size and schedule. In the past, steel pipe was designated as *standard* (S), *extra-strong* (X), and *double extra-strong* (XX). However, these designations have been replaced by a numerical rating. For example, schedule 40 now corresponds to a standard wall steel pipe in most cases.

The approximate schedule required can be found from equation 3.198.

$$\text{schedule} = \frac{(1000)(p)}{SE} \qquad 3.198$$

p is the operating pressure in psig; S is the allowable material stress in psi; E is the joint efficiency. A value of 6500 psi can be used for the product SE with low carbon steel in butt-welded lines and temperatures less than 650°F.

When copper is used as a pipe material, there is a potential for confusion, as there are two different sets of dimensions for copper pipe. Copper pipe in the K, L, and M categories is available in both annealed rolls ("tubing") and hardened straight lengths. Dimensions for such copper pipe, commonly referred to as *copper water tubing*, are given in Appendix G.

Type DWV copper drainage tube also is available. It is recommended for sanitary drainage installations above ground. The tube walls are thinner than type M, making it lighter and less expensive. It is strictly for non-pressure applications.

Copper and brass can also be formed into pipe with the dimensions given in Appendix H. Since the term "copper pipe" is ambiguous, the application must be used to determine the correct dimensions. Type L tubing in straight lengths is used principally in domestic and commercial plumbing because of its cost and the availability of soldered fittings. However, brass piping may be used with high-temperature water and corrosive fluids.

14 FLOW OF COMPRESSIBLE FLUIDS AND STEAM

Under certain conditions, Compressible fluids can be handled as incompressible flow. Specifically, a compressible gas, such as air or steam, can be treated as incom-

Table 3.13
Recommended Pipe Materials for Various Services

SERVICE		PIPE
REFRIGERANTS 12, 22, 500 and 502	Suction Line	Hard copper tubing, Type L* Steel pipe, standard wall Lap welded or seamless
	Liquid Line	Hard copper tubing, Type L* Steel pipe, standard wall Lap welded or seamless
	Hot Gas Line	Hard copper tubing, Type L* Steel pipe, standard wall Welded or seamless
CHILLED WATER		Plain or Galvanized steel pipe† Hard copper tubing†
CONDENSER OR MAKE-UP WATER		Galvanized steel pipe† Hard copper tubing†
DRAIN OR CONDENSATE LINES		Galvanized steel pipe† Hard copper tubing†
STEAM OR CONDENSATE		Steel pipe† Hard copper tubing†
HOT WATER		Steel pipe Hard copper tubing†

*Except for sizes 1/4″ and 3/8″ OD where wall thicknesses of 0.30 and 0.32 in. are required. Soft copper refrigeration tubing may be used for sizes 1 3/8″ OD and smaller. Mechanical joints must not be used with soft copper tubing in sizes larger than 7/8″ OD.

†Normally, standard wall steel pipe or Type M hard copper tubing is satisfactory for air conditioning applications. However, the piping material selected should be checked for the design temperature-pressure ratings.

pressible if the pressure drop along the pipe run is not excessive.

If the pressure drop, based on the entrance pressure, is less than 10%, the fluid properties can be evaluated at any known point long the pipe run. If the pressure drop is greater than 10% but less than 40%, use of the mid-point properties will yield reasonably close friction loss calculations. If the pressure drop is greater than 40%, exact compressible gas dynamics equations should be used.

Since the pressure drop is being used to determine if the pressure drop is excessive, several iterations may be necessary to determine the pressures by trial and error.

Appendix A
Properties of Water at Atmospheric Pressure

Temp. °F	(ρ) Density lbm/ft^3	(μ) Absolute Viscosity lbf-sec/ft^2	(ν) Kinematic Viscosity ft^2/sec	(T) Surface Tension lbf/ft	Vapor Pressure Head ft	(E) Bulk Modulus lbf/in^2
32	62.42	3.746 EE−5	1.931 EE−5	0.518 EE−2	0.20	293 EE3
40	62.43	3.229 EE−5	1.664 EE−5	0.514 EE−2	0.28	294 EE3
50	62.41	2.735 EE−5	1.410 EE−5	0.509 EE−2	0.41	305 EE3
60	62.37	2.359 EE−5	1.217 EE−5	0.504 EE−2	0.59	311 EE3
70	62.30	2.050 EE−5	1.059 EE−5	0.500 EE−2	0.84	320 EE3
80	62.22	1.799 EE−5	0.930 EE−5	0.492 EE−2	1.17	322 EE3
90	62.11	1.595 EE−5	0.826 EE−5	0.486 EE−2	1.61	323 EE3
100	62.00	1.424 EE−5	0.739 EE−5	0.480 EE−2	2.19	327 EE3
110	61.86	1.284 EE−5	0.667 EE−5	0.473 EE−2	2.95	331 EE3
120	61.71	1.168 EE−5	0.609 EE−5	0.465 EE−2	3.91	333 EE3
130	61.55	1.069 EE−5	0.558 EE−5	0.460 EE−2	5.13	334 EE3
140	61.38	0.981 EE−5	0.514 EE−5	0.454 EE−2	6.67	330 EE3
150	61.20	0.905 EE−5	0.476 EE−5	0.447 EE−2	8.58	328 EE3
160	61.00	0.838 EE−5	0.442 EE−5	0.441 EE−2	10.95	326 EE3
170	60.80	0.780 EE−5	0.413 EE−5	0.433 EE−2	13.83	322 EE3
180	60.58	0.726 EE−5	0.385 EE−5	0.426 EE−2	17.33	313 EE3
190	60.36	0.678 EE−5	0.362 EE−5	0.419 EE−2	21.55	313 EE3
200	60.12	0.637 EE−5	0.341 EE−5	0.412 EE−2	26.59	308 EE3
212	59.83	0.593 EE−5	0.319 EE−5	0.404 EE−2	33.90	300 EE3

Appendix B
Properties of Air at Atmospheric Pressure

Temp. °F	(ρ) Density lbm/ft^3	(ν) Kinematic Viscosity ft^2/sec	(μ) Absolute Viscosity lbf-sec/ft^2
0	0.0862	12.6 EE−5	3.28 EE−7
20	0.0827	13.6 EE−5	3.50 EE−7
40	0.0794	14.6 EE−5	3.62 EE−7
60	0.0763	15.8 EE−5	3.74 EE−7
68	0.0752	16.0 EE−5	3.75 EE−7
80	0.0735	16.9 EE−5	3.85 EE−7
100	0.0709	18.0 EE−5	3.96 EE−7
120	0.0684	18.9 EE−5	4.07 EE−7
250	0.0559	27.3 EE−5	4.74 EE−7

Appendix C
Viscosity of Water

Temperature (°F)	Absolute Viscosity	Kinematic Viscosity		
	Centipoise	Centistokes	SSU	ft^2/sec
32	1.79	1.79	33.0	0.00001931
50	1.31	1.31	31.6	0.00001410
60	1.12	1.12	31.2	0.00001217
70	0.98	0.98	30.9	0.00001059
80	0.86	0.86	30.6	0.00000930
85	0.81	0.81	30.4	0.00000869
100	0.68	0.69	30.2	0.00000739
120	0.56	0.57	30.0	0.00000609
140	0.47	0.48	29.7	0.00000514
160	0.40	0.41	29.6	0.00000442
180	0.35	0.36	29.5	0.00000385
212	0.28	0.29	29.3	0.00000319

Appendix D
Important Fluid Conversions

Multiply	By	To Obtain
cubic feet	7.4805	gallons
cfs	448.83	gpm
cfs	.64632	MGD
gallons	.1337	cubic feet
gpm	.002228	cfs
inches of mercury	.491	psi
inches of mercury	70.7	psf
inches of mercury	13.60	inches of water
inches of water	5.199	psf
inches of water	.0361	psi
inches of water	.0735	inches of mercury
psi	144	psf
psi	2.308	feet of water
psi	27.7	inches of water
psi	2.037	inches of mercury
psf	.006944	psi

Appendix E
Properties of Liquids

Liquid	Temp, °F	Specific Gravity*	Viscosity centistokes	Viscosity SSU	ft²/sec
Acetone	68	.792	.41		
Alcohol, ethyl	68	.789	1.52	31.7	1.65 EE−5
(C_2H_5OH)	104	.772	1.2	31.5	
Alcohol, methyl	68	.79			
(CH_3OH)	59		.74		
Ammonia	0	.662	.30		
Butane	−50		.52		
	30		.35		
	60	.584			
Castor Oil	68	.96			1110 EE−5
	104	.95	259–325	1200–1500	
	130		98–130	450–600	
Ethylene glycol	60	1.125			
	70		17.8	88.4	
Freon-11	70	1.49	21.1	.21	
Freon-12	70	1.33	21.1	.27	
Fuel oils, #1 to #6	60	.82–.95			
Fuel oil #1	70		2.39–4.28	34–40	
	100		−2.69	32–35	
Fuel oil #2	70		3.0–7.4	36–50	
	100		2.11–4.28	33–40	
Fuel oil #3	70		2.69–5.84	35–45	
	100		2.06–3.97	32.8–39	
Fuel oil #5A	70		7.4–26.4	50–125	
	100		4.91–13.7	42–72	
Fuel oil #5B	70		26.4–	125–	
	100		13.6–67.1	72–310	
Fuel oil #6	122		97.4–660	450–3000	
	160		37.5–172	175–780	
Gasoline (regular)	60	.728			.73 EE−5
	80	.719			.66 EE−5
	100	.710			.60 EE−5
Kerosene	60	.78–.82			
	68		2.17	35	
Jet Fuel	−30		7.9	52	
	60	.82			
Mercury	70	13.6	21.1	.118	
	100	13.6	37.8	.11	
Oils, SAE 5 to 150	60	.88–.94			
SAE-5W	0		1295 max	6000 max	
SAE-10W	0		1295–2590	6000–12000	
SAE-20W	0		2590–10350	12000–48000	
SAE-20	210		5.7–9.6	45–58	
SAE-30	210		9.6–12.9	58–70	
SAE-40	210		12.9–16.8	70–85	
SAE-50	210		16.8–22.7	85–110	
Salt Water (5%)	39	1.037			
	68		1.097	31.1	
Salt Water (25%)	39	1.196			
	60	1.19	2.4	34	

*Measured with respect to 60°F water

Appendix F
Dimensions of Welded and Seamless Steel Pipe

Nominal Diameter	Schedule	Outside Diameter	Wall Thickness	Internal Diameter	Internal Area	Internal Diameter	Internal Area
Inches	Schedule	Inches	Inches	Inches	Sq Inches	Feet	Sq Feet
$\frac{1}{8}$	40 (S)	0.405	0.068	0.269	0.0568	0.0224	0.00039
	80 (X)		0.095	0.215	0.0363	0.0179	0.00025
$\frac{1}{4}$	40 (S)	0.540	0.088	0.364	0.1041	0.0303	0.00072
	80 (X)		0.119	0.302	0.0716	0.0252	0.00050
$\frac{3}{8}$	40 (S)	0.675	0.091	0.493	0.1909	0.0411	0.00133
	80 (X)		0.126	0.423	0.1405	0.0353	0.00098
$\frac{1}{2}$	40 (S)	0.840	0.109	0.622	0.3039	0.0518	0.00211
	80 (X)		0.147	0.546	0.2341	0.0455	0.00163
	160		0.187	0.466	0.1706	0.0388	0.00118
	(XX)		0.294	0.252	0.499	0.0210	0.00035
$\frac{3}{4}$	40 (S)	1.050	0.113	0.824	0.5333	0.0687	0.00370
	80 (X)		0.154	0.742	0.4324	0.0618	0.00300
	160		0.219	0.612	0.2942	0.0510	0.00204
	(XX)		0.308	0.434	0.1479	0.0362	0.00103
1	40 (S)	1.315	0.133	1.049	0.8643	0.0874	0.00600
	80 (X)		0.179	0.957	0.7193	0.0798	0.00500
	160		0.250	0.815	0.5217	0.0679	0.00362
	(XX)		0.358	0.599	0.2818	0.0499	0.00196
$1\frac{1}{4}$	40 (S)	1.660	0.140	1.380	1.496	0.1150	0.01039
	80 (X)		0.191	1.278	1.283	0.1065	0.00890
	160		0.250	1.160	1.057	0.0967	0.00734
	(XX)		0.382	0.896	0.6305	0.0747	0.00438
$1\frac{1}{2}$	40 (S)	1.900	0.145	1.610	2.036	0.1342	0.01414
	80 (X)		0.200	1.500	1.767	0.1250	0.01227
	160		0.281	1.338	1.406	0.1115	0.00976
	(XX)		0.400	1.100	0.9503	0.0917	0.00660
2	40 (S)	2.375	0.154	2.067	3.356	0.1723	0.02330
	80 (X)		0.218	1.939	2.953	0.1616	0.02051
	160		0.344	1.687	2.235	0.1406	0.01552
	(XX)		0.436	1.503	1.774	0.1253	0.01232
$2\frac{1}{2}$	40 (S)	2.875	0.203	2.469	4.788	0.2058	0.03325
	80 (X)		0.276	2.323	4.238	0.1936	0.02943
	160		0.375	2.125	3.547	0.1771	0.02463
	(XX)		0.552	1.771	2.464	0.1476	0.01711
3	40 (S)	3.500	0.216	3.068	7.393	0.2557	0.05134
	80 (X)		0.300	2.900	6.605	0.2417	0.04587
	160		0.438	2.624	5.408	0.2187	0.03755
	(XX)		0.600	2.300	4.155	0.1917	0.02885
$3\frac{1}{2}$	40 (S)	4.000	0.226	3.548	9.887	0.2957	0.06866
	80 (X)		0.318	3.364	8.888	0.2803	0.06172
4	40 (S)	4.500	0.237	4.026	12.73	0.3355	0.08841
	80 (X)		0.337	3.826	11.50	0.3188	0.07984
	120		0.438	3.624	10.32	0.3020	0.07163
	160		0.531	3.438	9.283	0.2865	0.06447
	(XX)		0.674	3.152	7.803	0.2627	0.05419

Nominal Diameter Inches	Schedule	Outside Diameter Inches	Wall Thickness Inches	Internal Diameter Inches	Internal Area Sq Inches	Internal Diameter Feet	Internal Area Sq Feet
5	40 (S)	5.563	0.258	5.047	20.01	0.4206	0.1389
	80 (X)		0.375	4.813	18.19	0.4011	0.1263
	120		0.500	4.563	16.35	0.3803	0.1136
	160		0.625	4.313	14.61	0.3594	0.1015
	(XX)		0.750	4.063	12.97	0.3386	0.09004
6	40 (S)	6.625	0.280	6.065	28.89	0.5054	0.2006
	80 (X)		0.432	5.761	26.07	0.4801	0.1810
	120		0.562	5.501	23.77	0.4584	0.1650
	160		0.719	5.187	21.13	0.4323	0.1467
	(XX)		0.864	4.897	18.83	0.4081	0.1308
8	20	8.625	0.250	8.125	51.85	0.6771	0.3601
	30		0.277	8.071	51.16	0.6726	0.3553
	40 (S)		0.322	7.981	50.03	0.6651	0.3474
	60		0.406	7.813	47.94	0.6511	0.3329
	80 (X)		0.500	7.625	45.66	0.6354	0.3171
	100		0.594	7.437	43.44	0.6198	0.3017
	120		0.719	7.187	40.57	0.5989	0.2817
	140		0.812	7.001	38.50	0.5834	0.2673
	(XX)		0.875	6.875	37.12	0.5729	0.2578
	160		0.906	6.813	36.46	0.5678	0.2532
10	20	10.75	0.250	10.250	82.52	0.85417	0.5730
	30		0.307	10.136	80.69	0.84467	0.5604
	40 (S)		0.365	10.020	78.85	0.83500	0.5476
	60 (X)		0.500	9.750	74.66	0.8125	0.5185
	80		0.594	9.562	71.81	0.7968	0.4987
	100		0.719	9.312	68.11	0.7760	0.4730
	120		0.844	9.062	64.50	0.7552	0.4479
	140 (XX)		1.000	8.750	60.13	0.7292	0.4176
	160		1.125	8.500	56.75	0.7083	0.3941
12	20	12.75	0.250	12.250	117.86	1.0208	0.8185
	30		0.330	12.090	114.80	1.0075	0.7972
	(S)		0.375	12.000	113.10	1.0000	0.7854
	40		0.406	11.938	111.93	0.99483	0.7773
	(X)		0.500	11.750	108.43	0.97917	0.7530
	60		0.562	11.626	106.16	0.96883	0.7372
	80		0.688	11.374	101.61	0.94783	0.7056
	100		0.844	11.062	96.11	0.92183	0.6674
	120 (XX)		1.000	10.750	90.76	0.89583	0.6303
	140		1.125	10.500	86.59	0.87500	0.6013
	160		1.312	10.126	80.53	0.84383	0.5592
14 OD	10	14.00	0.250	13.500	143.14	1.1250	0.9940
	20		0.312	13.376	140.52	1.1147	0.9758
	30 (S)		0.375	13.250	137.89	1.1042	0.9575
	40		0.438	13.124	135.28	1.0937	0.9394
	(X)		0.500	13.000	132.67	1.0833	0.9213
	60		0.594	12.812	128.92	1.0677	0.8953
	80		0.750	12.500	122.72	1.0417	0.8522
	100		0.938	12.124	115.45	1.0104	0.8017
	120		1.094	11.812	109.58	0.98433	0.7610
	140		1.250	11.500	103.87	0.95833	0.7213
	160		1.406	11.188	98.31	0.93233	0.6827

Nominal Diameter Inches	Schedule	Outside Diameter Inches	Wall Thickness Inches	Internal Diameter Inches	Internal Area Sq Inches	Internal Diameter Feet	Internal Area Sq Feet
16 OD	10	16.00	0.250	15.500	188.69	1.2917	1.3104
	20		0.312	15.376	185.69	1.2813	1.2895
	30 (S)		0.375	15.250	182.65	1.2708	1.2684
	40 (X)		0.500	15.000	176.72	1.2500	1.2272
	60		0.656	14.688	169.44	1.2240	1.1767
	80		0.844	14.312	160.88	1.1927	1.1172
	100		1.031	13.938	152.58	1.1615	1.0596
	120		1.219	13.562	144.46	1.1302	1.0032
	140		1.438	13.124	135.28	1.0937	0.9394
	160		1.594	12.812	128.92	1.0677	0.8953
18 OD	10	18.00	0.250	17.500	240.53	1.4583	1.6703
	20		0.312	17.376	237.13	1.4480	1.6467
	(S)		0.375	17.250	233.71	1.4375	1.6230
	30		0.438	17.124	230.00	1.4270	1.5993
	(X)		0.500	17.000	226.98	1.4167	1.5762
	40		0.562	16.876	223.68	1.4063	1.5533
	60		0.750	16.500	213.83	1.3750	1.4849
	80		0.938	16.124	204.19	1.3437	1.4180
	100		1.156	15.688	193.30	1.3073	1.3423
	120		1.375	15.250	182.65	1.2708	1.2684
	140		1.562	14.876	173.81	1.2397	1.2070
	160		1.781	14.438	163.72	1.2032	1.1370
20 OD	10	20.00	0.250	19.500	298.65	1.6250	2.0739
	20 (S)		0.375	19.250	291.04	1.6042	2.0211
	30 (X)		0.500	19.000	283.53	1.5833	1.9689
	40		0.594	18.812	277.95	1.5677	1.9302
	60		0.812	18.376	265.21	1.5313	1.8417
	80		1.031	17.938	252.72	1.4948	1.7550
	100		1.281	17.438	238.83	1.4532	1.6585
	120		1.500	17.000	226.98	1.4167	1.5762
	140		1.750	16.500	213.83	1.3750	1.4849
	160		1.969	16.062	202.62	1.3385	1.4071
24 OD	10	24.00	0.250	23.500	433.74	1.9583	3.0121
	20 (S)		0.375	23.250	424.56	1.9375	2.9483
	(X)		0.500	23.000	415.48	1.9167	2.8852
	30		0.562	22.876	411.01	1.9063	2.8542
	40		0.688	22.624	402.00	1.8853	2.7917
	60		0.969	22.062	382.28	1.8385	2.6547
	80		1.219	21.562	365.15	1.7802	2.5358
	100		1.531	20.938	344.32	1.7448	2.3911
	120		1.812	20.376	326.92	1.6980	2.2645
	140		2.062	19.876	310.28	1.6563	2.1547
	160		2.344	19.312	292.92	1.6093	2.0342

Nominal Diameter		Outside Diameter	Wall Thickness	Internal Diameter	Internal Area	Internal Diameter	Internal Area
Inches	Schedule	Inches	Inches	Inches	Sq Inches	Feet	Sq Feet
30 OD	10	30.00	0.312	29.376	677.76	2.4480	4.7067
	(S)		0.375	29.250	671.62	2.4375	4.6640
	20 (X)		0.500	29.000	660.52	2.4167	4.5869
	30		0.625	28.750	649.18	2.3958	4.5082

S=Wall thickness, formerly designated "standard weight"

X=Wall thickness, formerly designated "extra strong"

XX=Wall thickness, formerly designated "double extra strong"

Actual wall thickness may vary slightly.

Extracted from American Standard Wrought Steel and Wrought Iron Pipe (ASA B36, 10—1959),
  The American Society of Mechanical Engineers.

APPENDIX G
Dimensions of Copper Water Tubing

CLASSIFICATION	NOM. TUBE SIZE (in.)	OUTSIDE DIAM (in.)	WALL THICK-NESS (in.)	INSIDE DIAM (in.)	TRANS-VERSE AREA (sq. in.)	SAFE WORKING PRESSURE (psi)
HARD	1/4	3/8	.025	.325	.083	1000
	3/8	1/2	.025	.450	.159	1000
	1/2	5/8	.028	.569	.254	890
	3/4	7/8	.032	.811	.516	710
	1	1 1/8	.035	1.055	.874	600
	1 1/4	1 3/8	.042	1.291	1.309	590
Type	1 1/2	1 5/8	.049	1.527	1.831	580
"M"	2	2 1/8	.058	2.009	3.17	520
250 psi	2 1/2	2 5/8	.065	2.495	4.89	470
Working	3	3 1/8	.072	2.981	6.98	440
Pressure	3 1/2	3 5/8	.083	3.459	9.40	430
	4	4 1/8	.095	3.935	12.16	430
	5	5 1/8	.109	4.907	18.91	400
	6	6 1/8	.122	5.881	27.16	375
	8	8 1/8	.170	7.785	47.6	375
HARD	3/8	1/2	.035	.430	.146	1000
	1/2	5/8	.040	.545	.233	1000
	3/4	7/8	.045	.785	.484	1000
	1	1 1/8	.050	1.025	.825	880
	1 1/4	1 3/8	.055	1.265	1.256	780
Type	1 1/2	1 5/8	.060	1.505	1.78	720
"L"	2	2 1/8	.070	1.985	3.094	640
250 psi	2 1/2	2 5/8	.080	2.465	4.77	580
Working	3	3 1/8	.090	2.945	6.812	550
Pressure	3 1/2	3 5/8	.100	3.425	9.213	530
	4	4 1/8	.110	3.905	11.97	510
	5	5 1/8	.125	4.875	18.67	460
	6	6 1/8	.140	5.845	26.83	430
HARD	1/4	3/8	.032	.311	.076	1000
	3/8	1/2	.049	.402	.127	1000
	1/2	5/8	.049	.527	.218	1000
	3/4	7/8	.065	.745	.436	1000
	1	1 1/8	.065	.995	.778	780
	1 1/4	1 3/8	.065	1.245	1.217	630
Type	1 1/2	1 5/8	.072	1.481	1.722	580
"K"	2	2 1/8	.083	1.959	3.014	510
400 psi	2 1/2	2 5/8	.095	2.435	4.656	470
Working	3	3 1/8	.109	2.907	6.637	450
Pressure	3 1/2	3 5/8	.120	3.385	8.999	430
	4	4 1/8	.134	3.857	11.68	420
	5	5 1/8	.160	4.805	18.13	400
	6	6 1/8	.192	5.741	25.88	400
SOFT	1/4	3/8	.032	.311	.076	1000
	3/8	1/2	.049	.402	.127	1000
	1/2	5/8	.049	.527	.218	1000
	3/4	7/8	.065	.745	.436	1000
	1	1 1/8	.065	.995	.778	780
	1 1/4	1 3/8	.065	1.245	1.217	630
Type	1 1/2	1 5/8	.072	1.481	1.722	580
"K"	2	2 1/8	.083	1.959	3.014	510
250 psi	2 1/2	2 5/8	.095	2.435	4.656	470
Working	3	3 1/8	.109	2.907	6.637	450
Pressure	3 1/2	2 5/8	.120	3.385	8.999	430
	4	4 1/8	.134	3.857	11.68	420
	5	5 1/8	.160	4.805	18.13	400
	6	6 1/8	.192	5.741	25.88	400

Appendix H
Dimensions of Brass and Copper Tubing

regular

pipe size in.	nominal dimensions in.			cross sectional area of bore sq. in.	lb per ft	
	O.D.	I.D.	wall		red brass	copper
1/8	.405	.281	.062	.062	.253	.259
1/4	.540	.376	.082	.110	.447	.457
3/8	.675	.495	.090	.192	.627	.641
1/2	.840	.626	.107	.307	.934	.955
3/4	1.050	.822	.114	.531	1.270	1.300
1	1.315	1.063	.126	.887	1.780	1.820
1 1/4	1.660	1.368	.146	1.470	2.630	2.690
1 1/2	1.900	1.600	.150	2.010	3.130	3.200
2	2.375	2.063	.156	3.340	4.120	4.220
2 1/2	2.875	2.501	.187	4.910	5.990	6.120
3	3.500	3.062	.219	7.370	8.560	8.750
3 1/2	4.000	3.500	.250	9.620	11.200	11.400
4	4.500	4.000	.250	12.600	12.700	12.900
5	5.562	5.062	.250	20.100	15.800	16.200
6	6.625	6.125	.250	29.500	19.000	19.400
8	8.625	8.001	.312	50.300	30.900	31.600
10	10.750	10.020	.365	78.800	45.200	46.200
12	12.750	12.000	.375	113.000	55.300	56.500

extra strong

pipe size in.	nominal dimensions in.			cross sectional area of bore sq. in.	lb per ft	
	O.D.	I.D.	wall		red brass	copper
1/8	.405	.205	.100	.033	.363	.371
1/4	.540	.294	.123	.068	.611	.625
3/8	.675	.421	.127	.139	.829	.847
1/2	.840	.542	.149	.231	1.230	1.250
3/4	1.050	.736	.157	.425	1.670	1.710
1	1.315	.951	.182	.710	2.460	2.510
1 1/4	1.660	1.272	.194	1.270	3.390	3.460
1 1/2	1.900	1.494	.203	1.750	4.100	4.190
2	2.375	1.933	.221	2.94	5.670	5.800
2 1/2	2.875	2.315	.280	4.21	8.660	8.850
3	3.500	2.892	.304	6.57	11.600	11.800
3 1/2	4.000	3.358	.321	8.86	14.100	14.400
4	4.500	3.818	.341	11.50	16.900	17.300
5	5.562	4.812	.375	18.20	23.200	23.700
6	6.625	5.751	.437	26.00	32.200	32.900
8	8.625	7.625	.500	45.70	48.400	49.500
10	10.750	9.750	.500	74.70	61.100	62.400

Appendix I
Specific Gravity of Hydrocarbons

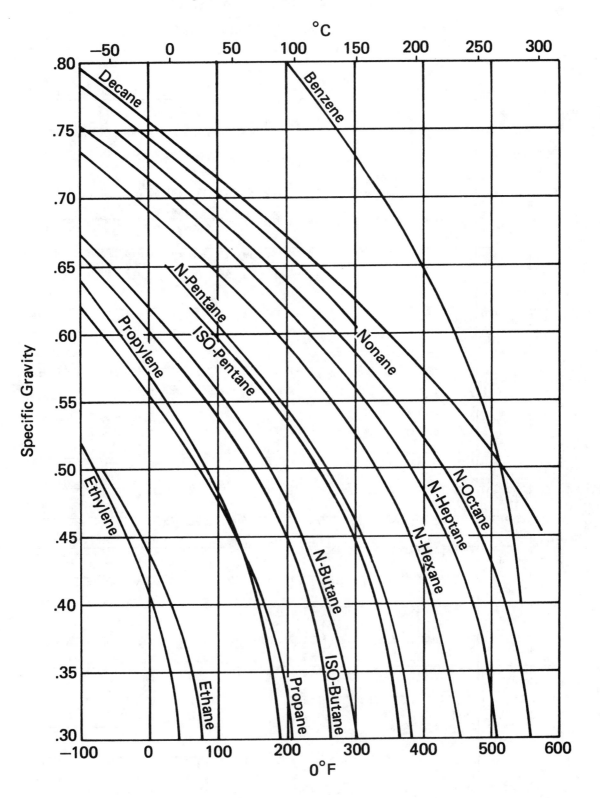

Appendix J: Equivalent Length of Straight Pipe for Various Fittings (feet)*

(turbulent flow only, for any fluid)

c.i. = cast iron

fittings			1/4	3/8	1/2	3/4	1	1 1/4	1 1/2	2	2 1/2	3	4	5	6	8	10	12	14	16	18	20	24
regular 90° ell	screwed	steel	2.3	3.1	3.6	4.4	5.2	6.6	7.4	8.5	9.3	11.0	13.0										
		c.i.										9.0	11.0										
	flanged	steel			0.92	1.2	1.6	2.1	2.4	3.1	3.6	4.4	5.9	7.3	8.9	12.0	14.0	17.0	18.0	21.0	23.0	25.0	30.0
		c.i.										3.6	4.8		7.2	9.8	12.0	15.0	17.0	19.0	22.0	24.0	28.0
long radius 90° ell	screwed	steel	1.5	2.0	2.2	2.3	2.7	3.2	3.4	3.6	3.6	4.0	4.6										
		c.i.										3.3	3.7										
	flanged	steel				1.1	1.3	1.6	2.0	2.3	2.7	2.9	3.4	4.2	5.0	5.7	7.0	8.0	9.0	9.4	10.0	11.0	12.0	14.0
		c.i.										2.8	3.4		4.7	5.7	6.8	7.8	8.6	9.6	11.0	11.0	13.0
regular 45° ell	screwed	steel	0.34	0.52	0.71	0.92	1.3	1.7	2.1	2.7	3.2	4.0	5.5										
		c.i.										3.3	4.5										
	flanged	steel			0.45	0.59	0.81	1.1	1.3	1.7	2.0	2.6	3.5	4.5	5.6	7.7	9.0	11.0	13.0	15.0	16.0	18.0	22.0
		c.i.										2.1	2.9		4.5	6.3	8.1	9.7	12.0	13.0	15.0	17.0	20.0
tee-line flow	screwed	steel	0.79	1.2	1.7	2.4	3.2	4.6	5.6	7.7	9.3	12.0	17.0										
		c.i.										9.9	14.0										
	flanged	steel			0.69	0.82	1.0	1.3	1.5	1.8	1.9	2.2	2.8	3.3	3.8	4.7	5.2	6.0	6.4	7.2	7.6	8.2	9.6
		c.i.										1.9	2.2		3.1	3.9	4.6	5.2	5.9	6.5	7.2	7.7	8.8
tee-branch flow	screwed	steel	2.4	3.5	4.2	5.3	6.6	8.7	9.9	12.0	13.0	17.0	21.0										
		c.i.										14.0	17.0										
	flanged	steel			2.0	2.6	3.3	4.4	5.2	6.6	7.5	9.4	12.0	15.0	18.0	24.0	30.0	34.0	37.0	43.0	47.0	52.0	62.0
		c.i.										7.7	10.0		15.0	20.0	25.0	30.0	35.0	39.0	44.0	49.0	57.0
180° return bend	reg. screwed	steel	2.3	3.1	3.6	4.4	5.2	6.6	7.4	8.5	9.3	11.0	13.0										
		c.i.										9.0	11.0										
	reg. flanged	steel			0.92	1.2	1.6	2.1	2.4	3.1	3.6	4.4	5.9	7.3	8.9	12.0	14.0	17.0	18.0	21.0	23.0	25.0	30.0
		c.i.										3.6	4.8		7.2	9.8	12.0	15.0	17.0	19.0	22.0	24.0	28.0
	long rad. flanged	steel				1.1	1.3	1.6	2.0	2.3	2.7	2.9	3.4	4.2	5.0	5.7	7.0	8.0	9.0	9.4	10.0	11.0	12.0	14.0
		c.i.										2.8	3.4		4.7	5.7	6.8	7.8	8.6	9.6	11.0	11.0	13.0
globe valve	screwed	steel	21.0	22.0	22.0	24.0	29.0	37.0	42.0	54.0	62.0	79.0	110.0										
		c.i.										65.0	86.0										
	flanged	steel			38.0	40.0	45.0	54.0	59.0	70.0	77.0	94.0	120.0	150.0	190.0	260.0	310.0	390.0						
		c.i.										77.0	99.0		150.0	210.0	270.0	330.0						
gate valve	screwed	steel	0.32	0.45	0.56	0.67	0.84	1.1	1.2	1.5	1.7	1.9	2.5										
		c.i.										1.6	2.0										
	flanged	steel								2.6	2.7	2.8	2.9	3.1	3.2	3.2	3.2	3.2	3.2	3.2	3.2	3.2	3.2
		c.i.										2.3	2.4		2.6	2.7	2.8	2.9	2.9	3.0	3.0	3.0	3.0
angle valve	screwed	steel	12.8	15.0	15.0	15.0	17.0	18.0	18.0	18.0	18.0	18.0	18.0										
		c.i.										15.0	15.0										
	flanged	steel			15.0	15.0	17.0	18.0	18.0	21.0	22.0	28.0	38.0	50.0	63.0	90.0	120.0	140.0	160.0	190.0	210.0	240.0	300.0
		c.i.										23.0	31.0		52.0	74.0	98.0	120.0	150.0	170.0	200.0	230.0	280.0
swing check valve	screwed	steel	7.2	7.3	8.0	8.8	11.0	13.0	15.0	19.0	22.0	27.0	38.0										
		c.i.										22.0	31.0										
	flanged	steel				3.8	5.3	7.2	10.0	12.0	17.0	21.0	27.0	38.0	50.0	63.0	90.0	120.0	140.0					
		c.i.										22.0	31.0		52.0	74.0	98.0	120.0						
coupling or union	screwed	steel	0.14	0.18	0.21	0.24	0.29	0.36	0.39	0.45	0.47	0.53	0.65										
		c.i.										0.44	0.52										
bell mouth inlet		steel	0.04	0.07	0.10	0.13	0.18	0.26	0.31	0.43	0.52	0.67	0.95	1.3	1.6	2.3	2.9	3.5	4.0	4.7	5.3	6.1	7.6
		c.i.										0.55	0.77		1.3	1.9	2.4	3.0	3.6	4.3	5.0	5.7	7.0
square mouth inlet		steel	0.44	0.68	0.96	1.3	1.8	2.6	3.1	4.3	5.2	6.7	9.5	13.0	16.0	23.0	29.0	35.0	40.0	47.0	53.0	61.0	76.0
		c.i.										5.5	7.7		13.0	19.0	24.0	30.0	36.0	43.0	50.0	57.0	70.0
re-entrant pipe		steel	0.88	1.4	1.9	2.6	3.6	5.1	6.2	8.5	10.0	13.0	19.0	25.0	32.0	45.0	58.0	70.0	80.0	95.0	110.0	120.0	150.0
		c.i.										11.0	15.0		26.0	37.0	49.0	61.0	73.0	86.0	100.0	110.0	140.0

*Source: The Hydraulic Institute

PRACTICE PROBLEMS: FLUIDS

Warm-ups (All water is 60°F unless stated otherwise.)

1. Calculate the kinematic viscosity of air at 80°F and 70 psia.

2. What is the absolute pressure if a gage reads 8.7 psi vacuum?

3. A mercury manometer is used in a water line and exhibits a 7″ level differential. What is the pressure differential across the taps?

4. A 10″ composition pipe is compressed by a tree root until its inside height is only 7.2 inches. What is its approximate hydraulic radius when flowing half full?

5. What is the Reynolds number of 70°F water flowing at 1.5 cfs through 1200 feet of 6″ diameter new schedule 40 steel pipe?

6. What is the friction loss in problem 5?

7. Water is carried in a pipe which changes gradually from 6″ at A to 18″ at B. B is 15 feet higher than A. Five cfs are flowing, and the respective pressures at A and B are 10 psia and 7 psia. What is the direction of flow?

8. A fire hose discharges at 50 fps and 45° from the horizontal. Neglecting air friction, what is the jet's maximum range?

9. A horizontal turbine reduces 100 cfs of water from 30 psig to 5 psi vacuum. Ignoring friction, what is the horsepower generated?

10. A pump discharges water at 12 fps through a 6″ line. The inlet is a section of 8″ line. The pump suction is at 5 psig below atmospheric. If the pump is 20 horsepower and is 70% efficient, what is the maximum height above the pump inlet that 14.7 psia water is available? Friction losses are 10 feet of water.

Concentrates

1. A blimp contains 10,000 pounds of hydrogen at 56°F and 30.2″ Hg. What is its lift if the hydrogen and air are in thermal and pressure equilibrium?

2. 500 gpm of 100°F water flow through 300 feet of 6″ schedule 40 pipe containing two 6″ elbows (flanged, steel), two full-open gate valves, a full-open 90° angle valve, and a back-flow limiter. If the discharge is 20 feet above the suction, find the pressure difference across the 300 feet.

3. Repeat problem 2 with 70°F air flowing at 60 fps.

4. Brine with a specific gravity of 1.2 is pumped at the rate of 2000 gpm through an inlet of 12″ and out an 8″ discharge at the same level. The inlet gage reads 6″ of mercury below atmospheric. The discharge gage reads 20 psig and is located 4 feet above the centerline of the

pump outlet. If the pump efficiency is 85%, what is the input power?

5. A 5-layer coil with mean radius of 1 foot is constructed of 2″ copper tubing. 200°F water flows at 10 fps. What is the friction loss?

6. A water jet 2″ in diameter has an absolute velocity of 40 fps as it strikes a curved blade moving away with velocity of 15 fps. The blade makes an angle of 60° with the horizontal. What is the reaction on the blade?

7. A 1/20 airplane model is tested in a wind tunnel at full velocity and temperature. What is the pressure?

8. 68°F castor oil is to be pumped at 1000 rpm in a pump. A model twice the actual pump size uses 68°F air. What should be the model speed for similarity?

9. What is the discharge of benzene through an $8''/3\frac{1}{2}''$ venturi ($C_d = .99$) if a mercury manometer measures 4″?

10. A sharp-edged ISA orifice is used in a water line of 12″ diameter. What is the smallest orifice if a 25 foot head loss across the orifice is not to be exceeded when 10 cfs of water flow? Figure 3.25 is applicable.

Timed (1 hour allowed for each)

1. Three reservoirs (A,B,C) are interconnected with a common junction at elevation 25 feet above some arbitrary reference point. The water surface levels for reservoirs A, B, and C are at elevations of 50, 40, and 22 feet, respectively. The pipe from reservoir A to the junction is 800 feet of 3″ pipe. The pipe from reservoir B to the junction is 500 feet of 10″ pipe. The pipe from reservoir C to the junction is 1000 feet of 4″ pipe. Determine the flow velocity and direction in each pipe. Assume $f = 0.02$ and disregard minor losses and velocity heads.

2. A cast iron pipe has an outside diameter of 24″ and a wall thickness of 0.75″. Water is flowing at 6 fps. (a) If a valve is closed instantaneously, what pressure will be created? (b) If the pipe is 500 feet long, over what length of time must the valve be closed to create a pressure equivalent to instantaneous closure?

3. An automobile manufacturer is introducing a new model with the following characteristics:

frontal area:	28 ft^2
weight:	3300 pounds
drag coefficient:	.42
rolling resistance:	1% of car weight
thermal engine efficiency:	28%
fuel heating value:	115,000 BTU/gal

(a) What is the fuel consumption at 55 mph based on only the drag and rolling resistance? (b) What is the percent increase in fuel consumption at 65 mph?

$P_{SIG} = 20 + 14.7$

RESERVED FOR FUTURE USE

4 HYDRAULIC MACHINES

Nomenclature

bhp	brake horsepower	hp
C_v	coefficient of velocity	–
d	diameter	inches
D	diameter	ft
ehp	electrical horsepower	hp
E	energy	ft-lbf/lbm
f	Darcy friction factor, or frequency	–, hertz
fhp	friction horsepower	hp
g	acceleration due to gravity (32.2)	ft/sec^2
g_c	gravitational constant (32.2)	$\dfrac{\text{lbm-ft}}{\text{lbf-sec}^2}$
h	head	ft
H	head added by pump	ft
L	pipe length	ft
m	mass	lbm
n	rotational speed	rpm
n_s	specific speed	rpm
n_{ss}	suction specific speed	rpm
NPSHA	net positive suction head available	ft
NPSHR	net positive suction head required	ft
p	pressure	psf
Q	flow quantity	gpm
s	slip	–
S.G.	specific gravity	–
T	temperature	°F
v	velocity	ft/sec
V	volume	ft^3
w	weight	lbf
whp	hydraulic ("water") horsepower	hp
z	height above datum	ft

Symbols

η	efficiency	–
ϵ	specific pipe roughness	ft
ρ	density	lbm/ft^3
β	blade angle	°
σ	cavitation coefficient	–

Subscripts

a	atmospheric
A	added by pump
d	discharge
f	friction
i	intake
j	jet
m	motor
n	nozzle
p	pump or pressure
s	suction
sd	static discharge
th	theoretical
ts	total static
T	turbine
v	velocity
vp	vapor pressure

1 INTRODUCTION

Pumps and turbines are the two types of hydraulic machines discussed in this chapter. Pumps convert mechanical energy into fluid energy. Turbines convert fluid energy into mechanical energy.

2 TYPES OF PUMPS

Pumps can be classified in several ways. The clearest categorization is based on the method by which pumping energy is transferred to the fluid. This approach separates pumps into *kinetic pumps* and *positive displacement pumps*.

The two most common forms of positive displacement pumps are *reciprocating action pumps* (which use pistons, plungers, diaphragms, or bellows) and *rotary action pumps* (using vanes, screws, lobes, or progressing cavities). Such pumps discharge a fixed volume for each stroke or revolution. Energy is added intermittently to the fluid flow.

Kinetic pumps rely on a transformation of kinetic energy to static pressure. Jet and ejector pumps fall into this category, but *centrifugal pumps* are the primary

Table 4.1
Characteristics of Kinetic and Displacement Pumps

Characteristic	Displacement	Kinetic
flow rate	low	high
pressure rise per stage	high	low
constant variable over operating range	flow rate	pressure rise
self-priming	yes	no
outlet stream	pulsing	steady
works with high-viscosity fluids	yes	no

examples of kinetic pumps. Pumps covered in this chapter are assumed to be centrifugal pumps.

Liquid flowing into the *suction side* (the *inlet*) of a centrifugal pump is captured by the *impeller* and thrown to the outside of the pump casing. Within the casing, the velocity imparted to the fluid by the impeller is converted into pressure energy.

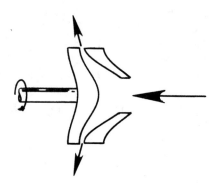

Figure 4.2 Centrifugal (Radial) Flow Pump

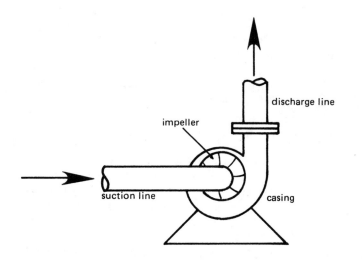

Figure 4.1 A Centrifugal Pump

3 TYPES OF CENTRIFUGAL PUMPS

Centrifugal pumps can be classified into three general categories according to the way the impeller imparts energy to the fluid. Each of these categories has its own range of specific speeds and appropriate applications.

Radial flow impellers impart energy primarily by centrifugal force. Liquid enters the impeller at the hub and flows radially to the outside of the casing. Single suction impellers have a specific speed less than 5000. Double suction impellers have a specific speed less than 6000.

Mixed flow impellers impart energy partially by centrifugal force and partially by axial force, since the vanes act partially as an axial compressor. Liquid enters the impeller at the hub and flows both radially and axially to discharge. Specific speeds of mixed flow pumps range from 4200 to 9000.

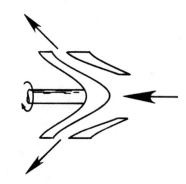

Figure 4.3 Mixed Flow Pump

Axial flow impellers impart energy to the fluid by acting as axial flow compressors. Fluid enters and exits along the axis of rotation. Specific speed is greater than 9000.

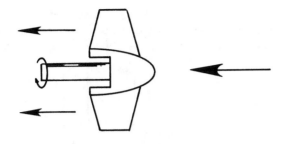

Figure 4.4 Axial Flow Pump

Radial flow and mixed flow centrifugal pumps can be designed for either single or double suction operation. In a *single suction pump*, fluid enters only one side of the impeller. In a *double suction pump*, fluid enters both sides of the impeller. Thus, for an impeller with a given specific speed, a greater flow rate can be expected from a double suction pump. In addition, a double suction pump has a lower NPSHR for a given flow than a single suction pump.

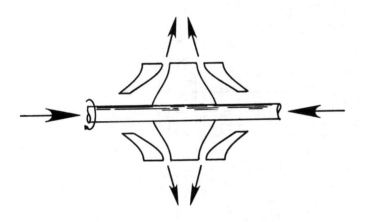

Figure 4.5 Radial Flow Pump (Double Suction)

A *multiple stage pump* consists of two or more impellers within a single casing. The discharge of one stage is the input of the next stage. In this manner, higher heads are achieved than would be possible with a single impeller.

4 PUMP AND HEAD TERMINOLOGY

Like most specialized subjects, the centrifugal pump field has developed its own terminology. It is essential that this terminology be understood, since its interpretation will often affect an installation's physical configuration.

All of the terms which follow are *head terms*, and as such, have units of feet. Of course, any head term can be converted to pressure by using equation 4.1.

$$p = \rho h \qquad 4.1$$

Friction head (h_f): The head required to overcome resistance to flow in the pipe, fittings, valves, entrances, and exits.

$$h_f = \frac{fL_e v^2}{2Dg_c} \qquad 4.2$$

Velocity head (h_v): The head of a fluid as a result of its kinetic energy.

$$h_v = \frac{v^2}{2g_c} \qquad 4.3$$

Atmospheric head (h_a): Atmospheric pressure converted to feet of fluid being pumped.

$$h_a = \frac{p_a}{\rho} \qquad 4.4$$

Pressure head (h_p): Pressure converted to feet of fluid being pumped.

$$h_p = \frac{p}{\rho} \qquad 4.5$$

Vapor pressure head (h_{vp}): Fluid vapor pressure converted to feet of fluid being pumped. Steam tables can be used to evaluate the vapor pressure of water. Figure 4.9 can be used with hydrocarbons.

$$h_{vp} = \frac{p_{vp}}{\rho} \qquad 4.6$$

Static suction head (h_s): The vertical distance in feet above the centerline of the inlet to the free level of the fluid source. If the free level of the fluid source is below the inlet, h_s will be negative. In this case, h_s is known as *static suction lift.*

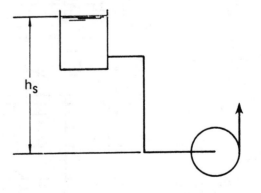

Figure 4.6 Static Suction Head

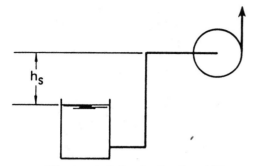

Figure 4.7 Static Suction Lift

Total (dynamic) suction head (H_s): The static suction head minus the friction head in the suction line (i.e., the total energy of the fluid entering the impeller). If h_s is negative (i.e., the free level of the fluid source is blow the inlet), then H_s will be negative. In this case, H_s is known as *total (dynamic) suction lift.*

$$H_s = h_s - h_{f(s)} \qquad 4.7$$

Static discharge head (h_{sd}): The vertical distance in feet above the pump centerline to the free level of the discharge tank or point of free discharge.

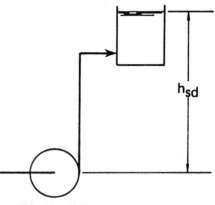

Figure 4.8 Static Discharge Head

Total (dynamic) discharge head (H_d): The static discharge head plus the discharge velocity head

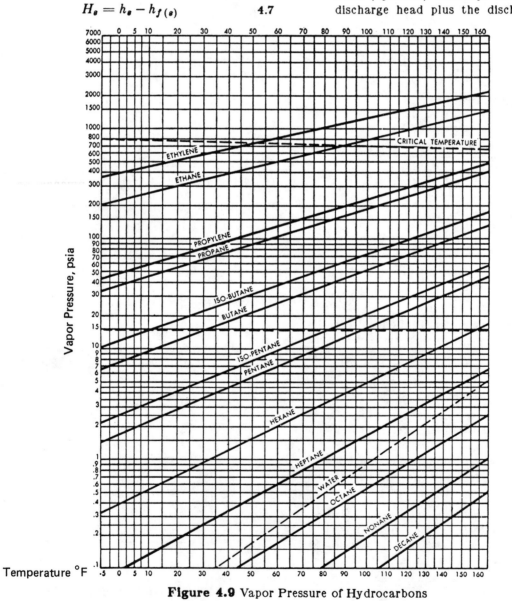

Figure 4.9 Vapor Pressure of Hydrocarbons

plus the friction head in the discharge line (i.e., the total energy of the fluid leaving the pump).

$$H_d = h_{sd} + h_{vd} + h_{f(d)} \qquad 4.8$$

$$= h_{sd} + \frac{v_d^2}{2g_c} + h_{f(d)} \qquad 4.9$$

Total static head (h_{ts}): The vertical distance in feet between the free level of the supply and either the point of free discharge or the free level of the discharge tank.

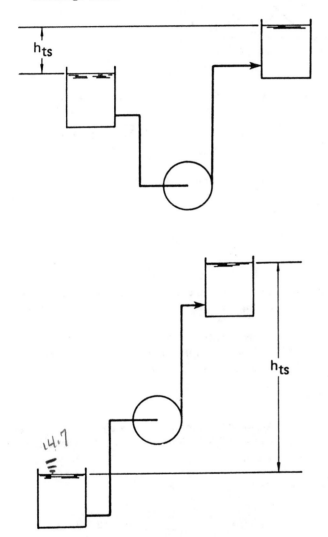

Figure 4.10 Total Static Head

Total (dynamic) head (H): The total discharge head less the total suction head.

$$H = H_d - H_s \qquad 4.10$$

$$= h_{sd} - h_s + \frac{v_d^2}{2g_c} + h_{f(s)} + h_{f(d)} \qquad 4.11$$

5 NET POSITIVE SUCTION HEAD AND CAVITATION

Liquid is not sucked into a pump. A positive head (normally atmospheric pressure) must push the liquid into the impeller (i.e., "flood" the impeller). *Net Positive Suction Head Required* (*NPSHR*) is the minimum fluid energy required at the inlet by the pump for satisfactory operation. NPSHR is usually specified by the pump manufacturer.[1] *Net Positive Suction Head Available* (*NPSHA*) is the actual fluid energy at the inlet.[2]

$$\text{NPSHA} = h_a + h_s - h_{f(s)} - h_{vp} \qquad 4.12$$

$$= h_{p(i)} + h_{v(i)} - h_{vp} \qquad 4.13$$

If NPSHA is less than NPSHR, the fluid will cavitate. *Cavitation* is the vaporization of fluid within the casing or suction line. If the fluid pressure is less than the vapor pressure, pockets of vapor will form. As vapor pockets reach the surface of the impeller, the local high fluid pressure will collapse them, causing noise, vibration, and possible structural damage to the pump.

Cavitation may be caused by any of the following conditions:

1. Discharge heads far below the pump's calibrated head at peak efficiency.

2. Suction lift higher or suction head lower than the manufacturer's recommendation.

3. Speeds higher than the manufacturer's recommendation.

4. Liquid temperatures (thus, vapor pressures) higher than that for which the system was designed.

The following steps can be used to check for cavitation:

step 1: Determine the minimum NPSHR for the given pump. This should be given as part of the pump performance data. NPSHR follows the Q^2 law. If NPSHR is known for one flow rate, it can be determined for another flow rate from equation 4.14.

$$\frac{\text{NPSHR}_2}{\text{NPSHR}_1} = \left(\frac{Q_2}{Q_1}\right)^2 \qquad 4.14$$

[1] If NPSHR is multiplied by the fluid density, it is known as *NIPR*, the *Net Inlet Pressure Required*. Similarly, NPSHA can be converted to *NIPA*.

[2] Equations 4.12 and 4.13 represent two totally different methods, both of which are correct, for calculating NPSHA. Equation 4.12 is based on the conditions at the fluid surface at the top of a tank. There is potential energy (the h_s term) but no kinetic energy. Equation 4.13 is based on the conditions at the immediate entrance to the pump. At that point, some of the potential head has been converted to velocity head.

step 2: Calculate NPSHA from either equation 4.12 or 4.13.

step 3: If NPSHA is greater than NPSHR, cavitation will not occur. A good safety margin is 2–3 feet of fluid. If NPSHA is insufficient, it should be increased or the NPSHR should be decreased. NPSHA can be increased by:

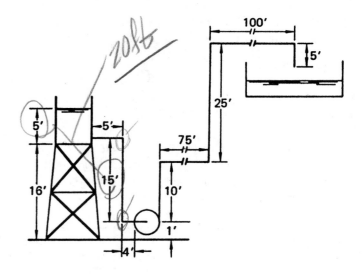

   a. increasing the height of the free fluid level of the supply tank

   b. reducing the distance and minor losses between the supply tank and the pump, or by using a larger pipe size

   c. reducing the temperature of the fluid

   d. pressurizing the supply tank

   e. **reducing the flow rate or velocity to reduce friction in the suction line**

NPSHR can be reduced by:

   a. placing a throttling valve in the discharge line (i.e., this will increase the total head, thereby reducing the capacity of the pump and driving its operating point into a region of lower NPSHR)

   b. using a double suction pump

step 1: Assume 60°F and 14.7 psia. From equation 4.4,

$$h_a = \frac{(14.7)(144)}{62.4} = 33.9 \text{ ft}$$

step 2: For 6″ schedule 40 steel pipe, $D = .505$ ft, $A = .201$ ft^2.

step 3:

$$v = \frac{Q}{A} = \frac{2}{.201} = 9.95 \text{ ft/sec}$$

step 4: The equivalent lengths of the pipe and flanged fittings are:

square entrance loss	1 × 16	=16
90° long radius elbows	2 × 5.7	=11.4
pipe run (5 + 15 + 4)		24
		51.4 ft

Applications which require very high NPSHR, such as boiler feed pumps needing 150 to 250 feet, should use booster pumps in front of the high NPSHR pumps. Such booster pumps are typically single stage, double suction pumps running at low speed. Their NPSHR can be 25 feet or less.

step 5: At 60°, the kinematic viscosity of water is 1.217 EE−5 ft^2/sec. The vapor pressure is .6 feet of water.

step 6: The Reynolds number is

$$N_{Re} = \frac{(.505)(9.95)}{1.217 \text{ EE}-5} = 4.13 \text{ EE5}$$

It is important to note that throttling the input line to a pump and venting or evacuating the receiving tank both work to increase cavitation. Throttling the input line increases the friction head term and decreases NPSHA. Evacuating the receiving tank increases the flow rate, which also increases the friction head term.

step 7: From the Moody friction factor chart, $f = .0165$, so

$$h_f = \frac{(.0165)(51.4)(9.95)^2}{(2)(.505)(32.2)} = 2.6 \text{ ft}$$

Example 4.1

2 cfs of water are pumped from a feed tank mounted on a platform to an open reservoir through 6″ schedule 40 ($\epsilon/D = .000293$) steel pipe. Determine the static suction head, total suction head, and NPSHA.

step 8: From equation 4.12,

NPSHA = 33.9 + 20 − 2.6 − .6 = 50.7 ft

6 PUMPING HYDROCARBONS AND OTHER LIQUIDS

NPSHR is specified by pump manufacturers for use with cold (85°F) water.[3] Minor variations in the water temperature do not change the NPSHR appreciably. Experiments have shown, however, that the NPSHR can be reduced from the cold water values when hydrocarbons are pumped. This reduction apparently is due to the slow vapor release of complex organic liquids. Figure 4.11 gives the percentage correction to be applied to NPSHR values for cold water. Notice that the vapor pressure at the pumping temperature must be known. This can be obtained from figure 4.9.

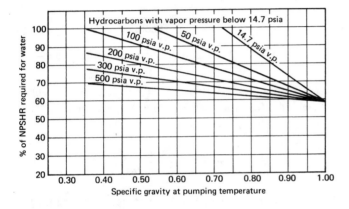

Figure 4.11 Hydrocarbon NPSHR Correction Factor

Head developed by a pump is independent of the liquid being pumped, although the required horsepower is dependent on the fluid specific gravity. Because of this independence, pump performance curves from water tests can be used with other Newtonian fluids (e.g., gasoline, alcohol, and saline solutions).

High viscosity fluids, however, result in decreased head and capacity, as well as in an increase in input horsepower when compared to the pumping of water. Therefore, corrections are required when using water-based pump curves with viscous fluids. The parameters that would be used with water curves are calculated from equations 4.15 through 4.17.

$$H_{\text{water}} = \frac{H_{\text{viscous}}}{C_H} \qquad 4.15$$

$$Q_{\text{water}} = \frac{Q_{\text{viscous}}}{C_Q} \qquad 4.16$$

$$\eta_{\text{water}} = \frac{\eta_{\text{viscous}}}{C_E} \qquad 4.17$$

No exact method for determining the correction factors, other than actual tests of an installation with both fluids, exists. Nevertheless, several sources have

[3]The term "cold *clear* water" is frequently used. This chapter omits the *clear* qualification.

produced correction factor charts based on experiments in limited viscosity and size ranges. One such chart is reproduced as Appendix E of this chapter.

Example 4.2

10°F liquid iso-butane is to be used with a centrifugal pump whose NPSHR is 12 psia with cold water. What NPSHR should be used with the butane?

From figure 4.9, the vapor pressure of 10°F iso-butane is approximately 15 psia. From Appendix I in chapter 3, the specific gravity at 10°F is approximately .58.

From figure 4.11, the intersection of .58 specific gravity and 15 psia is above the horizontal line. Therefore, the NPSHR is a full 12 psia.

7 RECIRCULATION

Cavitation is a high-volume problem. If a pump is forced to operate a flow rate higher than originally intended, one of the results could be cavitation.

Operating a pump at a flow rate much less than it was designed for, on the other hand, can result in *recirculation*. Recirculation of the fluid at both the impeller inlet and the pump outlet is possible. Such recirculation produces characteristics similar to cavitation. Specifically, vibration and noise are produced when the fluid energy is reduced through internal friction and fluid shear.

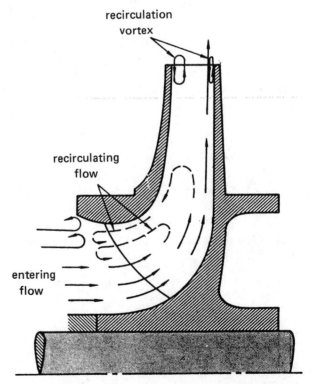

Figure 4.12 Low-Volume Recirculation

8 PUMPING POWER AND EFFICIENCY

The energy (head) added by a pump can be determined by evaluating Bernoulli's equation on either side of the pump. Writing Bernoulli's equation for the discharge and inlet conditions produces equation 4.18.

$$h_A = \frac{p_d}{\rho} - \frac{p_i}{\rho} + \frac{v_d^2}{2g_c} - \frac{v_i^2}{2g_c} + z_d - z_i \qquad 4.18$$

The work performed by a pump is a function of the total head and the mass of the liquid pumped in a given time. Pump output is measured in *hydraulic horsepower, whp*.[4] Relationships for finding the hydraulic horsepower are given in table 4.2.

Table 4.2
Hydraulic Horsepower Equations

ft^3/sec

	Q in gpm	$\dot{m}$ in lbm/sec	$\dot{V}$ in cfs
h_A is added head in feet	$\dfrac{h_A Q (S.G.)}{3956}$	$\dfrac{h_A \dot{m}}{550}$	$\dfrac{h_A \dot{V} (S.G.)}{8.814}$
p is added head in psf	$\dfrac{pQ}{2.468\ EE5}$	$\dfrac{p\dot{m}}{(34320)(S.G.)}$	$\dfrac{p\dot{V}}{550}$

The input horsepower delivered to the pump shaft is the *brake horsepower*.

$$bhp = \frac{whp}{\eta_p} \qquad \text{hydraulic horsepower} \qquad 4.19$$

The difference between hydraulic horsepower and brake horsepower is the power lost within the pump due to mechanical and hydraulic friction. This is referred to as *heat horsepower* or *friction horsepower* and is determined from equation 4.20.

$$fhp = bhp - whp \qquad 4.20$$

Electrical horsepower to the motor is

$$ehp = \frac{bhp}{\eta_m} \qquad 4.21$$

Overall efficiency is the pump efficiency multiplied by the motor efficiency:

$$\eta = (\eta_p)(\eta_m) = \frac{whp}{ehp} \qquad 4.22$$

Ideal pump efficiency is a function of the flow rate and specific speed. Figure 4.13 can be used if both quantities are known.

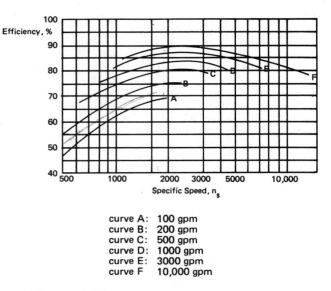

curve A: 100 gpm
curve B: 200 gpm
curve C: 500 gpm
curve D: 1000 gpm
curve E: 3000 gpm
curve F: 10,000 gpm

Figure 4.13 Efficiency versus Specific Speed

Larger horsepower pumps usually are driven by three-phase *induction motors*. The *synchronous speed* of such a motor is the speed of the rotating field.

$$n = \frac{(120)(f)}{\text{no. of poles}} \qquad 4.23$$

The frequency, f, is either 60 Hz (cycles per second) or 50 Hz, depending on the location of the installation.[5] The number of poles can be two or usually some multiple of four, but it must be an even number.

Table 4.3
Synchronous Speeds

Number of Poles	Synchronous Speed
2	3600
4	1800
6	1200
8	900
12	600
18	400
24	300
48	150

Induction motors do not run at their synchronous speeds. Rather, they run at slightly less than synchronous speed. The percentage deviation is known as the *slip*. Slip is seldom greater than 10%, and it is usually much less than that. 4% is a good estimate for evaluation studies.

$$s = \frac{n_{\text{synchronous}} - n_{\text{actual}}}{n_{\text{synchronous}}} \qquad 4.24$$

---

[4]This term also is known as *water horsepower*.

[5]50 Hz is used in Europe.

Of course, special gear or belt drives can be used with various reduction ratios to obtain any required operating speed. For belt drive applications, most motors are of the 1800 rpm variety.

Table 4.4 lists common motor sizes in horsepower, which should be greater than the brake horsepower requirement of the pump. (Do not select on basis of electrical horsepower.) It is always best to specify standard motor sizes when possible.

Table 4.4
Standard Motor Sizes (BHP)

$\frac{1}{8}, \frac{1}{6}, \frac{1}{4}, \frac{1}{3}$

.5, .75, 1, 1.5, 2, 3, 5, 7.5

10, 15, 20, 25, 30, 40, 50, 60

75, 100, 125, 150, 200, 250

Induction motors are usually specified in terms of their KVA (kilo-volt-amp) ratings. KVA is not the same as the motor power in kilowatts, although one can be derived from the other if the motor's power factor is known. Such power factors can range from .8 to .9, depending on the installation and motor size.

$$KVA \ rating = \frac{kilowatt \ power}{power \ factor} \qquad 4.25$$

$$kilowatt \ power = (.7457)ehp \qquad 4.26$$

Example 4.3

Recommend a 6-pole induction motor size for the pump in example 4.1. The friction loss in the discharge line is 12.0 feet. Neglect the electrical motor efficiency.

The Bernoulli equation can be used to find the head added by the pump.

$$0 + 0 + 20 + h_A = 0 + \frac{(9.95)^2}{2(32.2)} + 30 + 2.6 + 12$$

$$h_A = 26.14 \ ft$$

From table 4.2,

$$whp = \frac{(26.14)(2)(1)}{8.814} = 5.93 \ hp$$

The flow rate is

$$\frac{2}{.002228} = 897.7 \ gpm$$

Assuming a motor speed of 1200 rpm, equation 4.27 gives the specific speed

$$n_s = \frac{1200\sqrt{897.7}}{(26.14)^{.75}} = 3110 \ rpm$$

From figure 4.13, a pump efficiency of about 82% can be expected. So, the minimum motor horsepower would be

$$\frac{5.93}{.82} = 7.23$$

From table 4.4, choose a 7.5 hp or larger motor.

9 SPECIFIC SPEED

The capacity or flow rate of a centrifugal pump is governed by the impeller thickness. For a given impeller diameter, the deeper the vanes, the greater the capacity of the pump. For a desired flow rate or a desired discharge head, there will be one optimum impeller design. The impeller that is best for developing a high discharge pressure will have different proportions from an impeller designed to produce a high flow rate. The quantitative index of this optimization is called *specific speed* (n_s).

Specific speed is a function of the a pump's capacity, head, and rotational speed at peak efficiency. For a given pump and impeller configuration, the specific speed remains essentially constant over a range of flow rates and heads. Theoretically, specific speed is the speed in rpm at which a homologous pump would have to turn in order to put out 1 gpm at 1 foot total head. **(For double suction pumps, Q in equation 4.27 is not divided by 2.)**

$$n_s = \frac{n\sqrt{Q}}{(h_A)^{.75}} \qquad 4.27$$

Specific speed is used as a guide to selecting the most efficient pump type. Given a desired flow rate, pipeline geometry, and motor speed, n_s is calculated from equation 4.27. The type of impeller is chosen from table 4.5

Table 4.5
Specific Speed versus Impeller Types

Approximate Range of Specific Speed (rpm)	Impeller Type
500-1000	radial vane
2000-3000	Francis (mixed) vane
4000-7000	mixed flow
9000 and above	axial flow

Highest heads per stage are developed at low specific speeds. However, for best efficiency, a centrifugal pump's specific speed should be greater than 650 at its operating point. At low specific speeds, the impeller diameter is large with high mechanical friction and hydraulic losses. If the specific speed for a given set of conditions drops below 650, a multiple stage pump should be selected.[6]

As the specific speed increases, the ratio of the impeller diameter to the inlet diameter decreases. As this ratio decreases, the pump is capable of developing less head. Best efficiencies are usually obtained from pumps with specific speeds between 1500 and 3000. At specific speeds of 10,000 or higher, the pump is suitable for high flow rates but low discharge heads.

Other uses for specific speed are:

- If the pump and impeller are known, a maximum specific speed can be determined from table 4.5. This maximum specific speed can be translated into maximum values of rpm and flow rate, as well as a minimum value of total head added.

- If specific speed is known, an approximate pump efficiency can be found from figure 4.13.

- Specific speed limits have been established by the Hydraulic Institute.[7] These limits are presented in graphical form at the end of this chapter.

If NPSHR is substituted for total head in the expression for specific speed, a formula for *suction specific speed* results.

Suction specific speed is an index of the suction characteristics of the impeller.

$$n_{ss} = \frac{n\sqrt{Q}}{\text{NPSHR}^{.75}} \qquad 4.28$$

Ideally, n_{ss} should be approximately 7900 for single suction pumps and 11,200 for double suction pumps. (Unlike for specific speed, the value of Q in equation 4.28 should be halved for double suction pumps.) If these ideal values are assumed, equation 4.28 can be solved for approximate values of NPSHR.

$$\text{NPSHR}_{\text{single suction}} \approx (6.36\text{ EE}-6)\,n^{1.33}Q^{.67} \qquad 4.29$$

$$\text{NPSHR}_{\text{double suction}} \approx (3.99\text{ EE}-6)\,n^{1.33}Q^{.67} \qquad 4.30$$

---

[6]Relatively recent advances in *partial emission, forced vortex centrifugal pumps* allow operation down to specific speeds of 150. Such partial emission pumps use radial vanes and a single tangential discharge. Discharge is through a conical diffuser in which the energy conversion occurs. Partial emission pumps have been used for low flow, high head applications, such as petrochemical high-pressure cracking processes.

[7]Hydraulic Institute, 1230 Keith Building, Cleveland, OH 44115.

Example 4.4

A 3600 rpm pump ($Q = 150$ gpm) is used to increase fluid pressure from 35 psi to 220 psi. The pump adds negligible velocity or potential energy. If efficiency is to be maximized, how many stages should be used? Assume that each stage adds its proportionate share of head.

The head added is

$$h_A = \frac{(220-35)(144)}{62.4} = 426.9 \text{ ft}$$

From equation 4.27, the specific speed is

$$n_s = \frac{(3600)\sqrt{150}}{(426.9)^{.75}} = 470$$

From figure 4.13, the approximate efficiency is 45%.

Try a two stage pump. The head is split between the stages.

$$n_s = \frac{(3600)\sqrt{150}}{\left(\frac{426.9}{2}\right)^{.75}} = 790$$

From figure 4.13, the efficiency is approximately 60%.

This process continues until the specific speed reaches 2000, at which time the number of stages will be approximately seven. It is questionable, however, whether the cost of multi-staging is worthwhile in this low-volume application.

Example 4.5

A direct driven pump is to discharge 150 gpm against a 300 foot total head when turning at the fully-loaded speed of 3500 rpm. What type of pump should be selected?

Calculate the specific speed from equation 4.27.

$$n_s = \frac{3500\sqrt{150}}{(300)^{.75}} = 594.7 \text{ rpm}$$

From table 4.5, the pump should be a radial vane type. However, pumps with best efficiencies have n_s greater than 650. To increase the specific speed, the rotational speed can be increased or the total head can be decreased. Since 3600 rpm is the maximum practical speed for induction motors, the better choice would be to divide the total head between two stages (or to use two pumps in a series).

In a two stage system, the specific speed would be:

$$n_s = \frac{3500\sqrt{150}}{(150)^{.75}} = 1000$$

This is satisfactory for a radial vane pump.

10 AFFINITY LAWS–CENTRIFUGAL PUMPS

Most parameters (impeller diameter, speed, and flow rate) determining a pump's performance can vary. If the impeller diameter is held constant and the speed varied, the following ratios are maintained with no change of efficiency:

$$\frac{Q_2}{Q_1} = \frac{n_2}{n_1} \qquad 4.31$$

$$\frac{h_2}{h_1} = \left(\frac{n_2}{n_1}\right)^2 = \left(\frac{Q_2}{Q_1}\right)^2 \qquad 4.32$$

$$\frac{bhp_2}{bhp_1} = \left(\frac{n_2}{n_1}\right)^3 = \left(\frac{Q_2}{Q_1}\right)^3 \qquad 4.33$$

If the speed is held constant and the impeller size varied,

$$\frac{Q_2}{Q_1} = \frac{d_2}{d_1} \qquad 4.35$$

$$\frac{h_2}{h_1} = \left(\frac{d_2}{d_1}\right)^2 \qquad 4.36$$

$$\frac{bhp_2}{bhp_1} = \left(\frac{d_2}{d_1}\right)^3 \qquad 4.37$$

These relationships assume that the efficiencies of the larger and smaller pumps are the same. In reality, larger pumps will be more efficient than smaller pumps. Therefore, extrapolations to much larger or much smaller sizes should be avoided.

Equation 4.34 can be used to predict the efficiency of a larger or smaller pump based on homologous data.

$$\frac{1 - \eta_{\text{smaller}}}{1 - \eta_{\text{larger}}} = \left(\frac{d_{\text{larger}}}{d_{\text{smaller}}}\right)^{.2} \qquad 4.34$$

Example 4.6

A pump delivers 500 gpm against a total head of 200 feet operating at 1770 rpm. Changes have increased the total head to 375 feet. At what rpm should this pump be operated to achieve this new head at the same efficiency?

From equation 4.32,

$$n_2 = 1770\sqrt{\frac{375}{200}} = 2424$$

11 PUMP SIMILARITY

The performance of one pump can be used to predict the performance of a *dynamically similar (homologous)*

pump. This can be done by using equations 4.38 through 4.40.

$$\frac{n_1 d_1}{\sqrt{h_1}} = \frac{n_2 d_2}{\sqrt{h_2}} \qquad 4.38$$

$$\frac{Q_1}{d_1^2\sqrt{h_1}} = \frac{Q_2}{d_2^2\sqrt{h_2}} \qquad 4.39$$

$$\frac{bhp_1}{\rho_1 d_1^2 h_1^{1.5}} = \frac{bhp_2}{\rho_2 d_2^2 h_2^{1.5}} \qquad 4.40$$

$$\frac{Q_1}{n_1 d_1^3} = \frac{Q_2}{n_2 d_2^3} \qquad 4.41$$

$$\frac{bhp_1}{\rho_1 n_1^3 d_1^5} = \frac{bhp_2}{\rho_2 n_2^3 d_2^5} \qquad 4.42$$

$$\frac{n_1\sqrt{Q_1}}{(h_1)^{.75}} = \frac{n_2\sqrt{Q_2}}{(h_2)^{.75}} \qquad 4.43$$

These so-called *similarity laws* assume that both pumps

* operate in the turbulent region.
* have the same operating efficiency.
* operate with the same percentage of wide-open flow.

Similar pumps also will have the same specific speed and cavitation number.

Example 4.7

A 6″ pump operating at 1770 rpm discharges 1500 gpm of cold water (S.G.= 1.0) against an 80 foot head at 80% efficiency. A homologous 8″ pump operating at 1170 rpm is being considered as a replacement. What capacity and total head can be expected from the new pump? What would be the new power requirement?

From equation 4.38,

$$H_2 = \left[\frac{(8)(1170)}{(6)(1770)}\right]^2 (80) = 62.14 \text{ ft}$$

From equation 4.41,

$$Q_2 = \left[\frac{(1170)(8)^3}{(1770)(6)^3}\right](1500) = 2350 \text{ gpm}$$

$$whp_2 = \frac{(2350.3)(62.14)(1.0)}{3956} = 36.92 \text{ hp}$$

$$bhp_2 = \frac{36.92}{.8} = 46.1 \text{ hp}$$

12 THE CAVITATION NUMBER

Although it is difficult to predict when cavitation will occur, the *cavitation number (cavitation coefficient)* can be used in modeling and in comparing experimental results. The actual cavitation number, given by equation 4.44, is compared with the critical cavitation number from experimental results. If the critical cavitation number is larger, it is concluded that cavitation will result.

$$\sigma = \frac{p - p_{vp}}{\frac{\rho v^2}{2g_c}} = \frac{\text{NPSHA}}{h_A \text{ per stage}} \qquad 4.44$$

The two forms of equation 4.44 yield slightly different results. The first form is essentially the ratio of the net pressure available for collapsing a vapor bubble to the velocity pressure creating the vapor. The first form is useful in model experiments, whereas the second form is applicable to tests of production model pumps.

13 PUMP PERFORMANCE CURVES

Evaluating the performance of a pump is often simplified by examining a graphical representation of its operating characteristics. For a given impeller diameter and constant speed, the head added by a centrifugal pump will decrease as the flow rate increases. This is illustrated by figure 4.14. Other operating characteristics also vary with the flow rate. These can be presented on individual graphs. However, since the independent variable (flow rate) is the same for all, common practice is to plot all charcteristics together on a single graph. A pump *performance curve* is illustrated in figure 4.15.

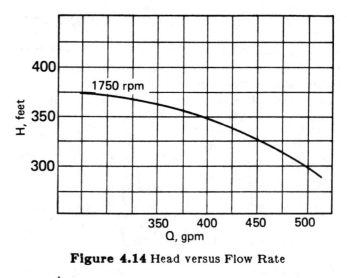

Figure 4.14 Head versus Flow Rate

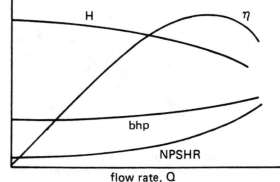

Figure 4.15 Pump Performance Curves

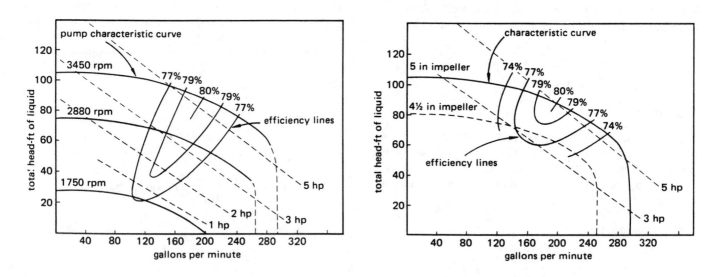

Figure 4.16　Two Types of Characteristic Curves

Figures 4.14 and 4.15 are for a pump with a fixed impeller diameter and rotational speed. The characteristics of a pump operated over a range of speeds are illustrated in figure 4.16. For maximum efficiency, the operating point should fall along the dotted line.

Manufacturers' performance curves show pump performance at a limited number of calibration speeds. The desired operating point can be outside the range of the published curves. It is then necessary to estimate a speed at which the pump would give the required performance. This is done by using the affinity laws, as illustrated in example 4.8.

Example 4.8

A pump with the 1750 rpm performance curve shown is required to pump 500 gpm at 425 feet total head. At what speed must this pump be driven to achieve the desired performance with no change in efficiency or impeller size?

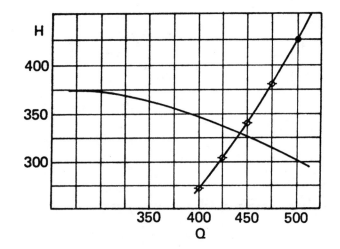

Then, from equation 4.31,

$$n_2 = 1750 \left(\frac{500}{440} \right) = 1989 \text{ rpm}$$

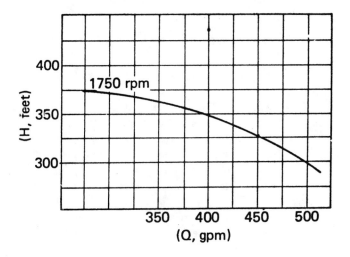

From equation 4.32, the quantity (H/Q^2) is constant for a pump with a given impeller size. In this case,

$$\frac{425}{(500)^2} = 1.7 \text{ EE} - 3$$

In order to apply an affinity equation, it is necessary to know the operating point on the 1750 rpm curve. To find the operating point, choose random values of Q and solve for H such that $(H/Q^2) = 1.7 \text{ EE} - 3$.

Q	H
475	383
450	344
425	307
400	272

These four points are plotted on the performance curve graph. The intersection of the constant efficiency line and the original 1750 rpm curve is at 440 gpm.

14 SYSTEM CURVES

A *system curve* graph can also be made from the resistance to flow of the piping system. This resistance varies with the square of the flow rate since h_f varies with v^2 in the Darcy friction formula.

$$\frac{H_1}{Q_1^2} = \frac{H_2}{Q_2^2} \qquad 4.45$$

Equation 4.45 is illustrated by figure 4.17, in which there is no static head (h_{ts}) to overcome.

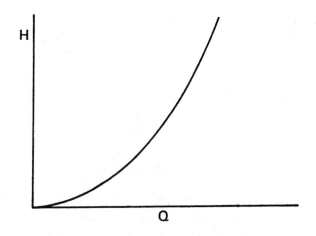

Figure 4.17 System Performance Curve
(Dynamic Losses Only)

When a static head (h_{ts}) exists in a system, the loss curve is displaced upward an amount equal to the static head. This is illustrated in figure 4.18.

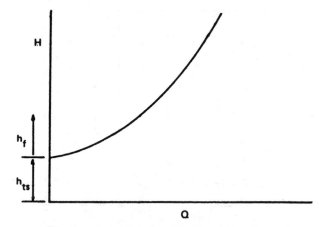

Figure 4.18 System Performance Curve

The intersection of the pump characteristic curve with the system curve defines the *operating point* as shown in figure 4.19.

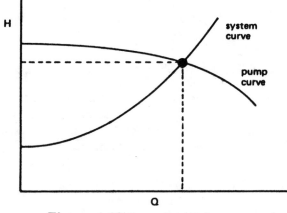

Figure 4.19 Operating Point

After a pump is installed, it may be desired to vary the pump's performance. If a valve is placed in the discharge line, the operating point may be moved along the performance curve by opening or closing the valve. This is illustrated in figure 4.20. (A throttling valve should never be placed in the intake line since that would reduce NPSHA.)

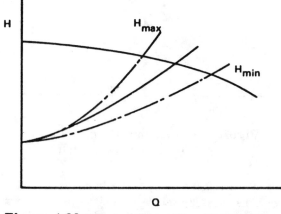

Figure 4.20 Effect of Throttling the Discharge

In most systems, the static head will vary as the feed tank is drained or as the discharge tank fills. The system head is then defined by a pair of parallel curves intersecting the performance curve. The two intercept points are the maximum and the minimum capacity requirements.

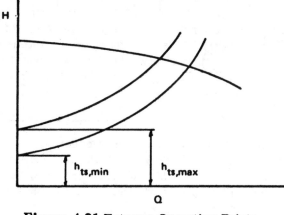

Figure 4.21 Extreme Operating Points

15 PUMPS IN SERIES AND PARALLEL

Parallel operation is obtained by having two pumps discharging into a common header. This type of connection is advantageous when the system demand varies greatly. A single pump providing total flow would have to operate far from its optimum efficiency at one point or another. With two pumps in parallel, one can be shut down during low demand. This allows the remaining pump to operate close to its optimum efficiency point.

Figure 4.22 illustrates that parallel operation increases the capacity of the system while maintaining the same total head.

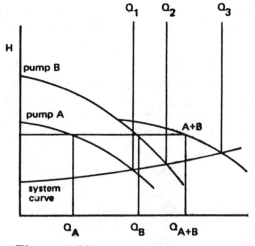

Figure 4.22 Pumps Operating in Parallel

The performance curve for a set of pumps in parallel can be plotted by adding the capacities of the two pumps at various heads. Capacity does not increase at heads above the maximum head of the smaller pump. Furthermore, a second pump will operate only when its discharge head is greater than the discharge head of the pump already running.

When the parallel performance curve is plotted with the system head curve, the operating point is the intersection of the system curve with the $A + B$ curve. With pump A operating alone, the capacity is given by Q_1. When pump B is added, the capacity increases to Q_3 with a slight increase in total head.

Series operation is achieved by having one pump discharge into the suction of the next. This arrangement is used primarily to increase the discharge head, although a small increase in capacity also results.

The performance curve for a set of pumps in series can be plotted by adding the heads of the two pumps at various capacities.

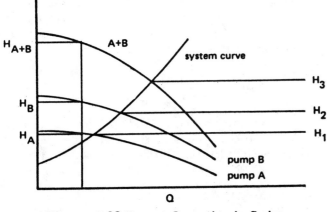
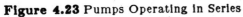

Figure 4.23 Pumps Operating in Series

16 IMPULSE TURBINES

As shown in figure 4.24, an *impulse turbine* converts the energy of a fluid stream into kinetic energy by use of a nozzle which directs the stream jet against the turbine blades. Impulse turbines are generally employed where the available head exceeds 800 feet.

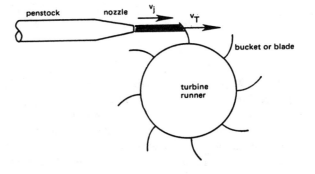

Figure 4.24 A Simple Impulse Turbine

The *total head available* to an impulse turbine is given by equation 4.46. (p is the pressure of the fluid at the nozzle entrance.)

$$H = \frac{p}{\rho} + \frac{v^2}{2g_c} - h_n \qquad 4.46$$

$$= h_s - \frac{fL_e v^2}{2Dg_c} - h_n \qquad 4.47$$

The *nozzle loss* is

$$h_n = \left(\frac{p}{\rho} + \frac{v^2}{2g_c}\right)(1 - C_v^2) \qquad 4.48$$

The velocity of the fluid jet is

$$v_j = C_v \sqrt{2gH} \qquad 4.49$$

The energy transmitted by each pound of fluid to the turbine runner is given by equation 4.50.

$$E = \frac{v_T(v_j - v_T)}{g_c}(1 - \cos\beta) \qquad 4.50$$

Multiplying the energy by the fluid flow rate gives an expression for the theoretical horsepower output of the turbine.

$$bhp_{th} = \frac{Q\rho(v_j - v_T)v_T(1 - \cos\beta)}{(2.47\ \text{EE5})g_c} \qquad 4.51$$

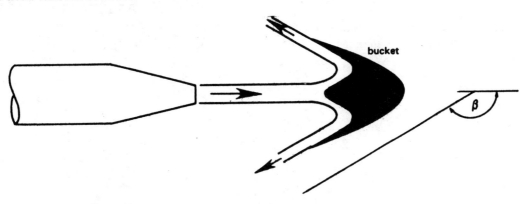

Figure 4.25 Turbine Blade Geometry

The actual output will be less than the theoretical output. Typical efficiencies range from 80% to 90%.

Example 4.9

A Pelton wheel impulse turbine developing 100 bhp (net) is driven by a water stream from an 8" schedule 40 penstock. Total head (before nozzle loss) is 200 feet. If the turbine runner is rotating at 500 rpm and its efficiency is 80%, determine the area of the jet, the flow rate, and the pressure head at the nozzle entrance ($C_v = 0.95$).

From equation 4.49, the jet velocity is

$$v_j = .95\sqrt{2(32.2)(200)} = 107.8 \text{ ft/sec}$$

From equation 4.48, the nozzle loss is

$$h_n = 200[1 - (0.95)^2] = 19.5 \text{ ft}$$

Using table 4.2, the flow rate is

$$\dot{V} = \frac{(8.814)(100)}{(200 - 19.5)(1)(0.8)} = 6.10 \text{ cfs}$$

The jet area is

$$A_j = \frac{6.10}{107.8} = 0.0566 \text{ ft}^2$$

The flow area of 8" schedule 40 pipe is .3474 ft^2.

The velocity at the nozzle entrance is

$$\frac{6.10}{0.3474} = 17.56 \text{ ft/sec}$$

The pressure head at the nozzle entrance is

$$200 - \frac{(17.56)^2}{(2)(32.2)} = 195.2 \text{ ft}$$

17 REACTION TURBINES

Reaction turbines are essentially centrifugal pumps in reverse. They are used when the total head is small, typically below 800 feet. However, their energy conversion efficiency is higher than for impulse turbines, typically 90%–95%.

Reaction turbines are classified in the same way as centrifugal pumps, according to the manner in which the impeller extracts energy from the fluid. Each of these types is associated with a range of specific speeds.

1. Centrifugal radial flow and mixed flow turbines are designed to operate most efficiently under heads of 80' to 600' with specific speeds ranging from 10 to 110. Radial flow turbines have the lowest specific speeds. Best efficiencies are found in turbines with specific speeds between 40 and 60.

2. Axial flow (propeller) turbines operate with specific speeds between 100 and 200 rpm and have the best efficiencies between 120 and 160.

$$n_s = \frac{n\sqrt{bhp}}{H^{1.25}} \qquad\qquad 4.52$$

Since reaction turbines are centrifugal pumps in reverse, all of the affinity and similarity relationships (equations 4.31 through 4.42) can be used when comparing homologous turbines.

Example 4.10

A reaction turbine develops 500 bhp. Flow through the turbine is 50 cfs. Water enters at 20 fps with a 100' pressure head. Elevation of the turbine above tailwater level is 10'. Find the effective head and turbine efficiency.

The effective (total) fluid head is

$$H = 100 + \frac{(20)^2}{2(32.2)} + 10 = 116.2 \text{ ft}$$

From table 4.2,

$$(whp)_{in} = \frac{(116.2)(50)(1)}{8.814} = 659.2 \text{ hp}$$

$$\eta_T = \frac{500}{659.2} = .758$$

Appendix A
Standard Atmosphere

Altitude ft	Temperature °R	Pressure psia	Altitude ft	Temperature °R	Pressure psia
0	518.7	14.696	35000	393.9	3.458
1000	515.1	14.175	36000	392.7	3.296
2000	511.6	13.664	37000	392.7	3.143
3000	508.0	13.168	38000	392.7	2.996
4000	504.4	12.692	39000	392.7	2.854
5000	500.9	12.225	40000	392.7	2.721
6000	497.3	11.778	41000	392.7	2.593
7000	493.7	11.341	42000	392.7	2.475
8000	490.2	10.914	43000	392.7	2.358
9000	486.6	10.501	44000	392.7	2.250
10000	483.0	10.108	45000	392.7	2.141
11000	479.5	9.720	46000	392.7	2.043
12000	475.9	9.347	47000	392.7	1.950
13000	472.3	8.983	48000	392.7	1.857
14000	468.8	8.630	49000	392.7	1.768
15000	465.2	8.291	50000	392.7	1.690
16000	461.6	7.962	51000	392.7	1.611
17000	458.1	7.642	52000	392.7	1.532
18000	454.5	7.338	53000	392.7	1.464
19000	450.9	7.038	54000	392.7	1.395
20000	447.4	6.753	55000	392.7	1.331
21000	443.8	6.473	56000	392.7	1.267
22000	440.2	6.203	57000	392.7	1.208
23000	436.7	5.943	58000	392.7	1.154
24000	433.1	5.693	59000	392.7	1.100
25000	429.5	5.452	60000	392.7	1.046
26000	426.0	5.216	61000	392.7	.997
27000	422.4	4.990	62000	392.7	.953
28000	418.8	4.774	63000	392.7	.909
29000	415.3	4.563	64000	392.7	.864
30000	411.7	4.362	65000	392.7	.825
31000	408.1	4.165			
32000	404.6	3.978			
33000	401.0	3.797			
34000	397.5	3.625			

Appendix B
Upper Limits of Specific Speeds
Single Stage, Single and Double Suction Pumps
Handling Clear Water at 85°F at Sea Level

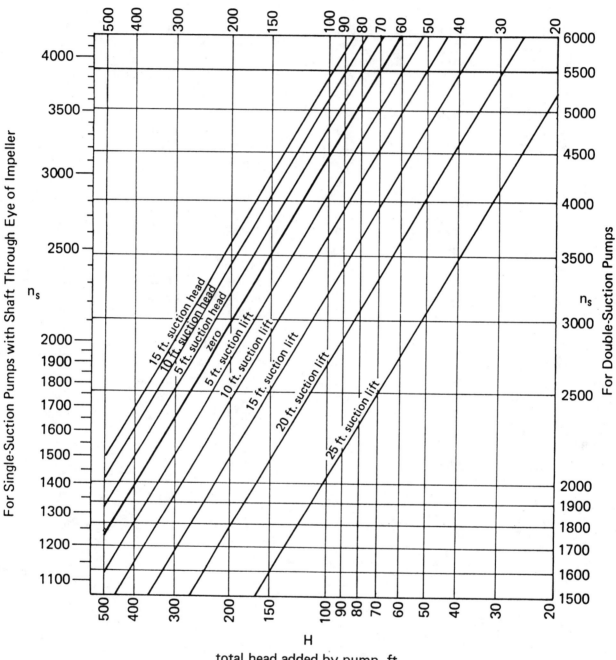

H

total head added by pump, ft

Appendix C
Upper Limits of Specific Speeds
Single Stage, Single Suction, Mixed and Axial Flow Pumps
Handling Clear Water at 85°F at Sea Level

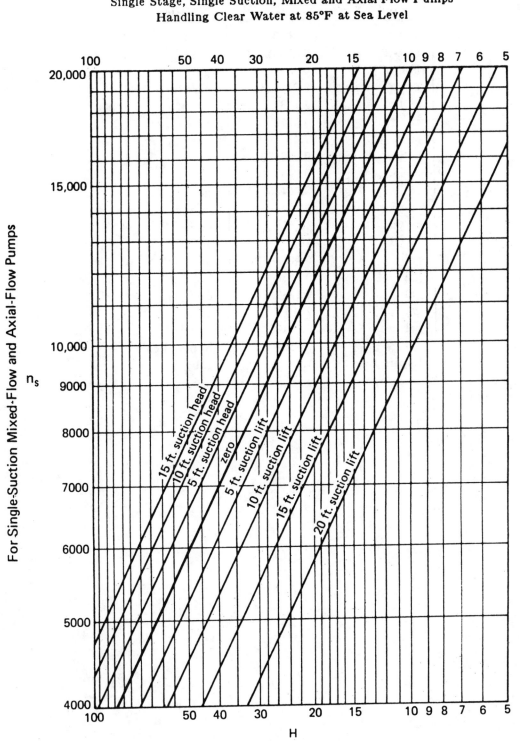

total head added by pump, ft

Appendix D
Volumetric Conversion Factors

Multiply	By	To Obtain
acre-ft	43,560	cu ft
acre-ft	325,851	gal
bbl (oil)	42	gal
cu ft	0.0000230	acre-ft
cu ft	1,728	cu in.
cu ft	0.0370	cu yd
cu ft	7.48	gal
cu in.	0.000579	cu ft
cu in.	0.0000214	cu yd
cu in.	0.00433	gal
cu yd	27	cu ft
cu yd	46,656	cu in.
cu yd	202	gal
gal	0.00000307	acre-ft
gal	0.0238	bbl (oil)
gal	0.1337	cu ft
gal	231	cu in.
gal	0.00495	cu yd
gal	0.8327	Imperial gal
Imperial gal	1.2	gal

Appendix E
Pump Performance Correction Factor Chart
(Prepared from tests on 1″ and smaller pumps)

Instructions for use: Start with the actual fluid flow rate on the bottom horizontal scale. Move vertically until the required head is reached on the diagonal head scale. Move horizontally to the left until the fluid viscosity is reached on the diagonal viscosity scale. Move upward and read the correction factors.

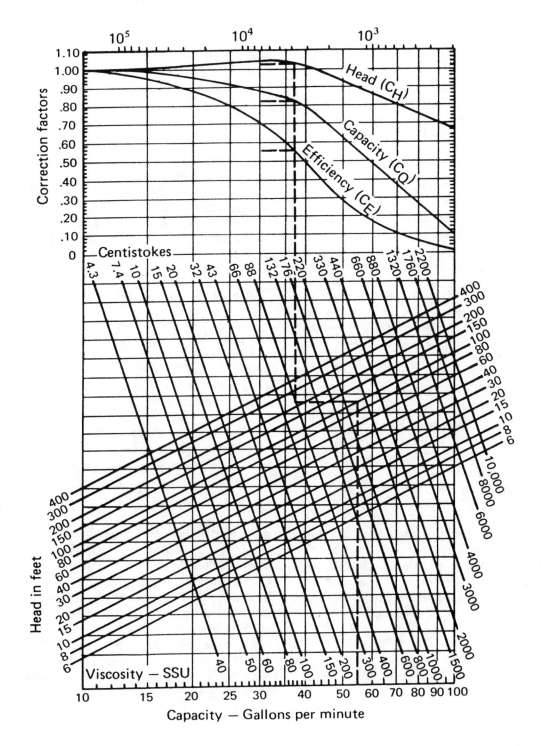

PRACTICE PROBLEMS: HYDRAULIC MACHINES

Warm-ups

1. Convert 72 gpm to cfs.

2. What horsepower is required to increase the pressure of 37 gpm of SAE 40 oil at 80°F to 40 psig?

3. A 1750 rpm pump is normally splined to a $\frac{1}{2}$ horse-power motor. What horsepower motor is required if the pump is to run at 2000 rpm?

4. What is the specific speed of a double-suction water pump adding 20 feet of head to 300 gps when turning at 900 rpm?

5. What is the maximum suggested specific speed for a 2-stage centrifugal pump adding 300 feet of head to water pulled through an inlet 10 feet below it?

Concentrates

1. What horsepower is required to pump 1.25 cfs of water from a settling tank through 700 feet (includes a 50 foot rise in elevation through two right-angle elbows) of 4″ schedule 40 pipe containing a gate valve and a check valve? The inlet pressure is 50 psig, and a working pressure of 20 psig at the end of the pipe is needed.

2. The relative velocity of a certain point on a marine propeller 8 feet below the surface of 68 °F sea water (2.5% salt by weight) is approximately 4.2 times the boat velocity. What is the practical maximum boat velocity?

3. Water (281°F, 80 psia) empties through 30 feet of $1\frac{1}{2}$″ pipe by a pump whose inlet and outlet are 20 feet below the surface of the water level when the tank is full. The pumping rate is 100 gpm, and the NPSHR is 10 feet for that rate. If the inlet line contains two gate valves and two long-radius elbows, and the discharge is into a 2 psig tank, when will the pump cavitate?

4. The inlet of a centrifugal pump is 7 feet above a water surface level. The inlet is 12 feet of 2″ pipe and contains one long-radius elbow and one check valve. The 2″ outlet is 8 feet above the surface level and contains two long-radius elbows in its 80 feet of length. The discharge is 20 feet above the surface. The following pump curve data is available. What is the flow rate at 70°F?

gpm	head	gpm	head
0	110	50	93
10	108	60	87
20	105	70	79
30	102	80	66
40	98	90	50

5. 8 MGD flow into the smooth steel pipe network shown below. Disregarding minor losses, calculate the quantity flowing in both branches and find the pressure drop between the inlet and the outlet.

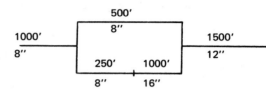

6. Water at 500 psig will be used to drive a 250 horse-power turbine at 1750 rpm against a backpressure of 30 psig. What type of turbine would you suggest? If the 4″ diameter jet discharging at 35 fps is deflected 80° by a single moving vane with velocity of 10 fps, what is the total force acting on the blade?

7. 80 gpm of 80 °F water/acid solution (S.G. = 1) are lifted 12 feet through a 2″ rubber hose into a brush used to clean the bottom of a tank filled with 8 feet of water. The hose is 50 feet long, and the pump is located between the water-acid source and the brush, at the top of the 12-foot high tank. What head is added by the pump?

Timed: (1 hour allowed for each)

1. A Francis hydraulic reaction turbine with 22″ diameter blades runs at 610 rpm and develops 250 horsepower when 25 cfs of water flow through it. The pressure head at the turbine entrance is 92.5 feet. The elevation of the turbine above the tailwater level is 5.26 feet. The inlet and outlet velocities are 12 fps. Find
    (a) The effective head
    (b) The turbine efficiency
    (c) The rpm at 225 feet effective head
    (d) The BHP at 225 feet effective head
    (e) The discharge in cfs at 225 feet effective head

2. A system has 100 radiative heaters with a maximum capacity of 10,000 BTUH each. Circulating water enters at 200°F and exits at 180°F. The heaters all are connected in a reverse-return circuit, making all pipe lengths equal. The equivalent line length is 420 feet, including all losses for valves, fittings, and bends. The building specification requires that the system piping be sized such that the pressure drop per equivalent foot of pipe is between .25″ and .65″ water. Curves for 3 possible pumps to supply the water to the system are given. Select the best pump for the job and determine the required pipe size for (a) a 1,000,000 BTUH system and (b) a 300,000 BTUH system.

feet of head

	pump number		
gpm	1	2	3
10	5.3	8.8	13.3
20	5.5	9.0	13.4
30	5.5	9.0	13.4
40	5.4	8.9	13.3
50	5.2	8.7	13.1
60	4.9	8.4	12.7
70	4.5	8.0	12.3
80	3.9	7.4	11.8
90	3.3	6.8	11.1
100	2.5	5.8	10.0
110	1.4	4.3	8.3

3. A 3-zone heating system uses hot water passing through the piping network shown. The heater increases the water temperature 20°F. All pipes are copper, type L.

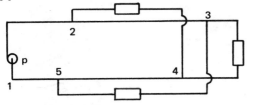

circuit	equivalent length (ft)	pipe diameter (in)	flow rate (gpm)
5–1–p–2	40	$2\frac{1}{2}$	60
2–4	70	$1\frac{1}{2}$	20
2–3	55	2	40
3–4	65	$1\frac{1}{2}$	20
3–5	60	$1\frac{1}{2}$	20
4–5	50	2	40

(a) What is the total head added by the pump?

(b) Assuming a pump efficiency of 45%, what size electric motor should be used?

(c) What is the heat flow rate into the water in BTUH?

4. A pump is used to transfer gasoline (specific gravity = .7, viscosity = 6 EE–6 ft^2/sec) from a tanker to a storage tank. The storage tank is open to the atmosphere and has a free surface 60 feet above the tanker's free surface. The piping consists of 500 feet of 3″ steel pipe with 6 flanged elbows and 2 gate valves. The pump has the following performance curve:

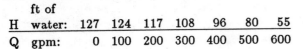

ft of H water:	127	124	117	108	96	80	55
Q gpm:	0	100	200	300	400	500	600

(a) What will be the flow in gpm?

(b) If both the pump and motor have efficiencies of 88% each, and if electricity costs $.045 per kw-hr, what is the total cost of operating the pump for one hour?

5. A pump station is used to fill a tank on a hill from a lake below. The pump is 12 feet above the lake, and the tank is 350 feet above the pump. The flow rate is 10,000 gallons per hour. The equivalent length between the lake and the tank is 7000 feet, including all fittings and bends. The total efficiency of the pump and motor combination is 70%.

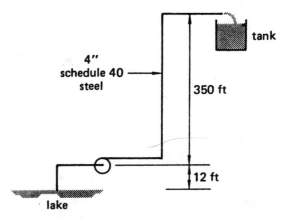

(a) What does it cost to operate the pump per hour at $.04 per kw-hr?

(b) What motor horsepower is required?

(c) If the equivalent length from the lake to the pump is 300 feet, what is the NPSHA to the pump?

6. A centrifugal pump has the curve shown at 1400 rpm. The pump will be bolted into an existing system with head requirements given by

$$H = 30 + 2Q^2 \qquad (H \text{ in feet, } Q \text{ in } ft^3/sec)$$

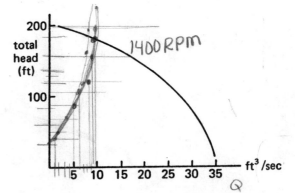

(a) What is the flow rate if the pump is turned at 1400 rpm?

(b) What is the input horsepower to the pump?

(c) What is the flow rate if the pump is turned at 1200 rpm?

RESERVED FOR FUTURE USE

5 FANS AND DUCTWORK

Nomenclature

a	short side, or interference factor	ft, –
A	duct area	ft^2
ACFM	actual flow rate	cfm
b	long side	ft
BHP	brake power	hp
c	minor loss coefficient	–
C	coefficient	–
d	diameter	inches
D	diameter	ft
f	Darcy friction factor	–
g	acceleration of gravity (32.2)	ft/sec^2
K	factor	–
L	duct length	ft
p	pressure	inches of water
P	power	ft-lbf/sec
m	mass flow rate	lbm/min
n	fan speed	rpm
r	radius	ft
R	ratio, or regain coefficient	–
Q	volume flow rate	cfm
SCFM	standard flow rate	cfm
v	flow velocity	fpm
x	distance from outlet	ft

Symbols

ρ	density	lbm/ft^3
η	efficiency	–

Subscripts

d	discharge, or density
e	equivalent, or entrainment
f	friction, or fan
m	main run
o	outlet
P	performance
s	static
t	total
v	velocity, or viscosity
var	variation between grilles

PART 1: Fans

1 TYPES OF FANS

A fan is essentially a pump used to increase the static and kinetic energies of an air flow. There are two main types of fans: axial flow fans and centrifugal fans.

Axial flow fans (propeller fans) develop static pressure by changing the air-flow velocity. They often are simple propellers mounted with small tip clearances in ducts. They are usually used when it is necessary to move large quantities (greater than 500,000 cfm) of air against low static pressure (less than 12 inches of water).

Axial flow fans can be organized further into tubeaxial and vaneaxial varieties. *Tubeaxial fans*, also known simply as *duct fans*, seldom are called upon to move air against more than three inches of water. They may have four to eight blades, and there will be very little separation between the blade tips and the surrounding duct. Tubeaxials can be identified by their hub diameters, which are less than 50% of the overall tip-to-tip fan diameter.

Vaneaxial fans can be distinguished from tubeaxial fans, since their hub diameters will be greater than 50% of the fan diameter. Furthermore, the fan assembly will have vanes downstream from the fan to straighten the air flow. Vaneaxial fans typically have as many as 24 blades, which may have airfoil cross sections. Vaneaxials are capable of moving air against pressures up to 12 inches of water.

Centrifugal fans can be used in installations requiring volumes less than 1,000,000 cfm and pressures less than 60 inches of water. They develop static pressure by increasing kinetic energy and by imposing a centrifugal force on the rotating air. Depending on the blade curvature, kinetic energy can be made greater (forward-curved blades) or less (backward-curved blades) than the tangential velocity of the impeller blades. For the same operating speed, backward-curved blade fans develop more pressure; forward-curved blade fans have a greater capacity (higher velocity) but require a larger scroll.

2 BASIC PRINCIPLES

Three parameters are important in the selection of fans: air horsepower, pressure developed, and operating efficiency. These three parameters vary with the volume flowing and, therefore, are usually presented graphically.

Moving or stationary air confined in a duct will exert a pressure perpendicular to the duct wall. The fan must supply this pressure, called the *static pressure*, p_s. Static pressure is usually measured in *inches of water* (*inches w.g.*).

$$\text{inches w.g.} = \frac{\text{psig}}{.0361} \qquad 5.1$$

The fan must also supply the velocity pressure, p_v.

$$v = 4005\sqrt{p_v} \qquad 5.2$$

$$p_v = \frac{\left(\frac{v}{60}\right)^2}{2g}(12)\left(\frac{.075}{62.4}\right) \approx \left(\frac{v}{4005}\right)^2 \qquad 5.3$$

The total pressure always decreases along the direction of flow.[1] However, the static pressure can increase with diameter increases. The loss in pressure due to friction should not be confused with the change in pressure due to diameter changes.

The *total pressure* is the sum of the static and velocity pressures. Were it not for friction, the total pressure would be constant along the length of the duct.

$$p_t = p_s + p_v \qquad 5.4$$

The efficiency of the fan can be found from the brake and air horsepowers.

$$\eta_f = \frac{\text{AHP}}{\text{BHP}} \qquad 5.5$$

The efficiency of most centrifugal fans will be .50 to .65, although values as high as .80 are possible.

---

[1]Total pressure is defined as the sum of the velocity pressure and the static pressure. Contributions from potential energy are ignored in duct problems.

The *static efficiency* is defined as

$$\eta_s = \eta_f\left(\frac{p_s}{p_t}\right) \qquad 5.6$$

The *air horsepower* is the theoretical horsepower required to drive a 100% mechanically efficient fan.

$$\text{AHP} = \frac{Qp_t}{6356} \qquad 5.7$$

The actual horsepower required will be

$$\text{BHP} = \frac{\text{AHP}}{\eta_f} \qquad 5.8$$

Typical fan characteristic curves for AHP, p_t (or p_s), and η_f (or η_s) are given in figure 5.1.

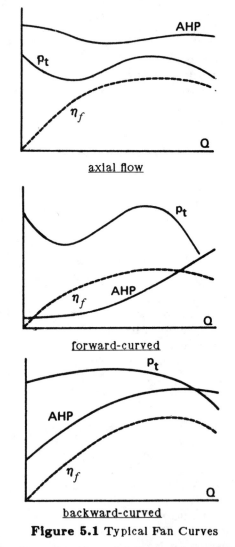

axial flow

forward-curved

backward-curved

Figure 5.1 Typical Fan Curves

The dips in p_t for the axial flow and centrifugal fans are characteristic. The highest efficiency to the right of the dip should be chosen as the operating point. In choosing between fan types, the following points should be noted:

- Forward-curving blades (centrifugal)

  · Maximum efficiency occurs near the point of maximum static pressure.

  · Power rises rapidly with increases in the delivery rate.

  · Motor overloading is possible if the duct losses are not calculated carefully.

  · Minimum sound occurs at maximum pressures.

- Backward-curving blades (centrifugal)

  · The fan may operate over a greater range without encountering unstable air.

  · Overloading is less likely.

  · The efficiency often is greater.

  · The fan is noisier than the forward-curving fan.

  · Minimum sound occurs at the highest efficiencies.

- Axial flow

  · Higher outlet velocities are possible than with centrifugals.

  · Overloading is less likely due to the flat power curve.

  · Minimum sound occurs at maximum efficiencies.

Example 5.1

A motor delivers 13.12 horsepower to a fan moving 27,000 cfm of 1800 fpm air against 2″ w.g. static pressure. What are the total and static efficiencies?

From equation 5.2,

$$p_v = \left(\frac{1800}{4005}\right)^2 = .2″ \text{ w.g.}$$

$$p_t = .2 + 2″ = 2.2″ \text{ w.g.}$$

From equation 5.7,

$$\text{AHP} = \frac{(27,000)(2.2)}{6356} = 9.35$$

$$\eta_t = \frac{9.35}{13.12} = .71$$

$$\eta_s = .71\left(\frac{2}{2.2}\right) = .65$$

3 SYSTEM CHARACTERISTICS

Air flowing through ducts encounters frictional resistance. Bernoulli's equation can be written (ignoring the gravitational head term) as

$$(p_s + p_v)_1 = (p_s + p_v)_2 + p_{loss} \qquad 5.9$$

The pressure loss varies approximately with the square of the velocity.[2]

$$p_{loss} \propto \frac{Lv^2}{D} \qquad 5.10$$

Since $Q = Av$, the pressure loss also varies according to Q^2. This relationship can be shown by plotting p_{loss} versus Q, resulting in the *system characteristic curve*. The curve will be flatter for lower resistance ductwork.

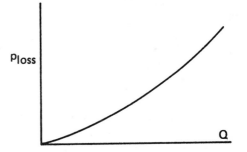

Figure 5.2 Typical System Curve

Given one point on the curve, the remainder of the curve (assuming turbulent flow) can be found from equation 5.11.

$$\frac{p_2}{p_1} = \left(\frac{Q_2}{Q_1}\right)^2 \qquad 5.11$$

The intersection of the fan and system characteristic curves determines the *operating point*, as shown in figure 5.3.

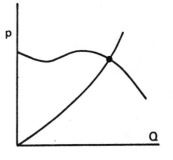

Figure 5.3 Operating Point

Most manufacturers provide fan rating tables similar to table 5.1 with their fans. Such a table may give rpm and BHP versus Q and p_s. For the given pressure required (column), the highest efficiency usually will be

[2]See equations 5.27 and 5.28.

In the middle third of the Q column. If the maximum efficiency point is not indicated (as with the underscoring in table 5.1), the actual efficiency should be calculated for each point using equation 5.8.

Table 5.1
Typical Fan Rating Table

Q	$p_s = 1''$ rpm	$p_s = 1''$ BHP	$p_s = 2''$ rpm	$p_s = 2''$ BHP
5000	440	1.20	617	2.67
10000	492	2.18	626	4.20
15000	600	4.06	706	6.45
20000	816	9.59	830	10.83

If the fan operation is evaluated from fan and system characteristic curves, the following design guides should be observed:

- The operating point should be to the right of the peak fan pressure to avoid pressure and volume fluctuations with accompanying noise and uneven motor loading.

- A fan with a steep pressure curve should be chosen to avoid wide volume changes with changes in the duct loss.

- The operating point should be as close as possible to the peak efficiency to minimize required horsepower.

Example 5.2

The friction pressure loss in a low-velocity system is 1.5" w.g. at 3500 cfm. What will be the flow rate if a 1000 rpm fan with the characteristics shown is used? Disregard velocity head and terminal pressure.

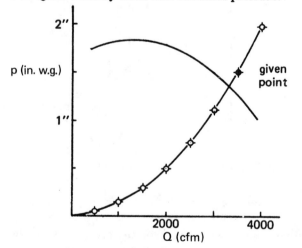

Using equation 5.11, the rest of the system curve is obtained:

Q	p
500	.03
1000	.12
1500	.28
2000	.49
2500	.77
3000	1.1
3500	1.50
4000	1.96

The system and fan characteristic curves intersect at 3300 cfm.

4 MODIFYING FAN PERFORMANCE

If the operating point is not as desired, several different steps can be taken to change it.

- A different fan can be used.

- Two fans in parallel can be used. For any value of p, the corresponding value of Q will be $Q_1 + Q_2$.

- Two fans in series can be used. For any value of Q, the corresponding pressure will be $p_1 + p_2$.

- The fan speed can be changed.[3]

- The fan size can be changed.

Predicting fan performance at a new speed can be accomplished with the *fan laws*. These three fan laws assume that the fan duct system and air density remain unchanged.

$$\frac{Q_2}{Q_1} = \frac{n_2}{n_1} \qquad 5.12$$

$$\frac{p_2}{p_1} = \left(\frac{n_2}{n_1}\right)^2 \qquad 5.13$$

$$\left(\frac{AHP_2}{AHP_1}\right) = \left(\frac{n_2}{n_1}\right)^3 \qquad 5.14$$

If two fans are different but geometrically similar (i.e., *homologous fans*), equations 5.15 through 5.20 can be used.

$$\frac{Q_A}{Q_B} = \left(\frac{D_A}{D_B}\right)^3 \left(\frac{n_A}{n_B}\right) = \left(\frac{D_A}{D_B}\right)^2 \sqrt{\frac{p_A}{p_B}} \sqrt{\frac{\rho_B}{\rho_A}} \quad 5.15$$

$$\frac{p_A}{p_B} = \left(\frac{D_A}{D_B}\right)^2 \left(\frac{n_A}{n_B}\right)^2 \left(\frac{\rho_A}{\rho_B}\right) \qquad 5.16$$

[3]Changing the fan speed will not change the relative position of the operating point with respect to the dip in the fan curve. The locus of peak points follows the Q^2 rule also. So, if an operating point is left of the peak point at a high rpm, it will be left of the peak point at lower speeds.

$$\frac{AHP_A}{AHP_B} = \left(\frac{D_A}{D_B}\right)^5 \left(\frac{n_A}{n_B}\right)^3 \left(\frac{\rho_A}{\rho_B}\right)$$

$$= \left(\frac{D_A}{D_B}\right)^2 \left(\frac{p_A}{p_B}\right)^{1.5} \sqrt{\frac{\rho_B}{\rho_A}} \qquad 5.17$$

$$= \left(\frac{Q_A}{Q_B}\right)\left(\frac{p_A}{p_B}\right) \qquad 5.18$$

$$\frac{n_A}{n_B} = \left(\frac{D_B}{D_A}\right)\sqrt{\frac{p_A}{p_B}}\sqrt{\frac{\rho_B}{\rho_A}}$$

$$= \sqrt{\frac{Q_B}{Q_A}}\left(\frac{p_A}{p_B}\right)^{.75}\left(\frac{\rho_B}{\rho_A}\right)^{.75} \qquad 5.19$$

$$\frac{D_A}{D_B} = \sqrt{\frac{Q_A}{Q_B}}\left(\frac{p_B}{p_A}\right)^{0.25}\left(\frac{\rho_A}{\rho_B}\right)^{0.25} \qquad 5.20$$

Use of the similarity laws is best confined to predicting the behavior of a larger fan from a smaller fan's performance, since the efficiency of the larger can be expected to be greater. The laws should not be used where there is a significant decrease in air density, although increases are allowed. Also, extrapolations to larger fans should be avoided when the ratio of the larger to smaller fan diameters, the ratio of speeds, or the product of the diameter and speed ratios exceeds 3.0.

Example 5.3

An 800 rpm fan uses 6.2 bhp while delivering 10,000 cfm against 2.25″ w.g. static pressure. If the fan is driven at 1400 rpm, what will be the required horsepower? If the static pressure is increased to 2.85″ w.g., what speed is required?

$$BHP = 6.2\left(\frac{1400}{800}\right)^3 = 33.2 \text{ hp}$$

$$n_2 = 800\sqrt{\frac{2.85}{2.25}} = 900 \text{ rpm}$$

5 FAN OPERATION WITH NON-STANDARD DENSITIES

Rating and performance charts are given for a standard air density of .075 lbm/ft³. Small variations in density due to normal temperature and humidity fluctuations need not be considered. However, if the system is to operate at elevated temperatures or reduced atmospheric pressures, corrections are necessary.

Fans are constant-volume devices. Thus, if the temperature or altitude is increased, or the pressure is decreased, the air volume moved will remain constant. However, the air mass will decrease. To select a fan from the standard rating tables, the following relationship can be used:

Density factor

$$K_d = \left(\frac{14.7}{p_{\text{actual}}}\right)\left(\frac{460 + {}^\circ F}{530}\right) \qquad 5.21$$

Viscosity factor (typically disregarded)

$$K_v = (\text{kinematic vicosity in}$$
$$\text{ft}^2/\text{sec}/1.63 \text{ EE} - 4)^{0.10} \qquad 5.22$$

$$SCFM = \frac{ACFM}{K_d} = \text{ standard ft}^3/\text{min} \qquad 5.23$$

$$(p_{\text{loss}})_{\text{actual}} = (p_{\text{loss}})_{\text{std}}\left(\frac{K_v}{K_d}\right) \qquad 5.24$$

$[(p_{\text{loss}})_{\text{std}} \text{ found using actual velocity}]$

$$(BHP)_{\text{actual}} = \frac{(BHP)_{\text{std}}}{K_d} \qquad 5.25$$

$$(p_s)_{\text{std}} = (p_s)_{\text{actual}}\frac{K_d}{K_v} \qquad 5.26$$

The values of ACFM (*actual cubic feet per minute*) and $(p_{\text{loss}})_{\text{std}}$ should be used to select a fan from rating tables. The table should be entered and the speed and BHP read. The speed will be correct, but the BHP should be modified with equation 5.25.

Example 5.4

A fan is chosen to move 18,000 scfm of air against .85″ w.g. using 4.2 horsepower. If the fan is moved to a 150°F environment, what will be the required horsepower and the new friction loss?

Assuming no change in local pressure, and disregarding the change in viscosity, the density factor is

$$K_d = \frac{460 + 150}{530} = 1.15$$

$$BHP = \frac{4.2}{1.15} = 3.65$$

$$p_{\text{loss}} = \frac{.85}{1.15} = .74″ \text{ w.g.}$$

Example 5.5

A 1500 fpm duct system at 5000 feet altitude delivers 39,000 acfm. The actual duct resistance at that altitude is 1.5″ w.g. Find the equivalent static pressure and the BHP asuming a 75% mechanical efficiency.

At 5000 feet altitude, the atmospheric pressure is 12.2 psia. Thus, the density correction factor is

$$K_d = \frac{14.7}{12.2} = 1.2$$

$$(p_s)_{\text{std}} = (1.5)(1.2) = 1.8″ \text{ w.g.}$$

Fans are constant volume devices. The velocity pressure is

$$p_{v,\text{std}} = p_{v,\text{actual}} = \left(\frac{1500}{4005}\right)^2 = 0.14$$

From equation 5.7, the actual air horsepower based on actual conditions is

$$AHP = \frac{(39,000)(1.5 + 0.14)}{6356} = 10.06 \text{ hp}$$

From equation 5.25, the standardized brake horsepower is

$$BHP = \frac{(1.2)(10.06)}{(.75)} = 16.1$$

PART 2: Duct Design

1 FRICTION LOSSES IN DUCTS

Friction losses can be calculated from the Moody equation.

$$p_{loss} = \frac{fLp_v}{D} \approx (.0270) \frac{L}{(d)^{1.22}} \left(\frac{v}{1000}\right)^{1.82} \qquad 5.27$$

$$\approx (3.9 \text{ EE} - 9)(v)^{2.43} \frac{L}{(Q)^{.61}} \qquad 5.28$$

In practice, equation 5.27 seldom is used. Figure 5.4 is based on equation 5.27 with a value of specific roughness equal to .0005, a standard density of .075 lbm/ft³, clean round galvanized metal ductwork, and approximately 40 joints per 100 feet. The chart can be used for temperatures between 50 °F and 90 °F. For operation outside this range, the pressure loss should be corrected with equation 5.29. K_v is usually taken as 1.0.

$$(p_{loss})_{actual} = p_{loss, \text{ fig. }5.4} \times \frac{K_v}{K_d} \qquad 5.29$$

Figure 5.4 is for use with standard round, galvanized ducts. Multiply the friction losses by the factors in table 5.2 for other materials. (Actual values are velocity dependent. Tables and charts exist for this purpose.)

Table 5.2
Multiplicative Factors for Non-Standard Ducts

Smooth ducts—no joints	.6–.95
Smooth concrete	1.1–1.4
Rough concrete/good brick	1.2–1.8

The *equivalent diameter* of a rectangular air duct with dimensions a and b, and aspect ratio less than 8.0 is

$$D_e = 1.3 \frac{(ab)^{.625}}{(a+b)^{.25}} \qquad 5.30$$

A round duct with diameter D_e will have the same friction and capacity as a square duct with dimensions a and b. Figure 5.4 can be used with D_e and flow rate to find the friction loss.

If the aspect ratio of the rectangular duct is known, a round duct can be converted to a rectangular duct of equal friction. The *aspect ratio*, which should be kept below 8.0 for ease of manufacturing, is

$$R = \frac{\text{longest side}}{\text{short side}} \qquad 5.31$$

The short side, a, is given by equation 5.32.

$$a = \frac{D_e(R+1)^{\frac{1}{4}}}{1.3(R)^{.625}} \qquad 5.32$$

Example 5.6

2000 cfm of air flow in a 13″ diameter duct. What is the velocity and the friction loss per 100 feet of duct?

Although the $Q = Av$ relationship could be used to find the velocity, it is expedient to use figure 5.4. By locating the intersection of the 2000 cfm and the 13″ lines, the velocity is found to be 2200 fpm.

Dropping straight down from the intersection point to the horizontal scale gives the friction loss as approximately .5″ w.g. per 100 feet.

Example 5.7

What size duct is required to carry 2000 cfm at 1600 fpm?

Figure 5.4 shows that a 15″ diameter duct is required.[4] The friction loss is approximately 0.23″ w.g. per 100 feet.

2 MINOR AND DYNAMIC LOSSES

Minor losses are fairly independent of air velocity and roughness. In the *loss coefficient method*, the losses are calculated as a percentage of the velocity pressure.

$$p = c\left(\frac{v}{4005}\right)^2 = cp_v \qquad 5.33$$

Typical values of c are given in table 5.3. Subscripts 1 and 2 refer to upstream and downstream, respectively. The coefficient c always should be used with the velocity at the point corresponding to its subscript.

The *equivalent length method* also can be used to calculate the friction of a bend or an elbow. As with equivalent lengths used in liquid flow problems, each obstruction produces a frictional loss equivalent to some length of duct. These lengths are given in multiples of the duct diameter in table 5.3.

---

[4] Any size duct can be manufactured. However, there are standard sizes available, and these sizes should be chosen to minimize cost. Generally, every whole-inch size up to 30″ diameter is available, although some odd-number sizes may be premium-priced. After 30″, sizes are available in 2″ increments.

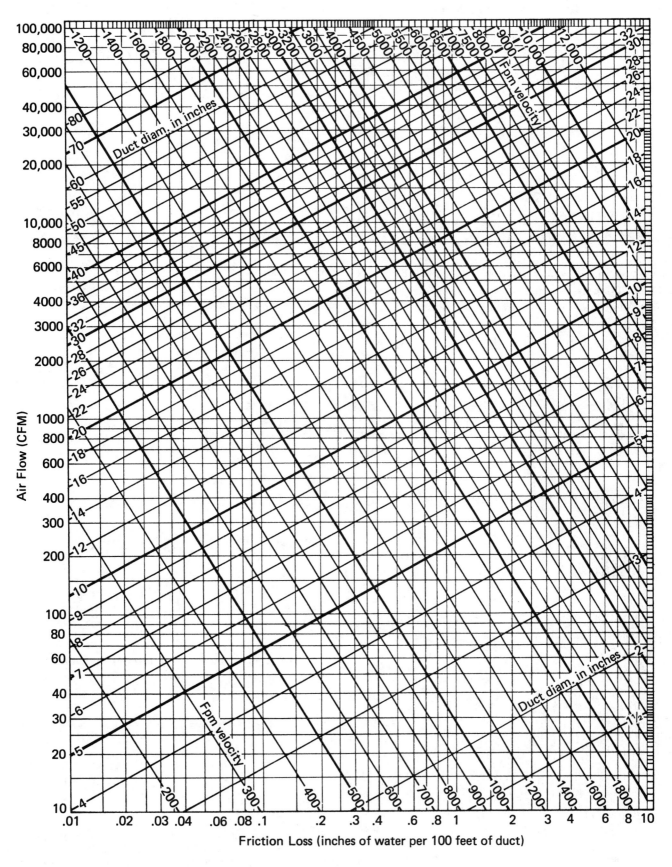

Figure 5.4 Friction Loss in Inches of Water per 100 Feet Standard Duct

Table 5.3
Minor Loss Coefficients

abrupt expansion	$\dfrac{A_1}{A_2} = 0$	$c_1 =$	1.0
	.1		.81
	.2		.64
	.3		.49
	.4		.36
	.5		.25
	.6		.16
	.7		.09
	.8		.04
	.9		.01
square-edged orifice with area A_0 at exit	$\dfrac{A_0}{A_1} = .2$	$c_0 =$	2.44
	.4		2.26
	.6		1.96
	.8		1.54
pipe of diameter E across duct of diameter D	$\dfrac{E}{D} = .1$	$c_1 =$	.2
	.25		.55
	.50		2.0
abrupt contraction	$\dfrac{A_2}{A_1} = .2$	$c_2 =$	.32
	.4		.25
	.6		.16
	.8		.06
90° smooth round elbow of radius r and diameter D	$\dfrac{r}{D} = .5$	$L_e =$	45D
	.75		23D
	1.0		17D
	1.5		12D
	2.0		10D
90° miter elbow		$L_e =$	65D

3 FRICTION LOSSES IN DIVIDED-FLOW FITTINGS

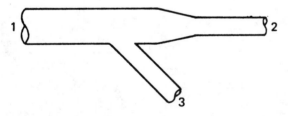

Figure 5.5 Duct with Take-off

If the friction effect of the take-off is ignored, the increase in static pressure due to the decrease in velocity at 2 is

$$\Delta p_{s,1-2} = \frac{v_1^2 - v_2^2}{(4005)^2} \qquad 5.34$$

This increase in pressure will be reduced by the friction and turbulence of the take-off. The amount of reduction

is typically between 10% and 25%. This reduction can be represented by multiplying by a coefficient less than 1.0

$$\Delta p_{s,1-2} = R\left(\frac{v_1^2 - v_2^2}{(4005)^2}\right) \qquad 5.35$$

Δp is known as the *static regain*, and R is the *static regain coefficient*. R has typical values of .75 to .90 for well-designed ducts without reducing sections.

The total pressure drop from 1 to 2 is

$$\Delta p_{t,1-2} = (1-R)\left(\frac{v_1^2 - v_2^2}{(4005)^2}\right) \qquad 5.36$$

The friction loss between 1 and 3 can be found from

$$\Delta p_{t,1-3} = c_b\left(\frac{v_1}{4005}\right)^2 \qquad 5.37$$

$$\Delta p_{s,1-3} = \left(\frac{v_3}{4005}\right)^2 - (1-c_b)\left(\frac{v_1}{4005}\right)^2 \qquad 5.38$$

Values of c_b are given in table 5.4.

Table 5.4
Approximate Values of c_b
(Round mains and branches only)

$\dfrac{v_3}{v_1}$	90°	60°	45°
.5	1.1	.8	.5
1.0	1.5	.8	.5
1.5	2.2	1.1	.9
2.0	3.0	2.9	2.8
2.5	4.3	3.3	3.2
3.0	5.6	5.2	4.9

4 LOW-VELOCITY DUCT DESIGN

Low velocity duct systems (up to 2500 fpm) are sized and designed by a variety of methods, some of which will be described in this chapter. General recommendations which apply to all duct designs are given here.

- Routes should be as direct as possible.

- Sudden changes in direction and velocity should be avoided.

- Turning vanes should be used whenever possible.

- Rectangular ducts should be as square as possible. Aspect ratios greater than 8:1 should be avoided, and 4:1 or less should be used whenever space permits.

- Smooth metal construction should be used.

- Since calculations are approximate, a fan with some excess capacity should be selected.

- To allow for leakage, the airflow through the fan should be 10% more than the sum of the outlet requirements.

- Dampers should be installed in all branches for balancing, even when the static-regain method is used in the design. Dampers should be installed as close as possible to the main duct in order to reduce noise.

- Nothing should be put in or through the ducts.

- For minimum fan horsepower and noise, the flow velocity should be as low as possible.

A. THE VELOCITY-REDUCTION METHOD

Ducts can be sized by the *velocity-reduction method,* a simple procedure in which the fan discharge is selected by judgment. Arbitrary reductions in velocity are made down the run, usually at branches and take-offs. This method requires great knowledge on the part of the designer and is not usually used except for estimating simple layouts.

Table 5.5 lists recommended maximum values of duct velocities for conventional low-velocity installations. The velocity-reduction method is summarized below.

step 1: Select a velocity, v, for each outlet.

step 2: Select the air flow requirements, Q, for each outlet.

step 3: Determine the duct size from $A = \dfrac{Q}{v}$.

step 4: By inspection, find the highest resistance duct. Calculate the static pressure drop in the run to size the fan.

Table 5.5
Recommended Maximum Duct Velocities (fpm)

Application	large supply ducts	small supply ducts	return ducts
residences	800	600	600
apartments/hotel bedrooms	1500	1100	1000
theaters	1600	1200	1200
private offices—deluxe		1100	800
private offices—average		1300	1000
general offices	2200	1400	1200
restaurants	1800	1400	1200
shops—small		1500	1200
department stores			
lower floors	2100	1600	1200
upper floors	1800	1400	1200

B. EQUAL-FRICTION METHODS

The *equal-friction per foot of length method* is superior to the velocity-reduction method and is used typically for low-velocity (i.e., less than 2500 fpm) systems in which velocity pressure (and possible regain) is minimal. A system thus designed still will require extensive dampering, however, since no attempt is made to equalize pressure drops in the branches.

The equal-friction method can be summarized as follows:

step 1: Select the main duct velocity, v_m, from table 5.5.

step 2: From v_m and Q_m, find the friction loss per foot from figure 5.4.

step 3: After each branch, reduce Q_m by the branch flow. Find the new velocity and duct size to keep the same friction loss per foot.

step 4: Calculate the equivalent length and the static pressure drop in the highest resistance duct. Decrease the pressure drop by any significant increase in velocity head.[5] Base the fan requirements on the static pressure drop plus the desired outlet pressure.

step 5: Compare the actual system pressure to the design system pressure, if known. If they are considerably different, repeat all steps with a different value of v_m.

step 6: Size shorter branch runs the same way— keeping the same friction rate and using dampers to equalize pressure.

A *combination method,* in which the main duct is sized by the equal-friction method and the branch ducts are sized to dissipate the additional friction is sometimes used. This is done by subtracting the desired outlet pressure from the pressure at the take-off to get the pressure which must be dissipated in the branch. This pressure is divided by the estimated equivalent length to find the drop per 100 feet and the duct size. Very high velocities should not be chosen, as excessive noise will result.

---

[5]It is not uncommon to ignore all changes in velocity head and the accompanying changes in static pressure head. This is in keeping with the philosophy that the equal-friction method is a low-velocity design method in which the kinetic energy available for conversion to static pressure is implicitly low. Such low-velocity systems are known as *conventional systems* to distinguish them from *high-velocity supply systems.*

Example 5.8

Size the theater duct system shown, using the equal-friction method.

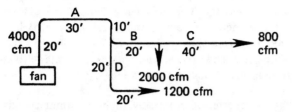

step 1: From table 5.5, select the velocity at A as 1600 fpm.

step 2: From figure 5.4, the duct diameter in section A is 21″. The friction loss per 100 feet is .15″ w.g.

step 3: The volume at B is 2800 cfm. From figure 5.4, with 2800 cfm and .15″ w.g. loss per 100 feet, the diameter is 18″, and the velocity is 1500 fpm. Similarly, the diameters at C and D are 11.5″ and 13″, respectively. The velocity at C is 100 fpm.

step 4: By inspection, the longest run is ABC. The equivalent length of this section is:

20′ fan to first bend

18′ first bend, assuming $\frac{r}{D} = 1.5$ from table 5.3

30′ first bend to second bend

18′ second bend, assuming $\frac{r}{D} = 1.5$

 70′ runs B and C

156′

Thus, the friction loss in the longest run, not counting the take-off loss, is

$$\frac{156}{100}(.15) = .23'' \text{ w.g.}$$

The take-off fitting loss, including any regain, with

$$\frac{V_B}{V_A} \approx 1 \text{ and } c_b = 1.5$$

is found from equation 5.37.

$$1.5\left(\frac{1600}{4005}\right)^2 = .24'' \text{ w.g.}$$

step 5: Assume a desired outlet pressure of .15″ w.g.

step 6: The net static pressure which the fan must supply is

$$.23 + .24 + .15 = .62'' \text{ w.g.}$$

This calculation does not include the static pressure regain between sections B and C. Static regain usually is not considered with the equal-friction method.

Example 5.9

A duct system consists of a long run and several smaller side ducts as shown. The friction loss in the longest run is .15″ w.g. Rather than rely on dampering to increase the pressure in duct A, it is desired to size duct A small enough to equalize the losses through increased velocity. The longest duct is sized with the equal friction method using a pressure drop of .2″ w.g. per 100 feet. Use the combination method to size duct A.

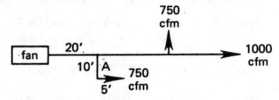

The pressure left to be dissipated in duct A is

$$.15 - \frac{20}{100}(.2) = .11$$

Assume that the equivalent length of both the branch take-off and the elbow is 15 feet. Then, the required loss per 100 feet in duct A is

$$(.11)\left(\frac{100}{10+5+15+15}\right) = .24'' \text{ w.g./100 feet}$$

Using figure 5.4 with 750 cfm and .24″ w.g. yields a velocity of 1250 fpm and a 10″ diameter.

A second iteration with better estimates of the take-off and elbow losses is possible now that an approximate velocity is known.

C. THE STATIC-REGAIN METHOD

If it is desired to have uniform static pressure at all branches and grilles without extensive dampering, it will be necessary to reduce the velocity at each branch. This reduction should be such that the recovery in static pressure (i.e., the *regain*) exactly offsets the friction loss in the succeeding section.

$$\Delta p_{e,1-2} = p_{f,2-3} \qquad 5.39$$

This procedure also is known as the *static-regain method* and the *total pressure method*. The static-regain method also can be used to recover the branch friction drops as long as the duct sizes remain reasonable.

p_f can be found analytically from equation 5.28, yielding an expression for v_2.

$$v_1^2 - v_2^2 = c(v_2)^{2.43}\frac{L_2}{(Q_2)^{.61}} \qquad 5.40$$

$$c = .0832 \text{ for } R = .75$$

$$c = .0693 \text{ for } R = .90$$

However, as v_2 appears on both sides of equation 5.40, it is easier to use figue 5.6. The (L/Q) ratio used in this figure is

$$\frac{L}{Q} \text{ ratio} = \frac{L_2}{(Q_2)^{.61}} \qquad 5.41$$

Usually, it is assumed that the regain will equal the friction loss. In that case, v_2 is read directly below the intersection of the (L/Q) and v_1 curves. However, figure 5.6 also can be used to determine a velocity which will increase or decrease the static pressure by some given amount. If a loss in static pressure is required, move to the right of the intersection point until the vertical distance between the two curves equals the desired loss. Then, drop down to read v_2. A gain is handled similarly, moving to the left.

The following is a summary of the static-regain method. This procedure assumes that the design is a long main run with numerous short-length distribution grilles. If the design consists of a main duct with take-offs and long branches, it will be necessary to include the es-timated branch friction drops, if any, in the value of $p_{s, start}$.

step 1: Assume an allowable outlet pressure variation, p_{var}, between the nearest and the farthest grilles. This is used to keep the area of the last sections from becoming unrealistically large in very long duct systems. The value of p_{var} is a matter of judgment, and .10″ w.g. may be used as a rule of thumb. However, $p_{var} = 0$ can be used to simplify calculations.

step 2: Assume an average grille discharge pressure, p_d. A good range is .15″ to .25″ w.g. (Use manufacturer's specifications if available.)

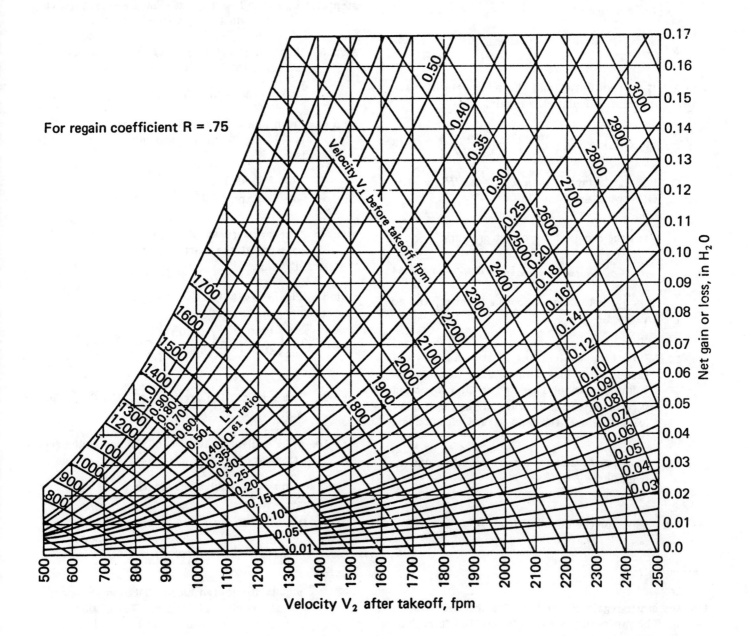

For regain coefficient R = .75

step 3: Assume a main run velocity, v_m, from table 5.5.

step 4: Size the main run from $A_m = \frac{Q_m}{v_m}$.

step 5: Find the equivalent length of duct from the fan to the first take-off. It may be necessary to make some assumptions about the bend radii.

step 6: From figure 5.4, find the friction loss in the main run up to the first take-off. This is Δp_{main}.

step 7: The design pressure at the nearest grille outlet is p_d. Therefore, the pressure at the start of the main run is

$$p_{s,\,\text{start}} = p_d + \Delta p_{\text{main}} \qquad 5.42$$

step 8: If the fan outlet and the main velocities are different, calculate the regain.

$$\Delta p_{\text{fan}} = R\left(\left(\frac{v_{\text{fan}}}{4005}\right)^2 - \left(\frac{v_m}{4005}\right)^2\right) \qquad 5.43$$

If v_m is larger than v_{fan}, use $R = 1.1$. Otherwise, use $R = .75$.

step 9: Find the required fan outlet pressure.

$$p_{s,\,\text{fan}} = p_{s,\,\text{start}} - \Delta p_{\text{fan}} \qquad 5.44$$

If $p_{s,\,\text{fan}}$ is too high, choose a lower v_m and go to step 4.

step 10: Find the equivalent length between grilles and take-offs. It may be necessary to make assumptions about sizes and radii. Proportion p_{var} along the grille run, giving the higher losses to longer runs and to runs with fittings.

step 11: Knowing L_e and Q_2 for each section, find v_2 from figure 5.6.

step 12: Solve for the duct size from $A = \frac{Q}{v}$.

Example 5.10

Use the static-regain method to size the duct system shown. The fan outlet is 1500 cfm at 1700 fpm. The equivalent length of the bend is 15 feet.

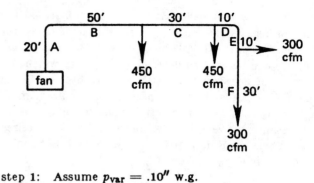

step 1: Assume $p_{\text{var}} = .10''$ w.g.

step 2: Assume an average $p_d = .25''$ w.g.

step 3: Assume $v_m = 1500$ fpm.

step 4: $A_m = \frac{1500}{1500} = 1$ sq. ft. For a round duct, the diameter will be approximately 14''.

step 5: The equivalent length of sections A and B, including the bend, is

$$20 + 15 + 50 = 85'$$

step 6: From figure 5.4, the friction loss per 100 feet is .20'' w.g. Thus, the loss in sections A and B is

$$\left(\frac{85}{100}\right).20 = .17'' \text{ w.g.}$$

step 7: The pressure at outlet 1 is .25'' w.g. The pressure at the start of run A is .25 + .17 = .42'' w.g.

step 8: The fan regain is

$$.75\left[\left(\frac{1700}{4005}\right)^2 - \left(\frac{1500}{4005}\right)^2\right] = .03'' \text{ w.g.}$$

step 9: The required fan pressure is .42 − .03 = .39'' w.g.

step 10: The equivalent lengths of the remaining sections are found:

section	equivalent length	Q
C	30	1050
D/E	10 + 10 + 15 = 35	600
F	30	300
	115' total	

p_{var} is distributed along CDEF in proportion to the equivalent length. Thus, .03 is distributed to sections C and F, and $\frac{D}{E}$ is given as .04'' w.g.

step 11: $L_e/(Q^{.61})$ is calculated for each section:

section	$L_e/Q^{.61}$
C	0.43
D/E	0.71
F	0.92

v_2 is found from figure 5.6, allowing for a loss in the static pressure between the v_1 and $L_e/Q^{.61}$ curves equal to the proportion of p_{var} calculated in step 10.

$$v_{2,C} = 1275 \text{ fpm}$$
$$v_{2,D/E} = 1050 \text{ fpm}$$
$$v_{2,F} = 850 \text{ fpm}$$

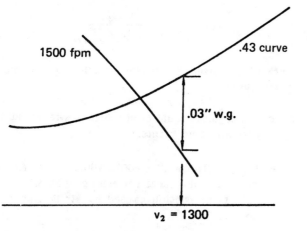

step 12: The required duct areas and diameters are:

$$A_C = \frac{1050}{1275} = 0.82 \qquad \text{or } 12'' \text{ diameter}$$

$$A_{D/E} = \frac{600}{1050} = 0.57 \qquad \text{or } 10'' \text{ diameter}$$

$$A_F = \frac{300}{850} = 0.35 \qquad \text{or } 8'' \text{ diameter}$$

5 DAMPERS

Dampers are included in all runs to balance the system. *Balancing* is the act of closing down dampers to equalize the friction losses in all ducts. Without dampering, the majority of the air flowing would escape out the closest grilles.

It is generally a good idea to install dampers even when more sophisticated design methods are used. Such dampers can be used for fine-tuning the installation.

Dampers can be motorized or operated manually. Figure 5.7 illustrates several common types of dampers.

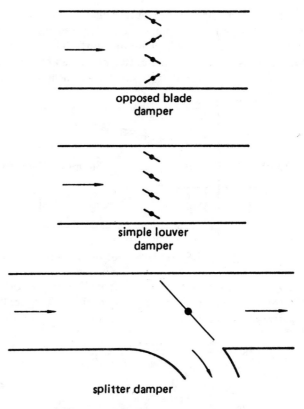

Figure 5.7 Types of Dampers

It is possible to design the outlet grilles to act as a damper. Such registers may be designed as a plate with multiple small holes through which the air must escape. The disadvantage of combining the tasks of air distribution and friction generation is that the noise created by the friction generation is projected directly into the room. It is better to have dampers installed close to the main supply and as far away from the grille as possible.

6 AIR DISTRIBUTION

An *outlet* is the general term used to describe any opening through which air enters the treated space. Although they are not strictly adhered to, the following definitions can be used to distinguish among outlets with different functions.

- grille: a decorative covering for an outlet

- diffuser: a functional grille guiding air direction

- register: a grille with an internal damper

The *gross area* or *core area* of the grille is its total cross sectional area. The total area of the openings in the grille constitutes the *free area* or *daylight area*.

When an outlet has been properly chosen, and the discharge pressure is available, the *terminal velocity* should be 50–75 fpm at the *distribution point*. This point is located a distance from the outlet known as the *throw*. This, and other concepts, are illustrated in figure 5.8.

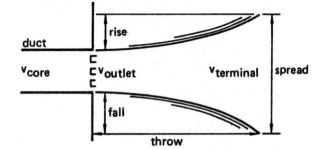

Figure 5.8 Air Distribution Terms

The outlet velocity can be found from the core velocity and the discharge coefficient of the outlet. C_d typically is between .7 and .9.

$$v_{outlet} = \frac{v_{core}A_{core}}{C_dA_{free}} = \frac{v_{core}}{C_dR_{f/c}} \qquad 5.45$$

$$R_{f/c} = \frac{A_{free}}{A_{gross}} \qquad 5.46$$

As it emerges from the outlet, duct air will entrain room air. This increase in moving air can increase the spread by 1 or 2 feet, even for straight outlets. The conservation of momentum laws can be used to predict the amount of entrained air.

$$m_o v_o + m_e v_e = (m_o + m_e) v_{final} \qquad 5.47$$

Since the velocity of the room air is initially zero, equation 5.47 becomes

$$m_o v_o = (m_o + m_e) v_{final} \qquad 5.48$$

$$v_o = R_e v_{final} \qquad 5.49$$

$$R_e = \frac{m_o + m_e}{m_o} \qquad 5.50$$

R_e is known as the *induction ratio* or the *entrainment ratio*.

The *centerline velocity* of distributed air emerging from a duct is approximately twice that of the average velocity across the entire flow. This centerline velocity can be predicted as a function of distance from the outlet if the approximate outlet constant, K_o, is known for the outlet.

$$v_x = \frac{K_o Q_o}{x\sqrt{A_c C_d R_{f/c}}} \qquad 5.51$$

Since the throw is defined roughly as the distance at which the average distribution velocity is 50 fpm, the throw can be calculated from equation 5.52.

$$throw = \frac{K_o Q_o}{100\sqrt{A_c C_d R_{f/c}}} \qquad 5.52$$

Values of K_o usually will be supplied by an outlet manufacturer.

The following rules should be considered when designing outlet types and placements.

- Increase throw 25%–50% when air is released along a wall or near the ceiling. This will counteract the friction caused by air flowing close to a surface.

- Select a throw approximately 75% of the distance from the outlet surface to the new normal surface. For example, the throw should be 75% of the floor-to-ceiling height for ceiling mounted outlets.

- For quiet operation, the outlet velocity should be in the 500–750 fpm range. High noise areas, such as stores and offices, however, might be able to tolerate 1000–1500 fpm outlets.

PART 3: Wind Power

1 TYPES OF WIND ENERGY COLLECTORS

Rotor-based wind energy collectors[6] can be classified in terms of the orientation of their rotational axes.

- Head-On Horizontal-Axis Rotors (Wind-Axis Rotors): The axis of rotation is parallel to the windstream.

- Cross-Wind Horizontal-Axis Rotors: The axis of rotation is horizontal to the earth and perpendicular to the windstream.

- Vertical-Axis Rotors: The axis of rotation is perpendicular to both the surface of the earth and the windstream.

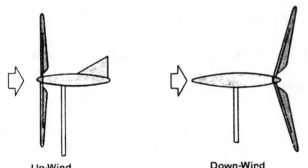

Up-Wind   Down-Wind

(a)   Head-On Horizontal-Axis

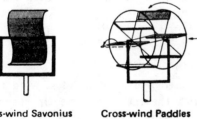

Cross-wind Savonius   Cross-wind Paddles

(b)   Cross-Wind Horizontal-Axis

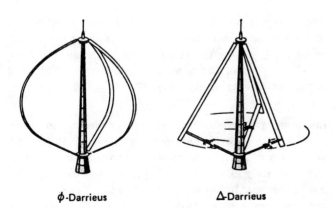

φ-Darrieus   Δ-Darrieus

[6]Another name commonly used is *Wind Energy Conversion System* (WECS).

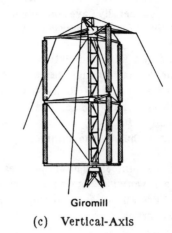

Giromill

(c)   Vertical-Axis

Figure 5.9 Types of Wind Energy Collectors

2 POWER CONTENT OF A WINDSTREAM

The total power available in a freely flowing windstream can be found from the momentum theory. The kinetic energy per unit volume is multiplied by the volumetric flow rate, resulting in a velocity-cubed term.

$$P = (Av)\left(\frac{\rho v^2}{2g}\right)\ \left(\frac{\text{ft-lb}}{\text{sec}}\right) \qquad 5.53$$

$$A = \pi\left(r_{\text{rotor}}^2\right) \qquad 5.54$$

As a result of this cubed term, the power densities of winds at sea level increase from approximately 5 watts per square foot at 10 mph to over 140 watts per square foot at 30 mph, and to approximately 650 watts per square foot at 50 mph.[7] At higher altitudes, the air density is less, so the wind power density is lower.

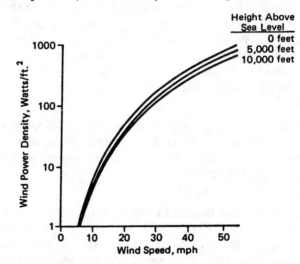

Figure 5.10 Windpower Density versus Wind Speed and Altitude

[7]Multiply ft-lbf/sec by 1.3558 to get watts.

Because of the need to overcome friction and inertia, WECS will not operate efficiently (or at all) at much less than 5 mph. Generally, average speeds above 10 mph are desired for maximum power conversion.

The wind at a given site frequently varies in direction, and its speed may change rapidly under gusting conditions. Its average velocity usually changes significantly with the changing seasons. In most locations, its amplitude often will be two or three times higher in the winter than in the summer.

Graphs showing the number of hours per year that the windstream, at a given site, attains a specified hourly mean wind speed are known as *annual average velocity duration* (AAVD) curves.

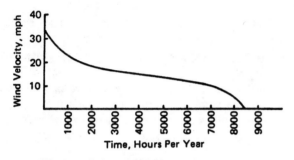

Figure 5.11 Typical AAVD Curve

Curves showing the distribution of annual average wind power per unit subtended area, as a function of windspeed, are called *annual average power density distribution* (AAPD) curves.

Because of the cubic relationship between wind power and wind velocity, and because wind gusts and is seldom steady, it is estimated that the actual wind power available at a given site can be two or three times that calculated on the basis of average annual wind speeds at that site. Therefore, depending on the responsiveness of a wind machine to these changes in wind speeds, the estimated performance of the machine can be conservative if based on annual average wind speeds.

The annual average wind *energy* density distribution is equal to the annual average *power* density distribution times the number of hours per year that the corresponding wind speeds occur.

Plots of the distribution of the annual average density of winds of various speeds at a given site show that most of the energy content of the wind is at speeds above the average wind speed. The contribution to the total annual average energy content of winds of all speeds is usually small for winds with speeds greater than approximately three times the average wind speed.

3 POWER EXTRACTION

A wind energy conversion system will be unable to generate power at the rate predicted by equation 5.53. Unfortunately, the extracted power is greatly affected by the type of wind turbine. Therefore, considerable experimentation is required to determine the operating characteristics of an actual installation.

None of the three methods for predicting output from a WECS presented here considers the mechanical efficiency of the drive or the electrical conversion efficiency of the generator. These two efficiencies should be included if the true electrical output from a WECS is needed.

A. THE POWER COEFFICIENT METHOD

The *power coefficient*, C_P, of a WECS is defined as the power actually delivered divided by the theoretical power from equation 5.53.

$$C_P = \frac{P_{actual}}{\left(\frac{A\rho v^3}{2g}\right)} \qquad 5.55$$

Alternately, the power delivered by the WECS can be determined if the power coefficient is known.

$$P_{actual} = C_P \left(\frac{A\rho v^3}{2g}\right) \qquad 5.56$$

The power coefficient of an ideal wind machine rotor varies with the ratio of blade tip speed to free-flow windstream speed and approaches the maximum of 0.593 when this ratio reaches a value of 5 or 6. This is illustrated in figure 5.12.

Experimental evidence indicates that two-bladed rotors of good aerodynamic design, running at high rotational speeds, (i.e., where the ratio of the blade-tip speed-to-free-flow speed of the windstream is 5 or 6), will have power coefficients as high as 0.47.

Likewise, a Darrieus rotor can be expected to reach a maximum power coefficient of approximately 0.35 at a ratio of peripheral speed-to-windstream speed of approximately 6. Other designs have been found to have lower maximum power coefficients. These maxima occur at lower speed ratios than the high-speed two-bladed horizontal-axis rotors and the Darrieus vertical-axis rotors.

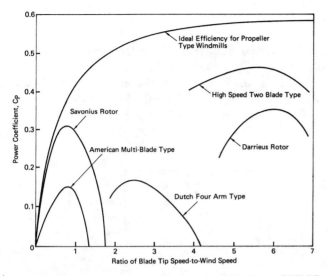

Figure 5.12 Power Coefficients for Various WECS Configurations

B. THE ENERGY CONSERVATION METHOD

The power extracted from the airflow can be determined from the conservation of energy equation if the pressure drop and the change in velocity are known. (p in equation 5.57 is measured in psi or psf.)

$$P_{actual} = Av\rho\left[\frac{p_2 - p_1}{\rho} + \frac{v_2^2 - v_1^2}{2g}\right] \qquad 5.57$$

C. MOMENTUM THEORY APPROACH

It is possible to make several simplifying assumptions in order to use conventional momentum calculations in the analysis of a wind-axis WECS. If the radial pressure gradient and the kinetic energy of the swirl component of velocity in the wake of the propeller are neglected, a stream tube containing all air affected by the WECS can be hypothesized, as in figure 5.13.

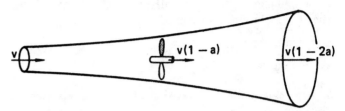

Figure 5.13 Stream Tube Used in Momentum Theory Approach

The air passing through the propeller is slowed according to the *interference factor, a*. The interference factor has a theoretical maximum value of $\frac{1}{3}$ when C_p is 0.593. The factor of 0.593 is known as the *Betz coefficient*. The interference factor can be derived from the power coefficient by using equation 5.58.

$$C_p = 4a(1-a)^2 \qquad 5.58$$

Using the interference factor with the ideal energy density of the air (equation 5.53) produces equation 5.59.

$$P_{actual} = 2a(1-a)^2 A\rho\left(\frac{v^3}{g}\right) \qquad 5.59$$

PRACTICE PROBLEMS: FANS AND DUCTWORK

Warm-ups

1. An aspect ratio of 4:1 is to be maintained in a rectangular duct with a diameter of 18″. What will be the equal-friction dimensions?

2. A fan moves 10,000 scfm against 4″ w.g. Draw the system curve.

3. A $\frac{1}{8}$ scale model fan tested at 300 rpm moves 40 cfm at standard conditions against $\frac{1}{2}$″ w.g. What horse-power will be required if a full-size fan operates outside at 5000 feet altitude at half the model's speed? Use standard atmospheric data and the fan similarity laws.

4. What is the friction loss through 750 feet of gal-vanized 20″ diameter ductwork flowing 6000 scfm if there are four round elbows with radius-to-diameter ratio of 1.5 and two 2″ diameter pipes passing perpen-dicularly through the center of the duct at two loca-tions?

5. What is the static regain in an 18″ duct flowing 1500 scfm which reduces to 14″ after a take-off of 400 scfm?

Concentrates

1. A fan in a theater delivers 1500 cfm to the system shown. Size the system with rectangular duct. Use an aspect ratio of 1.5, and allow 0.25″ w.g. at each outlet. Use the equal-friction method. Ignore take-off fitting losses. $r/D = 1.5$ for all elbows.

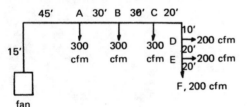

2. Use round duct with 1.25 r/D ratio elbows to size the system (in a theater) shown. Each outlet receives 300 cfm at 0.15″ w.g. Use the equal-friction method. Disregard fitting take-off losses. Assume the longest run has the largest equivalent friction loss.

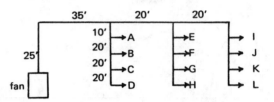

3. Repeat problem 2 with the static-regain method and a regain coefficient of .75.

Timed: (1 hour allowed for each)

1. Air with a density of .075 lbm/ft^3 flows through a 12″ diameter duct (section A). The Darcy friction factor is .02 everywhere in the system. The 4-piece 90° ells have a $\frac{r}{D}$ ratio of 1.5. The static regain coefficient for this system is .65. Use the static regain method to calculate the diameters of sections B and C. Specify where dampers should be placed (sections A, B, or C) and what friction loss should be associated with each.

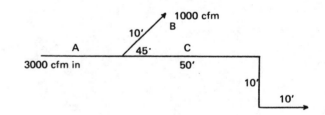

RESERVED FOR FUTURE USE

RESERVED FOR FUTURE USE

6 THERMODYNAMICS

PART 1: Properties of a Substance

Nomenclature

a	van der Waals' correction factor	atm-ft^6/pmole
A	area	ft^2
b	van der Waals' correction factor	ft^3/pmole
B	volumetric fraction	–
c	specific heat	BTU/lbm-°R
C	specific heat	BTU/pmole-°R
E	energy	BTU/lbm
g	local gravitational acceleration	ft/sec^2
g_c	gravitational constant (32.2)	lbm-ft/lbf-sec^2
G	gravimetric fraction	–
h	enthalpy	BTU/lbm
H	enthalpy	BTU/pmole
J	Joule's constant (778)	ft-lbf/BTU
k	ratio of specific heats, Boltzmann constant	–, J/°K
m	mass	lbm
$\dot{m}$	mass flow rate	lbm/sec
M	molecular weight, Mach number	lbm/pmole,–
n	number of moles, polytropic exponent	–
N	number of molecules	–
N_o	Avogadro's number (6.023 EE23)	molecules/gmole
p	pressure	lbf/ft^2
P	power	BTU/sec
Q	heat	BTU or BTU/lbm
R	specific gas constant	ft-lbf/lbm-°R
R*	universal gas constant (1545.33)	ft-lbf/pmole-°R
s	entropy	BTU/lbm-°R
S	entropy	BTU/pmole-°R
T	temperature	°R
u	internal energy	BTU/lbm

U	internal energy	BTU/pmole
v	velocity	ft/sec
V	volume	ft^3
W	work	BTU or BTU/lbm
x	quality, or mole fraction	–
z	height above datum	ft
Z	compressibility factor	–

Symbols

η	efficiency	–
ρ	density	lbm/ft^3
υ	specific volume	ft^3/lbm
ω	humidity ratio	–
μ	Joule-Thompson coefficient	°R-ft^2/lbf
ϕ	relative humidity	–
Φ	availability function	BTU/lbm

Subscripts

*	at sonic velocity
a	moist air
c	critical
f	saturated liquid
fg	vaporization
g	saturated vapor
k	kinetic
l	latent
m	mean
o	environment
p	potential, constant pressure, probable
r	ratio, reduced
rms	root-mean-squared
s	isentropic, sensible
sat	saturated
th	thermal
T	total
v	constant volume
w	water

1 PHASES OF A PURE SUBSTANCE

Thermodynamics is the study of a substance's energy-related properties. This study can be theoretical and based on derivations, or it can be result-oriented. This chapter is practical in its approach, and it provides appropriate background for the useful applications introduced in chapter 7.

The properties of a substance and the procedures used to determine those properties depend on the phase of the substance. It is convenient to distinguish between more than just the usual solid, liquid, and gas phases. Because they behave according to different rules, it is necessary to distinguish between the following phases and sub-phases.[1]

   solid—A solid does not take on the shape or volume of its container.

   subcooled liquid—If a liquid is not saturated (i.e., the liquid is not at its boiling point), it is said to be subcooled. 60°F water at standard atmospheric pressure is subcooled, as the addition of a small amount of heat will not cause vaporization.

   saturated liquid—A saturated liquid has absorbed as much heat energy as it can without vaporizing. Liquid water at standard atmospheric pressure and 212°F is an example of a saturated liquid.

   liquid-vapor mixture—A liquid and a vapor can coexist at the same temperature and pressure. This is called a two-phase, liquid-vapor mixture.

   saturated vapor—A vapor (e.g., steam at standard atmospheric pressure and 212°F) which is on the verge of condensing is said to be saturated.

   superheated vapor—A superheated vapor is one which has absorbed more heat than is needed merely to vaporize it. A superheated vapor will not condense when small amounts of heat are removed.

   ideal gas—A gas is a highly superheated vapor. If the gas behaves according to the ideal gas laws, it is called an ideal gas.

   real gas—A real gas does not behave according to the ideal gas laws.

   gas mixtures—Most gases mix together freely. Two or more pure gases together constitute a gas mixture.

   vapor/gas mixtures—Atmospheric air is an example of a mixture of several gases and water vapor.

These phases and sub-phases can be illustrated with a pure substance in the piston/cylinder arrangement shown in figure 6.1. The pressure in this system is

---

[1]Plasma and solids near absolute zero are not discussed in this chapter.

determined by the weight of the piston, which moves freely to permit volume changes.

In illustration (a), the volume is minimum. This is the solid phase. The temperature will rise as heat, Q, is added to the solid. This increase in temperature is accompanied by a small increase in volume. The temperature increases until the melting point is reached.

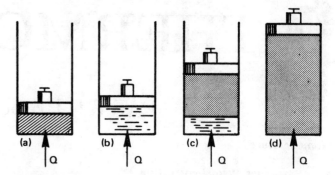

Figure 6.1 Phase Changes at Constant Pressure

The solid will begin to melt as heat is added to it at the melting point. The temperature will not increase until all of the solid has been turned into liquid. The liquid phase, with its small increase in volume, is illustrated by (b).

If the subcooled liquid continues to receive heat, its temperature will rise. This temperature increase continues until evaporation is imminent. The liquid at this point is said to be saturated. Any increase in heat energy will cause a portion of the liquid to vaporize. This is shown by (c), in which a liquid/vapor mixture exists.

When a substance exists as part liquid and part vapor at the saturation temperature, its *quality*, x, is defined as the fraction of the total mass which is vapor.

$$x = \frac{m_{\text{vapor}}}{m_{\text{vapor}} + m_{\text{liquid}}} \qquad 6.1$$

As with melting, evaporation occurs at constant temperature and pressure but with a very large increase in volume. The temperature cannot increase until the last drop of liquid has been evaporated, at which point the vapor is said to be saturated. This is shown by (d).

Additional heat will result in a high-temperature *superheated vapor*. This vapor may or may not behave according to the ideal gas laws, depending on the temperature.

2 DETERMINING PHASE

It is important to know which phase[2] a substance is in since most equipment is incompatible with some phases.

---

[2]The word *phase* is always used instead of *state*, which has a different meaning in thermodynamics. The state of a substance will change any time a property changes, even though the phase remains the same.

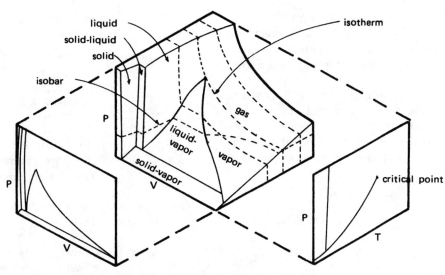

Figure 6.2 An Equilibrium Solid

For example, you cannot put ice through a turbine. Nor can you put gas through a centrifugal pump.

It is theoretically possible to develop a three-dimensional surface that predicts the substance's phase based on the properties of pressure, temperature, and specific volume. Such an *equilibrium solid* is illustrated in figure 6.2. Equilibrium solids are not of much value in quantitative problems.

If one property is held constant through a process, a two-dimensional projection of the equilibrium solid can be used. This projection is known as an *equilibrium diagram* or a *phase diagram*.

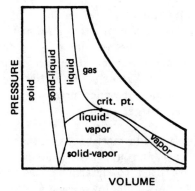

Figure 6.3 Phase Diagram

The important part of a phase diagram is limited to the liquid and vapor region. A general phase diagram showing this region and the bell-shaped dividing line (known as the *vapor dome*) is given in figure 6.4.

The vapor dome region can be drawn with many variables for the axes. For example, either temperature or pressure can be used for the vertical axis. Energy, volume, or entropy can be chosen for the horizontal axis.

However, the principles presented here apply to all combinations.

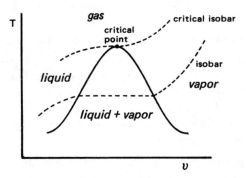

Figure 6.4 The Vapor Dome with Isobars

The left-hand part of the vapor dome separates the liquid phase from the liquid/vapor phase. This part of the line is known as the *saturated liquid line*. Similarly, the right-hand part of the line separates the liquid/vapor phase from the vapor phase. This line is called the *saturated vapor line*.

Lines of constant pressure (isobars) can be superimposed on the vapor dome. Each isobar is horizontal as it passes through the two-phase region, verifying that both temperature and pressure remain unchanged as a liquid vaporizes.

Notice that there is no dividing line between liquid and vapor at the top of the vapor dome. Far above the vapor dome, there is no distinction between liquids and gases as their properties are identical. The phase is assumed to be a gas.

The implied dividing line between liquid and gas is the isobar which passes through the top-most part of the vapor dome. This is known as the *critical isobar*, and the top-most point of the vapor dome is known as the *critical point.*[3] This critical isobar also provides a way to distinguish between a vapor and a gas. A substance below the critical isobar (but to the right of the vapor dome) is a vapor. Above the critical isobar, it is a gas.

Figure 6.5 illustrates a vapor dome for which pressure has been chosen as the vertical axis and enthalpy (*h*) has been chosen as the horizontal axis. The shape of the dome is essentially the same, but the lines of constant temperature (*isotherms*) have a different slope direction from isobars.

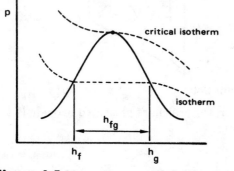

Figure 6.5 Vapor Dome with Isotherms

Figure 6.5 also illustrates the subscripting convention used to identify points on the saturation line. The subscript *f* (fluid) is used to indicate a saturated liquid. The subscript *g* (gas)[4] is used to indicate a saturated vapor. The subscript *fg* is used to indicate the difference in saturation properties.

The vapor dome is a good tool for illustration, but it cannot be used to determine a substance's phase. Such a determination must be made based on the substance's pressure and temperature.

For example, consider water in a container surrounded by 14.7 psia air. Water's boiling temperature (the *saturation temperature*) at 14.7 psia is 212°F. If water has a temperature lower than 212°F, say 85°F, we know that the water is a liquid. On the other hand, if the water's properties are 14.7 psia and 270°F, we know that it is a vapor.

This illustration is valid only for 14.7 psia water. Water at other pressures will have other boiling temperatures. (The lower the pressure, the lower the boiling temperature.) However, the rules given here follow directly from the previous example. The rules will become more meaningful as you progress through this chapter.

Rule 6.1 A substance is a subcooled liquid if its temperature is less than the saturation

temperature for the pressure to which it is exposed.

Rule 6.2 A substance is in the liquid-vapor region if its temperature is equal to the saturation temperature for the pressure to which it is exposed.

Rule 6.3 A substance is a superheated vapor if its temperature is greater than the saturation temperature for the pressure to which it is exposed.

The rules that follow can be stated using pressure as the determining variable.

Rule 6.4 A substance is a subcooled liquid if its pressure is greater than the saturation pressure for the temperature to which it is exposed.

Rule 6.5 A substance is in the liquid-vapor region if its pressure is equal to the saturation pressure for the temperature to which it is exposed.

Rule 6.6 A substance is a superheated vapor if its pressure is less than the saturation pressure for the temperature to which it is exposed.

3 PROPERTIES POSSESSED BY A SUBSTANCE

The thermodynamic *state* or condition of a substance is determined by its properties. *Intensive properties* are independent of the amount of substance present. Temperature, pressure, and stress are examples of intensive properties. *Extensive properties* are dependent on the amount of substance present. Examples are volume, strain, charge, and mass.

In this chapter, and in most books on thermodynamics, both lower case and upper case forms of the same characters are used to determine the basis for the properties. For example, lower case *h* is used to represent enthalpy in BTU's[5] per pound (BTU/lbm). Upper case *H* is used to represent enthalpy in BTU's per mole (BTU/pmole).

A. MASS: *m*

The mass of a substance is measured exclusively in English units in this chapter. Mass can be expressed in either pounds-mass (lbm) or pound-moles (pmole). A pound-mole is an amount of substance which has a mass

---

[3]The critical properties for water are 1165°R and 218.2 atmospheres. Critical properties for other substances are given in table 6.7.

[4]Although this book makes it a rule never to call a vapor a gas, this convention is not adhered to in the field of thermodynamics. The subscript *g* is standard for a saturated vapor.

[5]The *British Thermal Unit* is a measure of heat energy. It is approximately the energy given off by burning one wooden match.

In pounds equal to the molecular weight. For example, 18 lbm equals one pmole of water.

B. TEMPERATURE: T

Temperature is a thermodynamic property of a substance which depends on energy content. Heat energy entering a substance will increase the temperature of that substance. Normally, heat energy will flow only from a hot object to a cold object. If two objects are in thermal equilibrium (are at the same temperature), no heat will flow.

If two systems are in thermal equilibrium, they must be at the same temperature. If both systems are in equilibrium with a third, then all three are at the same temperature. This concept is known as the *Zeroth Law of Thermodynamics*.

The scales most commonly used for measuring temperature are the Fahrenheit and Celsius scales. The relationship between these two scales is:

$$T_{\circ F} = 32 + \left(\frac{9}{5}\right) T_{\circ C} \qquad 6.2$$

The absolute temperature scale defines temperature independently of the properties of any particular substance. This is unlike the Celsius and Fahrenheit scales, which are based on the freezing point of water. The absolute temperature scale should be used for all calculations unless a temperature difference is needed.

In the English system, the absolute scale is the *Rankine scale*.

$$T_{\circ R} = T_{\circ F} + 460^{\circ} \qquad 6.3$$
$$\Delta T_{\circ R} = \Delta T_{\circ F} \qquad 6.4$$

The absolute temperature scale in the SI system is the *Kelvin scale*.

$$T_{\circ K} = T_{\circ C} + 273^{\circ} \qquad 6.5$$
$$\Delta T_{\circ K} = \Delta T_{\circ C} \qquad 6.6$$

These four temperature scales are compared in figure 6.6.

---

[6] The name *Centigrade* is no longer correct.

C. PRESSURE: p

Pressure in thermodynamics problems can be given in psi, psf, or atmospheres. One *standard*[7] atmosphere is approximately 14.7 psia. Other pressure units include inches of water (407.1 inches of water equal one atmosphere), millimeters of mercury (760 millimeters of mercury equal one atmosphere), torr (760 torr equal one atmosphere), and bars (one bar equals one atmosphere). Torr and millimeters of mercury are essentially identical.

D. DENSITY: ρ

Density has been covered in chapter 4. As in that chapter, density will be given in lbm/ft^3 in this chapter. Density is the reciprocal of specific volume.

$$\rho = \frac{1}{v} \qquad 6.7$$

E. SPECIFIC VOLUME: v

Specific volume is the reciprocal of density. It is the volume that is occupied by one pound-mass of the substance. As such, its units are ft^3/lbm.

$$v = \frac{1}{\rho} \qquad 6.8$$

F. INTERNAL ENERGY: u and U

Internal energy encompasses all of the potential and kinetic energies of the atoms or molecules in a substance. Energies in the translational, rotational, and vibrational modes are included. Since this movement increases as the temperature increases, internal energy is a function of temperature. Internal energy does not depend on the process or path taken to reach a particular temperature.

Internal energy can be represented by a lower case u with units of BTU/lbm, or it can be represented by an upper case U with units of BTU/pmole. Of course, the relationship between u and U depends on the molecular weight of the substance.

$$U = Mu \qquad 6.9$$

---

[7] The term *STP (Standard Temperature and Pressure)* has several meanings. The standard pressure always is one atmosphere. The most common standard temperature is 32°F (0°C). However, 60°F, 68°F, and 70°F also are used in specific industrial situations.

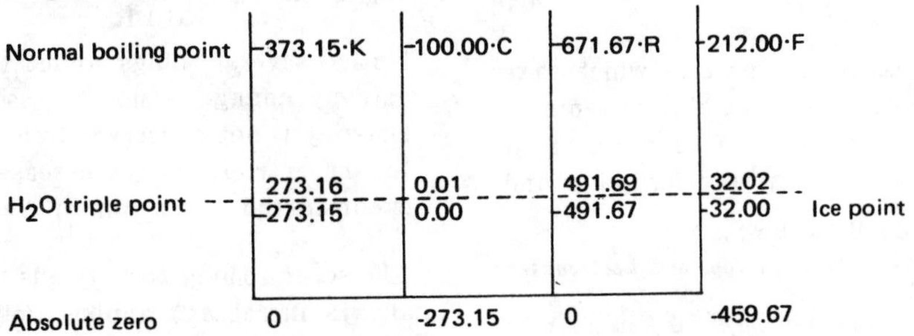

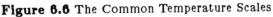

	Normal boiling point	373.15·K	100.00·C	671.67·R	212.00·F
H₂O triple point		273.16	0.01	491.69	32.02
		273.15	0.00	491.67	32.00	Ice point
Absolute zero		0	-273.15	0	-459.67

Figure 6.6 The Common Temperature Scales

Calculations in this book use u (the *specific internal energy*) exclusively. However, many tabulations of thermodynamic properties in other books are given in BTU/pmole (the *molar internal energy*). A conversion from a molar basis to a specific basis can be performed with equation 6.9.

G. ENTHALPY: h and H

Enthalpy[8] is a property which represents the total useful energy[9] in the substance. Useful energy consists of two parts—the internal energy and the *flow energy*.[10]

$$h = u + \frac{pv}{J} \qquad \text{(BTU/lbm)} \qquad 6.10$$

$$H = U + \frac{pV}{J} \qquad \text{(BTU/pmole)} \qquad 6.11$$

$$H = Mh \qquad 6.12$$

Enthalpy is defined as useful energy because, if the ambient conditions are proper, all of it can be used to perform useful tasks. It takes energy to increase the temperature of a substance. If that internal energy is recovered, it can be used to heat something else (e.g., to vaporize water in a boiler). Also, it takes energy to increase pressure and volume (as in blowing up a balloon). If pressure and volume are decreased, useful energy is given up.

The J term in equations 6.10 and 6.11 is known as *Joule's constant*. It has a value of 778 ft-lbf/BTU and is a conversion factor between ft-lbf and BTU. It is needed because internal energy (u) has the units of BTU, but the product of pressure and volume results in units of ft-lbf.

H. ENTROPY: s and S

Entropy[11] is a measure of energy which is no longer available to perform useful work within the current environment. Other definitions are frequently used (e.g., disorder of the system, randomness), but these alternate definitions are difficult to use in the framework of a numerical calculation.

The total unavailable energy in a system is equal to the summation of all unavailable energy inputs over the life of the system. That is,

$$s = \sum \Delta s \qquad 6.13$$

For an isothermal process (i.e., a process which takes place at a constant temperature) occurring at temperature T_o, the change in entropy is a function of the energy transfer. If Q is the energy transfer per pound-

mass of substance, the entropy change is given by equation 6.14.

$$\Delta s = \frac{Q}{T_o} \qquad 6.14$$

For processes that occur over a varying temperature, the entropy change must be found by integration.

$$\Delta s = \int ds = \int \frac{dQ}{T} \qquad 6.15$$

From equations 6.14–6.15, it can be seen that the units of entropy are BTU/lbm-°R. Entropy also can be given with units of BTU/pmole-°R, in which case a capital S would be used. Of course, s and S are related by the substance's molecular weight.

$$S = Ms \qquad 6.16$$

The concept of entropy and how it relates to unavailable energy is illustrated by the three identical planets in figure 6.7. The average temperatures of planets A, B, and C are 530°R, 520°R, and 510°R, respectively. All three planets are large, so small energy transfers among from them will not change the average temperature. Therefore, such energy transfers can be considered isothermal.

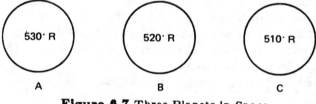

Figure 6.7 Three Planets in Space

Suppose that heat radiation transfers 1000 BTU/lbm from planet B to planet C. This transfer will occur naturally because planet B is hotter than planet C. The energy is gone from planet B, and it cannot be used. Furthermore, it cannot be recovered through a natural process because heat will not flow by itself from a cold object to a hot object.[12]

From equation 6.14, the entropy changes in planets B and C are:

$$\Delta s_B = \frac{(-1000)\ \text{BTU/lbm}}{520°R} = -1.92\ \text{BTU/lbm-°R}$$

$$\Delta s_C = \frac{(1000)\ \text{BTU/lbm}}{510°R} = 1.96\ \text{BTU/lbm-°R}$$

Notice several things about this transfer. First, the entropy change is not the same for the two planets. Entropy is not conserved in an energy transfer process. In fact, entropy always increases when the total universe is considered. (The sum of Δs_B and Δs_C is non-zero.)

The second thing to notice is that the phrase "entropy always increases" applies only to the universe as a

---

[8]The accent is on the second syllable: En-thal'-py.

[9]The older terms of *total heat*, *total energy*, and *heat content* are no longer recommended.

[10]Other names for the pv term are pv-*work* and *flow work*.

[11]The accent is on the first syllable: Eń-tro-py.

[12]This is one way of stating the *Second Law of Thermodynamics*.

whole. Localized collections of matter, such as planet B, can have a decrease in entropy if they lose energy.

Finally, notice that planet B can be brought back to its original condition if 1000 BTU/lbm are transferred to it from planet A. The transfer must be from planet A to planet B. It cannot be from planet C to planet B as heat will not flow naturally in that direction.

An analogous situation in which entropy is continually increasing and decreasing is water in a closed boiler/turbine installation. The entropy increases when the water is vaporized in the boiler. The entropy decreases when the steam is expanded through the turbine. When the water returns to the boiler, its entropy increases to its original value.

The Second Law of Thermodynamics can be related to entropy: *A natural process that starts in one equilibrium state and ends in another will go in the direction that causes the entropy of the system and the environment to increase.* In this form, the Second Law applies only to *irreversible processes*—those processes having a "natural direction."

Although all real-world processes result in an overall increase in entropy, it is possible to conceptualize processes that have a zero entropy change. Such processes are said to be *reversible* or *isentropic*. For a reversible process,

$$\Delta s = 0 \qquad 6.17$$

Processes which contain friction are never reversible. Other processes which are not as obviously irreversible are listed here.

- stirring a viscous liquid
- moving fluid coming to rest
- magnetic hysteresis
- ideal gas expanding into a vacuum
- throttling
- releasing a stretched spring
- chemical reactions
- diffusion of gases
- freezing a supercooled liquid
- condensation of vapor
- heat conduction

I. SPECIFIC HEAT: c and C

Heat energy is needed to cause a rise in temperature. Substances differ in the quantity of heat needed to produce a given temperature increase. The ratio of heat to the temperature change is called the *specific heat* of the substance, c.

$$c = \frac{Q}{m\Delta T} \qquad 6.18$$

Equation 6.18 can be rearranged to give the heat required to change the temperature of an object with mass m.

$$Q = mc\Delta T \qquad 6.19$$

The lower case c implies that the units are BTU/lbm-°R. The *molar specific heat* may be given, and its units are BTU/pmole-°R. A capital C is used to represent the molar specific heat.

$$C = Mc \qquad 6.20$$

Values of specific heat are listed in table 6.1.

Table 6.1
Approximate Specific Heats of Some Liquids and Solids
(BTU/lbm-°R)

Substance	T, °F	c
Aluminum, pure	100	.225
alloy (2024-T4)	−200-200	.151-.217
Ammonia	−50-100	1.00-1.16
Asbestos	70	.195
Brass, red	70	.093
Bronze	70	.082
Concrete	70	.21
Copper, pure	100	.094
Freon-12	−20-100	.214-.240
Gasoline	0-100	.465-.526
Glass	70	.18
Gold, pure	100	.031
Ice	32	.49
Iron, pure	70	.11
cast (4% C)	70	.10
Lead, pure	100	.031
Magnesium, pure	100	.24
Mercury	0-600	.033-.032
Oil, light	100-300	.46-.54
Silver, pure	100	.06
Steel, (1010)	70	.102
stainless (301)	70	.109
Tin, pure	200	.055
Titanium, pure	100	.13
Tungsten, pure	100	.032
Water	32	1.007
	100	.998
Wood, typical	70	.6
Zinc, pure	100	.088

For gases, the specific heat depends on the conditions under which the heat exchange occurs. The most common conditions are those of constant volume and constant pressure, designated by subscripts v and p, respectively.

$$Q = mc_v\Delta T \qquad \text{(constant volume)} \qquad 6.21$$
$$Q = mc_p\Delta T \qquad \text{(constant pressure)} \qquad 6.22$$

Values of c_p and c_v for common gases are given in table 6.4. Specific heats of solids and liquids vary only slightly with temperature. The mean specific heat is generally used for processes covering a large temperature range. c_p and c_v for solids and liquids are essentially the same.

$$Q = \dot{m}_w C_p \Delta T (R) = V \varsigma C_p \Delta T$$

J. RATIO OF SPECIFIC HEATS: k

For gases, the ratio of specific heats is defined by equation 6.23.

$$k = \frac{c_p}{c_v} \qquad 6.23$$

Values of k are given in table 6.4.

K. LATENT HEATS

Energy which changes the phase of a substance is distinguished from energy which produces a change only in temperature. Energy entering a substance, known as the *total heat*, may be divided into *latent* and *sensible* heats. Latent heat is that energy which produces a change in phase without causing a temperature change. Examples of latent effects are melting, vaporization, sublimation, and changes in crystalline form. Some typical values of latent energy for water are listed in table 6.2.

Table 6.2
Latent Heats for 14.7 psia Water

	BTU/lbm	cal/g	kcal/mole
Fusion (ice to water)	143.4	79.71	1.434
Vaporization (water to vapor)	970.3	539.55	9.703

The latent heat of vaporization is so important in thermodynamic calculations that it has its own symbol— h_{fg}. h_{fg} can be found tabulated in most listings of thermodynamic properties of liquids.

Sensible heat is the heat that changes temperature. The amount of heat required is dependent on the temperature and specific heat of a substance. Total heat and sensible heat are equal for substances which undergo a temperature change without a phase change.

Example 6.1

How much energy is required to vaporize 1 pound-mass of water which is originally at 75°F and standard atmospheric pressure?

The sensible heat required to raise the temperature of the water from 75°F to 212°F is

$$Q_s = mc(T_2 - T_1)$$
$$= (1)\ \text{lbm}\ (1)\ \frac{\text{BTU}}{\text{lbm-°R}}\ (212 - 75)\,°\text{F}$$
$$= 137\ \text{BTU}$$

The latent heat required to vaporize the water is found in table 6.2 to be 970.3 BTU/lbm.

Therefore, the total heat required is

$$Q = Q_s + Q_l = 137 + 970.3$$
$$= 1107.3\ \text{BTU}$$

4 USING TABLES TO FIND PROPERTIES

A. MOLLIER DIAGRAM

The *Mollier diagram* is a graph of enthalpy versus entropy for steam. It is particularly suitable for determining property changes between the superheated vapor and the liquid-vapor regions. For this reason, the Mollier diagram covers only a limited region, as indicated by the dashed lines in figure 6.8.

The Mollier diagram plots the enthalpy for a pound of steam as the ordinate and plots the entropy as the abscissa. Lines of constant pressure (isobars) slope upward from left to right. (In the low-pressure region at the right-hand side, dotted lines represent absolute pressure in inches of mercury and are convenient for exhaust steam calculations.) Below the saturation line, curves of constant moisture content slope down from left to right. Above the saturation line are lines of constant temperature and lines of constant superheat.

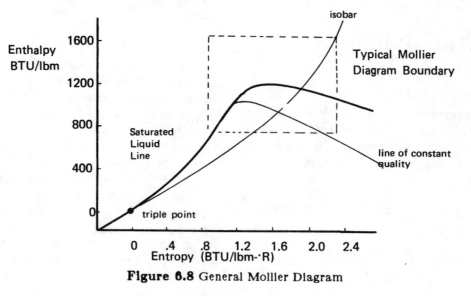

Figure 6.8 General Mollier Diagram

A complete Moller diagram is given at the end of this chapter.

Example 6.2

Find the following properties using the Mollier diagram. (a) enthalpy and entropy of steam at 700 psia and 1000°F, (b) enthalpy of steam at 1 in. Hg absolute and 80% quality, and (c) final temperature of steam throttled from 700 psia and 1000°F to 450 psia.

(a) Reading directly from the Mollier diagram at the end of this chapter: $h = 1510$ BTU/lbm, $s = 1.698$ BTU/lbm-°R.

(b) 80% quality is the same as 20% moisture. Reading at the intersection of 20% moisture and 1 in. Hg: $h = 885$ BTU/lbm.

(c) By definition, a throttling process does not change the enthalpy. This process is represented by a horizontal line to the right on the Mollier diagram. Starting at the intersection of 700 psia and 1000°F and moving horizontally to the right until 450 psia is reached defines the end point of the process. The final temperature is interpolated as 988°F.

B. STEAM TABLES

The information presented graphically by the Mollier diagram can be obtained with greater accuracy from *steam tables*.[13] The tables represent extensive tabulations of data for liquid and vapor phases of water.

The steam tables contain data about specific volume (v), enthalpy (h), internal energy (u), and entropy (s). These properties are functions of temperature, and appendix A is organized in this manner.

However, as shown in figure 6.4, there is a unique pressure associated with each temperature (i.e., there is only one horizontal isobar for each temperature). Since the pressure does not vary in even increments when temperature is changed, the second column of appendix A varies irregularly. A second property table, appendix B, is set up so that the pressure increments evenly.

In appendix A, the first column after temperature gives the corresponding saturation pressure. The next three columns give specific volume. The first of these gives the specific volume of the saturated liquid, v_f; the second column gives the increase in specific volume when the state changes from saturated liquid to saturated vapor, v_{fg}; the third column gives the specific volume of the saturated vapor, v_g.

---

[13]A more general name is *property tables*. Property tables are available in essentially the same format for all common working fluids.

The relationship between v_f, v_{fg}, and v_g is given by equation 6.24.

$$v_g = v_f + v_{fg} \qquad 6.24$$

The subsequent columns list the same data for enthalpy and entropy.

$$h_g = h_f + h_{fg} \qquad 6.25$$

$$s_g = s_f + s_{fg} \qquad 6.26$$

In appendix B, the first column after the pressure gives the corresponding saturation temperature. The next two columns give specific volume in a manner similar to that of appendix A, except that v_{fg} is not listed. When necessary, v_{fg} can be found by subtracting v_f from v_g. Appendix B also has a tabulation of internal energy.

C. SUPERHEAT TABLES

In the superheated region, pressure and temperature are independent properties. Therefore, for each pressure, a large number of temperatures is possible. Appendix C is a superheated steam table which gives the properties of specific volume, enthalpy, and entropy for various combinations of temperature and pressure.

D. COMPRESSED LIQUID TABLES

Water is only slightly compressible. For most thermodynamic problems, changes in properties for a liquid are negligible. In problems where the exact values are needed, appendix D should be used. This table gives the properties at the saturation state and the corrections to those properties for various pressures.

E. GAS TABLES

Gas tables are essentially superheat tables for which an assumption has been made about the pressure. For example, appendix F is a gas table for air at low pressures. "Low pressure" means less than several hundred psi pressure. However, reasonably good results can be expected even if pressures are higher.

Gas tables are indexed by temperature. That is, implicit in their use is the assumption that properties are functions of temperature only. Gas tables are not arranged in the same way as other property tables.

The *volume ratio* (v_r) and the *pressure ratio* (p_r) columns can be used when gases take part in an isentropic process. These two columns are not the specific volume or pressure. They are ratios with arbitrary references that make analysis of isentropic processes easier. Their use is illustrated in example 6.3 and is based on equations 6.27 and 6.28.

$$\frac{v_{r,1}}{v_{r,2}} = \frac{V_1}{V_2} \qquad (\Delta s = 0) \qquad 6.27$$

$$\frac{p_{r,1}}{p_{r,2}} = \frac{p_1}{p_2} \qquad (\Delta s = 0) \qquad 6.28$$

Entropy is not listed at all in appendix F. Instead a column of *entropy functions* (ϕ) with the same units as entropy is given. The entropy function is not the same as specific entropy, but it can be used to calculate the change in entropy as the gas goes through a process. This entropy change is calculated with equation 6.29.

$$s_2 - s_1 = \phi_2 - \phi_1 - \left(\frac{R}{J}\right) ln\left(\frac{p_2}{p_1}\right) \qquad 6.29$$

Example 6.3

Air is originally at 60°F and 14.7 psia. It is compressed isentropically to 86.5 psia. What is its new temperature and enthalpy?

From appendix F for 520°R, $p_{r,1} = 1.2147$. Using equation 6.28,

$$p_{r,2} = \frac{(1.2147)(86.5)}{14.7} = 7.148$$

Searching the p_r column of appendix F results in $T = 860°R$ and $h = 206.46$ BTU/lbm.

5 DETERMINING PROPERTIES

A. PROPERTIES OF SOLIDS

There are few mathematical relationships that predict the thermodynamic properties of solids. Properties, such as temperature, specific heat, and density, usually are known or stated in a problem. If the properties are not known or given, they must be found from tables. For example, table 6.1 gives specific heats of some solids.

The reference point for properties of solids is usually absolute zero temperature. That is, properties like enthalpy and entropy[14] are defined to be zero at 0°R. This is an arbitrary convention. The choice of reference point does not affect the *change* in properties between two temperatures.

B. PROPERTIES OF SUB-COOLED AND COMPRESSED LIQUIDS

A sub-cooled liquid has a temperature lower than the saturation temperature for the existing pressure. Unless the pressure of the liquid is very high (in which case appendix D should be used), the various thermodynamic properties can be considered to be functions only of the liquid's temperature.

Assuming that the properties are a function of temperature only, enthalpy, entropy, internal energy, and specific volume can be read directly from the property tables in the h_f, s_f, u_f, and v_f columns, respectively, regardless of the pressure. Use appendix A or B for water.

---
[14] Without regard to any arbitrary reference point, the Third Law of Thermodynamics says that *the absolute value of entropy of a perfect crystalline solid at absolute zero temperature is zero.*

If the liquid is compressed greatly, appendix D must be used. However, such extreme accuracy seldom is called for.

Example 6.4

What is the enthalpy of water at 30 psia and 240°F?

Although the substance is water, we do not know what phase the water is in. It could be liquid, vapor, or a combination. From appendix B, the saturation (boiling) temperature for 30 psia is 250.33°F. Since the water temperature is less than the boiling temperature, the water is a liquid.

As a liquid, the properties are functions of temperature only. From appendix A for 240°F, $h = 208.34$ BTU/lbm.

Example 6.5

A piston compresses 1 lbm of 300°F saturated water to 1000 psia. What is the specific volume?

From appendix A or D, the specific volume of the water in its original saturated state is .01745 ft³/lbm.

The difference in specific volumes for the saturated and compressed liquids is given by appendix D.

$$(v - v_f)\, EE5 = -6.9$$

To ease tabulation, this entry has been multiplied by EE5. The actual correction must be multiplied by EE−5 to recover the correct value.

$$v_2 = .01745 - 6.9\, EE - 5 = .01738\ ft^3/lbm$$

C. PROPERTIES OF SATURATED LIQUIDS

Either the temperature or the pressure of a saturated liquid must be given in order to identify its thermodynamic state. Knowing one defines the other, since there is a one-to-one relationship between saturation pressure and saturation temperature. The first two columns of appendix A and appendix B can be used to determine saturation temperatures and pressures.

Since the property tables are set up specifically for saturated substances, enthalpy, entropy, internal energy, and specific volume can be read directly from the h_f, s_f, u_f, and v_f columns, respectively. Density can be calculated as the reciprocal of the specific volume. The liquid's vapor pressure is the same as the saturation pressure listed in the table.

D. PROPERTIES OF SATURATED VAPORS

Properties of saturated vapors can be read directly from the property tables. The vapor's pressure or its temperature can be used to define its thermodynamic state. Enthalpy, entropy, internal energy, and specific volume can be read directly as h_g, s_g, u_g, and v_g, respectively.

E. PROPERTIES OF A LIQUID-VAPOR MIXTURE

If a thermodynamic property has a value between the saturated liquid and vapor values (i.e., h is between h_f and h_g), the substance will consist of a mixture of liquid and vapor. The quality of such a mixture can be calculated from equation 6.30, where h_f and h_{fg} are read from the property table.

$$x = \frac{h - h_f}{h_{fg}} \qquad 6.30$$

Similar equations can be written for entropy, internal energy, and specific volume.

Once the quality is known, it can be used to calculate the other thermodynamic properties by using equations 6.31 through 6.34.

$$h = h_f + x h_{fg} \qquad 6.31$$
$$s = s_f + x s_{fg} \qquad 6.32$$
$$u = u_f + x u_{fg} \qquad 6.33$$
$$v = v_f + x v_{fg} \qquad 6.34$$

Example 6.6

What is the specific volume of a 200°F steam mixture with a quality of 90%?

Using appendix A and equation 6.34,

$$v = .01663 + (.90)(33.62)$$
$$= 30.27 \text{ ft}^3/\text{lbm}$$

Example 6.7

What is the final enthalpy of steam which is expanded isentropically through a turbine from 100 psia and 500°F (superheated) to 3 psia?

From the superheat table (appendix C), $s_1 = 1.7085$ BTU/lbm-°R.

Since the expansion is isentropic (see equation 6.17), $s_2 = s_1$.

From appendix B for 3 psia vapor, the entropy of a saturated vapor (s_g) is 1.8863. Since s_2 is less than s_g, the expanded steam is in the liquid vapor region. The quality of the mixture can be found from equation 6.30.

$$x = \frac{s - s_f}{s_{fg}} = \frac{1.7085 - .2008}{1.6855} = .895$$

The final steam enthalpy can be found from equation 6.31.

$$h = 109.37 + (.895)(1013.2)$$
$$= 1016.2 \text{ BTU/lbm}$$

F. PROPERTIES OF A SUPERHEATED VAPOR

Unless a vapor is highly superheated, its properties should be found from a superheat table, such as appen-

dix C for water vapor. Since temperature and pressure are independent for a superheated vapor, both must be known in order to define the thermodynamic state.

If the vapor's temperature and pressure do not correspond to the superheat table entries, single or double interpolation will be required. Such interpolation can be avoided by using more complete tables, but where required, interpolation is standard practice.

Example 6.8

What is the enthalpy of water at 200 psia and 900°F?

It is not obvious which phase the water is in. From appendix B, the saturation (boiling) temperature for 200 psia water is 381.79°F. Since the water temperature exceeds the boiling temperature, the water exists as a vapor.

From appendix C, enthalpy can be read directly as 1476.2 BTU/lbm.

G. PROPERTIES OF AN IDEAL GAS

A gas can be considered *ideal* if its pressure is very low and the temperature is much higher than its critical temperature. By definition, ideal gases behave according to the various ideal gas laws.

The first of two specific laws that define the behavior of an ideal gas is *Boyle's law*. This law states that volume and pressure vary inversely when the temperature is held constant.

$$p_1 V_1 = p_2 V_2 \qquad 6.35$$

The second specific law is *Charles' law*, which states that the volume and temperature vary proportionally when the pressure is constant.

$$\frac{T_1}{V_1} = \frac{T_2}{V_2} \qquad 6.36$$

Equations 6.35 and 6.36 are derived from a general law that applies to ideal gases undergoing any process.

$$\frac{p_1 V_1}{T_1} = \frac{p_2 V_2}{T_2} \qquad 6.37$$

Avogadro's law states that equal volumes of different gases with the same temperature and pressure contain equal numbers[15] of molecules. For one mole of any gas, Avogadro's law can be reformulated as the *Equation of State*.

$$\frac{pV}{T} = R^* \qquad 6.38$$

R^* is known as the *universal gas constant*. It is *universal* because the same number can be used with any gas.

---

[15] It is frequently stated that *Avogadro's* number is approximately 6.023 EE23 molecules per mole. This is the correct value for a gram-mole (gmole). This chapter uses pound-moles (pmoles), so the number of molecules would be considerably larger.

Due to the different units which can be used with the variables p, V, and T, there are different values of R^*.

Table 6.3
Values of the Universal Gas Constant

1545.33 ft-lbf/pmole-°R
0.08206 atm-liter/gmole-°K
1.986 BTU/pmole-°R
1.986 cal/gmole-°K
8.314 joule/gmole-°K
0.730 atm-ft³/pmole-°R

Equation 6.38 can be modified to allow more than one mole of gas. If there are n moles, then

$$pV = nR^*T \qquad 6.39$$

The number of moles can be calculated from the substance's mass and molecular weight.

$$n = \frac{m}{M} \qquad 6.40$$

Equations 6.39 and 6.40 can be combined. R (no asterisk) is the *specific gas constant*. It is *specific* because it is valid only for a gas with a molecular weight of M.

$$pV = \frac{mR^*T}{M} = m\left(\frac{R^*}{M}\right)T \qquad 6.41$$
$$= mRT \qquad 6.42$$

Values of the specific gas constant and the molecular weight of various gases are contained in table 6.4.

Example 6.9

What mass of nitrogen is contained in a 2000 ft³ tank if the pressure and temperature are 14.7 psia and 70°F, respectively?

From table 6.4, the specific gas constant of nitrogen is 55.2 ft-lbf/lbm-°R. From equation 6.42,

GAS constant

$$m = \frac{pV}{RT} = \frac{(14.7)(144)(2000)}{(55.2)(460+70)} = 144.7 \text{ lbm}$$

Example 6.10

A 25 ft³ tank contains 10 lbm of an ideal gas with a molecular weight of 44. What is the pressure if the temperature is 70°F?

The specific gas constant is

$$R = \frac{R^*}{M} = \frac{1545.33 \text{ ft-lbf/pmole-}°R}{44 \text{ lbm/pmole}} = 35.1 \frac{\text{ft-lbf}}{\text{lbm-}°R}$$

From equation 6.42,

$$p = \frac{mRT}{V} = \frac{(10)(35.1)(460+70)}{25} = 7441 \text{ lbf/ft}^2$$

Values of h, u, and v for gases usually are read from gas tables, such as appendix F for air. Such tables are valid for gases under low pressure. The thermodynamic

Table 6.4
Approximate Properties of Gases

GAS	SYMBOL	MOLECULAR WEIGHT	R	c_p	c_v	k
Acetylene	C_2H_2	26.0	59.4	0.350	0.2737	1.30
Air	---------	29.0	53.3	0.24	0.1714	1.40
Ammonia	NH_3	17.0	91.0	0.523	0.4064	1.32
Argon	A	39.9	38.7	0.124	0.0743	1.67
Carbon dioxide	CO_2	44.0	35.1	0.205	0.1599	1.28
Carbon monoxide	CO	28.0	55.2	0.243	0.1721	1.40
Chlorine	Cl_2	70.9	21.8	0.115	0.0865	1.33
Ethane	C_2H_6	30.07	51.3	0.422	0.357	1.18
Ethylene	C_2H_4	28.0	55.1	0.40	0.3292	1.22
Freon (R-12)	CCl_2F_2	120.9	12.6	---------	---------	1.13
Helium	He	4.0	386.3	1.25	0.754	1.66
Hydrogen	H_2	2.0	766.8	3.42	2.435	1.41
Isobutane	C_4H_{10}	58.12	26.6	0.420	0.387	1.09
Krypton	Kr	82.9	18.6	---------	---------	1.67
Methane	CH_4	16.0	96.4	0.593	0.4692	1.32
Neon	Ne	20.18	76.4	0.248	0.151	1.64
Nitrogen	N_2	28.0	55.2	0.247	0.1761	1.40
Oxygen	O_2	32.0	48.3	0.217	0.1549	1.40
Propane	C_3H_8	44.09	35.0	0.404	0.360	1.12
Steam (see note)	H_2O	18.0	85.8	0.46	0.36	1.28
Sulfur dioxide	SO_2	64.1	24.0	0.154	0.1230	1.26
Xenon	Xe	130.2	11.9	---------	---------	1.67

Note: Values for steam are approximate and may be used for low pressures and high temperatures only.
R is in ft-lbf/lbm-°R. Both c_p and c_v are in BTU/lbm-°F.

properties are considered to be functions of temperature only.

Enthalpy can be related to the equation of state because both contain a pV term.

$$h = u + \frac{pv}{J} = u + \frac{RT}{J} \qquad 6.43$$

Furthermore, density is the reciprocal of specific volume, so the density of an ideal gas can be derived from the equation of state. If $m = 1$, then

$$p = \left(\frac{1}{v}\right)RT = \rho RT \qquad 6.44$$

Or

$$\rho = \frac{p}{RT} \qquad 6.45$$

The specific heats of an ideal gas are related to the gas constants. The following equations make it possible to find one specific heat if the other is known.

$$c_p - c_v = \frac{R}{J} \qquad 6.46$$

$$C_p - C_v = \frac{R^*}{J} \qquad 6.47$$

$$c_p = \frac{Rk}{J(k-1)} \qquad 6.48$$

$$C_p = \frac{R^* k}{J(k-1)} \qquad 6.49$$

$$k = \frac{c_p}{c_v} = \frac{C_p}{C_v} \qquad 6.50$$

Example 6.11

What is the enthalpy of air at 100°F and 50 psia?

Since the pressure is low (less than 300 psia), appendix F can be used. From the $T = 560°R$ line, the enthalpy is read directly as 133.86 BTU/lbm.

Statistical thermodynamics predicts the kinetic behavior of gas molecules. The *kinetic gas theory* results in an equation which gives a distribution of the number of gas molecules versus velocity. The *Maxwell-Boltzmann distribution* is illustrated in figure 6.9. The distribution law itself is given in equation 6.51, where k is the *Boltzmann constant*[16] and m is the mass of a molecule. Kinetic gas theory calculations traditionally are done in SI units. English units can be used, however, if all constants and units are consistent.

$$\frac{dN}{dv} = \frac{4N}{\sqrt{\pi}} (v^2) \left(\frac{m}{2kT}\right)^{3/2} \exp\left(\frac{-mv^2}{2kT}\right) \qquad 6.51$$

---

[16]The Boltzmann constant has a value of 1.3803 EE—23 J/molecule-°K. These units are the same as kg-m²/sec²-molecule-°K.

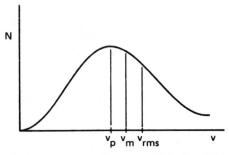

Figure 6.9 Maxwell-Boltzmann Velocity Distribution

The *most probable speed* of the molecules is

$$v_p = \sqrt{\frac{2kT}{m}} \qquad 6.52$$

The *average speed* is

$$v_m = 2\sqrt{\frac{2kT}{\pi m}} \qquad 6.53$$

The *root-mean-square speed* is

$$v_{\text{rms}} = \sqrt{\frac{3kT}{m}} \qquad 6.54$$

The three velocities illustrated in figure 6.9 and defined by equations 6.52, 6.53, and 6.54 are related.

$$\frac{v_m}{v_p} = 1.128 \qquad 6.55$$

$$\frac{v_{\text{rms}}}{v_p} = 1.225 \qquad 6.56$$

The thermodynamic property called *temperature* has a molecular interpretation derived from the kinetic gas theory. It can be shown that the root-mean-square velocity is related to the absolute temperature. That is,

$$T = \frac{m}{3k} (v_{\text{rms}})^2 \qquad 6.57$$

Since $\frac{1}{2}mv^2$ is the definition of kinetic energy, the mean translational kinetic energy of the molecule is proportional to the mean absolute temperature.

$$\frac{1}{2}m (v_{\text{rms}})^2 = \frac{3}{2}kT \qquad 6.58$$

Example 6.12

What are the kinetic energy and rms velocity for 275°K argon molecules?

From equation 6.58,

$$E_k = \left(\frac{3}{2}\right)(1.3803 \text{ EE} - 23)(275)$$
$$= 5.69 \text{ EE} - 21 \; J/\text{molecule}$$

The molecular mass of argon is its mass per mole divided by the number of molecules in a mole.

$$m = \frac{M}{N_o} = \frac{(39.9)g/\text{gmole}(.001)kg/g}{(6.023 \text{ EE23}) \text{ molecules/gmole}}$$
$$= 6.62 \text{ EE} - 26 \; kg/\text{molecule}$$

From equation 6.54, the rms velocity is

$$v\mathrm{rms} = \sqrt{\frac{(3)(1.3803\ EE - 23)(275)}{6.62\ EE - 26}}$$
$$= 414.7\ \mathrm{m/s}$$

H. PROPERTIES OF IDEAL GAS MIXTURES

A gas mixture consists of an aggregation of molecules of each gas component, the molecules of any single component being distributed uniformly and moving as if they alone occupied the space. As a consequence, the total pressure exerted by the mixture against the walls of its container is the sum of the *partial pressures* of the various gas components. The partial pressure of the components is the pressure that the gas would have if it occupied the total volume at the same temperature. The values of the partial pressure of the gases in a mixture are:

$$p_A = \frac{m_A R_A T}{V} \qquad 6.59$$

$$p_B = \frac{m_B R_B T}{V} \qquad 6.60$$

$$p_C = \frac{m_C R_C T}{V} \qquad 6.61$$

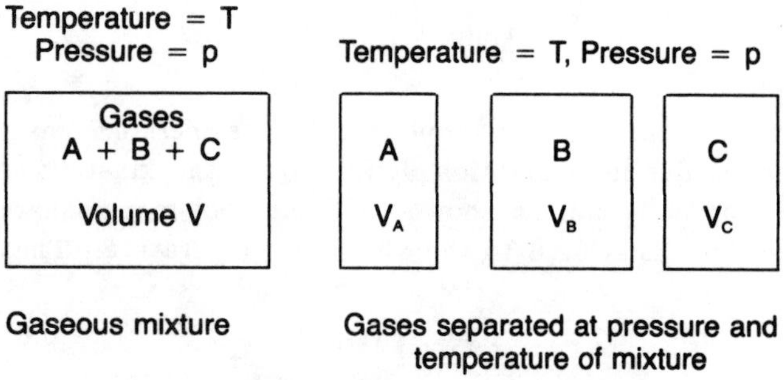

Figure 6.10 A Mixture of Ideal Gases

According to *Dalton's law*, the *total pressure* of the mixture is the sum of the partial pressures.

$$p = p_A + p_B + p_C \qquad 6.62$$

The ideal gas law can be used with one gas component and the entire mixture to calculate the partial pressure of the component.

$$\frac{p_A V}{pV} = \frac{n_A R^* T}{n R^* T} \qquad 6.63$$

The ratio (n_A/n) is known as the *mole fraction*.

$$x_A = \frac{n_A}{n} = \frac{n_A}{n_A + n_B + n_C} \qquad 6.64$$

Once the mole fraction is known for a component, it can be used with equation 6.63 to calculate the partial pressure. Since the V, R^*, and T terms cancel,

$$p_A = x_A p \qquad 6.65$$

The mole fraction is not the same as the *gravimetric fraction*, the ratio of component masses.

$$G_A = \frac{m_A}{m} = \frac{m_A}{m_A + m_B + m_C} \qquad 6.66$$

The partial pressures also can be found from the gravimetric fraction. However, the average specific gas constant of the mixture is needed.

$$p_A = G_A \left(\frac{R_A}{R_{\mathrm{mixture}}} \right) p \qquad 6.67$$

The *volumetric fraction* is defined as the ratio of a component's partial volume to the overall mixture volume. The partial volume is the volume the gas would occupy at the mixture temperature and pressure.

$$B_A = \frac{V_A}{V} = \frac{V_A}{V_A + V_B + V_C} \qquad 6.68$$

Since $pV = nR^*T$, equation 6.68 can be written as

$$B_A = \frac{n_A R^* T/p}{n R^* T/p} = \frac{n_A}{n} \qquad 6.69$$

Thus, the partial pressure ratio, the mole fraction, and the volumetric fraction are the same for ideal gas mixtures. The partial volume can be found from the volumetric fraction.

$$V_A = B_A V \qquad 6.70$$

Amagat's law states that the total volume of a mixture of non-reacting gases is equal to the sum of the partial gas volumes.

$$V = V_A + V_B + V_C \qquad 6.71$$

The assumed absence of intermolecular forces in an ideal gas mixture is responsible for the total pressure being the sum of the partial pressures of the various component gases. For the same reason, a mixture's internal energy and enthalpy are equal to the sum of its various components.

If the mixing is reversible and adiabatic, the entropy also will be equal to the sum of the individual entropies. This is an essential element of *Gibb's theorem*, which can be stated as follows: the total property (U, H, or S) of a mixture of ideal gases is the sum of the properties that the individual gases would have if each occupied the total mixture volume alone at the same temperature.

A summary of the composite gas properties is given in table 6.5.

Table 6.5
Summary of the Composite Gas Properties

Gravimetrically Weighted	Volumetrically Weighted
u h c_p c_v R s	M ρ

Example 6.13

0.14 lbm of octane vapor ($M = 114$) is mixed with 2 lbm of air in the manifold of the engine. The total pressure in the manifold is 12.5 psia, and the temperature is 520°R. Assume octane vapor behaves ideally. (a) What is the total volume of this mixture? (b) What is the partial pressure of the air in the mixture?

(a) The average molecular weight of air is

$$M = \frac{R^*}{R_{air}} = \frac{1545.33}{53.3} = 29.0$$

The number of moles of octane and air are

$$n_{oct} = \frac{0.14}{114} = .001228$$

$$n_{air} = \frac{2}{29} = .068966$$

From equation 6.39,

$$V = \frac{(.001228 + .068966)(1545.33)(520)}{(12.5)(144)}$$

$$= 31.34 \text{ ft}^3$$

(b) The mole fraction of air is

$$x_{air} = \frac{.068966}{.001228 + .068966} = .983$$

$$p_{air} = (.983)(12.5) = 12.29 \text{ psia}$$

I. PROPERTIES OF REAL GASES

Real gases do not meet the basic assumptions set forth for an ideal gas. The molecules of the real gas occupy a definite volume that is not negligible in comparison with the total volume of the gas. This is true especially for gases at low temperatures. Furthermore, real gases are subject to *Van der Waals' forces*, which are attractive forces existing between molecules.

A modification of the perfect gas laws accounts for molecular volumes and weakly-attractive intermolecular forces. *Van der Waals' equation of state* is:

$$(p + \frac{a}{V^2})(V - b) = nR^*T \qquad 6.72$$

For an ideal gas, the terms a and b are zero. When the spacing between molecules is close, such as at low temperatures, the molecules tend to attract each other and reduce the pressure exerted by the gas. The pressure is then corrected by the term (a/V^2). b is a constant dependent on the volume occupied by molecules in a dense state. Usually these corrections need to be applied only when the gas is below the critical pressure. Table 6.6 contains some typical correction factors.

Table 6.6
Van der Waals' Factors

Material	a (atm-ft⁶/pmole)	b (ft³/pmole)
air	345.2	.585
CO_2	926	.686
H_2	62.8	.427
O_2	348	.506
steam	1400	.488

The ideal gas equation also can be modified to predict real gas behavior by introducing a *compressibility factor*. The compressibility factor, Z, is a dimensionless constant that is dependent upon pressure, temperature, and type of gas.

$$pv = ZRT \qquad 6.73$$

Compressibility factors for each gas can be plotted against pressure and temperature. One diagram can be constructed for several gases if the *principle of corresponding states* is applied. This principle states that *all gases behave alike whenever they have the same reduced variables.* These *reduced variables* are the ratios of pressure, volume, and temperature to their corresponding critical values.

$$p_r = \frac{p}{p_c} \qquad 6.74$$

$$T_r = \frac{T}{T_c} \qquad 6.75$$

$$V_r = \frac{v}{v_c} \qquad 6.76$$

The critical properties needed to calculate the reduced properties can be found from table 6.7.

Example 6.14

What is the specific volume of carbon dioxide at 2680 psia and 300°F?

The absolute temperature is $(460+300) = 760°R$. From table 6.7, the critical temperature and pressure of carbon dioxide are 547.8°R and 1071.0 psia. From equations 6.74 and 6.75,

$$T_r = \frac{760}{547.8} = 1.39 \qquad p_r = \frac{2680}{1071} = 2.5$$

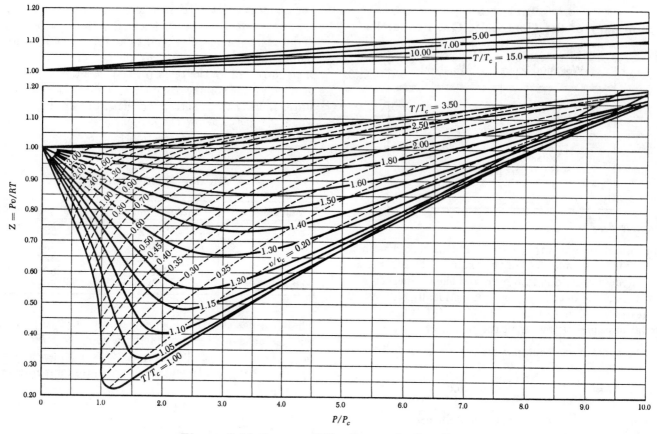

Figure 6.11 Compressibility Factors for Low Pressures

As originally presented by Professor Edward F. Obert and L.C. Nelson in "Generalized P-V-T Properties of Gases," ASME Transactions, 76, 1057 (1954).

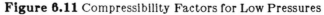

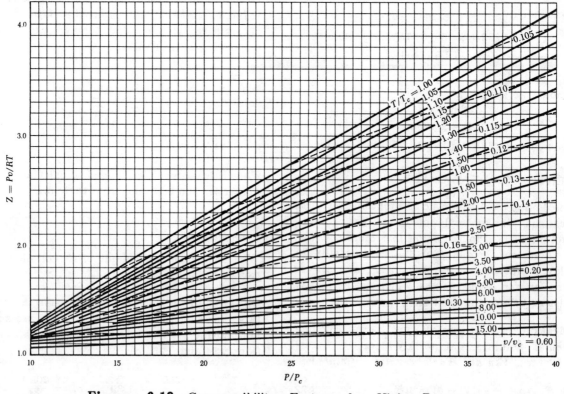

Figure 6.12 Compressibility Factors for High Pressures

Table 6.7
Approximate Critical Properties

Gas	Critical Temperature (°R)	Critical Pressure (psia)
air	235.8	547.0
ammonia	730.1	1639.0
argon	272.2	705.0
carbon dioxide	547.8	1071.0
carbon monoxide	242.2	508.2
chlorine	751.0	1116.0
ethane	549.8	717.0
ethylene	509.5	745.0
helium	10.0	33.8
hydrogen	60.5	188.0
mercury	2109.0	2646.0
methane	343.9	673.3
neon	79.0	377.8
nitrogen	227.2	492.5
oxygen	278.1	730.9
propane	666.3	617.0
sulfur dioxide	775.0	1141.0
water vapor	1165.4	3206.0
xenon	521.9	855.3

From figure 6.11, $Z = .75$. From equation 6.73,

$$v = \frac{(.75)(35.1)(760)}{(2680)(144)} = .0518 \text{ ft}^3/\text{lbm}$$

J. PROPERTIES OF ATMOSPHERIC AIR

Dry atmospheric air is a mixture of oxygen, nitrogen, and small amounts of carbon dioxide, water vapor, argon, and other inert gases. If all constituents except oxygen are grouped with the nitrogen, the air composition is as given in table 6.8. It is necessary to supply $(1/0.2315)$ or 4.32 pounds of air to obtain one pound of oxygen. The molecular weight of air is 28.9.

Table 6.8
Composition of Dry Atmospheric Air
(Rare inert gases included as N_2)

	% by weight	% by volume
Oxygen (O_2)	23.15	20.9
Nitrogen (N_2)	76.85	79.1

Moist atmospheric air can be considered a mixture of two ideal gases—air and water vapor. As a mixture, everything in the previous section applies. For example, Dalton's law is applicable.

$$p_a = p_{\text{air}} + p_w \qquad 6.77$$

Since the study of atmospheric air is so important, it has its own name: *psychrometrics*. Psychrometrics initially seems complicated by the three different definitions of temperature. The terms *dry bulb temperature*, *wet bulb temperature*, and *dewpoint temperature* are not interchangeable.

The dry bulb temperature is the temperature that a regular thermometer would measure if exposed to atmospheric air. The wet bulb temperature is the temperature of air which has gone through an *adiabatic saturation process*. The dewpoint temperature is the dry bulb temperature at which water starts to condense when cooled at a constant pressure.

The wet bulb temperature is always less than the dry bulb temperature unless the air is saturated, in which case the two temperatures are the same. An adiabatic saturation process can be achieved with a *psychrometer*, which is a regular thermometer with its bulb covered with wet cotton or gauze. If the thermometer is moved rapidly through the air (usually by twirling at the end of a string), the water in the gauze will evaporate, with the heat of vaporization coming from the air itself. The thermometer measures the temperature of the air, which has cooled because of the removal of the heat of vaporization.

For every temperature, there is a unique pressure at which water vapor is saturated. This pressure can be found from the second column of appendix A. If the partial pressure of the water vapor in atmospheric air is equal to this saturation pressure, the air is said to be *saturated*. (Actually, the water vapor, not the air, is saturated. This inconsistency, however, is part of psychrometrics.)

The amount of water vapor in the atmosphere can be measured by two different indexes. The *humidity ratio* (also known as *specific humidity*) is the ratio of water vapor and dry air masses.

$$\omega = \frac{m_w}{m_{\text{air}}} \qquad 6.78$$

Since $m = \rho V$, and since $V_w = V_{\text{air}}$, the humidity ratio also can be written as

$$\omega = \frac{\rho_w}{\rho_{\text{air}}} \qquad 6.79$$

Also, from the ideal gas equation of state, $m = \frac{pV}{RT}$, $V_w = V_{\text{air}}$, and $T_w = T_{\text{air}}$, so the ratio of masses is

$$\omega = \frac{R_{\text{air}} p_w}{R_w p_{\text{air}}} = \frac{53.3 p_w}{85.8 p_{\text{air}}} = .621 \frac{p_w}{p_{\text{air}}} \qquad 6.80$$

The other index of moisture content is the *relative humidity*. The relative humidity is the partial water vapor pressure divided by the saturation pressure.

$$\phi = \frac{p_w}{p_{\text{sat}}} \qquad 6.81$$

From the ideal gas equation of state, $\rho = \frac{p}{RT}$, so the relative humidity can be written as

$$\phi = \frac{\rho_w}{\rho_{\text{sat}}} \qquad 6.82$$

Combining equations 6.80 and 6.81 yields equation 6.83.

$$\phi = 1.61\omega \left(\frac{p_{\text{air}}}{p_{\text{sat}}}\right) \qquad \textbf{6.83}$$

It is possible to develop mathematical relationships for enthalpy, entropy, internal energy, and specific volume for atmospheric air. However, in practice this is almost never done. Psychrometric properties can be read directly from a *psychrometric chart*, such as figure 6.14.

A psychrometric chart is easy to use, despite its multiplicity of scales. The thermodynamic state is defined by specifying any two intersecting scales (e.g., dry bulb and wet bulb temperatures, or a temperature and a humidity). Once the air state has been located on the chart, all other properties can be read directly.

The psychrometric chart is useful also because many air conditioning processes follow along straight paths on the chart. The paths taken by air during typical air conditioning processes are shown in figure 6.13.

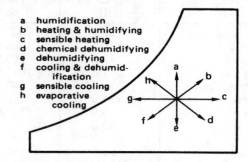

a   humidification
b   heating & humidifying
c   sensible heating
d   chemical dehumidifying
e   dehumidifying
f   cooling & dehumid-
       ification
g   sensible cooling
h   evaporative
       cooling

Figure 6.13 Air Conditioning Processes

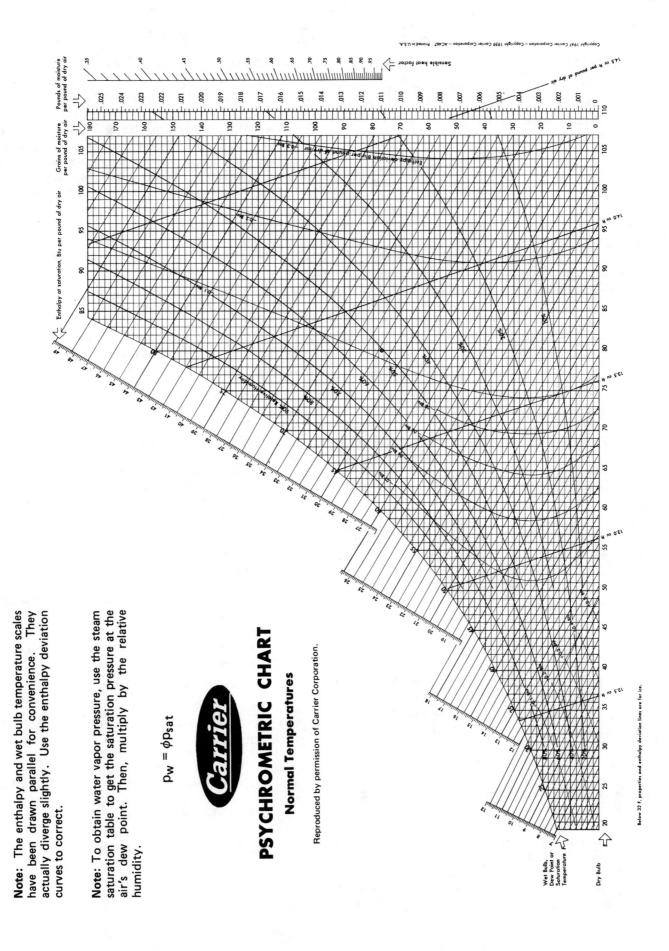

Note: The enthalpy and wet bulb temperature scales have been drawn parallel for convenience. They actually diverge slightly. Use the enthalpy deviation curves to correct.

Note: To obtain water vapor pressure, use the steam saturation table to get the saturation pressure at the air's dew point. Then, multiply by the relative humidity.

$$p_w = \phi p_{sat}$$

PSYCHROMETRIC CHART

Normal Temperatures

Reproduced by permission of Carrier Corporation.

PART 2: Changes in Properties

1 SYSTEMS

A thermodynamic *system* is defined as the matter enclosed within an arbitrarily but precisely defined volume. Everything external to the system is *surroundings* or *environment*. The surroundings and system are separated by the *system boundaries*. The region defined by the boundaries is known as a *control volume*. The surface of the control volume is known as a *control surface*.

If mass flows through a system, the system is an *open system*. Typical open systems are pumps, heat exchangers, and jet engines. Energy can enter or leave an open system. If no mass crosses the system boundary, the system is said to be a *closed system*. Closed systems need not have a constant volume. The gas compressed by a piston in a cylinder is an example of a closed system with a variable volume. Energy also can cross closed system boundaries.

An important type of open system is the *steady flow open system*. In a steady flow open system, matter must enter the system at the same rate as it leaves. Pumps, turbines, heat exchangers, boilers, and other thermodynamic devices can be assumed to be steady flow open systems.

2 TYPES OF PROCESSES

Changes in properties to working fluids in thermodynamic systems often depend on the type of process undergone by the fluids. This is particularly true of gases. Some of the most common processes are listed here.

- *constant pressure process*—also known as an *isobaric* process
- *constant volume process*—also known as an *isochoric* or *isometric* process
- *constant temperature process*—also known as an *isothermal process*
- *adiabatic process*—a process in which no energy (heat) crosses the system boundary. This is not the same as a constant temperature process. Adiabatic processes are categorized further into isentropic and throttling processes.
- *isentropic process*—a process in which no change in entropy occurs
- *throttling process*—a process in which no change in enthalpy occurs
- *polytropic process*—any process for which the working fluid properties can be predicted by the polytropic equation of state (equation 6.84). Generally, polytropic processes are limited to systems of ideal gases. n is the *polytropic exponent*, a property of the equipment, not of the

gas. For efficient air compressors, n typically is between 1.25 and 1.30.

$$p_1 (V_1)^n = p_2 (V_2)^n \qquad 6.84$$

A *reversible process* is one that is performed in such a way that, at the conclusion of the process, both the system and the local surroundings can be restored to their initial states. A process that does not meet these requirements is said to be *irreversible*. Some of the factors that render a real process irreversible are friction, unrestrained expansion of gases, heat transfer, and the mixing of two materials. A reversible process is implicitly isentropic.

3 APPLICATION OF THE FIRST LAW TO REAL EQUIPMENT

In order to apply thermodynamic principles to actual operating equipment, it is necessary to make assumptions about the processes and the equipment itself.

A. PUMPS

The purpose of a pump is to increase the total energy content of the fluid flowing through it. Pumps can be considered as adiabatic devices. If the pump inlet and outlet are the same size and at the same elevation, the kinetic and potential energy terms can be neglected. The *Steady Flow Energy Equation* (SFEE) (see equation 6.104) reduces to equation 6.85.

$$P = \dot{m}(h_2 - h_1) \qquad 6.85$$

B. TURBINES

Turbines can be thought of as pumps operating in reverse. A turbine extracts energy from the fluid as the fluid decreases in temperature and pressure. The energy extraction process is adiabatic. Changes in potential and kinetic energies can be neglected.

$$P = \dot{m}(h_2 - h_1) \qquad 6.86$$

C. HEAT EXCHANGERS AND FEEDWATER HEATERS

A heat exchanger transfers energy from one fluid to another. Since energy cannot be created or destroyed, the total energy content of both input streams must be the same as the total energy content of both output streams. Heat exchangers can be considered adiabatic. No work is done within a heat exchanger. The potential and kinetic energies of the fluids can be ignored. Therefore, from the SFEE,

$$h_2 - h_1 = 0 \qquad 6.87$$

FOR WATER $Q = V \rho c_p \Delta T$

This can be rewritten as equation 6.88.

$$\dot{m}_A h_{A,\text{in}} + \dot{m}_B h_{B,\text{in}} = \dot{m}_A h_{A,\text{out}} + \dot{m}_B h_{B,\text{out}} \quad 6.88$$

D. CONDENSERS

Condensers remove the heat of vaporization from fluids. This heat is transferred to the environment. It is possible to analyze a condenser as a heat exchanger. However, if the total heat removal is known, the SFEE can be written as equation 6.89.

$$Q = \dot{m}(h_2 - h_1) \quad 6.89$$

Since the heat flow is out of the condenser, both Q and $(h_2 - h_1)$ will be negative.

E. COMPRESSORS

Compressors can be evaluated using the assumptions for pumps.

F. VALVES

Flow through valves is adiabatic. No work is done on the fluid as it passes through a valve. If the potential energy changes are neglected, the SFEE reduces to equation 6.90.

$$h_1 + \frac{v_1^2}{2g_c J} = h_2 + \frac{v_2^2}{2g_c J} \quad 6.90$$

If the kinetic energy changes are neglected, $h_1 = h_2$. This is the definition of a *throttling* process.

Throttling is a constant temperature process for ideal gases. Real gases experience a temperature drop, however, upon throttling over normal temperatures. Whether or not a rise or drop in temperature occurs is dependent on the range of pressure and temperature over which the throttling occurs. There is one temperature at which no temperature changes occur upon throttling. This is called the *inversion temperature*.

Table 6.9
Approximate Inversion Temperatures

Gas	Temperature, °R
air	1085 (max)
argon	1301 (max)
carbon dioxide	2700 (max)
helium	45 (1 atm)
	72 (max)
hydrogen	364 (max)
nitrogen	1118 (max)

The *Joule-Thompson coefficient* is defined as the ratio of the change in temperature to the change in pressure when a gas is throttled. The Joule-Thompson coefficient is zero for an ideal gas.

$$\mu = \frac{\partial T}{\partial p} \quad 6.91$$

G. NOZZLES AND ORIFICES

If a high energy fluid is directed through a nozzle or orifice, its velocity will increase at the expense of the energy. This acceleration is adiabatic. No work is performed. The potential energy change can be neglected. Therefore, the SFEE reduces to equation 6.92.

$$h_1 + \frac{v_1^2}{2g_c J} = h_2 + \frac{v_2^2}{2g_c J} \quad 6.92$$

If the initial velocity is neglected, the exit velocity is

$$v_2 = \sqrt{2g_c J (h_1 - h_2)} \quad 6.93$$

4 PROPERTY CHANGES IN PROCESSES

One or more properties of a system will change when the system takes part in a process. For example, the temperature and pressure of a gas will increase if that gas is compressed in a cylinder by a piston. The remainder of this chapter is devoted to predicting the changes to the various properties.

A definite sign convention is used in calculating property changes. This convention is explained here, and it is consistent with the signs that are generated automatically by the formulas in this chapter.

- Q is positive if energy (heat) flows into the system.
- W is positive if the system does work on the surroundings.
- ΔH, ΔU, and ΔS are positive if they increase within the system.

In calculating changes to properties, it should be remembered that most formulas can be written in several forms. All equations can be written in terms of both *lbm* and *pmole* bases. The change in enthalpy for a perfect gas, for example, can be written as either equation 6.94 or 6.95.

$$\Delta h = c_p \Delta T \quad 6.94$$
$$\Delta H = C_p \Delta T \quad 6.95$$

It is not practical to list every relationship for property changes. You will have to use your knowledge of the behavior of the system to make substitutions into the various equations.

5 THE FIRST LAW OF THERMODYNAMICS

There is one basic principle that underlies all property changes as a system experiences a process: all energy must be accounted for. Energy that enters a system must either leave the system or be stored in some manner.

A. CLOSED SYSTEMS

This principle can be stated as the *First Law of Thermodynamics*: energy cannot be created or destroyed. The First Law can be written in differential form for *closed systems*.

$$dQ = dU + W \qquad 6.96$$

Most thermodynamic problems can be solved without resorting to differential calculus. The First Law can be written in finite terms.

$$Q = \Delta U + W \qquad 6.97$$

Equation 6.97 is the First Law formulation for closed systems. It says that heat entering a closed system can either increase the temperature (increase U) or be used to perform work (W) on the surroundings. Heat energy entering the system also can leak to the surroundings (a nonadiabatic closed system), but in that case, the term Q is understood to be the *net* heat entering the system.

Q will be negative if the net heat exchange to the system is a loss. ΔU will be negative if the internal energy of the system decreases. W will be negative if the surroundings do work on the system (e.g., a piston compressing gas in a cylinder). These signs are consistent with the convention used in this chapter.

Since Q and U contain units of BTU, the work term also must be expressed in BTU. If work is given in ft-lbf, Joule's constant must be incorporated into the First Law.

$$Q = \Delta U + \frac{W}{J} \qquad 6.98$$

B. OPEN SYSTEMS

The First Law of Thermodynamics can be written for open systems also, but more terms are required to account for the many energy forms that can change. If the mass flow rate is constant, the system is a steady flow system. The applicable First Law formulation is essentially the Bernoulli energy conservation equation extended to non-adiabatic processes. This equation is known as the *Steady Flow Energy Equation* (SFEE).

$$Q = \Delta U + \Delta E_p + \Delta E_k + W_{\text{flow}} + W_{\text{shaft}} \qquad 6.99$$

The terms in equation 6.99 can be illustrated by figure 6.15. Q is the net heat flow into the system. It can be supplied by furnace flame, electrical heating, nuclear reaction, or by any other method. Of course, if the system is known to be adiabatic, Q is zero.

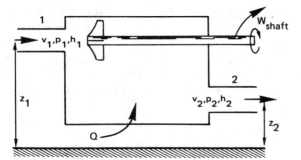

Figure 6.15 A Steady Flow Device

ΔU is the change in the internal energy of the system. It can be evaluated by using one of the many formulas presented in this chapter. It is seldom necessary to work with internal energy with a steady flow device.

ΔE_p is the change in *potential energy* of the fluid. This can be found on a per-pound basis from equation 6.100.

$$\Delta E_p = \frac{g(z_2 - z_1)}{J g_c} \qquad 6.100$$

Similarly, ΔE_k is the change in *kinetic energy*. This can be found on a per-pound basis from equation 6.101.

$$\Delta E_k = \frac{v_2^2 - v_1^2}{2 g_c J} \qquad 6.101$$

W_{flow} is the *flow work* (*pV work*) previously presented in the definition of enthalpy. Looking at figure 6.15, there is a pressure (p_2) at the exit of the steady flow device. This exit pressure opposes the entrance of the fluid. Therefore, the flow work term represents the work required to cause the flow into the system against the existing pressure. The flow work can be calculated on a per-pound basis from equation 6.102.

$$W_{\text{flow}} = \frac{p_2 v_2 - p_1 v_1}{J} \qquad 6.102$$

W_{shaft} is known as *shaft work*. This represents work that the steady flow device does on the surroundings. Its name is derived from the output shaft that almost always transmits the energy out of the system. For example, turbines and internal combustion engines have output shafts that perform useful tasks. Of course, W_{shaft} also can be negative, as it would be in the case of a pump or compressor.

Enthalpy was previously defined as the sum of internal energy and flow energy. Therefore, the internal energy change and flow work terms can be combined in an enthalpy change. That is,

$$\Delta h = u_2 - u_1 + \frac{p_2 v_2 - p_1 v_1}{J} \qquad 6.103$$

The complete formulation of the *Steady Flow Energy Equation* is given as equation 6.104. The shaft work term must be in units of ft-lbf/lbm. It does not represent the total work of the device. Similarly, Q is on a per-pound basis.

$$Q = h_2 - h_1 + \frac{v_2^2 - v_1^2}{2 g_c J} + \frac{g(z_2 - z_1)}{g_c J} + \frac{W_{\text{shaft}}}{J} \qquad 6.104$$

Both sides of equation 6.104 can be multiplied by $\dot{m}$ to obtain units of BTU/sec. If both sides are multiplied by $\dot{m} J$, the units become ft-lbf/sec.

Example 6.15

4 lbm/sec of steam enter a turbine with velocity of 65 ft/sec and enthalpy of 1350 BTU/lbm. The steam enters the condenser after being expanded to 1075

BTU/lbm and 125 ft/sec. The total heat loss from the turbine casing is 50 BTU/sec. What power is generated by this turbine?

Equation 6.104 can be used to solve this problem. Solving for the shaft work, including the mass flow term, and neglecting the potential energy change,

$$P = -50 + (4)(1350 - 1075) + (4)\left[\frac{(65)^2 - (125)^2}{(2)(32.2)(778)}\right]$$

$$= 1049 \text{ BTU/sec}$$

1049 BTU/sec corresponds to approximately 1500 horsepower.

A simple application of the first law involves the establishment of a thermal equilibrium between two substances. A thermal equilibrium is reached when all parts of a system are at the same temperature. Thermal equilibrium is achieved naturally (without the addition of external work) whenever two liquids, or a solid and a liquid with different temperatures, are mixed.

A thermal equilibrium problem can be solved with the first law. The entering energy comes from the heat given off by the cooling substance. Energy is stored by increasing the temperature of the warming substance. Since no work is added, the W term is zero.

$$(\text{Heat loss})_A = (\text{Heat gain})_B \qquad 6.105$$

$$Q_A = Q_B \qquad 6.106$$

The form of the equation for Q depends on the phases of the substances. If both substances are liquid or solid, equation 6.18 can be used.

$$m_A c_A (T_{1,A} - T_{2,A}) = m_B c_B (T_{2,B} - T_{1,B}) \qquad 6.107$$

For an open system, the m terms should be replaced by $\dot{m}$.

Example 6.16

A two pound block of steel ($c = .11$ BTU/lbm-°F) is removed from a furnace and quenched in a five pound aluminum ($c = .21$ BTU/lbm-°F) tank filled with 12 pounds of water. The water and the tank initially are in equilibrium at 75°F, and their temperature rises to 100°F after quenching. What is the initial steel temperature?

From equation 6.105, the heat lost by the steel is equal to the heat gained by the tank and the water.

$$m_s c_s (T_{s,1} - T_2) = (m_a c_a + m_w c_w)(T_2 - T_{w,1})$$

$$(2)(.11)(T_{s,1} - 100) = [(5)(.21) + (12)(1)](100 - 75)$$

$$T_{s,1} = 1583° \text{ F}$$

6 AVAILABILITY

The maximum possible work that can be obtained from a system is known as the system's *availability*. Both the first and the second law must be applied to determine this availability.

Refer to figure 6.15. Heat is rejected to an environment that is at a constant temperature, T_o. Work is done on the environment at a steady rate, $\dot{W}$. Assuming steady flow, and neglecting kinetic and potential energies, the first law can be written as

$$\dot{m}h_1 + \dot{Q} = \dot{m}h_2 - \dot{W} \qquad 6.108$$

Entropy is not a part of the first law. The leaving entropy, however, can be calculated from the entering entropy and the entropy production.

$$\dot{m}s_2 = \dot{m}s_1 + \frac{\dot{Q}}{T_o} \qquad 6.109$$

Since equations 6.108 and 6.109 both contain $\dot{Q}$, they can be combined. Since the second law of thermodynamics requires that $s_2 \geq s_1$, the combined equation can be written as an inequality.

$$\dot{W} \leq \dot{m}(h_1 - T_o s_1 - h_2 + T_o s_2) \qquad 6.110$$

This equation can be simplified by introducing the *steady flow availability function*, Φ. The maximum work output (availability) is

$$\dot{W}_{\text{max}} = \Phi_1 - \Phi_2 \qquad 6.111$$

$$\Phi = h - T_o s \qquad 6.112$$

If the equality in equation 6.110 holds, both the process within the control volume and the energy transfers between the system and the environment must be reversible. Maximum work output, therefore, will be obtained in a reversible process. The difference between the maximum and the actual work output is known as the *process irreversibility*.

Example 6.17

What is the maximum useful work which can be produced per pound of saturated steam which enters a steady flow system at 800 psia and leaves in equilibrium with the atmosphere at 70°F and 14.7 psia?

From the saturated steam table, $h_1 = 1198.6$ BTU/lbm, and $s_1 = 1.4153$ BTU/lbm-°R.

The final properties are obtained from the saturated steam table for water at 70°F. $h_2 = 38.04$ BTU/lbm, and $s_2 = .0745$ BTU/lbm-°R.

From equation 6.111 using $T_o = 530$°R, the availability is

$$W_{\text{max}} = [1198.6 - 530(1.4153)]$$

$$- [38.04 - 530(.0745)]$$

$$= 449.94 \text{ BTU/lbm}$$

7 PROPERTY CHANGES IN IDEAL GASES

For solids, liquids, and vapors, the changes in properties must be calculated the "hard way." That is, the change is found by subtracting the initial property value from the final property value. For ideal gases, however, the changes can be found directly, without knowing the initial and final property values.

Some of the methods available for finding property changes do not depend on the type of process experienced by the gas. For example, the equation of state can be applied to all problems.

$$pV = mRT \qquad 6.113$$

Similarly, the changes in enthalpy and internal energy can be found from equations 6.114 and 6.115, regardless of the process.

$$\Delta h = c_p \Delta T \qquad 6.114$$

$$\Delta u = c_v \Delta T \qquad 6.115$$

Changes in entropy can be calculated from equations 6.116 and 6.117.

$$\Delta s = c_p \, ln\left(\frac{T_2}{T_1}\right) - \frac{R}{J} \, ln\left(\frac{p_2}{p_1}\right) \qquad 6.116$$

$$= c_v \, ln\left(\frac{T_2}{T_1}\right) + \frac{R}{J} \, ln\left(\frac{v_2}{v_1}\right) \qquad 6.117$$

Equations 6.114 and 6.115 should not be confused with equation 6.19, which gives the energy (heat) transfer required to achieve a change in energy. The heat equation depends on the type of process.

$$\frac{Q}{m} = c_p \Delta T \quad \text{(constant pressure processes)} \quad 6.118$$

$$\frac{Q}{m} = c_v \Delta T \quad \text{(constant volume processes)} \quad 6.119$$

Relationships between the variables and properties are given in the following paragraphs for specific processes. It should be remembered that all of these specific equations can be written in terms of pounds and moles. For the sake of compactness, only the *lbm*-basis equations are listed. However, you can convert all equations to *pmole*-basis by substituting V for v, H for h, R[*] for R, etc. The sign conventions previously established apply to the following equations.

A. CONSTANT PRESSURE, CLOSED SYSTEMS

$$p_2 = p_1 \qquad 6.120$$

$$T_2 = T_1\left(\frac{v_2}{v_1}\right) \qquad 6.121$$

$$v_2 = v_1\left(\frac{T_2}{T_1}\right) \qquad 6.122$$

$$Q = h_2 - h_1 \qquad 6.123$$

$$= c_p(T_2 - T_1) \qquad 6.124$$

$$= c_v(T_2 - T_1) + p(v_2 - v_1) \qquad 6.125$$

$$u_2 - u_1 = c_v(T_2 - T_1) \qquad 6.126$$

$$= \frac{c_v p(v_2 - v_1)}{R} \qquad 6.127$$

$$= \frac{p(v_2 - v_1)}{k - 1} \qquad 6.128$$

$$W = p(v_2 - v_1) \qquad 6.129$$

$$= R(T_2 - T_1) \qquad 6.130$$

$$s_2 - s_1 = c_p \, ln\left(\frac{T_2}{T_1}\right) \qquad 6.131$$

$$= c_p \, ln\left(\frac{v_2}{v_1}\right) \qquad 6.132$$

$$h_2 - h_1 = Q \qquad 6.133$$

$$= c_p(T_2 - T_1) \qquad 6.134$$

$$= \frac{kp(v_2 - v_1)}{k - 1} \qquad 6.135$$

B. CONSTANT VOLUME, CLOSED SYSTEMS

$$p_2 = p_1\left(\frac{T_2}{T_1}\right) \qquad 6.136$$

$$T_2 = T_1\left(\frac{p_2}{p_1}\right) \qquad 6.137$$

$$v_2 = v_1 \qquad 6.138$$

$$Q = u_2 - u_1 \qquad 6.139$$

$$= c_v(T_2 - T_1) \qquad 6.140$$

$$u_2 - u_1 = Q \qquad 6.141$$

$$= c_v(T_2 - T_1) \qquad 6.142$$

$$= \frac{c_v v(p_2 - p_1)}{R} \qquad 6.143$$

$$= \frac{v(p_2 - p_1)}{k - 1} \qquad 6.144$$

$$W = 0 \qquad 6.145$$

$$s_2 - s_1 = c_v \, ln\left(\frac{T_2}{T_1}\right) \qquad 6.146$$

$$= c_v \, ln\left(\frac{p_2}{p_1}\right) \qquad 6.147$$

$$h_2 - h_1 = c_p(T_2 - T_1) \qquad 6.148$$

$$= \frac{kv(p_2 - p_1)}{k - 1} \qquad 6.149$$

C. CONSTANT TEMPERATURE, CLOSED SYSTEMS

$$p_2 = p_1\left(\frac{v_1}{v_2}\right) \qquad 6.150$$

$$T_2 = T_1 \qquad 6.151$$

$$v_2 = v_1\left(\frac{p_1}{p_2}\right) \qquad 6.152$$

$$Q = W \qquad 6.153$$

$$= T(s_2 - s_1) \qquad 6.154$$

$$= p_1 v_1 \, ln\left(\frac{v_2}{v_1}\right) \qquad 6.155$$

$$= RT \, ln\left(\frac{v_2}{v_1}\right) \qquad 6.156$$

$$u_2 - u_1 = 0 \qquad 6.157$$

$$W = Q \qquad 6.158$$

$$= p_1 v_1 \, ln\left(\frac{v_2}{v_1}\right) \qquad \text{6.159}$$

$$= RT \, ln\left(\frac{v_2}{v_1}\right) \qquad \text{6.160}$$

$$= RT \, ln\left(\frac{p_1}{p_2}\right) \qquad \text{6.161}$$

$$s_2 - s_1 = \frac{Q}{T} \qquad \text{6.162}$$

$$= R \, ln\left(\frac{v_2}{v_1}\right) \qquad \text{6.163}$$

$$= R \, ln\left(\frac{p_1}{p_2}\right) \qquad \text{6.164}$$

$$h_2 - h_1 = 0 \qquad \text{6.165}$$

D. ISENTROPIC, CLOSED SYSTEMS (REVERSIBLE ADIABATIC)

$$p_2 = p_1\left(\frac{v_1}{v_2}\right)^k \qquad \text{6.166}$$

$$= p_1\left(\frac{T_2}{T_1}\right)^{\frac{k}{k-1}} \qquad \text{6.167}$$

$$T_2 = T_1\left(\frac{v_1}{v_2}\right)^{k-1} \qquad \text{6.168}$$

$$= T_1\left(\frac{p_2}{p_1}\right)^{\frac{k-1}{k}} \qquad \text{6.169}$$

$$v_2 = v_1\left(\frac{p_1}{p_2}\right)^{\frac{1}{k}} \qquad \text{6.170}$$

$$= v_1\left(\frac{T_1}{T_2}\right)^{\frac{1}{k-1}} \qquad \text{6.171}$$

$$Q = 0 \qquad \text{6.172}$$

$$u_2 - u_1 = -W \qquad \text{6.173}$$

$$= c_v\,(T_2 - T_1) \qquad \text{6.174}$$

$$= \frac{c_v\,(p_2 v_2 - p_1 v_1)}{R} \qquad \text{6.175}$$

$$= \frac{p_2 v_2 - p_1 v_1}{k - 1} \qquad \text{6.176}$$

$$W = u_1 - u_2 \qquad \text{6.177}$$

$$= c_v\,(T_1 - T_2) \qquad \text{6.178}$$

$$= \frac{p_1 v_1 - p_2 v_2}{k - 1} \qquad \text{6.179}$$

$$= \frac{p_1 v_1}{k - 1}\left[1 - \left(\frac{p_2}{p_1}\right)^{\frac{k-1}{k}}\right] \qquad \text{6.180}$$

$$s_2 - s_1 = 0 \qquad \text{6.181}$$

$$h_2 - h_1 = c_p\,(T_2 - T_1) \qquad \text{6.182}$$

$$= \frac{k\,(p_2 v_2 - p_1 v_1)}{k - 1} \qquad \text{6.183}$$

E. POLYTROPIC, CLOSED SYSTEMS

$$p_2 = p_1\left(\frac{v_1}{v_2}\right)^n \qquad \text{6.184}$$

$$= p_1\left(\frac{T_2}{T_1}\right)^{\frac{n}{n-1}} \qquad \text{6.185}$$

$$T_2 = T_1\left(\frac{v_1}{v_2}\right)^{n-1} \qquad \text{6.186}$$

$$= T_1\left(\frac{p_2}{p_1}\right)^{\frac{n-1}{n}} \qquad \text{6.187}$$

$$v_2 = v_1\left(\frac{p_1}{p_2}\right)^{\frac{1}{n}} \qquad \text{6.188}$$

$$= v_1\left(\frac{T_1}{T_2}\right)^{\frac{1}{n-1}} \qquad \text{6.189}$$

$$Q = \frac{c_v\,(n - k)(T_2 - T_1)}{n - 1} \qquad \text{6.190}$$

$$u_2 - u_1 = c_v\,(T_2 - T_1) \qquad \text{6.191}$$

$$= \frac{p_2 v_2 - p_1 v_1}{n - 1} \qquad \text{6.192}$$

$$W = \frac{R\,(T_1 - T_2)}{n - 1} \qquad \text{6.193}$$

$$= \frac{p_1 v_1 - p_2 v_2}{n - 1} \qquad \text{6.194}$$

$$= \frac{p_1 v_1}{n - 1}\left[1 - \left(\frac{p_2}{p_1}\right)^{\frac{n-1}{n}}\right] \qquad \text{6.195}$$

$$s_2 - s_1 = \frac{c_v\,(n - k)}{n - 1}\left[ln\left(\frac{T_2}{T_1}\right)\right] \qquad \text{6.196}$$

$$h_2 - h_1 = c_p\,(T_2 - T_1) \qquad \text{6.197}$$

$$= \frac{n\,(p_2 v_2 - p_1 v_1)}{n - 1} \qquad \text{6.198}$$

F. ISENTROPIC, STEADY FLOW SYSTEMS

p_2, v_2, and T_2 are the same as for isentropic, closed systems.

$$Q = 0 \qquad \text{6.199}$$

$$W = h_1 - h_2 \qquad \text{6.200}$$

$$= c_p T_1\left[1 - \left(\frac{p_2}{p_1}\right)^{\frac{k-1}{k}}\right] \qquad \text{6.201}$$

$$u_2 - u_1 = c_v\,(T_2 - T_1) \qquad \text{6.202}$$

$$h_2 - h_1 = -W \qquad \text{6.203}$$

$$= c_p\,(T_2 - T_1) \qquad \text{6.204}$$

$$= \frac{k\,(p_2 v_2 - p_1 v_1)}{k - 1} \qquad \text{6.205}$$

$$s_2 - s_1 = 0 \qquad \text{6.206}$$

G. POLYTROPIC, STEADY FLOW SYSTEMS

p_2, v_2, and T_2 are the same as for polytropic, closed systems.

$$Q = \frac{c_v\,(n - k)(T_2 - T_1)}{n - 1} \qquad \text{6.207}$$

$$W = h_1 - h_2 \qquad \text{6.208}$$

$$= \frac{n c_v (1-k) T_1}{n-1} \left[1 - \left(\frac{p_2}{p_1} \right)^{\frac{n-1}{n}} \right] \qquad 6.209$$

$$u_2 - u_1 = c_v (T_2 - T_1) \qquad 6.210$$

$$\mathbf{h_2 - h_1 = -W} \qquad 6.211$$

$$= c_p (T_2 - T_1) \qquad 6.212$$

$$= \frac{n (p_2 v_2 - p_1 v_1)}{n-1} \qquad 6.213$$

$$s_2 - s_1 = \frac{c_v (n-k)}{n-1} \left[ln \left(\frac{T_2}{T_1} \right) \right] \qquad 6.214$$

H. THROTTLING, STEADY FLOW SYSTEMS

$$p_1 v_1 = p_2 v_2 \qquad 6.215$$

$$p_2 < p_1 \qquad 6.216$$

$$v_2 > v_1 \qquad 6.217$$

$$T_2 = T_1 \qquad 6.218$$

$$Q = 0 \qquad 6.219$$

$$W = 0 \qquad 6.220$$

$$u_2 - u_1 = 0 \qquad 6.221$$

$$h_2 - h_1 = 0 \qquad 6.222$$

$$s_2 - s_1 = R \, ln \left(\frac{p_1}{p_2} \right) \qquad 6.223$$

$$= R \, ln \left(\frac{v_2}{v_1} \right) \qquad 6.224$$

Example 6.18

Two pounds of hydrogen are cooled from 760°F to 660°F at constant volume. What heat is removed?

Equation 6.140 is on a per pound basis. The total heat for m pounds is $Q = m c_v (T_2 - T_1)$. c_v is found from **table 6.4 to be 2.435 BTU/lbm-°F.**

$$Q = (2)(2.435)(660 - 760)$$
$$= -487 \text{ BTU}$$

The minus sign is consistent with the convention that a heat loss is a negative heat flow.

Example 6.19

Four pounds of air initially are at 14.7 psia and 530°R. The air is compressed to 100 psia in a constant temperature, closed process. What heat is removed during the compression?

Either equation 6.155 or 6.156 could be used if the volumes were known. The volumes could be found from the equation of state, but it is not necessary to do so. Combining equations 6.152 and 6.156, and including the mass term,

$$Q = mRT \, ln \left(\frac{p_1}{p_2} \right)$$

$$= (4)(53.3)(530) \, ln \left(\frac{14.7}{100} \right)$$

$$= -216,650 \text{ ft-lbf}$$

Notice that (a) the pressure did not have to be converted to *psf* in the ratio, (b) the units are ft-lbf, not BTU, and (c) the heat flow is out of the gas.

8 CYCLES

A *cycle* is a process or series of processes that eventually brings the working fluid back to its original condition. Most cycles repeat over and over. For example, the same water can be vaporized in a boiler and expanded through a turbine. Also, a four-stroke internal combustion engine operates by repeating continually the compression-expansion-exhaust-intake processes.[17]

Although heat can be extracted and work performed by a single process, a cycle is necessary to obtain energy in useful quantities. Cycles are covered in greater detail in chapter 7.

The *Second Law of Thermodynamics* can be stated in several ways. The Kelvin-Planck statement is that it is impossible to operate an engine operating in a cycle that will have no other effect than to extract heat from a reservoir and turn it into an equivalent amount of work. That is, it is impossible to build a cyclic engine that will have a thermal efficiency of 100%.

The second law is not a contradiction of the first law. The first law does not preclude the possibility of converting heat entirely into work—it only denies the possibility of creating or destroying energy. The second law says that if some heat is converted entirely into work, some other energy must be lost to the surroundings.

The *thermal efficiency* of a process is defined as the ratio of useful work output to the supplied input energy.

$$\eta_{th} = \frac{\text{work output}}{\text{energy input}} = \frac{W}{Q_{in}} \qquad 6.225$$

The first law can be written as

$$Q_{in} = Q_{out} + W \qquad 6.226$$

Combining equations 6.225 and 6.226 produces equation 6.227.

$$\eta_{th} = 1 - \frac{Q_{out}}{Q_{in}} \qquad 6.227$$

If a heat engine could be built with a 100% thermal efficiency, Q_{out} would be zero. Such an engine is precluded by the second law. The most efficient engine cycle possible is the *Carnot cycle*. This cycle is discussed in detail in chapter 7.

[17] The definition of a cycle is not dependent on the same working fluid being used repeatedly, as it is with a boiler/turbine.

SPECIAL TOPIC
Property Changes for an Incompressible Liquid

With the exception of a fleeting reference to table D, this chapter assumes that the thermodynamic properties of a liquid are a function of the temperature only. That is, it has been assumed that enthalpy, entropy, internal energy, and specific volume are independent of pressure.

There are times, however, when it is necessary to evaluate the property changes of an incompressible liquid, no matter how small they may be. For example, the pump work must be evaluated for Rankine and other power cycles.

If it is assumed that the specific heat of the liquid is constant, then the following equations predict changes in the primary thermodynamic properties.

$$v_2 - v_1 = 0 \quad \text{(incompressible)}$$

$$u_2 - u_1 = c(T_2 - T_1)$$

$$s_2 - s_1 = c\ln\left(\frac{T_2}{T_1}\right)$$

$$h_2 - h_1 = c(T_2 - T_1) + \frac{v(p_2 - p_1)}{J}$$

These equations indicate that only enthalpy is a function of pressure. The internal energy and entropy are, indeed, functions of temperature only.

SPECIAL TOPIC
Variations in Specific Heats for Gases

It is frequently assumed in thermodynamic problems that the specific heats of a gas are constant. This is a valid assumption for ideal gases.

When greater accuracy is required, it can be assumed that the specific heats are a function of temperature only. If the gas goes through a process over a wide temperature range, it may be possible to use the specific heat at the average temperature.

The specific heats can be predicted as a function of temperature based on an equation of the form

$$c_p = a + bT + cT^2 + \frac{d}{\sqrt{T}}$$

Once c_p has been determined, c_v can be found from

$$c_p - c_v = \frac{R}{J}$$

(In these two equations, T is in °R, and the specific heats are expressed in BTU/lbm-°R. Accordingly, R is the specific gas constant.)

The coefficients a, b, c, and d are chosen to give good correlation with experimental data. Typical values are given here for various temperature ranges.

Specific heat varies somewhat with pressure, although this variation is small at high temperatures. If neces-sary, the variation can be estimated using the following procedure.

step 1: Determine the heat capacity at low pressures (*zero pressure specific heat*), C_{p0}, for the temperature of the gas.

step 2: Using the critical pressure, calculate the reduced pressure.

$$p_r = \frac{p}{p_c}$$

step 3: Using the critical temperature, calculate the reduced temperature.

$$T_r = \frac{T}{T_r}$$

step 4: If T_r is less than 1.2, C_p under pressure can be found from

$$C_p = C_{p0} + \frac{9p_r}{(T_r)^5} \quad \text{(BTU/pmole-°R)}$$

This equation is empirical. It can be used when the correction for pressure is less than 2.5 BTU/pmole-°R.

step 5: If T_r is more than 1.2, and the correction for pressure is no more than 1 or 2 BTU/pmole-°R, the value of C_p under pressure can be estimated from

$$C_p = C_{p0} + \frac{5p_r}{(T_r)^2} \quad \text{(BTU/pmole-°R)}$$

Gas	Temperature range (°R)	a	$b \times 10^5$	$c \times 10^9$	d	For 70° F			R/J
						c_p	c_v	k	
air	400 to 1200	0.2405	−1.186	20.1	0	0.240	0.171	1.40	0.0685
	1200 to 4000	0.2459	3.22	−3.74	−0.833				
CO	400 to 1200	0.2534	−2.35	26.88	0	0.249	0.178	1.40	0.0709
	1200 to 4000	0.2763	3.04	−3.89	−1.5				
CO_2	400 to 4000	0.328	3.2	−4.4	−3.33	0.199	0.154	1.29	0.0451
H_2	400 to 1000	2.853	145	−883	0	3.42	2.43	1.41	0.9850
	1000 to 2500	3.447	−4.7	70.3	0				
	2500 to 4000	2.841	45	−31.2	0				
H_2O	400 to 1800	0.4267	2.425	23.85	0	0.446	0.331	1.37	0.1102
	1800 to 4000	0.3275	14.67	−13.59	0				
N_2	400 to 1200	0.2510	−1.63	20.4	0	0.248	0.177	1.40	0.0709
	1200 to 4000	0.2192	4.38	−5.14	−0.124				
O_2	400 to 1200	0.213	0.188	20.3	0	0.220	0.158	1.39	0.0621
	1200 to 4000	0.340	−0.36	0.616	−3.19				
CH_4	400 to 1000	0.453	0.62	268.8	0	0.532	0.408	1.30	0.1238
	1000 to 4000	1.152	32.58	−41.29	−22.42				
C_8H_{18}	400 to 1200	0.0693	52.6	0	0	0.348	0.331	1.05	0.0174
$C_{12}H_{26}$	400 to 1200	0.0510	52.2	0	0	0.328	0.316	1.04	0.0117

By permission from C.O. Mackay, W.N. Barnard, and F.O. Ellenwood, *Engineering Thermodynamics* (Wiley, New York, 1957).

Appendix A
Saturated Steam: Temperatures

Temp. T,°F	Abs. Press. psia p	Specific Volume, ft³/lbm			Enthalpy, BTU/lbm			Entropy, BTU/lbm-°R			Temp. T,°F
		Sat. Liquid v_f	Evap. v_{fg}	Sat. Vapor v_g	Sat. Liquid h_f	Evap. h_{fg}	Sat. Vapor h_g	Sat. Liquid s_f	Evap. s_{fg}	Sat. Vapor s_g	
32°	0.08854	0.01602	3306	3306	0.00	1075.8	1075.8	0.0000	2.1877	2.1877	32°
35	0.09995	0.01602	2947	2947	3.02	1074.1	1077.1	0.0061	2.1709	2.1770	35
40	0.12170	0.01602	2444	2444	8.05	1071.3	1079.3	0.0162	2.1435	2.1597	40
45	0.14752	0.01602	2036.4	2036.4	13.06	1068.4	1081.5	0.0262	2.1167	2.1429	45
50	0.17811	0.01603	1703.2	1703.2	18.07	1065.6	1083.7	0.0361	2.0903	2.1264	50
60°	0.2563	0.01604	1206.6	1206.7	28.06	1059.9	1088.0	0.0555	2.0393	2.0948	60°
70	0.3631	0.01606	867.8	867.9	38.04	1054.3	1092.3	0.0745	1.9902	2.0647	70
80	0.5069	0.01608	633.1	633.1	48.02	1048.6	1096.6	0.0932	1.9428	2.0360	80
90	0.6982	0.01610	468.0	468.0	57.99	1042.9	1100.9	0.1115	1.8972	2.0087	90
100	0.9492	0.01613	350.3	350.4	67.97	1037.2	1105.2	0.1295	1.8531	1.9826	100
110°	1.2748	0.01617	265.3	265.4	77.94	1031.6	1109.5	0.1471	1.8106	1.9577	110°
120	1.6924	0.01620	203.25	203.27	87.92	1025.8	1113.7	0.1645	1.7694	1.9339	120
130	2.2225	0.01625	157.32	157.34	97.90	1020.0	1117.9	0.1816	1.7296	1.9112	130
140	2.8886	0.01629	122.99	123.01	107.89	1014.1	1122.0	0.1984	1.6910	1.8894	140
150	3.718	0.01634	97.06	97.07	117.89	1008.2	1126.1	0.2149	1.6537	1.8685	150
160°	4.741	0.01639	77.27	77.29	127.89	1002.3	1130.2	0.2311	1.6174	1.8485	160°
170	5.992	0.01645	62.04	62.06	137.90	996.3	1134.2	0.2472	1.5822	1.8293	170
180	7.510	0.01651	50.21	50.23	147.92	990.2	1138.1	0.2630	1.5480	1.8109	180
190	9.339	0.01657	40.94	40.96	157.95	984.1	1142.0	0.2785	1.5147	1.7932	190
200	11.526	0.01663	33.62	33.64	167.99	977.9	1145.9	0.2938	1.4824	1.7762	200
210°	14.123	0.01670	27.80	27.82	178.05	971.6	1149.7	0.3090	1.4508	1.7598	210°
212	14.696	0.01672	26.78	26.80	180.07	970.3	1150.4	0.3120	1.4446	1.7566	212
220	17.186	0.01677	23.13	23.15	188.13	965.2	1153.4	0.3239	1.4201	1.7440	220
230	20.780	0.01684	19.365	19.382	198.23	958.8	1157.0	0.3387	1.3901	1.7288	230
240	24.969	0.01692	16.306	16.323	208.34	952.2	1160.5	0.3531	1.3609	1.7140	240
250°	29.825	0.01700	13.804	13.821	218.48	945.5	1164.0	0.3675	1.3323	1.6998	250°
260	35.429	0.01709	11.746	11.763	228.64	938.7	1167.3	0.3817	1.3043	1.6860	260
270	41.858	0.01717	10.044	10.061	238.84	931.8	1170.6	0.3958	1.2769	1.6727	270
280	49.203	0.01726	8.628	8.645	249.06	924.7	1173.8	0.4096	1.2501	1.6597	280
290	57.556	0.01735	7.444	7.461	259.31	917.5	1176.8	0.4234	1.2238	1.6472	290
300°	67.013	0.01745	6.449	6.466	269.59	910.1	1179.7	0.4369	1.1980	1.6350	300°
310	77.68	0.01755	5.609	5.626	279.92	902.6	1182.5	0.4504	1.1727	1.6231	310
320	89.66	0.01765	4.896	4.914	290.28	894.9	1185.2	0.4637	1.1478	1.6115	320
330	103.06	0.01776	4.289	4.307	300.68	887.0	1187.7	0.4769	1.1233	1.6002	330
340	118.01	0.01787	3.770	3.788	311.13	879.0	1190.1	0.4900	1.0992	1.5891	340
350°	134.63	0.01799	3.324	3.342	321.63	870.7	1192.3	0.5029	1.0754	1.5783	350°
360	153.04	0.01811	2.939	2.957	332.18	862.2	1194.4	0.5158	1.0519	1.5677	360
370	173.37	0.01823	2.606	2.625	342.79	853.5	1196.3	0.5286	1.0287	1.5573	370
380	195.77	0.01836	2.317	2.335	353.45	844.6	1198.1	0.5413	1.0059	1.5471	380
390	220.37	0.01850	2.0651	2.0836	364.17	835.4	1199.6	0.5539	0.9832	1.5371	390
400°	247.31	0.01864	1.8447	1.8633	374.97	826.0	1201.0	0.5664	0.9608	1.5272	400°
410	276.75	0.01878	1.6512	1.6700	385.83	816.3	1202.1	0.5788	0.9386	1.5174	410
420	308.83	0.01894	1.4811	1.5000	396.77	806.3	1203.1	0.5912	0.9166	1.5078	420
430	343.72	0.01910	1.3308	1.3499	407.79	796.0	1203.8	0.6035	0.8947	1.4982	430
440	381.59	0.01926	1.1979	1.2171	418.90	785.4	1204.3	0.6158	0.8730	1.4887	440
450°	422.6	0.0194	1.0799	1.0993	430.1	774.5	1204.6	0.6280	0.8513	1.4793	450°
460	466.9	0.0196	0.9748	0.9944	441.4	763.2	1204.6	0.6402	0.8298	1.4700	460
470	514.7	0.0198	0.8811	0.9009	452.8	751.5	1204.3	0.6523	0.8083	1.4606	470
480	566.1	0.0200	0.7972	0.8172	464.4	739.4	1203.7	0.6645	0.7868	1.4513	480
490	621.4	0.0202	0.7221	0.7423	476.0	726.8	1202.8	0.6766	0.7653	1.4419	490
500°	680.8	0.0204	0.6545	0.6749	487.8	713.9	1201.7	0.6887	0.7438	1.4325	500°
520	812.4	0.0209	0.5385	0.5594	511.9	686.4	1198.2	0.7130	0.7006	1.4136	520
540	962.5	0.0215	0.4434	0.4649	536.6	656.6	1193.2	0.7374	0.6568	1.3942	540
560	1133.1	0.0221	0.3647	0.3868	562.2	624.2	1186.4	0.7621	0.6121	1.3742	560
580	1325.8	0.0228	0.2989	0.3217	588.9	588.4	1177.3	0.7872	0.5659	1.3532	580
600°	1542.9	0.0236	0.2432	0.2668	617.0	548.5	1165.5	0.8131	0.5176	1.3307	600°
620	1786.6	0.0247	0.1955	0.2201	646.7	503.6	1150.3	0.8398	0.4664	1.3062	620
640	2059.7	0.0260	0.1538	0.1798	678.6	452.0	1130.5	0.8679	0.4110	1.2789	640
660	2365.4	0.0278	0.1165	0.1442	714.2	390.2	1104.4	0.8987	0.3485	1.2472	660
680	2708.1	0.0305	0.0810	0.1115	757.3	309.9	1067.2	0.9351	0.2719	1.2071	680
700°	3093.7	0.0369	0.0392	0.0761	823.3	172.1	995.4	0.9905	0.1484	1.1389	700°
705.4	3206.2	0.0503	0	0.0503	902.7	0	902.7	1.0580	0	1.0580	705.4

THERMODYNAMICS

Appendix B
Saturated Steam: Pressures

Abs. Press. psia	Temp. T,°F	Specific Volume		Enthalpy, BTU/lbm			Entropy, BTU/lbm-°R			Internal Energy		Abs. Press psia
		Sat. Liquid	Sat. Vapor	Sat. Liquid	Evap	Sat. Vapor	Sat. Liquid	Evap.	Sat. Vapor	Sat. Liquid	Sat. Vapor	
p		v_f	v_g	h_f	h_{fg}	h_g	s_f	s_{fg}	s_g	u_f	u_g	p
1.0	101.74	0.01614	333.6	69.70	1036.3	1106.0	0.1326	1.8456	1.9782	69.70	1044.3	1.0
2.0	126.08	0.01623	173.73	93.99	1022.2	1116.2	0.1749	1.7451	1.9200	93.98	1051.9	2.0
3.0	141.48	0.01630	118.71	109.37	1013.2	1122.6	0.2008	1.6855	1.8863	109.36	1056.7	3.0
4.0	152.97	0.01636	90.63	120.86	1006.4	1127.3	0.2198	1.6427	1.8625	120.85	1060.2	4.0
5.0	162.24	0.01640	73.52	130.13	1001.0	1131.1	0.2347	1.6094	1.8441	130.12	1063.1	5.0
6.0	170.06	0.01645	61.98	137.96	996.2	1134.2	0.2472	1.5820	1.8292	137.94	1065.4	6.0
7.0	176.85	0.01649	53.64	144.76	992.1	1136.9	0.2581	1.5586	1.8167	144.74	1067.4	7.0
8.0	182.86	0.01653	47.34	150.79	988.5	1139.3	0.2674	1.5383	1.8057	150.77	1069.2	8.0
9.0	188.28	0.01656	42.40	156.22	985.2	1141.4	0.2759	1.5203	1.7962	156.19	1070.8	9.0
10	193.21	0.01659	38.42	161.17	982.1	1143.3	0.2835	1.5041	1.7876	161.14	1072.2	10
14.696	212.00	0.01672	26.80	180.07	970.3	1150.4	0.3120	1.4446	1.7566	180.02	1077.5	14.696
15	213.03	0.01672	26.29	181.11	969.7	1150.8	0.3135	1.4415	1.7549	181.06	1077.8	15
20	227.96	0.01683	20.089	196.16	960.1	1156.3	0.3356	1.3962	1.7319	196.10	1081.9	20
25	240.07	0.01692	16.303	208.42	952.1	1160.6	0.3533	1.3606	1.7139	208.34	1085.1	25
30	250.33	0.01701	13.746	218.82	945.3	1164.1	0.3680	1.3313	1.6993	218.73	1087.8	30
35	259.28	0.01708	11.898	227.91	939.2	1167.1	0.3807	1.3063	1.6870	227.80	1090.1	35
40	267.25	0.01715	10.498	236.03	933.7	1169.7	0.3919	1.2844	1.6763	235.90	1092.0	40
45	274.44	0.01721	9.401	243.36	928.6	1172.0	0.4019	1.2650	1.6669	243.22	1093.7	45
50	281.01	0.01727	8.515	250.09	924.0	1174.1	0.4110	1.2474	1.6585	249.93	1095.3	50
55	287.07	0.01732	7.787	256.30	919.6	1175.9	0.4193	1.2316	1.6509	256.12	1096.7	55
60	292.71	0.01738	7.175	262.09	915.5	1177.6	0.4270	1.2168	1.6438	261.90	1097.9	60
65	297.97	0.01743	6.655	267.50	911.6	1179.1	0.4342	1.2032	1.6374	267.29	1099.1	65
70	302.92	0.01748	6.206	272.61	907.9	1180.6	0.4409	1.1906	1.6315	272.38	1100.2	70
75	307.60	0.01753	5.816	277.43	904.5	1181.9	0.4472	1.1787	1.6259	277.19	1101.2	75
80	312.03	0.01757	5.472	282.02	901.1	1183.1	0.4531	1.1676	1.6207	281.76	1102.1	80
85	316.25	0.01761	5.168	286.39	897.8	1184.2	0.4587	1.1571	1.6158	286.11	1102.9	85
90	320.27	0.01766	4.896	290.56	894.7	1185.3	0.4641	1.1471	1.6112	290.27	1103.7	90
95	324.12	0.01770	4.652	294.56	891.7	1186.2	0.4692	1.1376	1.6068	294.25	1104.5	95
100	327.81	0.01774	4.432	298.40	888.8	1187.2	0.4740	1.1286	1.6026	298.08	1105.2	100
110	334.77	0.01782	4.049	305.66	883.2	1188.9	0.4832	1.1117	1.5948	305.30	1106.5	110
120	341.25	0.01789	3.728	312.44	877.9	1190.4	0.4916	1.0962	1.5878	312.05	1107.6	120
130	347.32	0.01796	3.455	318.81	872.9	1191.7	0.4995	1.0817	1.5812	318.38	1108.6	130
140	353.02	0.01802	3.220	324.82	868.2	1193.0	0.5069	1.0682	1.5751	324.35	1109.6	140
150	358.42	0.01809	3.015	330.51	863.6	1194.1	0.5138	1.0556	1.5694	330.01	1110.5	150
160	363.53	0.01815	2.834	335.93	859.2	1195.1	0.5204	1.0436	1.5640	335.39	1111.2	160
170	368.41	0.01822	2.675	341.09	854.9	1196.0	0.5266	1.0324	1.5590	340.52	1111.9	170
180	373.06	0.01827	2.532	346.03	850.8	1196.9	0.5325	1.0217	1.5542	345.42	1112.5	180
190	377.51	0.01833	2.404	350.79	846.8	1197.6	0.5381	1.0116	1.5497	350.15	1113.1	190
200	381.79	0.01839	2.288	355.36	843.0	1198.4	0.5435	1.0018	1.5453	354.68	1113.7	200
250	400.95	0.01865	1.8438	376.00	825.1	1201.1	0.5675	.9588	1.5263	375.14	1115.8	250
300	417.33	0.01890	1.5433	393.84	809.0	1202.8	0.5879	0.9225	1.5104	392.79	1117.1	300
350	431.72	0.01913	1.3260	409.69	794.2	1203.9	0.6056	0.8910	1.4966	408.45	1118.0	350
400	444.59	0.0193	1.1613	424.0	780.5	1204.5	0.6214	0.8630	1.4844	422.6	1118.5	400
450	456.28	0.0195	1.0320	437.2	767.4	1204.6	0.6356	0.8378	1.4734	435.5	1118.7	450
500	467.01	0.0197	0.9278	449.4	755.0	1204.4	0.6487	0.8147	1.4634	447.6	1118.6	500
550	476.94	0.0199	0.8424	460.8	743.1	1203.9	0.6608	0.7934	1.4542	458.8	1118.2	550
600	486.21	0.0201	0.7698	471.6	731.6	1203.2	0.6720	0.7734	1.4454	469.4	1117.7	600
650	494.90	0.0203	0.7083	481.8	720.5	1202.3	0.6826	0.7548	1.4374	479.4	1117.1	650
700	503.10	0.0205	0.6554	491.5	709.7	1201.2	0.6925	0.7371	1.4296	488.8	1116.3	700
750	510.86	0.0207	0.6092	500.8	699.2	1200.0	0.7019	0.7204	1.4223	498.0	1115.4	750
800	518.23	0.0209	0.5687	509.7	688.9	1198.6	0.7108	0.7045	1.4153	506.6	1114.4	800
850	525.26	0.0210	0.5327	518.3	678.8	1197.1	0.7194	0.6891	1.4085	515.0	1113.3	850
900	531.98	0.0212	0.5006	526.6	668.8	1195.4	0.7275	0.6744	1.4020	523.1	1112.1	900
950	538.43	0.0214	0.4717	534.6	659.1	1193.7	0.7355	0.6602	1.3957	530.9	1110.8	950
1000	544.61	0.0216	0.4456	542.4	649.4	1191.8	0.7430	0.6467	1.3897	538.4	1109.4	1000
1100	556.31	0.0220	0.4001	557.4	630.4	1187.8	0.7575	0.6205	1.3780	552.9	1106.4	1100
1200	567.22	0.0223	0.3619	571.7	611.7	1183.4	0.7711	0.5956	1.3667	566.7	1103.0	1200
1300	577.46	0.0227	0.3293	585.4	593.2	1178.6	0.7840	0.5719	1.3559	580.0	1099.4	1300
1400	587.10	0.0231	0.3012	598.7	574.7	1173.4	0.7963	0.5491	1.3454	592.7	1095.4	1400
1500	596.23	0.0235	0.2765	611.6	556.3	1167.9	0.8082	0.5269	1.3351	605.1	1091.2	1500
2000	635.82	0.0257	0.1878	671.7	463.4	1135.1	0.8619	0.4230	1.2849	662.2	1065.6	2000
2500	668.13	0.0287	0.1307	730.6	360.5	1091.1	0.9126	0.3197	1.2322	717.3	1030.6	2500
3000	695.36	0.0346	0.0658	802.5	217.8	1020.3	0.9731	0.1885	1.1615	783.4	972.7	3000
3206.2	705.40	0.0503	0.0503	902.7	0	902.7	1.0580	0	1.0580	872.9	872.9	3206.2

Appendix C
Superheated Steam

Abs. Press. psia (Sat. Temp.) (°F)		Temperature-Degrees Fahrenheit												
		200°	300°	400°	500°	600°	700°	800°	900°	1000°	1100°	1200°	1400°	1600°
1 (101.74)	v	392.6	452.3	512.0	571.6	631.2	690.8	750.4	809.9	869.5	929.1	988.7	1107.8	1227.0
	h	1150.4	1195.8	1241.7	1288.3	1335.7	1383.8	1432.8	1482.7	1533.5	1585.2	1637.7	1745.7	1857.5
	s	2.0512	2.1153	2.1720	2.2233	2.2702	2.3137	2.3542	2.3923	2.4283	2.4625	2.4952	2.5566	2.6137
5 (162.24)	v	78.16	90.25	102.26	114.22	126.16	138.10	150.03	161.95	173.87	185.79	197.71	221.6	245.4
	h	1148.8	1195.0	1241.2	1288.0	1335.4	1383.6	1432.7	1482.6	1533.4	1585.1	1637.7	1745.7	1857.4
	s	1.8718	1.9370	1.9942	2.0456	2.0927	2.1363	2.1767	2.2148	2.2509	2.2851	2.3178	2.3792	2.4363
10 (193.21)	v	38.85	45.00	51.04	57.05	63.03	69.01	74.98	80.95	86.92	92.88	98.84	110.77	122.69
	h	1146.6	1193.9	1240.6	1287.5	1335.1	1383.4	1432.5	1482.4	1533.2	1585.0	1637.6	1745.6	1857.3
	s	1.7927	1.8595	1.9172	1.9689	2.0160	2.0596	2.1002	2.1383	2.1744	2.2086	2.2413	2.3028	2.3598
14.696 (212.00)	v	......	30.53	34.68	38.78	42.86	46.94	51.00	55.07	59.13	63.19	67.25	75.37	83.48
	h	......	1192.8	1239.9	1287.1	1334.8	1383.2	1432.3	1482.3	1533.1	1584.8	1637.5	1745.5	1857.3
	s	......	1.8160	1.8743	1.9261	1.9734	2.0170	2.0576	2.0958	2.1319	2.1662	2.1989	2.2603	2.3174
20 (227.96)	v	......	22.36	25.43	28.46	31.47	34.47	37.46	40.45	43.44	46.42	49.41	55.37	61.34
	h	......	1191.6	1239.2	1286.6	1334.4	1382.9	1432.1	1482.1	1533.0	1584.7	1637.4	1745.4	1857.2
	s	......	1.7808	1.8396	1.8918	1.9392	1.9829	2.0235	2.0618	2.0978	2.1321	2.1648	2.2263	2.2834
40 (267.25)	v	......	11.040	12.628	14.168	15.688	17.198	18.702	20.20	21.70	23.20	24.69	27.68	30.66
	h	......	1186.8	1236.5	1284.8	1333.1	1381.9	1431.3	1481.4	1532.4	1584.3	1637.0	1745.1	1857.0
	s	......	1.6994	1.7608	1.8140	1.8619	1.9058	1.9467	1.9850	2.0212	2.0555	2.0883	2.1498	2.2069
60 (292.71)	v	......	7.259	8.357	9.403	10.427	11.441	12.449	13.452	14.454	15.453	16.451	18.446	20.44
	h	......	1181.6	1233.6	1283.0	1331.8	1380.9	1430.5	1480.8	1531.9	1583.8	1636.6	1744.8	1856.7
	s	......	1.6492	1.7135	1.7678	1.8162	1.8605	1.9015	1.9400	1.9762	2.0106	2.0434	2.1049	2.1621
80 (312.03)	v	......	......	6.220	7.020	7.797	8.562	9.322	10.077	10.830	11.582	12.332	13.830	15.325
	h	......	......	1230.7	1281.1	1330.5	1379.9	1429.7	1480.1	1531.3	1583.4	1636.2	1744.5	1856.5
	s	......	......	1.6791	1.7346	1.7836	1.8281	1.8694	1.9079	1.9442	1.9787	2.0115	2.0731	2.1303
100 (327.81)	v	......	......	4.937	5.589	6.218	6.835	7.446	8.052	8.656	9.259	9.860	11.060	12.258
	h	......	......	1227.6	1279.1	1329.1	1378.9	1428.9	1479.5	1530.8	1582.9	1635.7	1744.2	1856.2
	s	......	......	1.6518	1.7085	1.7581	1.8029	1.8443	1.8829	1.9193	1.9538	1.9867	2.0484	2.1056
120 (341.25)	v	......	......	4.081	4.636	5.165	5.683	6.195	6.702	7.207	7.710	8.212	9.214	10.213
	h	......	......	1221.4	1277.2	1327.7	1377.8	1428.1	1478.8	1530.2	1582.4	1635.3	1743.9	1856.0
	s	......	......	1.6287	1.6869	1.7370	1.7822	1.8237	1.8625	1.8990	1.9335	1.9664	2.0281	2.0854
140 (353.02)	v	......	......	3.468	3.954	4.413	4.861	5.301	5.738	6.172	6.604	7.035	7.895	8.752
	h	......	......	1221.1	1275.2	1326.4	1376.8	1427.3	1478.2	1529.7	1581.9	1634.9	1743.5	1855.7
	s	......	......	1.6087	1.6683	1.7190	1.7645	1.8063	1.8451	1.8817	1.9163	1.9493	2.0110	2.0683
160 (363.06)	v	......	......	3.008	3.443	3.849	4.244	4.631	5.015	5.396	5.775	6.152	6.906	7.656
	h	......	......	1217.6	1273.1	1325.0	1375.7	1426.4	1477.5	1529.1	1581.4	1634.5	1743.2	1855.7
	s	......	......	1.5908	1.6519	1.7033	1.7491	1.7911	1.8301	1.8667	1.9014	1.9344	1.9962	2.0535
180 (373.53)	v	......	......	2.649	3.044	3.411	3.764	4.110	4.452	4.792	5.129	5.466	6.136	6.804
	h	......	......	1214.0	1271.0	1323.5	1374.7	1425.6	1476.8	1528.6	1581.0	1634.1	1742.9	1855.2
	s	......	......	1.5745	1.6373	1.6894	1.7355	1.7776	1.8167	1.8534	1.8882	1.9212	1.9831	2.0404
200 (381.79)	v	......	......	2.361	2.726	3.060	3.380	3.693	4.002	4.309	4.613	4.917	5.521	6.123
	h	......	......	1210.3	1268.9	1322.1	1373.6	1424.8	1476.2	1528.0	1580.5	1633.7	1742.6	1855.0
	s	......	......	1.5594	1.6240	1.6767	1.7232	1.7655	1.8048	1.8415	1.8763	1.9094	1.9713	2.0287
220 (389.86)	v	......	......	2.125	2.465	2.772	3.066	3.352	3.634	3.913	4.191	4.467	5.017	5.565
	h	......	......	1206.5	1266.7	1320.7	1372.6	1424.0	1475.5	1527.5	1580.0	1633.3	1742.3	1854.7
	s	......	......	1.5453	1.6117	1.6652	1.7120	1.7545	1.7939	1.8308	1.8656	1.8987	1.9607	2.0181
240 (397.37)	v	......	......	1.9276	2.247	2.533	2.804	3.068	3.327	3.584	3.839	4.093	4.597	5.100
	h	......	......	1202.5	1264.5	1319.2	1371.5	1423.2	1474.8	1526.9	1579.6	1632.9	1742.0	1854.5
	s	......	......	1.5319	1.6003	1.6546	1.7017	1.7444	1.7839	1.8209	1.8558	1.8889	1.9510	2.0084
260 (404.42)	v	......	......	.....	2.063	2.330	2.582	2.827	3.067	3.305	3.541	3.776	4.242	4.707
	h	......	......	.....	1262.3	1317.7	1370.4	1422.3	1474.2	1526.3	1579.1	1632.5	1741.7	1854.2
	s	......	......	.....	1.5897	1.6447	1.6922	1.7352	1.7748	1.8118	1.8467	1.8799	1.9420	1.9995
280 (411.05)	v	......	......	.....	1.9047	2.156	2.392	2.621	2.845	3.066	3.286	3.504	3.938	4.370
	h	......	......	.....	1260.0	1316.2	1369.4	1421.5	1473.5	1525.8	1578.6	1632.1	1741.4	1854.0
	s	......	......	.....	1.5796	1.6354	1.6834	1.7265	1.7652	1.8033	1.8383	1.8716	1.9337	1.9912
300 (417.33)	v	......	......	.....	1.7675	2.005	2.227	2.442	2.652	2.859	3.065	3.269	3.674	4.078
	h	......	......	.....	1257.6	1314.7	1368.3	1420.6	1472.8	1525.2	1578.1	1631.7	1741.0	1853.7
	s	......	......	.....	1.5701	1.6268	1.6751	1.7184	1.7582	1.7954	1.8305	1.8638	1.9260	1.9835
350 (431.72)	v	......	......	.....	1.4923	1.7036	1.8980	2.084	2.266	2.445	2.622	2.798	3.147	3.493
	h	......	......	.....	1251.5	1310.9	1365.5	1418.5	1471.1	1523.8	1577.0	1630.7	1740.3	1853.1
	s	......	......	.....	1.5481	1.6070	1.6563	1.7002	1.7403	1.7777	1.8130	1.8463	1.9086	1.9663
400 (444.59)	v	......	......	.....	1.2851	1.4770	1.6508	1.8161	1.9767	2.134	2.290	2.445	2.751	3.055
	h	......	......	.....	1245.1	1306.9	1362.7	1416.4	1469.4	1522.4	1575.8	1629.6	1739.5	1852.5
	s	......	......	.....	1.5281	1.5984	1.6398	1.6842	1.7247	1.7623	1.7977	1.8311	1.8936	1.9513

THERMODYNAMICS

Appendix C—Continued
Superheated Steam

Abs. Press. psia (Sat. Temp.) (°F)		500°	550°	600°	620°	640°	660°	680°	700°	800°	900°	1000°	1200°	1400°	1600°
450 (456.28)	v	1.1231	1.2155	1.3005	1.3332	1.3652	1.3967	1.4278	1.4584	1.6074	1.7516	1.8928	2.170	2.443	2.714
	h	1238.4	1272.0	1302.8	1314.6	1326.2	1337.5	1348.8	1359.9	1414.3	1467.7	1521.0	1628.6	1738.7	1851.9
	s	1.5095	1.5437	1.5735	1.5845	1.5951	1.6054	1.6153	1.6250	1.6699	1.7108	1.7486	1.8177	1.8803	1.9381
500 (467.01)	v	0.9927	1.0800	1.1591	1.1893	1.2188	1.2478	1.2763	1.3044	1.4405	1.5715	1.6996	1.9504	2.197	2.442
	h	1231.3	1266.8	1298.6	1310.7	1322.6	1334.2	1345.7	1357.0	1412.1	1466.0	1519.6	1627.6	1737.9	1851.3
	s	1.4919	1.5280	1.5588	1.5701	1.5810	1.5915	1.6016	1.6115	1.6571	1.6982	1.7363	1.8056	1.8683	1.9262
550 (476.94)	v	0.8852	0.9686	1.0431	1.0714	1.0989	1.1259	1.1523	1.1783	1.3038	1.4241	1.5414	1.7706	1.9957	2.219
	h	1223.7	1261.2	1294.3	1306.8	1318.9	1330.8	1342.5	1354.0	1409.9	1464.3	1518.2	1626.6	1737.1	1850.6
	s	1.4751	1.5131	1.5451	1.5568	1.5680	1.5787	1.5890	1.5991	1.6452	1.6868	1.7250	1.7946	1.8575	1.9155
600 (486.21)	v	0.7947	0.8753	0.9463	0.9729	0.9988	1.0241	1.0489	1.0732	1.1899	1.3013	1.4096	1.6208	1.8279	2.033
	h	1215.7	1255.5	1289.9	1302.7	1315.2	1327.4	1339.3	1351.1	1407.7	1462.5	1516.7	1625.5	1736.3	1850.0
	s	1.4586	1.4990	1.5323	1.5443	1.5558	1.5667	1.5773	1.5875	1.6343	1.6762	1.7147	1.7846	1.8476	1.9056
700 (503.10)	v	.....	0.7277	0.7934	0.8177	0.8411	0.8639	0.8860	0.9077	1.0108	1.1082	1.2024	1.3853	1.5641	1.7405
	h	.....	1243.2	1280.6	1294.3	1307.5	1320.3	1332.8	1345.0	1403.2	1459.0	1513.9	1623.5	1734.8	1848.8
	s	.....	1.4722	1.5084	1.5212	1.5333	1.5449	1.5559	1.5665	1.6147	1.6573	1.6963	1.7666	1.8299	1.8881
800 (518.23)	v	.....	0.6154	0.6779	0.7006	0.7223	0.7433	0.7635	0.7833	0.8763	0.9633	1.0470	1.2088	1.3662	1.5214
	h	.....	1229.8	1270.7	1285.4	1299.4	1312.9	1325.9	1338.6	1398.6	1455.4	1511.0	1621.4	1733.2	1847.5
	s	.....	1.4467	1.4863	1.5000	1.5129	1.5250	1.5366	1.5476	1.5972	1.6407	1.6801	1.7510	1.8146	1.8729
900 (531.98)	v	.....	0.5264	0.5873	0.6089	0.6294	0.6491	0.6680	0.6863	0.7716	0.8506	0.9262	1.0714	1.2124	1.3509
	h	.....	1215.0	1260.1	1275.9	1290.9	1305.1	1318.8	1332.1	1393.9	1451.8	1508.1	1619.3	1731.6	1846.3
	s	.....	1.4216	1.4653	1.4800	1.4938	1.5066	1.5187	1.5303	1.5814	1.6257	1.6656	1.7371	1.8009	1.8595
1000 (544.61)	v	.....	0.4533	0.5140	0.5350	0.5546	0.5733	0.5912	0.6084	0.6878	0.7604	0.8294	0.9615	1.0893	1.2146
	h	.....	1198.3	1248.8	1265.9	1281.9	1297.0	1311.4	1325.3	1389.2	1448.2	1505.1	1617.3	1730.0	1845.0
	s	.....	1.3961	1.4450	1.4610	1.4757	1.4893	1.5021	1.5141	1.5670	1.6121	1.6525	1.7245	1.7886	1.8474
1100 (556.31)	v	.....	.....	0.4532	0.4738	0.4929	0.5110	0.5281	0.5445	0.6191	0.6866	0.7503	0.8716	0.9885	1.1031
	h	.....	.....	1236.7	1255.3	1272.4	1288.5	1303.7	1318.3	1384.3	1444.5	1502.2	1615.2	1728.4	1843.8
	s	.....	.....	1.4251	1.4425	1.4583	1.4728	1.4862	1.4989	1.5535	1.5995	1.6405	1.7130	1.7775	1.8363
1200 (567.22)	v	.....	.....	0.4016	0.4222	0.4410	0.4586	0.4752	0.4909	0.5617	0.6250	0.6843	0.7967	0.9046	1.0101
	h	.....	.....	1223.5	1243.9	1262.4	1279.6	1295.7	1311.0	1379.3	1440.7	1499.2	1613.1	1726.9	1842.5
	s	.....	.....	1.4052	1.4243	1.4413	1.4568	1.4710	1.4843	1.5409	1.5879	1.6293	1.7025	1.7672	1.8263
1400 (587.10)	v	.....	.....	0.3174	0.3390	0.3580	0.3753	0.3912	0.4062	0.4714	0.5281	0.5805	0.6789	0.7727	0.8640
	h	.....	.....	1193.0	1218.4	1240.4	1260.3	1278.5	1295.5	1369.1	1433.1	1493.2	1608.9	1723.7	1840.0
	s	.....	.....	1.3639	1.3877	1.4079	1.4258	1.4419	1.4567	1.5177	1.5666	1.6093	1.6836	1.7489	1.8083
1600 (604.90)	v	.....	.....	.....	0.2733	0.2936	0.3112	0.3271	0.3417	0.4034	0.4553	0.5027	0.5905	0.6738	0.7545
	h	.....	.....	.....	1187.8	1215.2	1238.7	1259.6	1278.7	1358.4	1425.3	1487.0	1604.6	1720.5	1837.5
	s	.....	.....	.....	1.3489	1.3741	1.3952	1.4137	1.4303	1.4964	1.5476	1.5914	1.6669	1.7328	1.7926
1800 (621.03)	v	.....	.....	.....	.....	0.2407	0.2597	0.2760	0.2907	0.3502	0.3986	0.4421	0.5218	0.5968	0.6693
	h	.....	.....	.....	.....	1185.1	1214.0	1238.5	1260.3	1347.2	1417.4	1480.8	1600.4	1717.3	1835.0
	s	.....	.....	.....	.....	1.3377	1.3638	1.3855	1.4044	1.4765	1.5301	1.5752	1.6520	1.7185	1.7786
2000 (635.82)	v	.....	.....	.....	.....	0.1936	0.2161	0.2337	0.2489	0.3074	0.3532	0.3935	0.4668	0.5352	0.6011
	h	.....	.....	.....	.....	1145.6	1184.9	1214.8	1240.0	1335.5	1409.2	1474.5	1596.1	1714.1	1832.5
	s	.....	.....	.....	.....	1.2945	1.3300	1.3564	1.3783	1.4576	1.5139	1.5603	1.6384	1.7055	1.7660
2500 (668.13)	v	.....	.....	.....	.....	.....	0.1484	0.1686	0.2294	0.2710	0.3061	0.3678	0.4244	0.4784	
	h	.....	.....	.....	.....	.....	1132.3	1176.8	1303.6	1387.8	1458.4	1585.3	1706.1	1826.2	
	s	.....	.....	.....	.....	.....	1.2687	1.3073	1.4127	1.4772	1.5273	1.6088	1.6775	1.7389	
3000 (695.36)	v	.....	.....	.....	.....	.....	.....	0.0984	0.1760	0.2159	0.2476	0.3018	0.3505	0.3966	
	h	.....	.....	.....	.....	.....	.....	1060.7	1267.2	1365.0	1441.8	1574.3	1698.0	1819.9	
	s	.....	.....	.....	.....	.....	.....	1.1966	1.3690	1.4439	1.4984	1.5837	1.6540	1.7163	
3206.2 (705.40)	v	.....	.....	.....	.....	.....	.....	.....	0.1583	0.1981	0.2288	0.2806	0.3267	0.3703	
	h	.....	.....	.....	.....	.....	.....	.....	1250.5	1355.2	1434.7	1569.8	1694.6	1817.2	
	s	.....	.....	.....	.....	.....	.....	.....	1.3508	1.4309	1.4874	1.5742	1.6452	1.7080	
3500	v	.....	.....	.....	.....	.....	.....	0.0306	0.1364	0.1762	0.2058	0.2546	0.2977	0.3381	
	h	.....	.....	.....	.....	.....	.....	780.5	1224.9	1340.7	1424.5	1563.3	1689.8	1813.6	
	s	.....	.....	.....	.....	.....	.....	0.9515	1.3241	1.4127	1.4723	1.5615	1.6336	1.6968	
4000	v	.....	.....	.....	.....	.....	.....	0.0287	0.1052	0.1462	0.1743	0.2192	0.2581	0.2943	
	h	.....	.....	.....	.....	.....	.....	763.8	1174.8	1314.4	1406.8	1552.1	1681.7	1807.2	
	s	.....	.....	.....	.....	.....	.....	0.9347	1.2757	1.3827	1.4482	1.5417	1.6154	1.6795	
4500	v	.....	.....	.....	.....	.....	.....	0.0276	0.0798	0.1226	0.1500	0.1917	0.2273	0.2602	
	h	.....	.....	.....	.....	.....	.....	753.5	1113.9	1286.5	1388.4	1540.8	1673.5	1800.9	
	s	.....	.....	.....	.....	.....	.....	0.9235	1.2204	1.3529	1.4253	1.5235	1.5990	1.6640	
5000	v	.....	.....	.....	.....	.....	.....	0.0268	0.0593	0.1036	0.1303	0.1696	0.2027	0.2329	
	h	.....	.....	.....	.....	.....	.....	746.4	1047.1	1256.5	1369.5	1529.6	1665.3	1794.5	
	s	.....	.....	.....	.....	.....	.....	0.9152	1.1622	1.3231	1.4034	1.5066	1.5839	1.6499	
5500	v	.....	.....	.....	.....	.....	.....	0.0262	0.0463	0.0880	0.1143	0.1516	0.1825	0.2106	
	h	.....	.....	.....	.....	.....	.....	741.3	985.0	1224.1	1349.3	1518.2	1657.0	1788.1	
	s	.....	.....	.....	.....	.....	.....	0.9090	1.1093	1.2930	1.3821	1.4908	1.5699	1.6369	

Appendix D
Compressed Water

Temperature, °F

Abs. Press., psia (Sat. Temp.)	Saturated Liquid		32	100	200	300	400	500	600	700
		P	0.08854	0.9492	11.526	67.013	247.31	680.8	1542.9	3093.7
		v_f	0.016022	0.016132	0.016634	0.017449	0.018639	0.020432	0.023629	0.03692
		h_f	0	67.97	167.99	269.59	374.97	487.82	617.0	823.3
		s_f	0	0.12948	0.29382	0.43694	0.56638	0.68871	0.8131	0.9905
200 (381.79)	$(v-v_f)$EE5		−1.1	−1.1	−1.1	−1.1				
	$(h-h_f)$		+0.61	+0.54	+0.41	+0.23				
	$(s-s_f)$EE3		+0.03	−0.05	−0.21	−0.21				
400 (444.59)	$(v-v_f)$EE5		−2.3	−2.1	−2.2	−2.8	−2.1			
	$(h-h_f)$		+1.21	+1.09	+0.88	+0.61	+0.16			
	$(s-s_f)$EE3		+0.04	−0.16	−0.47	−0.56	−0.40			
800 (518.23)	$(v-v_f)$EE5		−4.6	−4.0	−4.4	−5.6	−6.5	−1.7		
	$(h-h_f)$		+2.39	+2.17	+1.78	+1.35	+0.61	−0.05		
	$(s-s_f)$EE3		+0.10	−0.40	−0.97	−1.27	−1.48	−0.53		
1000 (544.61)	$(v-v_f)$EE5		−5.7	−5.1	−5.4	−6.9	−8.7	−6.4		
	$(h-h_f)$		+2.99	+2.70	+2.21	+1.75	+0.84	−0.14		
	$(s-s_f)$EE3		+0.15	−0.53	−1.20	−1.64	−2.00	−1.41		
1500 (596.23)	$(v-v_f)$EE5		−8.4	−7.5	−8.1	−10.4	−14.1	−17.3		
	$(h-h_f)$		+4.48	+3.99	+3.36	+2.70	+1.44	−0.29		
	$(s-s_f)$EE3		+0.20	−0.86	−1.79	−2.53	−3.32	−3.56		
2000 (635.82)	$(v-v_f)$EE5		−11.0	−9.9	−10.8	−13.8	−19.5	−27.8	−32.6	
	$(h-h_f)$		+5.97	+5.31	+4.51	+3.64	+2.03	−0.38	−2.5	
	$(s-s_f)$EE3		+0.22	−1.18	−2.39	−3.42	−4.57	−5.58	−4.3	
3000 (695.36)	$(v-v_f)$EE5		−16.3	−14.7	−16.0	−20.7	−30.0	−47.1	−87.9	
	$(h-h_f)$		+9.00	+7.88	+6.76	+5.49	+3.33	−0.41	−6.9	
	$(s-s_f)$EE3		+0.28	−1.79	−3.56	−5.12	−7.03	−9.42	−12.4	
4000	$(v-v_f)$EE5		−21.5	−19.2	−21.0	−27.5	−40.0	−64.5	−132.2	−821
	$(h-h_f)$		+11.88	+10.49	+9.03	+7.41	+4.71	−0.16	−10.0	−59.5
	$(s-s_f)$EE3		+0.29	−2.42	−4.74	−6.77	−9.40	−13.03	−19.3	−55.8
5000	$(v-v_f)$EE5		−26.7	−23.6	−26.0	−34.0	−49.6	−80.5	−169.3	−1017
	$(h-h_f)$		+14.75	+13.08	+11.30	+9.36	+6.08	+0.25	−12.1	−76.9
	$(s-s_f)$EE3		+0.22	−3.07	−5.92	−8.40	−11.74	−16.47	−25.3	−75.3

THERMODYNAMICS

Appendix E
Mollier Diagram For Water

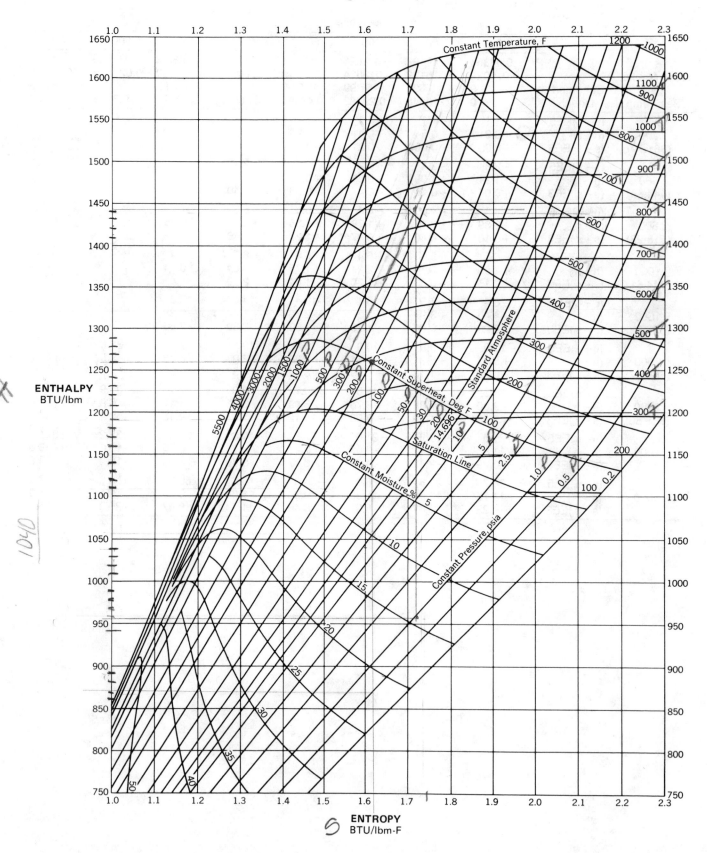

ENTHALPY
BTU/lbm

ENTROPY
BTU/lbm-F

Appendix F
Air Table

T°R	h BTU/lbm	p_r	u BTU/lbm	v_r	φ BTU/lbm-°R	T°R	h BTU/lbm	p_r	u BTU/lbm	v_r	φ BTU/lbm-°R
360	85.97	0.3363	61.29	396.6	0.50369	1460	358.63	50.34	258.54	10.743	0.84704
380	90.75	0.4061	64.70	346.6	0.51663	1480	363.89	53.04	262.44	10.336	0.85062
400	95.53	0.4858	68.11	305.0	0.52890	1500	369.17	55.86	266.34	9.948	0.85416
420	100.32	0.5760	71.52	270.1	0.54058	1520	374.47	58.78	270.26	9.578	0.85767
440	105.11	0.6776	74.93	240.6	0.55172	1540	379.77	61.83	274.20	9.226	0.86113
460	109.90	0.7913	78.36	215.33	0.56235	1560	385.08	65.00	278.13	8.890	0.86456
480	114.69	0.9182	81.77	193.65	0.57255	1580	390.40	68.30	282.09	8.569	0.86794
500	119.48	1.0590	85.20	174.90	0.58233	1600	395.74	71.73	286.06	8.263	0.87130
520	124.27	1.2147	88.62	158.58	0.59173	1620	401.09	75.29	290.04	7.971	0.87462
537	128.10	1.3593	91.53	146.34	0.59945	1640	406.45	78.99	294.03	7.691	0.87791
540	129.06	1.3860	92.04	144.32	0.60078	1660	411.82	82.83	298.02	7.424	0.88116
560	133.86	1.5742	95.47	131.78	0.60950	1680	417.20	86.82	302.04	7.168	0.88439
580	138.66	1.7800	98.90	120.70	0.61793	1700	422.59	90.95	306.06	6.924	0.88758
600	143.47	2.005	102.34	110.88	0.62607	1720	428.00	95.24	310.09	6.690	0.89074
620	148.28	2.249	105.78	102.12	0.63395	1740	433.41	99.69	314.13	6.465	0.89387
640	153.09	2.514	109.21	94.30	0.64159	1760	438.83	104.30	318.18	6.251	0.89697
660	157.92	2.801	112.67	87.27	0.64902	1780	444.26	109.08	322.24	6.045	0.90003
680	162.73	3.111	116.12	80.96	0.65621	1800	449.71	114.03	326.32	5.847	0.90308
700	167.56	3.446	119.58	75.25	0.66321	1820	455.17	119.16	330.40	5.658	0.90609
720	172.39	3.806	123.04	70.07	0.67002	1840	460.63	124.47	334.50	5.476	0.90908
740	177.23	4.193	126.51	65.38	0.67665	1860	466.12	129.95	338.61	5.302	0.91203
760	182.08	4.607	129.99	61.10	0.68312	1880	471.60	135.64	342.73	5.134	0.91497
780	186.94	5.051	133.47	57.20	0.68942	1900	477.09	141.51	346.85	4.974	0.91788
800	191.81	5.526	136.97	53.63	0.69558	1920	482.60	147.59	350.98	4.819	0.92076
820	196.69	6.033	140.47	50.35	0.70160	1940	488.12	153.87	355.12	4.670	0.92362
840	201.56	6.573	143.98	47.34	0.70747	1960	493.64	160.37	359.28	4.527	0.92645
860	206.46	7.149	147.50	44.57	0.71323	1980	499.17	167.07	363.43	4.390	0.92926
880	211.35	7.761	151.02	42.01	0.71886	2000	504.71	174.00	367.61	4.258	0.93205
900	216.26	8.411	154.57	39.64	0.72438	2020	510.26	181.16	371.79	4.130	0.93481
920	221.18	9.102	158.12	37.44	0.72979	2040	515.82	188.54	375.98	4.008	0.93756
940	226.11	9.834	161.68	35.41	0.73509	2060	521.39	196.16	380.18	3.890	0.94026
960	231.06	10.610	165.26	33.52	0.74030	2080	526.97	204.02	384.39	3.777	0.94296
980	236.02	11.430	168.83	31.76	0.74540	2100	532.55	212.1	388.60	3.667	0.94564
1000	240.98	12.298	172.43	30.12	0.75042	2150	546.54	233.5	399.17	3.410	0.95222
1020	245.97	13.215	176.04	28.59	0.75536	2200	560.59	256.6	409.78	3.176	0.95868
1040	250.95	14.182	179.66	27.17	0.76019	2250	574.69	281.4	420.46	2.961	0.96501
1060	255.96	15.203	183.29	25.82	0.76496	2300	588.82	308.1	431.16	2.765	0.97123
1080	260.97	16.278	186.93	24.58	0.76964	2350	603.00	336.8	441.91	2.585	0.97732
1100	265.99	17.413	190.58	23.40	0.77426	2400	617.22	367.6	452.70	2.419	0.98331
1120	271.03	18.604	194.25	22.30	0.77880	2450	631.48	400.5	463.54	2.266	0.98919
1140	276.08	19.858	197.94	21.27	0.78326	2500	645.78	435.7	474.40	2.125	0.99497
1160	281.14	21.18	201.63	20.293	0.78767	2550	660.12	473.3	485.31	1.9956	1.00064
1180	286.21	22.56	205.33	19.377	0.79201	2600	674.49	513.5	496.26	1.8756	1.00623
1200	291.30	24.01	209.05	18.514	0.79628	2650	688.90	556.3	507.25	1.7646	1.01172
1220	296.41	25.53	212.78	17.700	0.80050	2700	703.35	601.9	518.26	1.6617	1.01712
1240	301.52	27.13	216.53	16.932	0.80466	2750	717.83	650.4	529.31	1.5662	1.02244
1260	306.65	28.80	220.28	16.205	0.80876	2800	732.33	702.0	540.40	1.4775	1.02767
1280	311.79	30.55	244.05	15.518	0.81280	2850	746.88	756.7	551.52	1.3951	1.03282
1300	316.94	32.39	227.83	14.868	0.81680	2900	761.45	814.8	562.66	1.3184	1.03788
1320	322.11	34.31	231.63	14.253	0.82075	2950	776.05	876.4	573.84	1.2469	1.04288
1340	327.29	36.31	235.43	13.670	0.82464	3000	790.68	941.4	585.04	1.1803	1.04779
1360	332.48	38.41	239.25	13.118	0.82848	3500	938.40	1829.3	698.48	0.7087	1.09332
1380	337.68	40.59	243.08	12.593	0.83229	4000	1088.26	3280	814.06	0.4518	1.13334
1400	342.90	42.88	246.93	12.095	0.83604	4500	1239.86	5521	931.39	0.3019	1.16905
1420	348.14	45.26	250.79	11.622	0.83975	5000	1392.87	8837	1050.12	0.20959	1.20129
1440	353.37	47.75	254.66	11.172	0.84341	6000	1702.29	20120	1291.00	0.11047	1.25769
						6500	1858.44	28974	1412.87	0.08310	1.28268

Appendix G
Saturated Ammonia

Abstracted from Tables of *Thermodynamic Properties of Ammonia*, U.S. Dept. of Commerce, Bureau of Standards Circular No. 142, 1945.

BY TEMPERATURE

| Temp. (°F) | Press. (psia) | Specific Volume | | Enthalpy | | | Entropy | | | Temp. (°F) |
| | | Sat. liquid | Sat. vapor | Sat. liquid | Evap. | Sat. vapor | Sat. liquid | Evap. | Sat. vapor | |
T	p	v_f	v_g	h_f	h_{fg}	h_g	s_f	s_{fg}	s_g	T
−60	5.55	0.02278	44.73	−21.2	610.8	589.6	−0.0517	1.5286	1.4769	−60
−50	7.67	0.02299	33.08	−10.6	604.3	593.7	−0.0256	1.4753	1.4497	−50
−40	10.41	0.02322	24.86	0.0	597.6	597.6	0.0000	1.4242	1.4242	−40
−30	13.90	0.02345	18.97	10.7	590.7	601.4	0.0250	1.3751	1.4001	−30
−20	18.30	0.02369	14.68	21.4	583.6	605.0	0.0497	1.3277	1.3774	−20
−10	23.74	0.02393	11.50	32.1	576.4	608.5	0.0738	1.2820	1.3558	−10
0	30.42	0.02419	9.116	42.9	568.9	611.8	0.0975	1.2377	1.3352	0
5	34.27	0.02432	8.150	48.3	565.0	613.3	0.1092	1.2161	1.3253	5
10	38.51	0.02446	7.304	53.8	561.1	614.9	0.1208	1.1949	1.3157	10
20	48.21	0.02474	5.910	64.7	553.1	617.8	0.1437	1.1532	1.2969	20
30	59.74	0.02503	4.825	75.7	544.8	620.5	0.1663	1.1127	1.2790	30
40	73.32	0.02533	3.971	86.8	536.2	623.0	0.1885	1.0733	1.2618	40
50	89.19	0.02564	3.294	97.9	527.3	625.2	0.2105	1.0348	1.2453	50
60	107.6	0.02597	2.751	109.2	518.1	627.3	0.2322	0.9972	1.2294	60
70	128.8	0.02632	2.312	120.5	508.6	629.1	0.2537	0.9603	1.2140	70
80	153.0	0.02668	1.955	132.0	498.7	630.7	0.2749	0.9242	1.1991	80
86	169.2	0.02691	1.772	138.9	492.6	631.5	0.2875	0.9029	1.1904	86
90	180.6	0.02707	1.661	143.5	488.5	632.0	0.2958	0.8888	1.1846	90
100	211.9	0.02747	1.419	155.2	477.8	633.0	0.3166	0.8539	1.1705	100
110	247.0	0.02790	1.217	167.0	466.7	633.7	0.3372	0.8194	1.1566	110
120	286.4	0.02836	1.047	179.0	455.0	634.0	0.3576	0.7851	1.1427	120

BY PRESSURE

| Press. (psia) | Temp. (°F) | Specific Volume | | Enthalpy | | | Entropy | | | Press. (psia) |
| | | Sat. liquid | Sat. vapor | Sat. liquid | Evap. | Sat. vapor | Sat. liquid | Evap. | Sat. vapor | |
p	T	v_f	v_g	h_f	h_{fg}	h_g	s_f	s_{fg}	s_g	p
5	−63.11	0.02271	49.31	−24.5	612.8	588.3	−0.0599	1.5456	1.4857	5
10	−41.34	0.02319	25.81	−1.4	598.5	597.1	−0.0034	1.4310	1.4276	10
15	−27.29	0.02351	17.67	13.6	588.8	602.4	0.0318	1.3620	1.3938	15
20	−16.64	0.02377	13.50	25.0	581.2	606.2	0.0578	1.3122	1.3700	20
30	−0.57	0.02417	9.236	42.3	569.3	611.6	0.0962	1.2402	1.3364	30
40	11.66	0.02451	7.047	55.6	559.8	615.4	0.1246	1.1879	1.3125	40
50	21.67	0.02479	5.710	66.5	551.7	618.2	0.1475	1.1464	1.2939	50
60	30.21	0.02504	4.805	75.9	544.6	620.5	0.1668	1.1119	1.2787	60
80	44.40	0.02546	3.655	91.7	532.3	624.0	0.1982	1.0563	1.2545	80
100	56.05	0.02584	2.952	104.7	521.8	626.5	0.2237	1.0119	1.2356	100
120	66.02	0.02618	2.476	116.0	512.4	628.4	0.2452	0.9749	1.2201	120
140	74.79	0.02649	2.132	126.0	503.9	629.9	0.2638	0.9430	1.2068	140
170	86.29	0.02692	1.764	139.3	492.3	631.6	0.2881	0.9019	1.1900	170
200	96.34	0.02732	1.502	150.9	481.8	632.7	0.3090	0.8666	1.1756	200
230	105.30	0.02770	1.307	161.4	472.0	633.4	0.3275	0.8356	1.1631	230
260	113.42	0.02806	1.155	171.1	462.8	633.9	0.3441	0.8077	1.1518	260

Appendix H
Superheated Ammonia

Abstracted from Tables of *Thermodynamic Properties of Ammonia*, U.S. Dept. of Commerce, Bureau of Standards Circular No. 142, 1945.

Abs. press. psia (Sat. temp.)		Temperature—Degrees Fahrenheit											
		−40	−20	0	20	40	60	80	100	150	200	250	300
5 (−63.11)	v	52.36	54.97	57.55	60.12	62.69	65.24	67.79	70.33	76.68	......	......	......
	h	600.3	610.4	620.4	630.4	640.4	650.5	660.6	670.7	696.4	......	......	......
	s	1.5149	1.5385	1.5608	1.5821	1.6026	1.6223	1.6413	1.6598	1.7038	......	......	......
10 (−41.34)	v	......	26.58	28.58	29.90	31.20	32.49	33.78	35.07	38.26	41.45	......	......
	h	......	603.2	618.9	629.1	639.3	649.5	659.7	670.1	695.8	722.2	......	......
	s	......	1.4420	1.4773	1.4992	1.5200	1.5400	1.5593	1.5779	1.6222	1.6637	......	......
15 (−27.29)	v	......	18.01	18.92	19.82	20.70	21.58	22.44	23.31	25.46	27.59	......	......
	h	......	606.4	617.2	627.8	638.2	648.5	658.9	669.2	695.3	721.7	......	......
	s	......	1.4031	1.4272	1.4497	1.4709	1.4912	1.5108	1.5296	1.5742	1.6158	......	......
20 (−16.64)	v	......	......	14.09	14.78	15.45	16.12	16.78	17.43	19.05	20.66	......	......
	h	......	......	615.5	626.4	637.0	647.5	658.0	668.5	694.7	721.2	......	......
	s	......	......	1.3907	1.4138	1.4356	1.4562	1.4760	1.4950	1.5399	1.5817	......	......
25 (−7.96)	v	......	......	11.19	11.75	12.30	12.84	13.37	13.90	15.21	16.50	17.79	......
	h	......	......	613.8	625.0	635.8	646.5	657.1	667.7	694.1	720.8	748.0	......
	s	......	......	1.3616	1.3855	1.4077	1.4287	1.4487	1.4679	1.5131	1.5552	1.5948	......
30 (−0.57)	v	......	......	......	9.731	10.20	10.65	11.10	11.55	12.65	13.73	14.81	......
	h	......	......	......	623.5	634.6	645.5	656.2	666.9	693.5	720.3	747.5	......
	s	......	......	......	1.3618	1.3845	1.4059	1.4261	1.4456	1.4911	1.5334	1.5733	......
35 (5.89)	v	......	......	......	8.287	8.695	9.093	9.484	9.869	10.82	11.75	12.68	......
	h	......	......	......	622.0	633.4	644.4	655.3	666.1	692.9	719.9	747.2	......
	s	......	......	......	1.3413	1.3646	1.3863	1.4069	1.4265	1.4724	1.5148	1.5547	......
40 (11.86)	v	......	......	......	7.203	7.568	7.922	8.268	8.609	9.444	10.27	11.08	11.88
	h	......	......	......	620.4	632.1	643.4	654.4	665.3	692.3	719.4	746.8	774.6
	s	......	......	......	1.3231	1.3470	1.3692	1.3900	1.4098	1.4561	1.4987	1.5387	1.5766
50 (21.67)	v	......	......	......	......	5.988	6.280	6.564	6.843	7.521	8.185	8.840	9.489
	h	......	......	......	......	629.5	641.2	652.6	663.7	691.1	718.5	746.1	774.0
	s	......	......	......	......	1.3169	1.3399	1.3613	1.3816	1.4286	1.4716	1.5219	1.5500
60 (30.21)	v	......	......	......	......	4.933	5.184	5.428	5.665	6.239	6.798	7.348	7.892
	h	......	......	......	......	626.8	639.0	650.7	662.1	689.9	717.5	745.3	773.3
	s	......	......	......	......	1.2913	1.3152	1.3373	1.3581	1.4058	1.4493	1.4898	1.5281
70 (37.70)	v	......	......	......	......	4.177	4.401	4.615	4.822	5.323	5.807	6.287	6.750
	h	......	......	......	......	623.9	636.6	648.7	660.4	688.7	716.6	744.5	772.7
	s	......	......	......	......	1.2688	1.2937	1.3166	1.3378	1.3863	1.4302	1.4710	1.5095
80 (44.40)	v	......	......	......	......	......	3.812	4.005	4.190	4.635	5.063	5.487	5.894
	h	......	......	......	......	......	634.3	646.7	658.7	687.5	715.6	743.8	772.1
	s	......	......	......	......	......	1.2745	1.2981	1.3199	1.3692	1.4136	1.4547	1.4933
90 (50.47)	v	......	......	......	......	......	3.353	3.529	3.698	4.100	4.484	4.859	5.228
	h	......	......	......	......	......	631.8	644.7	657.0	688.7	714.7	743.0	771.5
	s	......	......	......	......	......	1.2571	1.2814	1.3038	1.3863	1.3988	1.4401	1.4789
100 (56.05)	v	......	......	......	......	......	2.985	3.149	3.304	3.672	4.021	4.361	4.695
	h	......	......	......	......	......	629.3	642.6	655.2	685.0	713.7	742.2	770.8
	s	......	......	......	......	......	1.2409	1.2661	1.2891	1.3401	1.3854	1.4271	1.4660
120 (66.02)	v	......	......	......	......	......	......	2.576	2.712	3.029	3.326	3.614	3.895
	h	......	......	......	......	......	......	638.3	651.6	682.5	711.8	740.7	769.6
	s	......	......	......	......	......	......	1.2386	1.2628	1.3157	1.3620	1.4042	1.4435
140 (74.79)	v	......	......	......	......	......	......	2.166	2.288	2.569	2.830	3.080	3.323
	h	......	......	......	......	......	......	633.8	647.8	679.9	709.9	739.2	768.3
	s	......	......	......	......	......	......	1.2140	1.2396	1.2945	1.3418	1.3846	1.4243
160 (82.64)	v	......	......	......	......	......	......	......	1.969	2.224	2.457	2.679	2.895
	h	......	......	......	......	......	......	......	643.9	677.2	707.9	737.6	767.1
	s	......	......	......	......	......	......	......	1.2186	1.2757	1.3240	1.3675	1.4076

Appendix H—Continued
Superheated Ammonia

Abstracted from Tables of *Thermodynamic Properties of Ammonia*, U.S. Dept. of Commerce, Bureau of Standards Circular No. 142, 1945.

Abs. press. psia (Sat. temp.)		Temperature—Degrees Fahrenheit											
		−40	−20	0	20	40	60	80	100	150	200	250	300
180 (89.78)	v	......	......	......	......	......	......	......	1.720	1.995	2.167	2.367	2.561
	h	......	......	......	......	......	......	......	639.9	674.6	705.9	736.1	765.8
	s	......	......	......	......	......	......	......	1.1992	1.2586	1.3081	1.3521	1.3926
200 (96.34)	v	......	......	......	......	......	......	......	......	1.740	1.935	2.118	2.295
	h	......	......	......	......	......	......	......	......	671.8	703.9	734.5	764.5
	s	......	......	......	......	......	......	......	......	1.2429	1.2935	1.3382	1.3791
220 (102.42)	v	......	......	......	......	......	......	......	......	1.564	1.745	1.914	2.076
	h	......	......	......	......	......	......	......	......	669.0	701.9	732.9	763.2
	s	......	......	......	......	......	......	......	......	1.2281	1.2801	1.3255	1.3668
240 (108.09)	v	......	......	......	......	......	......	......	......	1.416	1.587	1.745	1.895
	h	......	......	......	......	......	......	......	......	666.1	699.8	731.3	762.0
	s	......	......	......	......	......	......	......	......	1.2145	1.2677	1.3137	1.3554
260 (113.42)	v	......	......	......	......	......	......	......	......	1.292	1.453	1.601	1.741
	h	......	......	......	......	......	......	......	......	663.1	697.7	729.7	760.7
	s	......	......	......	......	......	......	......	......	1.2014	1.2560	1.3027	1.3349
280 (118.45)	v	......	......	......	......	......	......	......	......	1.184	1.339	1.478	1.610
	h	......	......	......	......	......	......	......	......	660.1	695.6	728.1	759.4
	s	......	......	......	......	......	......	......	......	1.1888	1.2449	1.2924	1.3350
300 (123.21)	v	......	......	......	......	......	......	......	......	1.091	1.239	1.372	1.496
	h	......	......	......	......	......	......	......	......	656.9	693.5	726.5	758.1
	s	......	......	......	......	......	......	......	......	1.1767	1.2344	1.2827	1.3257

Appendix I
Saturated Freon-12

Freon is a registered trademark of the E. I. du Pont de Nemours & Co., which originally tabulated this data.

BY TEMPERATURE

Temp. (°F)	Press. (psia)	Specific volume		Enthalpy			Entropy			Temp. (°F)
		Sat. liquid	Sat. vapor	Sat. liquid	Evap.	Sat. vapor	Sat. liquid	Evap.	Sat. vapor	
T	p	v_f	v_g	h_f	h_{fg}	h_g	s_f	s_{fg}	s_g	T
−60	5.37	0.01036	6.516	−4.20	75.33	71.13	−0.0102	0.1681	0.1783	−60
−50	7.13	0.01047	5.012	−2.11	74.42	72.31	−0.0050	0.1717	0.1767	−50
−40	9.32	0.0106	3.911	0.00	73.50	73.50	0.00000	0.17517	0.17517	−40
−30	12.02	0.0107	3.088	2.03	72.67	74.70	0.00471	0.16916	0.17387	−30
−20	15.28	0.0108	2.474	4.07	71.80	75.87	0.00940	0.16335	0.17275	−20
−10	19.20	0.0109	2.003	6.14	70.91	77.05	0.01403	0.15772	0.17175	−10
0	23.87	0.0110	1.637	8.25	69.96	78.21	0.01869	0.15222	0.17091	0
5	26.51	0.0111	1.485	9.32	69.47	78.79	0.02097	0.14955	0.17052	5
10	29.35	0.0112	1.351	10.39	68.97	79.36	0.02328	0.14687	0.17015	10
20	35.75	0.0113	1.121	12.55	67.94	80.49	0.02783	0.14166	0.16949	20
30	43.16	0.0115	0.939	14.76	66.85	81.61	0.03233	0.13654	0.16887	30
40	51.68	0.0116	0.792	17.00	65.71	82.71	0.03680	0.13153	0.16833	40
50	61.39	0.0118	0.673	19.27	64.51	83.78	0.04126	0.12659	0.16785	50
60	72.41	0.0119	0.575	21.57	63.25	84.82	0.04568	0.12173	0.16741	60
70	84.82	0.0121	0.493	23.90	61.92	85.82	0.05009	0.11692	0.16701	70
80	98.76	0.0123	0.425	26.28	60.52	86.80	0.05446	0.11215	0.16662	80
86	107.9	0.0124	0.389	27.72	59.65	87.37	0.05708	0.10932	0.16640	86
90	114.3	0.0125	0.368	28.70	59.04	87.74	0.05882	0.10742	0.16624	90
100	131.6	0.0127	0.319	31.16	57.46	88.62	0.06316	0.10268	0.16584	100
110	150.7	0.0129	0.277	33.65	55.78	89.43	0.06749	0.09793	0.16542	110
120	171.8	0.0132	0.240	36.16	53.99	90.15	0.07180	0.09315	0.16495	120

BY PRESSURE

Press. (psia)	Temp. (°F)	Specific volume		Enthalpy			Entropy			Press. (psia)
		Sat. liquid	Sat. vapor	Sat. liquid	Evap.	Sat. vapor	Sat. liquid	Evap.	Sat. vapor	
p	T	v_f	v_g	h_f	h_{fg}	h_g	s_f	s_{fg}	s_g	p
5	−62.5	0.01034	6.953	−4.73	75.56	70.83	−0.0115	0.1943	0.1788	5
10	−37.3	0.0106	3.662	0.54	73.28	73.82	0.00127	0.17360	0.17487	10
15	−20.8	0.0108	2.518	3.91	71.87	75.78	0.00902	0.16381	0.17283	15
20	−8.2	0.0109	1.925	6.53	70.74	77.27	0.01488	0.15672	0.17160	20
30	11.1	0.0112	1.324	10.62	68.86	79.48	0.02410	0.14597	0.17007	30
40	25.9	0.0114	1.009	13.86	67.30	81.16	0.03049	0.13865	0.16914	40
50	38.3	0.0116	0.817	16.58	65.94	82.52	0.03597	0.13244	0.16841	50
60	48.7	0.0117	0.688	18.96	64.69	83.65	0.04065	0.12726	0.16791	60
80	66.3	0.0120	0.521	23.01	62.44	85.45	0.04844	0.11872	0.16716	80
100	80.9	0.0123	0.419	26.49	60.40	86.89	0.05483	0.11176	0.16659	100
120	93.4	0.0126	0.350	29.53	58.52	88.05	0.06030	0.10580	0.16610	120
140	104.5	0.0128	0.298	32.28	56.71	88.99	0.06513	0.10053	0.16566	140
160	114.5	0.0130	0.260	34.78	54.99	89.77	0.06958	0.09564	0.16522	160
180	123.7	0.0133	0.228	37.07	53.31	90.38	0.07337	0.09139	0.16476	180
200	132.1	0.0135	0.202	39.21	51.65	90.86	0.07694	0.08730	0.16424	200
220	139.9	0.0138	0.181	41.22	50.28	91.50	0.08021	0.08354	0.16375	220

THERMODYNAMICS

Appendix J
Superheated Freon-12

Abs. press. psia (Sat. temp.)		−40	−20	0	20	40	60	80	100	150	200	250	300
		Temperature—Degrees Fahrenheit											
5 (−62.5)	v	7.363	7.726	8.088	8.450	8.812	9.173	9.533	9.893	10.79	11.69	......	......
	h	73.72	76.36	79.05	81.78	84.56	87.41	90.30	93.25	100.84	108.75	......	......
	s	0.1859	0.1920	0.1979	0.2038	0.2095	0.2150	0.2205	0.2258	0.2388	0.2513	......	......
10 (−37.3)	v	......	3.821	4.006	4.189	4.371	4.556	4.740	4.923	5.379	5.831	6.281	......
	h	......	76.11	78.81	81.56	84.35	87.19	90.11	93.05	100.66	108.63	116.88	......
	s	......	0.1801	0.1861	0.1919	0.1977	0.2033	0.2087	0.2141	0.2271	0.2396	0.2517	......
15 (−20.8)	v	......	2.521	2.646	2.771	2.895	3.019	3.143	3.266	3.571	3.877	4.191	......
	h	......	75.89	78.59	81.37	84.18	87.03	89.94	92.91	100.53	108.49	116.78	......
	s	......	0.17307	0.17913	0.18499	0.19074	0.19635	0.20185	0.20723	0.22028	0.23282	0.24491	......
20 (−8.2)	v	......	......	1.965	2.060	2.155	2.250	2.343	2.437	2.669	2.901	3.130	......
	h	......	......	78.39	81.14	83.97	86.85	89.78	92.75	100.40	108.38	116.67	......
	s	......	......	0.17407	0.17996	0.18573	0.19138	0.19688	0.20229	0.21537	0.22794	0.24005	......
25 (2.2)	v	......	......	......	1.712	1.793	1.873	1.952	2.031	2.227	2.422	2.615	......
	h	......	......	......	80.95	83.78	86.67	89.61	92.56	100.26	108.26	116.56	......
	s	......	......	......	0.17637	0.18216	0.18783	0.19336	0.19748	0.21190	0.22450	0.23665	......
30 (11.1)	v	......	......	......	1.364	1.430	1.495	1.560	1.624	1.784	1.943	2.099	......
	h	......	......	......	80.75	83.59	86.49	89.43	92.42	100.12	108.13	116.45	......
	s	......	......	......	0.17278	0.17859	0.18429	0.18983	0.19527	0.20843	0.22105	0.23325	......
35 (18.9)	v	......	......	......	1.109	1.237	1.295	1.352	1.409	1.550	1.689	1.827	......
	h	......	......	......	80.49	83.40	86.30	89.26	92.26	99.98	108.01	116.33	......
	s	......	......	......	0.16963	0.17591	0.18162	0.18719	0.19266	0.20584	0.21849	0.23069	......
40 (25.9)	v	......	......	......	......	1.044	1.095	1.144	1.194	1.315	1.435	1.554	......
	h	......	......	......	......	83.20	86.11	89.09	92.09	99.83	107.88	116.21	......
	s	......	......	......	......	0.17322	0.17896	0.18455	0.19004	0.20325	0.21592	0.22813	......
50 (38.3)	v	......	......	......	......	0.821	0.863	0.904	0.944	1.044	1.142	1.239	1.332
	h	......	......	......	......	82.76	85.72	88.72	91.75	99.54	107.62	116.00	124.69
	s	......	......	......	......	0.16895	0.17475	0.19040	0.18591	0.19923	0.21196	0.22419	0.23600
60 (48.7)	v	......	......	......	......	......	0.708	0.743	0.778	0.863	0.946	1.028	1.108
	h	......	......	......	......	......	85.33	88.35	91.41	99.24	107.36	115.54	124.29
	s	......	......	......	......	......	0.17120	0.17689	0.18246	0.19585	0.20865	0.22094	0.23280
70 (57.9)	v	......	......	......	......	......	0.553	0.642	0.673	0.750	0.824	0.896	0.967
	h	......	......	......	......	......	84.94	87.96	91.05	98.94	107.10	115.54	124.29
	s	......	......	......	......	......	0.16765	0.17399	0.17961	0.19310	0.20597	0.21830	0.23020
80 (66.3)	v	......	......	......	......	......	......	0.540	0.568	0.636	0.701	0.764	0.826
	h	......	......	......	......	......	......	87.56	90.68	98.64	106.84	115.30	124.08
	s	......	......	......	......	......	......	0.17108	0.17675	0.19035	0.20328	0.21566	0.22760
90 (73.6)	v	......	......	......	......	......	......	......	0.505	0.568	0.627	0.685	0.742
	h	......	......	......	......	......	......	......	90.31	98.32	106.56	115.07	123.88
	s	......	......	......	......	......	......	......	0.17443	0.18813	0.20111	0.21356	0.22554
100 (80.9)	v	......	......	......	......	......	......	......	0.442	0.499	0.553	0.606	0.657
	h	......	......	......	......	......	......	......	89.93	97.99	106.29	114.84	123.67
	s	......	......	......	......	......	......	......	0.17210	0.18590	0.19894	0.21145	0.22347
120 (93.4)	v	......	......	......	......	......	......	......	0.357	0.407	0.454	0.500	0.543
	h	......	......	......	......	......	......	......	89.13	97.30	105.70	114.35	123.25
	s	......	......	......	......	......	......	......	0.16803	0.18207	0.19529	0.20792	0.22000
140 (104.5)	v	......	......	......	......	......	......	......	......	0.341	0.383	0.423	0.462
	h	......	......	......	......	......	......	......	......	96.65	105.14	113.85	122.85
	s	......	......	......	......	......	......	......	......	0.17868	0.19205	0.20479	0.21701
160 (114.5)	v	......	......	......	......	......	......	......	......	0.318	0.335	0.372	0.408
	h	......	......	......	......	......	......	......	......	95.82	104.50	113.33	122.39
	s	......	......	......	......	......	......	......	......	0.17561	0.18927	0.20213	0.21444
180 (123.7)	v	......	......	......	......	......	......	......	......	0.294	0.287	0.321	0.353
	h	......	......	......	......	......	......	......	......	94.99	103.85	112.81	121.92
	s	......	......	......	......	......	......	......	......	0.17254	0.18648	0.19947	0.21187
200 (132.1)	v	......	......	......	......	......	......	......	......	0.241	0.255	0.288	0.317
	h	......	......	......	......	......	......	......	......	94.16	103.12	112.20	121.42
	s	......	......	......	......	......	......	......	......	0.16970	0.18395	0.19717	0.20970
220 (139.9)	v	......	......	......	......	......	......	......	......	0.188	0.232	0.254	0.282
	h	......	......	......	......	......	......	......	......	93.32	102.39	111.59	120.91
	s	......	......	......	......	......	......	......	......	0.16685	0.18142	0.19387	0.20753

PRACTICE PROBLEMS: THERMODYNAMICS

Warm-ups

1. What is the enthalpy of 92% quality, 250°F steam in BTU/pmole?

2. What is the availability of an isentropic process operating between 300 psia and 50 psia and starting with 95% quality steam?

3. Cast iron is heated from 80°F to 780°F. What heat is required?

4. What is the ratio of specific heats of air at 600°F?

5. What are the speeds of sound in 70°F air, steel, and water?

6. What is the density of 600°F, 1 atmosphere helium?

7. Calculate the isentropic efficiency of a process that expands dry steam from 100 psia to 3 psia and 90% quality.

8. Convert 470 BTU to kw-hrs.

9. Find the work done when 8 cubic feet of 180°F air are cooled to 100°F at a constant 14.7 psia.

10. A building requires 3 EE5 cubic feet of air per hour. The air is heated from 35°F to 75°F by water whose temperature decreases from 180°F to 150°F. How many gpm of water are required?

Concentrates

.46 PSIA

1. A 5000 kw steam turbine uses 200 psia steam with 100°F superheat and exhausts at 1″ Hg absolute. What is the water rate at full load? If the actual load is only 2500 kw, what is the loss in available energy?

2. A 10,000 kw steam turbine operates on 400 psia, 750°F dry steam and exhausts to 2″ Hg absolute. What is the adiabatic heat drop available for power production?

3. A 750 kw steam turbine has a water rate of 20 lbm/kw-hr when expanding 50°F superheat steam from 150 psig to 26″ Hg absolute. Find the quantity of 65°F cooling water required for a mixing condenser if the turbine operates straight condensing.

4. 332,000 lbm/hr of 81°F water enter a 2-pass closed feedwater heater with 1850 square feet of 5/8″ copper tubing. 4.45 psia steam from a turbine bleed is condensed from a saturated vapor to a saturated liquid. The heated water leaves at 150 °F with an enthalpy of 1100 BTU/lbm. (a) What is the overall transfer coefficient? (b) What is the steam extraction rate?

5. A 2-pass surface condenser constructed of 1″ B.W.G. tubing operates from a condensing turbine with a steam flow of 82,000 lbm/hr at 980 BTU/lbm and condenser

pressure of 1″ Hg absolute. Water is circulated at 8 fps through an equivalent length of 120 feet of extra strong 30″ steel pipe. A head loss of 6″ w.g. is incurred at the intake screens. (a) What is the head added by the pump? (b) If the water temperature increases 10°F, how many gallons of cooling water circulate per minute?

6. What is the cost of turning 100 lbm/hr of 60°F water into 14.7 psia (saturated) steam in an electric boiler if electricity is $.04 per kilowatt-hour and radiation losses are 35%?

7. What is the efficiency of a gas-fired boiler which delivers 250 lbm/hr of 98% dry steam at 25 psig if the feedwater is 60°F? The 80°F gas is at 4″ w.g. and has a heating value of 550 BTU per cubic foot at standard (industrial) conditions. The environment is at 30.2″ Hg. 13.5 cfm of gas are used.

8. A deLaval type, single-stage turbine with an 18″ diameter rotor operates at 12,000 rpm (which is 40% of the jet velocity). The jet stream is offset 20° from the rotor plane. If the exhaust is at 14.7 psia and the nozzle pressure is 100 psia, what is the enthalpy change of the steam?

9. Find the process efficiency of a boiler which evaporates 8.23 lbm of 120°F water per pound of coal fired, producing 100 psig saturated steam. The coal is 2% moisture by weight as fired, and dry coal is 5% ash. The dry coal has a heating value of 12,800 BTU/lbm. 12% of the coal is removed from the ash pit. Assume ash-pit coal has the same composition as unfired, dry coal.

10. What percent of the combustion heat is lost up the stack in 550°F stack gases if 26 lbm of 70°F air are used per pound of coal? The coal is 7% ash, 5% hydrogen, 5% oxygen, and 83% carbon. Assume a reasonable pit loss.

Timed (1 hour allowed for each)

1. A miniature air turbine is used to drive a small drill. Air enters the turbine at 140°F at the rate of 15 lbm/hr. The output of the turbine is .250 horsepower. The turbine exhaust is to 15 psia. The turbine flow is steady, and the process is adiabatic. The adiabatic efficiency of the process is 60%. For the following questions, use $c_v = .171$ BTU/lbm-°R, R= 53.35 ft-lbf/lbm-°R, and k= 1.4.

   (a) What is the air exhaust temperature?

   (b) What is the inlet air pressure?

   (c) What is the change in entropy through the turbine?

2. Water is used in an adiabatic desuperheater to cool steam. The conditions of the process are:

Hot steam in: 600°F, 200 psig, 1000 lbm/hr, negligible velocity

Water in: 82°F, 50 lbm/hr, negligible velocity

Cold steam out: 100 psia, 2000 fps

(a) What is the temperature and quality of the exiting steam?

(b) What is the entropy production across the device?

3. Xenon gas (20 psia, 70°F) is compressed to 3800 psia and 70°F by a compressor/heat exchanger combination. The compressed gas is stored in a 100 cubic foot rigid container initially charged with xenon gas at 20 psia. (a) What is the weight of the xenon gas originally stored in the tank? (b) What is the average flow rate of xenon gas in pounds per hour if the compressor charges the tank in exactly one hour? (c) If charging takes exactly one hour and electricity costs $.045 per kw-hr, what is the cost of charging the tank?

4. A 20 cubic foot tank is constructed of 40 pounds of steel with a specific heat of .11 BTU/lbm-°R and is placed in a large room containing air at 14.7 psia and 70°F. The tank is evacuated to 1 psia and 70°F. A valve is opened suddenly, allowing the tank to fill with room air. Find the air temperature inside the tank immediately after filling and before any heat transfer takes place into the room. Assume that the temperature of the steel surface and the air inside the tank are the same. Assume that the air entering the tank is turbulent and well-mixed.

5. In the figure shown, compressor A handles 600 cfm of 14.7 psia, 80°F air. The input conditions are the same for compressor B. Output C uses 100 cfm of 80 psia, 85°F air. Output D uses 120 cfm of 85 psia, 80°F air. Output E uses 8 pounds per minute of 85°F air. The storage tank contains 1000 psia, 90°F air. Assume that air is an ideal gas and calculate the input flow rate to compressor B.

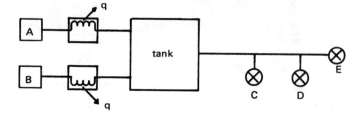

6. 300 cfm of 90°F air at 14.7 psia enter a compressor. The compressor discharges the air into a water-cooled heat exchanger. The compressed air at 90°F is stored at 300 psig in a 1000 ft³ tank. The tank feeds three air-driven tools with the flow and property characteristics listed. How long can the system run at the conditions given? (Assume that the compressor inlet and outlet conditions do not change.)

	tool #1	tool #2	tool #3
flow rate (cfm)	40	15	?
flow rate (lbm/min)	?	?	6
minimum pressure (psig)	90	50	80
temperature (°F)	90	85	80

7. A gas turbine air heater receives 540°F air at 100 psia and heats it to 1540°F. The outside temperature is 100°F. The pressure of the air drops 20 psia as it passes through the heater. What is the loss in available energy due to the pressure drop? Express your answer in percent.

RESERVED FOR FUTURE USE

RESERVED FOR FUTURE USE

7 VAPOR, COMBUSTION, REFRIGERATION, AND COMPRESSION CYCLES

PART 1: Vapor Cycles

Nomenclature for Vapor Cycles

h	enthalpy	BTU/lbm
J	Joule's constant (778)	ft-lbf/BTU
m	mass	lbm
p	pressure	psf
Q	heat	BTU/lbm
s	entropy	BTU/°R-lbm
T	temperature	°R
W	work	BTU/lbm
x	quality or bleed fraction	decimal
y	bleed fraction	decimal

Symbols

v	specific volume	ft^3/lbm
η	efficiency	decimal

Subscripts

comp	compressor
ext	external
id	ideal
int	internal
m	mechanical
s	isentropic
th	thermal
turb	turbine

1 DEFINITIONS

Air heater: See "air preheater."

Air preheater: A device that heats combustion air by recovering energy from stack gases. A convection preheater uses a conventional heat exchanger; a *regenerative preheater* uses a rotating drum which is alternately exposed to both gas flows.

Back-pressure turbine: A turbine that exhausts to a second turbine operating in a lower pressure range. Alternative definition: a turbine which exhausts to a pressure greater than atmospheric.

Bleed: A removal of partially expanded steam from a turbine.

Co-generation cycle: A cycle in which steam from an electricity generating process is used for subsequent processes, or vice-versa.

Condensing cycle: A cycle in which steam is condensed and returned to the boiler after expanding through a turbine.

Condition line: The locus of all states of steam during the expansion process, as in line d-e in figure 7.9.

Deaerator: A heat exchanger in which water is heated to a point where dissolved corrosive gases (primarily oxygen) are liberated.

Economizer: A heat exchanger that heats feedwater by exposing it to stack gases.

Evaporator: A closed heat exchanger that vaporizes untreated water at atmospheric pressure by high-pressure steam. The vaporized water is condensed and stored for use in the boiler.

Extraction heater: A feedwater heater using extracted steam as a heating source. Also, see "feedwater heater."

Extraction rate: The rate, usually expressed in BTU/hr, at which partially expanded steam is bled off from a turbine.

Extraction turbine: A turbine with one or more bleeds.

Feedwater heater: A device used to heat water from the condenser prior to pumping into the boiler. *Open heaters* mix extracted steam directly with feedwater. *Closed heaters* are conventional shell and tube heat exchangers, with the extracted steam typically being routed to the hot well.

Flue gas: See "stack gas."

Heat rate: For an electrical generating system, the heat rate is defined as the total energy input of the process in BTU/hr divided by the energy output in kilowatts. If the output is mechanical, the total energy is divided by the horsepower output.

High-pressure turbine: A turbine exhausting to a pressure greater than atmospheric.

Hot well: The lower portion of a condenser containing water in its liquid form.

Impulse turbine: A turbine with steam directed at moving blades through stationary nozzles.

Intercooling: Cooling of compressed air between gas turbine and compressor stages.

Low-pressure turbine: A turbine exhausting to a pressure lower than atmospheric.

Non-condensing cycle: A cycle in which steam is exhausted to the atmosphere or otherwise utilized without passing through a condenser.

Preheater: See "air preheater."

Reaction turbine: A turbine with steam discharging from moving nozzles.

Regenerator: A heat exchanger in a gas turbine used to preheat compressed air prior to combustion by exposure to exhaust gases.

Reheat factor: The ratio of actual work output in a multi-stage expansion to ideal work assuming a one-stage expansion.

Reheater: A section of the boiler used to reheat steam after partial expansion in a turbine.

Stack gases: Products of a combustion reaction, consisting of nitrogen, carbon dioxide, and water vapor.

Superposed turbine: See "back-pressure turbine."

Surface condenser: A heat exchanger in which the vapor and coolant do not mix.

Throttle: A series of valves designed to vary the amount of steam admitted to a turbine in accordance with varying loads. Also known as a "throttle valve."

Throttling valve: A device used to drop the pressure of a vapor without any significant change in enthalpy.

Topping turbine: See "back-pressure turbine."

Water rate: For electrical generating systems, the water rate is defined as the steam flow rate in pounds per hour divided by the output in kilowatts.[1] If the output is mechanical, the steam flow rate is divided by the horsepower output.

2 SYMBOLS FOR DRAWINGS

There are many different symbols for drawings in use. Those commonly used are noted here. Should a symbol that is not given be needed, a simple rectangular box labeled inside with the function provided usually is sufficient.

---

[1] The water rate for electrical systems simplifies to $3413/\Delta h_{turbine}$. Water rate is also known as *steam rate*.

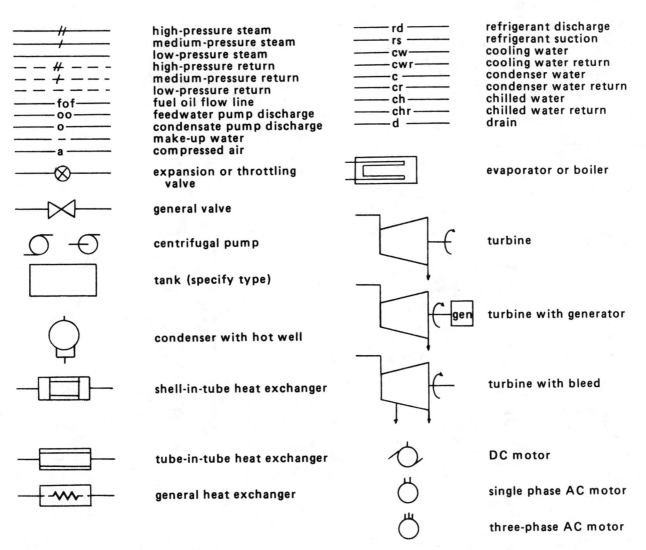

	high-pressure steam
	medium-pressure steam
	low-pressure steam
	high-pressure return
	medium-pressure return
	low-pressure return
fof	fuel oil flow line
oo	feedwater pump discharge
o	condensate pump discharge
	make-up water
a	compressed air

expansion or throttling valve

general valve

centrifugal pump

tank (specify type)

condenser with hot well

shell-in-tube heat exchanger

tube-in-tube heat exchanger

general heat exchanger

rd	refrigerant discharge
rs	refrigerant suction
cw	cooling water
cwr	cooling water return
c	condenser water
cr	condenser water return
ch	chilled water
chr	chilled water return
d	drain

evaporator or boiler

turbine

turbine with generator

turbine with bleed

DC motor

single phase AC motor

three-phase AC motor

3 TYPICAL EFFICIENCIES

Table 7.1 lists typical efficiencies which can be assumed in a problem when required.

4 POWER CYCLE EQUIPMENT

In power plants, many different pieces of equipment work together to extract energy from fuel. This section will investigate the more important equipment types.

A. BOILERS

Boilers, also known as *steam generators*, are of two general types: fire-tube boilers and water-tube boilers. *Fire-tube boilers* seldom are used today because of their limitations to low-pressure steam. Hot combustion gases pass through small diameter tubes and transfer heat energy to a surrounding water jacket known as a *drum*. The large diameter and typically-riveted construction of fire-tube drums limit the maximum pressure the drum can sustain. Pressures are limited to approximately 150

psia for riveted construction. Pressures as high as 400 psia are possible for welded construction.

A variation on the fire-tube boiler is the *horizontal return tubular (HRT) boiler*. In this boiler, the combustion gases pass both under and through each fire tube. It is essentially a two-pass heat exchanger.

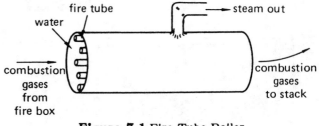

Figure 7.1 Fire-Tube Boiler

In a *water-tube boiler*, hot water passes through small diameter pipes as combustion gases pass around the pipes. There may be more than one drum.[2] The steam-

[2] A drum at the bottom of the boiler, which collects sediment, is called the *mud drum*.

Table 7.1
Approximate Full-Load Efficiencies

Device	Rating, or type	mechanical	isentropic	volumetric	electrical	thermal
steam turbine	EE3 kw	.98	.62–.66			
	EE4 kw	.98	.72–.78			
	EE5 kw	.98	.80–.82			
electrical generator	EE2 kva				.90	
	EE3 kva				.94	
	2 EE3 kva				.96	
	10 EE3 kva				.97	
	20 EE3 kva				.98	
	75 EE3 kva				.99	
electrical motor (AC)	1 hp				.80	
	10 hp				.85	
	50 hp				.90	
	EE3 hp				.95	
	5 EE3 hp				.96	
boiler	hand-fired					.50–.60
	chain-grate					.60–.70
	pulverized coal					.80–.90
	oil					.85–.90
pump, piston, water	1000 psi	.90		.71		
pump, centrifugal	200 gpm	.60–.70				
	500 gpm	.70–.75				
	1000 gpm	.75–.80				
	3000 gpm	.80–.85				
	10,000 gpm	.85–.87				
compressor, piston		.88–.93	.85–.93			
compressor, turbine	500 hp	.95	.62			
	5000 hp	.98	.76			
	10,000 hp	.99	.84			
Otto IC engine	4-stroke	.85		.90		.25–.30
Diesel IC engine		.85		.92		

collecting drum may be parallel to the tubes (a *longitudinal drum*), or it may be perpendicular to the tubes (a *cross drum*). Tubes may be straight (a *straight-tube boiler*) or bent (a *bent-tube boiler*). Water-tube boilers typically are limited to approximately 500 psia.

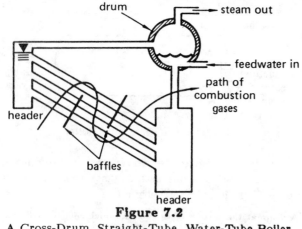

Figure 7.2
A Cross-Drum, Straight-Tube, Water-Tube Boiler

Boiler capacity commonly is given in pounds of steam per hour. This is not an exact determination of the thermodynamic output, however, since the enthalpies of the entering water and the leaving steam are not specified. It is better to use equation 7.1.

$$\dot{Q} = \dot{m}_{steam}(h_{steam} - h_{feed}) \qquad \textbf{7.1}$$

If the units of $\dot{m}_{steam}$ are lbm/hr, Q will be in BTU/hr. Dividing by 1000 or 1,000,000 will result in kB/hr and MB/hr, respectively.

At *maximum capacity*, $\dot{m}$ in equation 7.1 is as high as it can be. At *normal capacity*, the furnace efficiency is as high as it can be.

There are a number of archaic terms which are used infrequently. They are presented here for completeness. The mass flow rate is in lbm/hr in each case.

$$\text{boiler horsepower} = \frac{\dot{m}_{steam}(h_{steam} - h_{feed})}{(970.3)(34.5)} \quad 7.2$$

$$\text{factor of evaporation} = \frac{h_{steam} - h_{feed}}{970.3} \quad 7.3$$

$$\text{equivalent evaporation} = \frac{\dot{m}_{steam}(h_{steam} - h_{feed})}{970.3} \quad 7.4$$

The boiler efficiency is calculated from equation 7.5. It will range from 75% to 90%, depending on installation, although 85% to 88% is a normal range for efficient operations.

$$\eta_{boiler} = \frac{\dot{m}_{steam}(h_{steam} - h_{feed})}{\dot{m}_{fuel}(\text{heating value})} \quad 7.5$$

B. SUPERHEATERS

Superheaters generally are simple heat exchangers in the furnace which are used to increase the energy of the steam. If the tubes are exposed to the stack gases, the superheater is of the *convection type*. If it is exposed directly to the flames, it is of the *radiant type*. Because they have different lag times, both types can be used concurrently to maintain a constant superheat, regardless of boiler output and steam demand.

Other methods used to control superheat include the use of separately-fired superheaters, directable burner heads, and bypass dampers.

A *reheater* is a special type of superheater which adds energy to used steam.

C. DESUPERHEATERS

Desuperheaters and *attemperators* decrease the enthalpy of steam passing through them. Either the steam can be passed through a surface-type desuperheater, which is a closed heat exchanger surrounded by water, or the steam enthalpy can be decreased by water injection.

Even when it is not needed for normal operation, a desuperheater always should be installed across a high-pressure topping turbine. If the high pressure turbine is taken out of service, the high-pressure steam from the boilers can be desuperheated and routed around the high-pressure turbine to the low-pressure turbine.

D. ECONOMIZERS

An *economizer* is a water-tube heat exchanger exposed to stack gases. The economizer increases the temperature of water entering the boiler. The heat transfer takes place in the stack; it does not occur in the boiler.

External corrosion of the heat exchanger surface is avoided by keeping the temperature high enough to prevent acid formation. Internal corrosion is avoided by treating the feedwater.

The heat transfer may be through smooth steel tubes or cast iron tubes with fins. Regardless of the design, a forced draft usually is required because the stack gases are impeded by the economizer.

E. AIR PREHEATERS

A *preheater* increases the plant's operating efficiency by increasing the temperature of the combustion air. This is done by exposure to the stack gases. *Convective preheaters* are conventional heat exchangers which transfer energy through circular tubes or flat plates. A *regenerative preheater* uses a continually rotating drum which is exposed alternately to air and to stack gases.

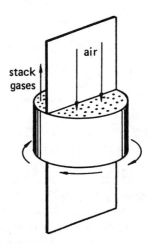

Figure 7.3 A Regenerative Air Preheater

F. EVAPORATORS

Evaporators are used to produce distilled water for boiler feed. The heating is provided by steam which is bled off from other parts of the plant. As the distilled water steam is produced, it is routed to a holding tank, an open feedwater heater, or a special condenser.

Single-effect evaporators use the distilled water steam produced by the evaporator. *Multiple-effect evaporators* pass the distilled water steam through other evaporators to recover and reuse several times the heat of evaporation. Up to four effects can be used in series in typical steam generating plants.

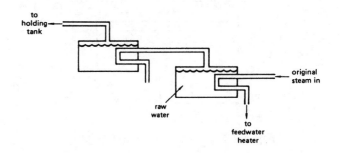

Figure 7.4 A Multiple-Effect Evaporator

G. DEAERATORS

Deaerators are baffled devices which remove dissolved gases from feedwater. This is done by breaking the water into small droplets in the baffles and then heating the droplets by exposure to high-temperature steam. The corrosive oxygen and carbon dioxide gases are vented to the atmosphere or are vacuum extracted.

Deaerating operations also can be performed in a feedwater heater.

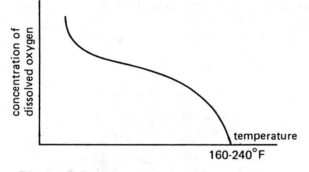

Figure 7.5 Typical Dissolved Oxygen Curve
(Water at atmospheric pressure)

H. FEEDWATER HEATERS

Feedwater heaters use bleed steam to increase the temperature of water entering the boiler. *Open heaters* (also known as *direct contact* and *direct mixing heaters*) physically mix bleed steam and condenser water.

Because they are essentially large boxes under steam pressure, open heaters usually are farthest from the boiler, where the pressure is low. In addition, they may be mounted quite high (i.e., 20 to 80 feet up) to provide the necessary NPSHA to boiler feed pumps below, which are handling water near the boiling point.

Open heaters may be vented to the atmosphere, or gases extracted may be removed by action of a vent condenser or a vacuum pump. Because open heaters are not highly pressurized, the output water temperature is approximately 212°F.

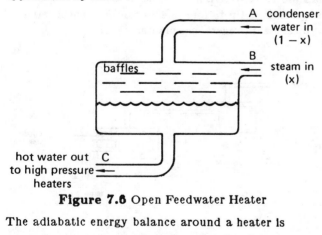

Figure 7.6 Open Feedwater Heater

The adiabatic energy balance around a heater is

$$(1-x)h_A + xh_B = h_C \qquad 7.6$$

If the three enthalpies are known, the *bleed fraction x* can be determined.

$$x = \frac{h_C - h_A}{h_B - h_A} \qquad 7.7$$

A *closed feedwater heater* is a heat exchanger which can operate at high or low pressures, although it almost always is used in high pressure locations close to the boiler. There is no mixing of water and steam in a closed feedwater heater, although the two flows may join at a later point after the steam has passed through a *drip pump* to raise its pressure.

In a closed feedwater heater (heat exchanger), the *terminal temperature difference* is the difference between the saturation temperature corresponding to the entering extraction steam pressure and the temperature of the water leaving the feedwater heater (but before combining with the drips). It typically varies between 5°F and 20°F. The *hot well depression*, also known as the *degrees of freedom*, is the difference in temperature between the saturation temperature corresponding to the heater pressure and the steam condensate in the heater. This typically varies between 5°F and 15°F.

There are various methods for disposing of the *drips*. They may be combined with the feedwater as in figure 7.7., or they may be returned to the hot well.

5 TYPICAL SYSTEM INTEGRATION

It is not possible to describe the numerous methods of combining the devices described in this chapter into a working system. The number of variations produced by multiplicity of elements (to provide both backup and increased efficiency) and by routing is enormous. Furthermore, it is not practical to include all of the various lines, valves, tanks, and other small details in a system illustration.

Figure 7.7 illustrates the main elements in a low-capacity power plant.

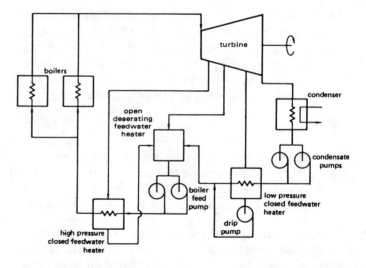

Figure 7.7 Typical Low-Capacity Installation

6 GENERAL VAPOR POWER CYCLES

The general vapor power cycle makes use of a boiler, a turbine, a condenser, and a boiler feed pump (compressor) as shown in figure 7.8.

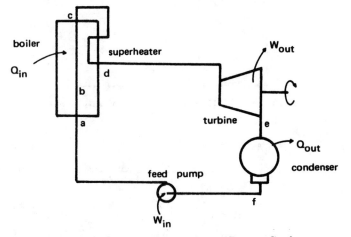

Figure 7.8 Generalized Vapor Power Cycle

The following processes take place in the vapor power cycle.

  a to b: Subcooled water is heated to the saturated fluid (subscript F) temperature in the boiler.

  b to c: Saturated water is vaporized in the boiler, producing saturated gas (subscript G).

  c to d: An optional superheating process increases the steam temperature while maintaining the pressure.

  d to e: Vapor expands through a turbine and does work as it decreases in temperature, pressure, and quality.

  e to f: Vapor is liquified in the condenser.

  f to a: Liquid water is brought up to the boiler pressure.

The expansion process between states d and e is essentially adiabatic. Ideal turbine expansion also is isentropic. Isentropic expansion is described by a vertical line downward on the Mollier diagram, as shown in figure 7.9.

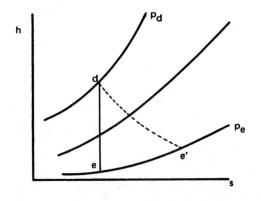

Figure 7.9 Single Stage Turbine Expansion

If the flow is steady, if changes in potential and kinetic energies are insignificant (as they are in a turbine), and if the expansion is isentropic, the energy extracted per pound of steam is

$$W_{id} = h_d - h_e \qquad 7.8$$

If the expansion process is not isentropic, entropy will increase and h'_e will be higher than h_e. The actual energy extracted is

$$W' = h_d - h'_e \qquad 7.9$$

If friction losses are deducted, the turbine shaft work can be found.

$$W_{turb} = W' - W_{friction} = \eta_{mech}W' \qquad 7.10$$

The isentropic efficiency is defined as

$$\eta_s = \frac{W'}{W_{id}} \qquad 7.11$$

The overall turbine efficiency is

$$\eta_{turb} = \frac{W_{turb}}{W_{id}} = \eta_s\eta_{mech} \qquad 7.12$$

Inasmuch as the frictional losses are usually very small (see table 7.1), the isentropic and turbine efficiencies are essentially identical, and

$$W_{turb} = W' = \eta_s W_{id} = \eta_s (h_d - h_e) \qquad 7.13$$

$$\eta_{turb} = \eta_s \qquad 7.14$$

Power plant output may be qualified by the terms "thermal," "electrical," "gross," or "net." If a plant is classified as 1,000,000 BTUH *thermal*, this indicates the energy transfer to the feedwater. 1,000,000 kw *electrical* indicates the generator output.

Some of the electrical output is used to drive auxiliary devices such as pumps and motors. The electrical output is *gross* before the auxiliary loads are removed and *net* after.

Example 7.1

Steam is expanded from 700°F and 200 psia to 5 psia through an 87% efficient turbine. What is the final enthalpy of the steam?

Refer to figure 7.9. From the superheated steam tables,

$$h_d = 1373.6 \text{ BTU/lbm}$$
$$s_d = 1.7232 \text{ BTU/lbm-°R}$$

Proceed as if the turbine is 100% efficient. From the saturated steam tables for 5 psia,

$$s_F = .2347 \qquad h_F = 130.13$$
$$\text{and}$$
$$s_{FG} = 1.6094 \qquad h_{FG} = 1001.0$$

Since it is assumed that expansion is isentropic (100% efficient), $s_e = s_d = 1.7232$. The quality at point e can be found from

$$x_e = \frac{s_e - s_F}{s_{FG}} = \frac{1.7232 - .2347}{1.6094} = .92$$

Now that the quality is known, the enthalpy can be found.

$$h_e = h_F + xh_{FG} = 130.13 + (.92)(1001.0) = 1051.1$$

However, this value of h_e was found assuming isentropic expansion through the turbine. Since the turbine is capable of extracting only 87% of the ideal energy, the actual value of h_e is

$$h_e' = h_d - \eta_{turb}(h_d - h_e)$$
$$= 1373.6 - (.87)(1373.6 - 1051.1)$$
$$= 1093.0 \text{ BTU/lbm}$$

Example 7.2

Repeat example 7.1 using the Mollier diagram.

h_d is read directly from the Mollier diagram at the intersection of 700°F and 200 psia. $h_d \approx 1375$. Greater accuracy is possible with larger Mollier diagrams.

h_e is found by dropping straight down (which keeps entropy constant) to the 5 psia line. h_e is read as approximately 1050. h_e' is calculated as in example 7.1.

If the expansion is multiple stage, or if the turbine has a bleed at an intermediate pressure, p_m, the expansion process for each stage or bleed will be illustrated by figure 7.10.

In the first stage, the ideal and actual outputs per pound of steam are

$$W_{id,1} = h_d - h_m \qquad \qquad 7.15$$
$$W_1' = h_d - h_m' = \eta_{s,1}(W_{id,1}) \qquad 7.16$$

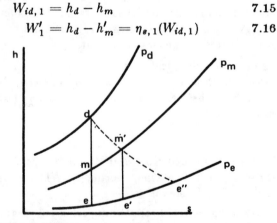

Figure 7.10 Two-stage Turbine Expansion

For the case of a bleed at pressure p_m, if $y\%$ of the original pound of steam expands to pressure p_e, the additional ideal and actual outputs are

$$W_{id,2} = y(h_m' - h_e') \qquad \qquad 7.17$$
$$W_2' = y(h_m' - h_e'') = \eta_{s,2} y(h_m' - h_e') \qquad 7.18$$

The actual work done per pound of steam is

$$W_{turb} = W_1' + W_2' = h_d - h_m' + y(h_m' - h_e'') \quad 7.19$$

A similar analysis can be made for a pump. The ideal and the actual work inputs are shown by the lines (f to a) and (f to a'), respectively, in figure 7.11.

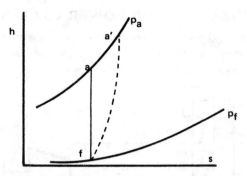

Figure 7.11 One-stage Pump Compression

The pump relationships are

$$W_{id} = h_a - h_f \qquad \qquad 7.20$$
$$W' = h_a' - h_f = \frac{W_{id}}{\eta_s} \approx \frac{v_f(p_a - p_f)}{J\eta_s} \qquad 7.21$$
$$\eta_s = \frac{h_a - h_f}{h_a' - h_f} \qquad \qquad 7.22$$
$$\eta_{pump} = \eta_{mech}\eta_s \qquad \qquad 7.23$$

Since W_{id} is so small—only a few BTU/lbm—the pump mechanical efficiency often is ignored, and η_{pump} is taken as η_s.

The thermal efficiency of an entire power cycle with single-stage expansion is

$$\eta_{th} = \frac{Q_{in} - Q_{out}}{Q_{in}} = \frac{W_{turb} - W_{pump}}{Q_{in}}$$
$$= \frac{(h_d - h_e') - (h_a' - h_f)}{h_d - h_a'} \qquad 7.24$$

Example 7.3

A boiler feed pump increases the pressure of 14.7 psia, 90°F water to 150 psia. What is the final water temperature if the pump has an isentropic efficiency of 80%?

Refer to figure 7.11. Since the properties of a liquid are essentially independent of pressure, the properties of 90°F water can be read from the saturated steam table.

$$h_f = 57.99 \text{ BTU/lbm}$$
$$v_f = 0.01610 \text{ ft}^3/\text{lbm}$$

The enthalpy of the boiler feedwater (point a on figure 7.11) is equal to the enthalpy at point f plus the energy put into the water by the pump. Assuming the water is incompressible, the specific volumes at points a and f are the same.

$$h_a = h_f + \frac{v_f(p_a - p_f)}{J}$$
$$= 57.99 + \frac{(.01610)(150 - 14.7)(144)}{778}$$
$$= 57.99 + .40 = 58.39$$

This calculation assumes that the pump is capable of isentropic compression. Because of the pump's inefficiency, not all of the .40 BTU/lbm enthalpy increase goes into raising the pressure. Therefore, to get to 150 psia, more than .40 BTU/lbm must be added to the water. The actual enthalpy at point a is

$$h'_a = 57.99 + \frac{.40}{.80} = 58.49 \text{ BTU/lbm}$$

The enthalpy was increased by .5 BTU/lbm. Assuming an average specific heat of 1.0 BTU/lbm-°F, the temperature also was increased by .5°F. The final temperature is 90.5°F.

7 THE CARNOT CYCLE

The Carnot cycle is an ideal power cycle which is impractical to implement. However, its work output sets the maximum attainable from any heat engine, as evidenced by the isentropic (reversible) processes between states (d and a) and (b and c) in figure 7.12. The working fluid in a Carnot cycle is irrelevant. As with most other cycles, it is necessary to have property tables for the working fluid if the cycle is to be evaluated.

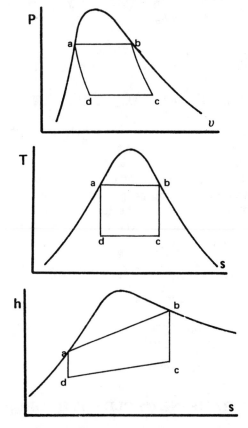

Figure 7.12 The Carnot Cycle

The processes involved are

a to b: isothermal expansion of saturated fluid to saturated gas

b to c: isentropic expansion

c to d: isothermal compression

d to a: isentropic compression

The properties at the various states can be found from the following solution methods. The capital letters "F" and "G" stand for saturated fluid and saturated gas, respectively. They do not correspond to states f and g on any cycle diagram. The Carnot cycle can be evaluated by working around, finding T, p, x, h, and s at each node.

at a: From the property table for T_{high}, read p_a, h_a, and s_a for a saturated fluid.

at b: $T_b = T_a$; $p_b = p_a$; $x = 1$; h_b read from the table as h_G; s_b read as s_G.

at c: Either p_c or T_c will be given. Read p_c from the T_c line on the property table or vice versa; $x_c = (s_b - s_F)/s_{FG}$; $h_c = h_F + x_c(h_{FG})$; $s_c = s_b$.

at d: $T_d = T_c$; $p_d = p_c$; $x_d = (s_a - s_F)/s_{FG}$; $h_d = h_F + x_d(h_{FG})$; $s_d = s_a$.

The turbine and compressor work terms are

$$W_{\text{turb}} = h_b - h_c \qquad \text{7.25}$$
$$W_{\text{comp}} = h_a - h_d \qquad \text{7.26}$$

The heat flows into and out of the system are

$$Q_{\text{in}} = T_{\text{high}}(s_b - s_a) = h_b - h_a \qquad \text{7.27}$$
$$Q_{\text{out}} = T_{\text{low}}(s_c - s_d) = T_{\text{low}}(s_b - s_a) = h_c - h_d \qquad \text{7.28}$$

The thermal efficiency of the entire cycle is

$$\eta_{th} = \frac{Q_{\text{in}} - Q_{\text{out}}}{Q_{\text{in}}} = \frac{W_{\text{turb}} - W_{\text{comp}}}{Q_{\text{in}}} \qquad \text{7.29}$$

$$= \frac{(h_b - h_c) - (h_a - h_d)}{h_b - h_a} = \frac{T_{\text{high}} - T_{\text{low}}}{T_{\text{high}}} \qquad \text{7.30}$$

If isentropic efficiencies for the pump and turbine are given, proceed as follows. Calculate all properties assuming that the efficiencies are 100%. Then, modify h_c and h_a as given in equations 7.31 and 7.32. Use the new values to find the actual thermal efficiency of the cycle.

$$h'_c = h_b - \eta_{\text{turb}}(h_b - h_c) \qquad \text{7.31}$$

$$h'_a = h_d + \frac{h_a - h_d}{\eta_{\text{comp}}} \qquad \text{7.32}$$

$$W'_{\text{turb}} = h_b - h'_c \qquad \text{7.33}$$

$$W'_{\text{comp}} = h'_a - h_d \qquad \text{7.34}$$

8 THE BASIC RANKINE CYCLE

The basic Rankine cycle is similar to the Carnot cycle except that the compression process occurs in the liquid region. The Rankine cycle is closely approximated in actual steam turbine plants. The efficiency of the Rankine cycle is lower than that of a Carnot cycle operating between the same temperature limits because the mean temperature at which heat is added to the system is lower than T_{high}. The piping diagram and property plots are shown in figures 7.13 and 7.14.

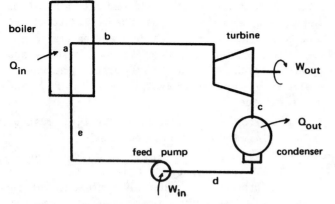

Figure 7.13 Basic Rankine Piping Diagram

The processes used in the basic Rankine cycle are

   a to b: vaporization in the boiler
   b to c: adiabatic expansion in the turbine
   c to d: condensation
   d to e: adiabatic compression to boiler pressure
   e to a: heating to fluid saturation temperature

The properties at each point can be found from the following procedure. The capital letters "F" and "G" refer to "saturated fluid" and "saturated gas," respectively. They do not correspond to locations f and g on any diagram. Usually T_{high} and T_{low} are given. The procedure is to work around the cycle, finding T, p, x, h, and s at each node.

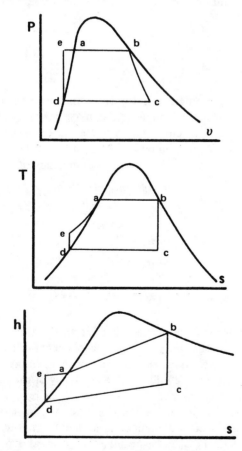

Figure 7.14 Basic Rankine Property Plots

at a: From the property table for T_{high}, read p_a, h_a, and s_a for a saturated fluid.

at b: $T_b = T_a$; $p_b = p_a$; $x = 1$; h_b read from the table as h_G; s_b read as s_G.

at c: Either p_c or T_c will be given. Read p_c from the T_c line on the property table or vice versa; $x_c = (s_b - s_F)/s_{FG}$; $h_c = h_F + x_c(h_{FG})$; $s_c = s_b$.

at d: $T_d = T_c$; $p_d = p_c$; $x_d = 0$; h_d read as h_F; s_d read as s_F; v_d read as v_F.

at e: $p_e = p_a$; $s_e = s_d$; $h_e = h_d + W_{pump} = h_d + v_d(p_e - p_d)/J$ (watch units); T_e found as the saturation temperature for a fluid with enthalpy equal to h_e.

The work and heat flow terms are

$$W_{turb} = h_b - h_c \qquad 7.35$$

$$W_{pump} = h_e - h_d \approx \frac{v_d(p_e - p_d)}{J} \qquad 7.36$$

$$Q_{in} = h_b - h_e \qquad 7.37$$

$$Q_{out} = h_c - h_d \qquad 7.38$$

The thermal efficiency of the entire cycle is

$$\eta_{th} = \frac{Q_{in} - Q_{out}}{Q_{in}} = \frac{W_{turb} - W_{pump}}{Q_{in}}$$
$$= \frac{(h_b - h_c) - (h_e - h_d)}{h_b - h_e} \qquad 7.39$$

If isentropic efficiencies for the pump and the turbine are given, calculate all properties as if these efficiencies were 100%. Then use the following relationships to modify h_c and h_e. Use the new values to recalculate the thermal efficiency.

$$h'_c = h_b - \eta_{turb}(h_b - h_c) \qquad 7.40$$

$$h'_e = h_d + \frac{h_e - h_d}{\eta_{pump}} \qquad 7.41$$

$$W'_{turb} = h_b - h'_c \qquad 7.42$$

$$W'_{pump} = h'_e - h_d \qquad 7.43$$

$$Q'_{in} = h_b - h'_e \qquad 7.44$$

9 THE RANKINE CYCLE WITH SUPERHEAT

Superheating occurs when heat in excess of that required to produce saturated vapor is added to the water. Superheat is used to raise the vapor above the critical temperature, to raise the mean effective temperature at which heat is added, and to keep the expansion primarily in the vapor region to reduce wear on the turbine blades. A maximum practical metallurgical limit on superheat is 1150°F. The piping and property plots are shown in figures 7.15 and 7.16, respectively.

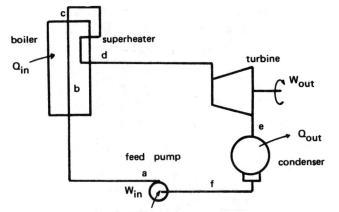

boiler

Q_{in}

superheater

turbine

W_{out}

feed pump

Q_{out}

condenser

W_{in}

Figure 7.15 Rankine with Superheat Piping Diagram

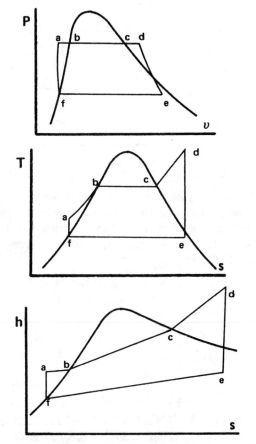

Figure 7.16 Rankine with Superheat Property Plots

The processes in the Rankine cycle with added superheat are similar to the basic Rankine cycle.

a to b: heating water to the saturation temperature in the boiler

b to c: vaporization of water in the boiler

c to d: superheating steam in the superheater region of the boiler

d to e: adiabatic expansion in the turbine

e to f: condensation

f to a: adiabatic compression to boiler pressure

The properties at each point can be found from the following procedure. The subscripts "F" and "G" refer

to "saturated fluid" and "saturated gas," respectively. They do not correspond to any points on the property plot diagram.

at a: Point a is covered below. Start the analysis at point b.

at b: From the property table for T_b or p_b, read h_b and s_b for a saturated fluid.

at c: $T_c = T_b$; $p_c = p_b$; $x = 1$; h_c read as h_G; s_c read as s_G.

at d: T_d usually is known; $p_d = p_c$; h_d read from superheat tables with T_d and p_d known. Same for s_d and v_d.

at e: T_e usually is known; p_e read from saturated table for T_e; $x_e = (s_e - s_F)/s_{FG}$; $h_e = h_F + x_e(h_{FG})$; $s_e = s_d$. If point e is in the superheated region, the Mollier diagram should be used to find the properties at point e.

at f: $T_f = T_e$; $p_f = p_e$; $x = 0$; h_f, s_f, and v_f read as h_F, s_F, and v_F from the saturated table.

at a: $p_a = p_b$; $h_a = h_f + v_f(p_a - p_f)/J$ (watch units); $s_a = s_f$. T_a is equal to the saturation temperature for a liquid with enthalpy equal to h_a.

The work and heat flow terms are

$$W_{\text{turb}} = h_d - h_e \qquad 7.45$$

$$W_{\text{pump}} = v_f(p_a - p_f) = h_a - h_f \qquad 7.46$$

$$Q_{\text{in}} = h_d - h_a \qquad 7.47$$

$$Q_{\text{out}} = h_e - h_f \qquad 7.48$$

The thermal efficiency of the entire cycle is

$$\eta_{th} = \frac{Q_{\text{in}} - Q_{\text{out}}}{Q_{\text{in}}} = \frac{W_{\text{turb}} - W_{\text{pump}}}{Q_{\text{in}}}$$

$$= \frac{(h_d - h_a) - (h_e - h_f)}{h_d - h_a} \qquad 7.49$$

If the Rankine cycle gives efficiencies for the turbine and the pump, calculate all quantities as if those efficiencies were 100%. Then modify h_e and h_a prior to recalculating the thermal efficiency.

$$h'_e = h_d - \eta_{\text{turb}}(h_d - h_e) \qquad 7.50$$

$$h'_a = h_f + \frac{h_a - h_f}{\eta_{\text{pump}}} \qquad 7.51$$

$$W'_{\text{turb}} = h_d - h'_e \qquad 7.52$$

$$W'_{\text{pump}} = h'_a - h_f \qquad 7.53$$

$$Q'_{\text{in}} = h_d - h'_a \qquad 7.54$$

10 RANKINE CYCLE WITH SUPERHEAT AND REHEAT—THE REHEAT CYCLE

Reheat is used to increase the mean effective temperature at which heat is added without producing significant expansion in the liquid-vapor region. The analysis given assumes that $T_d = T_f$, as is usually the case. It is possible, however, that the two temperatures will be different.

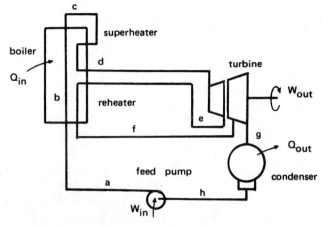

Figure 7.17 Reheat Cycle Piping Diagram

The properties at each point can be found from the following procedure.

at a: Point a is covered below. Start the analysis at point b.

at b: From the vapor table for T_b, read p_b, h_b, and s_b as a saturated fluid.

at c: $T_c = T_b$; $p_c = p_b$; $x_c = 1$; h_c read as h_G; s_c read as s_G.

at d: T_d usually is known; $p_d = p_c$; h_d read from superheat tables with T_d and p_d known. Same for s_d and v_d.

at e: p_e usually is known; T_e read from property table for p_e; $s_e = s_d$; $x_e = (s_e - s_F)/s_{FG}$; $h_e = h_F + x_e(h_{FG})$. (Use the Mollier diagram if superheated.)

at f: T_f usually is known; $p_f = p_e$; h_f read from superheat tables with T_f and p_f known. Same for s_f and v_f.

at g: T_g usually is known; p_g read from property tables for T_g; $s_g = s_f$; $x_g = (s_g - s_F)/s_{FG}$; $h_g = h_F + x_g(h_{FG})$. (Use the Mollier diagram if superheated.)

at h: $T_h = T_g$; $p_h = p_g$; $x_h = 0$; h_h, s_h, and v_h read as h_F, s_F, and v_F.

at a: $p_a = p_b$; $h_a = h_h + v_h(p_a - p_h)/J$ (watch units); $s_a = s_h$. T_a is the saturation temperature for a liquid with enthalpy equal to h_a.

The work and heat flow terms are

$$W_{\text{turb}} = (h_d - h_e) + (h_f - h_g) \qquad 7.55$$

$$W_{\text{pump}} = v_h(p_a - p_h) = h_a - h_h \qquad 7.56$$

$$Q_{\text{in}} = (h_d - h_a) + (h_f - h_e) \qquad 7.57$$

$$Q_{\text{out}} = h_g - h_h \qquad 7.58$$

The thermal efficiency of the entire cycle is

$$\eta_{th} = \frac{(h_d - h_a) + (h_f - h_e) - (h_g - h_h)}{(h_d - h_a) + (h_f - h_e)} \qquad 7.59$$

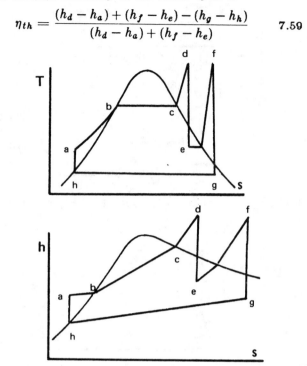

Figure 7.18 Reheat Cycle Property Plot

If the reheat cycle is specified with isentropic efficiencies for the pump and the turbine, calculate all of the preceding quantities as if the efficiencies were 100%. Then modify h_e, h_g, and h_a before recalculating the thermal efficiency.

$$h'_e = h_d - \eta_{\text{turb}}(h_d - h_e) \qquad 7.60$$

$$h'_g = h_f - \eta_{\text{turb}}(h_f - h_g) \qquad 7.61$$

$$h'_a = h_h + \frac{h_a - h_h}{\eta_{\text{pump}}} \qquad 7.62$$

$$W'_{\text{turb}} = (h_d - h'_e) + (h_f - h'_g) \qquad 7.63$$

$$W'_{\text{pump}} = (h'_a - h_h) \qquad 7.64$$

$$Q'_{\text{in}} = (h_d - h'_a) + (h_f - h'_e) \qquad 7.65$$

11 THE RANKINE CYCLE WITH REGENERATION—THE REGENERATIVE CYCLE

If the mean effective temperature at which heat is added can be increased, the overall thermal efficiency of the cycle will be improved. This can be accomplished by raising the temperature at which the condensed fluid enters the boiler.

In the regenerative cycle, portions of the steam in the turbine are withdrawn at various points. Heat is transferred from this bleed stream to the feedwater coming from the condenser. Although only two bleeds are used in the following analysis, seven or more exchange locations can be used in a large installation. The regenerative cycle always involves superheating, although conceptually it need not.

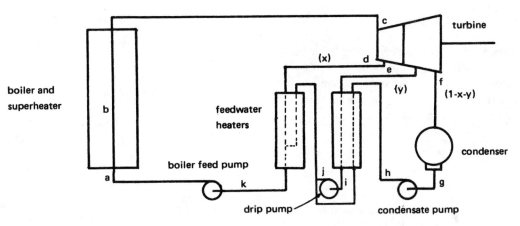

Figure 7.19 Regenerative Cycle with Two Feedwater Heaters

In the analysis, x is the first bleed fraction, and y is the second bleed fraction.

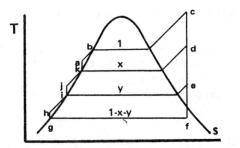

Figure 7.20 Regenerative Cycle Property Plot

The heat exchange may be by direct mixing of the bleed and the feedwater in an *open heater*, or by use of a conventional heat exchanger known as a *closed heater*. These feedwater heaters are illustrated in figures 7.21 and 7.22.

In the open heater with x% bleed and $(1 - x)$% feedwater, the following energy balance can be used to find the bleed fraction if the enthalpies are known. The open heater is assumed to be an adiabatic device.

$$(1 - x)h_2 + xh_1 = h_3 \qquad 7.66$$

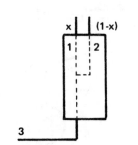

Figure 7.21 Open Feedwater Heater

If the heater is closed, a special *drip pump* can be used. This drip pump is in addition to the condensate and boiler feed pumps. There are various possibilities for disposing of the drips, including mixing with the feedwater (as in figures 7.19 and 7.22) or piping back to

the hot well. For a closed heater with y% bleed, the adiabatic energy balance is given by equation 7.67.

$$(1 - x - y)h_5 + y(h_4) + W_{\text{drip pump}} = (1 - x)h_8 \qquad 7.67$$

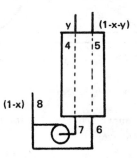

Figure 7.22 Closed Feedwater Heater

Because it is small, the drip pump work can be omitted for first approximations. Equation 7.67 can be solved for y once x is known. For that reason, it is best to start with the heater nearest the boiler when evaluating the regenerative cycle.

The *terminal temperature difference* for a closed heater is defined as $TTD = T_{\text{sat},p_4} - T_6$. If the heater is ideal, TTD is zero.

The work and heat flow terms for the regenerative cycle are

$$W_{\text{turb}} = (h_c - h_d) + (1 - x)(h_d - h_e)$$
$$+ (1 - x - y)(h_e - h_f) \qquad 7.68$$
$$W_{\text{pumps}} = (h_a - h_k) + y(h_j - h_i)$$
$$+ (1 - x - y)(h_h - h_g) \qquad 7.69$$
$$Q_{\text{in}} = h_c - h_a \qquad 7.70$$
$$Q_{\text{out}} = (1 - x - y)(h_f - h_g) \qquad 7.71$$

The thermal efficiency of the entire cycle is

$$\eta_{th} = \frac{Q_{\text{in}} - Q_{\text{out}}}{Q_{\text{in}}} = \frac{W_{\text{turb}} - W_{\text{pumps}}}{Q_{\text{in}}}$$
$$= \frac{(h_c - h_a) - (1 - x - y)(h_f - h_g)}{h_c - h_a} \qquad 7.72$$

12 THE BINARY CYCLE

The binary cycle utilizes two different fluids, usually mercury and steam, achieving conditions unobtainable with a single working fluid. Operation of the binary cycle is essentially two Rankine cycles, and Rankine procedures should be used to evaluate it.

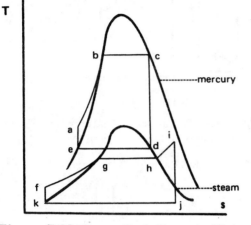

Figure 7.23 Binary Cycle Property Plot

The processes in a binary cycle are

   a to b: Mercury is heated to its saturation temperature.

   b to c: Mercury is vaporized in the boiler.

   c to d: Mercury expands adiabatically in the turbine.

   d to e: Mercury condenses in the combination condenser-boiler.

   e to a: Mercury is compressed adiabatically.

   f to g: Water is heated to its saturation temperature.

   g to h: Water vaporizes in a boiler.

   h to i: Steam is superheated.

   i to j: Steam expands adiabatically in the turbine.

   j to k: Water condenses.

   k to f: Water is compressed adiabatically.

If x pounds of mercury flow for every pound of steam, the thermal efficiency of the binary cycle is

$$\eta_{th} = \frac{W_{\text{turbines}} - W_{\text{pumps}}}{Q_{\text{in}}}$$

$$= \frac{x\,(h_c - h_d) + (h_i - h_j) - x\,(h_a - h_e) - (h_f - h_k)}{x\,(h_c - h_a) + (h_i - h_f)}$$

7.73

13 MAGNETOHYDRODYNAMICS

If an ionized gas flows through a magnetic field, an electric field is generated. The electric field can be used as a potential source for generating useful electricity. This concept is employed in a magnetohydrodynamic (MHD) generator.

In an MHD generator, a high-temperature, electrically-conducting plasma is passed through a nozzle and directed through a duct. The plasma passes through a toroidal coil or between two plates which contain the magnetic field. The electric field will be generated perpendicularly to the magnetic field.

Very high temperatures (2000°C to 2500°C) are required to achieve the desired plasma properties. This is achieved through the combustion of fossil fuels.

The high-temperature gas is doped ("seeded") with salts of easily-ionized elements, such as potassium or cesium, to obtain the necessary ionization (less than 1%). These

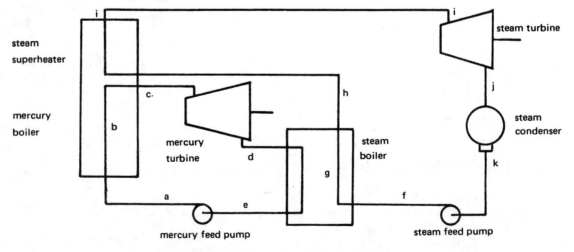

Figure 7.24 Binary Cycle Piping Diagram

alkali metal ions provide a secondary benefit by combining with sulfur in the fossil fuel combustion products. Since the sulfur compounds (such as potassium sulfate) can be recovered in traps or electrostatic precipitators, sulfur emissions would be minimal.

The required magnetic field strength is high (50,000 gauss).

The following assumptions generally are made in evaluating simple MHD problems.

(1) The gas velocity is much less than the speed of light.

(2) Collisions determine the fluid behavior. (Electrostatic interaction between the particles is negligible.)

(3) No plasma oscillations occur.

(4) Gas particles travel in straight lines, not in spirals.

(5) Capacitance effects are negligible.

Under the above assumptions, Ohm's law can be written as

$$\mathbf{I} = \sigma(\mathbf{E} + \mathbf{v} \times \mathbf{B}) \qquad 7.74$$

In equation 7.74, $\mathbf{I}$ is the current vector, σ is the gas conductivity (reciprocal of resistivity), $\mathbf{E}$ is the generated electrical field vector, $\mathbf{v}$ is the plasma velocity vector, and $\mathbf{B}$ is the magnetic field intensity vector.

It is possible (and desirable) to recover some of the heat energy from the generator gas in a *high-temperature air heater* (HTAH). This heat energy can be used in a conventional steam generating plant. Such an arrangement is illustrated in figure 7.25.

14 NUCLEAR STEAM GENERATORS

The moderator and method used to carry off heat from the reactor to the turbine characterize the nuclear reactor type. Once the heat has been transferred to water for steam generation, either directly or indirectly, the performance analysis follows the same steps that were used to evaluate the various Rankine cycles.

A. LIGHT-WATER MODERATED REACTORS (LWR)

Included in this category are the typical pressurized water reactor (PWR) and the boiling water reactor (BWR). Light water is circulated around the fuel rod bundles in each design. In a *pressurized water reactor*, high pressure water from the reactor carries away the reaction heat. This hot water passes through a heat exchanger (the steam generator), but it never passes through a turbine.

The primary (reactor) system is a closed loop which typically is operated in the 2000 psia to 2500 psia range. Therefore, the water temperature in this primary system is approximately 500°F.

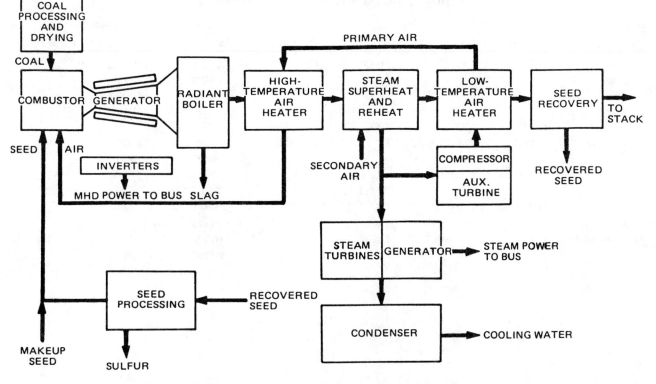

Figure 7.25 A Combined MHD/Steam Generating System

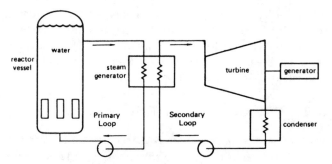

Figure 7.26 A Pressurized Water Reactor

In a *boiling water reactor*, steam is generated directly within the core. It is passed through a series of separators and driers before entering the turbine. The pressure in a typical BWR is approximately 1000 psia.

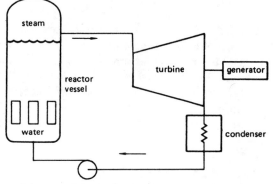

Figure 7.27 A Boiling Water Reactor

B. HEAVY-WATER REACTORS (HWR)

Reactors fueled with natural uranium (no enrichment) commonly are referred to as the CANDU type. They are similar to light-water PWR's in concept, although larger cores result in higher capital costs.

C. HIGH TEMPERATURE GAS-COOLED REACTORS (HTGR)

Helium appears to be the best coolant for HTGR's, although air and carbon dioxide also have been used. There is a closed-loop primary system in which helium at 1300°F–1500°F is circulated by turbine pumps.[3] The secondary system is similar to the PWR with typical steam properties of 1000°F and 1450 psia.

D. OTHER REACTOR TYPES

Other reactor types include *organic liquid cooled reactors* (OCR) and *liquid metal cooled reactors* (LMR). Sodium is the liquid metal most frequently used as a coolant in liquid metal cooled reactors. Bismuth and mixtures of sodium and potassium also are used. Reactors using liquid metals and operating as breeders are known as *liquid metal fast breeder reactors* (LMFBR).

---

[3]If the coolant gas temperature is kept below 1000°F, the reactor may be known as a *gas cooled reactor* (GCR).

Table 7.2
Typical Nuclear Reactor Characteristics

Reactor Type	Abbrev.	Fuel	Coolant	Moderator	Primary Coolant Conditions	Steam Conditions	Cycle Efficiency
Pressurized water	PWR	UO_2	H_2O	H_2O	2300 psi 610°F	780 psi 514°F	33%
Boiling water	BWR	UO_2	H_2O	H_2O	1000 psi satur. temp.	1000 psi 545°F	33%
Gas-cooled	GCR	UO_2	CO_2	Graphite			
High-temp gas-cooled	HTGR	$(Th-U^{235})C_2$	He	Graphite	700 psi 1430°F	2400 psi 950°F	39%
Organic-cooled	OCR	U-Mo	HB-40 Terphenyl	D_2O	100 psi 750°F		35%
Heavy water	HWR	UO_2	D_2O	D_2O	1300 psi 570°F		31%
Fast breeder	LMFBR	(U-Pu)C	Na-K	–	100 psi 1025°F	2415 psi 950°F	28%

PART 2: Combustion Cycles

Nomenclature for Combustion Cycles

c	specific heat	BTU/lbm-°R
h	enthalpy	BTU/lbm
HHV	higher heating value	BTU/lbm
k	adiabatic exponent (c_p/c_v)	–
m	mass	lbm
p	pressure	psf
Q	heat flow	BTU/lbm
R	compression ratio	–
s	entropy	BTU/lbm-°R
T	temperature, or torque	°R, ft-lbf
V	actual volume	ft³
W	work	BTU/lbm

Symbols

η	efficiency	–
ρ	density	lbm/ft³
υ	specific volume	ft³/lbm

Subscripts

a	air
a/f	air/fuel
c	cut-off
f	fuel
p	constant pressure
th	thermal
v	constant volume

Combustion power cycles differ from vapor power cycles in that combustion products cannot be returned to their initial conditions for reuse. Due to the difficulties of working with mixtures of fuel and air, combustion power cycles often are analyzed as air-standard cycles. The cycle is said to be *air-standard* if it is analyzed as a closed system using ideal air as a working fluid and with the heat of combustion added without regard to source. Actual engine efficiencies may be 10% to 50% lower than calculated for air-standard analyses. However, if excess air is used in combustion, there may be fair agreement.

15 FUELS

Fuels for combustion power cycles usually are gasoline or some form of diesel fuel oil. (Heating fuels, such as coal, natural gas, and Bunker-C oil, are covered in another chapter.) Commonly needed properties are listed in table 7.3, along with the properties of several other fuels.

A. GASOLINE

The heating value of gasoline varies only slightly with composition. It generally can be taken as 20,200 BTU/lbm, although 20,300 can be used for high octane aviation fuel. The variation is less than ±1.5%. If necessary, the higher heating value can be calculated from the Baumé specific gravity.

$$HHV = 18,320 + 40(°Baumé - 10) \text{ in BTU/lbm} \quad 7.75$$

The *volatility* of gasoline is the percentage of the fuel that will evaporate at a given temperature. (Since gasoline is a mixture of hydrocarbons, it will not all evaporate at a single temperature.) Typical specifications call for 10% volatility at 167°F, 50% at 221°F, and 90% at 275°F. Low volatility will result in hard starting and poor engine operation at low temperatures.

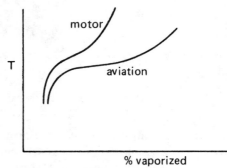

Figure 7.28 Typical Volatility Curves

The *octane number* is a measure of knock resistance. The measure is based on comparison, performed in a standardized one cylinder engine, with the burning of isooctane and n-heptane. N-heptane ("normal"), C_7H_{16}, is rated 0 and produces violent knocking in the test engine. Isooctane, C_8H_{18}, is rated 100 and produces relatively knock-free operation. The percentage blend by volume of these fuels that matches the performance of the gasoline is the octane rating.

B. DIESEL FUELS

The heating value of diesel fuel can be determined from its specific gravity.[4]

$$HHV = 22,320 - 3780(SG)^2 \text{ in BTU/lbm} \quad 7.76$$

The *cetane number* is used as an index of the fuel's ignition capabilities. Like the octane rating, the cetane rat-

---

[4]This specific gravity is with respect to 60°F water. In fact, this equation can be used with fair precision to predict the higher heating value of any hydrocarbon fuel, including gasoline.

Table 7.3
Properties of Common Fuels

	Gasoline	Iso-Octane	Propane	Ethanol	No. 1 Diesel	No. 2 Diesel
Chemical formula	—	C_8H_{18}	C_3H_8	C_2H_5OH	—	—
Molecular weight	≈ 126	114.2	44	46	≈ 170	≈ 184
Carbon % by weight	—	84	82	52	—	—
Hydrogen % by weight	—	16	18	13	—	—
Oxygen % by weight	—	—	—	35	—	—
Heating value						
Higher BTU/lbm	20,260	20,590	21,646	12,800	19,240	19,110
Lower BTU/lbm	18,900	19,160	19,916	11,500	18,250	18,000
BTU/gal (lower)	116,485	111,824	81,855	76,152	133,332	138,110
Latent heat of						
Vaporization BTU/lbm	142	141	147	361	115	105
Specific gravity	.68 (to .74)	.692	.493	.794	.876	.920
Research octane	85–94	100	112	106		—
Motor octane	77–86	100	97	89	10–30	
Cetane number	10 to 20	—	—	−20 to 8	≈ 45	—
Stoichiometric						
Mass A/F ratio	14.7	15.1	—	9.0	—	—
Distillation						
Temperature (°F)	90–410	211	—	173	340–560	—
Flammability						
Limits (volume percent)	1.4 to 7.6	—	—	4.3 to 19	—	—

ing is determined by comparision with standard fuels. Cetane, $C_{16}H_{32}$, has a cetane number of 100. N-methyl-napthalene, $C_{11}H_{10}$, has a cetane number of zero.

A cetane number of approximately 30 is required for satisfactory operation of a low-speed diesel engine. High-speed engines, such as those used in modern cars, require cetane numbers of 45 or more.

Since the octane number measures the tendency not to pre-ignite, and the cetane number measures the tendency to ignite, there is an approximate relationship between the two numbers.

$$\text{cetane no.} \approx \frac{104 - \text{octane no.}}{2.75} \qquad 7.77$$

The *pour point* number of a diesel fuel refers to the viscosity of the fuel. A fuel with a pour point of 10°F will flow freely above that temperature. A fuel with a high pour point will thicken in cold weather.

The *cloud point* refers to the wax crystals which cloud diesel fuel at lower temperatures. The cloud point should be 20°F or higher. Below that temperature, the engine will not run well.

C. ALCOHOL FUELS

Both methanol and ethanol are available for combustion engines. *Methanol*, also known as *methyl alcohol*, is produced from coal, wood, and organic garbage. *Ethanol*, also known as *ethyl alcohol* and *grain alcohol*, is distilled from grain, sugar cane, potatoes, and other agricultural products containing various amounts of sugars, starches, and cellulose. Although methanol works as well as ethanol, only ethanol can be produced in large quantities from farm by-products.

Alcohol is water-soluble. The concentration of alcohol is measured by its proof, where 200 proof is a pure alcohol. (180 proof is 90% alcohol and 10% water.) 190 proof is a typical maximum concentration limited by the economics of distillation.

Gasohol is a mixture of approximately 90% gasoline and 10% alcohol.

Use of high percentages of alcohol in gasoline engines may require the following modifications: enlarge carburetor jets, advance timing, preheat fuel in cold weather, line tanks to prevent rusting, use alcohol-resistant gaskets.

Alcohol is a poor substitute for diesel fuel, since alcohol's cetane number is low—from −20 to +8. Straight injection of alcohol results in poor performance and heavy knock. Attempts to adapt diesel engines to alcohol use have centered on the use of expensive additives, mixing 30% ethanol with 70% diesel, and carburating ethanol into the air instead of using direct injection. The latter method appears to be the most promising.

16 THE AIR-STANDARD CARNOT CYCLE

The T-s diagram (figure 7.29) is the same for any Carnot cycle. However, isothermal expansion and compression do not occur at constant pressure as they do in a vapor power cycle. The Carnot air-standard cycle is not a practical engine cycle, but it does represent an upper maximum on efficiency due to the isentropic processes.

The air-standard Carnot cycle is described by the following processes.

   a to b: isentropic compression

   b to c: isothermal expansion power stroke with compression ignition and metered fuel injection

   c to d: isentropic expansion

   d to a: isothermal compression

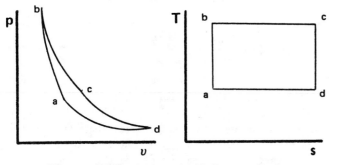

Figure 7.29 Air-Standard Carnot Cycle

The p, V, and T properties can be found from the isentropic and isothermal relationships in chapter 6. The compression ratio is defined as

$$R = \frac{V_a}{V_b} = \frac{V_d}{V_c} \qquad 7.78$$

The thermal efficiency of the process is

$$\eta_{th} = \frac{T_b - T_a}{T_b} = \frac{T_c - T_d}{T_c} = 1 - \frac{1}{R^{k-1}} \qquad 7.79$$

17 THE AIR-STANDARD OTTO CYCLE

The Otto cycle is a four-stroke cycle. Four processes together constitute the cycle: intake, compression,

power, and exhaust. These four processes require two revolutions of the crank to be completed. Therefore, each cylinder contributes one power stroke every other revolution.

$$\frac{\text{no. power strokes}}{\text{per minute}} = \frac{(\text{no. cylinders})(\text{rpm})}{2} \qquad 7.80$$

The actual Otto cycle is not a true cycle because it doesn't close on itself. That is, the exhaust gases are eliminated from the system, not brought back to their original conditions. For ease of analysis, the Otto cycle is evaluated on an air-standard basis. Actual engine efficiency will be approximately 50% of the efficiency calculated from the air-standard analysis.

The air-standard Otto cycle is less efficient than the Carnot cycle operating within the same temperature limits. However, equal efficiencies will be obtained if the compression ratios are made equal. This will require the Otto engine to operate at higher temperatures than the Carnot cycle. The processes in the cycle are shown in figure 7.30.

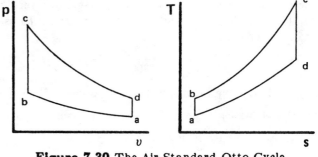

Figure 7.30 The Air-Standard Otto Cycle

The Otto cycle comprises the following processes.

   a to b: adiabatic compression

   b to c: constant volume heat addition

   c to d: adiabatic expansion

   d to a: constant volume heat rejection

It is useful to note the following relationships.

$$R = \frac{V_a}{V_b} = \frac{V_d}{V_c} \qquad 7.81$$

$$\frac{T_d}{T_c} = \frac{T_a}{T_b} \qquad 7.82$$

The p, V, and T relationships for the isentropic and constant volume ideal gas processes from chapter 6 can be used to analyze the Otto cycle.[5] The work and heat flow terms are

[5] Use of the temperature-related equations for ideal gases assumes that the specific heats are constant during the four processes. They actually vary greatly over the wide temperature extremes encountered. Therefore, air tables should be used whenever possible.

$$Q_{in} = c_v(T_c - T_b) \qquad\qquad\qquad 7.83$$

$$Q_{out} = c_v(T_d - T_a) \qquad\qquad\qquad 7.84$$

$$W_{out} = c_v(T_c - T_d) \qquad\qquad\qquad 7.85$$

$$W_{in} = c_v(T_b - T_a) \qquad\qquad\qquad 7.86$$

$$\eta_{th} = \frac{Q_{in} - Q_{out}}{Q_{in}} = \frac{W_{out} - W_{in}}{Q_{in}} = 1 - \frac{1}{R^{k-1}} \quad 7.87$$

$$= 1 - \left(\frac{T_d}{T_c}\right) = 1 - \left(\frac{T_a}{T_b}\right) \qquad 7.88$$

Equation 7.87 indicates that the thermal efficiency can be increased by increasing k and R. Increasing k can be done only by substituting another inert gas, such as helium, for the atmospheric nitrogen. Increases in R are limited to approximately 10:1 by fuel detonation. (Sealing does not become a problem until compression ratios of 16:1 are exceeded.)

Performance of an internal combustion engine operating on the Otto cycle at full throttle also can be analyzed on a macroscopic basis by the "PLAN" formula. It is necessary to know the size of the engine (bore ✕ stroke) in inches, the number of engine cycles per minute, and the operating pressures. Then the horsepower output can be found from equation 7.89. It is important to use the units given.

$$hp = \frac{PLAN}{33000} \qquad\qquad\qquad 7.89$$

hp is the engine output in horsepower

P (or MEP) is the mean effective pressure in psig

L is the stroke length in feet

A is the piston area in square inches

N is the number of power strokes per minute, equal to $\dfrac{(2n)(\text{no. cylinders})}{(\text{no. strokes per cycle})}$

n is the engine speed in rpm

A *prony brake* applies a frictional resistance to a rotating engine. The torque force and its moment arm can be used to calculate the brake horsepower.

$$BHP = \frac{2\pi(r)(F)(rpm)}{33,000} \qquad\qquad 7.90$$

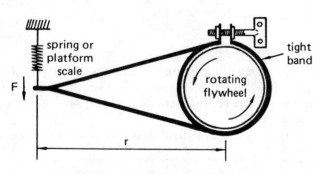

Figure 7.31 A Prony Brake

rF is the *brake torque*, so

$$T = \frac{(33,000)(BHP)}{2\pi(rpm)} = \frac{(5252)(BHP)}{rpm} \qquad 7.91$$

The output also can be determined from the specific fuel consumption (SFC in lbm/hp-hr).

$$hp = \frac{\text{actual fuel flow}}{SFC} \qquad\qquad 7.92$$

Several of the parameters may be given as either *brake* or *indicated*, such as BHP and IHP, BMEP and IMEP, or BSFC and ISFC. The term "brake" refers to the actual performance at the output shaft as installed in an operating environment. The term "indicated" refers to a test of the engine under frictionless conditions. It is important that all terms in equations 7.89 and 7.90 be consistent with regard to "brake" and "indicated."[6]

Various efficiencies can be calculated.

$$\eta_{th} = \frac{2545}{(SFC)(\text{heating value})}$$

$$= \frac{(2545)(hp)}{(\text{fuel consumption})(\text{heating value})} \qquad 7.93$$

$$\eta_{mechanical} = \frac{BHP}{IHP} \qquad\qquad 7.94$$

$$\eta_{volumetric} = \frac{(\text{air intake in ft}^3/\text{hr})}{(30)(rpm)(\text{engine displacement in ft}^3)} \; 7.95$$

$$\eta_{ideal} = 1 - \frac{1}{R^{k-1}} \qquad\qquad 7.96$$

$$\eta_{relative} = \frac{\eta_{th}}{\eta_{ideal}} \qquad\qquad 7.97$$

Since a lower atmospheric density decreases the available oxygen per intake stroke, engine output decreases with altitude. A numerical approach for determining the output under new conditions of altitude, pressure, or temperature is given in the following steps. It is assumed that engine speed is constant.

step 1:   Let 1 and 2 be the lower and higher altitudes, respectively.

step 2:   Calculate the frictionless horsepower.

$$IHP_1 = \frac{BHP_1}{\eta_{mech,\,1}} \qquad\qquad 7.98$$

step 3:   Calculate the friction horsepower, which is assumed to be constant at constant speed.

$$FHP = IHP_1 - BHP_1 \qquad\qquad 7.99$$

step 4:   **Find the air densities ρ_{a1} and ρ_{a2} from the data in appendix C of chapter 8.**

---

[6] If the horsepower is calculated from the actual pressure and area, but ignoring the friction losses in the engine and the drive train, equation 7.89 predicts the *indicated* horsepower.

step 5: Calculate

$$IHP_2 = IHP_1 \left(\frac{\rho_{a2}}{\rho_{a1}} \right) \qquad \text{7.100}$$

step 6: Calculate

$$BHP_2 = IHP_2 - FHP \qquad \text{7.101}$$

step 7: Calculate

$$\eta_{\text{mech}, 2} = \frac{BHP_2}{IHP_2} \qquad \text{7.102}$$

step 8: The volumetric air flow rates are the same.

$$\dot{V}_{a2} = \dot{V}_{a1} \qquad \text{7.103}$$

step 9: The original air and fuel rates are

$$\dot{m}_{f1} = (BSFC_1)(BHP_1) \qquad \text{7.104}$$
$$\dot{m}_{a1} = (R_{a/f})(\dot{m}_{f1}) \qquad \text{7.105}$$
$$V_{a1} = \frac{\dot{m}_{a1}}{\rho_{a1}} \qquad \text{7.106}$$

step 10: The new air mass flow rate is

$$\dot{m}_{a2} = \dot{V}_{a2}\rho_{a2} \quad \text{(see step 8)} \qquad \text{7.107}$$

step 11: For engines with metered injection, $\dot{m}_{f2} = \dot{m}_{f1}$. For engines with carburetors,

$$\dot{m}_{f2} = \frac{\dot{m}_{a2}}{R_{a/f}} \qquad \text{7.108}$$

step 12: The new fuel consumption is

$$BSFC_2 = \frac{\dot{m}_{f2}}{BHP_2} \qquad \text{7.109}$$

Example 7.4

A 10″ bore, 18″ stroke single-cylinder engine runs at 200 rpm. The gross weight on a prony brake is 140 pounds, the tare weight is 25 pounds, and the arm length is 66 inches. The indicator card shows an area of 1.20 in² with an overall length of 3 inches. The spring scale was 200 psi/inch. What is the mechanical efficiency of this engine?

First find the indicated horsepower.

The mean effective pressure is

$$p = \left(\frac{1.20}{3} \right)(200) = 80 \text{ psi}$$

$$L = \frac{18}{12} = 1.5 \text{ ft}$$
$$A = \frac{\pi}{4}(10)^2 = 78.54 \text{ in}^2$$
$$N = \frac{200}{2} = 100 \text{ power strokes per minute}$$

From equation 7.89, the indicated horsepower is

$$IHP = \frac{(80)(1.5)(78.54)(100)}{33,000} = 28.56 \text{ hp}$$

Now find the brake horsepower.

$$r = \frac{66}{12} = 5.5 \text{ ft}$$
$$F = 140 - 25 = 115 \text{ lbf}$$

From equation 7.90, the brake horsepower is

$$BHP = \frac{(2\pi)(5.5)(115)(200)}{33,000} = 24.09$$

The mechanical efficiency is

$$\eta = \frac{24.09}{28.56} = .843$$

Example 7.5

An internal combustion engine is being evaluated on the basis of an air-standard Otto cycle. The suction pressure is 14.7 psia. The temperature of the incoming air is 140°F. The maximum pressure and temperature in the cycle are 340 psia and 2150°F, respectively.

Find the pressure and temperature at the end of the compression stroke. Determine the cycle efficiency.

The ideal gas laws will be used. They require that the temperatures be expressed in °R.

$$T_a = 140 + 460 = 600°R$$
$$T_c = 2150 + 460 = 2610°R$$

a to b:

$$p_b = 14.7 \left(\frac{T_b}{600} \right)^{\frac{1.4}{1.4-1}} = 14.7 \left(\frac{T_b}{600} \right)^{3.5}$$

b to c:

$$T_b = 2610 \left(\frac{p_b}{340} \right) = 7.676\, p_b$$

Now combine these two results and solve for p_b.

$$p_b = 14.7 \left(\frac{7.676\, p_b}{600} \right)^{3.5}$$
$$p_b = 152.5 \text{ psia}$$

Then,

$$T_b = 7.676(152.5) = 1170.6$$

From equation 7.88,

$$\eta_{th} = 1 - \frac{600}{1170.6} = .487$$

Example 7.6

An Otto engine operates with 15% clearance. (Clearance is the ratio of clear volume to the swept volume.) Atmospheric air at 14.0 psia and 120°F is used. The fuel/air ratio is .06. Fuel with a higher heating value of 19,500 BTU/lbm is available.

Find the temperature and pressure of the mixture at all four endpoints.

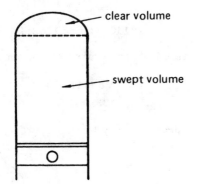

Refer to figure 7.30.

at a:
$$V_a = V_{\text{swept}} + V_{\text{clear}} = 1.15 V_{\text{swept}}$$
$$p_a = 14.0 \text{ psia}$$
$$T_a = 120°F = 580°R$$

From the air table at 580°R, $v_{ra} = 120.7$, and $p_{ra} = 1.78$.

at b:
$$V_b = .15 V_{\text{swept}}$$
$$R = \frac{V_a}{V_b} = \frac{1.15}{.15} = 7.67$$

For an isentropic process,

$$\frac{V_a}{V_b} = \frac{v_{ra}}{v_{rb}}$$

So,

$$v_{rb} = \frac{120.7}{7.67} = 15.74$$

Finding this pressure ratio in the air table yields

$$T_b = 1273°R$$
$$u_b = 222.72 \text{ BTU/lbm}$$
$$p_{rb} = 29.93$$
$$p_b = 14\left(\frac{29.93}{1.78}\right) = 235.4 \text{ psia}$$

at c: The heat added by the fuel per pound of air is

$$q = (.06)(19,500) = 1170 \text{ BTU/lbm}$$

For a closed system, with $W = 0$, the first law of thermodynamics says $Q = \Delta U$.

$$u_c = u_b + q = 222.72 + 1170 = 1392.72 \text{ BTU/lbm}$$

(Notice that this is not $h_c = h_b + q$.)

Finding this value of internal energy in the air tables produces

$$T_c = 6420°R$$
$$v_{rc} = .08685$$
$$p_{rc} = 27381$$

For an ideal gas in a constant volume process,

$$p_c = 235.4\left(\frac{6420}{1273}\right) = 1187.2 \text{ psia}$$

(p_r cannot be used to predict point c because the process to point c is not isentropic.)

at d: Since the process from c to d is isentropic,

$$v_{rd} = v_{rc}(R) = (.08685)(7.67) = .6661$$

From the air table for this volume ratio,

$$T_d = 3565°R$$
$$p_{rd} = 1981.9$$
$$p_d = (1187.2)\left(\frac{1981.9}{27381}\right) = 85.9 \text{ psia}$$

18 ACTUAL OTTO ENGINE

The actual pressure-volume relationship will be considerably different from the plot shown in figure 7.30. A *drum indicator* can be used to investigate and plot the p-V relationship. These plots are known as *indicator drawings* or *indicator cards*.

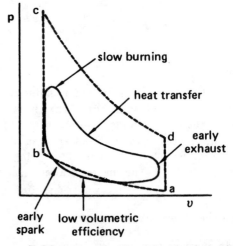

Figure 7.32 Indicator Card for an Otto Cycle

Figure 7.32 illustrates a typical indicator card for an Otto cycle. The causes for the deviation from the ideal cycle (shown by a dotted line) are given. Due to the heat transfer in the a-b and c-d processes, the lines are not true adiabats. Rather, these two processes are polytropic with a polytropic exponent of approximately 1.32.

Due to the numerous deviations from the ideal cycle, the actual Otto cycle will have a thermal efficiency of approximately 50% of the ideal efficiency.

An indicator card can be used to determine the mean effective pressure required in the PLAN formula, equation 7.89. The area of the indicator diagram is determined, and that area is divided by the overall width of the plot. The resulting number then is multiplied by the indicator spring constant (or scale).

19 THE AIR-STANDARD DIESEL CYCLE

The air-standard diesel cycle uses the heat of compression to ignite the air/fuel mixture. Spark plugs are not used. Fuel is injected directly into the combustion area in a critically-timed, high pressure operation. Diesel engines can operate as 4-stroke or 2-stroke devices, depending on the application.[7]

The air-standard diesel cycle is less efficient than the Otto cycle, given the same compression ratio. However, it is more efficient than the Otto cycle with the same pressure. Figure 7.33 illustrates the Diesel cycle.

The processes in the diesel cycle are

a to b: adiabatic compression

b to c: constant pressure heat addition

c to d: adiabatic expansion

d to a: constant volume heat rejection

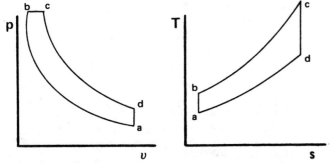

Figure 7.33 The Air-Standard Diesel Cycle

The following definitions commonly are used with the diesel cycle. (V_c is known as the *cut-off volume*.)

---

[7]The 2-stroke and 4-stroke diesel cycles are analyzed the same way, since the four processes on the p-V diagram do not correspond to the four strokes. The main difference is the number of power strokes per revolution, which is one for the 2-stroke engine.

$$R = \frac{V_a}{V_b} \quad \text{(compression ratio)} \qquad 7.110$$

$$R_c = \frac{V_c}{V_b} = \frac{T_c}{T_b} \qquad 7.111$$

The p, V, and T properties can be evaluated readily by using the ideal gas process equations presented in chapter 6. The work and heat flow terms are

$$Q_{\text{in}} = c_p(T_c - T_b) \qquad 7.112$$

$$Q_{\text{out}} = c_v(T_d - T_a) \qquad 7.113$$

$$W_{\text{in}} = c_v(T_b - T_a) \qquad 7.114$$

$$W_{\text{out}} = c_v(T_c - T_d) + (c_p - c_v)(T_c - T_b) \qquad 7.115$$

$$\eta_{th} = \frac{Q_{\text{in}} - Q_{\text{out}}}{Q_{\text{in}}} = \frac{W_{\text{out}} - W_{\text{in}}}{Q_{\text{in}}}$$

$$= 1 - \frac{T_d - T_a}{k(T_c - T_b)} \qquad 7.116$$

20 THE AIR-STANDARD DUAL CYCLE

The air-standard dual cycle is a compromise between the diesel cycle and the Otto cycle, with an intermediate efficiency. It most closely approximates the operation of an actual internal combustion engine. The dual cycle is illustrated in figure 7.34.

The processes of the dual cycle are

a to b: adiabatic compression

b to c: constant volume compression

c to d: constant pressure heat addition

d to e: adiabatic expansion

e to a: constant volume heat rejection

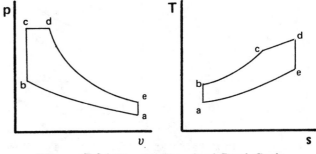

Figure 7.34 The Air-Standard Dual Cycle

The following definitions are used to describe the dual cycle.

$$R = \frac{V_a}{V_b} \quad \text{(compression ratio)} \qquad 7.117$$

$$R_c = \frac{V_d}{V_c} \quad \text{(cut-off ratio)} \qquad 7.118$$

$$R_p = \frac{p_c}{p_b} \quad \text{(pressure ratio)} \qquad 7.119$$

21 GAS TURBINE BRAYTON CYCLE (THE AIR-STANDARD JOULE CYCLE)

Strictly speaking, the Brayton cycle is an internal combustion cycle. It differs from previously discussed cycles in that each process is carried out in a different location, air flow and fuel injection are steady, and air-standard calculations are realistic since a large air/fuel ratio is used to keep combustion temperatures below metallurgical limits.

Figure 7.35 illustrates the physical arrangement of components used to achieve the Brayton cycle. Almost all installations drive the compressor from the turbine. (Approximately 75% of the turbine power is required to drive the high-efficiency compressor.) The actual arrangement differs from the air-standard property plot shown in figure 7.36 in that the exhaust products exiting at point d are not cooled and returned to the compressor at point a.

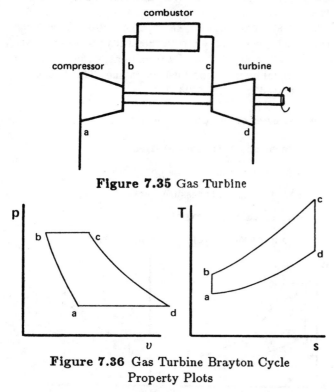

Figure 7.35 Gas Turbine

Figure 7.36 Gas Turbine Brayton Cycle Property Plots

The processes in the air-standard Brayton cycle follow.

a to b: adiabatic compression in the compressor

  b to c: constant pressure heat addition in the combustor

c to d: adiabatic expansion in the turbine

  d to a: constant pressure heat rejection to the sink

Combustors are said to be *open* if the incoming air and combustion products flow to the turbine. A high percentage, 30% to 60%, of excess air is needed to cool the gases. Depending on the turbine construction details, the temperature of the air entering the turbine will be between 1200°F and 1800°F. Turbojet and turboprop engines typically use *open combustors*.

A *closed combustor* is a heat exchanger. Air flowing to the turbine is not combined with the combustion gases. This allows any type of fuel to be used, but the bulk of the heat exchanger limits closed combustors to stationary power plants and pipeline pumping stations.

Relationships for steady flow systems from chapter 6 can be used to evaluate the p, V, and T properties at points a, b, c, and d. Constant values of k, c_p, and c_v can be assumed if air is considered ideal. An air table also can be used if the isentropic efficiencies are known. Usually, it is assumed that p_b and p_c are equal.

The work and heat flow terms are

$$Q_{in} = c_p(T_c - T_b) = h_c - h_b \qquad 7.120$$

$$W_{turb} = c_p(T_c - T_d) = h_c - h_d \qquad 7.121$$

$$W_{comp} = c_p(T_b - T_a) = h_b - h_a \qquad 7.122$$

$$\eta_{th} = \frac{Q_{in} - Q_{out}}{Q_{in}} = \frac{W_{turb} - W_{comp}}{Q_{in}}$$

$$= \frac{(h_c - h_b) - (h_d - h_a)}{h_c - h_b} \qquad 7.123$$

If the gas is ideal so that c_p is constant, then

$$\eta_{th} = \frac{(T_c - T_b) - (T_d - T_a)}{T_c - T_b} \qquad 7.124$$

Example 7.7

A gas turbine compressor takes 14.7 psia, 80°F incoming air and isentropically compresses it through a pressure ratio of 4.5. The turbine inlet conditions are 64 psia and 2200°R. The turbine's pressure ratio is 25%, and the expansion efficiency is 85%. What is the thermal efficiency if the exhaust is to the atmosphere? Use the ideal gas laws.

at a: $T_a = 80°F = 540°R$

       $p_a = 14.7$ psia

at b: $T_b = 540(4.5)^{\frac{1.4-1}{1.4}} = 830\ °R$

       $p_b = 4.5(14.7) = 66.15$ psia

at c: $T_c = 2200°R$

       $p_c = 64$ psia

at d: $p_d = \frac{64}{4} = 16$ psia

If the expansion had been isentropic, the temperature would have been

$$T_d = 2200\left(\frac{1}{4}\right)^{\frac{1.4-1}{1.4}} = 1480.5°R$$

The actual temperature is

$$T'_d = 2200 - .85(2200 - 1480.5) = 1588.4°R$$

From equation 7.124, the thermal efficiency is

$$\eta_{th} = \frac{(2200 - 830) - (1588.4 - 540)}{2200 - 830} = .235$$

Example 7.8

Repeat example 7.7 using the air table.

at a: $T_a = 540°R$

$p_a = 14.7$ psia

$h_a = 129.06$ BTU/lbm

$p_{ra} = 1.3860$

at b: Since the a-b process is isentropic, the pressure ratio can be used to predict the properties at point b.

$$p_{rb} = (4.5)(p_{ra}) = (4.5)(1.3860) = 6.237$$

Searching the air table for this pressure ratio,

$T_b = 827.6°R$

$h_b = 198.5$ BTU/lbm

$p_b = (4.5)(14.7) = 66.15$ psia

at c: $T_c = 2200°R$

$p_c = 64$ psia

$h_c = 560.59$ BTU/lbm

$p_{rc} = 256.6$

at d: $$p_{rd} = \frac{256.6}{4} = 64.15$$

Searching the air tables,

$T_d = 1554.6°R$

$h_d = 383.66$

However, the expansion efficiency is 85%. Therefore,

$$h_d = 560.59 - .85(560.59 - 383.66) = 410.2 \text{ BTU/lbm}$$

Using equation 7.123,

$$\eta_{th} = \frac{(560.59 - 198.5) - (410.2 - 129.06)}{560.59 - 198.5} = .224$$

22 THE AIR-STANDARD BRAYTON CYCLE WITH REGENERATION

Regeneration typically is used to improve the efficiency of the Brayton cycle. Regeneration involves transferring some of the heat from the exhaust products to the air in the compressor. The transfer occurs in a regenerator, which is usually a cross-flow heat exchanger. There is no effect on turbine work, compressor work, or net output. However, the cycle is more efficient since less heat is added. Of course, T_b cannot be greater than T_f. Similarly, T_c cannot be greater than T_e.

The physical arrangement and property plots are shown in figures 7.37 and 7.38.

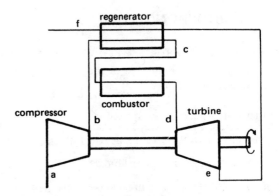

Figure 7.37 Brayton with Regeneration

The processes involved are

a to b: adiabatic compression

b to c: constant pressure heat addition in regenerator

c to d: constant pressure heat addition in combustor

d to e: adiabatic expansion

e to f: constant pressure heat removal in regenerator

f to a: constant pressure heat removal in the sink

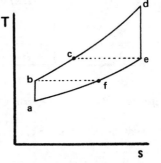

Figure 7.38 Brayton with Regeneration Property Plot

If $T_c = T_e$, the regenerator is said to be 100% efficient. Otherwise, the *regenerator efficiency* is calculated from equation 7.125.[8]

$$\eta_{regen} = \frac{h_c - h_b}{h_e - h_b} \qquad 7.125$$

---

[8] Regenerator efficiency rarely exceeds 75%.

The thermal efficiency of the air-standard Brayton cycle with regeneration is

$$\eta_{th} = \frac{W_{out} - W_{in}}{Q_{in}} = \frac{(h_d - h_e) - (h_b - h_a)}{h_d - h_c} \qquad 7.126$$

If air is considered to be an ideal gas, temperatures can be substituted for enthalpies in equations 7.125 and 7.126.

23 THE ERICSSON CYCLE

The Ericsson cycle offers the best chance of achieving an efficiency approaching that of the Carnot cycle. The processes shown in figure 7.39 are

a to b: isothermal compression

b to c: constant pressure heat addition

c to d: isothermal expansion

d to a: constant pressure heat rejection

The efficiency of the Ericsson cycle is equal to that of the Carnot cycle if reversible regeneration is used to transfer heat from the (d to a) process to the (b to c) process. Isothermal processes can be approximated by reheating and intercooling. The constant-pressure processes can be approached with counter-flow heat exchangers. Modifications to the Brayton cycle, including regeneration, intercooling, and reheat, approximate the Ericsson cycle.

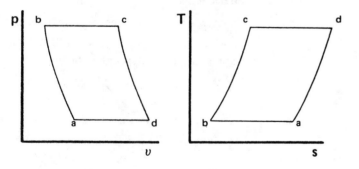

Figure 7.39 The Ericsson Cycle

The work and heat flow terms for the cycle as shown in figure 7.41 are

$$W_{out} = W_{c-d} \qquad\qquad 7.127$$
$$W_{in} = W_{a-b} \qquad\qquad 7.128$$
$$Q_{in} = Q_{b-c} + Q_{c-d} \qquad 7.129$$

If a reversible regenerator is used such that $Q_{b-c} = 0$,

$$\eta_{th} = \frac{T_{high} - T_{low}}{T_{high}} \qquad 7.130$$

24 BRAYTON CYCLE WITH REGENERATION, INTERCOOLING, AND REHEATING

Multiple compression and expansion can be used to improve the efficiency of the Brayton cycle still further. Physical limitations usually preclude more than two stages of intercooling and reheat. This section assumes only one stage of each. The physical arrangement is shown in figure 7.40.

The processes are

a to b: adiabatic compression

b to c: cooling at constant pressure (usually back to T_a)

c to d: adiabatic compression

d to f: constant pressure heat addition

f to g: adiabatic expansion

g to h: reheating at constant pressure in combustor or reheater (usually back to T_f)

h to i: adiabatic expansion

i to a: constant pressure heat rejection

Calculation of the work and heat flow terms and of the thermal efficiency is similar to that in the previous cycles, except that there are two W_{turb}, W_{comp}, and Q_{in} terms. If efficiencies for the compressor, turbine, and regenerator are given, the following relationships are required.

$$p'_b = p_b \qquad\qquad 7.131$$

$$h'_b = h_a + \frac{h_b - h_a}{\eta_{comp}} \qquad 7.132$$

$$p'_d = p_d \qquad\qquad 7.133$$

$$h'_d = h_c + \frac{h_d - h_c}{\eta_{comp}} \qquad 7.134$$

$$p'_e = p_e \qquad\qquad 7.135$$

$$h'_e = h_d + \eta_{regen}(h_i - h_d) \qquad 7.136$$

$$p'_g = p_g \qquad\qquad 7.137$$

$$h'_g = h_f - \eta_{turb}(h_f - h_g) \qquad 7.138$$

$$p'_i = p_i \qquad\qquad 7.139$$

$$h'_i = h_h - \eta_{turb}(h_h - h_i) \qquad 7.140$$

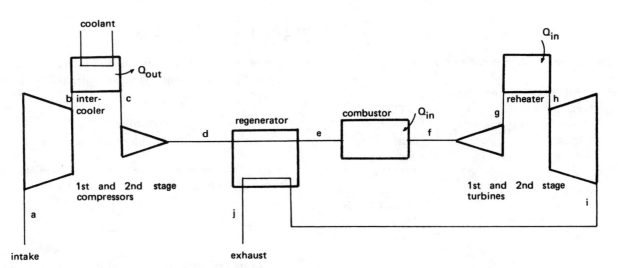

Figure 7.40 Augmented Brayton Cycle

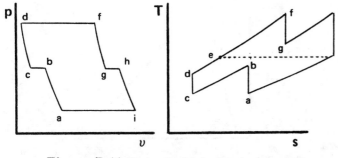

Figure 7.41 Property Plots for Augmented Brayton Cycle

25 THE STIRLING CYCLE

The Stirling cycle, shown in figure 7.42, can have a thermal efficiency equal to the Carnot efficiency if a reversible regenerator is used to transfer heat from the (c to d) process to the (a to b) process. The isothermal processes are possible with reheating and intercooling, but the steady flow constant volume processes are not. The processes are

 a to b: constant volume heat addition

 b to c: isothermal expansion with heat addition

 c to d: constant volume heat rejection

 d to a: isothermal compression with heat rejection

The Stirling cycle must be analyzed with the aid of the ideal gas process relationships in chapter 6. Enthalpy relationships cannot easily be used.

$$W_{\text{out}} = W_{b-c} = Q_{\text{in}} \qquad 7.141$$

$$W_{\text{in}} = W_{d-a} = Q_{\text{out}} \qquad 7.142$$

$$Q_{\text{in}} = Q_{a-b} + Q_{b-c} \qquad 7.143$$

If a reversible regenerator is used so that Q_{a-b} is recovered from the (c to d) process (i.e., $Q_{a-b} = 0$),

$$\eta_{th} = \frac{T_{\text{high}} - T_{\text{low}}}{T_{\text{high}}} \qquad 7.144$$

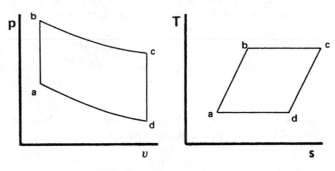

Figure 7.42 The Stirling Cycle

PART 3: Refrigeration Cycles

Nomenclature for Refrigeration Cycles

COP	coefficient of performance	–
h	enthalpy	BTU/lbm
k	adiabatic exponent (c_p/c_v)	–
m	mass	lbm
p	pressure	psf
Q	heat flow	BTU/lbm
R	compression ratio	–
s	entropy	BTU/lbm-°R
T	temperature	°R
W	work	BTU/lbm

26 GENERAL REFRIGERATION CYCLES

Refrigeration cycles are essentially power cycles in reverse. It is necessary to do work on the refrigerant in order to get it to discharge energy to the high-temperature sink. A *heat pump* also operates on a refrigeration cycle. The only difference between a refrigerator and a heat pump is the purpose of each. The refrigerator's main purpose is to cool a low temperature area and to reject the absorbed heat to a high temperature area. The heat pump's main purpose is to reject the absorbed heat to a high temperature area, having obtained that heat from a low temperature area.

The *coefficient of performance* (COP) of a device operating on a refrigeration cycle is defined as the ratio of useful heat transferred to the work input. If Q_{in} is the heat absorbed from the low temperature area, the COP for a refrigerator is

$$\text{COP}_{\text{refrig}} = \frac{Q_{in}}{Q_{out} - Q_{in}} = \frac{Q_{in}}{W_{in}} \qquad 7.145$$

The COP for a heat pump includes the desired heating effect of the compressor work input.

$$\text{COP}_{\text{heat pump}} = \frac{Q_{in} + W_{in}}{Q_{out} - Q_{in}} = \frac{Q_{in} + W_{in}}{W_{in}} \qquad 7.146$$
$$= \text{COP}_{\text{refrig}} + 1$$

The *capacity* of a refrigerator is the rate at which heat is removed expressed in *tons*. A ton is 12,000 BTU/hr or 200 BTU/min. The ton is derived from the heat flow required to melt a ton of ice in 24 hours. The relationship between horsepower, COP, and the capacity of the refrigerator in tons is given by equation 7.147.

$$\text{COP} = \frac{(4.715)(\text{tonnage})}{\text{horsepower}} \qquad 7.147$$

A new term used to evaluate the performance of cooling devices is the *energy efficiency ratio* (EER) as defined in equation 7.148.

$$\text{EER} = \frac{\text{cooling in BTU/hr}}{\text{input power in watts}} \qquad 7.148$$

In general, the required refrigerant flow in lbm/hr can be found from equation 7.149.

$$\dot{m} = \frac{Q_{in}}{h_c - h_b} \qquad 7.149$$

27 THE REVERSED CARNOT REFRIGERATION CYCLE

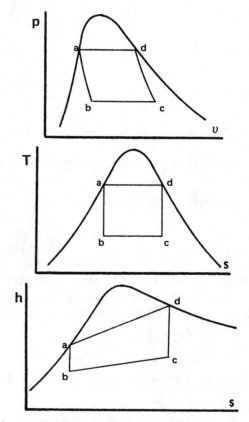

Figure 7.43 The Carnot Refrigeration Cycle

The reversed Carnot cycle can be constructed if a condensable vapor, such as ammonia or Freon, is used. The cycle's main disadvantage is the complex linkwork required for the (a to b) process. The cycle's processes, shown in figure 7.43, are

a to b: isentropic expansion behind a piston

b to c: isothermal heat addition (vaporization)

c to d: isentropic compression

d to a: isothermal cooling (condensation)

The solution method is reversed but identical in concept to the Carnot power cycle. The coefficients of performance are

$$\text{COP}_{\text{refrig}} = \frac{T_{\text{low}}}{T_{\text{high}} - T_{\text{low}}} \qquad \textbf{7.150}$$

$$\text{COP}_{\text{heat pump}} = \frac{T_{\text{high}}}{T_{\text{high}} - T_{\text{low}}} = \text{COP}_{\text{refrig}} + 1$$
$$\textbf{7.151}$$

28 THE VAPOR COMPRESSION CYCLE

In the vapor compression cycle, an irreversible expansion through a throttling valve replaces the isentropic expansion behind a piston in the reversed Carnot cycle.

The processes illustrated in figure 7.44 are

a to b: irreversible, isenthalpic expansion

b to c: isothermal vaporization

c to d: adiabatic compression

d to a: isothermal condensation

Compression of saturated or superheated vapor is said to be *dry compression* and is favored over *wet compression* due to wear in the compressor. For that reason, the refrigerator usually is designed so that the refrigerant leaves the evaporator either saturated or slightly super-heated, as shown in figure 7.45.

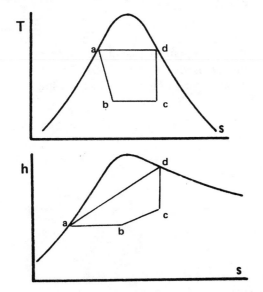

Figure 7.44 Wet Vapor Compression Cycle

If the refrigerant is saturated when leaving the evaporator at point c, the following solution method can be used.

at a: saturated liquid; $h_a = h_b$; $p_a = p_d$

at b: $T_b = T_c$; $h_b \approx h_a$

at c: saturated gas; $T_c = T_b$; $s_c = s_d$

at d: h_d found from the refrigerant table, given s_d and either p_d or T_d; $s_d = s_c$

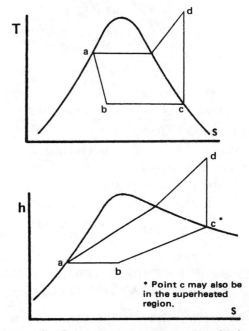

* Point c may also be in the superheated region.

Figure 7.45 Dry Vapor Compression Cycle

Example 7.9

A refrigeration cycle uses Freon-12 to remove 10,000 BTU's per hour. The refrigerant leaves the evaporator saturated at 0°F. The pressure of the refrigerant entering the condenser is 120 psia. The refrigerant leaves the condenser as a saturated liquid. The isentropic efficiency of the compressor is 80%.

Find the temperature out of the compressor, the refrigerant flow rate, the compressor work, and the coefficient of performance as a refrigerator.

Use figure 7.45 and the Freon-12 data from chapter 6.

at c:

$$T_c = 0°F \text{ (given)}$$
$$p_c = 23.87 \text{ psia}$$
$$h_c = 78.21 \text{ BTU/lbm}$$
$$s_c = .17091$$

at d:

$$p_d = 120 \text{ psia (given)}$$
$$s_d = s_c = .17091 \text{ (isentropic compression)}$$
$$h_d = 90.8 \text{ (interpolating in the 120 psia}$$
$$\text{superheat table, assuming}$$
$$\text{isentropic compression)}$$

$$h'_d = 78.21 + \left(\frac{90.8 - 78.21}{.8}\right) = 93.95$$

$T'_d = 129°F$ (interpolating the 120 psia

superheat table, knowing h'_d

at a:

$$p_a = 120 \text{ psia}$$
$$T_a = 93.4°F$$
$$h_a = 29.53$$

at b:

$$h_b = h_a = 29.53$$

The refrigerant flow rate is

$$\frac{\text{total heat removed}}{\Delta h \text{ per pound Freon}} = \frac{10,000 \text{ BTU/hr}}{(78.21 - 29.53) \text{ BTU/lbm}}$$
$$= 205.4 \text{ lbm/hr}$$

The compressor work is

$$205.4(93.95 - 78.21) = 3233 \text{ BTU/hr}$$

The coefficient of performance as a refrigerator is

$$\text{COP} = \frac{Q_{\text{removed}}}{W_{\text{in}}} = \frac{78.21 - 29.53}{93.95 - 78.21} = 3.09$$

29 THE AIR REFRIGERATION CYCLE (REVERSED BRAYTON CYCLE)

The air refrigeration cycle is not very common because of its high power consumption. However, air as a refrigerant is non-flammable, readily available, and very safe. Therefore, the air refrigeration cycle sees considerable use in aircraft air-conditioning and gas liquefication. The processes shown in figure 7.46 are

a to b: isentropic expansion in a turbine

b to c: constant pressure heat addition

c to d: isentropic compression

d to a: constant pressure cooling

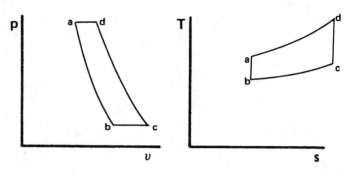

Figure 7.46 The Air Refrigeration Cycle

If air is considered to be an ideal gas, the ideal gas formulas from chapter 6 can be used to find the properties at each point. In that instance, the COP will be

$$\text{COP} = \frac{T_c - T_b}{(T_d - T_a) - (T_c - T_b)} \qquad 7.152$$

If the (a to b) and (c to d) processes are both isentropic, the following equation can be used to find the coefficient of performance.

$$\text{COP} = \frac{1}{R^{\frac{k-1}{k}} - 1} \qquad 7.153$$

$$R = \frac{p_{\text{high}}}{p_{\text{low}}} \qquad 7.154$$

If it cannot be assumed that air is an ideal gas, an air table must be used to find the COP. Enthalpy can then be substituted for temperature in equation 7.152.

PART 4: Compression Cycles

Nomenclature for Compression Cycles

BHP	brake horsepower	hp
c	clearance $= V_d/(V_b - V_d)$	decimal
h	enthalpy	BTU/lbm
k	ratio of specific heats	–
n	polytropic exponent	–
p	pressure	psf
Q	heat	BTU/lbm
R	compression ratio	–
s	entropy	BTU/lbm-°R
T	temperature	°R
V	volume	ft^3
W	work	BTU/lbm

Symbols

η efficiency

Subscripts

m   mechanical
s   isentropic
v   volumetric

30 RECIPROCATING, SINGLE-STAGE COMPRESSOR WITH ZERO CLEARANCE

The following processes describe the zero-clearance compressor illustrated in figure 7.47.

   a to b: suction at constant pressure (intake valve open)

   b to c: compression following polytropic relationships

   c to d: discharge at constant delivery pressure (outlet valve open)

   d to a: not a closed cycle, so the gas properties do not return to their original state.

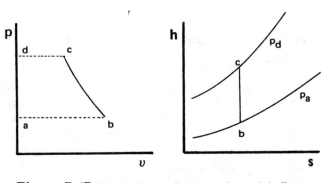

Figure 7.47 Single-Stage Compression with Zero Clearance

The dotted lines on figure 7.47 indicate the cylinder conditions, not the gas. The work done per cycle depends on the polytropic exponent, since the (b to c) process is assumed to be polytropic. For efficient air compressors, n is between 1.25 and 1.30. All properties can be analyzed using the closed system polytropic process equations given in chapter 6.

In addition to the work of compression in the (b to c) process, the motor must supply power to overcome friction and power to discharge the air in the (c to d) process. the brake work can be found from equation 7.155.

$$\text{brake work} = \frac{W_{\text{compression}}}{\eta_m \eta_s} \qquad 7.155$$

31 RECIPROCATING SINGLE-STAGE COMPRESSOR WITH CLEARANCE

V_d in figure 7.48 represents the *clearance volume* of the compressor. The gases remaining in the clearance volume after the discharge valve closes at top-dead-center (TDC) are known as the *residual gases*. The residual gases expand along with the new gas during the intake stroke to reduce the capacity per stroke. (The dotted line in figure 7.48 is for the residual gases, not the main charge.)

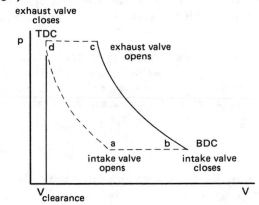

Figure 7.48 Single-Stage Compressor with Clearance

The *percentage clearance* (or just *clearance*) depends on the *swept volume* (*piston displacement*, $V_b - V_d$). (Swept volume is not the volume of the free gas.)

$$\text{percent clearance} = 100 \times \frac{\text{clearance volume}}{\text{swept volume}}$$

$$= 100 \times \frac{V_d}{V_b - V_d} \qquad 7.156$$

Although clearance affects the volumetric efficiency, for two compressors with the same gas flow rate, clearance does not affect the required power. The power requirement depends only on the mass of gas passing through the compressor. During steady state compressor operation, the mass of gas entering the cylinder equals the mass of air discharged, and clearance is not considered.

The parameters affecting performance are the polytropic exponent, n, the compression ratio, R, and the volumetric efficiency, η_v. For convenience, the *polytropic exponent*, n, for expansion is usually assumed to be the same as for compression. (However it is a simple matter to consider different values.) If the compression is isentropic, then the ratio of specific heats should replace the polytropic exponent (i.e., $n = k$.)

The *compression ratio*, R, is defined by equation 7.157. The compression ratio for compressors is a ratio of pressures (not a ratio of volumes as it is for internal combustion engines).

$$R = \frac{p_d}{p_a} = \frac{p_c}{p_b} \qquad 7.157$$

The *volumetric efficiency*, η_v, is the ratio of the actual mass of gas compressed to the mass of gas in the swept volume. Volumetric capacity is evaluated at the inlet conditions.

$$\eta_v = \left.\frac{\text{actual mass of gas compressed}}{\text{mass of gas in swept volume}}\right|_{\text{inlet conditions}}$$

$$= \frac{V_b - V_a}{V_b - V_d} = 1 - (R^{1/n} - 1)c \qquad 7.158$$

As with the zero clearance case, the work per cycle can be evaluated with the aid of air tables or the ideal gas process equations in chapter 6.

The mass flow rate through the compressor can be determined from the mass per stroke and the number of strokes per minute. Assuming an ideal gas and evaluating the gas at inlet conditions, the ideal mass per stroke (i.e., in the swept volume) is

$$m_{\text{ideal}} = \frac{p_a V}{R T_a} = \frac{p_a(V_b - V_d)}{R T_a} \qquad 7.159$$

The relationship between pressures and volumes are predicted by the ideal gas laws.

$$\frac{V_a}{V_d} = \left(\frac{p_d}{p_a}\right)^{1/n}$$

$$\frac{V_b}{V_c} = \left(\frac{p_c}{p_b}\right)^{1/n} \qquad 7.160$$

Assuming an ideal gas, the work of compression per stroke is

$$W_{bc} = h_c - h_b = c_p(T_c - T_b)$$

$$= \frac{n}{1-n}(p_c V_c - p_b V_b)$$

$$= \frac{n p_b V_b}{1-n}\left[\left(\frac{p_c}{p_b}\right)^{(n-1)/n} - 1\right] \qquad 7.161$$

The net work per cycle is

$$W_{\text{net}} = W_{bc} - W_{da} = h_b - h_a$$

$$= c_p(T_b - T_a) \qquad 7.162$$

The actual power requirement is

$$P = \eta_v \dot{m}_{\text{ideal}} W_{\text{net}}$$

32 RECIPROCATING MULTI-STAGE COMPRESSORS WITH CLEARANCE

In the multi-stage reciprocating compressor, the partially compressed gas is withdrawn, cooled, and compressed further. Desirable results include a reduction in required power, a decrease in working temperature, and a decrease in valve and ring loading due to the reduced pressure differential. The term *perfect intercooling* refers to the case when $T_b = T_d$ and $P_c = P_d$ (see figure 7.49). The discharge from an *intercooler* usually will be approximately 20 °F higher than the jacket water temperature.

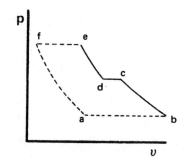

Figure 7.49 Multi-Stage Compressor with Clearance

Analysis of the multi-stage compressor is similar to that of the single-stage compressor.

The minimum work requirement occurs when all stage pressure ratios are the same. In that case,

$$\frac{p_c}{p_b} = \frac{p_e}{p_d} \qquad 17.163$$

Since $p_c = p_d$,

$$p_c^2 = p_e p_b \qquad 17.164$$

33 CENTRIFUGAL (DYNAMIC) COMPRESSION

The analysis of dynamic compressors is not exceptionally difficult. It can be handled by the open system (steady flow) process relationships in chapter 6. One added consideration is that of kinetic energy, which may have to be included in the calculations. Although this can be done adequately by using the steady flow energy equation (SFEE), it is most expedient to use isentropic gas flow tables.

Appendix A
Energy, Work, and Power Conversions

Multiply	By	To Get
BTU	3.929 EE−4	hp-hrs
BTU	778.3	ft-lbf
BTU	2.930 EE−4	kw-hrs
BTU	1.0 EE−5	therms
BTU/hr	.2161	ft-lbf/sec
BTU/hr	3.929 EE−4	hp
BTU/hr	0.2930 EE−3	kw
ft-lbf	1.285 EE−3	BTU
ft-lbf	3.766 EE−7	kw-hrs
ft-lbf	5.051 EE−7	hp-hrs
ft-lbf/sec	4.6272	BTU/hr
ft-lbf/sec	1.818 EE−3	hp
ft-lbf/sec	1.356 EE−3	kw
hp	2545.0	BTU/hr
hp	550	ft-lbf/sec
hp	.7457	kw
hp-hr	2545.0	BTU
hp-hr	1.976 EE6	ft-lbf
hp-hr	.7457	kw-hrs
kw	1.341	hp
kw	3412.9	BTU/hr
kw	737.6	ft-lbf/sec

Appendix B
Physical Properties of Pipe Materials for High Temperature Service (Welded Construction)

ASTM	Grade	Manufacture	ANSI Power Piping Code		ASME Boiler Code		Recommended use
			Temp. limit (°F)	Allowable stress at recommended temp. limit (p.s.i.)	Temp. limit (°F)	Allowable stress at recommended temp. limit (p.s.i.)	
A120..	...	Seamless...	450...	9600	Not approved...		Steam under 125 psig and 450°F.
		Butt welded.	450...	5700			Water under 175 psig. Threaded water pipe under 100 psig if temp. is above 220°F.
A53...	A...	Seamless...	775...	10,700 at 750°F.	800 ...	10,700 at 750°F.	Temps under 750°F.
	B...	Seamless...	775...	12,900 at 750°F.	800...	12,950 at 750°F.	Steam under 600 psig.
	A...	Elec. resistance weld.	750...	9,100...	800...	9,100 at 750°F.	Elec. resistance welded threaded water pipe under 100 psig if temp. is above 220°F
	B...	Elec. resistance weld.	750...	11,000...	800...	11,000 at 750°F.	Elec. resistance welded threaded steam pipe under 250 psig. Grade A only for close coiling, cold bending, and forge welding.
		Butt welded.	650...	6,750...	650...	6,750...	Temp. under 650°F. Steam under 250 p.s.i. Water under 400 p.s.i. Threaded water pipe under 100 psig if temp. is above 220°F. Do not use for flanging.
A106..	A...	Seamless...	775...	10,700 at 750°F.	800...	10,700 at 750°F.	Temp. under 750°F.
	B...	Seamless...	775...	12,950 at 750°F.	800...	12,950 at 750°F.	Grade A only for close coiling, cold bending, and forge welding.
A135..	A...	Elec. resistance weld.	750...	9,100...	800...	9,100 at 750°F.	Temp. under 750°F.
	B...	Elec. resistance weld.	750...	11,000...	800...	11,000 at 750°F.	Grade A only for flanging and bending.

ASTM	Grade	Manufacture	ANSI Power Piping Code		ASME Boiler Code		Recommended use
			Temp. limit (°F)	Allowable stress at recommended temp. limit (p.s.i.)	Temp. limit (°F)	Allowable stress at recommended temp. limit (p.s.i.)	
A155..	...	Elec. fusion weld.	775...	8,700–13,250 at 750°F. depending on chem. analysis.	Not approved...		Temp. under 750°F.
A134..	A-D..	Elec. fusion weld.	650...	8,800–10,100...	Not approved...		Temp. under 650°F. Steam under 250 psig. Water under 400 psig.
A139..	A...	Elec. fusion weld.	750...	8,300	Not approved...		Temp. under 750°F.
	B...	Elec. fusion weld.	750...	9,950			
A335..	P1...	C-1/2 Mo...	875...	13,150 at 850°F.	850...	13,150...	Temp. under 850°F.
	P2...	1/2 Cr-1/2 Mo.	975...	10,000 at 950°F.	1,000.	10,000 at 950°F.	Temp. under 950°F.
	P12..	1 Cr-1/2 Mo.	1,100.	7,500 at 1,000°F.	1,100.	7,500 at 1,000°F.	Temp. under 1,000°F.
	P11..	1.25 Cr-1/2 Mo.	1,100.	7,800 at 1,000°F.	1,100.	7,800 at 1,000°F.	Temp. under 1,000°F.
	P22..	2.25 Cr-1 Mo.	1,150.	5,800 at 1,050°F.	1,200.	5,800 at 1,050°F.	Temp. under 1,050°F.
	P21..	3 Cr-1 Mo.	1,150.	4,000 at 1,100°F.	1,200.	4,000 at 1,100°F.	Temp. under 1,100°F.
	P5...	5 Cr-1/2 Mo.	1,200.	3,300 at 1,100°F.	1,200.	3,300 at 1,100°F.	Temp. under 1,100°F.
	P9...	9 Cr-1/2 Mo.	...	...	1,200.	3,300 at 1,100°F.	Temp. under 1,100°F.

PRACTICE PROBLEMS: POWER CYCLES AND REFRIGERATION

Warm-ups

1. A steam cycle operates between 650°F and 100°F. What is the maximum possible efficiency?

2. What is the coefficient of performance of a Carnot heat pump operating between 700°F and 40°F?

3. What is the efficiency of a steam Carnot cycle operating between 650°F and 100°F if the turbine and compressor (pump) have isentropic efficiencies of .9 and .8, respectively?

4. An ammonia refrigerator operates on the Carnot cycle between 90°F and 5°F. What is the COP, horsepower per ton, and energy efficiency ratio?

5. What is the heating value of an oil with specific gravity of 40° API?

6. A 10″ × 18″ 2-cylinder, 4-stroke engine operates with a mean effective pressure of 95 psig at 200 rpm. If the actual developed torque is 600 ft-lbf, what is the friction horsepower?

7. A positive displacement air compressor with 7% clearance discharges 48 pounds of air per minute to 65 psia. If the polytropic exponent is 1.33, how many pounds of air are compressed each minute?

8. The air/fuel mixture enters a cylinder of a 3.1″ × 3.8″ internal combustion engine at 100 fps when the engine speed is 4000 rpm. The valve opens at TDC and closes at 40° past BDC. If the volumetric efficiency is 65%, what is the effective area of the intake valve?

9. 10 cubic feet per second of 200 psia, 1500°F air expand isentropically to 50 psia. What is the enthalpy change, the new temperature, and the new volume?

10. Atmospheric air at 500°F is compressed in a centrifugal compressor to 6 atmospheres with a 65% isentropic process efficiency. Find the work done, the final temperature, and the increase in entropy.

Concentrates

1. A steam Rankine cycle operates with 100 psia saturated steam which is reduced to 1 atmosphere through expansion in an 80% efficient turbine. The working fluid is 80°F liquid at 1 atmosphere upon entering the 60% efficient pump. What is the cycle thermal efficiency?

2. 500 psia steam is superheated to 1000°F before expanding through a 75% efficient turbine to 5 psia. No subcooling occurs, and the pump work is negligible compared to the 200,000 kw generated. What quantity of steam is required? What heat is rejected in the condenser?

3. A turbine and a pump are 88% and 96% efficient, respectively. The cycle operates between 60°F and 600°F with superheated 600 psia steam. Steam is reheated when its pressure drops to 200 psia through expansion. What is the thermal efficiency?

4. The cycle in problem 3 is modified to include a bleed of 270°F steam (within the second expansion) for feedwater heating in a closed feedwater heater. The **terminal temperature difference is 6 °F. Neglecting drip and condensate pump work, what is the thermal efficiency?**

5. The air in an Otto cycle engine enters an 11 ft^3 cylinder at 80°F and 14.2 psia. After the compression stroke (compression ratio of 10) 160 BTU's are added. What are the thermal efficiency and temperature at the end of the heat addition?

6. A 4.25″ × 6″ six cylinder standard diesel runs at 1200 rpm while consuming 28 pounds of fuel per hour. When the air/fuel ratio is 15, there is 13.7% CO_2 in the exhaust. However, the actual CO_2 is 9%. At one atmosphere and 70°F, what is the volumetric efficiency? Assume the CO_2 percentages are on a dry basis.

7. At 60°F and 14.7 psia, a fully loaded 1000 BHP diesel runs at 2000 rpm. The air/fuel ratio is 23, and BSFC is .45 lbm/BHP-hr. The mechanical efficiency is a constant 80%, independent of the operating altitude. What are the BHP and BSFC at 5000 feet altitude?

8. In an air-standard gas turbine, 60°F and 14.7 psia air enters a compressor and is compressed through a volume ratio of 5. Air enters the turbine at 1500°F and expands to 14.7 psia. If the efficiencies of the compressor and the turbine are .83 and .92, respectively, what is the thermal efficiency?

9. A 65% efficient regenerator is added to problem 8. What is the thermal efficiency?

10. 191,000 lbm/hr of 635°F gases flow through a 20 feet wide boiler stack whose front and back plates are 5′10″ apart. An integral economizer is being designed to heat water from 212°F to 285°F by dropping the stack gases to 470°F. Layers of 24 tubes (.957″ i.d., 1.315″ o.d., 20′ long) will be placed on a 2.315″ pitch in horizontal banks. How many 24-tube layers are required if $U_0 = 10$ BTU/hr-ft^2-°F?

Timed (1 hour allowed for each)

1. An internal combustion engine running on gasoline operates at 4600 rpm. The engine is 4-stroke and a V-8. The work required to compress the gas mixture is 1200 ft-lbf, and the work done by the gas in expansion is 1500 ft-lbf per piston for one complete cycle. The input energy is 1.27 BTU per piston per cycle. The engine displacement is 265 cubic inches. Neglecting friction, find the (a) indicated horsepower, (b) thermal efficiency, (c) gallons per hour of gasoline consumed, and (d) specific fuel consumption.

2. A reheat cycle is operated as shown. (a) What is the adiabatic efficiency of the high-pressure turbine? (b) What is the process thermal efficiency?

at 1: 800°F, 900 psia
at 2: 200 psia, 1270 BTU/lbm
at 3: 800°F, 190 psia
at 4: 50 psia, 1280 BTU/lbm
at 5: 2″ Hg absolute, 1075 BTU/lbm
at 6: 69.73 BTU/lbm
at 8: 250.2 BTU/lbm, .0173 ft³/lbm
at 9: 253.1 BTU/lbm
pump work (6 to 7) .15 BTU/lbm

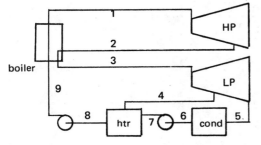

3. For each of the following processes, find the thermal efficiency or the COP.

(a) A jet engine compressor. Intake at −10°F and 8 psia; exhaust at 315°F and 40 psia.

(b) An ideal heat pump using refrigerant-12. Operating between 35.7 psia and 172.4 psia.

(c) An actual refrigeration cycle using Freon-12. Input power of 585 watts; heat absorption of 450 BTUH.

(d) A steam turbine power cycle producing 600 MW. Condenser load of 3.07 EE9 BTUH.

(e) A heat engine running on differential sea water temperature. Surface temperature at 82°F; temperature at −1200 ft is 40°F.

4. An internal combustion engine normally is fuel injected with gasoline (C_8H_{18}, lower heating value of 23,200 BTU/lbm). It is desired to switch to alcohol (lower heating value of 11,930 BTU/lbm, C_2H_5OH)

with a minimum of changes to the engine. Assume that the indicated and mechanical efficiencies remain unchanged and that the engine runs with a stoichiometric air/fuel ratio. (a) If the power output is to be unchanged, what must be the percent change in specific fuel consumption? (b) If the fuel injection velocity is to be unchanged, what is the percent change in jet size? (c) If no changes are made to the engine, what will be the percent change in power output?

5. A mixture of carbon dioxide and helium in an engine undergoes the following processes.

1 to 2: compression and heat removal
2 to 3: constant volume heating
3 to 1: isentropic expansion

point	temperature	pressure
1	520°R	14.7 psia
2	1240	?
3	1600	568.6

Find the gravimetric analysis of the fuel and the work done during the isentropic process. Draw the T-s and the p-V diagrams.

6. When the pressure is 14.7 psia and the temperature is 80°F, a diesel engine has the full-throttle characteristics listed. What are the corresponding quantities if the engine is operated at 12.2 psia and 60°F?

BHP: 200
BSFC: .48 lbm/hp-hr
air/fuel ratio: 22
mechanical efficiency: 86%

7. Waste steam previously used only for heating cold water is to be expanded through a low-pressure turbine with an isentropic efficiency of 60% and a mechanical efficiency of 96%. The steam then will flow through a mixing heater. A pressure drop of 5 psi occurs through the heater. The heater output must remain 180°F, but the output (6) may decrease. Find the shaft work output of the added turbine. Find the flow rate in pounds per hour at point (6), which discharges hot water at 180°F and 20 psia.

ORIGINAL USE OF WASTE STEAM
T_1 = 400°F T_2 = 70°F T_3 = 180°F
p_1 = 100 psia p_2 = 60 psia p_3 = 20 psia
   m_3 = 2000 lb/hr

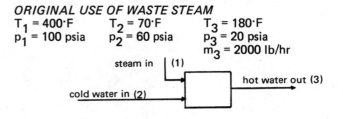

PROPOSED USE OF WASTE STEAM

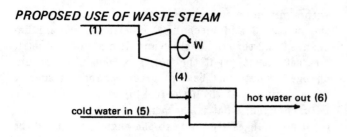

8. A gas turbine operating on the Brayton cycle (8:1 pressure ratio) develops 6000 bhp at 7000 feet altitude while consuming .609 pounds of fuel per hp-hour and 50,000 cfm of air. The ambient conditions at 7000 feet are 35°F and 12 psia. The turbine efficiency is 80%, and the compressor efficiency is 85%. The fuel has a lower heating value of 19,000 BTU/pound, and the fuel mass is small compared to the air mass. The turbine receives combustor gases at 1800°F. The air inlet filter area is 254 ft². If the turbine is moved to sea level (70°F, 14.7 psia) and the combustor temperature remains the same, find the new BHP and SFC.

RESERVED FOR FUTURE USE

RESERVED FOR FUTURE USE

8 COMPRESSIBLE FLUID DYNAMICS

1 DEFINITIONS OF TERMS

Adiabatic: A term used to describe a process in which there is no heat transfer.

Beta ratio: The ratio of throat to pipe diameters.

Chamber property: The value of a gas property in a location where the velocity is assumed to be zero.

Choked flow: A flow which occurs when some part of the device is operating at $M = 1$. A maximum possible flow.

Compressible fluid: A fluid capable of exhibiting volume changes when exposed to pressure fluctuations.

Critical flow: The flow at a point where $M = 1$.

Critical property: The property at a point where $M = 1$.

Hypersonic flow: The flow with M above 5.

Impact property: See "Stagnation property."

Isentropic: A term used to describe a constant entropy process.

Mach number: The ratio of the actual speed at a point to the speed of sound at that point.

Sonic flow: Flow with $M = 1$.

Stagnation property: The value of a gas property (density, pressure, temperture, etc.) obtained when the velocity of flow is reduced to zero in an isentropic process.

Static property: The value of a gas property measured by an observer moving with the flow.

Subsonic flow: Flow with M less than 1.

Supersonic flow: Flow with M between 1 and 5.

Total property: See "stagnation property."

Venturi: A subsonic converging/diverging nozzle.

2 INTRODUCTION

A high-velocity gas is defined as one with a velocity in excess of 200 ft/sec. There are several reasons why high-velocity gases must be handled differently from low-velocity gases.

- The Bernoulli equation assumes a constant density, and it cannot account for the conversion of internal energy to kinetic energy in an expanding gas. This conversion results in a lower temperature and a higher velocity than would be predicted by the Bernoulli equation.

- Density changes complicate the continuity equation, often resulting in more than one unknown.

- Pressure fluctuations and other local disturbances cannot be transmitted in flows with velocities greater than the speed of sound.

3 PROPERTIES OF AIR AND OTHER GASES

A. SPECIFIC HEATS

For most purposes, the specific heats for air can be taken as

$$c_p = 0.24 \text{ BTU/lbm-°R}$$
$$c_v = 0.1714 \text{ BTU/lbm-°R}$$

Values for other gases are given in chapter 6. The values of c_p and c_v are related by the following relationships.

$$c_p - c_v = \frac{R}{J} \qquad 8.1$$

$$c_p = \frac{Rk}{J(k-1)} \qquad 8.2$$

B. RATIO OF SPECIFIC HEATS

A required parameter in almost all equations is the ratio of specific heats, c_p/c_v. In most problems, this ratio is assumed to be constant. Since exact solution methods are too difficult to be done by hand, either the assumption of constant k should be made, or a value of k based on the average temperature of the process should be used.

Table 8.1
Values of k versus T for Air

T°R	k	T°R	k	T°R	k
500	1.400	1300	1.361	2200	1.322
600	1.399	1400	1.355	2400	1.317
700	1.396	1500	1.349	2600	1.313
800	1.392	1600	1.344	2800	1.309
900	1.387	1700	1.339	3000	1.306
1000	1.381	1800	1.335	3500	1.301
1100	1.375	1900	1.331	4000	1.298
1200	1.368	2000	1.328	4500	1.296
				5000	1.294

k is remarkably similar for gases with similar structures. For monatomic gases (He, Ar, Ne, Kr, etc.) k is exactly 1.666. For diatomic gases (N_2, O_2, H_2, CO, NO, and air) it is approximately 1.4 with slight temperature variations. For triatomic gases (H_2O or CO_2) k varies between 1.3 and 1.33. The value of k is less than 1.3 for more complex gases.

C. DENSITY

In this chapter, density is given in lbm/ft^3, the reciprocal of specific volume. Density can be found readily from the equation of state.

$$\rho = \frac{p}{RT} = \frac{1}{v} \qquad 8.3$$

D. VISCOSITY

The absolute viscosity of air is nearly independent of pressure. Within the temperature range of 300°R to 900°R the viscosity in lbf-sec/ft^2 can be found from the *Sutherland formula*.

$$\mu = (3.02 \text{ EE} - 7)\left(\frac{T}{392}\right)^{.76} \qquad 8.4$$

Between 180°R and 3400°R equation 8.5 should be used.

$$\mu = (2.27 \ \mathrm{EE} - 8)\left(\frac{T^{1.5}}{T + 198.6}\right) \qquad 8.5$$

Of course, the kinematic viscosity is affected by both pressure and temperature changes, since the density varies with these properties. The kinematic viscosity is defined as

$$\nu = \frac{\mu g_c}{\rho} \qquad 8.6$$

E. SPECIFIC GAS CONSTANT

The specific gas constant, R, is derived from the ideal gas constant and the gas molecular weight, both of which do not vary with temperature, pressure, or volume. For air, the value of R is 53.3 ft-lbf/lbm-°R.

$$R = \frac{R^*}{MW} \qquad 8.7$$

F. REYNOLDS NUMBER

The Reynolds number is a dimensionless number used to determine the fluid flow regime. It is given by equation 8.8.

$$N_{Re} = \frac{\rho v L}{g_c \mu} \qquad 8.8$$

For a gas flowing in a round duct, L is the inside duct diameter.

G. SONIC VELOCITY

The speed of sound in an ideal gas is essentially a function of the static temperature.[1]

$$c = \sqrt{k g_c R T} \qquad 8.9$$

For air, assuming $k = 1.4$, equation 8.9 becomes

$$c = 49.0\sqrt{T} \qquad 8.10$$

4 THE GENERAL FLOW EQUATION

Consider a compressible ideal gas with constant heat capacity, c_p, flowing in a duct with varying area. The flow can be assumed to be one-dimensional as long as the cross section varies slowly along the length. An adiabatic open-system energy balance, known as the *General Flow Equation* (GFE), can be written on a one pound basis, as in equation 8.11.

$$\frac{v_1^2}{2g_c} + z_1 + Jh_1 = \frac{v_2^2}{2g_c} + z_2 + Jh_2 \qquad 8.11$$

---

[1] In this chapter, property variables such as p (pressure) and T (temperature) without subscripts refer to the static condition. Total or stagnation properties will have the subscript T.

Since the potential energy changes, Δz, are minimal, this can be rewritten as

$$\frac{v_2^2 - v_1^2}{2g_c} = J(h_1 - h_2) \qquad 8.12$$

$$v_2 = \sqrt{2g_c J(h_1 - h_2) + v_1^2} \qquad 8.13$$

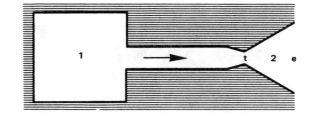

Figure 8.1 Reservoir Flow Locations

Point 1 often is a location where stagnation properties prevail, as in a chamber or reservoir, and thus, $v_1 = 0$. Point 1 can be assumed to be representative of the stagnation properties as long as $v_1 \leq .1v_2$. Substituting values of g_c and J produces equation 8.14.

$$v_2 = 223.8\sqrt{h_1 - h_2} \qquad 8.14$$

Other simple substitutions can be made to rewrite the GFE.

$$v_2 = \sqrt{2g_c J c_p (T_1 - T_2)} = \sqrt{\frac{2g_c R k (T_1 - T_2)}{k - 1}}$$
$$= \sqrt{\frac{2g_c k}{k - 1}\left[\frac{p_1}{\rho_1} - \frac{p_2}{\rho_2}\right]} \qquad 8.15$$

5 ISENTROPIC FLOW PARAMETERS

If the gas flow is adiabatic and reversible, it is said to be isentropic. The change in entropy for an isentropic flow is zero. As a practical matter, completely isentropic flow does not exist. Some steady state flow processes, however, proceed with little increase in entropy and are considered to be isentropic. The irreversible effects are accounted for by various correction factors, such as nozzle and discharge coefficients.

In isentropic flow, total pressure, total temperature, and total density remain constant, regardless of the flow area or the velocity. This is not to imply, however, that the static properties do not change. From equation 8.15,

$$v_2^2 = \frac{2g_c R k (T_T - T_2)}{k - 1} \qquad 8.16$$

Dividing both sides of 8.16 by $k g_c R T_2$ gives

$$\frac{v_2^2}{k g_c R T_2} = \frac{2\left(\frac{T_T}{T_2} - 1\right)}{k - 1} \qquad 8.17$$

But kg_cRT is the square of the speed of sound at the conditions at point 2, and the left hand side is M_2^2. Rearranging produces

$$\frac{T_T}{T_2} = \tfrac{1}{2}(k-1)M_2^2 + 1 \qquad 8.18$$

Example 8.1

Air flows isentropically from a large tank at 70°F through a converging/diverging nozzle and is expanded to supersonic velocities. What is the gas temperature at a point where the Mach number is 2.5? What is the actual velocity where $M = 2.5$?

$$\frac{T_T}{T_2} = \tfrac{1}{2}(1.4-1)(2.5)^2 + 1 = 2.25$$
$$T_T = 70° + 460° = 530°R$$
$$T_2 = \frac{530}{2.25} = 236°R = -224°F$$
$$v_2 = 2.5\sqrt{(1.4)(32.2)(53.3)(236)} = 1883 \text{ fps}$$

Similar results can be derived for the ratio of (p_T/p_2) and (ρ_T/ρ_2). The subscript "2" usually is omitted as being understood.

$$\frac{p_T}{p} = \left[\tfrac{1}{2}(k-1)M^2+1\right]^{\frac{k}{(k-1)}}$$
$$= \left[\frac{T_T}{T}\right]^{\frac{k}{(k-1)}} \qquad 8.19$$
$$\frac{\rho_T}{\rho} = \left[\tfrac{1}{2}(k-1)M^2+1\right]^{\frac{1}{(k-1)}}$$
$$= \left(\frac{T_T}{T}\right)^{\frac{1}{(k-1)}} \qquad 8.20$$

In order to design a nozzle capable of expanding a gas to some given velocity, it is necessary to be able to calculate the area perpendicular to the flow. Since the reservoir cross sectional area is an unrelated variable, it is not possible to develop a ratio for (A_T/A) as was done for temperature, pressure, and density. The usual choice for a reference area is the area at which the gas velocity is (or could be) sonic. This area is designated as A^*. The mass balance at this critical point is

$$A^* v^* \rho^* = Av\rho \qquad 8.21$$

Various substitutions can be made in equation 8.21 to arrive at the following relationship.

$$\frac{A}{A^*} = \frac{1}{M}\left[\frac{\tfrac{1}{2}(k-1)M^2+1}{\tfrac{1}{2}(k-1)+1}\right]^{\frac{k+1}{2(k-1)}} \qquad 8.22$$

Equation 8.22 can be plotted versus M with the results shown in figure 8.2. Notice that, as long as M is less than unity, the area must decrease in order for the

velocity to increase. However, if M exceeds unity, the area must increase in order for the velocity to increase.

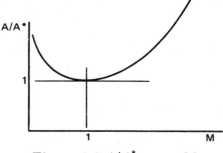

Figure 8.2 A/A^* versus M

In addition, equation 8.23, derived from the continuity formulas, gives the ratio of mass flow rate to the critical area. The formula is useful when you know the flow rate and want the critical area.

$$\frac{\dot{m}}{A^*} = \frac{\rho_T\sqrt{kg_cRT_T}}{\left[\tfrac{1}{2}(k-1)+1\right]^{\frac{k+1}{2(k-1)}}} \qquad 8.23$$

A final ratio is that of (v/c^*). This ratio is useful if you know the velocity and the total properties but need to find M. It is, of course, possible to solve for the local temperature and the speed of sound, but this is tedious.

$$\frac{v}{c^*} = \sqrt{\frac{\tfrac{1}{2}(k+1)M^2}{\tfrac{1}{2}(k-1)M^2+1}} \qquad 8.24$$

By now, it should be apparent that the algebraic burden of solving a gas dynamics problem is tremendous. The easiest way to solve such problems is with gas tables, given at the end of this chapter. The notation and symbols in this table, or in any other table that might be available, should be studied carefully to avoid confusion.

Example 8.2

The Mach number is .8 at a point in a nozzle where the area is 1.5 in². What is the area at the throat if sonic velocity is achieved there? Air is flowing in the nozzle.

From the isentropic flow table, $A/A^* = 1.0382$ at $M = .8$. Therefore, the throat area is

$$A_t = \frac{1.5}{1.0382} = 1.445 \text{ in}^2$$

Example 8.3

For the nozzle described in example 8.2, what is the area at a point where the Mach number is .4?

In this case, the $M = 1.0$ point will be used as a reference point. For $M = .4$, $A/A^* = 1.5901$.

$$A_{M=.4} = \frac{(1.5901)(1.5)}{1.0382} = 2.30 \text{ in}^2$$

Example 8.4

At a point where $T = 180°$ F, the velocity in a nozzle is 740 fps. What is the velocity in the throat where $M = 1$?

The speed of sound at the given point is

$$c_1 = \sqrt{(1.4)(32.2)(53.3)(460 + 180)} = 1240.1 \text{ fps}$$
$$M_1 = \frac{740}{1240} = .6$$

At $M = .6$, $v/c^* = .6348$. Therefore,

$$c_2 = \frac{740}{.6348} = 1165.7 \text{ fps}$$

Example 8.5

A frictionless, adiabatic duct receives 10 lbm/sec of air from a reservoir at 200°F and 30 psia. Somewhere in the duct, a velocity of 1400 ft/sec is attained. What are the cross-sectional area, temperature, pressure, and Mach number at this point?

In the reservoir,

$$\rho = \frac{p}{RT} = \frac{(30)(144)}{(53.3)(660)} = .1228 \text{ lbm/ft}^3$$

For a point where $M = 1$, the isentropic flow table gives

$$\frac{T}{T_T} = .8333 \quad \text{or} \quad T^* = (.8333)(660) = 550°R$$
$$\frac{\rho}{\rho_T} = .6339 \quad \text{or} \quad \rho^* = (.6339)(.1228) = .0778$$

From equation 8.10, the sonic velocity at the throat is

$$c^* = 49\sqrt{550} = 1149.2 \text{ fps}$$
$$A^* = \frac{\dot{m}}{\rho v} = \frac{10}{(.0778)(1149.2)} = .1118 \text{ ft}^2$$

Then,

$$\frac{v}{c^*} = \frac{1400}{1149.2} = 1.218$$

Searching the (v/c^*) column for 1.218 gives $M = 1.28$. For this Mach number, read

$$\frac{T}{T_T} = .7532 \quad \frac{p}{p_T} = .3708 \quad \frac{A}{A^*} = 1.058$$

So,

$$T = .7532\,(660) = 497.1°R$$
$$p = .3708\,(30) = 11.12 \text{ psia}$$
$$A = (1.058)(.1118) = .1183 \text{ ft}^2$$

Although it would seem that the gas exit velocity could increase without bound simply by increasing the exit area, this is not so. Velocity (kinetic energy) can be increased only at the expense of the temperature (internal energy). Assuming that the specific heat is constant and that the gas does not liquefy, the maximum attainable velocity is

$$v_{\max} = \sqrt{2Jg_c c_p T_T} \qquad \qquad 8.25$$

Also, it would seem that the velocity (and thus the mass flow rate) could be made to increase without bound by increasing the reservoir pressure accordingly. This is also false, as is shown in figure 8.3.

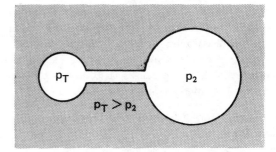

Figure 8.3 Reservoir Discharging to Environment

Figure 8.3 illustrates a high pressure reservoir discharging into a low pressure reservoir or local environment at absolute pressure, p_2. As p_T is increased, the mass flow will increase, as shown in figure 8.4. However, the increase will stop when $p_2/p_T = .5283$. If the assumptions of isentropic, one-dimensional, steady flow are made, the ratio of $p_2/p_T = .5283$ occurs when $M = 1$ in the duct. This ratio is known as the *critical pressure ratio*. It can be found for any gas from equation 8.26.

$$R_{cp} = \left(\frac{2}{k+1}\right)^{\frac{k}{k-1}} \qquad \qquad 8.26$$

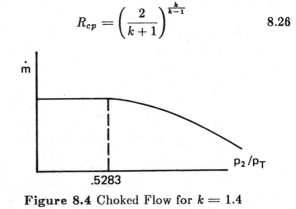

Figure 8.4 Choked Flow for $k = 1.4$

The velocity cannot be increased indefinitely by increasing the reservoir pressure. The pressure fluctuations required to increase the gas velocity travel at sonic velocity. However, if the gas in the duct is traveling at sonic velocity, the fact that a lower pressure region exists to the right of the duct cannot be transmitted to the source of the air. Such a flow is said to be *choked* because no additional mass can be made to flow without changing other upstream conditions.

The actual pressure decrease from p_T to p_2 when the flow is choked occurs across a *shock wave*, a process that is neither isentropic nor one-dimensional.

6 SHOCK WAVES

Small pressure disturbances are propagated in compressible fluids at the speed of sound. These waves commonly are known as *sonic waves*. Stronger waves separating radical changes in fluid properties are known as *shock waves*. These waves move at velocities faster than the local speed of sound.

The strength of a shock wave is measured in the change in Mach number across it. Actually, the only difference between a shock wave and a sonic wave is the strength. As a shock wave gets weaker, the change in Mach number across the shock decreases to 0, at which time the shock wave becomes a sonic wave.

The properties on either side of a shock wave are given in table 8.2.

Table 8.2
Properties Across a Shock Wave

before the shock	after the shock
supersonic	subsonic
low temperature	high temperature
low pressure	high pressure

The second law of thermodynamics limits shock waves to situations where the fluid velocity (or relative fluid flow) is from a supersonic region to a subsonic region. Such a *compression shock* results in an entropy increase.

Normal shock waves are those in which the fluid flow is perpendicular to the wave. This is the only kind of shock that occurs in one-dimensional flow. Another type of shock, the *oblique shock*, can occur with two-dimensional flow. Oblique shocks form at the leading edges of supersonic airframes causing sonic booms.

Note the following points about normal shocks.

- The flow across a shock is adiabatic, and the total temperature remains unchanged.

- The flow is not isentropic, and the total pressure decreases across the shock wave.

- The stagnation density increases across a shock.

- Shock waves are very thin, approximately 4 EE−7 inches at standard conditions. They become thicker as pressure decreases. For engineering purposes, they are usually assumed to be of negligible thickness.

- Momentum is conserved across a shock.

The following relationships can be used to calculate the properties on either side of a normal shock. The x subscript refers to upstream (supersonic) conditions; the y subscript refers to downstream (subsonic) conditions.

$$T_{Tx} = T_{Ty} \qquad\qquad 8.27$$

$$M_y^2 = \frac{M_x^2 + \frac{2}{k-1}}{\left[\frac{2k}{k-1}\right] M_x^2 - 1} \qquad\qquad 8.28$$

$$\frac{T_y}{T_x} = \frac{[1 + \frac{1}{2}(k-1)M_x^2](\frac{2k}{k-1}M_x^2 - 1)}{\left[\frac{(k+1)^2}{2(k-1)}\right] M_x^2} \qquad 8.29$$

$$\frac{p_y}{p_x} = \left[\frac{2k}{k+1}\right] M_x^2 - \frac{k-1}{k+1} \qquad 8.30$$

$$\frac{\rho_y}{\rho_x} = \left(\frac{p_y}{p_x}\right)\left(\frac{T_x}{T_y}\right) \qquad\qquad 8.31$$

$$\frac{p_{Ty}}{p_{Tx}} = \left[\frac{\frac{1}{2}(k+1)M_x^2}{1 + \frac{1}{2}(k-1)M_x^2}\right]^{\frac{k}{k-1}} \times$$
$$\left[\left(\frac{2k}{k+1}\right)M_x^2 - \frac{k-1}{k+1}\right]^{\frac{1}{1-k}}$$
$$8.32$$

$$\frac{p_{Ty}}{p_x} = \left[\frac{1}{2}(k+1)M_x^2\right]^{\frac{k}{k-1}} \times$$
$$\left[\left(\frac{2k}{k+1}\right)M_x^2 - \frac{k-1}{k+1}\right]^{\frac{1}{1-k}} \qquad 8.33$$

$$s_2 - s_1 = c_p ln\left(\frac{T_x}{T_y}\right) - R\, ln\left(\frac{p_x}{p_y}\right) \qquad 8.34$$

As in the isentropic flow case, the solution of shock wave problems will be facilitated by the use of normal shock tables, such as those at the end of this chapter.

Example 8.6

A shock wave occurs at a point where $M = 2.35$. If the static pressure before the shock is 73.96 psia, what are the static and total pressures after the shock? What is the Mach number after the shock?

From the normal shock tables for $M_x = 2.35$, the Mach number after the shock is .5286. Also,

$$\frac{p_y}{p_x} = 6.276 \quad \text{and} \quad \frac{p_x}{p_{Ty}} = .131$$

Therefore,

$$p_y = (73.96)(6.276) = 464.17 \text{ psia}$$
$$p_{Ty} = \frac{73.96}{.131} = 564.6 \text{ psia}$$

Example 8.7

The air temperature and pressure in an isolated desert prior to an explosive test are 70°F and 14.7 psia, respectively. At the center of the blast, the pressure is raised to 2.0 atmospheres. What is the shock wave velocity? What are the pressure, temperature, and velocity behind it?

Even though the expansion is spherical, the normal shock ratios can be used to a good approximation. Let x be the subscript for the still air into which the shock wave is expanding and let y be the subscript for the gas inside the expanding high-pressure region.

From the normal shock tables for $(p_y/p_x) = 2$, read $M = 1.36$. At these conditions, the speed of sound in air is 1128 ft/sec, so the shock travels at $(1128)(1.36) = 1534$ ft/sec. Behind the shock wave, the pressure is 2.0 $(14.7) = 29.4$ psia. Also, for $M = 1.36$, $T_y/T_x = 1.229$. So, $T_y = (530)(1.229) = 651.4°$R. Since $M_y = .757$,

$$v_y = .757(49.0)\sqrt{651.4} = 947 \text{ ft/sec}$$

7 SUPERSONIC FLOW PAST WEDGES AND CONES

Supersonic flow past wedges and cones creates shock waves which generally are attached to the tip and inclined at an angle which varies with the incident Mach number. In the case of the wedge, the shock wave is in the form of two inclined planes; for the cone, the shock wave has the form of a concentric outer cone.

A. WEDGES

If the bow shock is attached to the wedge, as it will be if the wedge semi-vertex angle δ is less than the *critical angle* found from table 8.3, the shock wave is straight and the flow downstream of the shock is parallel to the wedge surface. The conditions on either side of the shock can be found from the normal shock tables if $M_1 \sin \theta$ is used in place of M_x in the first column. If the shock is not attached, the shock tables cannot be used.

Table 8.3
Approximate Maximum Semi-Vertex Angles for Attached Shock Waves for Wedges and Cones
$k = 1.4$

M_1	wedge	cone	M_1	wedge	cone
1.0	0°	0°	2.6	31	47
1.2	4	19	2.8	32.5	48
1.4	9	27.5	3.0	34	49
1.6	14.5	33	3.2	35.5	50
1.8	19	37.5	3.4	36.5	51
2.0	23	41	3.6	37.5	51.7
2.2	26	43	3.8	38	52.2
2.4	28.5	45	4.0	39	52.7
			∞		45

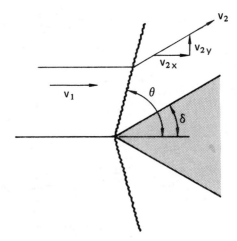

Figure 8.5 An Oblique Shock

θ is given by the formula

$$\tan \theta = \frac{v_1 - v_2 \cos \delta}{v_2 \sin \delta} \qquad 8.35$$

Unfortunately, there is no convenient exact formula for finding θ from M_1 and δ, although cubic solutions and iterative processes exist. However, the value of θ can be found graphically from figure 8.6.

In some instances, two shock waves and angles are theoretically possible, but only the *weaker shock* wave (smaller angle θ) attached to an isolated convex body is thought to exist. The Mach number after the shock is given by equation 8.36.[2]

$$M_2 = \frac{M_y \text{ from shock table}}{\sin (\theta - \delta)} \qquad 8.36$$

B. CONES

Shock waves attached to cones also are difficult to handle by exact methods. If θ is known, conditions behind the shock wave also can be determined from the shock tables using $M_1 \sin \theta$ for the first column value.

Example 8.8

A wedge with a semi-vertex angle, δ, of 10° is traveling through air at $M = 2$. Find the shock angle, the Mach number after the shock, the ratio of static pressures before and after the shock, and the ratio of static temperatures before and after the shock.

From table 8.3, the maximum semi-vertex angle for a wedge traveling at $M = 2$ is 23°. Therefore, the shock will be attached.

From figure 8.6, the semi-vertex shock angle will be approximately 40°.

---

[2]The velocity after an oblique shock wave does not necessarily drop below the sonic velocity, although the normal component of velocity will. Therefore, M_2 can be greater than 1.

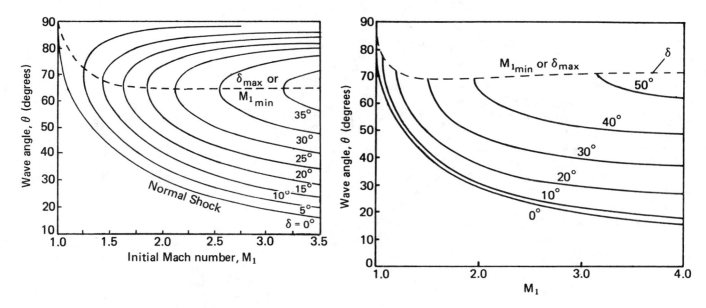

Figure 8.6 Shock Angle for (a) Wedges and (b) Cones $k = 1.4$

From the normal shock tables for $M_g = (2)(\sin 40°) = 1.29$,

$$\frac{p_x}{p_y} = \frac{1}{1.7748} = 0.563$$

$$\frac{T_x}{T_y} = \frac{1}{1.1846} = 0.844$$

$$M_2 = \frac{.79108}{\sin(40° - 10)} = 1.58$$

8 NOZZLES AND DIFFUSERS

Figure 8.2 gives the area ratio for a supersonic nozzle. If the Mach number is to increase steadily in supersonic flow, the cross-sectional area-distance relationship must be exactly as shown. This plot is for a converging-diverging nozzle, named a *deLaval nozzle* after its inventor who used it to obtain supersonic steam in early turbines.

It might seem possible to change the Mach number (or the area) by any amount over any chosen distance. However, if the rate of change, dA/dx, becomes too great, the assumptions of one-dimensional flow become invalid. Usually, the *converging section* has a steeper angle (known as the *convergent angle*) than the *diverging section*. If the *diverging angle* is too great, a normal shock may form in that part of the nozzle.

Consider the arrangement shown in figure 8.7 Here a high-pressure reservoir with a constant total pressure is connected through a converging-diverging nozzle to a variable low-pressure reservoir. Below the figure, the ratio of static pressure to total pressure (p_x/p_T) is plotted as a function of the location. Notice that p_e does not always equal p_2, although the exit gases must eventually reach p_2 through a shock wave or rarefaction.

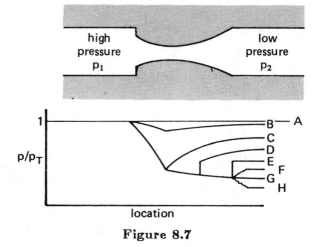

Figure 8.7

case A: If $p_2 = p_T$, there will be no flow.

case B: If p_2 is lowered to slightly less than p_T, flow begins. As long as the flow in the divergent section is subsonic (that is, M found in step 4 is less than 1.0), the exit pressure will be equal to p_2. Gas properties at any point in the nozzle can be determined from the isentropic flow table in the following manner.

step 1: Calculate the ratio p_2/p_T.

step 2: Locate (p_2/p_T) in the isentropic flow table and determine the exit Mach number and the ratio A/A^*. (At this point, there is no sonic flow because A_t is too large. A^* is the theoretical area the throat would have if there were sonic flow.)

step 3: Calculate $(A_t/A^*) = (A/A^*)(A_t/A_e)$

step 4: Locate (A_t/A^*) in the table and read the value of M_t and p_t/p_T.

step 5:   Calculate p_t from (p_t/p_T) and p_T.

step 6:   Repeat steps 3, 4, and 5 for all points for which the area is known.

case C: If p_2/p_T is lowered to the critical pressure ratio (.5283 for air), sonic velocity will be achieved in the throat. The gas properties at any point in the nozzle can be found from the procedure given in case B. Of course, $p_e = p_2$, and the exit velocity is subsonic.

case D: If p_2/p_T is lowered slightly below the critical ratio, the mass flow rate will remain the same since the flow is choked. However, since p_2 is lower, v_2 must be higher, even supersonic in some places. For the flow to be supersonic everywhere in the divergent section, the ratio p_e/p_T must be equal to the isentropic ratio associated with A_e/A^*. If the pressure ratio exceeds this value, the pressure in the divergent section will be too high for complete supersonic flow, so the flow will become sonic somewhere in the nozzle via a normal shock wave. Again, p_e will equal p_2. Given A_t, A_e, and p_e/p_{Tx}, the position in the divergent section where the shock occurs can be found from the following procedure.

step 1:   Assume some value of M_1 at which the shock wave occurs.

step 2:   Read p_{Ty}/p_{Tx} from the shock tables.

step 3:   Calculate $(p_e/p_{Ty}) = (p_e/p_{Tx})(p_{Tx}/p_{Ty})$.

step 4:   Find the value of p_e/p_{Ty} in the shock table and read M_e and (A/A^*).

step 5:   Calculate $(A_e/A_t)(p_{Ty}/p_{Tx})$.

step 6:   Compare the results of step 5 with the value of (A/A^*) from step 4. If these results differ, repeat with another value of M_1. (The step 5 value decreases as M_1 increases.)

step 7:   Once the value of M_1 has been found, read from the flow tables (A/A^*) and calculate the area at which the shock occurs.

case E: The shock wave will remain in the divergent section until p_2/p_{Tx} is equal to the pressure ratio given for the actual A_e/A_t, at which time the shock will stand at the exit.

cases F, G, and H: If p_2/p_{Tx} is equal to the pressure ratio given for the actual A_e/A_t, there will be no shock wave at all. This is case G. If the pressure ratio is slightly greater than this, a shock wave will form outside the nozzle (case F), but this will not change the nozzle conditions. If the pressure ratio is smaller than at

G, a *rarefaction* will occur outside the nozzle (case H).

The behavior of isentropic flow through a nozzle can be summarized.

- If $M < 1$ in a converging section, the velocity will increase as the area decreases. The maximum possible velocity is $M = 1$, which will occur in the throat, if it occurs anywhere.

- If $M > 1$ in a converging section, the velocity will decrease as the area decreases. The minimum possible velocity is $M = 1$, which will occur in the throat, if it occurs anywhere.

- If $M > 1$ in a diverging section, then $M = 1$ has been achieved in the throat, and the velocity will increase as the area increases.

- If $M < 1$ in a diverging section, then $M = 1$ has not been achieved in the throat, and the velocity will decrease as the area increases.

- If $M = 1$ in the throat, the velocity may or may not remain supersonic everywhere in the diverging section. If the ambient pressure is greater than the design pressure, there will be a normal shock somewhere within the nozzle.

A nozzle which causes the gas to behave as in case H is said to be an *under-expanding nozzle*. The fluid is discharged at a pressure greater than the ambient conditions because the exit area is too small. Expansion is incomplete in the nozzle, and the expansion continues outside the nozzle.

An *over-expanding nozzle* has an exit area too large for the ambient conditions (**case F**). With such a nozzle, the gas pressure drops below the ambient conditions and then returns to it by way of a rarefaction.

The effect of under-expansion and over-expansion is to reduce the gas exit velocity. In fixed steam nozzles, this results in a decrease in available energy for the turbine. In the case of propulsion systems, this decrease produces a proportional decrease in thrust.

However, a given percentage of over-expansion (based on the ratio of actual to theoretical exit areas) can reduce the available energy as much as 10 times the reduction from the same percentage of under-expansion. For that reason, nozzles are sometimes designed 10% to 20% too small to ensure under-expansion under light or partial loads.

Example 8.9

An attitude-adjustment jet uses high-pressure gas ($k = 1.4$, molecular weight = 21.0 lbm/pmole) generated by catalytic action in a combustion chamber. The chamber conditions are 450 psia and 4700°R. The gas is expanded to 2.97 psia at the nozzle exit, where it exits at supersonic velocity.

Find the sonic velocity in the throat, the ratio A_e/A_t, and the exit velocity.

Assume that the chamber properties can be used as the total properties. For a gas with $k = 1.4$, $T/T_T = .8333$ where $M = 1$. Therefore,

$$T_{throat} = (.8333)(4700) = 3916.5°R$$

The throat velocity is

$$c^* = \sqrt{(32.2)(1.4)\left(\frac{1545}{21}\right)(3916.5)} = 3604 \text{ fps}$$

At the exit,

$$\frac{p}{p_T} = \frac{2.97}{450} = 0.0066$$

Searching the isentropic flow table in the supersonic region for this value of p/p_T indicates that the exit Mach number is 4.0. At that point, $A/A^* = 10.7187$, and $T/T_T = .2381$. So, the exit velocity is

$$v_e = 4c_e = (4)\sqrt{(32.2)(1.4)\left(\frac{1545}{21}\right)(.2381)(4700)}$$
$$= 7706 \text{ fps}$$

Example 8.10

A supersonic wind tunnel is constructed so that the back pressure can be varied during the testing of a nozzle. The wind tunnel is fed from a reservoir with total properties of 1000 psia and 1000°R. A nozzle in the wind tunnel has a throat area of 1.5 in² and an exit area of 3.457 in².

Find the highest back pressure which will allow the flow to remain supersonic throughout the entire length of the nozzle.

The desired operating conditions (i.e., that of no shock wave in the nozzle) can be satisfied by case E, where there is a shock wave at or just out of the exit. Therefore, this example is *not* correctly solved by assuming case G and finding the design pressure.

If the velocity is supersonic in the diverging section, $M = 1$ has been achieved in the throat. The area ratio is

$$\frac{A_e}{A_t} = \frac{3.457}{1.5} = 2.3$$

Searching the A/A^* column for this value, the exit Mach number is found to be $M_e = 2.35$. At the exit, $p/p_t = .07396$, so the exit pressure before the shock wave is

$$p_{ex} = (.07396)(1000) = 73.96 \text{ psia}$$

This pressure is the design pressure for the nozzle and corresponds to case G. However, if there is a shock wave at the exit, the pressure after the shock will be higher.

Example 8.6 describes a case where the upstream Mach number is 2.35 and the upstream pressure is 73.96. In that example, the pressure after the shock is 464.17 psia. This is the maximum back pressure which can satisfy the desired operating conditions.

Example 8.11

A nozzle is tested with air flowing. The back pressure is adjustable. The total pressure upstream of the nozzle is 100 psia. There is a shock wave standing at the exit. The Mach number is 3.0 just before the shock.

Find the back pressures that will cause the shock wave at the exit to disappear completely.

There are two ways the shock wave can disappear. If the back pressure is decreased to the design pressure (case G), the shock wave will move out of the nozzle and will dissipate. However, if the back pressure is increased, the shock wave will move upstream and vanish at the throat (case C).

At $M = 3$, $p/p_t = .02722$. Therefore, the design pressure for this nozzle is

$$(.02722)(100) = 2.722 \text{ psia}$$

At $M = 3$, $A/A^* = 4.235$. Locating this value in the subsonic part of the isentropic flow tables gives $M = .138$. The shock wave will vanish, and the flow will be subsonic everywhere in the nozzle. At $M = .138$, $p/p_T = .987$. The back pressure will be $(.987)(100) = 98.7$ psia.

9 ADIABATIC FLOW WITH FRICTION IN CONSTANT AREA DUCTS

Unless the duct is very short, friction with high velocity flow cannot be ignored. High velocity gas flow can be assumed to be adiabatic if the pipe is insulated or if the flow is rapid. This type of flow is known as *Fanno flow*. Figure 8.8 illustrates an arrangement for Fanno flow.

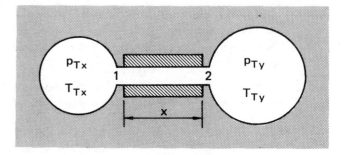

Figure 8.8 Fanno Flow

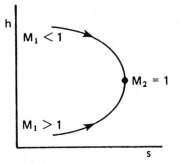

Figure 8.9 Fanno Line

The *Fanno line* is a plot of enthalpy versus entropy, derived from the energy conservation law, continuity equation, and the change in entropy relationships. Figure 8.9 illustrates the Fanno line, showing how velocity tends towards $M_2 = 1$, regardless of M_1.

With Fanno flow and an initially subsonic flow at point 1, the pressure drops as x increases. Since the density also drops, the velocity increases.[3] However, since the pressure drop is proportional to the square of the velocity, the rate of change of pressure (dp/dx) increases. Eventually, as p_{Ty} is lowered sufficiently, the velocity in the duct becomes sonic, after which no increase in velocity or mass flow can occur. Lowering p_{Ty} further will have no effect on the choked flow.[4]

Because the flow is adiabatic, the total temperature remains constant. Thus, equation 8.18 can be used to find the temperature at any point in the duct. The ratio of static pressure is given by equation 8.37.

$$\frac{p_2}{p_1} = \frac{\rho_2}{\rho_1}\left(1 + \left[\tfrac{1}{2}(k-1)M_1^2\right]\left[1 - \left(\frac{\rho_1}{\rho_2}\right)^2\right]\right) \quad 8.37$$

For round ducts, the *hydraulic diameter*, D_H, is equal to the inside diameter. $x_{\max}$ is the maximum distance required for M to become unity.[5]

$$\frac{4fx_{\max}}{D_H} = \left(\frac{1-M^2}{kM^2}\right) + \left(\frac{k+1}{2k}\right) \times$$
$$\quad ln\left(\frac{(k+1)M^2}{2\left[1 + \tfrac{1}{2}(k-1)M^2\right]}\right) \quad 8.38$$

The length of duct required to change M_1 to any other value, M_2, is

$$x = \frac{D_H}{4f}\left[\left(\frac{4fx_{\max}}{D_H}\right)_1 - \left(\frac{4fx_{\max}}{D_H}\right)_2\right] \quad 8.39$$

---

[3] If the velocity in the duct is initially supersonic, the velocity will decrease, approaching $M = 1$.

[4] Notice that the total pressure is not constant in Fanno flow. However, since the flow is adiabatic, the total temperature does not change.

[5] The *Fanning friction factor*, f, also is known as the *coefficient of friction*, C_f. Multiplying the Fanning friction factor by 4, as has been done in figure 8.10, converts it to the *Darcy friction factor*.

The static gas properties at any point are related to the sonic properties according to equations 8.40, 8.41, and 8.42.

$$\frac{p}{p^*} = \frac{1}{M}\sqrt{\frac{k+1}{2\left[1 + \tfrac{1}{2}(k-1)M^2\right]}} \quad 8.40$$

$$\frac{T}{T^*} = \frac{k+1}{2\left[1 + \tfrac{1}{2}(k-1)M^2\right]} \quad 8.41$$

$$\frac{p_{Tx}}{p_T^*} = \frac{1}{M}\left(\frac{2}{k+1}\left[1 + \tfrac{1}{2}(k-1)M^2\right]\right)^{\frac{k+1}{2(k-1)}} \quad 8.42$$

These equations are useful only in the simplest of problems. Tabulated or graphed values (as in figure 8.10) are required in most situations.

With Fanno flow and an initially sonic flow at point 1, the friction causes the pressure, density, and temperature to increase and the velocity to decrease. The frictional effects cause the fluid to reach $M = 1$ at the exit.

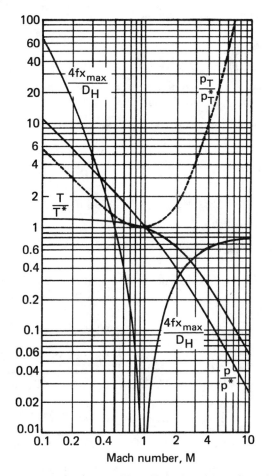

Figure 8.10 Fanno Flow Relationships, $k = 1.4$

Example 8.12

Gas ($k = 1.4$) enters an adiabatic, constant-area 0.3 foot diameter duct at 12 psia and $M = .3$. If the

Fanning friction factor is .003, what is the Mach number and pressure 50 feet downstream?

From figure 8.10 or equation 8.38, the value of $4fx_{max}/D_H$ is found to be 5.3.

50 feet downstream,

$$\frac{4fx}{D_H} = \frac{(4)(.003)(50)}{.3} = 2.0$$

From equation 8.39,

$$\left(\frac{4fx_{max}}{D_H}\right)_2 = 5.3 - 2.0 = 3.3$$

Solving for M_2 from equation 8.38 is not easy, but M_2 can be found from figure 8.10. It is approximately .35. **At this speed, p/p^* (from the chart or equation 8.40) is** approximately 3.09. At the entrance, p/p^* is 3.62, or

$$p^* = \frac{12}{3.62} = 3.32$$

Therefore, $p_2 = (3.32)(3.09) = 10.26$ psia.

Example 8.13

A duct ($D_i = 2.0$ inches, $f = .005$) receives air at $M = 2$. The total pressure and temperature at the entrance are 78.25 psia and 1080°R.

(a) Find the temperature and pressure at the entrance.

(b) Find the total temperature, total pressure, static temperature, and static pressure at a point where $M = 1.75$.

(c) Find the distance between the points where $M = 2$ and $M = 1.75$.

(a) From the isentropic flow table for $M = 2$,

$$T_1 = (.5556)(1080) = 600°R$$
$$p_1 = (.1278)(78.25) = 10 \text{ psia}$$

(b) Although it isn't known if the duct is long enough for choked flow ($M = 1$) to exist, that point still can be used as a reference point. The following values are read from a Fanno flow table. (Figure 8.10 can be used also.)

	$M = 2.0$	$M = 1.75$
$\frac{p}{p^*}$	.40825	.49295
$\frac{p_T}{p_T^*}$	1.6875	1.3865
$\frac{T}{T^*}$	.66667	.74419
$\frac{4fL}{D}$	.30499	.22504

$T_{T1} = T_{T2} = 1080°R$ since the flow is adiabatic.

$$T_2 = T_1\left(\frac{T_2}{T^*}\right)\left(\frac{T^*}{T_1}\right) = (600)\left(\frac{.74419}{.66667}\right) = 670°R$$

$$p_2 = p_1\left(\frac{p_2}{p^*}\right)\left(\frac{p^*}{p_1}\right) = (10)\left(\frac{.49295}{.40825}\right) = 12.1 \text{ psia}$$

$$p_{T2} = p_{T1}\left(\frac{p_{T2}}{p_T^*}\right)\left(\frac{p_T^*}{p_{T1}}\right) = (78.25)\left(\frac{1.3865}{1.6875}\right)$$
$$= 64.3 \text{ psia}$$

(c) At $M = 2$, the distance to reach $M = 1$ is

$$L_1 = \frac{(.30499)(2)}{(4)(.005)} = 30.5 \text{ inches}$$

At $M = 1.75$, the distance to reach $M = 1$ is

$$L_2 = \frac{(.22504)(2)}{(4)(.005)} = 22.5 \text{ inches}$$

The distance between points 1 and 2 is

$$L = L_1 - L_2 = 30.5 - 22.5 = 8 \text{ inches}$$

10 ISOTHERMAL FLOW WITH FRICTION

As the Mach number approaches unity, an infinite heat-transfer rate is required to keep the flow isothermal. That is why the gas flow is closer to adiabatic than to isothermal. A notable exception is encountered with long-distance pipelines where the earth supplies the needed heat to keep the flow isothermal.[6]

For any given pressure drop, the mass flow rate is given by the *Weymouth equation*.

$$\dot{m} = \sqrt{\frac{(p_1^2 - p_2^2)D^5 g_c(\pi/4)^2}{4fxRT}} \qquad 8.43$$

The Fanning friction factor, f, in the Weymouth equation is assumed to be given by equation 8.44.

$$f = \frac{.00349}{(D)^{.333}} \qquad 8.44$$

Example 8.14

A 40 inch (inside diameter) pipe is placed into service carrying natural gas between pumping stations 75 miles apart. The gas leaves the pumping stations at 650 psia, but the pressure drops to 450 psia by the time it reaches the next station. The gas temperature remains constant at 40°F. Assume the gas is methane (MW=16).

[6] The low velocities (less than 20 ft/sec) typical of natural gas pipelines hardly seem appropriate in this chapter. However, the gas experiences a density change along the length of pipe.

Use the Bernoulli equation and the discharge conditions to calculate the flow rate. Compare the Bernoulli flow rate with the flow rate predicted by the Weymouth equation.

The pipe diameter and area are

$$D = \frac{40}{12} = 3.333$$

$$A = \frac{\pi}{4}(3.333)^2 = 8.73 \text{ ft}^2$$

The pressures in psf are

$$p_1 = (650)(144) = 9.36 \text{ EE4 psf}$$

$$p_2 = (450)(144) = 6.48 \text{ EE4 psf}$$

$$\rho_1 = \frac{p}{RT} = \frac{9.36 \text{ EE4}}{(1545/16)(40 + 460)}$$

$$= 1.938 \text{ lbm/ft}^3$$

(a) The entire pressure drop is due to friction. Neglecting the low kinetic energy and the changes to potential energy, the Bernoulli equation based on the pump discharge conditions is

$$\frac{9.36 \text{ EE4}}{1.938} = \frac{6.48 \text{ EE4}}{1.938} + h_f$$

$$h_f = 14,861 \text{ ft}$$

Assume $f_{\text{Darcy}} = 0.015$. Then,

$$h_f = \frac{fLv^2}{2Dg}$$

$$14,861 = \frac{(0.015)(75)(5280)v^2}{(2)(3.333)(32.2)}$$

$$v = 23.17 \text{ ft/sec}$$

The mass flow rate is

$$\dot{m} = \rho v A = (1.938)(23.17)(8.73) = 392 \text{ lbm/sec}$$

(b) The Fanning friction factor from equation 8.44 is

$$f = \frac{.00349}{(3.333)^{.333}} = .00234$$

From equation 8.43,

$$\dot{m} = \sqrt{\frac{[(9.36 \text{ EE4})^2 - (6.48 \text{ EE4})^2](3.333)^5(32.2)\left(\frac{\pi}{4}\right)^2}{(4)(0.00234)(75)(5280)(1545/16)(460 + 40)}}$$

$$= 456 \text{ lbm/sec}$$

11 FRICTIONLESS FLOW WITH HEATING

If heat is added to a fluid flowing without friction through a constant-area duct, the flow is said to be a *Rayleigh flow*.[7] The heat comes from chemical reactions, phase changes, electrical heating, or external sources. (Frictional heating is not Rayleigh flow.) The

[7]This is known as *diabatic flow*—flow *with* heating.

usual conservation of energy, mass, and momentum equations hold. The following ratios of static to critical values can be calculated and plotted. (See figure 8.11.)

$$\dot{q} = c_p(T_{T2} - T_{T1}) \qquad 8.45$$

$$\frac{T}{T^*} = \left(\frac{1+k}{1+kM^2}\right)^2 M^2 \qquad 8.46$$

$$\frac{T_T}{T_T^*} = \frac{2(k+1)M^2[1 + \frac{1}{2}(k-1)M^2]}{(1+kM^2)^2} \qquad 8.47$$

$$\frac{p}{p^*} = \frac{1+k}{1+kM^2} \qquad 8.48$$

$$\frac{p_T}{p_T^*} = \left(\frac{2}{k+1}\right)^{\frac{k}{k-1}}\left(\frac{1+k}{1+kM^2}\right) \times$$

$$[1 + \frac{1}{2}(k-1)M^2]^{\frac{k}{k-1}} \qquad 8.49$$

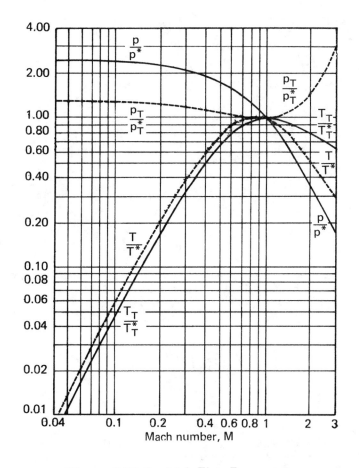

Figure 8.11 Rayleigh Flow Parameters

For Rayleigh flow, the Mach number approaches unity for both subsonic and supersonic flow. From the T/T^* curve, it is apparent that the maximum temperature occurs at $M = \sqrt{k}$. Also, the total pressure always decreases, even while the heat is being added. The *Rayleigh line* is plotted in figure 8.12.

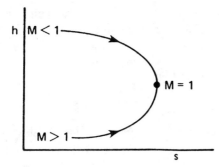

Figure 8.12 Rayleigh Line
(Arrows shown are for heating. The directions should be reversed for cooling.)

Example 8.15

Air enters a constant area duct at 50°F and is heated at the rate of 150 BTU/lbm by an external flame. If the inlet Mach number is .3, what are the exit temperature, exit Mach number, and exit stagnation temperature?

From the isentropic flow table,

$$T_{T1} = \frac{(460 + 50)}{.9825} = 519°R$$

The exit stagnation temperature is

$$T_{T2} = T_{T1} + \frac{\text{heat}}{c_p} = 519 + \frac{150}{.24} = 1144°R$$

From figure 8.11 at $M_1 = .3$, $(T_{T1}/T_T^*) = .35$. And since $(T_{T2}/T_{T1}) = \frac{1144}{519} = 2.20$,

$$\frac{T_{T2}}{T_T^*} = (2.20)(.35) = .77$$

From figure 8.11 for $(T_{T2}/T_T^*) = .77$, $M_2 = .56$, and $(T_2/T^*) = .87$. At the entrance, $M = .3$, and $(T_1/T^*) = .41$, so

$$T_2 = (.87)\left(\frac{1}{.41}\right)(460 + 50) = 1082°R$$

12 NOZZLE AND ROCKET PERFORMANCE CALCULATIONS

The flow rate of a nozzle can be found from the exit (or from any other location) properties.

$$\dot{m} = \rho_e v_e A_e \qquad 8.50$$

The ideal flow rate is given by equation 8.51.

$$\dot{m}_{\text{ideal}} = \frac{\dot{m}_{\text{actual}}}{C_d} \qquad 8.51$$

The ideal value of the flow rate should be used as the basis of nozzle design. Typical values of C_d vary around .97 and .98.

The force (*thrust*) exerted is

$$F = \left(\frac{\dot{m}v_e}{g_c}\right) + A_e(p_e - p_a) = \left(\frac{\dot{m}v_{\text{eff}}}{g_c}\right) \qquad 8.52$$

The *characteristic exhaust velocity* is defined as

$$v_{\text{char}} = \frac{g_c p_T A_t}{\dot{m}} = \frac{v_{\text{eff}}}{C_f} = \frac{I_{sp}g}{C_f} \qquad 8.53$$

C_f is known as the *coefficient of thrust.*

$$C_f = \frac{F}{p_T A_t} \qquad 8.54$$

The *effective exhaust velocity* is defined as

$$v_{\text{eff}} = \frac{F g_c}{\dot{m}} = I_{sp}g \qquad 8.55$$

The *specific impulse* is the thrust in lbf divided by the fuel consumption rate in lbm/sec.

$$I_{sp} = \sqrt{\left(\frac{2J}{g_c}\right)(h_T - h_e)} \qquad 8.56$$

The *total impulse* is the thrust multiplied by the time the thrust is applied.

$$I_T = (\text{fuel mass})I_{sp} = Ft \qquad 8.57$$

The *characteristic length* of the combustion chamber is

$$L^* = \frac{V_c}{A_t} \qquad 8.58$$

13 HIGH-VELOCITY GAS FLOW MEASUREMENTS WITH PITOT TUBES

Equations are given in chapter 3 for finding the velocity and mass flow of a subsonic gas by using a pitot tube. The compressibility effects were handled by using the compressibility factor. Exact solutions for the case when the initial Mach number is less than one are given by the following equations. (p_T is the stagnation pressure. p_x is the static pressure.)

$$M = \sqrt{\frac{2\left[\left(\frac{p_T}{p_x}\right)^{\frac{(k-1)}{k}} - 1\right]}{k - 1}} \qquad 8.59$$

or

$$v = \sqrt{\frac{2kRTg_c}{(k-1)}\left[\left(\frac{p_T}{p_x}\right)^{\frac{(k-1)}{k}} - 1\right]} \qquad 8.60$$

Of course, the mass flow is given by

$$\dot{m} = A\rho v \qquad 8.61$$

If the flow is supersonic, it cannot be brought to rest isentropically at the pitot tube inlet. Such an isentropic process would require a converging/diverging nozzle. Thus, a non-isentropic transition must occur at the Mach number of the original flow. The pressure read from a supersonic pitot tube is not the upstream stagnation pressure, but the downstream stagnation pressure. The shock tables have the value of (p_x/p_{Ty}) tabulated so that the incident Mach number can be read directly.

Example 8.16

A pitot tube in air measures an impact pressure of 20 psia. The accompanying static pressure is 4.5 psia. What is the Mach number?

Equation 8.59 could be used, but it is easier to use the isentropic flow tables. The ratio of static to total pressure is

$$\frac{p}{p_T} = \frac{4.5}{20} = .225$$

Searching the isentropic flow table for this value, the Mach number appears to be 1.63. However, since this is greater than one, there is a normal shock before the pitot tube.

Searching the p_x/p_{Ty} column, $M = 1.75$.

14 STEAM FLOW

There are few exact relationships for use with steam, since it is not an ideal gas. However, equation 8.62 can be used.

$$v_2 = \sqrt{2g_c J(h_1 - h_2) + v_1^2} \qquad 8.62$$

The other equations in this chapter cannot be used unless a reasonably accurate value of k (or c_p) is available for the steam. Figure 8.13 can be used for this purpose.[8]

Since steam expansion generally is not isentropic, the exit velocity is less than that calculated from equation 8.62. Isentropic efficiencies run between .85 and .95. This efficiency is defined by equation 8.63, which also defines the *coefficient of velocity*. C_v typically varies from .95 to .99 for superheated steam, although it can be considerably less for wet steam.

---

[8]Figure 8.13 is for superheated steam. k, for saturated steam up to approximately 500 psia, varies between 1.12 and 1.14.

$$\eta_n = \frac{(h_1 - h_2)_{\text{actual}}}{(h_1 - h_2)_{\text{ideal}}} = \left(\frac{v_{\text{actual}}}{v_{\text{ideal}}}\right)^2 = (C_v)^2 \qquad 8.63$$

The *discharge coefficient* is the ratio of actual mass flow to the ideal mass flow. The actual discharge is reduced from the ideal because of jet contraction and non-isentropic expansion.

$$C_d = C_c C_v = \frac{\dot{m}_{\text{actual}}}{\dot{m}_{\text{ideal}}} \qquad 8.64$$

Values of C_d for well designed steam nozzles vary from .95 to .98, although the discharge coefficient can exceed 1.00 when the steam becomes supersaturated during expansion.

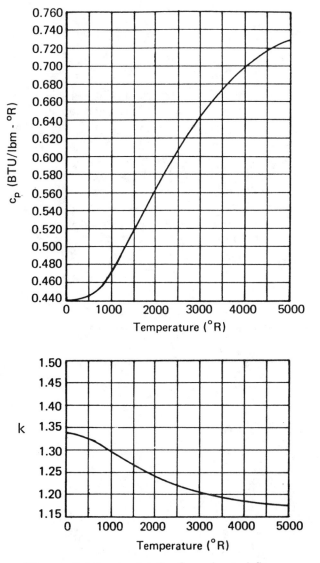

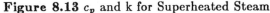

Figure 8.13 c_p and k for Superheated Steam

Several common design techniques used for steam follow.

- The critical pressure ratio can be found from equation 8.30 and an average value of k. However, the ratio usually is taken as .5457 for highly superheated steam and .57 for super-saturated steam.[9]

- If the flow is sonic at the throat, and if the throat conditions are known, the ratio of mass flow to throat area is

$$\frac{m}{A^*} = \rho_t c = \rho_t \sqrt{kgRT_t} \qquad 8.65$$

Experiments with steam have shown that, for common inlet pressures, sonic velocity will be achieved if the back pressure is 54% to 58% of the inlet pressure. (This corresponds to the p/p_T value at $M = 1$.) Therefore, assuming $R = 85.8$, equation 8.65 can be written in terms of the inlet conditions.

$$\frac{\dot{m}}{A^*} \approx 33.0 \rho_i \sqrt{T_i} \quad (k = 1.1)$$
$$\approx 35.0 \rho_i \sqrt{T_i} \quad (k = 1.3)$$

- A second method for finding A^* is to use an average value of k in equation 8.23. Of course, the properties (including the density) at the throat can be used to find A^* exactly. Assuming sonic flow at the throat, this would require finding the throat static pressure by multiplying the inlet pressure by the critical pressure ratio. Given the throat pressure and the original entropy, the throat state can be located in the superheated tables or on a Mollier diagram.

- The exit area can be found from the mass flow rate and the exit conditions. In a single-

---

[9] Because of the rapidity of the expansion process, the steam may not actually condense out when its enthalpy drops below the saturation value. This temporary condition is known as su-

persaturation. The effect of supersaturation is to decrease the energy actually extracted from the steam by a few percent. The degree of supersaturation is the ratio of the steam pressure to the saturation pressure corresponding to the steam temperature.

stage turbine, the turbine back pressure can be used to approximate the nozzle exit pressure. However, modern turbines are multi-staged. That is, the steam will pass through a series of blades and passageways before reaching the condenser. The nozzle exit pressure in such multi-staged turbines must be specified.

Example 8.17

Steam at 200 psia and 500°F enters an insulated nozzle with negligible velocity. It is expanded to 20 psia and 98% quality.

Find the exit velocity, the nozzle efficiency, and the coefficient of velocity.

From the Mollier diagram,

$$h_1 = 1269 \text{ BTU/lbm}$$
$$h'_2 = 1137$$

If the expansion had been isentropic (a straight drop on the Mollier diagram), the exit enthalpy would have been approximately 1080 BTU/lbm.

From equation 8.62,

$$v_2 = \sqrt{(2)(32.2)(1269 - 1137)(778)} = 2571 \text{ fps}$$

The efficiency and the coefficient of velocity are defined by equation 8.63.

$$\eta = \frac{1269 - 1137}{1269 - 1080} = .70$$
$$C_v = \sqrt{.70} = .84$$

Appendix A
Isentropic Flow Factors For $k = 1.4$

M = Mach number, P = pressure, TP = total pressure, D = density, TD = total density, T = temperature, TT = total temperature, A = area at a point, A* = throat area for M = 1, v = velocity, and c* = speed of sound at throat.

M	P/TP	D/TD	T/TT	A/A*	V/C*	M	P/TP	D/TD	T/TT	A/A*	V/C*
.00	1.0000	1.0000	1.0000	-------	.0000	.51	.8374	.8809	.9506	1.3212	.5447
.01	.9999	1.0000	1.0000	57.8737	.0110	.52	.8317	.8766	.9487	1.3034	.5548
.02	.9997	.9998	.9999	28.9420	.0219	.53	.8259	.8723	.9468	1.2865	.5649
.03	.9994	.9996	.9998	19.3005	.0329	.54	.8201	.8679	.9449	1.2703	.5750
.04	.9989	.9992	.9997	14.4815	.0438	.55	.8142	.8634	.9430	1.2549	.5851
.05	.9983	.9988	.9995	11.5914	.0548	.56	.8082	.8589	.9410	1.2403	.5951
.06	.9975	.9982	.9993	9.6659	.0657	.57	.8022	.8544	.9390	1.2263	.6051
.07	.9966	.9976	.9990	8.2915	.0766	.58	.7962	.8498	.9370	1.2130	.6150
.08	.9955	.9968	.9987	7.2616	.0876	.59	.7901	.8451	.9349	1.2003	.6249
.09	.9944	.9960	.9984	6.4613	.0985	.60	.7840	.8405	.9328	1.1882	.6348
.10	.9930	.9950	.9980	5.8218	.1094	.61	.7778	.8357	.9307	1.1767	.6447
.11	.9916	.9940	.9976	5.2992	.1204	.62	.7716	.8310	.9286	1.1656	.6545
.12	.9900	.9928	.9971	4.8643	.1313	.63	.7654	.8262	.9265	1.1551	.6643
.13	.9883	.9916	.9966	4.4969	.1422	.64	.7591	.8213	.9243	1.1451	.6740
.14	.9864	.9903	.9961	4.1824	.1531	.65	.7528	.8164	.9221	1.1356	.6837
.15	.9844	.9888	.9955	3.9103	.1639	.66	.7465	.8115	.9199	1.1265	.6934
.16	.9823	.9873	.9949	3.6727	.1748	.67	.7401	.8066	.9176	1.1179	.7031
.17	.9800	.9857	.9943	3.4635	.1857	.68	.7338	.8016	.9153	1.1097	.7127
.18	.9776	.9840	.9936	3.2779	.1965	.69	.7274	.7966	.9131	1.1018	.7223
.19	.9751	.9822	.9928	3.1123	.2074	.70	.7209	.7916	.9107	1.0944	.7318
.20	.9725	.9803	.9921	2.9635	.2182	.71	.7145	.7865	.9084	1.0873	.7413
.21	.9697	.9783	.9913	2.8293	.2290	.72	.7080	.7814	.9061	1.0806	.7508
.22	.9668	.9762	.9904	2.7076	.2398	.73	.7016	.7763	.9037	1.0742	.7602
.23	.9638	.9740	.9895	2.5968	.2506	.74	.6951	.7712	.9013	1.0681	.7696
.24	.9607	.9718	.9886	2.4956	.2614	.75	.6886	.7660	.8989	1.0624	.7789
.25	.9575	.9694	.9877	2.4027	.2722	.76	.6821	.7609	.8964	1.0570	.7883
.26	.9541	.9670	.9867	2.3173	.2829	.77	.6756	.7557	.8940	1.0519	.7975
.27	.9506	.9645	.9856	2.2385	.2936	.78	.6691	.7505	.8915	1.0471	.8068
.28	.9470	.9619	.9846	2.1656	.3043	.79	.6625	.7452	.8890	1.0425	.8160
.29	.9433	.9592	.9835	2.0979	.3150	.80	.6560	.7400	.8865	1.0382	.8251
.30	.9395	.9564	.9823	2.0351	.3257	.81	.6495	.7347	.8840	1.0342	.8343
.31	.9355	.9535	.9811	1.9765	.3364	.82	.6430	.7295	.8815	1.0305	.8433
.32	.9315	.9506	.9799	1.9218	.3470	.83	.6365	.7242	.8789	1.0270	.8524
.33	.9274	.9476	.9787	1.8707	.3576	.84	.6300	.7189	.8763	1.0237	.8614
.34	.9231	.9445	.9774	1.8229	.3682	.85	.6235	.7136	.8737	1.0207	.8704
.35	.9188	.9413	.9761	1.7780	.3788	.86	.6170	.7083	.8711	1.0179	.8793
.36	.9143	.9380	.9747	1.7358	.3893	.87	.6106	.7030	.8685	1.0153	.8882
.37	.9098	.9347	.9734	1.6961	.3999	.88	.6041	.6977	.8659	1.0129	.8970
.38	.9052	.9313	.9719	1.6587	.4104	.89	.5977	.6924	.8632	1.0108	.9058
.39	.9004	.9278	.9705	1.6234	.4209	.90	.5913	.6870	.8606	1.0089	.9146
.40	.8956	.9243	.9690	1.5901	.4313	.91	.5849	.6817	.8579	1.0071	.9233
.41	.8907	.9207	.9675	1.5587	.4418	.92	.5785	.6764	.8552	1.0056	.9320
.42	.8857	.9170	.9659	1.5289	.4522	.93	.5721	.6711	.8525	1.0043	.9406
.43	.8807	.9132	.9643	1.5007	.4626	.94	.5658	.6658	.8498	1.0031	.9493
.44	.8755	.9094	.9627	1.4740	.4729	.95	.5595	.6604	.8471	1.0021	.9578
.45	.8703	.9055	.9611	1.4487	.4833	.96	.5532	.6551	.8444	1.0014	.9663
.46	.8650	.9016	.9594	1.4246	.4936	.97	.5469	.6498	.8416	1.0008	.9748
.47	.8596	.8976	.9577	1.4018	.5038	.98	.5407	.6445	.8389	1.0003	.9832
.48	.8541	.8935	.9560	1.3801	.5141	.99	.5345	.6392	.8361	1.0001	.9916
.49	.8486	.8894	.9542	1.3595	.5243	1.00	.5283	.6339	.8333	1.0000	1.0000
.50	.8430	.8852	.9524	1.3398	.5345						

COMPRESSIBLE FLUID DYNAMICS

Appendix B
Isentropic Flow and Normal Shock Parameters $k = 1.4$

(X refers to upstream conditions. Y refers to downstream conditions.)
(For example, PX/TPY is the ratio of static pressure before the shock wave to total pressure after the shock wave.)

M	P/TP	D/TD	T/TT	A/A*	V/C*	MY	PY/PX	DY/DX	TY/TX	TPY/TPX	PX/TPY
1.00	.5283	.6339	.8333	1.0000	1.0000	1.0000	.1000 EE+1	.1000 EE+1	.1000 EE+1	1.0000	.5283
1.10	.4684	.5817	.8052	1.0079	1.0812	.9118	.1245 EE+1	.1169 EE+1	.1065 EE+1	.9989	.4689
1.20	.4124	.5311	.7764	1.0304	1.1583	.8422	.1513 EE+1	.1342 EE+1	.1128 EE+1	.9928	.4154
1.30	.3609	.4829	.7474	1.0663	1.2311	.7860	.1805 EE+1	.1516 EE+1	.1191 EE+1	.9794	.3685
1.40	.3142	.4374	.7184	1.1149	1.2999	.7397	.2120 EE+1	.1690 EE+1	.1255 EE+1	.9582	.3280
1.50	.2724	.3950	.6897	1.1762	1.3646	.7011	.2458 EE+1	.1862 EE+1	.1320 EE+1	.9298	.2930
1.60	.2353	.3557	.6614	1.2502	1.4254	.6684	.2820 EE+1	.2032 EE+1	.1388 EE+1	.8952	.2628
1.70	.2026	.3197	.6337	1.3376	1.4825	.6405	.3205 EE+1	.2198 EE+1	.1458 EE+1	.8557	.2368
1.80	.1740	.2868	.6068	1.4390	1.5360	.6165	.3613 EE+1	.2359 EE+1	.1532 EE+1	.8127	.2142
1.90	.1492	.2570	.5807	1.5553	1.5861	.5956	.4045 EE+1	.2516 EE+1	.1608 EE+1	.7674	.1945
2.00	.1278	.2300	.5556	1.6875	1.6330	.5774	.4500 EE+1	.2667 EE+1	.1687 EE+1	.7209	.1773
2.10	.1094	.2058	.5313	1.8369	1.6769	.5613	.4978 EE+1	.2812 EE+1	.1770 EE+1	.6742	.1622
2.20	.9352 EE−1	.1841	.5081	2.0050	1.7179	.5471	.5480 EE+1	.2951 EE+1	.1857 EE+1	.6281	.1489
2.30	.7997 EE−1	.1646	.4859	2.1931	1.7563	.5344	.6005 EE+1	.3085 EE+1	.1947 EE+1	.5833	.1371
2.40	.6840 EE−1	.1472	.4647	2.4031	1.7922	.5231	.6553 EE+1	.3212 EE+1	.2040 EE+1	.5401	.1266
2.50	.5853 EE−1	.1317	.4444	2.6367	1.8257	.5130	.7125 EE+1	.3333 EE+1	.2137 EE+1	.4990	.1173
2.60	.5012 EE−1	.1179	.4252	2.8960	1.8571	.5039	.7720 EE+1	.3449 EE+1	.2238 EE+1	.4601	.1089
2.70	.4295 EE−1	.1056	.4068	3.1830	1.8865	.4956	.8338 EE+1	.3559 EE+1	.2343 EE+1	.4236	.1014
2.80	.3685 EE−1	.9463 EE−1	.3894	3.5001	1.9140	.4882	.8980 EE+1	.3664 EE+1	.2451 EE+1	.3895	.9461 EE−1
2.90	.3165 EE−1	.8489 EE−1	.3729	3.8498	1.9398	.4814	.9645 EE+1	.3763 EE+1	.2563 EE+1	.3577	.8848 EE−1
3.00	.2722 EE−1	.7623 EE−1	.3571	4.2346	1.9640	.4752	.1033 EE+2	.3857 EE+1	.2679 EE+1	.3283	.8291 EE−1
3.10	.2345 EE−1	.6852 EE−1	.3422	4.6573	1.9866	.4695	.1104 EE+2	.3947 EE+1	.2799 EE+1	.3012	.7785 EE−1
3.20	.2023 EE−1	.6165 EE−1	.3281	5.1210	2.0079	.4643	.1178 EE+2	.4031 EE+1	.2922 EE+1	.2762	.7323 EE−1
3.30	.1748 EE−1	.5554 EE−1	.3147	5.6286	2.0278	.4596	.1254 EE+2	.4112 EE+1	.3049 EE+1	.2533	.6900 EE−1
3.40	.1512 EE−1	.5009 EE−1	.3019	6.1837	2.0466	.4552	.1332 EE+2	.4188 EE+1	.3180 EE+1	.2322	.6513 EE−1
3.50	.1311 EE−1	.4523 EE−1	.2899	6.7896	2.0642	.4512	.1412 EE+2	.4261 EE+1	.3315 EE+1	.2129	.6157 EE−1
3.60	.1138 EE−1	.4089 EE−1	.2784	7.4501	2.0808	.4474	.1495 EE+2	.4330 EE+1	.3454 EE+1	.1953	.5829 EE−1
3.70	.9903 EE−2	.3702 EE−1	.2675	8.1691	2.0964	.4439	.1580 EE+2	.4395 EE+1	.3596 EE+1	.1792	.5526 EE−1
3.80	.8629 EE−2	.3355 EE−1	.2572	8.9506	2.1111	.4407	.1668 EE+2	.4457 EE+1	.3743 EE+1	.1645	.5247 EE−1
3.90	.7532 EE−2	.3044 EE−1	.2474	9.7990	2.1250	.4377	.1758 EE+2	.4516 EE+1	.3893 EE+1	.1510	.4987 EE−1
4.00	.6586 EE−2	.2766 EE−1	.2381	10.7187	2.1381	.4350	.1850 EE+2	.4571 EE+1	.4047 EE+1	.1388	.4747 EE−1
4.10	.5769 EE−2	.2516 EE−1	.2293	11.7147	2.1505	.4324	.1944 EE+2	.4624 EE+1	.4205 EE+1	.1276	.4523 EE−1
4.20	.5062 EE−2	.2292 EE−1	.2208	12.7916	2.1622	.4299	.2041 EE+2	.4675 EE+1	.4367 EE+1	.1173	.4314 EE−1
4.30	.4449 EE−2	.2090 EE−1	.2129	13.9564	2.1732	.4277	.2140 EE+2	.4723 EE+1	.4532 EE+1	.1080	.4120 EE−1
4.40	.3918 EE−2	.1909 EE−1	.2053	15.2099	2.1837	.4255	.2242 EE+2	.4768 EE+1	.4702 EE+1	.9948 EE−1	.3938 EE−1
4.50	.3455 EE−2	.1745 EE−1	.1980	16.5622	2.1936	.4236	.2346 EE+2	.4812 EE+1	.4875 EE+1	.9170 EE−1	.3768 EE−1
4.60	.3053 EE−2	.1597 EE−1	.1911	18.0178	2.2030	.4217	.2452 EE+2	.4853 EE+1	.5052 EE+1	.8459 EE−1	.3609 EE−1
4.70	.2701 EE−2	.1464 EE−1	.1846	19.5828	2.2119	.4199	.2560 EE+2	.4893 EE+1	.5233 EE+1	.7809 EE−1	.3459 EE−1
4.80	.2394 EE−2	.1343 EE−1	.1783	21.2637	2.2204	.4183	.2671 EE+2	.4930 EE+1	.5418 EE+1	.7214 EE−1	.3319 EE−1
4.90	.2126 EE−2	.1233 EE−1	.1724	23.0671	2.2284	.4167	.2784 EE+2	.4966 EE+1	.5607 EE+1	.6670 EE−1	.3187 EE−1
5.00	.1890 EE−2	.1134 EE−1	.1667	25.0000	2.2361	.4152	.2900 EE+2	.5000 EE+1	.5800 EE+1	.6172 EE−1	.3062 EE−1
5.10	.1683 EE−2	.1044 EE−1	.1612	27.0696	2.2433	.4138	.3018 EE+2	.5033 EE+1	.5997 EE+1	.5715 EE−1	.2945 EE−1
5.20	.1501 EE−2	.9620 EE−2	.1561	29.2833	2.2503	.4125	.3138 EE+2	.5064 EE+1	.6197 EE+1	.5297 EE−1	.2834 EE−1
5.30	.1341 EE−2	.8875 EE−2	.1511	31.6491	2.2569	.4113	.3260 EE+2	.5093 EE+1	.6401 EE+1	.4913 EE−1	.2730 EE−1
5.40	.1200 EE−2	.8197 EE−2	.1464	34.1748	2.2631	.4101	.3385 EE+2	.5122 EE+1	.6610 EE+1	.4560 EE−1	.2631 EE−1
5.50	.1075 EE−2	.7578 EE−2	.1418	36.8690	2.2691	.4090	.3512 EE+2	.5149 EE+1	.6822 EE+1	.4236 EE−1	.2537 EE−1
5.60	.9643 EE−3	.7012 EE−2	.1375	39.7402	2.2748	.4079	.3642 EE+2	.5175 EE+1	.7038 EE+1	.3938 EE−1	.2448 EE−1
5.70	.8663 EE−3	.6496 EE−2	.1334	42.7974	2.2803	.4069	.3774 EE+2	.5200 EE+1	.7258 EE+1	.3664 EE−1	.2364 EE−1
5.80	.7794 EE−3	.6023 EE−2	.1294	46.0500	2.2855	.4059	.3908 EE+2	.5224 EE+1	.7481 EE+1	.3412 EE−1	.2284 EE−1
5.90	.7021 EE−3	.5590 EE−2	.1256	49.5075	2.2905	.4050	.4044 EE+2	.5246 EE+1	.7709 EE+1	.3179 EE−1	.2208 EE−1
6.00	.6334 EE−3	.5194 EE−2	.1220	53.1798	2.2953	.4042	.4183 EE+2	.5268 EE+1	.7941 EE+1	.2965 EE−1	.2136 EE−1
6.10	.5721 EE−3	.4829 EE−2	.1185	57.0772	2.2998	.4033	.4324 EE+2	.5289 EE+1	.8176 EE+1	.2767 EE−1	.2067 EE−1
6.20	.5173 EE−3	.4495 EE−2	.1151	61.2102	2.3042	.4025	.4468 EE+2	.5309 EE+1	.8415 EE+1	.2584 EE−1	.2002 EE−1
6.30	.4684 EE−3	.4187 EE−2	.1119	65.5899	2.3084	.4018	.4614 EE+2	.5329 EE+1	.8658 EE+1	.2416 EE−1	.1939 EE−1
6.40	.4247 EE−3	.3904 EE−2	.1088	70.2274	2.3124	.4011	.4762 EE+2	.5347 EE+1	.8905 EE+1	.2259 EE−1	.1880 EE−1
6.50	.3855 EE−3	.3643 EE−2	.1058	75.1343	2.3163	.4004	.4912 EE+2	.5365 EE+1	.9156 EE+1	.2115 EE−1	.1823 EE−1
6.60	.3503 EE−3	.3402 EE−2	.1030	80.3227	2.3200	.3997	.5065 EE+2	.5382 EE+1	.9411 EE+1	.1981 EE−1	.1768 EE−1
6.70	.3187 EE−3	.3180 EE−2	.1002	85.8049	2.3235	.3991	.5220 EE+2	.5399 EE+1	.9670 EE+1	.1857 EE−1	.1716 EE−1
6.80	.2902 EE−3	.2974 EE−2	.9758 EE−1	91.5935	2.3269	.3985	.5378 EE+2	.5415 EE+1	.9933 EE+1	.1741 EE−1	.1667 EE−1
6.90	.2646 EE−3	.2785 EE−2	.9504 EE−1	97.7017	2.3302	.3979	.5538 EE+2	.5430 EE+1	.1020 EE+2	.1634 EE−1	.1619 EE−1
7.00	.2416 EE−3	.2609 EE−2	.9259 EE−1	104.1429	2.3333	.3974	.5700 EE+2	.5444 EE+1	.1047 EE+2	.1535 EE−1	.1573 EE−1
7.10	.2207 EE−3	.2446 EE−2	.9024 EE−1	110.9309	2.3364	.3968	.5864 EE+2	.5459 EE+1	.1074 EE+2	.1443 EE−1	.1530 EE−1
7.20	.2019 EE−3	.2295 EE−2	.8797 EE−1	118.0799	2.3393	.3963	.6031 EE+2	.5472 EE+1	.1102 EE+2	.1357 EE−1	.1488 EE−1
7.30	.1848 EE−3	.2155 EE−2	.8578 EE−1	125.6046	2.3421	.3958	.6200 EE+2	.5485 EE+1	.1130 EE+2	.1277 EE−1	.1448 EE−1
7.40	.1694 EE−3	.2025 EE−2	.8367 EE−1	133.5200	2.3448	.3954	.6372 EE+2	.5498 EE+1	.1159 EE+2	.1202 EE−1	.1409 EE−1
7.50	.1554 EE−3	.1904 EE−2	.8163 EE−1	141.8415	2.3474	.3949	.6546 EE+2	.5510 EE+1	.1188 EE+2	.1133 EE−1	.1372 EE−1
7.60	.1427 EE−3	.1792 EE−2	.7967 EE−1	150.5849	2.3499	.3945	.6722 EE+2	.5522 EE+1	.1217 EE+2	.1068 EE−1	.1336 EE−1
7.70	.1312 EE−3	.1687 EE−2	.7777 EE−1	159.7665	2.3523	.3941	.6900 EE+2	.5533 EE+1	.1247 EE+2	.1008 EE−1	.1302 EE−1
7.80	.1207 EE−3	.1589 EE−2	.7594 EE−1	169.4030	2.3546	.3937	.7081 EE+2	.5544 EE+1	.1277 EE+2	.9510 EE−2	.1269 EE−1
7.90	.1111 EE−3	.1498 EE−2	.7417 EE−1	179.5114	2.3569	.3933	.7264 EE+2	.5555 EE+1	.1308 EE+2	.8982 EE−2	.1237 EE−1
8.00	.1024 EE−3	.1414 EE−2	.7246 EE−1	190.1094	2.3591	.3929	.7450 EE+2	.5565 EE+1	.1339 EE+2	.8488 EE−2	.1207 EE−1
8.10	.9449 EE−4	.1334 EE−2	.7081 EE−1	201.2148	2.3612	.3925	.7638 EE+2	.5575 EE+1	.1370 EE+2	.8025 EE−2	.1177 EE−1
8.20	.8723 EE−4	.1260 EE−2	.6921 EE−1	212.8461	2.3632	.3922	.7828 EE+2	.5585 EE+1	.1402 EE+2	.7592 EE−2	.1149 EE−1
8.30	.8060 EE−4	.1191 EE−2	.6767 EE−1	225.0221	2.3652	.3918	.8020 EE+2	.5594 EE+1	.1434 EE+2	.7187 EE−2	.1122 EE−1
8.40	.7454 EE−4	.1126 EE−2	.6617 EE−1	237.7622	2.3671	.3915	.8215 EE+2	.5603 EE+1	.1466 EE+2	.6806 EE−2	.1095 EE−1
8.50	.6898 EE−4	.1066 EE−2	.6472 EE−1	251.0862	2.3689	.3912	.8412 EE+2	.5612 EE+1	.1499 EE+2	.6449 EE−2	.1070 EE−1

Appendix C
Standard Atmosphere

Altitude ft	Temperature °R	Pressure psia	Altitude ft	Temperature °R	Pressure psia
0	518.7	14.696	35000	393.9	3.458
1000	515.1	14.175	36000	392.7	3.296
2000	511.6	13.664	37000	392.7	3.143
3000	508.0	13.168	38000	392.7	2.996
4000	504.4	·12.692	39000	392.7	2.854
5000	500.9	12.225	40000	392.7	2.721
6000	497.3	11.778	41000	392.7	2.593
7000	493.7	11.341	42000	392.7	2.475
8000	490.2	10.914	43000	392.7	2.358
9000	486.6	10.501	44000	392.7	2.250
10000	483.0	10.108	45000	392.7	2.141
11000	479.5	9.720	46000	392.7	2.043
12000	475.9	9.347	47000	392.7	1.950
13000	472.3	8.983	48000	392.7	1.857
14000	468.8	8.630	49000	392.7	1.768
15000	465.2	8.291	50000	392.7	1.690
16000	461.6	7.962	51000	392.7	1.611
17000	458.1	7.642	52000	392.7	1.532
18000	454.5	7.338	53000	392.7	1.464
19000	450.9	7.038	54000	392.7	1.395
20000	447.4	6.753	55000	392.7	1.331
21000	443.8	6.473	56000	392.7	1.267
22000	440.2	6.203	57000	392.7	1.208
23000	436.7	5.943	58000	392.7	1.154
24000	433.1	5.693	59000	392.7	1.100
25000	429.5	5.452	60000	392.7	1.046
26000	426.0	5.216	61000	392.7	.997
27000	422.4	4.990	62000	392.7	.953
28000	418.8	4.774	63000	392.7	.909
29000	415.3	4.563	64000	392.7	.864
30000	411.7	4.362	65000	392.7	.825
31000	408.1	4.165			
32000	404.6	3.978			
33000	401.0	3.797			
34000	397.5	3.625			

RESERVED FOR FUTURE USE

PRACTICE PROBLEMS: COMPRESSIBLE FLUID DYNAMICS

Warm-ups

1. 150°F air at 10 psia flows at 750 fps through a converging section into a chamber with a back pressure of 5.5 psia. What is the pressure, temperature, and Mach number at the throat?

2. A blunt bullet passes through STP air at 2000 fps. What are the stagnation values of temperature, pressure, and enthalpy at the bullet face?

3. 4.5 lbm/sec of 160 psia, 240°F air expands through a converging/diverging nozzle to 20 psia. What are the throat and exit areas? What are the throat and exit Mach numbers?

4. What is the maximum semi-vertex shock cone angle produced by a conical projectile with a semi-vertex angle of $\delta = 30°$ traveling at 2700 fps in 60°F atmospheric air?

5. A large tank containing 100 psia, 80°F air exhausts into a converging/diverging nozzle with a 1 square inch throat area. What are the area, mass flow rate, and temperature at a point where the Mach number is 2?

6. What is the air velocity behind a shock wave if the Mach number is 2 and the temperature is 500°R in front of it?

7. Air with a total pressure of 100 psia and a total temperature of 70°F flows through a converging/diverging nozzle. Find the temperature, pressure, and Mach number in the supersonic portion if $A/A^* = 1.555$.

8. What is the smallest area to which a 1 square foot tube could be reduced if 20 lbm/sec of 10 psia (static value) and 40°F (total value) air flows?

9. A supersonic pitot tube in a 1.38 psia (static) wind tunnel measures a 20 psia total pressure behind a shock wave formed at its entrance. What is the Mach number in the tunnel?

10. A boiler produces 100 psia, 800°F steam. A nozzle expands the steam to 60 psia. What is its velocity?

Concentrates

1. Air flows through a converging section. The conditions at some point are 50 psia, 1000°R, 600 fps, and area of .1 ft^2. Find the stagnation temperature, Mach number, critical exit area, critical pressure, and critical temperature.

2. An 85% efficient transonic nozzle expands 3 lbm/sec of 300 fps steam from 200 psia and 600°F to 80 psia. What is the throat area?

Timed (1 hour allowed for each)

1. 35,200 pounds per hour of steam flow through 95 feet of 3″ I.D. insulated pipe with a Fanning friction factor of .012. Although the average pressure is 200 psia, the exit pressure is 115 psia (at 540°F). Will the flow be maintained if 30 feet are added to the length of the pipe?

2. 3600 pounds per hour of air (160 psia, 660°R, $k = 1.4$) flow through a converging/diverging adiabatic nozzle. The nozzle has a 90% overall efficiency and a 95% throat efficiency. The diverging angle is 6° (total angle). The exit pressure is 14.7 psia. The length of the converging section is 5% of the diverging section. (a) Draw and dimension this nozzle. (b) What is the throat area?

RESERVED FOR FUTURE USE

COMBUSTION

Nomenclature

A	% ash by weight	%
c_p	specific heat at constant pressure	BTU/lbm-°F
FC	% fixed carbon by weight	%
HV	heating value	BTU/unit of fuel
HHV	higher heating value	BTU/unit of fuel
LHV	lower heating value	BTU/unit of fuel
m	mass	lbm
M	% bed moisture of coal by weight	%
q	heat	BTU/lbm
S	% sulfur by weight	%
T	temperature	°F
VM	volatile matter by weight	%
W	weight of gas per pound of fuel	lbm/lbm

Symbols

ν	SSU viscosity	seconds
η	efficiency	%
[ ]	gravimetric fraction	decimal
( )	volumetric fraction	decimal

1 DEFINITIONS

Anthracite: A clean, dense, hard coal, comparatively difficult to ignite, but which burns uniformly and smokelessly with a short flame.

Ash: Noncombustible, solid mineral matter.

Bituminous coal: A coal that varies in composition but which generally has a higher volatile content than anthracite, starts easily, and burns freely with a long flame. Smoke and soot are possible if this coal is improperly fired.

Coke: The carbonaceous residue containing ash and sulfur created by burning coal with little or no air. The lighter coal constituents are vaporized, while heavier hydrocarbons crack and form carbon.

Fixed carbon: The combustible residue (not all of which is carbon) remaining in coal when all volatile matter has been driven off.

Flash point: The temperature at which an oil spontaneously ignites in the presence of sparks or flame.

Flue gas: The hot gases resulting from combustion, consisting primarily of nitrogen, carbon dioxide, water, and small quantities of carbon monoxide and sulfur dioxide. If water vapor is removed, it is known as *dry stack gas*.

Fuel: A compound containing hydrogen and carbon in elemental form or compounds.

Gravimetric analysis: An analysis by weight.

Heating value: The heat generated during complete combustion of a fuel.

Ignition temperature: The temperature at which more heat is generated by the combustion reaction than is lost to the surroundings, after which combustion becomes self-sustaining.

Lignite: A coal of woody structure, very high in moisture with low heating value. It normally ignites slowly due to moisture content, breaks apart when burning, and burns with little smoke or soot.

Moisture: Water content of fuels. *Bed moisture* of coal indicates moisture level when removed from the earth. *As received moisture* indicates moisture level of a coal before being dried or burned.

Primary air: Air that is mixed with a fuel to initiate and sustain the combustion reaction.

Proximate analysis: A gravimetric analysis of a fuel broken down into moisture, volatile matter, fixed carbon, ash, and sulfur. Sulfur may be included with ash.

Secondary air: Air which combines with the flue gas.

SSU viscosity: Kinematic viscosity as determined on a Saybolt viscometer.

Stack gas: See "flue gas."

Stoichiometric combustion: (Also known as ideal, perfect, theoretical combustion.) A reaction in which all fuel is burned with the theoretically correct amount of air. Neither unburned fuel nor free oxygen is present in the stack gases. CO_2 is at a maximum value in the combustion products.

Sulfur: A common constituent of lower grade fuels. *Organic sulfur* is combined with hydrogen and carbon in compounds. *Pyritic sulfur* is elemental.

Ultimate analysis: A gravimetric analysis of a fuel broken down into the individual elements. Moisture is broken up into hydrogen and oxygen.

Volatile matter: That portion of a fuel which is driven off as a gas or a vapor when the fuel is heated according to standard tests. It consists mainly of organic gases such as methane.

2 FORMATION OF COMPOUNDS

The sum of the valences must be zero if a neutral compound is to form. For example, H_2O is a valid compound since the two hydrogen atoms have a total positive valence of $2 \times 1 = +2$. The oxygen ion has a valence of -2. These valences sum to zero.

On the other hand, $NaCO_3$ is not a valid compound formula. The sodium (Na) has a valence of $+1$. However, the carbonate radical[1] has a valence of -2. The correct sodium carbonate molecule is Na_2CO_3.

In order to evaluate whether or not a compound formula is valid, it is necessary to know the valences of the interacting atoms. Although some atoms have more than one possible valence, most do not. The valences of some common ions and radicals are given in table 9.1.

---

[1]A *radical* is a charged group of atoms that combines as a single ion.

Table 9.1
Valences of Ions and Radicals

Name	Symbol	Valence
acetate	$C_2H_3O_2$	-1
aluminum	Al	$+3$
ammonium	NH_4	$+1$
barium	Ba	$+2$
boron	B	$+3$
borate	BO_3	-3
bromine	Br	-1
calcium	Ca	$+2$
carbon	C	$+4, -4$
carbonate	CO_3	-2
chlorate	ClO_3	-1
chlorine	Cl	-1
chlorite	ClO_2	-1
chromate	CrO_4	-2
chromium	Cr	$+2, +3, +6$
copper	Cu	$+1, +2$
dichromate	Cr_2O_7	-2
fluorine	F	-1
gold	Au	$+1, +3$
hydrogen	H	$+1$
hydroxide	OH	-1
hypochlorite	ClO	-1
iron	Fe	$+2, +3$
lead	Pb	$+2, +4$
lithium	Li	$+1$
magnesium	Mg	$+2$
mercury	Hg	$+1, +2$
nickel	Ni	$+2, +3$
nitrate	NO_3	-1
nitrite	NO_2	-1
nitrogen	N	$-3, +1, +2, +3, +4, +5$
oxygen	O	-2
permanganate	MnO_4	-1
phosphate	PO_4	-3
phosphorus	P	$-3, +3, +5$
potassium	K	$+1$
silicon	Si	$+4, -4$
silver	Ag	$+1$
sodium	Na	$+1$
sulfate	SO_4	-2
sulfite	SO_3	-2
sulfur	S	$-2, +4, +6$
tin	Sn	$+2, +4$
zinc	Zn	$+2$

3 CHEMICAL REACTIONS

During chemical reactions, chemical bonds between atoms are broken, and new bonds are formed. *Reactants* either are converted to simpler products or are synthesized into more complex products. There are five common types of chemical reactions.

- *direct combination or synthesis*: This is the simplest type of reaction where two elements or compounds combine directly to form a compound.

$$2H_2 + O_2 \mapsto 2H_2O$$
$$SO_2 + H_2O \mapsto H_2SO_3$$

- *decomposition*: Bonds uniting a compound are disrupted by heat or another energy source to yield simpler compounds or elements.

$$2HgO \mapsto 2Hg + O_2$$
$$H_2CO_3 \mapsto H_2O + CO_2$$

- *single displacement*: This type of reaction is identified by one element and one compound as the reactants.

$$2Na + 2H_2O \mapsto 2NaOH + H_2$$
$$2KI + Cl_2 \mapsto 2KCl + I_2$$

- *double decomposition*: These are reactions characterized by having two compounds as reactants and forming two new compounds.

$$AgNO_3 + NaCl \mapsto AgCl + NaNO_3$$
$$H_2SO_4 + ZnS \mapsto H_2S + ZnSO_4$$

- *oxidation-reduction (Redox)*: These reactions involve oxidation of one substance and reduction of another. In the example, calcium loses electrons and is oxidized; oxygen gains electrons and is reduced.

$$2Ca + O_2 \mapsto 2CaO$$

The coefficients of the chemical symbols represent the number of molecules taking part in the reaction. Since matter cannot be destroyed in a chemical reaction, the number of atoms of each element must be equal on both sides of the arrow. This is the principle used in balancing chemical equations.

The coefficients also can represent the number of *moles*[2] taking part in a reaction. *Avogadro's law* states that the number of atoms in a mole of any substance is the same (6.023 EE23 molecules/gmole). Therefore, it is necessary only to multiply all of the coefficients in a chemical equation by 6.023 EE23 to show that the coefficients represent moles.

---

[2]A *mole* of a substance has a mass equal to its atomic or molecular weight. Mass can be measured in grams (the *gmole*), kilograms (the *kgmole*), or in pounds (the *pmole*). Avogadro's number is valid only when the mass of a mole is measured in grams.

If the reactants and the products are gases, the coefficients also represent the number of volumes taking part in the reaction. This also is the result of Avogadro's law, which states that equal numbers of gas molecules occupy equal volumes. (Such an interpretation is valid for water vapor only if the temperature is high.)

Example 9.1

Balance the reaction equation between aluminum and sulfuric acid.

$$Al + H_2SO_4 \mapsto Al_2(SO_4)_3 + H_2$$

step 1:   Since there are two aluminums on the right, multiply Al by 2.

$$2Al + H_2SO_4 \mapsto Al_2(SO_4)_3 + H_2$$

step 2:   Since there are three sulfate radicals (SO_4) on the right, multiply H_2SO_4 by 3.

$$2Al + 3H_2SO_4 \mapsto Al_2(SO_4)_3 + H_2$$

step 3:   Now there are six hydrogens on the left, so multiply H_2 by 3 to balance the equation.

$$2Al + 3H_2SO_4 \mapsto Al_2(SO_4)_3 + 3H_2$$

4 STOICHIOMETRY

Stoichiometry is the study of the proportions in which elements combine into compounds.

Stoichiometric problems are known as *weight and proportion* problems because their solutions use simple ratios to determine the weight of reactants required to produce given amounts of products. The procedure for solving these problems is essentially the same, regardless of the reaction.

step 1:   Write and balance the chemical equation.

step 2:   Calculate the molecular weight of each compound or element in the equation.

step 3:   Multiply the molecular weights by their respective coefficients and write the products under the formulas.

step 4:   Write the given weight data under the molecular weights calculated in step 3.

step 5:   Fill in missing information by using simple ratios.

Example 9.2

Caustic soda (NaOH) is made from sodium carbonate (Na_2CO_3) and slaked lime ($Ca(OH)_2$). How many pounds of caustic soda can be made from 2000 pounds of sodium carbonate?

The balanced chemical equation is

$$Na_2CO_3 + Ca(OH)_2 \mapsto 2NaOH + CaCO_3$$

MW's:	106	74	2×40	100
given:	2000		X	

The ratio used is

$$\frac{NaOH}{Na_2CO_3} = \frac{80}{106} = \frac{X}{2000}$$
$$X = 1509 \text{ pounds}$$

5 EMPIRICAL FORMULA DEVELOPMENT

A relationship between the atomic weights, combining masses, and the chemical formula can be developed for *binary compounds*—compounds constructed from two elements.

For example, suppose elements A and B combine to form compound A_mB_n according to the reaction

$$mA + nB \mapsto A_mB_n \qquad 9.1$$

Then, if w_A and w_B are the masses of A and B which combine together,

$$\frac{w_A}{w_B} = \left(\frac{m}{n}\right)\left(\frac{A.W._A}{A.W._B}\right) \qquad 9.2$$

Solving for m and n gives the formula. Solving for w_A and w_B gives the combining masses. If, in solving for m and n, one of the values is assumed to be one, the other often will come out to be some simple fraction. In such a case, multiply both the assumed value and the derived value by the smallest integer which clears the fraction. For example, $A_1B_{1.5}$ would not be allowed since only a whole atom can combine. The cleared formula would be A_2B_3.

Example 9.3

How many grams of carbon are required to form 18.7 grams of methane?

The chemical reaction is

$$C + 2H_2 \mapsto CH_4$$

In this example, $m = 1$, $n = 4$, $A.W._A = 12$, $A.W._B = 1$.

w_A is unknown, and $w_B = 18.7 - w_A$.

$$\frac{w_A}{18.7 - w_A} = \left(\frac{1}{4}\right)\left(\frac{12}{1}\right)$$
$$w_A = 14.0 \text{ grams}$$

A *tertiary compound* (three elements) formula can be analyzed in the following manner.

step 1: Convert all combining weights into percentages by dividing each by the sum of the combining weights.

step 2: Divide the percentages by the atomic weight of each of the respective elements. Call the result Y_i.

step 3: Divide all Y_i by the smallest Y_i. Call the result X_i. One of the X_i (corresponding to the element with the smallest Y_i) should be 1.

step 4: If all of the X_i are integers, the formula is $A_{X1}B_{X2}C_{X3}$. Otherwise, clear the fraction as before.

Example 9.4

An alcohol is analyzed and the following gravimetric analysis is recorded: carbon 37.5%, hydrogen 12.5%, oxygen 50%. What is the alcohol?

step 1: The percentages are given.

step 2: Divide the gravimetric analysis by the atomic weight.

$$C : \frac{37.5}{12} = 3.125$$
$$H : \frac{12.5}{1} = 12.5$$
$$O : \frac{50}{16} = 3.125$$

step 3: The smallest Y is 3.125. Dividing each by 3.125 gives

$$C : \frac{3.125}{3.125} = 1$$
$$H : \frac{12.5}{3.125} = 4$$
$$O : \frac{3.125}{3.125} = 1$$

step 4: The basic formula is CH_4O. The alcohol is CH_3OH (methyl alcohol).

In example 9.4, there is insufficient information in the chemical analysis to determine whether the formula is CH_4O or CH_3OH. (It might be that the compound analyzed behaves as if it possesses a hydroxyl (OH) radical.) Similarly, chemical properties would have to be used to distinguish between ethyl alcohol (C_2H_5OH) and dimethyl ether ($(CH_3)_2O$). Different arrangements of the same atoms are known as *isomers*.

Table 9.2
Common Combustion Reactions
(multiply oxygen by 3.78 to get nitrogen in air)

substance	molecular symbol	reaction equation (excluding nitrogen)
carbon (to CO)	C	$2C + O_2 \mapsto 2CO$
carbon (to CO_2)	C	$2C + 2O_2 \mapsto 2CO_2$
sulfur (to SO_2)	S	$S + O_2 \mapsto SO_2$
sulfur (to SO_3)	S	$2S + 3O_2 \mapsto 2SO_3$
carbon monoxide	CO	$2CO + O_2 \mapsto 2CO_2$
methane	CH_4	$CH_4 + 2O_2 \mapsto CO_2 + 2H_2O$
acetylene	C_2H_2	$2C_2H_2 + 5O_2 \mapsto 4CO_2 + 2H_2O$
ethylene	C_2H_4	$C_2H_4 + 3O_2 \mapsto 2CO_2 + 2H_2O$
ethane	C_2H_6	$2C_2H_6 + 7O_2 \mapsto 4CO_2 + 6H_2O$
hydrogen	H_2	$2H_2 + O_2 \mapsto 2H_2O$
hydrogen sulfide	H_2S	$2H_2S + 3O_2 \mapsto 2H_2O + 2SO_2$
propane	C_3H_8	$C_3H_8 + 5O_2 \mapsto 3CO_2 + 4H_2O$
n-butane	C_4H_{10}	$2C_4H_{10} + 13O_2 \mapsto 8CO_2 + 10H_2O$
octane	C_8H_{18}	$2C_8H_{18} + 25O_2 \mapsto 16CO_2 + 18H_2O$
olefin series	C_nH_{2n}	$2C_nH_{2n} + 3nO_2 \mapsto 2nCO_2 + 2nH_2O$
paraffin series	C_nH_{2n+2}	$2C_nH_{2n+2} + (3n + 1)O_2 \mapsto 2nCO_2 + (2n + 2)H_2O$

6 COMBUSTION REACTIONS

Combustion reactions involving oxygen, combustible elements, and compounds in fuels occur according to fixed chemical principles.

- Specific compounds form in fixed combinations when two or more reactants combine.

- The mass of each element in the reactants must equal the mass of that element in the products (conservation of matter).

- Chemical compounds form from elements combining in fixed weight relationships (*law of combining weights*).

Table 9.2 lists combustible elements and compounds formed in typical combustion reactions.

7 TYPES OF FUELS AND BURNING SYSTEMS

A. INTRODUCTION

Hydrocarbon fuels generally are classified by phase—solid, liquid, or gas. Each type of fuel requires combustion equipment designed to mix reactants properly and to transfer the heat of combustion efficiently. Gaseous fuels can be burned in premix or diffusion burners. Liquid fuel burners atomize or vaporize the fuel prior to mixing with air. Solid fuel combustion equipment heats the fuel (in order to vaporize sufficient volatiles for ignition and to sustain combustion), holds the fuel

long enough for complete combustion, and contains and removes the combustion ash.

Selection of fuels and burning systems is based on availability and dependability of the fuel supply, convenience of use and storage, economy, and cleanliness. These factors must be combined with combustion equipment selection considerations such as cost, operating and service requirements, and ease of control.

B. LUMP-COAL FURNACES

Thorough mixing of fuel and air in lump-coal furnaces is the responsibility of the *stoker*. Stoking is the act of adding fuel and distributing it (initially and regularly repeated) in such a manner that air can reach the burning coal pieces. In low-volume home furnaces and early fire-tube boilers, stoking was accomplished by hand.

Stoking in lump-coal installations is the responsibility of automatic furnace equipment. Such equipment can be divided into two types, depending on how the air reaches the coal. In *overfeed stokers*, coal is placed above the air flow. In *underfeed stokers*, coal is placed under the air flow.

Figure 9.1 illustrates several types of overfeed stokers. Figure 9.1 (a) is a *sprinkler stoker*, which uses an impeller to spread coal pieces over a horizontal grate. Figure 9.1 (b) is an *inclined grate stoker* with a mechanical ram feed. The coal moves down the inclined grate under the action of gravity, or the grate vibrates. The latter case is known as a *vibrating grate stoker*.

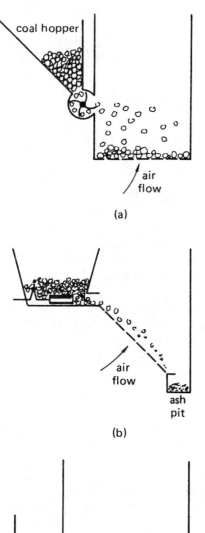

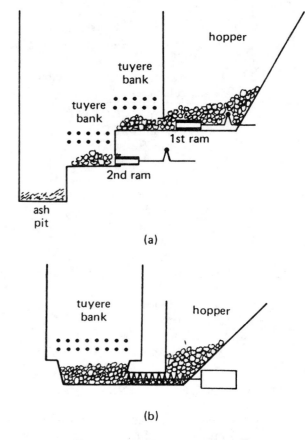

Figure 9.2 Underfeed Stokers

Figure 9.1 Overfeed Stokers

Figure 9.1 (c) illustrates the common *traveling grate stoker*. A 3″ to 7″ layer of coal is spread on the grate. Air passes through both bottom and top grates, or it enters from between the two grates (i.e., a *side feed stoker*).

Figure 9.2 illustrates two underfeed stokers. Figure **9.2** (a) is a *ram-feed stoker*, which pushes coal from the hopper into the combustion area. Figure 9.2 (b) is a *screw-fed stoker*. In both cases, the combustion area is known as a *retort* (hence the name *retort stoker*). Air is injected through ports in the furnace walls. These ports are known as *tuyeres*.[3]

C. PULVERIZED COAL FURNACES

Coal in high-volume installations frequently is pulverized prior to use. Then the finely-ground coal is suspended in a gaseous atmosphere while burning. Determining the two major categories of *pulverized coal furnaces* depends on when the coal is pulverized. With the *bin system*, coal is dried and pulverized, then stored in bins for later use. An air transport network can be used to transfer the powdered coal to the furnace when needed.

Now that pulverizing equipment is reliable, *direct firing furnaces* pulverize coal on demand. There is no shortage of pulverized coal. Direct firing furnaces react quickly to variations in demand.

D. FLUIDIZED BED FURNACES

The desire to make efficient use of "scrap fuels" has led to the development of *fluidized bed furnaces*.[4] Solid fuel is turned into a turbulent, fluid-like mass near the bottom of the combustion chamber by mixing the fuel

---

[3]Tuyeres is pronounced "twee-yer." Both one- and two-syllable pronunciations are used.

[4]*Scrap fuels* include oil shale, tar sands, green wood, seed and rice hulls, biomass refuse, peat, tire shreddings, and shingle/roofing waste. Of course, petroleum, coal, and other bituminous wastes also can be burned.

with a *bed material* and blowing air through the mixture at a controlled rate.

The bed material can be any solid substance that is not consumed. Sand, crushed limestone, and crushed dolomite are possibilities. The mixture of fuel and bed becomes fluidized and assumes free-flowing properties when the air flow through the mixture reaches minimum *fluidizing velocity*.

Fluidized bed combustion using limestone as the bed material can reduce pollutants considerably. One reason is that rapid and complete combustion in a well-controlled, low-temperature environment reduces nitrogen oxides. Also, sulfur dioxide is absorbed by the limestone sorbent, forming stable calcium sulfate.

8 IGNITION TEMPERATURES

Table 9.3 lists approximate ignition temperatures for common fuels.

Table 9.3
Approximate Minimum Ignition Temperatures

Fuel	Ignition Temperature, °F
charcoal	650
fixed carbon	
bituminous	770
semi-bituminous	870
anthracite	850–1100
sulfur	470
carbon monoxide	1170–1300
methane	1260–1380
acetylene	760–820
ethylene	990–1120
ethane	990–1120
hydrogen	1060–1170
propane	950–1080
n-butane	890–1020

9 SOLID FUELS

Solid fuels include wood, coals, coke, and various waste products from industrial and agricultural operations.

Coals consist of volatile matter, fixed carbon, moisture, ash, and sulfur. *Volatile matter* is driven off as a vapor when the coal is heated. *Fixed carbon* is the combustible residue left after the volatile matter is burned off. *Moisture* is present in coal as free water and water of hydration. *Sulfur* is an undesirable constituent of coal. Oxides of sulfur contribute to air pollution and cause corrosion of combustion equipment. However, sulfur contributes to coal heat content and should be considered in heat production.

Fuel components are known from a proximate or an ultimate analysis. A *proximate analysis* gives the percent by weight of moisture, volatile matter, fixed carbon, and ash (sulfur combined or given separately). The *ultimate analysis* also gives percent by weight of constituents but as the individual elements of carbon, hydrogen, sulfur, nitrogen, and moisture. Moisture can be broken into hydrogen and oxygen in an ultimate analysis of coal, in which case all oxygen is assumed locked up as water. For every percent oxygen, 1/8 percent hydrogen also is considered locked up as water. The remaining hydrogen is considered combustible. In an ultimate analysis of a fuel oil, oxygen is considered part of dissolved air.

Coal is burned efficiently in a particular furnace only if it is a uniform size. Therefore, it is not unexpected that coal will be sized according to certain standards. *Run-of-mine* or *run-of-the-mine* is coal as mined. *Lump coal* typically is in the 1″ to 6″ range. *Nut coal* is smaller, *pea coal screenings* are even smaller, and *fines* are dust.

Unfortunately, determining size ranges depends on who is categorizing. For example, pea coal might mean 1/4″ in one area and 1/2″ in another. Because of the variations and different trade names used, ASME recommends that screen sizes be used when grading coal.

Coke is produced by heating coal in the absence of oxygen. The heavy hydrocarbons *crack*, meaning that the hydrogen is driven off, leaving only the carbon. Coke is the carbonaceous residue containing ash and sulfur. It burns smokeless and typically is used in blast furnaces.

Breeze is coke smaller than 5/8″. It is not suitable for use in blast furnaces, but some steam boilers have been adapted to use breeze.

Char is produced from coal in a low-temperature (i.e., 900°F) process. The volatile matter is removed, but there is little cracking. This process is used to solidify tars, bitumins, and some gases.

Bagasse[5] is the fibrous part of sugar cane and other agricultural wastes after crushing. The fuel is approximately 50% water. The heating capacity is a low 3400 BTU/lbm.

Table 9.4 gives the source and compositions of selected coal samples.

Table 9.5 lists the heating value and the ultimate analysis for some woods.

---

[5]Pronounced "buh-gas'"

Table 9.4
Selected U.S. Coals

			Proximate Analysis, % (Coal As Received)						Ultimate Analysis, % (Dry, Ash Free)			
No.	State	County	M	VM	FC	A	S	HV (BTU)	C	H_2	O_2	N_2
1	PA	Schuylkill	2.0	1.8	86.2	10.0	0.79	13,070	93.9	2.1	2.3	0.3
2	PA	Lackawanna	2.0	6.3	79.7	12.0	0.60	13,000	93.5	2.6	2.3	0.9
3	VA	Montgomery	3.0	10.5	66.5	20.0	0.61	11,800	90.7	4.2	3.3	1.0
4	WV	McDowell	3.0	16.3	75.7	5.0	0.73	14,420	90.4	4.8	2.7	1.3
5	PA	Westmoreland	3.0	30.3	55.7	11.0	1.80	13,130	85.0	5.4	5.8	1.7
6	KY	Letcher; Pike	3.0	34.4	56.6	6.0	0.72	13,800	85.2	5.4	7.0	1.6
7	OH	Jefferson	6.0	34.8	49.2	10.0	2.44	12,450	82.0	5.5	7.7	1.7
8	IL	Saline; Perry	10.0	31.7	48.3	10.0	1.6	11,610	80.6	5.4	10.3	1.7
9	UT	Carbon; Emery	8.0	36.6	43.4	12.0	0.56	11,480	80.3	5.7	11.7	1.6
10	IA	Polk	13.9	36.9	35.2	14.0	6.15	10,244	75.8	7.7	26.0	1.2
11	CO	Weld; Boulder	24.0	30.2	40.8	5.0	0.36	9,200	75.0	5.1	17.9	1.5
12	WY	Campbell	24.0	30.0	36.0	10.0	0.33	8,450	74.1	5.1	18.7	1.3
13	ND	McLean; Morton	40.0	27.6	23.4	9.0	1.42	6,330	72.4	4.7	18.6	1.5

Table 9.5
Physical and Chemical Properties of Wood

Wood	Density, lbm/ft^3 air		Gross heating value, BTU/lbm	Ultimate Analysis, % (dry)			
	dried	green	(kiln dried)	C	H_2	O_2	ash
Ash, white	42	47	8,210	49.73	6.93	43.04	0.30
Birch, white	38	51	7,958	49.77	6.49	43.45	0.29
Fir	27	52	8,285	52.32	6.42	41.23	0.03
Oak, black	42	61	7,530	48.78	6.09	44.98	0.15
red	45	65	7,988	49.49	6.62	43.74	0.15
white	48	59	8,112	50.44	6.59	42.73	0.24
Pine, pitch	36	54	10,420	59.00	7.19	32.68	1.12
white	27	39	8,176	52.55	6.08	41.25	0.12
yellow	29	49	8,836	52.60	7.02	40.07	0.31

10 LIQUID FUELS

Liquid fuels commonly are lighter hydrocarbon products refined from crude petroleum oil. They include liquified petroleum gases (LPG), gasoline, kerosene, jet fuel, diesel fuels, and light heating oils. The level of refinement of liquid petroleum fuels determines fuel composition, ignition temperature, flash point, viscosity, and heating value.

Specifications for various grades of fuel oils are based on requirements of different types of burners. Fuel oils are classified as *distillate oils* (lighter petroleum products) and *residual fuel oils* (heavier oils).

- Grade No. 1: A light distillate with high volatility, used in vaporizing type burners; highest in cost/gallon.

- Grade No. 2: A distillate oil heavier in viscosity and API gravity than No. 1, used in pressure atomizing burners; in common use domestically and in medium capacity industrial burners.

- Grade No. 4: Light residual oil or heavy distillate used in burners designed to atomize oils of higher viscosities.

- Grade No. 5L (Light): A residual oil heavier than No. 4; may require preheating for pumping and burning.

- Grade No. 5H (Heavy): A residual oil more viscous than No. 5L requiring preheating.

- Grade No. 6: Also known as *Bunker C oil*; frequently used in industrial applications;

requires preheating for pumping and additional heating for burning; lowest in cost/gallon.

Tables 9.6 and 9.7 list typical properties of fuel oils.

Table 9.6
Properties of Fuel Oils

Grade No.	Weight, lbm/gallon	Heating value BTU/gallon
1	6.675–6.95	132,000–137,000
2	6.960–7.296	137,000–141,000
4	7.396–7.787	143,100–148,000
5L	7.686–7.94	146,800–150,000
5H	7.89–8.08	149,400–152,000
6	8.053–8.488	151,300–155,900

Table 9.7
Fuel Oil Grade vs. Firing Rate

Firing rate, gph	Recommended Grade
up to 25	No. 2
25–35	No. 2, No. 4
35–50	No. 2, No. 4
	No. 5 (Light)
	No. 5 (Heavy)
50–100	No. 5 (Heavy)
	No. 6

Fuel oil burner designs are based on oil atomizing viscosities according to table 9.8.

Table 9.8
Burner Type and Atomizing Viscosity

Burner type	Atomizing viscosity SSU
pressure	30–70
mechanical	35–150
low pressure air atomizing	80–90
steam/high pressure air atomizing	150–250
rotary cup	150–300
sonic	150–300

In handling fuel oils, suction pipes for No. 5 and No. 6 oils should not exceed 100 feet of equivalent length without a booster pump to prevent pump cavitation.

Specifications for various grades of *diesel oil* are based on characteristics similar to those of fuel oils.

- Grade No. 1 Diesel: A distillate oil for high-speed engines in service requiring frequent speed and load changes.

- Grade No. 2 Diesel: A distillate oil of lower volatility for engines in industrial and heavy mobile service.

- Grade No. 4 Diesel: More viscous distillate oils with blends of residual oils for use in medium speed engines under sustained loads.

Property specifications for No. 1, No. 2, and No. 4 diesel and fuel oils are identical except that diesel fuels can be specified by cetane number. *Cetane number* is a measure of the ignition quality of a fuel.

11 GASEOUS FUELS

Various gaseous fuels are used as energy sources, but most applications are limited to natural gas and liquefied petroleum gases. *Natural gas* is a mixture of methane (55 to 95%), higher hydrocarbons (primarily ethane), and noncombustible gases. Typical heating values range from 950 to 1100 BTU/ft^3 at industrial STP (30 inches Hg and 60°F). *Liquefied petroleum gases* are available as butane, propane, and mixtures of the two. At atmospheric pressure, propane boils at −40°F, while butane boils at 32°F.

There are a number of manufactured gases which can be used where available.

- coke-oven gas: Approximately 17% of the coal heated to form coke can be recovered. This gas is largely hydrogen.

- blast furnace gas: The gas discharged from blast furnaces is approximately 55% nitrogen and 20% carbon monoxide.

- water gas: Steam passing through burning coke will produce carbon monoxide and hydrogen gas.

- enriched water gas, carbureted water gas: This is water gas which has been mixed with blast furnace gas, or gas produced from oil cracked by spraying onto hot bricks.

- producer gas: This gas is produced by burning coal in an oxygen deficient atmosphere (as in burning coal seams underground instead of mining the coal). The gas is high in carbon monoxide.

Gas burners can be natural draft or forced draft. *Natural draft* burners rely on chimney draft to draw off combustion gases. A fan is used only to control combustion air. *Forced draft* burners also use the fan to move products through the burner; combustion occurs under pressure.

Compensation for altitudes above 1000 feet should be used since burners are constant volume devices (regardless of air density). Derating of burner heat output suggested by A.G.A. is 4% per 1000 feet of altitude. Higher volume fans can be used to compensate for loss. Heating value of gases also decreases with altitude (from approximately 1000 BTU/ft^3 at sea level to approximately 830 BTU/ft^3 at 5000 feet).

12 STOICHIOMETRIC COMBUSTION

Atmospheric air is a mixture of oxygen, nitrogen, and small amounts of carbon dioxide, water vapor, argon, and other inert gases. If all constituents except oxygen are grouped with the nitrogen, the air composition is as given in table 9.9. It is necessary to supply by weight (1/0.2315) or 4.32 pounds of air to obtain one pound of oxygen. The molecular weight of air is 28.9.

Table 9.9
Composition of Air
(Rare inert gases included as N_2)

	% by weight	% by volume
oxygen (O_2)	23.15	20.9
nitrogen (N_2)	76.85	79.1
ratio of nitrogen to oxygen	3.32	3.78

Stoichiometric or *theoretical, air* is the exact quantity of air necessary to provide the required amount of oxygen for complete combustion of a fuel. Stoichiometric air requirements usually are stated on the basis of pounds for solid and liquid fuels, and on the basis of cubic feet for gaseous fuels. The approximations in equations 9.3 through 9.5 are based on ultimate analysis.

$$\text{lbm air/lbm fuel} \atop \text{(solid fuels)} = 34.34\left(\frac{[C]}{3} + \left([H] - \frac{[O]}{8}\right) + \frac{[S]}{8}\right)$$
9.3

$$\text{lbm air/lbm fuel} \atop \text{(liquid fuels)} = 34.34\left(\frac{[C]}{3} + [H] + \frac{[S]}{8}\right) - [O]$$ 9.4

$$\text{ft}^3\text{ air/ft}^3\text{ fuel} \atop \text{(gaseous fuels)} = \sum_i k_i\,(\%i)$$
9.5

where i is the component gas index and k is a constant from table 9.10.

$$\text{lbm air/lbm fuel} \atop \text{(gaseous fuels)} = \sum_i j_i\,(\%i)$$
9.6

where j is a constant from table 9.10.

Table 9.10
Coefficients for Equations 9.5 and 9.6

gas, i	k_i	j_i
CO	2.39	2.47
H_2	2.39	34.34
CH_4	9.53	17.27
C_2H_6	16.86	16.12
C_3H_8	23.82	15.70
C_4H_{10}	30.97	15.49
C_2H_2	11.91	13.30
C_2H_4	14.29	14.81
H_2S	7.15	6.10
O_2	−4.78	−4.32

Example 9.5

For coal sample No. 2 from table 9.4, find the theoretical air requirements.

$$\frac{\text{lbm air}}{\text{lbm fuel}} = 34.34\left\{\frac{.935}{3} + \left[.026 - \frac{.023}{8}\right] \right.$$
$$\left. + \left[\frac{1 - (.935 + .026 + .023 + .009)}{8}\right]\right\} = 11.53$$

Example 9.6

Find the stoichiometric air requirements per pound of No. 1 fuel oil if the composition is 84% C, 15.3% H_2, 0.3% S and 0.4% N_2.

$$\frac{\text{lbm air}}{\text{lbm fuel}} = 34.34\left\{\frac{.84}{3} + .153 + \frac{.003}{8}\right\} = 14.88$$

Example 9.7

How many cubic feet of air are required to burn a natural gas that is 86.92% methane, 7.95% ethane, 2.16% propane, and 0.16% butane?

$$\frac{\text{ft}^3\text{ air}}{\text{ft}^3\text{ fuel}} = .8692\,(9.53) + .0795\,(16.86) + .0216\,(23.82)$$
$$+ .0016\,(30.97) = 10.19$$

The theoretical air required also can be found by standard analysis of the chemical combustion equation. Table 9.11 frequently is used to find stoichiometric air based on the chemical equation.

Example 9.8

Calculate the theoretical air required to burn CO to CO_2 at 90°F.

From table 9.11 for complete combustion of CO to CO_2, the required air is $1.88\ N_2 + 0.5\ O_2 = 2.38$ ft^3 of air which yields 1.0 ft^3 of CO_2.

Table 9.11
Consolidated Combustion Data

Fuel	FOR ONE MOLE OF FUEL					FOR ONE CU. FT. OF FUEL				FOR ONE POUND OF FUEL					Units of fuel
	AIR		Other products			AIR		Other products		AIR		Other products			
	O_2	N_2	CO_2	H_2O	SO_2	O_2	N_2	CO_2	H_2O	O_2	N_2	CO_2	H_2O	SO_2	
C carbon	1.0	3.76	1.0							.0833	.313	.0833			Moles
	379	1425	379							31.6	118.8	31.6			Cu. ft
	32.0	105	44.0							2.67	8.78	3.67			Pounds
H_2† hydrogen	0.5	1.88		1.0		.00132	.00496		.00264	.250	.940		.5		Moles
	189.5	712		379*		.5	1.88		1.0*	94.8	356		189.5*		Cu. ft
	16.0	52.6		18		.0422	.139		.0475	8.0	26.3		9.0		Pounds
S sulfur	1.0	3.76			1.0					.0312	.1176			.0312	Moles
	379	1425			379					11.84	44.6			11.84	Cu. ft
	32.0	105			64					1.0	3.29			2.0	Pounds
CO carbon monoxide	.5	1.88	1.0			.00132	.00496	.00264		.179	.0672	.0357			Moles
	189.5	712	379			.5	1.88	1.0		6.77	25.4	13.53			Cu. ft
	16.0	52.6	44.0			.0422	.139	.116		.571	1.88	1.57			Pounds
CH_4 methane	2.0	7.52	1.0	2.0		.00528	.0198	.00264	.00528	.125	.470	.0625	.125		Moles
	758	2850	379	758*		2.0	7.52	1.0	2.0*	47.4	178	23.7	47.4*		Cu. ft
	64.0	210	44	36		.169	.556	.116	.0950	4.0	13.17	2.75	2.25		Pounds
C_2H_2 acetylene	2.5	9.40	2.0	1.0		.0066	.0248	.00528	.00264	.0962	.362	.0769	.0385		Moles
	947	3560	758*	379*		2.5	9.40	2.0	1.0*	36.4	137	29.15	14.58*		Cu. ft
	80.0	263	88	18		.211	.694	.232	.0475	3.08	10.13	3.38	.692		Pounds
C_2H_4 ethylene	3.0	11.29	2.0	2.0		.00792	.0298	.00528	.00528	.1071	.403	.0714	.0713		Moles
	1137	4280	758	758*		3.0	11.29	2.0	2.0*	40.6	153	27.1	27.1*		Cu. ft
	96.0	316	88.0	36.0		.253	.834	.232	.0950	3.43	11.29	3.14	1.286		Pounds
C_2H_6 ethane	3.5	13.17	2.0	3.0		.00923	.0347	.00528	.0079	.1167	.439	.0667	.10		Moles
	1326	4990	758	1137*		3.5	13.17	2.0	3.0*	44.2	166.3	25.3	37.9*		Cu. ft
	112	369	88	54		.296	.972	.232	.1425	3.73	12.29	2.93	1.8		Pounds

This table gives the required combustion air and products for common combustibles burned with theoretical air requirements. Air and products are given in moles, cubic feet, and pounds (see the right-hand column) for one mole of fuel, one cubic foot of fuel, and one pound of fuel.

* The volumes shown for H_2O apply only when the combustion products are at such high temperatures that all of the H_2O is a gas.

† Varying assumptions for molecular weight introduce slight inconsistencies in the values of air and combustion products from the burning of hydrogen. The true molecular weight is 2.02; however, the approximate value of 2.00 is used in figuring the air and combustion products.

NOTE: These volumes are at 60°F psia. To obtain volumes for T°F, multiply by (T+460)/520.

The multiplicative factor is

$$\frac{90 + 460}{520} = 1.057$$

or $1.057\,(1.88\,N_2 + 0.5\,O_2) = 1.057\,(2.38) = 2.52$ ft^3 of air per cubic foot of CO.

13 EXCESS AIR

Complete combustion occurs when all fuel is completely burned; excess air usually is necessary to accomplish this. Excess air is expressed as a percentage of the theoretical air. Rule-of-thumb values for excess air are given in table 9.12, although actual values vary widely with equipment and installation.

Example 9.9

Find the percent CO_2 by volume in the stack gases with 20% excess air for propane gas.

step 1: The chemical equation is

$$C_3H_8 + 5O_2 \mapsto 3CO_2 + 4H_2O$$

step 2: From table 9.9, N_2 is 79% by volume in air. The number of volumes of N_2 carried to the flue gas with the O_2 is

$$\frac{79\% \; N_2}{21\% \; O_2} = \frac{X}{5 \text{ volumes } O_2}$$
$$X = 18.8 \text{ volumes } N_2$$

Table 9.12
Typical Amounts of Excess Air

Fuel	Type of Furnace or Burner	Excess Air, % by weight
Pulverized coal	Completely water-cooled furnace for slag-tap or dry-ash removal	15–20
	Partially water-cooled furnace for dry-ash removal	15–40
Crushed coal	Cyclone furnace—pressure or suction	10–15
Coal	Stoker-fired, forced-draft, chain grate	15–50
	Stoker-fired, forced-draft, underfeed	20–50
	Stoker-fired, natural-draft	50–65
Fuel oil	Oil burners, register-type	5–10
	Multiple burners and flat-flame	10–20
Natural gas	Register-type	5–10
	Multifuel burners	7–12
Blast-furnace gas	Intertube nozzle-type burners	15–18

step 3: The total volume of flue gas with stoichimetric combustion is

$$18.8N_2 + 3CO_2 + 4H_2O = 25.8 \text{ ft}^3/\text{ft}^3 \text{ fuel}$$

step 4: The % CO_2 is

$$\frac{3 \text{ ft}^3 CO_2}{25.8 \text{ ft}^3 \text{ flue gas}} = 11.6\% CO_2$$

step 5: For 20% excess air, the volume of N_2 is 120%(18.8) = 22.6 ft³. The excess O_2 carried to flue gas is 20%(5) = 1.0 ft³.

step 6: The total volume of flue gas with 20% excess air is

$$22.6N_2 + 1.0O_2 + 3CO_2 + 4H_2O = 30.6 \text{ ft}^3$$

step 7: The % CO_2 by volume with 20% excess air is

$$\frac{3 \text{ ft}^3 CO_2}{30.6 \text{ ft}^3 \text{ flue gas}} = 9.8\%$$

The effects of decreasing air, while keeping fuel flow constant, are given here.

- Furnace temperature increases since there is less cooling air.
- Stack temperature change depends on the physical construction and heat transfer coefficients. The change is either positive or negative.
- Flue gas decreases in quantity.
- Heat loss decreases.
- Furnace efficiency increases.

Incomplete combustion is both inefficient and dangerous. In addition to carbon monoxide, other products may include toxic alcohols, ketones, and aldehydes. Principle causes of incomplete combustion are

- Low combustion temperature or cold furnace
- Poor air supply
- Smothering from improperly vented stack gases
- Insufficient fuel/air mixing

Smoke is the obvious result of incomplete combustion. A slight *haze* is normal and indicates good firing conditions.

14 HEAT OF COMBUSTION

The heating value of a fuel can be determined experimentally in a calorimeter, or it can be estimated from its chemical analysis and the heating value of the constituents. The analysis percentage (gravimetric or volumetric analysis) is multipled by the heating value per unit (pound or ft³) for each combustible element, and all values are added together. Sulfur is included if its amount is known. Only hydrogen that is not locked up with oxygen as water is included. The correct percentage of free hydrogen is given by equation 9.7.

(In formulas to follow, gravimetric fractions are enclosed in brackets [ ] while volumetric fractions are enclosed in parentheses ( ).)

$$[H_2]_{\text{free}} = [H_2] - \frac{[O_2]}{8} \qquad 9.7$$

Example 9.10

Find the heating value for coal sample No. 1 from table 9.4. (Use appendix A to obtain the heating values of each component.)

The ultimate analysis is

$$[N_2] = .003, \quad [C] = .939$$
$$[H_2] = .021, \quad [O_2] = .023$$
$$[H_2]_{\text{free}} = .021 - \left(\frac{.023}{8}\right) = .018$$

Element	Heating Value of Element	% Element	Estimated Heating Value
C	14,093 BTU/lb	.939	13,233 BTU/lbm
H_2 free	60,958 BTU/lb	.018	1,097 BTU/lbm
			14,330 BTU/lbm

The *higher* or *gross heating value* of a fuel includes the heat of vaporization of water vapor formed during combustion. The *lower* or *net heating value* assumes that all the products of combustion remain gaseous. The higher heating value can be calculated from the lower heating value according to the following formulas.

$$\text{HHV} = \text{LHV} + 970 \ (\text{lbm of } H_2O \text{ produced}) \qquad 9.8$$
$$\text{HHV} = \text{LHV} + 8,730 \ (\text{lbm } H_2 \text{ burned to } H_2O) \qquad 9.9$$
$$\text{HHV} = \text{LHV} + 1,092 \ (\text{lbm } O_2 \text{ burned to } H_2O) \qquad 9.10$$
$$\text{HHV} = \text{LHV} + 17,460 \ (\text{moles } H_2 \text{ burned to } H_2O) \ 9.11$$

Dulong's formula (equation 9.12) gives 2% to 3% accuracy of high heating value for coals compared to calorimeter values.

$$\text{HV} = 14,093\,[C] + 60,958\left([H_2] - \frac{[O_2]}{8}\right) + 3983\,[S] \qquad 9.12$$

Equation 9.12 makes the following assumptions.

- None of the oxygen is in the form of carbonates.
- There is no free oxygen.
- The hydrogen and carbon are not locked together in the form of hydrocarbons.
- The carbon is amorphous (not graphitic).
- Sulfur burns to SO_2, not to SO_3.
- Sulfur is not in the form of sulfates.

Example 9.11

Using equation 9.12, find the heating value of coal sample No. 1 from table 9.4. Disregard sulfur content.

$$\text{HV} = 14,093\,[.939] + 60,958\left([.021] - \frac{[.023]}{8}\right)$$
$$= 14,338 \ \text{BTU/lbm}$$

15 ENTHALPY OF FORMATION

The *enthalpy of formation* is the energy absorbed or released by an element or a compound taking part in a reaction. The enthalpy of formation is assigned a value of zero for elements in their free states.

Combustion reactions which give off energy during the formation of compounds are known as *exothermic reactions*. In this case, enthalpies of formation will be negative. Many exothermic reactions begin spontaneously. On the other hand, *endothermic reactions* require the application of heat or other energy to start the reaction.

Table 9.13 contains enthalpies of formation for several elements and compounds.[6] The enthalpy of formation depends on the phase of the compound. Compounds in table 9.13 are solid unless indicated to be gases (g) or liquids (l).

The enthalpy of formation of a compound formed in a chemical reaction is found by summing the enthalpies of reaction over all products and subtracting the sum of enthalpies of reaction over all reactions.

Example 9.12

What is the heat of stoichiometric combustion of gaseous methane and oxygen at 25°C?

Balance the chemical equation and obtain enthalpies of reaction for each element and compound.

$$CH_4 + 2O_2 \mapsto 2H_2O + CO_2$$
$$(-17.90) + 2(0) \mapsto 2(-57.80) + (-94.05)$$

Subtract the reactants' enthalpy sum from the products' enthalpy sum.

$$2(-57.80) - 94.05 - (-17.90) = -191.75\,\frac{\text{kcal}}{\text{mole-methane}}$$

The negative sign indicates the reaction is exothermic.

Since the enthalpy of formation taken from table 9.13 is for gaseous water (g), the heating value obtained from this calculation is the lower heating value. If the value for liquid water were used $(-68.32$ instead of $-57.80)$, the higher heating value would be obtained.

---

[6]Multiply kcal/gmole by (1800/molecular weight) to get BTU/lbm.

Table 9.13
Standard Enthalpy of Formation
for Various Compounds
in kcal/gmole at 25°C

element/compound	ΔH_f
Al (g)	75.0
Al_2O_3	−399.09
C (graphite)	0.00
C (g)	171.70
C_2 (g)	234.7
CO (g)	−26.42
CO_2 (g)	−94.05
CH (g)	142.1
CH_2 (g)	95
CH_3 (g)	32.0
CH_4 (g)	−17.90
C_2H_2 (g)	54.19
C_2H_4 (g)	12.50
C_2H_6 (g)	−20.24
CCl_4 (g)	−25.5
$CHCl_4$ (g)	−24
CH_2Cl_2 (g)	−21
CH_3Cl (g)	−19.6
CS_2 (g)	27.55
COS (g)	−32.80
$(CH_3)_2S$ (g)	−8.98
CH_3OH (g)	−48.08
C_2H_5OH (g)	−56.63
$(CH_3)_2O$ (g)	−44.3
C_3H_6 (g)	9.0
C_6H_{12} (g)	−29.98
C_6C_{10} (g)	−1.39
C_6H_6 (g)	19.82
Fe	0.0
Fe (g)	99.5
Fe_2O_3	−196.8
Fe_3O_4	−267.8
H_2 (g)	0.00
H_2O (g)	−57.80
H_2O (l)	−68.32
H_2O_2 (g)	−31.83
H_2S (g)	−4.82
N_2 (g)	0.00
NO (g)	21.60
NO_2 (g)	8.09
NO_3 (g)	13
NH_3 (g)	−11.04
O_2 (g)	0.00
O_3 (g)	34.0
S	0.00
SO (g)	1.4
SO_2 (g)	−70.96
SO_3 (g)	−94.45

The heating value can be converted to BTU/lbm and compared to the value given in appendix A. (See footnote 6.) The molecular weight of methane is 16, so

$$LHV = \frac{(191.75)(1800)}{16} = 21,572 \text{ BTU/lbm}$$

16 FLUE GAS ANALYSIS

To control avoidable heat losses in combustion effectively, continuous measurement and control of combustion air are necessary. The amount of combustion air can be determined from flue gas analysis and from metering combustion air or flue gas rate. A quantitative determination of total air supplied to a combustion process requires a complete flue gas analysis for percentages of CO_2, O_2, CO, and N_2 by volume. The *Orsat gas analysis* apparatus is used to find the volumetric percentages of these flue gases.

The *actual air* taking part in combustion should be estimated from equation 9.13.

$$\frac{\text{lbm actual air}}{\text{lbm fuel}} = \frac{3.04\,(N_2)[C]}{(CO_2)+(CO)} \qquad 9.13$$

[C] is the gravimetric percent of carbon in the fuel. (N_2), (CO_2), and (CO) are from the Orsat analysis.

Example 9.13

Calculate the actual air used.

Coal analysis, dry: [C]= 75%, [ash]= 8%, $[H_2]$ + $[O_2]/8 = 17\%$.

Flue gas analysis: $(CO_2) = 12.6\%$, $(CO)= 1.0\%$, $(O_2) = 6.2\%$, $(N_2) = 80.2\%$

From equation 9.13,

$$\frac{\text{lbm actual air}}{\text{lbm fuel}} = \frac{(3.04)(.802)(.75)}{.126 + .01}$$
$$= 13.4 \text{ lbm/lbm fuel}$$

The percent excess air can be approximated from an Orsat analysis with equation 9.14.

$$\% \text{ excess air} = \frac{100\,((O_2) - 0.5\,(CO))}{0.264\,(N_2) - [(O_2) - 0.5\,(CO)]} \qquad 9.14$$

Example 9.14

Find the % excess air in example 9.13.

$$\% \text{ excess air} = \frac{100\,[.062 - (.5)(.01)]}{(.264)(.802) - [.062 - (.5)(.01)]}$$
$$= 36.8\%$$

The maximum theoretical % CO_2 can be approximated from an Orsat analysis with equation 9.15.

$$(CO_2)_{theoretical} = \frac{100\,(CO_2)_{actual}}{1 - 4.76\,(O_2)} \qquad 9.15$$

Example 9.15

Find the % theoretical CO_2 in example 9.13.

$$(CO_2)_{theoretical} = \frac{100\,(0.126)}{1 - 4.76\,(0.062)} = 17.9\%$$

The pounds of dry flue gas produced per pound of solid fuel can be estimated from an Orsat analysis.

$$\frac{lbm\ flue\ gas}{lbm\ solid\ fuel} =$$
$$\frac{\{11\,(CO_2) + 8\,(O_2) + 7\,\{(CO) + (N_2)\}\}\left\{[C] + \left(\frac{[S]}{1.833}\right)\right\}}{3\,((CO_2) + (CO))}$$
$$9.16$$

Example 9.16

Find the amount of flue gas in example 9.13.

$$\frac{lbm\ flue\ gas}{lbm\ solid\ fuel} =$$
$$\frac{\{11\,(.126) + 8\,(.062) + 7\,\{(.01) + (.802)\}\}\{.75\}}{3\,((.126) + (.01))} = 13.9$$

17 COMBUSTION HEAT AND LOSSES

Combustion heat not transferred to the heating medium (e.g., a boiler) is lost from the system in the following ways.

- In the *dry* flue gases,

$$q_1 = mc_p(T_{gas} - T_{air}) \qquad 9.17$$

  m is the mass of dry flue gas per pound of fuel from equation 9.16.

  c_p is the mean specific heat of the flue gases at constant pressure. This varies from 0.242 to 0.254 BTU/lbm-°F for flue gas temperatures from 300°F to 1,000°F. (Properties of nitrogen frequently are assumed for dry flue gas.)

- In the vapor formed as part of the products of combustion,

$$q_2 = 9[H_2](h_g - h_f) \qquad 9.18$$

  $[H_2]$ is the fraction of hydrogen in the fuel by weight.

h_g is the enthalpy of superheated steam at the flue gas exit temperature and partial pressure of the water vapor. (The saturation enthalpy often is used.)

h_f is the enthalpy of saturated water at the air entrance temperature.

- In the water vapor originally in the combustion air,

$$q_3 = \omega m(h_g - h'_g) \qquad 9.19$$

ω is the humidity ratio of the combustion air.

h'_g is the enthalpy of saturated steam at the air entrance temperature.

- In incomplete combustion,

$$q_4 = \frac{(HHV_{CO_2} - HHV_{CO})[C](CO)}{(CO_2) + (CO)}$$
$$\approx \frac{9746\,[C](CO)}{(CO_2) + (CO)} \qquad 9.20$$

[C] is the mass of carbon in the fuel per pound of fuel. (CO) and (CO_2) are from the Orsat analysis.

- In unburned carbon left in the fuel ash,

$$q_5 = \frac{14{,}093\,m_a[C_a]}{m_f} \qquad 9.21$$

C_a is the percentage combustible carbon in the ash.

m_a is the mass of the ash.

m_f is the mass of the fuel.

- Radiation and other losses (q_6) usually are ignored.

Flue gas loss is the sum of all the above, but for gas- and oil-fired equipment, incomplete combustion and unburned carbon losses are omitted.

18 COMBUSTION EFFICIENCY

The overall thermal efficiency of the combustion reaction is defined as

$$\eta_{thermal} = \frac{useful\ heat\ (100)}{heating\ value\ of\ fuel}$$
$$= \frac{100\,(HV - (q_1 + q_2 + q_3 + q_4 + q_5 + q_6))}{HV}$$
$$9.22$$

19 MAXIMUM THEORETICAL COMBUSTION TEMPERATURE

The maximum increase in temperature can be assumed to be the result of all liberated heat being absorbed by the combustion products.

$$\Delta T = \frac{\text{heat of combustion}}{(\text{mean specific heat}) \times \left(\begin{array}{c}\text{mass of stack gas}\\ \text{per pound of fuel}\end{array}\right)}$$

9.23

The mean specific heat is a gravimetric weighted average of the values of c_p for all combustion products.

20 POLLUTANTS AND IMPURITIES IN COAL

The three primary undesirable coal constituents are moisture, mineral matter (ash), and sulfur in various forms.

A. FREE MOISTURE

Although a moisture content around 5% by weight is reported to be beneficial in some mechanically-fired boilers, moisture generally is undesirable. *Free moisture* in coal increases its weight and transportation costs, increases its cost to the consumer, and decreases the combustion heat available to the boilers.

B. MINERAL MATTER

Strictly speaking, *mineral matter* is the noncombustible material in the coal, whereas *ash* is the inorganic residue remaining after fuel combustion. However, this naming convention is not always followed, and "ash" can be reported as a component of the unburned fuel.

If the mineral matter is part of the vegetation which subsequently becomes coal, it is known as *intrinsic mineral matter*. This distinguishes it from *adventitious mineral matter*, which is clay and other mineral matter picked up with or mixed with the coal.

The finely-powdered ash that covers combustion grates protects the grates from temperature extremes. If the ash has a low fusion point, it will melt and form hard *clinkers*. In extreme cases, it can adhere to the grates. 2200°F is considered a low fusion point, and 2600°F is considered a high fusion point. Ashes with high fusion points are known as *refractory ashes*.

The actual fusion point depends on the composition of the ash. Ash is primarily a mixture of *silica* (SiO_2), *alumina* (Al_2O_3), and *ferric oxide* (Fe_2O_3).[7] The relative proportions of each of these compounds will determine the fusion point, with higher percentages of ferric oxide reducing the fusion point. The fusion points of pure alumina and pure silica are in the 2700°F to 2800°F range.

Fly ash is carried out of the general combustion area by the flue gas. Fly ash may be deposited on boiler tubes, stack walls, economizer tubes, and on other devices in the stack system. It also may be discharged from the stack if not removed. As little as 10% of the total ash content might be recovered in the ash pit.

Fly ash can be removed from the flue gas by *dust catchers* and *gas washers*, which can be used in series for increased efficiency. Dust catchers can be mechanical separators (e.g., screens, baffle traps, and cyclone centrifugal-type collectors) or *electrostatic precipitators*. Electrostatic precipitators have a high efficiency—between 80% and 95% of all ash is collected. However, in addition to using some of a power plant's electrical energy, they are large and expensive.

C. SULFUR

Several forms of sulfur are present in coal. Its primary form is as *pyrite* (FeS_2). Iron sulfate and *gypsum* ($CaSO_4 \cdot 2H_2O$) also can be found in smaller amounts. Sulfur in elemental or pyritic forms oxidizes to sulfur dioxide. *Sulfur trioxide* can be formed under certain conditions. *Sulfurous acid* (H_2SO_3) and *sulfuric acid* (H_2SO_4) are produced when oxides of sulfur react with

---

[7] *Calcium oxide* (CaO), *magnesium oxide* (*magnesia*) (MgO), *titanium oxide* (*titania*) (TiO_2), *ferrous oxide* (FeO), *sulfur trioxide* (SO_3), and *alkalies* (Na_2O and K_2O) also are ashes which can be present in smaller percentages.

Table 9.14
Approximate Specific Heats, c_p(BTU/lbm-°R) at One Atmosphere

Gas	\multicolumn{8}{c}{Temperature, °R}
	500	1000	1500	2000	2500	3000	4000	5000
nitrogen	.248	.255	.270	.284	.294	.301	.310	.315
carbon dioxide	.196	.251	.282	.302	.314	.322	.332	.339
carbon monoxide	.248	.257	.274	.288	.298	.304	.312	.316
oxygen	.218	.236	.253	.264	.271	.276	.286	.294
water vapor	.444	.475	.519	.566	.609	.645	.696	.729
hydrogen	3.39	3.47	3.52	3.63	3.77	3.91	4.14	4.30
sulfur dioxide	.15	.16	.18	.19	.20	.21	.23	

moisture in the flue gas. Both of these acids are actively corrosive.

$$SO_2 + H_2O \mapsto H_2SO_3$$
$$SO_3 + H_2O \mapsto H_2SO_4$$

21 ATMOSPHERIC POLLUTANTS

The primary pollutants are smoke, sulfur compounds, fly ash, carbon monoxide, and oxides of nitrogen. Water vapor and carbon dioxide are not considered pollutants.

Smoke and carbon monoxide generally result from incomplete combustion. Therefore, the generation of these pollutants can be minimized by proper furnace monitoring. Oxides of nitrogen include *nitric oxide* (NO), *nitrogen dioxide* (NO_2), and *nitrous oxide* (N_2O). These nitrogen compounds form in small amounts at temperatures high enough for the nitrogen gas (N_2) to dissociate. Oxides of nitrogen are associated with the formation of *smog*.

22 DEW POINT OF FLUE GAS MOISTURE

Dalton's law of partial pressures can be used to predict the dew point of water in the flue gas. This law states that the partial pressure of the water vapor can be predicted from the mole fraction of the water vapor in the stack gas.

$$\text{partial pressure} = \left(\begin{array}{c}\text{water vapor}\\\text{mole fraction}\end{array}\right)\left(\begin{array}{c}\text{stack gas}\\\text{pressure}\end{array}\right) \quad 9.24$$

Once the water vapor partial pressure is known, the dew point temperature can be found from the steam tables as the saturation temperature corresponding to the partial pressure. The dew point with coals typically is around 100°F.

Of course, the entering air also can have moisture in it. This entering moisture should be added to the water vapor from combustion when calculating the mole fraction. More water vapor raises the dew point.

23 DRAFT AND STACK DESIGN

The amount of air that a furnace receives is determined by the furnace draft. *Draft* is the static pressure in the furnace or at any other point being evaluated. Some furnaces operate on *natural draft*, which implies that the air flows through the furnace and out the stack due to the pressure differential caused by reduced densities in the stack.

Induced drafting relies on an *induced draft fan* in the stack. Such a fan injects air into the stack after combustion.[8] The injected air draws the combustion products through the stack system.

Air also can be pushed into the furnace system by a *forced draft fan* located before the furnace. Forced draft fans are run at relatively high speeds (1200 or 1800 rpm) with direct-drive motors. Two or more fans are used in parallel to provide for efficient operation at low furnace demand.

The *draft loss* is the friction loss in the stack and the boiler. It usually is specified in inches of water. If the pressure at any point in the stack system reaches atmospheric pressure, a condition of *balanced draft* is said to have been achieved.

A positive pressure (e.g., 2 to 10 inches w.g.) is created by forced draft fans. This positive pressure should be reduced to a very small negative pressure by passing through the air heater, the ducts, and the windbox system. The negative pressure in the furnace serves to keep combustion gases from leaking into the furnace and boiler areas. The air pressure continues to drop as it passes through the boiler, the economizer, the air heater, and the pollution control equipment. The final negative pressure of 2 to 10 inches w.g. may or may not require the use of an induced draft fan.

Stack effect (*chimney action*) can be relied on to draw combustion products partially up the stack.[9] The stack effect uses the difference in air and stack gas densities to produce a pressure difference.

If a few assumptions are made, the *stack effect, SE,* can be calculated. Such calculations are necessary when sizing the stack. One assumption is that the pressure in the stack is approximately equal to the atmospheric pressure. The density difference is due to only the temperature difference. The problem is simplified further by assuming that the stack gas and the atmospheric air have the same composition (and thus the same molecular weight, gas content, etc.). This makes it unnecessary to know the actual condition of the stack gas.

Once the average temperature of the stack gas (at the stack midpoint) is known, the stack effect can be calculated from the stack height, z, and the average density. Be sure to use consistent units with equation 9.25.

$$SE = z\left(\rho_{\text{air}} - \rho_{\text{gas}}\right) = \frac{p_{\text{air}}z}{R_{\text{air}}}\left(\frac{1}{T_{\text{air}}} - \frac{1}{T_{\text{gas}}}\right) \quad 9.25$$

---

[8]Induced draft fans are located after dust collectors and precipitators in order to reduce the abrasive effects of fly ash. They are run at slower speeds than forced draft fans in order to reduce the abrasive effects further.

---

[9]The higher the chimney, the greater the stack effect. However, the stack effect in modern plants is reduced greatly by the friction and cooling effects of economizers, air heaters, and precipitators. Therefore, the function of a stack is primarily to carry the combustion products a sufficient distance upward to dilute the pollutants. Modern stacks seldom are built higher than 200 feet.

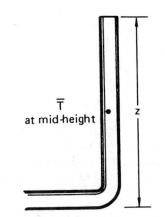

Figure 9.3 Stack Effect Calculation

Often the induced draft is less than 80% of ideal. Therefore, a coefficient K is introduced into equation 9.25.

$$SE' = K(SE) \qquad 9.26$$

The theoretical head of stack gases is

$$h = \frac{SE'}{\rho_{gas}} = z\left(\frac{T_{gas}}{T_{air}} - 1\right) \qquad 9.27$$

The ideal *stack velocity* is

$$v_{ideal} = \sqrt{2gh} = \sqrt{2gz\left(\frac{T_{gas}}{T_{air}} - 1\right)} \qquad 9.28$$

The actual velocity in the stack is 30% to 50% of the ideal velocity. This reduction is achieved by the introduction of the *coefficient of velocity*, C_v.

$$v_{actual} = C_v v_{ideal} \qquad 9.29$$

The stack gas area can be found from equation 9.30.

$$A_{stack} = \frac{Q_{gas}}{v_{actual}} \qquad 9.30$$

Example 9.17

The air surrounding a stack is 70°F and 14.7 psia. The average stack gas temperature is 400°F. The stack is 80 feet tall measured with respect to the ash pit elevation. What is the draft in inches of water?

$$T_{air} = 70 + 460 = 530°R$$
$$T_{gas} = 400 + 460 = 860°R$$

From equation 9.25,

$$SE = \frac{(14.7)(144)(80)}{53.3}\left[\frac{1}{530} - \frac{1}{860}\right] = 2.3 \text{ psf}$$

Converting to inches of water,

$$SE = \frac{(2.3)(12)}{62.4} = .44 \text{ inches w.g.}$$

24 FEEDWATER IMPURITIES

A. INTRODUCTION

Water impurities cause scaling and corrosion in boiler systems. Impurities in *make-up water* taken from local sources include dissolved solids, dissolved gases, and suspended matter. Industrial wastes also may be present.

Much of make-up water treatment involves the removal of dissolved minerals. Table 9.15 lists some of the minerals found in water.[10] The calcium and magnesium compounds are particularly troublesome. Their presence, alone or in combination, in large concentrations causes the water to become "hard." *Hard water* containing relatively insoluble calcium and magnesium compounds causes scale and other deposits in boiler systems.

Table 9.15
Dissolved Minerals in Water

Formula	Name	Common Name
$Ca(HCO_3)_2$	calcium bicarbonate	limestone
$Mg(HCO_3)_2$	magnesium bicarbonate	dolomite
$CaSO_4$	calcium sulfate	gypsum
$MgSO_4$	magnesium sulfate	epsom salt
$NaCl$	sodium chloride	table salt

Water also contains oxygen, nitrogen, and carbon dioxide in small amounts. The nitrogen is inert, and it does not need to be considered in make-up water purification. However, both the oxygen and the carbon dioxide need to be removed. Although the solubility of oxygen in water decreases with temperature, water at high pressures can hold large quantities of oxygen. High temperature dissolved oxygen readily attacks pipe and boiler metal. Carbon dioxide combines with water to form *carbonic acid*.

B. WATER QUALITY STANDARDS

Water entering the boiler to replace liquid which has been evaporated is known as *feedwater*. Water which has been added to replace losses is known as *make-up water*. The purity of the feedwater returned to the boiler (after condensing) depends on the purity of the make-up water, since impurities continually build up.

The purity requirements depend on the boiler equipment. Steam pressure is particularly significant. A low pressure firetube boiler may be able to tolerate higher hardness levels. Most modern high pressure boilers, however, require the removal of virtually all impurities.

[10]Water also can include small quantities of iron, manganese, fluorides, aluminum, nitrates, and phosphates.

Table 9.16
ABMA Maximum Limits for Boiler Water

Boiler Pressure (psig)	Total Solids (ppm)	Alkalinity (ppm)	Suspended Solids (ppm)	Silica* (ppm)
0-300	3500	700	300	125
301-450	3000	600	250	90
451-600	2500	500	150	50
601-750	2000	400	100	35
751-900	1500	300	60	20
901-1000	1250	250	40	8
1001-1500	1000	200	20	2.5
1501-2000	750	150	10	1.0
Over 2000	500	100	5	0.5

*Silica limits based on limiting silica in steam to 0.02-0.03 ppm.

Table 9.16 lists limits specified by the American Boiler Manufacturer's Association (ABMA) for solids and alkalinity.

C. WATER CHEMISTRY

Water impurities generally are expressed in *parts per million (ppm)*. One ppm of calcium carbonate ($CaCO_3$) requires one weight of calcium carbonate in one million weights of water. The older units of *grains per gallon (gpg)* might be encountered also.[11] *Unit equivalents per million (epm)* also are used in water analyses. This unit can be derived by dividing the concentration in ppm by the equivalent weight of the substance.[12]

Quantities of substances in water reactions are expressed "as $CaCO_3$." To simplify the calculation of chemical dosages, all chemical results are based on the molecular weight of calcium carbonate. Using this system, one ppm of one chemical (expressed as $CaCO_3$) reacts with one ppm of another chemical (expressed as $CaCO_3$), even though the combining weights are different. Appendix D gives the "as $CaCO_3$" equivalents of commonly required compounds.

Example 9.18

Soda ash (Na_2CO_3) combines with calcium sulfate ($CaSO_4$) in a treatment to remove sulfate hardness. How many ppm (as $CaCO_3$) of soda ash react with 20 ppm (as $CaCO_3$) of calcium sulfate? How many pounds of each compound are involved in removing one pmole of calcium sulfate?

By definition, 20 ppm (as $CaCO_3$) of soda ash react with 20 ppm (as $CaCO_3$).

[11] One grain equals 1/7000th of a pound. Multiply gpg by 17.1 to get ppm.

[12] Equivalent weight is the molecular weight divided by the valence.

From appendix D, the molecular weight of calcium sulfate is 136.2. One mole of soda ash is required. The molecular weight of soda ash is 106 (also from appendix D). Therefore, 106 pounds of soda ash react with 136.2 pounds of calcium sulfate.

Determining amounts of softening chemicals requires knowledge of the amounts of each water impurity. Unfortunately, water analysis tests frequently report water purity as *hardness* and *alkalinity*. Figure 9.4 can be used to calculate the quantities of each compound from hardness and alkalinity.

Figure 9.4 uses the following terms.

- total hardness: Total calcium and magnesium compounds.

- calcium: Total calcium hardness in the form of calcium bicarbonate.

- methyl orange alkalinity: Alkalinity as determined with methyl orange indicator. Either bicarbonates with carbohydrates or carbonates with hydrates. (Bicarbonates and hydrates cannot exist together.)

- hydrates: Calcium, magnesium, or sodium hydroxides.

- Sodium alkalinity: sodium hydroxide, carbonate, and phosphate.

D. WATER SOFTENING

1 Lime and Soda Ash Softening

Water softening can be accomplished with lime and soda ash to precipitate calcium and magnesium ions from the solution. *Lime treatment* has added benefits of iron removal and clarification. Practical limits of precipitation softening are 30 mg/l of $CaCO_3$ and 10 mg/l of $Mg(OH)_2$ (as $CaCO_3$) because of intrinsic solubilities.

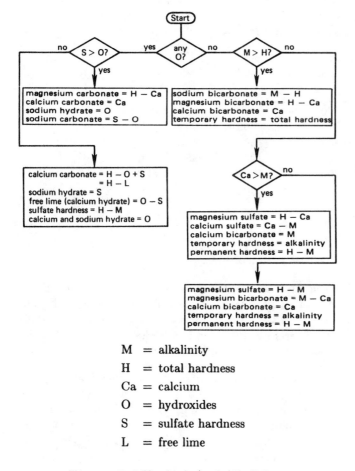

M = alkalinity

H = total hardness

Ca = calcium

O = hydroxides

S = sulfate hardness

L = free lime

Figure 9.4 Hardness and Alkalinity
(All results expressed as $CaCO_3$)

Lime (CaO) is available as granular *quicklime* (90% CaO, 10% MgO) or *hydrated lime* (68% CaO, the rest water). Both forms are *slaked* prior to use, which means that water is added to form a lime slurry in an exothermic reaction.

$$CaO + H_2O \mapsto Ca(OH)_2 + \text{heat}$$

Soda ash usually is available as 98% pure sodium carbonate (Na_2CO_3).

FIRST STAGE TREATMENT: In the *first stage treatment*, lime added to water reacts with any free carbon dioxide to form calcium carbonate precipitate.

$$CO_2 + Ca(OH)_2 \mapsto CaCO_3 \downarrow + H_2O$$

Next the lime reacts with calcium bicarbonate.

$$Ca(HCO_3)_2 + Ca(OH)_2 \mapsto 2CaCO_3 \downarrow + 2H_2O$$

Any magnesium hardness also is removed at this time.

$$Mg(HCO_3)_2 + Ca(OH)_2 \mapsto CaCO_3 \downarrow$$
$$+ 2H_2O + MgCO_3$$

To remove the soluble $MgCO_3$, the pH must be above 10.8. This is accomplished by adding an excess of approximately 35 mg/l of CaO or 50 mg/l of $Ca(OH)_2$.

$$MgCO_3 + Ca(OH)_2 \mapsto CaCO_3 \downarrow + Mg(OH)_2 \downarrow$$

The amount of lime required in the first stage softening is given by equation 9.32. Note that "alkalinity" includes both $Ca(HCO_3)_2$ and $Mg(HCO_3)_2$ contributions. If slaked lime is used, substitute 74 for 56 in equation 9.31.

$$
\begin{aligned}
\text{mg/l of} \atop \text{pure CaO} &= \frac{56}{44}\left(\text{mg/l of} \atop CO_2\right) + \frac{56}{100}\left(\text{mg/l of} \atop \text{alkalinity}\right) \\
&= \frac{56}{44}\left(\text{mg/l of} \atop CO_2\right) + \frac{56}{40.1}\left(\text{mg/l of} \atop Ca^{++}\right) \\
&\quad + \frac{56}{24.3}\left(\text{mg/l of} \atop Mg^{++}\right)
\end{aligned}
\qquad 9.31
$$

SECOND STAGE TREATMENT: The *second stage treatment* removes calcium *noncarbonate hardness* (sulfates and chlorides). This requires soda ash for precipitation.

$$CaSO_4 + Na_2CO_3 \mapsto CaCO_3 \downarrow + Na_2SO_4$$

Magnesium noncarbonate hardness needs both lime and soda ash.

$$MgSO_4 + Ca(OH)_2 \mapsto Mg(OH)_2 \downarrow + CaSO_4$$
$$CaSO_4 + Na_2CO_3 \mapsto CaCO_3 \downarrow + Na_2SO_4$$

The amount of soda ash required is

$$
\text{mg/l of} \atop Na_2CO_3 = \frac{106}{100}\left(\text{mg/l of noncarbonate hardness} \atop \text{as } CaCO_3\right)
\qquad 9.32
$$

Excess soda ash leaves sodium ions in the water. Luckily, noncarbonate hardness requiring soda ash is a small part of total hardness.

If hardness (as $CaCO_3$) and *alkalinity* (as $CaCO_3$) are the same, there are no SO_4^{--}, Cl^-, or NO_3^- ions present. (That is, there is no noncarbonate hardness.) If hardness is greater than alkalinity, noncarbonate hardness is present. If hardness is less than alkalinity, all hardness is carbonate hardness, and the extra HCO_3^- comes from other sources (such as $NaHCO_3$). Figure 9.4 ilustrates this in greater detail.

Example 9.19

How much slaked lime (90% pure), soda ash, and carbon dioxide are required to reduce the hardness of the water evaluated below to zero using the lime-soda ash process? Neglect the fact that this process cannot readily produce zero hardness. Base your answer on stoichiometric considerations.

total hardness: 250 mg/l as $CaCO_3$

alkalinity: 150 mg/l as $CaCO_3$

carbon dioxide: 5 mg/l

From equation 9.31, the pure slaked lime requirement is

$$CO_2: \left(\frac{74}{44}\right)(5) = \quad 8.4$$
$$\text{alkalinity:} \left(\frac{74}{100}\right)(150) = 111.0$$
$$\text{excess:} = \underline{\quad 50.0} \ (\text{required to raise pH to 10.8})$$
$$\text{Total: } 169.4 \text{ mg/l}$$

The actual requirement of 90% pure slaked lime is

$$\frac{169.4}{.9} = 188.2 \text{ mg/l}$$

The noncarbonate hardness is $(250 - 150) = 100$ mg/l. The soda ash requirement is given by equation 9.32.

$$\left(\frac{106}{100}\right)(100) = 106 \text{ mg/l}$$

2 Ion Exchange Method

In the *ion exchange process* (also known as *zeolite process* or *base exchange method*) water is passed through a filter bed of exchange material. This exchange material is known as *zeolite*. Ions in the insoluble exchange material are displaced by ions in the water. When the exchange material is spent, it is regenerated with a rejuvenating solution such as sodium chloride (salt). Typical flow rates are 6 gpm per ft^2 of filter bed. The processed water has a zero hardness. Since water with zero hardness may not be needed, some water can be bypassed around the unit.

There are three types of ion exchange materials. *Greensand* (glauconite) is a natural substance which is mined and treated with manganese dioxide. *Siliceous-gel zeolite* is an artificial solid used in small volume deionizer columns. *Polystyrene resins* also are synthetic. Polystyrene resins currently dominate the softening field.

During operation, the calcium and magnesium ions are removed according to the following reaction, in which R is the zeolite anion. The resulting sodium compounds are soluble.

$$\left\{\begin{matrix} Ca \\ Mg \end{matrix}\right\} \left\{\begin{matrix} (HCO_3)_2 \\ SO_4 \\ Cl_2 \end{matrix}\right\} + Na_2R \mapsto Na_2 \left\{\begin{matrix} (HCO_3)_2 \\ SO_4 \\ Cl_2 \end{matrix}\right\} + \left\{\begin{matrix} Ca \\ Mg \end{matrix}\right\} R$$

Typical characteristics of an ion exchange unit are listed here.

- exchange capacity: 3000 grains·hardness/ft^3 zeolite for natural; 5000–30,000 (20,000 typical) for synthetic.

- flow rate: 2 to 6 gpm/ft^3

- backwash flow: 5 to 6 gpm/ft^2

- salt dosage: 5 to 20 pounds/ft^3. Alternatively, .3 to .5 pounds of salt per 1000 grains of hardness removed.

- brine contact time: 25 to 45 minutes

Ion exchange does not reduce total solids, alkalinity, or silica.

Example 9.20

A plant uses water with a total hardness of 200 mg/l. The designed discharge hardness is 50 mg/l. If an ion exchange unit is used, what is the *bypass factor*?

Since the water passing through the ion exchange unit is reduced to zero hardness,

$$(1 - x)\,0 + x\,(200) = 50$$
$$x = \text{bypass factor} = .25$$

Appendix A
Heats of Combustion (1 atm and 60°F)
(Also, see page 7-18)

No.	Substance	Formula	Molecular Weight	Cu. ft. per Lbm	Heat of Combustion BTU PER CU. FT. Gross (high)	Net (low)	BTU PER LBM. Gross (high)	Net (low)
1.	Carbon	C	12.01				14,093	14,093
2.	Hydrogen	H_2	2.016	187.723	325	275	60,958	51,623
3.	Oxygen	O_2	32.000	11.819				
4.	Nitrogen	N_2	28.016	13.443				
5.	Carbon Monoxide	CO	28.01	13.506	322	322	4,347	4,347
6.	Carbon Dioxide	CO_2	44.01	8.548				
PARAFFIN SERIES (Alkanes)								
7.	Methane	CH_4	16.041	23.565	1,013	913	23,879	21,520
8.	Ethane	C_2H_6	30.067	12.455	1,792	1,641	22,320	20,432
9.	Propane	C_3H_8	44.092	8.365	2,590	2,385	21,661	19,944
10.	n-Butane	C_4H_{10}	58.118	6.321	3,370	3,113	21,308	19,680
11.	Isobutane	C_4H_{10}	58.118	6.321	3,363	3,105	21,257	19,629
12.	n-Pentane	C_5H_{12}	72.144	5.252	4,016	3,709	21,091	19,517
13.	Isopentane	C_5H_{12}	72.144	5.252	4,008	3,716	21,052	19,478
14.	Neopentane	C_5H_{12}	72.144	5.252	3,993	3,693	20,970	19,396
15.	n-Hexane	C_6H_{14}	86.169	4.398	4,762	4,412	20,940	19,403
OLEFIN SERIES (Alkenes and Alkynes)								
16.	Ethylene	C_2H_4	28.051	13.412	1,614	1,513	21,644	20,295
17.	Propylene	C_3H_6	42.077	9.007	2,336	2,186	21,041	19,691
18.	n-Butene	C_4H_8	56.102	6.756	3,084	2,885	20,840	19,496
19.	Isobutene	C_4H_8	56.102	6.756	3,068	2,869	20,730	19,382
20.	n-Pentene	C_5H_{10}	70.128	5.400	3,836	3,586	20,712	19,363
AROMATIC SERIES								
21.	Benzene	C_6H_6	78.107	4.852	3,751	3,601	18,210	17,480
22.	Toluene	C_7H_8	92.132	4.113	4,484	4,284	18,440	17,620
23.	Xylene	C_8H_{10}	106.158	3.567	5,230	4,980	18,650	17,760
MISCELLANEOUS GASES								
24.	Acetylene	C_2H_2	26.036	14.344	1,499	1,448	21,500	20,776
25.	Naphthalene	$C_{10}H_8$	128.162	2.955	5,854	5,654	17,298	16,708
26.	Methyl alcohol	CH_3OH	32.041	11.820	868	768	10,259	9,078
27.	Ethyl alcohol	C_2H_5OH	46.067	8.221	1,600	1,451	13,161	11,929
28.	Ammonia	NH_3	17.031	21.914	441	365	9,668	8,001
29.	Sulfur	S	32.06				3,983	3,983
30.	Hydrogen Sulfide	H_2S	34.076	10.979	647	596	7,100	6,545
31.	Sulfur Dioxide	SO_2	64.06	5.770				
32.	Water Vapor	H_2O	18.016	21.017				
33.	Air		28.9	13.063				
34.	Octane	C_8H_{18}		— See page 7-18				

Appendix B
Atomic Weights of Elements Referred to Carbon (12)

Element	Symbol	Atomic Weight	Element	Symbol	Atomic Weight
Actinium	Ac	(227)	Mercury	Hg	200.59
Aluminum	Al	26.9815	Molybdenum	Mo	95.94
Americium	Am	(243)	Neodymium	Nd	144.24
Antimony	Sb	121.75	Neon	Ne	20.183
Argon	Ar	39.948	Neptunium	Np	(237)
Arsenic	As	74.9216	Nickel	Ni	58.71
Astatine	At	(210)	Niobium	Nb	92.906
Barium	Ba	137.34	Nitrogen	N	14.0067
Berkelium	Bk	(249)	Osmium	Os	190.2
Beryllium	Be	9.0122	Oxygen	O	15.9994
Bismuth	Bi	208.980	Palladium	Pd	106.4
Boron	B	10.811	Phosphorus	P	30.9738
Bromine	Br	79.909	Platinum	Pt	195.09
Cadmium	Cd	112.40	Plutonium	Pu	(242)
Calcium	Ca	40.08	Polonium	Po	(210)
Californium	Cf	(251)	Potassium	K	39.102
Carbon	C	12.01115	Praseodymium	Pr	140.907
Cerium	Ce	140.12	Promethium	Pm	(145)
Cesium	Cs	132.905	Protactinium	Pa	(231)
Chlorine	Cl	35.453	Radium	Ra	(226)
Chromium	Cr	51.996	Radon	Rn	(222)
Cobalt	Co	58.9332	Rhenium	Re	186.2
Copper	Cu	63.54	Rhodium	Rh	102.905
Curium	Cm	(247)	Rubidium	Rb	85.47
Dysprosium	Dy	162.50	Ruthenium	Ru	101.07
Einsteinium	Es	(254)	Samarium	Sm	150.35
Erbium	Er	167.26	Scandium	Sc	44.956
Europium	Eu	151.96	Selenium	Se	78.96
Fermium	Fm	(253)	Silicon	Si	28.086
Fluorine	F	18.9984	Silver	Ag	107.870
Francium	Fr	(223)	Sodium	Na	22.9898
Gadolinium	Gd	157.25	Strontium	Sr	87.62
Gallium	Ga	69.72	Sulfur	S	32.064
Germanium	Ge	72.59	Tantalum	Ta	180.948
Gold	Au	196.967	Technetium	Tc	(99)
Hafnium	Hf	178.49	Tellurium	Te	127.60
Helium	He	4.0026	Terbium	Tb	158.924
Holmium	Ho	164.930	Thallium	Tl	204.37
Hydrogen	H	1.00797	Thorium	Th	232.038
Indium	In	114.82	Thulium	Tm	168.934
Iodine	I	126.9044	Tin	Sn	118.69
Iridium	Ir	192.2	Titanium	Ti	47.90
Iron	Fe	55.847	Tungsten	W	183.85
Krypton	Kr	83.80	Uranium	U	238.03
Lanthanum	La	138.91	Vanadium	V	50.942
Lead	Pb	207.19	Xenon	Xe	131.30
Lithium	Li	6.939	Ytterbium	Yb	173.04
Lutetium	Lu	174.97	Yttrium	Y	88.905
Magnesium	Mg	24.312	Zinc	Zn	65.37
Manganese	Mn	54.9380	Zirconium	Zr	91.22
Mendelevium	Md	(256)			

Appendix C
The Periodic Table of Elements

The number of electrons in filled shells is shown in the column at the extreme left; the remaining electrons for each element are shown immediately below the symbol for each element. Atomic numbers are enclosed in brackets. Atomic weights (rounded, based on Carbon-12) are shown above the symbols. Atomic weight values in parentheses are those of the isotopes of longest half-life for certain radioactive elements whose atomic weights cannot be precisely quoted without knowledge of origin of the element.

METALS   NON-METALS   TRANSITION METALS

periods	shells	I A	II A	III B	IV B	V B	VI B	VII B	VIII	VIII	VIII	I B	II B	III A	IV A	V A	VI A	VII A	O
1	0	1.0079 H[1] 1																1.0079 H[1] 1	4.0026 He[2] 2
2	2	6.941 Li[3] 1	9.0122 Be[4] 2											10.81 B[5] 3	12.011 C[6] 4	14.007 N[7] 5	16.000 O[8] 6	19.000 F[9] 7	21.179 Ne[10] 8
3	2,8	22.99 Na[11] 1	24.305 Mg[12] 2											26.982 Al[13] 3	28.086 Si[14] 4	30.974 P[15] 5	32.06 S[16] 6	35.453 Cl[17] 7	39.948 Ar[18] 8
4	2,8	39.098 K[19] 8,1	40.08 Ca[20] 8,2	44.956 Sc[21] 9,2	47.90 Ti[22] 10,2	50.941 V[23] 11,2	51.996 Cr[24] 13,1	54.938 Mn[25] 13,2	55.847 Fe[26] 14,2	58.933 Co[27] 15,2	58.70 Ni[28] 16,2	63.546 Cu[29] 18,1	65.38 Zn[30] 18,2	69.72 Ga[31] 18,3	72.59 Ge[32] 18,4	74.922 As[33] 18,5	78.96 Se[34] 18,6	79.904 Br[35] 18,7	83.80 Kr[36] 18,8
5	2,8,18	85.468 Rb[37] 8,1	87.62 Sr[38] 8,2	88.906 Y[39] 9,2	91.22 Zr[40] 10,2	92.906 Nb[41] 12,1	95.94 Mo[42] 13,1	(97) Tc[43] 14,1	101.07 Ru[44] 15,1	102.906 Rh[45] 16,1	106.4 Pd[46] 18	107.868 Ag[47] 18,1	112.41 Cd[48] 18,2	114.82 In[49] 18,3	118.69 Sn[50] 18,4	121.75 Sb[51] 18,5	127.60 Te[52] 18,6	126.905 I[53] 18,7	131.30 Xe[54] 18,8
6	2,8,18	132.905 Cs[55] 18,8,1	137.33 Ba[56] 18,8,2	* [57-71]	178.49 Hf[72] 32,10,2	180.948 Ta[73] 32,11,2	183.85 W[74] 32,12,2	186.207 Re[75] 32,13,2	190.2 Os[76] 32,14,2	192.22 Ir[77] 32,15,2	Pt[78] 32,17,1	196.967 Au[79] 32,18,1	200.59 Hg[80] 32,18,2	204.37 Tl[81] 32,18,3	207.2 Pb[82] 32,18,4	208.980 Bi[83] 32,18,5	(209) Po[84] 32,18,6	(210) At[85] 32,18,7	(222) Rn[86] 32,18,8
7	2,8,18,32	(223) Fr[87] 18,8,1	226.025 Ra[88] 18,8,2	† [89-103]	Rf[104] 32,10,2	Ha[105] 32,11,2	[106] 32,12,2	[107]	[108]										

* LANTHANIDE SERIES

138.906 La[57] 18,9,2	140.12 Ce[58] 20,8,2	140.908 Pr[59] 21,8,2	144.24 Nd[60] 22,8,2	(145) Pm[61] 23,8,2	150.4 Sm[62] 24,8,2	151.96 Eu[63] 25,8,2	157.25 Gd[64] 25,9,2	158.925 Tb[65] 27,8,2	162.50 Dy[66] 28,8,2	164.930 Ho[67] 29,8,2	167.26 Er[68] 30,8,2	168.934 Tm[69] 31,8,2	173.04 Yb[70] 32,8,2	174.97 Lu[71] 32,9,2

† ACTINIDE SERIES

(227) Ac[89] 18,9,2	232.038 Th[90] 18,10,2	231.036 Pa[91] 20,9,2	238.029 U[92] 21,9,2	237.048 Np[93] 23,8,2	(244) Pu[94] 24,8,2	(243) Am[95] 25,8,2	(247) Cm[96] 25,9,2	(247) Bk[97] 26,9,2	(251) Cf[98] 28,8,2	(254) Es[99] 29,8,2	(257) Fm[100] 30,8,2	(258) Md[101] 31,8,2	(255) No[102] 32,8,2	(260) Lr[103] 32,9,2

Appendix D
Data Used in Water Chemistry

Cations	Ion Formula	Ionic Equivalent Weight	Weight	Multiplying Factor $CaCO_3 = 100$ Substance to $CaCO_3$ equivalent $CaCO_3$ equivalent	to substance
Aluminum	Al^{+3}	27.0	9.0	5.56	0.18
Ammonium	NH_4^+	18.0	18.0	2.78	0.36
Calcium	Ca^{+2}	40.1	20.0	2.50	0.40
Hydrogen	H^+	1.0	1.0	50.00	0.02
Ferrous Iron	Fe^{+2}	55.8	27.9	1.79	0.56
Ferric Iron	Fe^{+3}	55.8	18.6	2.69	0.37
Magnesium	Mg^{+2}	24.3	12.2	4.10	0.24
Manganese	Mn^{+2}	54.9	27.5	1.82	0.55
Potassium	K^+	39.1	39.1	1.28	0.78
Sodium	Na^+	23.0	23.0	2.18	0.46
Cupric	Cu^{+2}	63.6	31.8	1.57	0.64
Cuprous	Cu^{+3}	63.6	21.2	2.36	0.42

Anions					
Bicarbonate	HCO_3^-	61.0	61.0	0.82	1.22
Carbonate	CO_3^{-2}	60.0	30.0	1.67	0.60
Chloride	Cl^-	35.5	35.5	1.41	0.71
Fluoride	F^-	19.0	19.0	2.66	0.38
Nitrate	NO_3^-	62.0	62.0	0.81	1.24
Hydroxide	OH^-	17.0	17.0	2.94	0.34
Phosphate (tribasic)	PO_4^{-3}	95.0	31.7	1.58	0.63
Phosphate (dibasic)	HPO_4^{-2}	96.0	48.0	1.04	0.96
Phosphate (monobasic)	$H_2PO_4^-$	97.0	97.0	0.52	1.94
Sulfate	SO_4^{-2}	96.1	48.0	1.04	0.96
Sulfite	SO_3^{-2}	80.1	40.0	1.25	0.80

Compounds	Formula	Molecular Equivalent Weight	Weight	Substance to $CaCO_3$ equivalent $CaCO_3$ equivalent	to substance
Aluminum hydroxide	$Al(OH)_3$	78.0	26.0	1.92	0.52
Aluminum sulfate	$Al_2(SO_4)_3$	342.1	57.0	0.88	1.14
Alumina	Al_2O_3	102.0	17.0	2.94	0.34
Sodium Aluminate	$Na_2Al_2O_4$	164.0	27.3	1.83	0.55
Calcium bicarbonate	$Ca(HCO_3)_2$	162.1	81.1	0.62	1.62
Calcium carbonate	$CaCO_3$	100.1	50.1	1.00	1.00
Calcium chloride	$CaCl_2$	111.0	55.5	0.90	1.11
Calcium hydroxide (pure)	$Ca(OH)_2$	74.1	37.1	1.35	0.74
Calcium hydroxide (90%)	$Ca(OH)_2$	—	41.1	1.22	0.82
Calcium sulfate (anhydrous)	$CaSO_4$	136.2	68.1	0.74	1.36
Calcium sulfate (gypsum)	$CaSO_4 \cdot 2H_2O$	172.2	86.1	0.58	1.72
Calcium phosphate	$Ca_3(PO_4)_2$	310.3	51.7	0.97	1.03
Disodium phosphate	$Na_2HPO_4 \cdot 12H_2O$	358.2	119.4	0.42	2.39
Disodium phosphate (anhydrous)	Na_2HPO_4	142.0	47.3	1.06	0.95

	Formula	Molecular Weight	Equivalent Weight	Substance to CaCO$_3$ equivalent	CaCO$_3$ equivalent to substance
Ferric oxide	Fe$_2$O$_3$	159.6	26.6	1.88	0.53
Iron oxide (magnetic)	Fe$_3$O$_4$	321.4	—	—	—
Ferrous sulfate (copperas)	FeSO$_4$·7H$_2$O	278.0	139.0	0.36	2.78
Magnesium oxide	MgO	40.3	20.2	2.48	0.40
Magnesium bicarbonate	Mg(HCO$_3$)$_2$	146.3	73.2	0.68	1.46
Magnesium carbonate	MgCO$_3$	84.3	42.2	1.19	0.84
Magnesium chloride	MgCl$_2$	95.2	47.6	1.05	0.95
Magnesium hydroxide	Mg(OH)$_2$	58.3	29.2	1.71	0.58
Magnesium phosphate	Mg$_3$(PO$_4$)$_2$	263.0	43.8	1.14	0.88
Magnesium sulfate	MgSO$_4$	120.4	60.2	0.83	1.20
Monosodium phosphate	NaH$_2$PO$_4$·H$_2$O	138.1	46.0	1.09	0.92
Monosodium phosphate (anhydrous)	NaH$_2$PO$_4$	120.1	40.0	1.25	0.80
Metaphosphate	NaPO$_3$	102.0	34.0	1.47	0.68
Silica	SiO$_2$	60.1	30.0	1.67	0.60
Sodium bicarbonate	NaHCO$_3$	84.0	84.0	0.60	1.68
Sodium carbonate	Na$_2$CO$_3$	106.0	53.0	0.94	1.06

Compounds

	Formula	Molecular Weight	Equivalent Weight	Substance to CaCO$_3$ equivalent	CaCO$_3$ equivalent to substance
Sodium chloride	NaCl	58.5	58.5	0.85	1.17
Sodium hydroxide	NaOH	40.0	40.0	1.25	0.80
Sodium nitrate	NaNO$_3$	85.0	85.0	0.59	1.70
Sodium sulfate	Na$_2$SO$_4$	142.0	71.0	0.70	1.42
Sodium sulfite	Na$_2$SO$_3$	126.1	63.0	0.79	1.27
Tetrasodium EDTA	(CH$_2$)$_2$N$_2$(CH$_2$COONa)$_4$	380.2	95.1	0.53	1.90
Trisodium phosphate	Na$_3$PO$_4$·12H$_2$O	380.2	126.7	0.40	2.53
Trisodium phosphate (anhydrous)	Na$_3$PO$_4$	164.0	54.7	0.91	1.09
Trisodium NTA	(CH$_2$)$_3$N(COONa)$_3$	257.1	85.7	0.58	1.71

Gases

	Formula	Molecular Weight	Equivalent Weight	Substance to CaCO$_3$ equivalent	CaCO$_3$ equivalent to substance
Ammonia	NH$_3$	17	17	2.94	0.34
Carbon dioxide	CO$_2$	44	22	2.27	0.44
Hydrogen	H$_2$	2	1	50.00	0.02
Oxygen	O$_2$	32	8	6.25	0.16
Hydrogen sulfide	H$_2$S	34	17	2.94	0.34

Acids

	Formula	Molecular Weight	Equivalent Weight	Substance to CaCO$_3$ equivalent	CaCO$_3$ equivalent to substance
Carbonic	H$_2$CO$_3$	62.0	31.0	1.61	0.62
Hydrochloric	HCl	36.5	36.5	1.37	0.73
Phosphoric	H$_3$PO$_4$	98.0	32.7	1.53	0.65
Sulfuric	H$_2$SO$_4$	98.1	49.1	1.02	0.98

PRACTICE PROBLEMS: COMBUSTION

Warm-ups

1. A compound is analyzed. What is the simplest molecular formula if the gravimetric analysis is 40% C, 6.7% H, and 53.3% O?

2. Assuming that the furnace efficiency is 50%, how many kilograms of water can be heated from 15°C to 95°C by burning 200 standard liters of methane? The heating value of methane is 24,000 BTU/lbm.

3. 15 pounds of propane (C_3H_8) are burned in air each hour stoichiometrically. How many cubic feet of dry CO_2 are formed after cooling to 70°F and 14.7 psia?

4. How many pounds of nitrogen pass through a furnace that is burning 4000 cfh of methane? Assume 30% excess air (15 psia, 100°F) is needed for complete combustion.

5. How many pounds of air are required to burn one pound of a fuel which is 84% carbon, 15.3% hydrogen, .3% sulfur, and .4% nitrogen? Assume complete combustion.

6. Propane (C_3H_8) is burned with 20% excess air. What percent of CO_2 by weight will be in the flue gas?

Concentrates

1. Coal contains 80% carbon, 4% hydrogen, and 2% oxygen. The stack gases are 12% carbon dioxide, 1% carbon monoxide, 7% oxygen, and 80% nitrogen by volume. Approximately how many pounds of air are required to burn one pound of coal under these conditions?

2. Coal has the following composition: 75% carbon, 5% hydrogen, 3% oxygen, and 2% nitrogen by weight. What are the theoretical and actual temperatures of the combustion products? 60°F air is available.

3. A fuel oil has the following ultimate analysis: 85.43% carbon, 11.31% hydrogen, 2.7% oxygen, .22% nitrogen, .34% sulfur. What volumes of 600°F wet and dry stack gases will be produced with 60% excess air? What will be the percent of carbon dioxide with 60% excess air?

4. 15,395 pounds of coal are burned, and 2816 pounds of ash containing 20.9% carbon are recovered. The original ultimate analysis was: 51.45% carbon, 4.02% hydrogen, 7.28% oxygen, .93% nitrogen, 3.92% sulfur, 16.69% ash, 15.71% moisture. How much air was supplied per pound of fuel if 13.3 pounds of dry gases are formed per pound burned?

5. A coal containing 65% carbon requires 9.45 pounds of theoretical air per pound. During combustion, 3% of the coal is lost in the refuse. The flue gas analysis is 9.5% carbon dioxide, 9.0% oxygen, 81.5% nitrogen. What was the percent excess air over theoretical?

6. What heat loss is experienced due to the formation of carbon monoxide if coal is burned with a 3% carbon loss? The coal has the following composition: 67.34% carbon, 4.28% sulfur, 4.43% hydrogen. 4.91% oxygen, 1.08% nitrogen. The stack gases (by volume) are: 15.5% carbon dioxide, 1% oxygen, 1.6% carbon monoxide, 81.9% nitrogen.

7. A natural gas is 93% methane, 3.4% nitrogen, .45% carbon monoxide, 1.82% hydrogen, .25% ethylene, .18% hydrogen sulfide, .35% oxygen, and .22% carbon dioxide by volume. What is the gas density? What are the theoretical air requirements? What will be the wet and dry stack gas analyses with 40% excess air? Use 1 atmosphere and 60°F.

8. Coal (78.42% carbon, 5.56% hydrogen, 8.25% oxygen, 1.09% nitrogen, 1.0% sulfur, and 5.68% ash) with 1.91% moisture and having a heating value of 14,000 BTU/lbm is consumed with a loss of 7.03% in refuse and ash. The ash contains 31.5% carbon. Air at 67°F wet bulb and 73°F dry bulb is supplied. The 575°F stack gases are 80.08% nitrogen, 14.0% carbon dioxide, 5.5% oxygen, and .42% carbon monoxide. 11.12 pounds of water at 1 atmosphere are evaporated per pound of dry coal burned. What is the complete heat balance? Saturated water enters the boiler at 212°F. The coal is originally at 73°F.

9. A town's water supply has the following hypothetical ion concentrations. (a) What is the total hardness in mg/l (as $CaCO_3$)? (b) How much lime ($Ca(OH)_2$) and soda ash are required to remove the carbonate hardness?

Ca^{++}	80.2 mg/l	CO_3^{--}	0
Na^+	46.0 mg/l	Mg^{++}	24.3 mg/l
NO_3^-	0	Fl^-	0
Cl^-	85.9 mg/l	SO_4^{--}	125 mg/l
CO_2	19 mg/l	Fe^{++}	1.0 mg/l
Al^{+++}	0.5 mg/l	HCO_3^-	185 mg/l

10. The following concentrations of inorganic compounds are found during a routine analysis of a city's water supply.

$$Ca(HCO_3)_2 = 137 \text{ mg/l (as } CaCO_3)$$
$$CO_2 = 0 \text{ mg/l}$$
$$MgSO_4 = 72 \text{ mg/l (as } CaCO_3)$$

(a) How many pounds of lime ($Ca(OH)_2$) and soda ash (Na_2CO_3) are required to soften one million gallons of this water to 100 mg/l if 30 mg/l of excess lime is required for a complete reaction? (b) How many pounds of salt would be required if a zeolite process with the following characteristics is used?

exchange capacity: 10,000 grains hardness/ft^3

salt requirement: .5 pound/1000 grains hardness removed

Timed (1 hour allowed for each)

1. A coal-fired utility boiler has the following characteristics.

coal feed rate: 15,300 lbm/hr

electric power rating: 17,000 kw (max)

generator efficiency: 95%

steam generator efficiency: 86%

refuse removal rate: 410 lbm/hr

30% carbon, 0% sulfur in refuse

ultimate fuel analysis: 76.56% carbon, 7.7% oxygen, 2.44% sulfur, 6.1% silicon, 1.7% nitrogen, 5.5% hydrogen.

(a) Assume all sulfur and carbon is either burned or removed as ash. Determine the emission rate of solid particulates in lbm/hr.

(b) Determine sulfur dioxide production in lbm/hr.

(c) 225 cubic feet per second of river water are diverted for cooling. What is the temperature rise?

(d) The EPA has set a limit of 0.1 lbm particulate per million BTU per hour fuel feed. If stack gas particulate collectors are required to meet the limit, what efficiency must the collection system have?

2. 250 SCFM of C_3H_8 are mixed with an oxidizer consisting of 60% O_2 and 40% N_2 by volume in a proportion allowing 40% excess oxygen by weight. Maximum velocity for the two reactants when combined is 400 fpm at 14.7 psia and 80°F. Maximum velocity for the exhaust is 800 fpm at 8 psia and 460°F. Size the inlet pipes and the stack. Find the actual flow of oxygen, actual stack gas volume, and the exhaust dew point.

3. An industrial process needs heated gas at 3600°R and 14.7 psia. It is proposed that propane (C_3H_8) be burned in a mixture of nitrogen and oxygen. After passing through the process, gas will be exhausted through a duct, being cooled slowly to 100°F and 14.7 psia before exhausting.

(a) Determine the proportions by weight of the oxygen and the nitrogen to be used. Assume stoichiometric combustion.

(b) Determine the amount of liquid water (per mole of methane), if any, present in the exhaust. The following data are available.

Enthalpy of formation (BTU/pmole @ reference temperature)

$$C_3H_8\ (g): \quad \Delta H_f = +28,800$$
$$CO_2\ (g): \quad \Delta H_f = -169,300$$
$$H_2O\ (g): \quad \Delta H_f = -104,040$$

Enthalpy increase (from reference temperature to 3600°R, in BTU/pmole)

H_2O:	31,658
CO_2:	39,791
N_2:	24,471

RESERVED FOR FUTURE USE

RESERVED FOR FUTURE USE

10 HEAT TRANSFER

Nomenclature

a	diffusivity	ft^2/hr
A	area	ft^2
c	specific heat	BTU/lbm-°R
d	diameter	inches
D	diameter	ft
E	emissive power	BTU/hr-ft^2
f	friction factor	–
F_c	correction factor	–
g	acceleration due to gravity (4.17 EE8)	ft/sec^2 or ft/hr^2
G	mass flow per unit area (v ρ)	lbm/sec-ft^2
h	film coefficient	BTU/hr-ft^2-°R
k	thermal conductivity	BTU-ft/hr-ft^2-°R
L	thickness, or (characteristic) length	ft
$\dot{m}$	mass flow	lbm/sec
n	exponent	–
p	pressure	psf
q	heat flow[1]	BTU/hr
Q	flow rate	ft^3/hr
r	radius	ft
R	fouling factor	hr-°F-ft^2/BTU
S	spacing	ft
t	wall thickness, or time	ft, hr
T	temperature	°R
U	overall coefficient of heat transfer	BTU/hr-ft^2-°R
v	velocity	ft/sec
x	distance	ft
z	depth	ft

Symbols

α	absorptivity	–
β	volumetric coefficient of thermal expansion	%/°R or %/°F
γ	temperature coefficient of thermal conductivity	%/°R or %/°F
ϵ	emissivity	–

[1]BTUH also is used as an abbreviation for BTU per hour.

ρ	density, reflectivity, or resistivity	lbm/ft^3, –, or ohm-in
σ	electrical conductivity, or Stefan-Boltzmann constant (.1713 EE–8)	–, or BTU/hr-ft^2-°R^4
τ	transmissivity	–
ν	kinematic viscosity	ft^2/sec
μ	absolute fluid viscosity	lbm/ft-sec
$\mathcal{E}$	effectiveness	–

Subscripts

a	arrangement
b	black body, or boiling
e	electrical, or emissive
f	fouling
g	gray
H	hydraulic
i	inner
l	liquid
L	longitudinal
m	logarithmic mean
o	outer, or reference, or overall
p	constant pressure
r	radiation
s	surface
t	time t, transverse, or total
v	vapor
∞	far distance

1 INTRODUCTION

In the mid-1800's, researchers noticed a relationship between thermal and electrical conductivities. In 1900, the *electron gas hypothesis* was proposed. According to this hypothesis, thermal and electrical conductance are due to the free valence electrons moving through the metallic crystalline lattice. Insulating solids, which have few free electrons, conduct heat by the agitation of atoms vibrating about their equilibrium points in the lattice. This mode of transfer is several orders of magnitude less efficient than conduction by free electrons.

Although conductivity in perfect crystals would be independent of the temperature, real solids exhibit conductivities which vary with changes in temperature. Conductivity at absolute zero is zero, and for the first few degrees above absolute zero, it increases with increases in temperature.

At common temperatures, *thermal conductivity* (k) in solids varies according to equation 10.1.

$$k(T) = k_o(1 + \gamma T) \qquad 10.1$$

Tabulated values of γ are not common, having been replaced with tabulations of k itself versus T for common materials. In most calculations, the average thermal conductivity (conductivity at the arithmetic mean temperature) is used, and no other attention is paid to variations in conductivity with temperature.

In liquids, heat is transmitted by longitudinal vibrations, similar to sound waves. Conductivity in water and aqueous solutions increases with increases in temperature up to approximately 250°F and then gradually decreases. Conductivity decreases with increased concentrations of aqueous solutions, as it does with most other liquids. Conductivity increases with increases in pressure. Of the nonmetallic liquids, water is the best thermal conductor.

The *net transport theory* can be used to explain heat conduction through gases. Hot molecules move faster than cold molecules, traveling to cold areas with greater frequency than cold molecules travel to hot areas. Conductivity in gases increases almost linearly with increases in temperature, but it is fairly independent of pressure in common ranges.

Table 10.1 lists representative thermal conductivities for several substances. All the conductivities except hydrogen were evaluated at 32°F. Hydrogen was evaluated at 100°F. Notice that BTU-ft/hr-ft²-°R is the same as BTU/hr-°R-ft. These units are not the same as BTU-in/hr-°R-ft², which also is used.[2]

Table 10.1
Typical Thermal Conductivities
(BTU-ft/hr-ft²-°R)

Material	k
silver	242
copper	224
aluminum	117
brass	56
steel 1%C	27
lead	20
ice	1.3
glass	.63
concrete	.5
water	.32
hydrogen	.11
fiberglass	.03
cork	.025
oxygen	.016
air	.014

2 STEADY STATE CONDUCTION THROUGH PLANE SURFACES

Conduction, the flow of heat through solids, is given by *Fourier's law.*[3]

$$q = kA \frac{dT}{dL} \qquad 10.2$$

If the heat transmission is steady, and both k and A are constant, heat flow through a single slab of thickness L is given by equation 10.3.

$$q = \frac{kA\Delta T}{L}$$
$$\Delta T = T_1 - T_2 \qquad 10.3$$

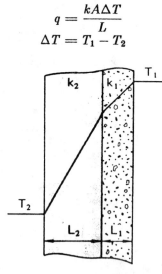

Figure 10.1 A Composite Slab Wall

---

[2]Multiply cal-cm/sec-°K-cm² by 241.9 to get BTU-ft/hr-°R-ft².

[3]Fourier's law usually is written with a minus sign to indicate that the heat flow is opposite the direction of the thermal gradient. This chapter assumes that the direction of heat flow can be determined by inspection.

For composite sandwiched materials, the heat flow due to conduction, as shown in figure 10.1, is

$$q = \frac{A\Delta T}{\sum \frac{L_i}{k_i}} \qquad 10.4$$

To account for the thermal resistance of films on exposed surfaces without having to measure the film thickness, the film thermal resistance is given by a *film coefficient, h*.[4] The heat flow through a film is

$$q = hA\Delta T \qquad 10.5$$

Table 10.2
Typical Film Coefficients
(BTU/hr-ft^2-°F)

Film Material	h
No change in phase	
air, still	1.65
air, with 7.5 mph wind (summer)	4.00
air, with 15 mph wind (winter)	6.00
water	150 to 2000
other gases	3 to 50
gasoline, kerosene, alcohol, and other organic solvents	60 to 500
oils	10 to 120
Condensing	
steam	1000 to 3000
organic solvents	150 to 500
light oils	200 to 400
heavy oils	20 to 50
ammonia	500 to 1000
Evaporating	
water	800 to 2000
organic solvents	100 to 300
light oils	150 to 300
heavy oils	10 to 50
ammonia	200 to 400
Freon-12	100 to 600

The heat flow through a composite sandwich with films is

$$q = \frac{A\Delta T}{\sum \frac{L_i}{k_i} + \sum \frac{1}{h_j}} = UA\Delta T \qquad 10.6$$

The *overall conductivity* is[5]

$$U = \frac{1}{\sum \frac{L_i}{k_i} + \sum \frac{1}{h_j}} \qquad 10.7$$

---

[4] The film coefficient commonly is written as $\overline{h}$, the overbar indicating an *average* film coefficient over the entire area of heat transfer. This chapter omits the overbar.

[5] The overall conductivity also is known as the *overall coefficient of heat transfer* and the *heat transmittance*.

The temperature at any point within a wall, composite or not, can be found if the heat transfer rate, q, is known. The thermal resistance is calculated up to the point whose temperature is wanted. Then equation 10.6 is solved for ΔT. Since this method requires knowing q, it may be necessary to work the problem iteratively, particularly if k varies greatly with temperature in the material being investigated.

Example 10.1

A 100 square foot wall consists of four inches of red brick, one inch of pine, and 1/2 inch of plasterboard in series. Evaluate the heat loss in a 15 mph wind if the external and internal temperatures are 30°F and 72°F, respectively. Use the following data.

material	t	k	h
brick	.333	.38	
pine	.083	.06	
plaster	.042	.30	
air, inside			1.65
air, outside			6.00

From equation 10.7, the overall conductivity is

$$\frac{1}{U} = \frac{.333}{.38} + \frac{.083}{.06} + \frac{.042}{.30} + \frac{1}{1.65} + \frac{1}{6.00} = 3.17$$
$$U = .315 \text{ BTU/hr-ft}^2\text{-°F}$$

The heat flow is

$$q = (.315)(100)(72 - 30) = 1323 \text{ BTU/hr}$$

3 CONDUCTION THROUGH CURVED SURFACES

Equations 10.3 and 10.6 require a uniform path length and a constant area. In a situation where the heat flow is through a changing area, the *logarithmic mean area* should be used.

$$A_m = \frac{A_o - A_i}{ln\left(\frac{A_o}{A_i}\right)} \qquad 10.8$$

However, if $A_o/A_i \leq 2$, an approximation of $A = \frac{1}{2}(A_o + A_i)$ will result in a maximum error of 4%, with q being too high.

A. HOLLOW CYLINDER

The radial heat flow out of an uninsulated hollow cylinder without films (neglecting the heat flow out of the cylinder ends) is

$$q = \frac{2\pi k L \Delta T}{ln\left(\frac{r_o}{r_i}\right)} \qquad 10.9$$

$$A_m = \frac{2\pi L (r_o - r_i)}{ln\left(\frac{r_o}{r_i}\right)} \qquad 10.10$$

B. INSULATED PIPE

Equation 10.11 can be used for an insulated pipe with films. (All dimensions must be in feet.)

$$q = \frac{2\pi L \Delta T}{\dfrac{1}{r_a h_a} + \dfrac{ln\left(\frac{r_b}{r_a}\right)}{k_{pipe}} + \dfrac{1}{r_b h_b} + \dfrac{ln\left(\frac{r_c}{r_b}\right)}{k_{in}} + \dfrac{1}{r_c h_c}} \qquad 10.11$$

h VALUES ON PAGE 10-3

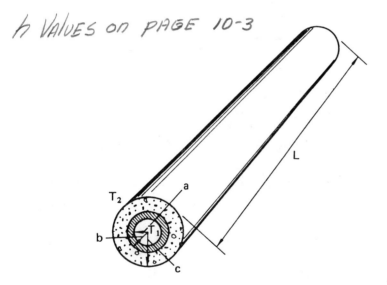

Figure 10.2 An Insulated Pipe

If the wall and the layers are very thin, or if the radii are large so that $A \approx A_a \approx A_c$, the effects of curvature can be ignored, and equation 10.6 can be used.

The addition of insulation to a bare pipe or wire also increases the surface area. Therefore, adding insulation up to the $\boxed{critical\ thickness}$ actually will increase the heat loss above bare-pipe levels. This critical radius usually is very small, and it is most relevant in the cases of thin wires or capillaries. The *critical radius* is

$$r_{critical} = \frac{k_{insulation}}{h} \qquad 10.12$$

Example 10.2

Liquid oxygen at $-290°F$ is stored in a 5' inside diameter, 20' long cylindrical stainless steel tank covered with one foot of powdered diatomaceous silica (average thermal conductivity of .022 BTU/ft-hr-°F). The environment temperature is 70 °F, and the wind is 15 mph. The tank walls are 3/8" thick. Compare the radial heat gain to the liquid oxygen using equations 10.6 and 10.11. Disregard heat gain through the ends. Use the following data.

material	t	k	h
stainless	.031	28.0	
silica	1.0	.022	
air, outside			6.0
oxygen, inside			∞

$3/8'' = .031'$. The necessary radii are

$$r_a = \frac{5}{2} = 2.5 \text{ ft}$$

$$r_b = \frac{5 + (2)(.031)}{2} = 2.53 \text{ ft}$$

$$r_c = \frac{5 + (2)(.031) + (2)(1)}{2} = 3.53 \text{ ft}$$

Equation 10.11 gives the exact solution as

$$q = \frac{2\pi (20)(70 + 290)}{\dfrac{ln\left(\frac{2.53}{2.50}\right)}{28.0} + \dfrac{ln\left(\frac{3.53}{2.53}\right)}{.022} + \dfrac{1}{(3.53)(6.0)}}$$

$$= 2980 \text{ BTU/hr}$$

If the effects of curvature are ignored, equation 10.6 predicts the heat loss based on the outside area.

$$q = \frac{2\pi (3.53)(20)(70 + 290)}{\dfrac{.031}{28.0} + \dfrac{1.0}{.022} + \dfrac{1}{6}} = 3500 \text{ BTU/hr}$$

C. SPHERE

Spheres have the largest volume-to-surface area ratio, so they are used in situations where heat transmission is to be kept at a minimum. Heat transmission through spheres, however, cannot be reduced indefinitely by the addition of great amounts of insulation, whereas such a reduction is possible for flat or cylindrical walls. The heat flow out of a hollow, uninsulated sphere is given by equation 10.13.[6]

$$q = \frac{4\pi k \Delta T (r_i r_o)}{(r_o - r_i)} \qquad 10.13$$

---

[6] Use equation 10.13 for insulated spheres by ignoring the thermal resistance of the tank's material. Be sure to check the Biot modulus to see if transient methods should be used.

4 TRANSIENT HEAT FLOW

Special Nomenclature

A_s	surface area	ft^2
h	average surface film coefficient	BTU/hr-ft^2-°R
k	body thermal conductivity	BTU-ft/hr-ft^2-°R
L	characteristic length $= V/A_s$	ft
q_t	heat transfer at time t	BTU/hr
t	time	hr
ΔT	$T_o - T_\infty$	°R
T_o	initial body temperature	°R
T_t	body temperature at time t	°R
T_∞	constant environment temperature	°R
V	body volume	ft^3
ρ	body density	lbm/ft^3

If the temperature difference between a body and its surroundings is not too large, the total rate of heat transfer by conduction, convection, and radiation is approximately proportional to the temperature difference. This empirical relationship is known as *Newton's law of cooling*. The temperature at time t is

$$T_t = T_\infty + (T_o - T_\infty)e^{-rt} \qquad 10.14$$

r is a constant dependent on the situation, usually found by solving equation 10.15 for r given some temperature versus time information. The time to achieve a drop in temperature is

$$t = -\left(\frac{1}{r}\right)ln\left(\frac{T_t - T_\infty}{T_o - T_\infty}\right) \qquad 10.15$$

Example 10.3

Water in a 5″ diameter, 10′ long tank cools from 200°F to 190°F in 1.25 hours. The ambient temperature is 75°F. How long will it take for the water to cool from 200°F to 125°F?

The thickness of this tank is small, and it can be assumed that the internal resistance is small. Sufficient information is given to calculate the rate constant.

From equation 10.14,

$$190 = 75 + (200 - 75)e^{-r(1.25)}$$
$$r = .0667 \ 1/hr$$

Now that r is known, t can be found from either equation 10.14 or 10.15.

$$125 = 75 + (200 - 75)e^{-.0667t}$$
$$t = 13.74 \text{ hours}$$

A simplified approach is possible if the body temperature is homogeneous throughout the cooling process. This will be true if the external resistance is much greater than the internal resistance. To check, calculate the *Biot number* (also known as the *Biot modulus*), N_{Bi}. If N_{Bi} is less than .10, the assumption is good, and the so-called *lumped parameter method* can be used to approximate the transient heat flow (with an error of less than 5%).

$$N_{Bi} = \frac{hL}{k} \qquad 10.16$$

The *Fourier number*, also known as *relative time*, is also needed in transient heat flow problems. (a is known as the *thermal diffusivity* of the material. L is redefined when used with charts such as figure 10.3.)

$$N_{Fo} = \frac{kt}{\rho c_p L^2} = \frac{at}{L^2} \qquad 10.17$$

$$a = \frac{k}{\rho c_p} \qquad 10.18$$

Using the lumped parameter method, the temperature at time t is

$$T_t = T_\infty + \Delta T e^{-N_{Bi}N_{Fo}} \qquad 10.19$$

The heat flow at time t is

$$q_t = hA_s\Delta T e^{-N_{Bi}N_{Fo}} \qquad 10.20$$

The analogy of a capacitor discharging through a resistor often is used to describe lumped parameter performance. This analogy has given rise to the terms of thermal capacitance and thermal resistance.

The *thermal capacitance (capacity)* is

$$C_e = c_p\rho V \qquad \text{(in BTU/°R)} \qquad 10.21$$

The *thermal resistance* (reciprocal of conductance) is

$$R_e = \frac{1}{hA_s} \qquad \left(\text{in } \frac{\text{hr-°R}}{\text{BTU}}\right) \qquad 10.22$$

The thermal *time constant* of the thermal gradient decay is given by equation 10.23. The heat content of the source will be reduced by 63% in one time constant (i.e., reduced to 37% of its original value).

$$\text{time constant} = \frac{c_p\rho L}{h} \qquad \text{(in hr)} \qquad 10.23$$

These analogous variables can be used to rewrite equations 10.19 and 10.20. The temperature at time t is

$$T_t = T_\infty + \Delta T e^{-\frac{t}{R_e C_e}} \qquad 10.24$$

The heat flow at time t is

$$q_t = hA_s\Delta T e^{-\frac{t}{R_e C_e}} \qquad 10.25$$

Example 10.4

A .03125″ diameter copper wire is heated by a short circuit to 300°F before a slow-blow fuse burns out. If the still air temperature is 100°F, how long will it take for the wire to drop to 120°F?

material	k	h	c	ρ
copper	224		.091	558
still air		1.65		

The characteristic length of the wire is

$$L = \frac{V}{A_s} = \frac{\pi r^2 l}{2\pi r l} = \frac{r}{2} = \frac{.03125}{4(12)} = .00065 \text{ ft}$$

The Biot number is

$$N_{Bi} = \frac{hL}{k} = \frac{(1.65)(.00065)}{224} = .00000479$$

Since the Biot number is less than .1, the internal thermal resistance can be ignored. The Fourier number is

$$N_{Fo} = \frac{kt}{\rho c_p L^2} = \frac{224\,t}{(558)(.091)(.00065)^2} = 10,441,000\,t$$

The temperature response, from equation 10.19, is

$$T_t = 100 + 200\,e^{[-(.00000479)(10,441,000t)]}$$
$$= 100 + 200\,e^{(-50t)}$$

But $T_t = 120$, so

$$.1 = e^{(-50t)}$$
$$ln(.1) = -50t$$
$$t = .046 \text{ hours}$$

If the Biot modulus is greater than .10, a graphical approach is necessary. Graphs to aid in the evaluation of transient heat flow problems are available only for the simplest of shapes (e.g., spheres, infinite slabs, and cylinders).

Figure 10.3 can be used with spheres. (The quantity r_o is the outside radius. It is not the characteristic length. The quantity k/hr_o is not the reciprocal of the Biot modulus, but the Biot modulus must be calculated to determine if it is necessary to use this figure.) Similar charts are available for other shapes.

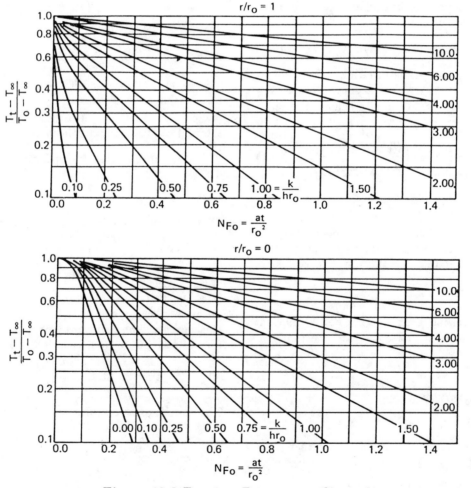

Figure 10.3 Transient Temperature Charts for Spheres

Example 10.5

Whole peaches at 80°F are to be cooled in 14.7 psia, 10°F air by convection prior to being completely frozen. The peaches are placed on trays with spacing far enough apart so that they do not affect each other. How long will it take for the peach centers to cool to 40°F? Assume the peaches are homogeneous three inch diameter spheres with the following properties.

conductivity $= .4$ BTU/hr-ft-°F
film coefficient $= 5.8$ BTU/hr-ft²-°F
specific heat $= 1.0$ BTU/lbm-°F
thermal diffusivity $= .00625$ ft²/hr

Check the Biot modulus to see if this problem can be solved with the lumped parameter method. The characteristic length is

$$L = \frac{V}{A_s} = \frac{\frac{4}{3}\pi r^3}{4\pi r^2} = \frac{r}{3} = \frac{3}{(2)(12)(3)} = .0417$$

$$N_{Bi} = \frac{hL}{k} = \frac{(5.8)(.0417)}{.4} = .605$$

Since .605 is greater than .10, figure 10.3 must be used.

$$\frac{T_t - T_\infty}{T_o - T_\infty} = \frac{40 - 10}{80 - 10} = .43$$

$$r_o = \frac{3}{(2)(12)} = .125 \text{ ft}$$

$$\frac{k}{hr_o} = \frac{.4}{(5.8)(.125)} = .55$$

Since the center temperature is given, use the $r/r_o = 0$ chart in figure 10.3.

The Fourier modulus (based on r_o) is found to be approximately .3.

From equation 10.17, using r_o for L,

$$t = \frac{r_o^2 N_{Fo}}{a} = \frac{(.125)^2(.3)}{.00625} = .75 \text{ hr}$$

5 CONDUCTION WITH INTERNAL HEAT GENERATION

Some solids generate their own heat. Examples are electric heating elements, furnace coal beds, and nuclear fuel rods. Only the simplest cases can be handled analytically. The following analysis assumes a steady state operation and uniform heat generation in a homogeneous material.

A. FLAT PLATE

Consider a flat plate which is infinite in two out of three dimensions, has a thickness of 2L, and generates heat

internally at the rate of q^* BTU/ft³-hr, which is dissipated from two sides. The temperature at layer x is given by equation 10.26. (x is measured from the closest surface and cannot exceed L.) T_s is the surface temperature.

$$T_x = \left(\frac{-q^*}{2k}\right)x^2 + \left(\frac{q^*L}{k}\right)x + T_s \qquad 10.26$$

$$T_{center} - T_s = \frac{q^*L^2}{2k} \qquad 10.27$$

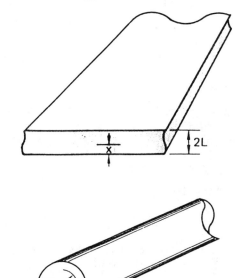

Figure 10.4 Infinite Plate and Long Bar

B. LONG BAR

$$T_r = T_s + \left(\frac{q^*r^2}{4k}\right)\left[1 - \left(\frac{x}{r}\right)^2\right]$$

$$= T_{center} - \left(\frac{x^2 q^*}{4k}\right) \qquad 10.28$$

$$T_s = T_{center} - \left(\frac{r^2 q^*}{4k}\right) \qquad 10.29$$

Example 10.6

A $\frac{1}{4}''$ copper plate is surrounded by 80°F coolant. The resistivity of the copper is .68 EE−6 ohm-inch. If a potential of .01 volt is applied across the faces, what is the center temperature?

Consider a $\frac{1}{4}'' \times 6.93' \times 6.93'$ lump (which has a volume of 1 ft³). The resistance of such a lump is

$$R = \rho\left(\frac{L}{A}\right) = \frac{(.68 \text{ EE} - 6) \text{ ohm-in } (.25) \text{ in}}{(12) \text{ in/ft } (12) \text{ in/ft } (6.93)^2 \text{ ft}^2}$$

$$= 2.46 \text{ EE} - 11 \text{ ohms}$$

The power dissipation in the lump is

$$q^* = \frac{V^2}{R} = \frac{(.01)^2}{2.46 \, EE-11}$$
$$= 4.07 \, EE6 \text{ watts/ft}^3$$
$$= 1.39 \, EE7 \text{ BTU/hr-ft}^3$$

From equation 10.27, the center temperature is

$$80 + \frac{(1.39 \, EE7)\left[\frac{0.25}{(2)(12)}\right]^2}{(2)(224)} = 83.4 \text{ °F}$$

6 FINNED HEAT RADIATORS

Special Nomenclature

A	uniform cross sectional area (annulus for pipes) ft^2	
h	**average film coefficient**	BTU/hr-ft^2-°R
h$_e$	film coefficient for fin end	BTU/hr-ft^2-°R
m	$\sqrt{hP/kA}$	ft^{-1}
P	**fin perimeter (circumference)** $(2 \times (b+t))$ for rectangular fin	ft
T$_s$	temperature at base	°R
T$_\infty$	temperature of surrounding fluid	°R

If a uniform temperature across the face of the fin is assumed, equations 10.32 through 10.35 can be used

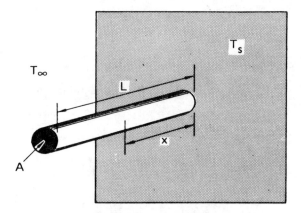

Figure 10.5 A Finite Rod Fin

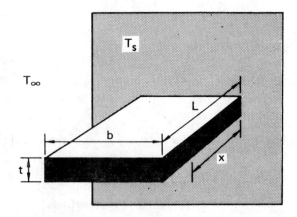

Figure 10.6 A Finite Rectangular Fin

to calculate the heat transfer and the temperature distribution along the fin length. (The arguments of the hyperbolic functions must be expressed in radians.)

The *fin efficiency* is the ratio of the actual heat loss from the fin to the ideal heat loss assuming the entire fin is at T_s.

$$\eta = \frac{q}{hA_s(T_s - T_\infty)}$$
$$= \frac{\tanh(Lm)}{Lm} \qquad 10.30$$

If hA/Pk is greater than 1, the fin will serve as insulation, and it will not increase the heat transfer.

If h is not known, the fluid properties should be evaluated at the film temperature.

$$T_{\text{film}} = \tfrac{1}{2}\left[\tfrac{1}{2}(T_s + T_{x=L}) + T_\infty\right] \qquad 10.31$$

A. INFINITE ROD

$$T_x - T_\infty = (T_s - T_\infty)e^{-mx} \qquad 10.32$$
$$q = \sqrt{hPkA}\,(T_s - T_\infty) \qquad 10.33$$

B. FINITE ROD WITH ADIABATIC TIP

$$T_x - T_\infty = (T_s - T_\infty)\left(\frac{\cosh[m(L-x)]}{\cosh(mL)}\right) \qquad 10.34$$

$$q = \sqrt{hPkA}\,(T_s - T_\infty)\tanh(mL) \qquad 10.35$$

C. RECTANGULAR FINS

Use the rod formula with $P = 2(b+t)$.

Example 10.7

A $\frac{1}{2}'' \times \frac{1}{2}'' \times 10''$ rod ($k = 80$ BTU/hr-ft-°F) has one end held at a constant 300°F. The still air temperature is 80°F. What is the temperature 3" from the base if the average film coefficient is 1.65?

Since the tip area is only 1.2% of the total surface area, the adiabatic case (equation 10.34) will be used. Let $P = 2(.5 + .5)/12 = .167$ ft. The area is $A = (.5)^2/144 = .00174$. Then

$$m = \sqrt{\frac{(1.65)(.167)}{(80)(.00174)}} = 1.41$$

$$T_3 = 80 + 220\left[\frac{\cosh\left[\left(\frac{7}{12}\right)(1.41)\right]}{\cosh\left[\left(\frac{10}{12}\right)(1.41)\right]}\right] = 248.4°F$$

7 TWO-DIMENSIONAL CONDUCTION

Many important heat transfer problems cannot be solved with the assumption of one-dimensional heat flow. Unfortunately, there are only a few problems that have closed-form solutions. More complex problems require iterative, graphical, or numerical methods to solve.

A. BURIED SPHERE

If a sphere of diameter D and temperature T_2 is buried a distance z (as measured from the sphere's center), the heat transfer through the soil (conductivity k) is given by equation 10.36. T_1 is the soil's surface temperature.

$$q = \left[\frac{2\pi D}{1 - \left(\frac{D}{4z}\right)}\right] k(T_2 - T_1) \qquad 10.36$$

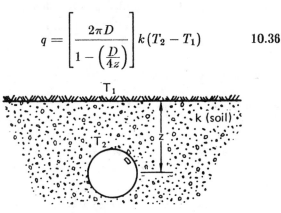

Figure 10.7 A Buried Sphere

B. BURIED PIPE

If an isothermal pipe is buried, its heat transfer can be found from equation 10.37 if both D and z are much smaller than the pipe length. If L is not known, solve for q/L.

$$q = \left[\frac{2\pi L}{\text{arccosh}\left(\frac{2z}{D}\right)}\right] k(T_2 - T_1) \qquad 10.37$$

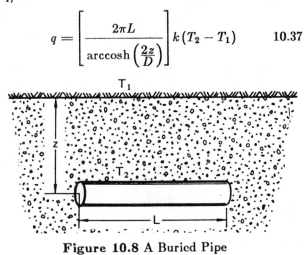

Figure 10.8 A Buried Pipe

8 FORCED CONVECTION

A. INTRODUCTION

Unlike conduction, convective heat transfer depends on the movement of a fluid. *Forced convection* results from a fan, a pump, or a relative motion driven flow. If the flow is over a flat surface, the fluid near the surface will experience laminar flow since the surface fluid particles

adhere. The velocity distribution at the surface is approximately parabolic.

The *hydrodynamic boundary layer* (HBL) consists of all flow less than 99% of the free-flow velocity. This layer comprises the major thermal resistance. Since the fluid is essentially stationary, heat transfer through the HBL is by conduction.

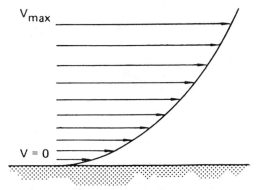

Figure 10.9 A Film Boundary Layer

Newton's law of convection gives the heat transfer for both natural and forced convection. T_∞ is the temperature of the surroundings.

$$q = hA(T_s - T_\infty) \qquad 10.38$$

B. TURBULENT FLOW IN PIPES

Finding the average film coefficient is the major part of any convection problem. For forced convection, the *Nusselt equation* can be used to find the inside film coefficient for turbulent fluid flow in round, horizontal pipes.[7] The thermal resistance of the pipe itself usually is ignored, and c_p and k refer to the fluid properties. Equation 10.39 is restricted to $(N_{Re} > 10,000)$, $(.7 < N_{Pr} < 100)$, and $(L/D > 60)$. All fluid properties are evaluated at the bulk temperature.

$$\frac{hD}{k} = .0225(N_{Re})^{.8}(N_{Pr})^n \qquad 10.39$$

$$N_{Re} = \frac{DG}{\mu} = \frac{vD}{\nu} \qquad 10.40$$

$$N_{Pr} = \frac{c_p\mu}{k} \qquad 10.41$$

$n = .3$ for heat flow out of pipe
   $(T_{\text{bulk}} > T_{\text{wall}})$
$= .4$ for heat flow into pipe
   $(T_{\text{bulk}} < T_{\text{wall}})$

If there is a large change in viscosity due to the heat transfer process, as would be the case with heating

---

[7]Equation 10.39 predicts the actual h with an accuracy of only $\pm 20\%$. Therefore, the coefficient 0.0225 frequently is rounded to **0.023**.

oils or other viscous fluids, equation 10.39 should be modified.

$$\frac{hD}{k} = 0.023\,(N_{Re})^{.8}(N_{Pr})^{.333}\left(\frac{\mu_{\text{bulk}}}{\mu_{\text{wall}}}\right)^{.14} \qquad 10.42$$

Two good approximations are available for air and water. For turbulent air between 0°F and 240°F flowing in a pipe, equation 10.43 can be used. (T is in °F in equation 10.44.)

$$h = \frac{C(3600G)^{.8}}{(D)^{.2}} \qquad 10.43$$

$$C = .00351 + .000001583T \qquad 10.44$$

For turbulent water between 40°F and 220°F flowing in a pipe of diameter d inches at T (in °F),

$$h = \frac{150\,(1 + .011T)(v)^{.8}}{(d)^{.2}} \qquad 10.45$$

Problems that require finding the pipe length to accomplish a given change in temperature can be solved directly from equation 10.46. (The 3600 term assumes v is in feet per second. Other variables also must have consistent units.)

$$L = \frac{\rho D_i^2 c_p (T_{\text{out}} - T_{\text{in}})(3600)\,v}{4U_o D_o \Delta T_m} \qquad 10.46$$

Example 10.8

What is the length of a two inch tube (2.00″ I.D., 2.125″ O.D.) needed to raise the temperature of water flowing in the pipe at 30 fps from 70 °F to 130 °F? The tube wall temperature is a constant 170 °F. (Assume $U_o = h_i$.)

The water properties are evaluated at the mean (bulk) temperature of $\frac{1}{2}(70 + 130) = 100°F$.

$$\rho = 62.0 \ \text{lbm/ft}^3$$
$$c_p = .998 \ \text{BTU/lbm-°}F$$
$$\nu = .74 \ \text{EE} - 5 \ \text{ft}^2/\text{sec}$$
$$k = .364 \ \text{BTU/hr-ft-°}F$$
$$N_{Pr} = 4.52$$

From equation 10.66, the logarithmic mean temperature is

$$\Delta T_m = \frac{100 - 40}{ln\left(\frac{100}{40}\right)} = 65.5°$$

The Reynolds number is

$$N_{Re} = \frac{vD}{\nu} = \frac{(30)\left(\frac{2}{12}\right)}{.74 \ \text{EE} - 5} = 6.76 \ \text{EE5 \ (turbulent)}$$

From equation 10.39,

$$h = \frac{(.364)(.0225)(6.76 \ \text{EE5})^{.8}(4.52)^{.4}}{\left(\frac{2}{12}\right)}$$

$$= 4144 \ \frac{\text{BTU}}{\text{hr-ft}^2\text{-°F}}$$

From equation 10.46,

$$L = \frac{(62)\left(\frac{2}{12}\right)^2(.998)(130 - 70)(3600)(30)}{(4)(4144)\left(\frac{2.125}{12}\right)(65.5)}$$

$$= 57.9 \ \text{ft}$$

C. FLOW OVER SINGLE CYLINDERS

Equations 10.47 and 10.48 give the average value of the film coefficient for use with the entire cylinder area. The pressure is assumed to be one atmosphere. The film properties should be evaluated at the temperature $\frac{1}{2}(T_{\text{wall}} + T_\infty)$. Values of C_1 and C_2 are given in table 10.3.

For air, the Hilbert/Morgan equation is

$$\frac{hD}{k} = C_1(N_{Re})^n \qquad 10.47$$

For other fluids,

$$\frac{hD}{k} = C_2(N_{Re})^n(N_{Pr})^{.333} \qquad 10.48$$

$$N_{Re} = \frac{v_\infty D}{\nu} \qquad 10.49$$

$$q = hA(T_{\text{wall}} - T_\infty) \qquad 10.50$$

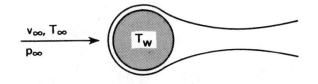

Figure 10.10 A Single Cylinder in Cross Flow

Table 10.3

N_{Re}	C_1	C_2	n
1–4	.891	.989	.330
4–40	.821	.911	.385
40–4000	.615	.683	.466
4000–40000	.174	.193	.618
40000–250000	.0239	.0266	.805

D. TUBE BUNDLES IN CROSS FLOW

Tube bundles can be in-line or staggered as shown in figure 10.11. Notice the different definitions of S_t in the diagrams.

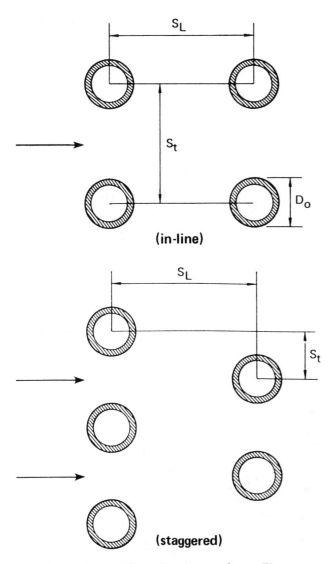

Figure 10.11 Tube Bundles in Cross Flow

Since the fluid velocity is not constant in the exchanger, the value to be used in the calculation of the Reynolds number must be defined carefully. The velocity used should be the maximum velocity.

$$v_{\max} = \frac{Q}{A_{\text{minimum}}} \qquad 10.51$$

$$
\begin{aligned}
A_{\text{minimum}} &= L_{\text{tube}} \times \quad S_t - D_o \qquad \text{(in-line)}\\
&= L_{\text{tube}} \times \text{ minimum} \left\{ \begin{array}{c} 2S_t - D_o \\ \sqrt{S_t^2 + S_L^2} - D_o \end{array} \right\}\\
&\qquad\qquad\qquad\qquad \text{(staggered)} \quad 10.52
\end{aligned}
$$

The heat transfer coefficient for exchangers with 10 or more transverse rows can be found from the *Colburn*

equation if the Reynolds number is greater than 5000 and great accuracy is not required. The properties (with a subscript "f") should be evaluated at $\frac{1}{2}(T_{\text{wall}} + T_\infty)$. C is equal to .33 for staggered bundles and is equal to .26 for in-line bundles.

$$\frac{hD_o}{k} = C\,(N_{Re})^{.6}(N_{Pr})^{.3} \qquad 10.53$$

$$N_{Re} = \frac{G_{\max}D_o}{\mu_f} = \frac{v_{\max}\rho D_o}{\mu_f} \qquad 10.54$$

If there are fewer than 10 transverse rows, multiply the value of the film coefficient, h, by the correction factor given in table **10.4**.

The turbulent flow pressure drop across N rows, in psf, is given by equation 10.55. f is an empirical friction factor from table 10.5.[8] The exponent .14 in equation 10.55 is for turbulent flow. Replace it with .25 for laminar flow.

$$\Delta p = \frac{f\,(3600 G_{\max})^2 N}{\rho\,(2.09\ \text{EE8})}\left(\frac{\mu_{\text{wall}}}{\mu_{\text{bulk}}}\right)^{.14} \qquad 10.55$$

Table 10.5
Approximate Friction Factors

N_{Re}	In-Line	Staggered
10	1.8–5.1	2.1–6.5
100	.24–.55	.40–.90
1000	.075–.13	.15–.20
10,000	~.10	~.10
100,000	~.08	~.08
1,000,000	~.05	~.05

9 NATURAL CONVECTION

If heat is removed by fluid acting under the influence of a density gradient, the heat removal mechanism is called *natural convection*.[9]

An alternate form of the Nusselt equation is used to find the film coefficient for natural convection. As is usually done, the film temperature is assumed to be $\frac{1}{2}(T_{\text{wall}} + T_\infty)$.

$$\frac{hL}{k} = C\,(N_{Gr}N_{Pr})^n \qquad \text{for } N_{Pr} > .6 \qquad 10.56$$

[8]Data in table 10.5 is based on limited testing. If greater accuracy is required, f can be predicted analytically by other methods. Interpolation between 1000 and 10,000 is not recommended since the transition to turbulent flow occurs in that range.

[9]The equation also can be used for cooling.

Table 10.4

	1	2	3	4	5	6	7	8	9	10
Staggered	.68	.75	.83	.89	.92	.95	.97	.98	.99	1.0
In-line	.64	.80	.87	.90	.92	.94	.96			

N_{Pr} is the dimensionless Prandtl number from equation 10.41. N_{Gr} is the dimensionless Grashoff number. (g and μ must have the same unit of time to make N_{Gr} dimensionless.)

$$N_{Gr} = \frac{L^3 \rho^2 \beta \Delta T g}{\mu^2} \qquad 10.57$$

$$\Delta T = T_s - T_\infty \qquad 10.58$$

L in equation 10.57 is the *characteristic length* of the object. Table 10.6 defines the characteristic length, n and C, for various configurations.

Table 10.7 contains simplified equations for 14.7 psia air. These simplifications can be derived readily from the standard Nusselt equation for natural convection. They are not approximations.

Example 10.9

A horizontal 20' long, 4" O.D. pipe carries 300°F steam through a 100°F room. What is the convective heat loss? (Assume the pipe wall temperature is 300°F.)

This is an example of natural convection. The fluid properties are evaluated at $\frac{1}{2}(300 + 100) = 200°F$.

The characteristic length is the pipe diameter.

$$L = \frac{4}{12} \doteq .333 \text{ ft}$$

The temperature gradient is

$$\Delta T = 300 - 100 = 200°F$$

Fluid properties to evaluate the Grashoff number are found in this chapter's appendixes.

$$k = .0174 \text{ BTU-ft/ft}^2\text{-hr-}°F$$
$$\rho = .060 \text{ lbm/ft}^3$$
$$\mu = 1.44 \text{ EE}{-}5 \text{ lbm/ft-sec}$$
$$\beta = 1.52 \text{ EE}{-}3 \text{ 1/}°F$$
$$N_{Pr} = .72$$

The Grashoff number is

$$N_{Gr} = \frac{(.333)^3 (.060)^2 (1.52 \text{ EE} - 3)(200)(32.2)}{(1.44 \text{ EE} - 5)^2}$$
$$= 6.28 \text{ EE6}$$
$$(N_{Gr} N_{Pr}) = (6.28 \text{ EE6})(.72) = 4.52 \text{ EE6}$$

Table 10.6
Parameters for Natural Convection

Configuration	L	$N_{Gr} N_{Pr}$	C	n
vertical plate		EE4 to EE9	.59	.25
or cylinder	height	EE9 to EE12	.13	.33
horizontal cylinder	diameter	EE3 to EE9	.53	.25
		EE9 to EE12	.126	.33
hot surface facing up, cold surface facing down	$1/2(L_1 + L_2)$ or (diameter of circular plate)	EE5 to 2EE7	.54	.25
		2EE7 to EE10	.14	.33
thin horizontal or vertical wires	diameter	EE−7 to EE0	1.0	.1

Table 10.7
Natural Film Coefficients (h) for Air

Arrangement	$N_{Gr} N_{Pr}$	Simplified Equation
vertical plate	EE4 to EE9	$.29 (\Delta T/L)^{.25}$
or cylinder	EE9 to EE12	$.19 (\Delta T)^{.33}$
horizontal cylinder	EE3 to EE9	$.27 (\Delta T/L)^{.25}$
	EE9 to EE12	$.18 (\Delta T)^{.33}$
hot surface facing up, cold surface facing down	EE5 to 2EE7	$.27 (\Delta T/L)^{.25}$
	2EE7 to 3EE10	$.22 (\Delta T)^{.33}$

From equation 10.56,

$$h = \frac{.0174}{.333}(.53)(4.52 \text{ EE6})^{.25} = 1.28 \text{ BTU/hr-ft}^2\text{-}°R$$
$$q = \pi(.333)(20)(1.28)(300 - 100) = 5356 \text{ BTU/hr}$$

10 FILM COEFFICIENT WITH OUTSIDE CONDENSING VAPOR

Special Nomenclature

c	specific heat of the liquid	BTU/lbm-°F
D	tube diameter	ft
g	acceleration due to gravity (4.17 EE8)	ft/hr^2
h_{fg}	latent heat of condensation	BTU/lbm
k	thermal conductivity of the liquid	BTU/hr-ft-°F
T_s	surface temperature	°F
T_{sv}	saturated vapor temperature	°F
μ_l	liquid viscosity	lbm/hr-ft
ρ_l	liquid density	lbm/ft^3
ρ_v	vapor density	lbm/ft^3

When a vapor condenses on a surface, the condensate forms a thermal resistance to heat transfer. As the condensate falls or flows from the surface, it will remove thermal energy which has entered it from the surface.

With *filmwise condensation*, a continuous film covers the entire surface. The film flows downward over the pipe under the action of gravity. If the surface contains impurities or irregularities which prevent wetting, the film will be discontinuous, a condition known as *dropwise condensation*. Equation 10.59 predicts the film coefficient only for filmwise condensation. Coefficients for dropwise condensation can be 4 to 8 times larger than the filmwise coefficient because the film is thinner, providing a smaller thermal resistance.

Equation 10.59 can be used to calculate a conservative estimate of the film coefficient for horizontal 1″ to 3″ tubes with outside condensation.[10] (Watch the units of μ. They should be lbm/ft-hr.) The approximate range of h is 2000 to 4000.

$$h = 0.725\left[\frac{\rho_l(\rho_l - \rho_v)gh'_{fg}(k)^3}{D\mu_l(T_{sv} - T_s)}\right]^{.25} \qquad 10.59$$
$$q = Ah(T_{sv} - T_s) \qquad 10.60$$

In the absence of information about the actual surface temperature of the outside condensing tube, assume that $(T_{sv} - T_s)$ is between 5°F and 40°F.

The fluid properties should be evaluated at $\frac{1}{2}(T_s + T_{sv})$. However, $\rho_v = 0$ can be assumed as a first approximation.

[10] This equation also is known as the *Colburn equation*.

h'_{fg} is just the latent heat of vaporization (condensation) unless there is significant subcooling of the condensate. In the case of such subcooling, h'_{fg} can be calculated from equation 10.61.

$$h'_{fg} = h_{fg} + \left(\tfrac{3}{8}\right)c(T_{sv} - T_s) \qquad 10.61$$

If the vapor is condensing from a superheated condition, or if the quality of the vapor surrounding the cooling surface is less than 1.0, equation 10.59 still can be used. T_{sv} is the correct value, however, not the superheat temperature.

In the case of a group of horizontal layers of tubes, the condensate will drop from one layer to another. Equation 10.59 can be modified for n layers of tubes, each directly above the other, by inserting n into the denominator of the bracketed quantity. This produces a conservative estimate of the film coefficient.

Example 10.10

A 1″ O.D., $\frac{7}{8}$″ I.D. brass tube is surrounded by 7.5 psia, 92% quality steam. What is the outside film coefficient if the wall temperature is 60°F?

The saturation temperature for 7.5 psia steam is 180°F. The film properties are evaluated at the average of the wall and saturation temperatures.

$$T_{\text{film}} = \tfrac{1}{2}(60 + 180) = 120°F$$
$$k_{120} = .372 \text{ BTU-ft/hr-ft}^2\text{-}°R$$
$$\mu_{120} = (.392 \text{ EE}-3)(3600) = 1.41 \text{ lbm/hr-ft}$$
$$\rho_{l,120} = \frac{1}{v_{120}} = \frac{1}{.01620} = 61.73 \text{ lbm/ft}^3$$
$$\rho_{v,120} = \frac{1}{203.27} = .005 \approx 0$$
$$h_{fg} = 990.2 \text{ BTU/lbm} \quad (\text{at } 180°F)$$
$$T_{sv} = 180°F$$
$$D = \frac{1}{12} = .0833 \text{ ft}$$
$$g = 4.17 \text{ EE8 ft/hr}^2$$

From equation 10.59,

$$h = 0.725$$
$$\times\left[\frac{(61.73)(61.73 - 0)(4.17 \text{ EE8})(990.2)(.372)^3}{(.0833)(1.41)(180 - 60)}\right]^{.25}$$
$$= 1123 \text{ BTU/hr-ft}^2\text{-}°R$$

11 FILM COEFFICIENT WITH BOILING

If a fluid is vaporizing around a hot tube, the boiling can occur in three distinctly different ways. If the tube temperature is near the saturating temperature, there will be little or no bubble formation, and the heat transfer will be essentially convective. This is known as *pool boiling*.

At temperatures somewhat greater than the saturating temperature, *nucleate boiling* begins with the formation of vapor bubbles. When the surface temperature is much greater (i.e., 200°F higher or more) than the saturation temperature, a film of vapor will cover the surface. This is known as *film boiling*.

The *Bromley equation* (equation 10.62) can be used for laminar film boiling around horizontal tubes up to $\frac{1}{2}$ inch in diameter. No movement of fluid past the tubes is allowed. The flow is assumed to be laminar when the flow path around the tube is short. All properties should be evaluated at the saturation temperature.

$$h_b = .62 \left[\frac{\rho_v(\rho_l - \rho_v)gh'_{fg}k_v^3}{D\mu_v(T_s - T_{sv})} \right]^{.25} \quad \textbf{10.62}$$

$$h'_{fg} = h_{fg} + .4\,c_{pv}(T_s - T_{sv}) \quad \textbf{10.63}$$

Since the tube is likely to be at a high temperature, the effects of radiation are likely to be significant. Not only does the radiation heat transfer have to be included, but the radiation itself increases the film thickness, reducing h and the convective heat transfer.

Equation 10.64 can be solved iteratively for the overall combined film coefficient. (See equation 10.88 for a definition of h_r.)

$$h_o = h_b \left(\frac{h_b}{h_o} \right)^{.333} + h_r \quad \textbf{10.64}$$

The total heat loss in film boiling is

$$q = h_o A (T_s - T_{sv}) \quad \textbf{10.65}$$

12 HEAT EXCHANGERS

Heat exchangers employ two moving fluids—one hot and one cold. Primarily through forced convection, heat is transferred to the cold fluid.

There are several types of *tubular exchangers*, also known as *shell and tube heat exchangers*.

- Single pass: In a single pass exchanger, each fluid is exposed to the other fluid only once. The temperature difference between the two fluids is not constant, and the logarithmic mean temperature difference (equation 10.66) must be used.

The single pass exchanger can be *parallel flow* (if both fluids flow in the same direction) or *counterflow* (if the fluids flow in opposite directions). Counterflow is more efficient since the temperature gradient is essentially constant.

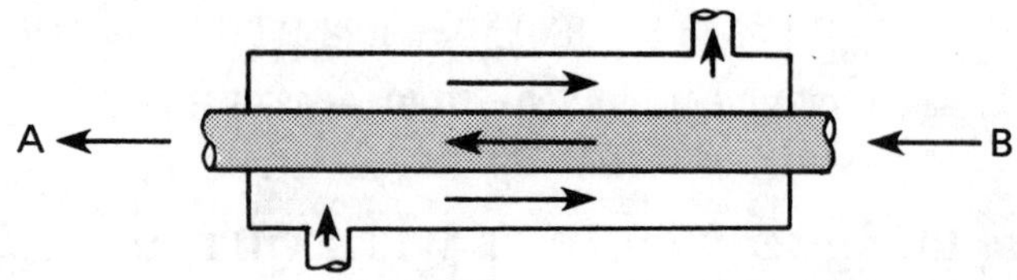

Figure 10.12 Single Pass Counterflow Exchanger

- Multiple pass: For increased efficiency, most heat exchangers provide for more than a single pass through the tubes, and the shell fluid is routed around baffles. An *x-y exchanger* will have *x* shell passes and *y* tube passes.

- Crossflow: In the crossflow exchanger, one fluid moves perpendicularly to the other. Crossflow exchangers can have both fluids unmixed, a typical situation when fluids are constrained within tubes and passageways, or one or both fluids may be mixed by forcing around pipes, baffles, or passages. If a fluid is mixed, its temperature is essentially uniform across the outlet, although the flow direction may be nonhomogeneous.

Since the temperature gradient is not uniform along the length of the exchanger, the *logarithmic mean temperature difference* must be used when calculating the heat transfer in heat exchangers. ΔT_A and ΔT_B are the temperature differences at the left end, A, and the right end, B, regardless of the flow directions (i.e., regardless of whether the fluids move in counterflow or parallel flow).

$$\Delta T_m = \frac{\Delta T_A - \Delta T_B}{ln\left(\frac{\Delta T_A}{\Delta T_B}\right)} \quad \textbf{10.66}$$

The heat transfer in an exchanger can be calculated from equation 10.67.

$$q = UA\Delta T_m \quad \textbf{10.67}$$

For multiple pass and crossflow heat exchangers, a correction factor for ΔT_m is required. ΔT_m is calculated as if the fluids are in counterflow. (When one of the fluids does not change temperature, as in condensing/evaporating problems, $F_c = 1$.) Values of F_c are given in figure 10.13.

$$q = F_c UA\Delta T_m \quad \textbf{10.68}$$

The *overall conductance*, U, can be specified for either inside tube area or outside tube area use. In heat exchanger problems, U is usually given for use with A_o. Of course, the heat transfer is independent of whether A_o or A_i is used.

$$U_i A_i = U_o A_o = \frac{q}{\Delta T_m} \qquad 10.69$$

The values of U can be calculated from the film coefficients and the pipe material conductivities.

$$\frac{1}{U_i} = \frac{1}{h_i} + \frac{r_i}{k} ln\left(\frac{r_o}{r_i}\right) + \frac{r_i}{r_o h_o} \qquad 10.70$$

$$\frac{1}{U_o} = \frac{1}{h_o} + \frac{r_o}{k} ln\left(\frac{r_o}{r_i}\right) + \frac{r_o}{r_i h_i} \qquad 10.71$$

The middle term (with the natural logarithm) is sometimes approximated by (t/k_{tube}). However, if the thermal resistance of the tube is kept at all, it is not much more difficult to calculate the exact value. If the thermal conductivity of the tube, k, is high, the middle term can be omitted. If the middle term is omitted,

$$\frac{1}{U_o} \approx \frac{1}{h_o} + \frac{r_o}{r_i h_i} \approx \frac{1}{h_o} + \frac{1}{h_i} \qquad 10.72$$

Unless the walls are clean, *fouling factors* should be incorporated into the thermal resistance to account for corrosion, scale, and dirt. Various values of the fouling factor, R, are given in table 10.8. These have been recommended by the Tubular Exchanger Manufacturer's Association. The overall coefficient is given by equation 10.73. Typical values are given in table 10.9.

$$\frac{1}{U_{o,dirty}} = \frac{1}{h_o} + \frac{r_o}{k} ln\left(\frac{r_o}{r_i}\right) + R_{fo} + \left(R_{fi} + \frac{1}{h_i}\right)\left(\frac{r_o}{r_i}\right)$$

$$\approx \frac{1}{h_o} + \frac{1}{h_i} + R_{fo} + R_{fi} \qquad 10.73$$

Table 10.8
Typical Fouling Factors
hr-°F-ft^2/BTU

Fluid	Below 125°F	Above 125°F
seawater	.0005	.001
distilled water	.0005	.0005
treated boiler feedwater	.001	.001
city or well water	.001	.002
untreated cooling tower water	.002	.002
hard water	.003	.005
air	.002	
diesel exhaust	.01	
clean steam	.0005	
oil-bearing steam	.001	

Example 10.11

A single pass heat exchanger is used to cool 10 gpm of water from 55°F to 45°F by using 10 gpm of 25% solution calcium chloride brine ($c_p = .68$, $\rho = 77.0$) initially at 10°F. The exchanger is built of schedule 40 $1\frac{1}{4}''$ steel pipe for the tube and $2\frac{1}{2}''$ pipe for the shell, with the brine flowing in the tube. How long should the exchanger be? Assume that the brine-side coefficient is 265 BTU/ft^2-hr-°F and that the water-side coefficient is 246 BTU/ft^2-hr-°F. Assume counterflow operation.

The mass flow rates are

$$\dot{m}_{water} = \frac{10}{7.48}(60)(62.4) = 5005 \text{ lbm/hr}$$

$$\dot{m}_{brine} = \frac{10}{7.48}(60)(77) = 6176 \text{ lbm/hr}$$

Table 10.9
Values of U
BTU/hr-ft^2-°F

Heating Applications		Clear Surface U		With Normal Fouling	
hot side	cold side	natural convection	forced convection	natural convection	forced convection
steam	watery solution	250–500	300–550	100–200	150–275
steam	light oils	50–70	110–140	40–45	60–110
steam	medium lube oils	40–60	110–130	35–40	50–100
steam	Bunker C or #6 oil	20–40	70–90	15–30	60–80
steam	air or gases	2–4	5–10	1–3	4–8
hot water	watery solution	115–140	200–250	70–100	110–160
Cooling Applications					
cold side	hot side				
water	watery solution	110–135	195–245	65–95	105–155
water	medium lube oil	8–12	20–30	5–8	10–20
water	air or gases	2–4	5–10	1–3	4–8
freon/ammonia	watery solution	35–45	60–90	20–35	40–60
calcium or sodium brine	watery solution	100–120	175–200	50–75	80–125

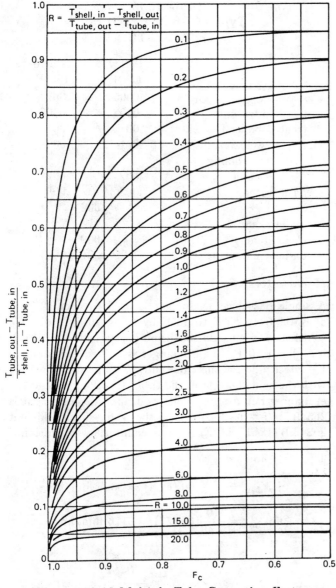

$$R = \frac{T_{shell,\,in} - T_{shell,\,out}}{T_{tube,\,out} - T_{tube,\,in}}$$

(vertical axis) $\dfrac{T_{tube,\,out} - T_{tube,\,in}}{T_{shell,\,in} - T_{tube,\,in}}$

F_c

Figure 10.13 Multiple Tube Correction Factor
(1 shell pass, 2 or more tube passes)

The heat transfer is found from the temperature gain.

$$q = (5005)(1.00)(55 - 45) = 50,050 \text{ BTU/hr}$$

The brine final temperature is

$$T_{b,2} = 10 + \frac{(50,050)}{(.68)(6176)} = 21.9°F$$

The logarithmic mean temperature difference is

$$\Delta T_m = \frac{(45 - 10) - (55 - 21.9)}{ln\left(\frac{35}{33.1}\right)} = 34.0$$

The pipe dimensions are

	$1\frac{1}{4}''$	$2\frac{1}{2}''$
I.D.	1.38	2.47
O.D.	1.66	2.88

The overall coefficient can be calculated from equation 10.73 assuming $k_{pipe} = 35$, $R_{fo} = .0005$, and $R_{fi} = .001$.

$$\frac{1}{U_o} = \frac{1}{246} + \frac{\frac{.83}{12}}{35} ln\left(\frac{1.66}{1.38}\right) + .0005$$
$$+ \left(.001 + \frac{1}{265}\right)\left(\frac{1.66}{1.38}\right) = .01067$$

$$U_o = 93.7$$

Then, since $q = UA\Delta T_m$,

$$50,050 = 93.7\left(\frac{1.66}{12}\right)\pi L(34.0)$$

$$L = 36.2 \text{ ft}$$

13 ANNULAR FLOW

Shell and tube heat exchangers are an example of annular flow. It is necessary to treat annular flow differently when calculating friction losses and film coefficients.

If it is necessary to calculate the friction loss through the shell, the *hydraulic diameter* should be used when calculating the Reynolds number and the friction head loss.

$$D_H = \frac{4 \text{ (area in flow)}}{\text{wetted perimeter}} = \text{I.D.}_{shell} - \text{O.D.}_{tube} \quad 10.74$$

When calculating the film coefficient on the outside of the tube, the *characteristic diameter* should be used in the Nusselt equation.

$$D = \frac{4 \text{ (area in flow)}}{\text{heated perimeter}} = \frac{(\text{I.D.})^2_{shell} - (\text{O.D.})^2_{tube}}{\text{O.D.}_{tube}}$$
$$10.75$$

14 HEAT EXCHANGER EFFECTIVENESS

Some problems that must be solved by tedious iterative methods can be handled easily with the *NTU (number of transfer units) method*. This method calculates the *effectiveness* (equal to q_{actual}/q_{ideal}) of the heat exchanger as the ratio of actual heat transfer to the maximum possible transfer. (The maximum possible transfer can occur only if the heat exchanger has an infinite length in which to extract thermal energy from the hot fluid.)

The first step is to determine the fluid capacity rates, $C = \dot{m}c$, of the two fluids. The smaller capacity rate is designated as C_{min}. Then, if the cold fluid has the minimum capacity rate,

$$\varepsilon = \frac{C_{hot}(T_{hot,in} - T_{hot,out})}{C_{min}(T_{hot,in} - T_{cold,in})} \quad 10.76$$

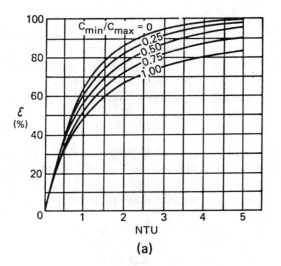

(a)

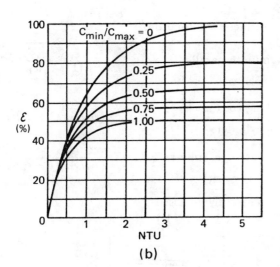

(b)

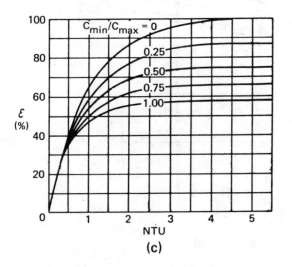

(c)

Figure 10.14 Heat Exchanger Effectiveness

(a) single pass counterflow, (b) single pass parallel flow,
(c) one shell pass, even number of tube passes

If the hot fluid has the minimum capacity rate,

$$\mathcal{E} = \frac{C_{cold}(T_{cold,out} - T_{cold,in})}{C_{min}(T_{hot,in} - T_{cold,in})} \qquad 10.77$$

If the effectiveness is known, the heat transfer can be found from equation 10.78. (Notice that the temperature difference is the difference of two entering temperatures.)

$$q = \mathcal{E} C_{min}(T_{hot,in} - T_{cold,in}) \qquad 10.78$$

The problem simplifies to finding the effectiveness. If the outlet temperatures are not known, equations 10.76 and 10.77 cannot be used. However, an alternate method using the number of transfer units (NTU) is available.

$$NTU = \frac{U_o A_o}{C_{min}} = \frac{U_i A_i}{C_{min}} \qquad 10.79$$

It is not very difficult to calculate the effectiveness. Nevertheless, the graphical approach is the most expedient, particularly if NTU is unknown. Figure 10.14 can be used for this purpose.

15 RADIATION

Thermal radiation is emitted from bodies in the .1 to 100 micron range. The total amount of radiation per unit area and time is called the *emissive power*, E, which depends on the temperature and the surface characteristics of the radiating body.

Thermal radiation hitting a body can be absorbed, reflected, or transmitted, with the total of the three equaling the incident energy. Thus the radiation conservation law is

$$\alpha + \rho + \tau = 1 \qquad 10.80$$

Kirchhoff's law states that for any two bodies in thermal equilibrium, the ratio of emissive powers to absorptivities is the same.

$$\left(\frac{E}{\alpha}\right)_1 = \left(\frac{E}{\alpha}\right)_2 \qquad 10.81$$

Since *absorptivity*, α, cannot exceed unity, Kirchhoff's law places an upper limit on emitted energy. Bodies that radiate at this upper limit are known as *black bodies* (ideal radiators) and have $\alpha = 1$. A black body never can be achieved in practice. In theory, a black body absorbs all incident energy; it also emits the maximum possible when acting as a source. E_b will be used to designate the emission of a black body.

Real bodies cannot radiate at the maximum level, and the ratio of E/E_b is called the *emissivity*, ϵ. (Notice that emissivity does not appear in the conservation law.) For a black body, both emissivity and absorptivity equal 1. However, emissivity equals absorptivity for any body at thermal equilibrium.

The *Stefan-Boltzmann law*, also known as the *fourth-power law*, gives the total emission from a black body over all wavelengths.

$$E_b = \sigma T^4 \qquad\qquad 10.82$$

$$\sigma = .1713\ EE - 8\ \frac{BTU}{hr\text{-}ft^2\text{-}^\circ R^4} \qquad 10.83$$

If a real body exhibits a constant emissivity, regardless of wavelength, the body is said to be a *gray body*. Usually, real bodies are assumed to be gray bodies. Then the Stefan-Boltzmann law is written as

$$E_g = \epsilon_g \sigma T^4 \qquad\qquad 10.84$$

The net heat transfer by radiation between two bodies at temperatures T_1 and T_2 can be found from equation 10.85.

$$E = \sigma F_e F_a [(T_1)^4 - (T_2)^4] \qquad 10.85$$

In equation 10.85, F_a is an *arrangement factor* (also known as the *geometric shape factor* or just *shape factor*), which depends on the spatial arrangement of the objects. For a hot radiator completely surrounded by an enclosure, or for infinite parallel planes, $F_a = 1$.

F_e is the *emissivity factor*, which depends on the emissivities of both surfaces. For a completely enclosed body, F_e is just the emissivity of the enclosed body. Other values are given in table 10.10. Sometimes F_a and F_e are combined into the *configuration factor*, F_{12}.[11]

$$F_{12} = F_e F_a \qquad\qquad 10.86$$

---

[11] This also is known as a *view factor* and *gray-body shape factor*.

The configuration factor F_{12} is the fraction of diffuse radiation leaving object 1 that will fall on object 2. Conversely, F_{21} is the fraction from object 2 falling on object 1. Area 1 must be used with F_{12}; area 2 must be used with F_{21}. Whether F_{12} or F_{21} is used depends on which is easier to evaluate. Then

$$A_1 F_{12} = A_2 F_{21} \qquad\qquad 10.87$$

Generally, it is difficult to evaluate configuration factors for complex arrangements of objects and surfaces. However, some of the simpler cases have been solved, and their graphical solutions are available.

Table 10.10
Emissivity Factors

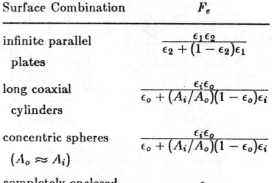

Surface Combination	F_e
infinite parallel plates	$\dfrac{\epsilon_1 \epsilon_2}{\epsilon_2 + (1 - \epsilon_2)\epsilon_1}$
long coaxial cylinders	$\dfrac{\epsilon_i \epsilon_o}{\epsilon_o + (A_i/A_o)(1 - \epsilon_o)\epsilon_i}$
concentric spheres $(A_o \approx A_i)$	$\dfrac{\epsilon_i \epsilon_o}{\epsilon_o + (A_i/A_o)(1 - \epsilon_o)\epsilon_i}$
completely enclosed body $(A_o >> A_i)$	ϵ_i

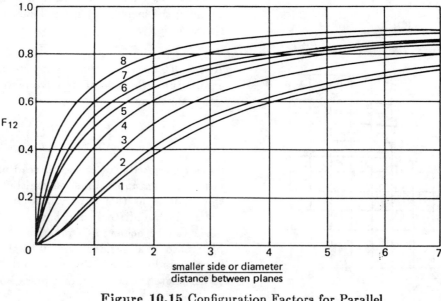
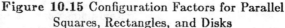

Figure 10.15 Configuration Factors for Parallel Squares, Rectangles, and Disks

curves 1, 2, 3, 4: direct radiation between planes
curves 5, 6, 7, 8: planes connected by nonconducting but reradiating wall.
curves 1, 5: disks

curves 2, 6: squares
curves 3, 7: rectangles (2:1 aspect ratio)
curves 4, 8: long, narrow rectangles

16 COMBINED HEAT TRANSFER

When heat is transferred by radiation and convection, it is convenient to define the *radiant heat transfer coefficient*.

$$h_r = \frac{F_e F_a \sigma (T_1^4 - T_2^4)}{T_1 - T_3} = \frac{q_r}{A_1(T_1 - T_3)} \qquad 10.88$$

T_1 is the hot body temperature, T_2 is the surrounding wall temperature for radiation purposes, and T_3 is the fluid environment temperature for convection purposes. The total heat transfer is

$$q = h_t A_1 (T_1 - T_3) \qquad 10.89$$
$$h_t = h + h_r \qquad 10.90$$

If T_2 and T_3 are not the same, the heat transfer also can be calculated separately. Trial and error may be required to determine T_2 or T_3 based on a known heat flow.

Example 10.12

A 1′ diameter heating duct (emissivity of .8 and 200°F surface temperature) carries air through a basement with 40°F air and 0°F walls. What is the heat loss per unit area of duct?

Since the surface temperature is known, this problem can be solved by superposition. The heat losses from convection and radiation can be calculated independently and added together.

The film temperature is assumed to be $\frac{1}{2}(200 + 40) = 120°F$. Interpolating the data at the end of this chapter, the Prandtl and Grashoff numbers can be determined.

$$N_{Pr,120} = .72$$

$$\left(\frac{g\beta\rho^2}{\mu^2}\right)_{120} = 1.58 \text{ EE6}$$

$$N_{Gr} = (1)^3(1.58 \text{ EE6})(200 - 40) = 2.53 \text{ EE8}$$

Use a simplified equation from table 10.7.

$$h = .27\left(\frac{200 - 40}{1}\right)^{.25} = .96 \text{ BTU/hr-ft}^2\text{-°}F$$

The heat loss is

$$\frac{q}{A} = h\Delta T = .96(200 - 40) = 153.6 \text{ BTU/hr-ft}^2$$

The radiation loss is calculated directly. (Notice that temperatures are expressed in °R in the radiation equation.)

$$\frac{q}{A} = (.1713 \text{ EE} - 8)(.8)[(660)^4 - (460)^4]$$
$$= 198.7 \text{ BTU/hr-ft}^2$$

The total heat loss is the sum of these two losses.

$$\frac{q_{total}}{A} = 153.6 + 198.7 = 352.3 \text{ BTU/hr-ft}^2$$

Example 10.13

A 1′ diameter heating duct with emissivity of .8 loses 700 BTU/hr per square foot of surface area. If the surrounding air is 40°F and the walls are 0°F, what is the duct temperature?

It is not practical to solve for the duct temperature directly in this type of problem. Not only does the unknown temperature appear in both the convection and radiation heat loss equations, but the duct surface temperature also is a parameter of the Nusselt equation, used to calculate the film coefficient.

The simplest approach is to choose a random value for the duct temperature and then calculate what the heat transfer would be. If the calculated heat transfer is equal to 700 BTU/hr, it can be concluded that the temperature chosen is correct. In all likelihood, several iterations will be necessary to bracket and close in on the correct temperature.

Assume that the duct temperature is 250°F and that the convection is laminar. Then the convective film coefficient on the outside of the duct can be calculated from a simplified equation in table 10.7.

$$h = .27\left(\frac{250 - 40}{1}\right)^{.25} = 1.03$$
$$\frac{q}{A} = (1.03)(250 - 40) = 216.3 \text{ BTU/hr-ft}^2$$

The radiation loss is found from equation 10.85.

$$\frac{q}{A} = (.1713 \text{ EE} - 8)(.8)[(710)^4 - (460)^4]$$
$$= 286.9 \text{ BTU/hr-ft}^2$$

The total loss from radiation and convection, assuming that the duct temperature is 250°F, is 503.2 BTU/hr-ft². Since the total loss is 700, the duct temperature must be higher. The solution would continue by trial and error.

Example 10.14

A tubular thermocouple with emissivity of .8 is used to measure the temperature of a gas flowing in a pipe with 350° walls. The apparent temperature of the gas, as measured by the thermocouple, is 850°F. Assume that the convective film coefficient for the thermocouple is 27 BTU/hr-ft²-°F. Find the true gas temperature. Neglect conduction.

Assume that the thermocouple is small compared to the pipe diameter, so that $F_a = 1$. Also, assume that the thermocouple temperature is constant so that

$$\left(\frac{q}{A}\right)_{\text{gain by convection}} = \left(\frac{q}{A}\right)_{\text{loss by radiation}}$$

Since the thermocouple loses heat by radiation, its temperature will be less that of the gas.

$$h\left(T_{\text{gas}} - T_{\text{probe}}\right) = \sigma\epsilon\left(T_{\text{probe}}^4 - T_{\text{walls}}^4\right)$$

Expressing the temperatures in °R,

$$T_{\text{probe}} = 850 + 460 = 1310°R$$
$$T_{\text{walls}} = 350 + 460 = 810°R$$

Then the thermal equilibrium equation is

$$27\left(T_{\text{gas}} - 1310\right) = (.1713\ EE - 8)(.8)[(1310)^4 - (810)^4]$$

The gas temperature is found directly to be 1437.6°R.

17 NOCTURNAL RADIATION

It is possible to measure the effective sky temperature on a cold, clear night. Measurements of radiation away from the earth indicate that the effective sky temperature is approximately 410°R. Since this is below the freezing point of water, it is possible for freezing (e.g., of citrus fruit or puddles of standing water) to occur even when the air temperature is above 32°F.

Example 10.15

A pond on a clear, cold night has a convective film coefficient of 5. The air temperature is 60°F. What is the water temperature if the pond's temperature remains constant throughout the night?

With an effective sky temperature of 410°R, the pond will be losing heat through radiation. Since the pond temperature is constant, it must be gaining heat from the air through convection. Therefore the pond temperature is less than 60°F.

Expressing the temperatures in °R,

$$T_{\text{sky}} = 410°R$$
$$T_{\text{air}} = 520°R$$

The equilibrium equation is

$$\left(\frac{q}{A}\right)_{\text{gain by convection}} = \left(\frac{q}{A}\right)_{\text{loss by radiation}}$$

The emissivity of the pond water is found to be .96 from appendix F.

$$5\left(520 - T_{\text{pond}}\right) = (.1713\ EE - 8)(.96)[(T_{\text{pond}})^4 - (410)^4]$$

Trial and error is required to solve for T_{pond}.

$$T_{\text{pond}} \approx 508°R = 48°F$$

Appendix A
Properties of Metals and Alloys

Material	k(BTU/hr-ft-°F) 32°F	212°F	572°F	932°F	c(BTU/lbm-°F) 32°F	ρ(lbm/ft³) 32°F
aluminum	117	119	133	155	.208	169
brass (70% Cu, 30% Zn)	56	60	66		.092	532
bronze (75% Cu, 25% Sn)	15				.082	540
cast iron, plain	33	31.8	27.7	24.8	.11	474
cast iron, alloy	30	28.3	27		.10	455
copper, pure	224	218	212	207	.091	558
iron, pure	35.8	36.6			.104	491
lead	20.1	19	18		.030	705
magnesium	91	92			.232	109
silver	242	238			.056	655
18-8 stainless (304)	8.0	9.4	10.9	12.4	.11	488
18-8 stainless (347)	8.0	9.3	11.0	12.8	.11	488
steel, mild (1% C)	26.5	26	25	22	.11	490
tin	36	34			.054	456
zinc	65	64	59		.091	446

HEAT TRANSFER

Appendix B
Properties of Non-metals

Material	Average temperature (°F)	k BTU/hr-ft-°F	c_p BTU/lbm-°F	ρ lbm/ft^3
asbestos	32	.087	.25	36
	392	.12		
brick, building	70	.38	.20	106
brick, fire-clay	392	.58	.20	144
	1832	.95		
brick, Kaolin				
insulating	932	.15		27
	2102	.26		
firebrick	392	.05		19
	1400	.11		
concrete, stone	70	.54	.20	144
with 10% moisture	70	.70		140
diatomaceous earth				
powdered	100	.030	.21	14
	300	.036		
	600	.046		
glass, window	70	.45	.2	170
glass wool (fine)	20	.022		
	100	.031		1.5
	200	.043		
glass wool (packed)	20	.016		
	100	.022		6.0
	200	.029		
ice	32	1.28	.46	57
magnesia, 85%, molded pipe	32	.032		17
covering $(T < 600°F)$	200	.037		
molded pipe covering,	400	.051		26
diatomaceous silica	1600	.088		
sand, dry	68	.20		95
sand, with 10% water	68	.60		100
soil, dry	70	.20	.44	
soil, wet	70	1.5		
wood, oak				
perpendicular to grain	70	.12	.57	51
parallel to grain	70	.20	.57	
wood, pine				
perpendicular to grain	70	.06	.67	31
parallel to grain	70	.14	.67	

Appendix C
Properties of Saturated Water

T °F	ρ lbm/ft^3	c_p BTU/lbm-°F	μ lbm/ft-sec	ν ft^2/sec	k BTU/hr-ft-°F	N_{Pr}	β 1/°F	$\frac{g\beta\rho^2}{\mu^2}$ 1/ft^3-°F
32	62.4	1.01	1.20 EE−3	1.93 EE−5	.319	13.7	−0.37 EE−4	
40	62.4	1.00	1.04 EE−3	1.67 EE−5	.325	11.6	.20 EE−4	2.3 EE6
50	62.4	1.00	.88 EE−3	1.40 EE−5	.332	9.55	.49 EE−4	8.0 EE6
60	62.3	.999	.76 EE−3	1.22 EE−5	.340	8.03	.85 EE−4	18.4 EE6
70	62.3	.998	.658 EE−3	1.06 EE−5	.347	6.82	1.2 EE−4	34.6 EE6
80	62.2	.998	.578 EE−3	.93 EE−5	.353	5.89	1.5 EE−4	56.0 EE6
90	62.1	.997	.514 EE−3	.825 EE−5	.359	5.13	1.8 EE−4	85.0 EE6
100	62.0	.998	.458 EE−3	.740 EE−5	.364	4.52	2.0 EE−4	118 EE6
150	61.2	1.00	.292 EE−3	.477 EE−5	.384	2.74	3.1 EE−4	440 EE6
200	60.1	1.00	.205 EE−3	.341 EE−5	.394	1.88	4.0 EE−4	1.11 EE9
250	58.8	1.01	.158 EE−3	.269 EE−5	.396	1.45	4.8 EE−4	2.14 EE9
300	57.3	1.03	.126 EE−3	.220 EE−5	.395	1.18	6.0 EE−4	4.00 EE9
350	55.6	1.05	.105 EE−3	.189 EE−5	.391	1.02	6.9 EE−4	6.24 EE9
400	53.6	1.08	.091 EE−3	.170 EE−5	.381	.927	8.0 EE−4	8.95 EE9
450	51.6	1.12	.080 EE−3	.155 EE−5	.367	.876	9.0 EE−4	12.1 EE9
500	49.0	1.19	.071 EE−3	.145 EE−5	.349	.87	10.0 EE−4	15.3 EE9
550	45.9	1.31	.064 EE−3	.139 EE−5	.325	.93	11.0 EE−4	17.8 EE9
600	42.4	1.51	.058 EE−3	.137 EE−5	.292	1.09	12.0 EE−4	20.6 EE9

Appendix D
Properties of Air at One Atmosphere

T °F	ρ lbm/ft^3	c_p BTU/lbm-°F	μ lbm/ft-sec	ν ft^2/sec	k BTU/hr-ft-°F	N_{Pr}	β 1/°F	$\frac{g\beta\rho^2}{\mu^2}$ 1/ft^3-°F
0	.086	.239	1.110 EE$-$5	.130 EE$-$3	.0133	.73	2.18 EE$-$3	4.2 EE6
32	.081	.240	1.165 EE$-$5	.145 EE$-$3	.0140	.72	2.03 EE$-$3	3.16 EE6
100	.071	.240	1.285 EE$-$5	.180 EE$-$3	.0154	.72	1.79 EE$-$3	1.76 EE6
200	.060	.241	1.440 EE$-$5	.239 EE$-$3	.0174	.72	1.52 EE$-$3	.850 EE6
300	.052	.243	1.610 EE$-$5	.306 EE$-$3	.0193	.71	1.32 EE$-$3	.444 EE6
400	.046	.245	1.75 EE$-$5	.378 EE$-$3	.0212	.689	1.16 EE$-$3	.258 EE6
500	.0412	.247	1.890 EE$-$5	.455 EE$-$3	.0231	.683	1.04 EE$-$3	.159 EE6
600	.0373	.250	2.000 EE$-$5	.540 EE$-$3	.0250	.685	.943 EE$-$3	.106 EE6
700	.0341	.253	2.14 EE$-$5	.625 EE$-$3	.0268	.690	.862 EE$-$3	70.4 EE3
800	.0314	.256	2.25 EE$-$5	.717 EE$-$3	.0286	.697	.794 EE$-$3	49.8 EE3
900	.0291	.259	2.36 EE$-$5	.815 EE$-$3	.0303	.705	.735 EE$-$3	36.0 EE3
1000	.0271	.262	2.47 EE$-$5	.917 EE$-$3	.0319	.713	.685 EE$-$3	26.5 EE3
1500	.0202	.276	3.00 EE$-$5	1.47 EE$-$3	.0400	.739	.510 EE$-$3	7.45 EE3
2000	.0161	.286	3.54 EE$-$5	2.14 EE$-$3	.0471	.753	.406 EE$-$3	2.84 EE3
2500	.0133	.292	3.69 EE$-$5	2.80 EE$-$3	.051	.763	.338 EE$-$3	1.41 EE3
3000	.0114	.297	3.86 EE$-$5	3.39 EE$-$3	.054	.765	.289 EE$-$3	.815 EE3

Appendix E
Properties of Steam at One Atmosphere

T °F	ρ lbm/ft^3	c_p BTU/lbm-°F	μ lbm/ft-sec	ν ft^2/sec	k BTU/hr-ft-°F	N_{Pr}	β 1/°F	$\frac{g\beta\rho^2}{\mu^2}$ 1/ft^3-°F
212	.0372	.451	.870 EE−5	.234 EE−3	.0145	.96	1.49 EE−3	.877 EE6
300	.0328	.456	1.00 EE−5	.303 EE−3	.0171	.95	1.32 EE−3	.459 EE6
400	.0288	.462	1.13 EE−5	.395 EE−3	.0200	.94	1.16 EE−3	.243 EE6
500	.0258	.470	1.265 EE−5	.490 EE−3	.0228	.94	1.04 EE−3	.139 EE6
600	.0233	.477	1.420 EE−5	.610 EE−3	.0257	.94	.943 EE−3	82 EE3
700	.0213	.485	1.555 EE−5	.725 EE−3	.0288	.93	.862 EE−3	52.1 EE3
800	.0196	.494	1.70 EE−5	.855 EE−3	.0321	.92	.794 EE−3	34.0 EE3
900	.0181	.50	1.810 EE−5	.987 EE−3	.0355	.91	.735 EE−3	23.6 EE3
1000	.0169	.51	1.920 EE−5	1.13 EE−3	.0388	.91	.685 EE−3	17.1 EE3
1200	.0149	.53	2.14 EE−5	1.44 EE−3	.0457	.88	.603 EE−3	9.4 EE3
1400	.0133	.55	2.36 EE−5	1.78 EE−3	.053	.87	.537 EE−3	5.49 EE3
1600	.0120	.56	2.58 EE−5	2.14 EE−3	.061	.87	.485 EE−3	3.38 EE3
1800	.0109	.58	2.81 EE−5	2.58 EE−3	.068	.87	.442 EE−3	2.14 EE3
2000	.0100	.60	3.03 EE−5	3.03 EE−3	.076	.86	.406 EE−3	1.43 EE3
2500	.0083	.64	3.58 EE−5	4.30 EE−3	.096	.86	.338 EE−3	.603 EE3
3000	.0071	.67	4.00 EE−5	5.75 EE−3	.114	.86	.289 EE−3	.293 EE3

Appendix F
Emissivities of Various Surfaces

Wavelength and average temperature

Material	microns °F	9.3 100°	5.4 500°	3.6 1000°	1.8 2500°	.6 solar
aluminum, polished		.04	.05	.08	.19	
aluminum, oxidized		.11	.12	.18		
aluminum (24-ST), weathered		.4	.32	.27		
aluminum, surface roofing		.22				
aluminum, anodized		.94	.42	.60	.34	
brass, polished		.10	.10			
brass, oxidized		.61				
brick, red		.93				.7
brick, fire clay		.9		.7	.75	
brick, silica		.9		.75	.84	
brick, magnesite refractory		.9			.4	
chromium, polished		.08	.17	.26	.40	.49
copper, polished		.04	.05	.18	.17	
copper, oxidized		.87	.83	.77		
enamel, white		.9				
glass		.9				
ice (at 32°F)		.97				
iron, polished		.06	.08	.13	.25	.45
iron, cast, oxidized		.63	.66	.76		
iron, galvanized, new		.23			.42	.66
iron, galvanized, dirty		.28			.90	.89
iron, oxide		.96		.85		.74
iron, molten					.3-.4	
magnesium		.07	.13	.18	.24	.30
paper, white		.95		.82	.25	.28
paint, aluminized lacquer		.65	.65			
paint, lacquer, black or white		.96	.98			
paint, lampblack		.96	.97		.97	.97
paint, white (ZnO)		.95		.91		.18
paint, enamel, white		.9				
stainless steel, 18-8, polished		.15	.18	.22		
stainless steel, 18-8, weathered		.85	.85	.85		
steel tube, oxidized			.8			
steel plate, rough		.94	.97	.98		
Tungsten filament		.03			.18	.35 (6000°F)
water		.96				
wood		.93				
zinc, polished		.02	.03	.04	.06	.46
zinc, galvanized sheet		.25				

Appendix G
Important Conversion Factors

Multiply	By	To obtain
BTU	1054.8	watt-sec
BTU/ft²-hr	3.152	watt/m²
BTU-ft/hr-ft²-°F	1.73	J/sec-m-°C
BTU-in/sec-ft²-°F	518.87	J/sec-m-°C
BTU/sec	1054.8	watt
cal	4.184	J
cal/cm²-min	697.33	watt/m²
cal/g-°C	1.0	BTU/lbm-°F
cal/sec	4.184	watt
cal/sec-cm-°C	241.9	BTU-ft/hr-ft²-°F
cal/sec-cm²-°C	7373	BTU/hr-ft²-°F
ft-lbf	1.356	J
ft-lbf/sec	1.356	watt
horsepower	745.7	watt
J/g-°C	.239	BTU/lbm-°F
kcal	3.968	BTU
kcal/hr-m-°C	.672	BTU-ft/hr-ft²-°F
kcal/hr-m²-°C	.2048	BTU/hr-ft²-°F
kw-hr	3.6 EE6	J
watt	3.413	BTU/hr
watt/cm²	10,000	watt/m²
watt/cm-°C	57.79	BTU-ft/hr-ft²-°F
watt-hr	3600	J
watt-sec	1.0	J

PRACTICE PROBLEMS: HEAT TRANSFER

Warm-ups

1. Experiments have shown that k varies with T according to the relationship $k = .030(1 + .0015T)$ with T in °F. What is the value of k that should be used with a process that operates in the $T = 150°F$ to $T = 350°F$ range?

2. What is the heat flow through a 1' thick oven wall ($k = .038$ BTU-ft/ft^2-hr-°F) with a 350°F gradient?

3. A fluid in a tank is maintained at 85°F by an immersed steam radiator (190°F in, 160°F out). What is the ΔT_m?

4. The fluid in problem 3 is raised from 85°F to 110°F. At what temperature should you evaluate (a) the fluid film coefficient, (b) the steam film coefficient?

5. Fuel oil is heated from 55°F to 87°F by stack gases which cool from 350°F to 270°F in a heat exchanger. What is ΔT_m?

6. What is the density of 87% wet steam at 50 psia?

7. What is the viscosity of 100°F water in lbm/hr-ft?

8. A 6" thick furnace wall has a 3" square peephole. If the interior temperature is 2200°F and the surroundings are 70°F, what is the heat loss through the open peephole?

9. What is the overall coefficient of heat transfer of a 4" thick wall ($k = 13.9$ BTU-ft/ft^2-hr-°F) in a 10 mph wind?

10. Light oil is heated from 95°F to 105°F in a .6" diameter pipe. Its average velocity is 2 fps. What is the Reynolds number?

Concentrates

1. A composite wall is made up of three inches of material A (exposed to 1000°F), five inches of material B, and six inches of material C (exposed to 200°F). The mean values of k for materials A, B, and C are .06, .5, and .8 BTU-ft/ft^2-hr-°F, respectively. What are the temperatures at the two interfaces?

2. A pipe (3.5" I.D., 4.0" O.D., and 100' long) carries 350°F air through a 50°F environment. Two inches of insulation ($k = .05$ BTU-ft/ft^2-hr-°F) covers the pipe. Flow is laminar. What is the heat loss?

3. A bare horizontal pipe (4" O.D.) carries saturated steam at 300 psia through a 70°F environment at atmospheric pressure. If the flow rate is 5000 pounds per hour, what decrease in quality will occur in the first 50 feet?

4. A bare horizontal wire (.6" O.D.) dissipates two watts per foot of length, cooled by free convection at 60°F. What is the wire's surface temperature if the film temperature is 100°F?

5. A tubular feedwater heater is being designed to heat 2940 lbm/hr of water from 70°F to 190°F. Saturated steam at 134 psia is condensing on the outer tube surface. The tubes are copper, 1" O.D., 0.9" I.D., and 8' long. The water velocity is 3 fps. What surface area is required?

6. The heat supply of a large building is turned off at 5:00 p.m. when the interior temperature is 70°F. The capacity of the building and its contents is 100,000 BTU/°F, and the conductance is 6500 BTU/hr-°F. What is the interior temperature at 1:00 a.m. if the outdoor temperature is a constant 40°F?

7. A hot air duct is painted with white lacquer and carries air through a room with air temperature of 80°F and wall temperature of 70°F. The duct is 9" in diameter and has a surface temperature of 200°F. What is the rate of heat transfer?

8. A uranium dioxide fuel rod ($k = 1.1$ BTU-ft/ft^2-hr-°F, O.D.$= 0.4"$) is clad with 0.020 inches of stainless steel and produces 4 EE7 BTU/hr per cubic foot. A coolant at 500°F surrounds the clad rod. If the film coefficient is 10,000 BTU/ft^2-hr-°F, what is the centerline temperature of the rod?

9. Two long pieces of 1/16" copper wire are soldered together (end to end, butt connection) with a 450°F iron. If the air temperature is 80°F, what is the minimum rate of heat application to keep the junction at 450°F? Assume a unit surface conductance of 3 BTU/ft^2-hr-°F.

10. A glass thermometer (O.D.$= 0.35"$, 100°F) is inserted perpendicularly to a 100 fps airflow at 150°F. What is the unit convective conductance between the air and the thermometer?

Timed (1 hour allowed for each)

1. A cold storage unit is constructed with walls shown. (a) Find the heat flow in BTU/ft²-hr. (b) What is the temperature of the aluminum foil?

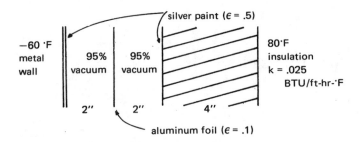

2. A white, uninsulated rectangular (18″ wide, 12″ high) duct runs through a 50′ long room with walls and contents at 70°F. The combined convection and radiation outside film coefficient is 2 BTU/ft²-hr-°F. 100°F air enters at 800 fpm at one end of the room. (a) What is the heat transfer to the room? (b) What is the temperature of the air in the duct at the other side of the room? (c) What is the friction pressure drop in the duct in inches of water?

3. Steel ball bearings varying in diameter from $\frac{1}{4}''$ to $1\frac{1}{2}''$ and with an initial temperature of 1800°F are individually quenched in a 110°F oil bath. The steel balls are to be removed when their temperature is 250°F. Assume a film coefficient of 56 BTU/ft²-hr-°F. (a) Derive an equation for the time the bearing balls should be in the oil bath. The equation should be a linear function of the diameter. State your assumptions. (b) What is the time constant for the cooling process?

4. A surface feedwater heater is to be designed to heat 500,000 pounds of water from 200°F to 390°F per hour. The heater is to be straight condensing. Dry saturated steam at 400°F is to be used as the heating medium. Drains leave as saturated liquid at 400°F. The tubes are 7/8″ O.D. with 1/16″ walls. The heater is to have two water passes, and it has been estimated that the overall coefficient of heat transfer is 700 BTU/hr-°F-ft². Specify the number and the effective length of the tubes. Assume a water velocity of 5 fps through the tubes.

5. The temperature of a gas in a duct with 600°F walls is evaluated with a .5″ diameter thermocouple probe. The flow rate is $G = 3480$ lbm/ft²-hr, and the velocity is 400 fpm. The film coefficient on the probe is given empirically as

$$h = \frac{0.024\,(G^{.8})}{(D^{.4})}$$

where G is the flow rate in lbm/ft²-hr and D is the thermocouple diameter in feet. The emissivity of the thermocouple is .8. (a) If the actual temperature is 300°F, what is the thermocouple reading? (b) If the thermocouple reading indicates that the gas temperature is 300°F, what is the gas temperature?

6. Air is flowing at 500 cfm through a 12″ diameter uninsulated duct, 50′ long, with emissivity of .28. The air entering the duct is 45°F, and an engineer states that the air leaving the duct will be 55°F. The duct goes through a room whose content and walls are at 80°F. Consider both convection and radiation to prove or disprove the engineer's statement.

7. A mild steel pipe (4.25″ O.D., 4.0″ I.D., 35′ long) is painted with dull gray lead (oil-base) paint on the outside. The pipe carries 200 cfm of 500°F, 25 psia air through a 70°F room. The conditions of the air at the end of the pipe are 350°F and 15 psia. (a) What is the overall coefficient of heat transfer? (b) Using theoretical or empirical methods, what is the expected coefficient of heat transfer? (c) Explain the possible reasons for differences between the actual and the expected coefficients of heat transfer.

8. A single pass heat exchanger is tested in the clean condition and is found to heat 70°F water to 140°F when flowing 100 gpm. The hot side uses 230°F steam. The tube's inner surface area is 50 square feet. After being used in the field for several months, the exchanger heats 100 gpm of 70°F water to 122°F. What is the fouling factor in units of hr-ft²-°F/BTU?

9. A semiconductor device is modeled as an upright circular cylinder, .75″ in diameter and 1.5″ high. The device emits 5 watts and is cooled by a combination of natural convection and radiation. The surface emissivity is .65. It can be assumed that the base is insulated and transmits no heat. The local environment is 14.7 psia air at 75°F. (a) What is the surface temperature? (b) What are the percentages of the heat loss that are convection and radiation?

RESERVED FOR FUTURE USE

HVAC— HEATING, VENTILATING, AND AIR CONDITIONING

PART 1: Heating Load

Nomenclature

a	thermal coefficient for air space	BTU/ft^2-°F-hr
A	area	ft^2
B	air leakage	ft^3/ft-hr
DD	degree days	°F-day
e	emissivity	–
E	effective space emissivity	–
F	heat loss coefficient	BTU/hr-ft-°F
h	film coefficient	BTU/ft^2-°F-hr
h_{fg}	latent heat of vaporization	BTU/lbm
HV	heating value of fuel	BTU/unit
k	thermal conductivity	BTU-ft/ft^2-°F-hr
L	length	ft
N	number of days in the heating season	days
NAC	number of air changes per hour	1/hr
p	slab perimeter	ft
q	heat transfer	BTU/hr
Q	volume flow	ft^3/hr
R	thermal resistance	ft^2-°F-hr/BTU
T	temperature	°F
U	overall coefficient of heat transfer	BTU/ft^2-°F-hr
v	velocity	mph
x	thickness	ft
y	distance from building mid-height	ft

Symbols

ρ	density	lbm/ft^3
ω	humidity ratio	lbm/lbm
η	efficiency	–

Subscripts

a	air
c	combustion
d	design
e	effective
i	indoor design
o	outdoor design
w	walls, water

Heating load calculations consist of (a) finding the heat lost through the walls, the roof, and the floor; (b) finding the heat required to warm cold air entering the heated space (infiltration losses); and (c) finding the heat required to warm any forced air ventilation. The basic procedure is essentially the same for all problems.

step 1: Determine the winter *outside design conditions* for the location. Generally, this requires finding temperature, wind direction, and wind speed.[1]

step 2: Select the *inside design conditions*. 75°F dry bulb with 50% relative humidity often is used regardless of season. This temperature is for the breathing line, three to five feet above the floor. The temperature variation with height is approximately .75°F per foot. For maximum comfort, the relative humidity should be kept between 20% and 60%.

---

[1]There are several ways to determine the outside design temperature. If a distribution of minimum annual temperatures is available, the outside design temperature can be taken as a value which will not be exceeded a desired percentage of the time (i.e., five years out of 100). As a rough estimate, the outside design temperature also can be chosen as 15 degrees more than the lowest temperature ever recorded.

Table 11.1
Recommended Inside Design Conditions
for General Comfort

	Dry Bulb, °F	Relative Humidity, %	Temperature Swing, °F
Summer			
Deluxe accommodations	74–76	50–45	—
Commercial practice	77–79	50–45	+2 to +4
Winter			
With humidification	74–76	35–30	−3 to −4
No humidification	75–77	—	−4

Table 11.1 gives recommendations for inside design conditions. The values in the table are only general guidelines. Certain applications will require modifications. For example, generally it is acceptable to allow slightly higher temperatures (2 or 3 degrees more) in the summer and slightly lower temperatures (2 or 3 degrees less) in the winter for retail shops and factories.

If the room is to be maintained at two different temperatures during different parts of the day, the average fuel consumption can be found from the duration-weighted average of the two temperatures. To size the heating equipment, however, the higher temperature should be used.

step 3: Determine the temperatures in all adjoining unheated rooms, such as attics, large closets, and basements.

step 4: Determine the coefficient of heat transmission, U, for all heat loss surfaces, including glass, outside walls, roof, floor, and walls which adjoin unheated rooms. If it is necessary to calculate U values from k values, film coefficients should be included. A value of 1.65 typically is used for an inside (still air) film coefficient. The outside film coefficient in a 15 mph wind is approximately 6.0.

$$U = \frac{1}{\left(\frac{1}{h_{\text{in}}}\right) + \left(\frac{1}{h_{\text{out}}}\right) + \sum\left(\frac{1}{a}\right) + \sum\left(\frac{x}{k}\right)} \qquad 11.1$$

Values of a, the *thermal conductivity of an air space*, depend on the space width, emissivities of both sides, orientation, mean temperature, and temperature differential. Representative values for a 50°F mean temperature differential are given in table 11.2. Values at other temperatures and differentials are within 20%.

In table 11.2, E is the *effective space emissivity*.

$$\frac{1}{E} = \left(\frac{1}{e}\right)_{\text{side 1}} + \left(\frac{1}{e}\right)_{\text{side 2}} - 1 \qquad 11.2$$

Values of U often are read directly from tables.[2] Such tables usually assume a 15 mph external wind. If the wind varies from 15 mph, a multiplicative factor should be applied from table 11.3.

step 5: Determine the area for each heat loss surface.

step 6: Calculate the heat loss from each heat loss surface.

$$q = UA(T_i - T_o) \qquad 11.3$$

Table 11.2
Typical Thermal Conductances of Air Spaces, a
BTU/ft²-hr-°F

orientation of air space	heat flow direction	space thickness	value of E .05	.20	.50	.82
horizontal	up	.75–4.00″	.41	.55	.82	1.11
vertical	up	.75–4.00″	.28	.42	.69	.99
horizontal	down	.75″	.28	.42	.69	.98
		1.50″	.18	.31	.58	.87
		4.00″	.11	.25	.52	.81

Table 11.3
Wind Correction Factors

wind velocity (mph)

Value of U at 15 mph	0	5	10	20	25	30
.310	.88	.95	.98	1.01	1.02	1.03
.500	.82	.93	.97	1.02	1.03	1.04
.700	.76	.90	.96	1.02	1.04	1.05
.900	.72	.88	.95	1.03	1.05	1.07
1.100	.67	.85	.94	1.04	1.06	1.08
1.300	.64	.83	.94	1.05	1.08	1.10

Since the ground temperature usually is higher than the winter air temperature, a higher temperature should be used to find the heat loss through floors. A 5°F difference frequently is assumed for conduction. If *slab edge* data is available for the floor, use the floor perimeter rather than the area with the actual values of T_i and T_o.

$$q = pF(T_i - T_o) \qquad 11.4$$

F varies between .81 BTU/hr-°F per foot of exposed edge for uninsulated edges to .55 BTU/hr-°F-ft for insulated edges.

step 7: Find the sensible heat required for *infiltration losses* from cracks around doors and windows. The losses will depend on the opening size, wind speed, and temperature difference. The specific heat, c_p, of air is taken as .24 BTU/lbm-°F.

$$q_a = 0.24Q\rho(T_i - T_o) \approx (0.018)Q(T_i - T_o) \qquad 11.5$$

Q can be found from the crack method or the air-change method. For the *crack method*, the infiltration air is calculated as

$$Q = BL \qquad 11.6$$

Representative values of B are given in table 11.4. In the absence of other information, choose the wind velocity as 15 mph. Since air entering through cracks on the windward side must leave through cracks on the leeward side, only half the total *crack length* is used for entire buildings. However, the amount of crack length used also depends on the building orientation, and the crack length should never be less than half the total length. The usual rules are given here.

- 1 exposed wall—Use total crack length
- 2 exposed walls—Use wall with greater crack length

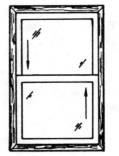 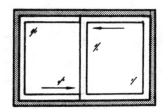
double-hung vertical (wood sash)   double-hung horizontal, or sliding (metal sash)

residential casement (metal sash)   industrial (horizontal) pivoted (metal sash)

architectural projected, top-hung, or ventilating sash (metal sash)   hollow metal vertical pivot (metal sash)

Figure 11.1 Types of Windows

- 3 exposed walls—Use crack length from 2 walls
- 4 exposed walls—Use one-half total crack length

For buildings over 100 feet high, it is not reasonable to assume that the wind velocity is uniform across the entire face. The wind velocity used in table 11.4 is the *effective wind velocity* calculated from equation 11.7.

$$v_e = \sqrt{v_o^2 - 1.75y} \qquad 11.7$$

In equation 11.7, y is measured from the mid-height of the building. y is positive above the mid-height

---

[2] Some tables give values of U. Others list values of *thermal resistance*, $R = 1/U$. The units indicate whether U or R values are provided. To further complicate the problem, thermal resistance values per inch of thickness frequently are given. The total R value can be obtained by multiplying such a *per-unit* resistance by the thickness of the material.

Table 11.4
Typical Infiltration Coefficients—ft³/hr-ft

| | | effective wind velocity, mph | | | | | |
		5	10	15	20	25	30
Double-hung wood sash windows, unlocked	average window, 1/16" crack, and 3/64" clearance, without						
	weather-stripping	7	21	39	59	80	104
	weather-stripped	4	13	24	36	49	63
Double-hung metal windows	no weather-stripping						
	locked	20	45	70	96	125	154
	unlocked	20	47	74	104	137	170
	weather-stripped, unlocked	6	19	32	46	60	76
Rolled section, steel sash window	residential casement						
	1/64" crack	6	18	33	47	60	74
	1/32" crack	14	32	52	76	100	128

and negative below. By convention, v_e is used on the windward side, but $(v_e - v_d)$ is used on the leeward side.

Equation 11.7 can be used to calculate the infiltration through windows at any height. If the total windward infiltration is required, it will be necessary to calculate and sum the infiltration through windows at all levels.

If the *air-change method* is used, Q is found from the room volume and the number of changes per hour. Recommended changes per hour can be found from tables. As an approximation for residential infiltration, assume one replacement per hour, which is based on windows on one side. For windows on two sides, assume $1\frac{1}{2}$ changes per hour. For windows on three sides, use two changes per hour.

For buildings with forced ventilation, the recommended air changes per hour can be found from table 11.5.

Table 11.5
Recommended Air Changes Per Hour

room type	NAC
boiler rooms	15–30
engine rooms	40–60
plant buildings, offices, restaurants, and assembly halls	6–12
warehouses	2–6

Heat also is required to warm any moisture that is brought in with the infiltrated air.[3]

[3] If the outside design temperature is less than 32°F, there will be little or no moisture in the incoming air. In that case, it is unnecessary to calculate heating requirements for incoming moisture.

$$q_w = m_w c_p \Delta T = \rho Q \omega c_p (T_i - T_o)$$
$$\approx (.075)(Q)(\omega)(.45)(T_i - T_o) \qquad 11.8$$
$$\approx (.0338)\omega Q(T_i - T_o)$$

The total heat required to warm incoming air is

$$q_t = q_a + q_w \qquad 11.9$$

However, the best way to calculate q_t is to use enthalpies from the psychrometric chart. If h_i and h_o are taken from the psychrometric chart,

$$q_t = m_a(h_i - h_o) \qquad 11.10$$

step 8: Find the latent heat required to add moisture to the air from infiltration.

$$q = h_{fg}Q\rho(\omega_i - \omega_o) \approx 79.5Q(\omega_i - \omega_o) \qquad 11.11$$

step 9: If the ventilation requirements have not been included in the infiltration losses, repeat steps 7 and 8 to find the sensible and latent heat requirements for the ventilation air.

step 10: An *exposure allowance* of 15% often is added to the totals of latent and sensible heat losses for unknown or severe climatic conditions.

step 11: Since the sky is assumed overcast, there is no solar heating contribution.

step 12: The heat supplied by permanent machinery and lighting should be subtracted from the required heat.[4]

[4] It may be unrealistic to assume that all heat generated by lights enters the room. If the air space above the lights in a dropped ceiling (the *plenum area*) is not directly conditioned, it is proper to assume that only 60% of the heating from lights enters the occupied space.

$$q_{\text{motors}} = \frac{(2545)(hp)}{\eta} \qquad 11.12$$

$$q_{\text{lights}} = (3413)(kw) \qquad 11.13$$

If a motor is not actually in the room, but the output shaft is, do not divide by the efficiency. For fluorescent lights, multiply the kilowatts by 1.2 to account for ballast heating. Unless the room is reasonably permanently occupied, the heating load should not be reduced by human heat output. Under no circumstances should the heating load be reduced to a point where the inside temperature would be 40°F or lower in the absence of heat added by external sources.

step 13: The *heating load* is the sum of all heat losses minus all heat sources. q can be converted to *therms* by dividing by 100,000.

If the building is heated only during the day, add 10% to the required heat capacity to allow for start-ups. If the building is left cool for extended periods, add 25% capacity.

The heating load calculated from the prior steps can be used to size the furnace. The furnace must be capable of keeping the building at the inside design temperature on the coldest days of the heating season. However, not all days in the heating season will have temperatures that cold, so the heating load q cannot be used directly to calculate the cost of heating throughout the cold season.

step 14: If the *degree days* are known for the location, the maximum heating load can be converted to an average heating requirement. This average heating requirement can be used to calculate heating costs.

Each day whose 24-hour average temperature is less than 65°F will accumulate $(T_{\text{ave}} - 65)$ degree days. The sum of these degree day terms over the entire heating season (which is different for each site) is the total degree days, DD, for that location. The average temperature over the heating season can be calculated from the degree days if the length of the heating season is known.

$$DD = N(65 - T_{\text{ave}}) \qquad 11.14$$

The heating cost can be calculated from the heating load and the degree days. It is reasonable to assume that the combustion efficiency in equation 11.15 is 60%–75%. The units of fuel consumption depend on the units of HV.[5]

$$\frac{\text{fuel}}{\text{consumption}} = \frac{24q(DD)}{(T_i - T_o)(HV)\eta_c} \qquad 11.15$$

Equation 11.15 does not include the cost of operating pumps, fans, stokers, and other mechanical or electrical devices.

---

[5] Equation 11.15 appears to imply that the lower T_o is, the lower the fuel consumption will be. This is not true. The temperature difference in the denominator actually cancels the same temperature difference used to calculate q. Thus, q is put on a "per degree" basis. The average temperature difference used in the calculation of DD converts the "per degree" heat loss to an average heat loss.

PART 2: Ventilation

Nomenclature

C	concentration	parts per million (ppm)
k	safety factor	–
LEL	lower explosive limit	parts per hundred (pph)
MW	molecular weight	lbm/pmole
SG	specific gravity	–
t	time	hrs
TLV	threshold limit value	parts per million (ppm)
V	volume	ft^3

Subscripts

r	room

1 RULES FOR ROOM DESIGN

The following general rules apply to room design for ventilation purposes.

- The building orientation should not be chosen on the basis of wind direction since that direction cannot be guaranteed.

- Ventilation inlets should be unobstructed.

- Ventilation paths should be unobstructed to allow for complete mixing with room air.

- Inlet and outlet openings should be of nearly equal areas.

- Direct short circuits between air inlets and outlets in the ventilation areas should be avoided.

- If natural ventilation is to be used, the air outlet should be higher than the air inlet.

2 PURPOSES OF VENTILATION

A. TO PROVIDE OXYGEN

The metabolic oxygen requirements for various grades of work in cubic feet per hour are given in table 11.6.

Table 11.6
Pure Oxygen Requirements
(cfh)

very light work	less than 1.06
light	1.06–2.12
moderate	2.12–3.18
heavy	3.18–4.24
very heavy	4.24–5.30
unduly heavy	5.30

The oxygen concentration should not be allowed to fall below 12% by volume under any circumstances. In most cases, infiltration will meet all oxygen requirements.

B. TO REMOVE CARBON DIOXIDE

The CO_2 concentration should not exceed 5% under any circumstances. A common equation for finding the time for a 3% build-up of CO_2 is given by equation 11.16. The air will become noticeably stale in about one fourth of this time.

$$t = \frac{(0.04)V_r}{\text{no. people in room}} \qquad 11.16$$

Although states have laws regarding ventilation requirements, a minimum airflow of 5 cfm per sedentary adult is recommended, and $7\frac{1}{2}$ cfm is preferred. In areas with excess activity and smoking, the air flow should be 25 cfm minimum and 40 cfm preferred. 15 cfm often is used as a design standard.

C. TO REMOVE ODORS

The air flow required to remove body odors depends on room size and activity. In the winter season, for sedentary adults without air conditioning, the minimum air supply to remove odors is given in table 11.7.

Table 11.7
Air Flow to Remove Odors

air space per person	cfm
100 ft^3	25
200	17
300	12
400	9
500	7

Values should be increased 50% to allow for moderate physical activity. If 15 cfm is used as a design standard, odors should not be a problem. When ventilation is controlled by odor removal, oxygen and carbon dioxide requirements also will be met. If the air is partly recirculated, no less than 4 cfm per person of outside air should be introduced.

D. TO REMOVE HEAT

Removal of sensible body heat is the controlling factor in ventilation requirements. If body heat removal is accomplished, all of the other needs will be met. The sensible heat output for sedentary adults is given in table 11.13.

The ventilation requirements can be calculated from the heat generation rate and equation 11.5. ΔT in equation

11.5 is the air temperature rise after being introduced into the populated space.

Usually, the sensible heat air flow requirements far exceed the latent heat air flow requirements, and ventilation can be calculated on that basis alone. However, when large moisture sources are present and it is desired to limit the humidity rise to $\Delta\omega$ grains/lbm, the ventilation can be found from equation 11.17 (which is based on $h_{fg, 70°F} \approx 1050$ BTU/lbm).

$$Q \approx \frac{(90)(\text{latent load, BTU/hr})}{\Delta\omega \text{ grains/lbm}}$$

$$\approx \frac{(93,300)(\text{latent load, lbm/hr})}{\Delta\omega \text{ grains/lbm}} \qquad 11.17$$

The higher of the two ventilation requirements (as determined from latent and sensible loads) should be used. Do not add them.

E. FOR TOXICITY DILUTION

Toxicity dilution refers to the dilution of contaminated air for reduced biological hazard and explosion control. *Dilution ventilation* is less effective than outright removal by exhaust ventilation. The required air volume cannot be too great or air velocities will become unreasonable. Also it is assumed that the toxicity is low, that the workers are not too close to the source, and that evolution is uniform and steady. This method generally is limited to organic liquids and low-toxicity solvents since fumes and dusts seldom are removed successfully by toxicity dilution.

step 1: Determine the volume of solvent evaporated in pints or pounds per hour.

step 2: Determine a safety factor, k, between 3 and 10. Choose the safety factor based on toxicity, uniformity of vapor evolution, and effectiveness of proposed ventilation system. Start with $k = 0$.

I. slightly toxic (TLV > 500 ppm)   add 1 to k
   moderately toxic (TLV 100–500 ppm) add 2 to k
   highly toxic (TLV < 100 ppm)   add 3 to k

II. uniform vapor evolution   add 1 to k
   non-uniform vapor evolution add 2 to k
   non-uniform and unpredictable add 3 to k

III. good ventilation   add 1 to k
   fair ventilation add 2 to k
   poor ventilation add 3 to k

step 3: Calculate the dilution ventilation requirements.

$$Q = \frac{(4.0 \text{ EE8})(SG)(\text{pints/hr})\,k}{(MW)(TLV)} \qquad 11.18$$

If the evaporation is measured in pounds per hour, replace the numerical coefficient in the numerator with (3.9 EE8). The volumes assume 70°F and one atmosphere. For other operating conditions (as in an oven), convert the volumes using the perfect gas law.

step 4: If there is no human exposure and if only reduced explosion hazard is desired, calculate the ventilation from equation 11.19.

$$Q = \frac{(4.0 \text{ EE4})(SG)(\text{pints/hr})\,k'}{(MW)(LEL)(B)} \qquad 11.19$$

$k' = 4$ for good ventilation and continuous operation

$k' = 10$ for good ventilation and batch or recirculating ovens, or if ventilation is poor

$B = 1$ for operation up to 250°F

$B = .7$ for operation above 250°F

Values of TLV and LEL are given in table 11.8. When two or more hazardous substances are present, the mixture threshold value is exceeded when equation 11.20 exceeds 1.0.

$$\left(\frac{C}{TLV}\right)_1 + \left(\frac{C}{TLV}\right)_2 + \cdots \qquad 11.20$$

The additive nature implied by equation 11.20 is assumed unless specific knowledge to the contrary is known (e.g., two solvents which act biologically independently). If the air must be breathed, calculate the air flow rates for each substance and add the rates together.

It is recommended that dilution ventilation not be used for carbon tetrachloride, chloroform, or gasoline.

Table 11.8
Approximate Values of TLV and LEL (1976)[6]

substance	MW	SG	LEL(pph)	TLV(ppm)
acetic acid	60.05	1.049	5.40	10
acetone	58.08	.792	2.55	1000
ammonia	17.03	.597	15.50	25
benzene	78.11	.879	1.40	25
butane	58.12	2.085	1.86	500
carbon monoxide	28.10	.968	12.5	50
carbon tetrachloride	153.84	1.595	*	10
chlorine	70.91	3.214	–	1
chloroform	119.39	1.478	*	25
ethyl alcohol (ethanol)	46.07	.789	3.28	1000
ethyl ether	74.12	.713	–	400
formaldehyde	30.03	.815	7.0	2
gasoline	86	.68 – .74	1.3	–
hydrogen chloride	36.47	1.268	–	5
methyl alcohol (methanol)	32.04	.792	6.72	200
methyl acetate	74.08	.928	3.15	200
methylene chloride	84.94	1.336		500
naptha (coal tar)	106.16	.85	–	500
octane	114.22	.703	.95	400
pentane	72.15	.626	1.4	500
propane	44.09	1.554	2.12	
toluene (toluol)	92.13	.866	1.27	100
vinyl chloride	62.50	.908	4	200
xylene	106.16	.881	1.0	100

* not flammable

----
[6]LEL and TLV values are subject to change and regulation by many government agencies. Values in table 11.8 are not necessarily correct for every situation.

PART 3: Mathematical Psychrometrics

Nomenclature

ADP	apparatus dew point	°F
BF	bypass factor	–
h	enthalpy	BTU/lbm
m	mass	lbm
MW	molecular weight	lbm/pmole
n	number of moles	–
p	pressure	psf
q	heat, or heat flow	BTU, or BTU/hr
Q	air flow rate	cfh
R	specific gas constant	ft-lbf/°R-lbm
R^*	universal gas constant (1545.33)	ft-lbf/°R-pmole
s	entropy	BTU/lbm-°R
SHR	sensible heat ratio	–
T	temperature	°R or °F
V	volume	ft^3
x	mole fraction	–

Symbols

ρ	density	lbm/ft^3
η	efficiency	–
ϕ	relative humidity	–
υ	specific volume	ft^3/lbm
ω	humidity ratio	lbm/lbm[7]
μ	degree of saturation	–

Subscripts

a	air
co	output of conditioner
db	dry bulb
dp	dew point
exh	exhaust
i	ith component, or inside
l	latent
o	outside
s	sensible
sat	saturation
t	total
w	water , or water vapor
wb	wet bulb
ws	water vapor at saturation

1 BASIC DEFINITIONS

Degree of Saturation, μ: The ratio of actual humidity ratio to saturated humidity ratio at the same temperature and total pressure. Also known as the *percentage humidity*.

Dew Point Temperature, T_{dp}: The temperature of saturated air at a given humidity ratio.

[7]Divide (grains/lbm) by 7000 to obtain (lbm/lbm).

Dry Bulb Temperature, T_{db}: The true temperature of still, moist air.

Humidity Ratio, ω: The ratio of water mass to the mass of dry air in a moist air sample. Also known as *specific humidity* and *mixing ratio*.

Latent Heat: The enthalpy content of added or condensed water vapor.

Mole Fraction, x: The ratio of the number of moles of substance i to the total number of moles in the sample. Same as volumetric fraction.

Relative Humidity, ϕ: The ratio of partial pressure of the water vapor in the sample to the saturation pressure at the same temperature.

Sensible Heat: The enthalpy content of the air in a sample.

Total Heat: The enthalpy content of the air and the water vapor. The sum of the latent and sensible heats.

Wet Bulb Temperature, T_{wb}: The temperature of adiabatically saturated air at the same pressure.

2 AIR AS A PERFECT GAS

Atmospheric air can be thought of as a mixture of two perfect gases, dry air (subscript *a*) and water vapor (subscript *w*). According to *Dalton's rule*,

$$V_t = \frac{n_a R^* T}{p_a} = \frac{n_w R^* T}{p_w} = \frac{(n_a + n_w) R^* T}{p_w + p_a} \qquad 11.21$$

$$p_w = \left(\frac{n_w}{n_t}\right) p_t = x_w p_t \qquad 11.22$$

$$p_a = \left(\frac{n_a}{n_t}\right) p_t = x_a p_t \qquad 11.23$$

From the definition of humidity ratio,

$$\omega = \left(\frac{m_w}{m_a}\right) = \left(\frac{n_w}{n_a}\right)\left(\frac{MW_w}{MW_a}\right) = .622\left(\frac{n_w}{n_a}\right)$$
$$= .622\left(\frac{x_w}{x_a}\right) \qquad 11.24$$

$$\frac{n_w}{n_a} = \frac{n_w}{(n_t - n_w)} = \frac{p_w}{(p_t - p_w)} \qquad 11.25$$

$$\omega = \frac{.622 p_w}{(p_t - p_w)} = \frac{.622 x_w}{(x_t - x_w)} \qquad 11.26$$

Using the mole fraction/pressure relationship, the *relative humidity* is defined.

$$\phi = \frac{n_w}{n_{ws}} = \frac{p_w}{p_{ws}} = \frac{\mu}{1 - \left[(1 - \mu)\frac{p_{ws}}{p_t}\right]} \qquad 11.27$$

Notice that $\phi = \mu$ only for completely saturated air and for completely dry air.

$$\mu = \frac{\omega}{\omega_{\text{sat}}} \qquad 11.28$$

Using the ideal gas law, the volume of moist air per pound of dry air is

$$v = \frac{R_a T}{p_a} = \frac{R_a T}{p_t - p_w} = (1 + 1.6078\omega)\frac{R_a T}{p_t} \qquad 11.29$$

A practical expression for enthalpy is

$$h_t = h_a + \omega h_w \qquad 11.30$$
$$h_a = .241T \qquad 11.31$$
$$h_w = .444T + 1061 \qquad 11.32$$

Notice that enthalpy of air usually is measured with respect to 0°F when psychrometric studies are made. Therefore T is in °F in equations 11.31 and 11.32.

The partial pressure of water vapor is approximated by equation 11.33. All pressures are in psig; temperatures are in °F.

$$p_w = p_{\text{sat, wb}} - \frac{(p_t - p_{\text{sat, wb}})(T_{db} - T_{wb})}{2800 - 1.3T_{wb}} \qquad 11.33$$

3 PSYCHROMETRIC CHARTS

Since the humidity ratio, ω, can have any value between 0 and ω_{sat}, the *degree of saturation* can be used to predict the properties of moist air in a manner similar to the use of quality with steam.

$$h = h_a + \mu h_{a,\text{ sat}} \qquad 11.34$$
$$s = s_a + \mu s_{a,\text{ sat}} \qquad 11.35$$
$$v = v_a + \mu v_{a,\text{ sat}} \qquad 11.36$$

The use of these formulas requires *moist air tables*. Such tables exist; however, corrections are required for use above 150°F, and the entropy correlation is poor even below 150°F. Therefore, graphical solutions are preferred.

Psychrometric charts are graphical representations of the moist air tables with the corrections built in. Different charts are required for different pressures.

There is one confusing point about psychrometric charts. Some of the properties are given "per pound of dry air." This unit basis incorrectly implies that the water vapor's contribution is not included. For example, if the enthalpy of air with water vapor is found to be 28.0 BTU per pound of dry air, the energy content of the water vapor *has* been included. However, to get the energy of a sample, 28.0 would be multiplied by the

mass of the dry air only, not by the combined water and air masses.

Example 11.1

In a room during the summer, air is at 75°F dry bulb and 50% relative humidity. Find its (a) wet bulb temperature, (b) humidity ratio, (c) enthalpy, (d) dew point temperature, (e) specific volume, (f) vapor pressure, and (g) degree of saturation.

Locate the point where the 75°F vertical line intersects the 50% humidity curve. All other values can be read directly from the chart.

(a) Follow the diagonal line up to the left until it intersects the wet bulb temperature scale. $T_{wb} = 62.6°F$.

(b) Follow the horizontal line to the right until it intersects the humidity ratio scale. Read $\omega = 64.8$ grains of moisture per pound of dry air.

(c) Finding the enthalpy is different on different charts. Some charts use the same diagonal lines for wet bulb temperature and humidity. Corrections are required in that case. Other charts use two alignment scales in conjunction with a straightedge. $h = 28.1$ BTU per pound of dry air.

(d) Follow the horizontal line until it intersects the dew point temperature scale. Read 55.1°F.

(e) Interpolate between diagonal specific volume lines. Read 13.68 cubic feet per pound of dry air.

(f) From the steam tables for a dew point temperature of 55.1°F, the vapor pressure is $p_{\text{sat}} = 0.214$ psia.

(g) The humidity ratio at 75°F and saturation is 131.5. Therefore,

$$\mu = \frac{64.8}{131.5} = .49$$

4 TYPICAL AIR CONDITIONING PROCESSES

All air conditioning processes are governed by conservation equations. These equations can be valuable in finding some portions of the solution. However, graphical solutions from the psychrometric chart usually are adequate.

- conservation of mass
  $$\sum (m_a + m_w) = \text{constant} \qquad 11.37$$
- conservation of energy
  $$\sum m_i h_i = \text{constant} \qquad 11.38$$
- conservation of water vapor
  $$\sum m_i \omega_i = \text{constant} \qquad 11.39$$

A. SENSIBLE COOLING

With sensible cooling, there is no change in the water content. The wet and dry bulb temperatures both decrease. Since the humidity ratio does not change, the process is represented by moving horizontally to the left on the psychrometric chart. This line is known as the *condition line*.

$$q_{removed} = m_a(h_1 - h_2) \approx m_a(.24 + .45\omega)(T_1 - T_2) \quad 11.40$$

The *bypass factor* is defined as

$$BF = \frac{T_{2,db} - T_{coil}}{T_{1,db} - T_{coil}} \quad 11.41$$

The bypass factor can be found graphically as the ratio of the distance $(T_2 - T_{coil})$ on the psychrometric chart to the distance $(T_1 - T_{coil})$.

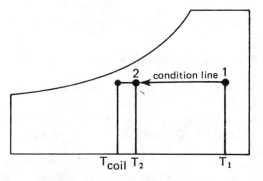

Figure 11.2 Sensible Cooling

The bypass factor can be thought of as the percentage of the air which does not come into contact with the cooling coil. If that interpretation is made, the remaining air (which does contact the cooling coil) is assumed to be reduced to the average coil temperature.

Typical bypass factors for coil cooling and heating equipment is given in table 11.9. Generally, the larger the installed unit, the smaller the bypass factor.

Table 11.9
Typical Equipment Bypass Factors

Installation	BF
residential	.35
small retail shop	.25
bank	.15
department store	.10

B. SENSIBLE HEATING

There is no change in the dew point since no moisture is absorbed. Wet and dry bulb temperatures both increase. The condition line is a horizontal line to the right on the psychrometric chart. As with sensible cooling,

$$q_{added} = m_a(h_2 - h_1) \approx m_a(.24 + .45\omega)(T_2 - T_1) \quad 11.42$$

The bypass factor is

$$BF = \frac{T_{coil} - T_2}{T_{coil} - T_1} \quad 11.43$$

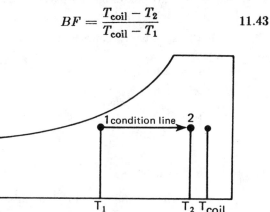

Figure 11.3 Sensible Heating

C. ADIABATIC MIXING OF TWO AIR STREAMS

Two ducts carrying moist air are shown joining in figure 11.4.

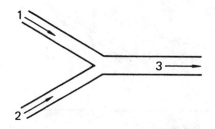

Figure 11.4 Adiabatic Mixing

The conservation laws can be written for adiabatic mixing.

$$m_{a1} + m_{a2} = m_{a3} \quad 11.44$$
$$m_{a1}h_1 + m_{a2}h_2 = m_{a3}h_3 \quad 11.45$$
$$m_{a1}\omega_1 + m_{a2}\omega_2 = m_{a3}\omega_3 \quad 11.46$$

Eliminating m_{a3},

$$\frac{m_{a1}}{m_{a2}} = \frac{h_2 - h_3}{h_3 - h_1} = \frac{\omega_2 - \omega_3}{\omega_3 - \omega_1} \quad 11.47$$

However, h and ω are linear scales on the psychrometric chart. Therefore, it is possible to use straight line proportions in the same ratio as m_{a1}/m_{a2}. To do so, draw a straight line between the two points (1 and 2) representing the input streams. The final mixture (3) will be on that line. Use the *lever rule* (on the basis of air masses) to locate the mixture point.[8]

---

[8]Because the water vapor adds little to the mixture volume, the air volumes can be approximated by the total mixture volumes.

Example 11.2

5000 cfm of 40°F dry bulb, 35°F wet bulb air are mixed with 15,000 cfm of 75°F dry bulb, 50% relative humidity air. Find the final dry bulb temperature.

First locate the two points on the psychrometric chart and draw a line between them.

Then read the specific volumes and calculate the dry air masses.

$$v_1 = 12.65$$
$$v_2 = 13.68$$
$$m_{a1} = Q_1 \rho_1 = \frac{5000}{12.65} = 395 \text{ lbm/min}$$
$$m_{a2} = \frac{15,000}{13.68} = 1096$$

The total air flow is

$$m_{a3} = 395 + 1096 = 1491 \text{ lbm/min}.$$

Now use the lever rule to locate point 3.

$$\frac{m_{a2}}{m_{a1} + m_{a2}} = \frac{1096}{1491} = .735$$

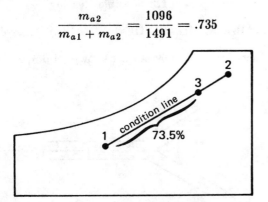

Because m_{a2} is greater than m_{a1}, point 3 will be closer to point 2 than to point 1. In fact, point 3 will be 73.5% of the distance from point 1 to point 2. Now it is necessary to measure the distance between points 1 and 2. (A centimeter scale or an inch scale with 1/10″ graduations is useful for this.) Multiplying that separation distance by .735 will locate point 3.

The dry bulb temperature is found to be approximately 66°F.

D. COOLING AND DEHUMIDIFICATION (COIL)

If a cooling coil temperature is below the incoming air's dew point, moisture will condense on the coil surface. The effective coil temperature in this instance is known as the *apparatus dew point*, *ADP*, determined from the intersection of the condition line and the saturation line. The ADP is the temperature to which the air would be cooled if 100% of the air contacted the coil. Thus, ADP and *effective coil temperature* are the same.

The water mass condensing on the coil will be

$$m_w = m_a(\omega_1 - \omega_2) \qquad 11.48$$

The total energy removed from the air is

$$q_t = m_a(h_1 - h_2) \qquad 11.49$$

The latent and sensible losses are calculated from the heat of vaporization evaluated at the average temperature.

$$q_l = m_a(\omega_1 - \omega_2)h_{fg} \qquad 11.50$$
$$q_s = q_t - q_l \qquad 11.51$$

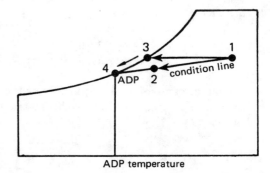

Figure 11.5 Cooling and Dehumidification

The air experiences sensible cooling from point 1 to point 3, after which it follows the saturation line down from point 3 to point 4, condensing water as it goes. Since some of the air does not contact the coil, the final condition of the air is actually at point 2.

For convenience, it is assumed that a straight line can be drawn between point 1 and the ADP, and that the final condition of the air will be along that straight line. Point 2 can be located if the coil bypass factor is known.

$$BF = \frac{T_{db,2} - ADP}{T_{db,1} - ADP} = \frac{\text{line segment } 4-2}{\text{line segment } 4-1} \qquad 11.52$$

The resulting air can be assumed to be BF% original air with (1−BF)% air at the apparatus dew point.

Water condenses out at a range of temperatures from 3 to 4. Usually, it is assumed that water leaving the coil is at the dew point temperature of the leaving air, $T_{dp,2}$.

The *sensible heat factor (sensible heat ratio)*, *SHR*, is the slope of the condition line.[9] The air masses cancel in equation 11.53.

$$SHR = \frac{q_s}{q_t} = \frac{m_a[(h_1 - h_2) - (\omega_1 - \omega_2)h_{fg}]}{m_a(h_1 - h_2)} \qquad 11.53$$

---

[9]The slope is not taken in the normal sense (i.e., y/x distances). Rather, the slope is defined by a sensible heat ratio scale built into the psychrometric chart. An alignment mark (usually a small circle) may appear in the center of the psychrometric chart for use with the scale. Lines determined from this point and scale need to be translated upwards or downwards to get them to pass through the appropriate points.

Example 11.3

A coil has a bypass factor of .2 and an apparatus dew point of 55°F. Air enters the coil at 85°F dry bulb and 69°F wet bulb. What are the sensible and latent heat losses? What is the sensible heat ratio?

The dry bulb temperature of the air leaving the coil will be

$$T_{2,db} = 55 + .2\,(85 - 55) = 61$$

This locates point 2.

Reading from the psychrometric chart, $h_1 = 33.1$, and $h_2 = 25.6$. The total energy loss is

$$q_t = 33.1 - 25.6 = 7.5 \text{ BTU/lbm dry air}$$

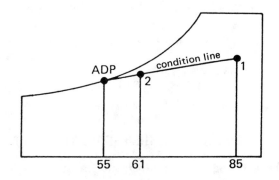

Evaluating h_{fg} at 60°F is an approximation for the average heat of vaporization. This is 1059.9 BTU/lbm moisture.

Reading from the chart,

$$\omega_1 = 81 \text{ grains/lbm}$$
$$\omega_2 = 69$$
$$q_l = \frac{(81 - 69)}{7000}\,(1059.9) = 1.84 \text{ BTU/lbm dry air}$$
$$q_s = 7.5 - 1.84 = 5.66$$
$$SHR = \frac{5.66}{7.5} = .75$$

Example 11.4

Find the sensible heat ratio for example 11.3 without calculating the sensible and latent losses.

First locate the ADP and point 1 on the psychrometric chart. Draw a straight line between them.

Draw a line parallel to the condition line from the reference point (the small circle near the center of the chart) to the sensible heat ratio scale. Read SHR= .75. (This example was chosen such that the condition line very nearly passes through the reference point. Most problems will require moving the parallel line much farther to reach the reference point.)

E. ADIABATIC MIXING OF MOIST AIR AND HOT WATER

If air is humidified by injecting steam or by passing the air through a hot water spray, the enthalpy of the air will increase. The conservation of energy equation determines the leaving air enthalpy.

$$m_a h_{a,\,in} + m_w h_w = m_a h_{a,\,out} \qquad 11.54$$

The required steam or hot water enthalpy can be calculated from equation 11.55.

$$h_w = \frac{m_a(h_{a,\,out} - h_{a,\,in})}{m_w} \qquad 11.55$$

The conservation of mass can be used to calculate the change in humidity ratio.

$$m_a \omega_{in} + m_w = m_a \omega_{out} \qquad 11.56$$
$$\omega_{out} - \omega_{in} = \frac{m_w}{m_a} \qquad 11.57$$

Eliminating m_w from equations 11.55 and 11.57,

$$\frac{\Delta h}{\Delta \omega} = \frac{m_a(h_{a,\,out} - h_{a,\,in})}{m_a(\omega_{out} - \omega_{in})} \qquad 11.58$$

Equation 11.58 gives the slope of the condition line. It defines the direction of the humidification process.[10]

Since water must be heated continually in this process, equilibrium never is reached. So this is not the same as passing air through a cold water spray.

Figure 11.6 illustrates that the condition line will be "above" the line of constant enthalpy that radiates from the incoming point. However, even though heat is added to the water, the air temperature can drop (as in case A), stay the same, or increase (as in case B).

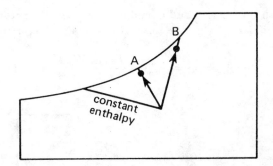

Figure 11.6 Heating and Humidification

The concept of bypass factors is not used with adiabatic saturation processes. Rather, the complement of a bypass factor is used, and this is known as the *satura-*

---

[10]The slope is based on the enthalpy and humidity scales. This slope is not the same as the SHR slope, nor is it a pure y/x scale slope.

tion efficiency or the *humidification efficiency*. Dry bulb temperatures are used in equation 11.59.

$$\eta_{sat} = 1 - BF = \frac{T_{a,\,in} - T_{a,out}}{T_{a,\,in} - T_w} \qquad 11.59$$

When the primary purpose of a recirculating spray is to cool the circulating water (e.g., a *cooling tower*), the water cooling efficiency is based on the water temperatures. This efficiency typically is in the 50% to 70% range.

$$\eta_w = \frac{T_{w,\,in} - T_{w,\,out}}{T_{w,\,in} - T_{a,wb,\,in}} \qquad 11.60$$

The lowest temperature to which the water can be cooled by purely evaporative means is the wet bulb temperature of the entering air. The *range* is the actual difference between leaving and entering water temperatures. The *approach* is defined as the difference between the leaving water temperature and the entering air wet bulb temperature. Water can be cooled 5°F to 15°F with approaches in the 5°F to 10°F range, typical of air conditioning installations.

F. COOLING AND HUMIDIFICATION

Cooling and humidification (also known as *adiabatic saturation* and *evaporative cooling*) occurs when air is passed through a continuously recirculating cold water spray, as in an *air washer*. This is a constant enthalpy process, since any evaporization that occurs requires heat to be drawn from the air. Because the removed heat goes into the recirculating water, the water temperature is raised to the wet bulb temperature of the incoming air, after which it is constant. Problems of this type assume that the recirculation has been going on for some time, so that the water already is at the wet bulb air temperature.

As with the heating and humidification case, the *saturation efficiency* takes the place of the bypass factor concept.

$$\eta_{sat} = \frac{T_{a,in} - T_{a,out}}{T_{a,in} - T_w} \qquad 11.61$$

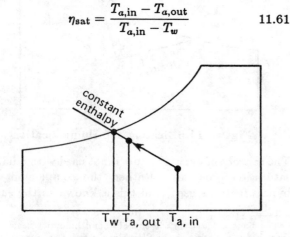

Figure 11.7 Cooling and Humidification

The *humidification load* is defined as the pounds of water evaporated per hour.

Of course, the air should be kept from freezing when its temperature drops. Therefore, the entering wet bulb temperature should be above 35°F. Since the leaving air will be cold, reheating may be necessary before returning washed air to the conditioned space.

Example 11.5

Air at 90°F dry bulb and 65°F wet bulb enters a spray type air conditioner. The air leaves at 90% relative humidity. The recirculated water to the spray is 65°F. What are the dry bulb and dew point temperatures of the leaving air?

When the spray water is the same temperature as the wet bulb of the entering air, the evaporative cooling process is along a line of constant enthalpy (constant wet bulb). Draw a line from the 90°F dry bulb line to the 90% relative humidity line along the 65°F wet bulb line. At the intersection with the 90% relative humidity line, read $T_{db} = 67.1°F$ and $T_{dp} = 64.0°F$.

G. COOLING AND DEHUMIDIFICATION (SPRAY)

If the entering water temperature in an air washer is below the entering wet bulb temperature, both the dry bulb and wet bulb temperatures will drop. This can occur during the start-up of an air washer used for humidification, or the water can be kept intentionally chilled.

If the leaving water temperature is below the entering dew point, dehumidification will result and the air will give up heat to the water. The final water temperature will depend on heat pick-up and water quantity. All air temperatures fall, and some moisture condenses.

The *performance factor* is defined as

$$1 - \frac{T_{a,wb,out} - T_{w,out}}{T_{a,wb,in} - T_{w,in}} \qquad 11.62$$

H. HEATING AND DEHUMIDIFICATION

Air passing through a solid or a liquid *adsorbant*, such as silica gel or activated alumina, will decrease in humidity ratio.[11] As the moisture is removed, chemical heat is evolved, as well as heat of vaporization being liberated.

Since heat is being generated, this is not an adibatic process.

---

[11] The correct term for a substance which collects water on its surface is *adsorbant*. By virtue of their great porosity, adsorbants have large surface areas. The attractive forces on the surfaces of these solids cause a thin layer of condensed water to form. Adsorbants can be reactivated by heating.

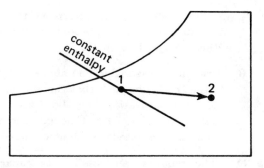

Figure 11.8 Heating and Dehumidification

5 THE RECIRCULATING AIR BYPASS PROBLEM

Figure 11.9 illustrates a common configuration of a conditioned space, an air conditioner, and a bypass duct. The bypass duct can be part of the air conditioner (as opposed to a separate duct), but that does not change the solution procedure given here.

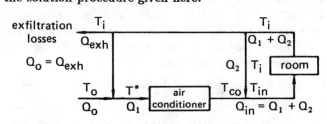

Figure 11.9 Recirculating Air Bypass

The solution to this type of problem is given on a step-by-step basis. It is necessary to know some things about the room, such as the sensible and latent heat loads, and the conditions of the outside and inside air. Careful attention needs to be paid to the subscripts, particularly "1" and "i," which are similar in appearance. Some modifications to the solution procedure may be required, depending on the known information.

step 1: Locate the indoor and outdoor design conditions on the psychrometric chart and draw a line between the two points. This line represents the mixture of two moist air streams combining prior to entering the air conditioner. (see point * on figure 11.9). Read h_i, h_o, and v_o.

The percentage of Q_o and Q_i will determine the actual mixture point, *. Assuming that the densities are approximately the same, the air masses and the volumes are proportional. From the lever rule,

$$\frac{T_{db}^* - T_{i,db}}{T_{o,db} - T_{i,db}} = \frac{\text{line segment}(i - *)}{\text{line segment}(i - o)} = \frac{Q_o}{Q_1} \qquad 11.63$$

Unfortunately, Q_1 may not be known at this point.

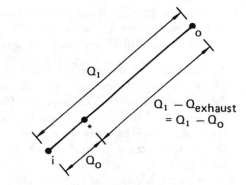

Figure 11.10 Step 1

step 2: Calculate the sensible heat ratio from the sensible and latent room loads. This ratio sometimes is called the *room sensible heat ratio*, *RSHR*, because heat associated with ventilation air is not included. (Do not include ventilation air in calculating sensible and latent heat loads.)

$$RSHR = \frac{q_s}{q_s + q_l} \qquad 11.64$$

step 3: Draw a line with slope RSHR (based on the sensible heat ratio scale) through point *i*. This condition line is the locus of points which could represent the air entering the conditioned space.

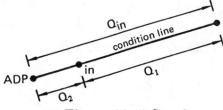

Figure 11.11 Step 3

The line drawn in step 3 represents the mixing of two air streams: Q_1 from the conditioner at the ADP, and Q_2 at T_i. The percentage of each will determine the "in" point and the system bypass factor according to the following lever rule.

$$\begin{aligned}(BF)_{\text{system}} &= \frac{\text{line segment}(ADP\text{-}in)}{\text{line segment}(ADP\text{-}i)}\\ &= \frac{T_{in,db} - T_{ADP,db}}{T_{i,db} - T_{ADP,db}}\\ &= \frac{Q_2}{Q_1 + Q_2} \qquad 11.65\end{aligned}$$

Unfortunately, either or both of Q_1 and Q_2 may be unknown.

To complete this step, extend the condition line to the left until it intersects the saturation line. This intersection point defines the ADP, which should be greater than 32°F.

If the condition line intersects the saturation line above 32°F, cooling and dehumidification are required. If the intersection is below 32°F, or if the condition line does not intersect the saturation line at all, ice will form on the coils and reheating will be necessary. (See step 10.)

step 4: The selection of the properties of the output air or the output flow rate are related. One determines the other. The larger the $(T_i - T_{in})$ temperature difference, the lower the flow rate required. However, since very large values of this temperature difference would require extremely efficient mixing in the room, and because a very low value of T_{in} probably would be uncomfortable for occupants, a 15–20°F temperature difference should not be exceeded.

Unless either T_{in} or Q_{in} is known, arbitrarily select T_{in}, which is the dry bulb temperature of the air entering the conditioned space. (Notice that T_i is not the same as T_{in}.) Remember to keep $T_i - T_{in}$ in the 15 °F to 20 °F range. (If Q_{in} is known, use equation 11.66 to calculate T_{in}.)

step 5: Locate the point with the chosen $T_{in, \, db}$ on the RSHR condition line. (This is point "in.") This establishes the ratio of Q_1 and Q_2 from equation 11.65.

step 6: Calculate Q_{in} into the room based on the room sensible heat load and the acceptable temperature rise.

$$Q_{in} = \frac{(55.3) \, q_s}{T_i - T_{in}} = Q_1 + Q_2 \qquad 11.66$$

step 7: Calculate Q_2 and Q_1.

$$Q_2 = (BF_{system})(Q_{in}) \qquad 11.67$$
$$Q_1 = Q_{in} - Q_2 \qquad 11.68$$

step 8: Locate point * using the lever rule (equation 11.63 from step 1).

As an alternative, draw a line through point "in" with a slope equal to the grand sensible heat ratio, GSHR.

$$GSHR = \frac{q_{s, \, room} + q_{s, \, ventilation \, air}}{q_{t, \, room} + q_{t, \, ventilation \, air}} \qquad 11.69$$

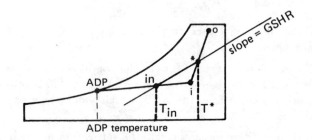

Figure 11.12 Step 8

step 9: If the condition line intersects the saturation line above 32°F, the solution is complete. Otherwise, go to step 10.

step 10: Draw any reasonable line above and parallel to the (in-*) line such that the new line intersects the saturation line above 32°F. Check that $(T_i - T_{in}) < 20$°F. Calculate the new circulation rate based on the new line.

step 11: The required air conditioner capacity (in BTU/hr) is

$$q_t = q_{s,room} + q_{l,room} + (h_o - h_i) Q_o \rho_o$$
$$\approx q_{s,room} + q_{l,room} + 4.5 (cfm)_o (h_o - h_i) \qquad 11.70$$

Example 11.6

A conditioned room with partial recirculation is to be kept at 75°F db, 62.5°F wb.[12] The local outside environment is 94°F db, 78°F wb. The sensible space load is 160,320 BTU/hr. The latent load from occupants and infiltration, but excluding ventilation, is 19,210 BTU/hr. Based on the occupancy rate, 1275 cfm of ventilation air are required.

Find (a) the temperature of the air entering the room.

(b) the volume of air passing through the room.

(c) the condition of the air entering the conditioner.

(d) the required apparatus dew point.

(e) the system bypass ratio.

This is a standard recirculating air bypass problem. Refer to figure 11.9.

step 1: Plot points i and o on the psychrometric chart.

step 2: Calculate the room sensible heat ratio.
$$RSHR = \frac{160,320}{160,320 + 19,210} = .89$$

step 3: Draw a line with slope .89 through point i.

step 4: Choose $T_{in} = 56$°F db.

step 5: Locate 56°F db on the condition line. For point "in" read

$$T_{in, \, wb} = 54.6°F$$

Calculate the system bypass ratio by measuring the line segments. The lengths of these lines will depend on the psychrometric chart, but the ratio given by equation 11.65 is approximately .12. Therefore, $BF_{system} = .12$.

---

[12]It is impossible to "keep" a room at a particular temperature. Cold air enters the room, and warm air is removed. The inside design conditions, therefore, are interpreted as the temperature of the air removed from the room.

step 6: Calculate the air flow into the room.

$$Q_{in} = \frac{(55.3)(160,320)}{(75-56)} = 466,600 \text{ cfh}$$

step 7:
$$Q_2 = (.12)(466,600) = 56,000 \text{ cfh}$$
$$Q_1 = 466,600 - 56,000 = 410,600$$

step 8: Locate point * by using equation 11.63.

$$\frac{(1275)(60)}{410,610} = .19$$

The line segment lengths will depend on the psychrometric chart. Moving 19% of the way up the $(i-o)$ line towards point o locates point *.

$$T_{db}^* = 78.5°F$$
$$T_{wb}^* = 66°F$$

PART 4: Cooling Load

Nomenclature

A	area	ft^2
C	coefficient	–
F$_{shg}$	solar heat gain factor	BTU/hr-ft^2
h	film coefficient, or enthalpy	BTU/hr-°F-ft^2 or BTU/lbm
HP	horsepower	hp
p	pressure	inches water
q	heat	BTU/hr
Q	air flow	ft^3/hr
T	temperature	°F
U	overall thermal conductance	BTU/hr-°F-ft^2

Symbols

η	efficiency	–
ρ	density	lbm/ft^3
ω	humidity ratio	lbm/lbm

Subscripts

f	fan
i	indoor design
o	outdoor design
s	shading
t	total
te	equivalent temperature

1 BASIC PROCEDURE

Finding the *cooling load* (the *air conditioning load*) is similar to finding the heating load, with the added consideration of solar heat gain and sensible/latent differences.

step 1: Determine the indoor design conditions.

step 2: Determine the summer outdoor design conditions.

step 3: Determine the required ventilation.

step 4: Find the instantaneous solar heat gain.[13] Of course, the instantaneous heat absorption is not the same as the instantaneous cooling load due to the time lag of thermal conductance. For that reason, the cooling load will not be constant with time of day (as was assumed with the heating load).

The instantaneous heat gain through windowless exterior walls and roofs is approximated by equation 11.71.

$$q = UA\Delta T_{te} \qquad 11.71$$

[13]The instantaneous heat *gain* is the heat which enters the conditioned space. It is not the same as the instantaneous heat *absorption* by the building. This naming convention is somewhat arbitrary, and it is not rigidly adhered to.

Table 11.10
Typical Total Equivalent Temperature Differences
Latitudes: 0° to 50° N
hottest weather and full sun exposure
(Includes radiation, convection, and conduction.
Assumes daily swing of 20°F and $T_o - T_i = 20°F$.)

	12 am	2 pm	4 pm	6 pm	8 pm
ROOFS					
light—1″ wood nominal	81	94	84	56	20
light—2″ concrete	69	85	83	63	32
light—2″ wood nominal	54	77	85	75	48
medium—2″ wood, 4″					
rock wool, ½″ plaster	16	32	49	61	63
heavy—4″ concrete	46	65	74	68	49
heavy—6″ concrete	32	49	61	63	54
WALLS (light colored)					
frame wall facing E	28	20	23	21	17
facing S	22	31	30	22	16
facing W	14	20	38	46	33
facing N	13	18	21	21	21
4″ brick or stone and					
frame facing E	30	28	20	22	20
facing S	10	21	28	28	20
facing W	10	15	20	35	41
facing N	9	13	17	20	19

ΔT_{te} is the *total equivalent temperature difference*, which is dependent on the construction type, geographical location, time of day, and wall orientation. It is read from extensive tabulations, a sample of which is given in table 11.10.

The instantaneous heat gain through transparent areas such as windows and skylights not in direct sunlight is

$$q = UA(T_o - T_i) \qquad 11.72$$

If the window is in direct sunlight, a portion of the incident solar radiation will be transmitted through the glass.[14] The instantaneous heat gain is

$$q = A(C_s F_{shg} + U(T_o - T_i)) \qquad 11.73$$

Both F_{shg} and C_s must be found from tables. If the windows have shades, curtains, or overhangs, additional corrections are needed.

Table 11.11
Typical Shading Coefficients

glass type	thickness	C_s
single		
regular sheet	$\frac{3}{32}''$, $\frac{1}{8}''$	1.00
regular plate	$\frac{1}{4}''$	.95
regular plate	$\frac{3}{8}''$	.91
heat absorbing sheet	$\frac{7}{32}''$	.71
heat absorbing plate*	$\frac{1}{4}''$	.67
	$\frac{3}{8}''$	.57
gray sheet	$\frac{1}{4}''$	.86
gray plate	$\frac{1}{4}''$	.70
double (up to $\frac{1}{2}''$ air space)		
both regular sheet	$\frac{3}{32}''$, $\frac{1}{8}''$	.90
both regular plate*	$\frac{1}{4}''$	.83

*Plate glass is sheet glass which has been ground and polished.

Example 11.7

A room in a building at 40° north latitude has 30 square feet of unshaded window facing west. (The area is based on sash dimensions.) The window is $\frac{1}{4}''$ sheet glass. What is the maximum afternoon solar heat gain? At which time does it occur?

From table 11.12,

$$F_{shg,max} = 216 \text{ BTU/hr-ft}^2 \text{ at 4 pm}$$

---

[14]*Fenestration* is the term used to describe the arrangement of windows or other openings transparent to solar radiation.

From table 11.11, the shading coefficient is .95. Therefore, the heat gain from equation 11.73 is

$$q = (30)(.95)(216) = 6156 \text{ BTU/hr}$$

Table 11.12
Typical Solar Heat Gain Factors
40° North latitude, July 21
BTU/hr-ft^2

time a.m.	time p.m.	N	E	S	W	horizontal
5	7	0	1	0	0	0
6	6	37	137	10	10	31
7	5	30	204	20	19	88
8	4	28	216	29	26	145
9	3	32	194	52	31	194
10	2	35	146	80	35	231
11	1	37	81	102	37	255
12	12	38	41	109	41	262
		N	W	S	E	horizontal

p.m. orientation

step 5: Find the heat gain from interior walls separating the conditioned space from facilities such as kitchens and boiler rooms. (Floors usually are ignored in cooling load calculations.)

$$q = UA(T_{\text{adjacent}} - T_i) \qquad 11.74$$

step 6: Find the heat gain from exposed air conditioning ducts.

step 7: Find the heat gain due to interior sources.

- people: Use table 11.13, which is based on a 75°F room dry bulb. For 80°F, the total heat remains the same, but the sensible heat is decreased approximately 20%, with the latent heat being adjusted accordingly. The "adjusted" column refers to a normal mix of men, women, and children for the application listed. It is assumed that the heat gain for an adult female is 85% of the adult male rate; the heat gain for a child is taken as 75% of the adult male rate.

- lighting:

$$q = 3.413 \text{ (watts)} \begin{Bmatrix} 1.0 \text{ for incandescent} \\ 1.2 \text{ for flourescent} \end{Bmatrix} \qquad 11.75$$

- power equipment:

$$q = \frac{2545(HP)}{\eta} \qquad 11.76$$

Table 11.13
Heat Gain from Occupants (BTU/hr)

activity	total heat adult males	total heat adjusted	sensible heat	latent heat
seated at rest	390	330	225	105
moderately active office work	475	450	250	200
standing, light work, walking slowly	550	450	250	200
walking 3 mph, moderately heavy work, dancing	1000	1000	375	625
heavy work	1500	1450	580	870

If the motor is not in the room, but the output shaft is, do not divide by the efficiency. Typical efficiencies are listed in table 11.14.

Table 11.14
Typical Motor Efficiencies

motor hp	efficiency
1/8	.55
1	.80
10 and up	.90

If the air conditioning fan and motor specifications are known, the heat from them also can be included.

$$q_{\text{fan}} = 2545\,(1 - \eta_{\text{fan}})(HP)_{\text{motor}} \qquad 11.77$$

$$q_{\text{motor}} = \frac{2545\,(1 - \eta_{\text{motor}})(HP)_{\text{motor}}}{\eta_{\text{motor}}} \qquad 11.78$$

The heat equivalent of the friction from air movement in the air conditioning duct between two points is

$$q_{\text{air movement}} = (0.4)(\Delta p_t)(cfm) \qquad 11.79$$

Since the equipment usually is not known in advance, it is reasonable to add 5–10% of the sensible load as an allowance for the fan and some duct.

step 8: Calculate the cooling load required to bring outside air for ventilation to the desired indoor conditions.

$$q_t = Q\rho\Delta h = 60\,(cfm)\,\rho\Delta h \approx 4.5\,(cfm)\,\Delta h$$
$$11.80$$

$$q_{\text{sensible}} = Q\rho(0.24 + 0.45\omega)\Delta T \approx 1.085\,(cfm)\Delta T$$
$$11.81$$

$$q_{\text{latent}} = 1076 Q\rho\Delta\omega \approx 4840\,(cfm)\,\Delta\omega$$
$$11.82$$

step 9: Determine any sensible heat losses due to cooling surfaces.

The *energy efficiency ratio* is used as a type of efficiency. EER usually is greater than 1.0.

$$EER = \frac{\text{cooling (BTU/hr)}}{\text{input power (watts)}} \qquad 11.83$$

2 AVERAGE COOLING COSTS

As with the heating load, the cooling load can be used to find the average cost of cooling per season. The approximate cost per season (or over any other period for which the *summer degree days* are known) is given by equation 11.84.[15]

$$\text{cost} = \frac{24\,(\text{cost/BTU})(\text{cooling load})(DD)}{T' - 70} \qquad 11.84$$

$$T' = T_o - \tfrac{1}{2}\,(\text{average summer daily range}) \qquad 11.85$$

---

[15] The summer degree days are not the same as winter degree days. Tables of summer degree days are far less common than of winter degree days.

PART 5: Solar Energy

Nomenclature

A	area	ft^2
c_p	specific heat	BTU/lbm-°F
E	incident solar radiation	BTU/hr-ft^2
F_R	heat removal factor	–
$\dot{m}$	mass flow rate	lbm/hr
q	heat	BTU/hr
T	temperature	°F
U_L	collection loss coefficient	BTU/hr-ft^2-°F

Symbols

α	plate absorptivity	–
η	efficiency	–
τ	cover transmittance	–

Subscripts

w	water

1 INTRODUCTION

The two basic categories of solar heating systems are passive systems and active systems. This chapter is applicable only to active systems and only to the two types discussed: rock and water thermal storage.

2 ACTIVE SYSTEMS

Active systems rely on mechanical components to collect and deliver heat. Air or liquid is heated in a *solar collector* and then is transported with the aid of fans or pumps. Solar collectors, also referred to as *collector panels*, usually are mounted on the roof or on the south wall of a house. Also, they can be mounted on their own supporting structure entirely separate from the house for better exposure to the sun.

For North American latitudes, where the winter sun is low in the southern sky, collectors should face south. However, a deviation of 20° from true south will not substantially reduce performance. Collectors also should be tilted toward the sun.

The combination of facing south and tilted toward the sun allows the collectors to receive a maximum amount of solar radiation.

The angle or tilt of the collectors toward the sun depends on the latitude of the location. A good rule of thumb for determining the best tilt angle is to add

15° to the latitude.[16] However, in some areas where snow covers the ground most of the winter, collectors mounted on the south wall of a structure can receive almost as much solar radiation as collectors mounted at the latitude plus 15° because of the high amount of radiation that reflects off the snow and onto the collector surface.

In addition to solar collectors, components used with active systems include a storage unit, a backup heating unit, a delivery system, and controls.

A. SOLAR COLLECTORS

Of the many solar collectors currently available for space and household water heating, the *flat-plate collector* is most often used. Flat-plate collectors are categorized by the type of heat-transfer fluid used—air or liquid. Internal components may vary, but a typical flat-plate collector is a shallow rectangular box covered with clear glass or plastic (*glazing*) to trap solar energy, an *absorber plate* usually made of steel, copper, or aluminum, an air pocket between the glazing and the absorber, and a layer of insulation behind the absorber to retain the trapped solar energy within the collector box.

The absorber plate may be flat, corrugated, or grooved, and usually is either painted black to increase absorption or treated with a surface coating (*selective surface*) which absorbs more solar radiation and emits less thermal radiation than ordinary paints. Some collectors use two sheets of glass or plastic (*double glazing*) to reduce further the heat loss from within the collector.

A solar collector works on the *greenhouse effect*. The window glass admits the sun's rays (shortwave radiation) where they strike the interior surfaces. These surfaces absorb the radiation, become warm, and lose this heat (longwave radiation) to the surrounding air. Because glass does not easily transmit longwave radiation, a large amount of heat is trapped.

The trapped heat then is absorbed by the air or by the liquid heat-transfer fluid as it passes by the absorber plate. Heat absorbed by the heat-transfer fluid then can be delivered for immediate use within the house or to a storage unit for later use.

---

[16]If the purpose of the collector is to heat water for bathing, it is not uncommon to tilt the collector at an angle equal to the latitude. If the collector is to heat an occupied space, however, the tilt angle should be the latitude plus 15°. This larger angle is required in order to collect maximum energy during the winter when the sun is lower in the sky.

B. STORAGE UNITS

Because solar radiation is intermittent, heat from solar collectors must be "stored" for use at night or during cloudy periods. Storage units commonly used with active solar heating systems contain either rocks or water.

A large bin or container filled with rocks is the best type of storage to use with air systems. The rocks absorb heat from collector-heated air blown into the storage bin by a small fan or blower. As it is needed, the stored heat can be blown from the storage unit through the delivery system ducts to the house.

The rocks in a storage bin should range in size from $\frac{1}{2}$ to $1\frac{1}{2}$ inches in diameter and should be washed to eliminate dust or other matter that may interrupt air flow. Rounded granite river rocks are excellent for use in a rock storage bin.

In liquid solar systems, a large insulated tank of water serves the same purpose as the rocks. Water, like rock, is inexpensive and can retain heat for long periods. Liquid storage tanks are made of several different materials. Glass-lined steel tanks probably are the best but are expensive. A concrete tank can be poured on site; however, the tank must be lined with thick polyethylene plastic to resist high temperatures and to protect against leaks.

Another type of tank that works well is a lightweight, noncorrosive fiberglass tank. Some fiberglass tanks are designed for solar heating systems but are slightly more expensive than other types. Whatever type is used, it must be heavily insulated to reduce heat loss.

C. BACKUP HEATING

Most solar heating systems are designed to provide from 40% to 80% of annual yearly heating needs (*solar fraction*) because it is less expensive to purchase a small amount of heat using a backup heating system than it is to build a solar system large enough to supply the total amount. A solar system sized to provide 100% of the energy needs would be expensive and oversized. (However, the cost of a system that will supply less than 30-40% of yearly heating needs seldom is justified.)

D. DELIVERY SYSTEM

A *delivery system* is required in both solar and conventional heating systems. The materials used in solar delivery systems are much the same as those used in conventional delivery systems. Solar systems employing air as the heat-transfer fluid use well insulated ducts and small blowers or fans to distribute heated air throughout the system.

Liquid solar systems use conventional plumbing techniques and materials such as piping, valves, and small pumps to control and distribute heated liquid.

Solar heating systems operate more efficiently at lower temperatures than most conventional heating systems. As a result, the delivered heat usually is warm rather than hot. The most effective method for delivering warm air is with a forced-air delivery system.

Air ducts used with forced-air delivery systems can deliver heat from either the solar system or a backup hot-air furnace. However, they must be larger than ducts used with conventional forced-air heating because a larger amount of solar heated air must be delivered to make up for its lower temperature. Forced-air delivery is well-established, and factory-made units called *air handlers* are available for solar systems and contain everything needed except ducts.

When a *heat exchanger* is used to transfer heat from liquid to air, forced-air is the most effective delivery system for liquid type systems. The heat exchanger is installed in the return air duct to the furnace, allowing it to operate also as a preheater when the backup system is operating.

3 A COMPARISON OF LIQUID AND AIR SYSTEMS

A. LIQUID SYSTEMS

Because liquid heating by conventional means has been used for years and the technology is well developed, liquid solar systems have been preferred over air solar systems. Since liquid has a higher heat capacity than air, it is a compact storage material.

Where space is limited, liquid pipes have an inherent advantage over larger air ducts. The disadvantages of liquid systems include the expense of leak-proof storage facilities and piping, the dangers of freezing and boiling, and corrosion problems.

There are several ways to avoid freezing problems. Liquid systems used in cold climates can be designed so the collectors drain at night to isolate the liquid from freezing temperatures. In another design, a slow circulation rate through the collector system is maintained during freezing temperatures to prevent substantial ice formation.

A third, more popular method, is to lower the freezing point of the liquid by adding antifreeze. However, most antifreezes break down and become corrosive if not replaced periodically. In addition, if the solar system also heats the household water supply, a heat exchanger must be used to keep the antifreeze from mixing with drinking or bath water.

B. AIR SYSTEMS

An increasing number of homes today are using **air solar heating systems** successfully. One advantage of air systems is the freedom from hazards associated with corrosion, freezing, boiling, and liquid leakage. Air is used in the collector-to-storage loop and in the storage-to-rooms loop, allowing for some savings in component costs. Also, air systems have an additional operating mode: direct heat delivery from the collectors to the building without heat exchange or storage.

One disadvantage of air systems is the size of the ductwork and the storage units, which often are quite bulky. Flow noise and blowing air also can be a nuisance. In some cases, the power requirement for blowers or fans is higher than for pumps used with liquid systems. A common problem is leaky ductwork, which can seriously degrade performance, but it is not easily detected.

4 COLLECTOR PERFORMANCE

A. DETERMINING AVAILABLE SOLAR ENERGY

Approximately 429 BTU/ft^2-hr strikes the earth's atmosphere.[17,18] However, reflection and absorption losses reduce this to 170–300 BTU/ft^2-hr that actually reach the earth's surface. Furthermore, the position of the sun, the clouds, and the atmospheric purity all affect the energy reaching the collector. Since most collectors rely on direct radiation, reflected and diffuse radiation components are not counted in the available solar energy.

The sun's position in the sky varies with time of day, time of year, and the latitude of the observer. The sun is highest in the sky at the start of summer, June 21. It is lowest at the start of winter, December 21. These dates often are used in solar calculations.[19]

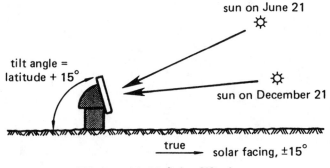

Figure 11.13 Solar Window

---

[17]Multiply *Langleys* by 3.69 to get BTU/ft^2. Multiply Langleys/min by 221 to get BTU/ft^2-hr.

[18]This is known as the *solar constant*.

[19]Actual values of incident solar radiation should be determined from tables or charts.

B. COLLECTOR EFFICIENCY

Solar collector efficiency is the percentage of the intercepted solar radiation that is converted to heat and is delivered to the building. There are many factors which affect this efficiency, including construction, type of fluid circulating, temperatures of the fluid and the surrounding air, and intensity of the solar radiation. Typical efficiencies are in the 70%–80% range when the collector and air temperatures are the same.

$$\eta_{collector} = \frac{\text{actual useful energy collected}}{\text{solar radiation striking the collector}} \qquad 11.86$$

As the temperature of the collector increases above the ambient air temperature, the collector itself loses heat to the air. This decreases the efficiency. The rate of heat decrease varies with different types of collectors. For example, a single glazed collector will experience a faster drop-off in efficiency than will a double glazed unit.

Collector efficiency usually is expressed as a function of the fluid inlet and the outdoor air temperature.

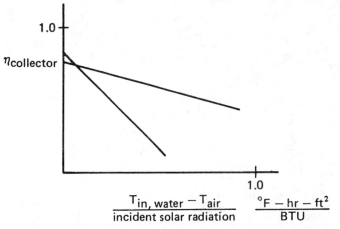

Figure 11.14 Typical Solar Collector Efficiency

The maximum efficiency can be read from the efficiency chart with a zero temperature difference. This maximum efficiency is equal to

$$\eta_{max} = F_R \tau \alpha \qquad 11.87$$

F_R is the *heat removal factor*. $\tau\alpha$ is the cover *transmittance plate-absorptivity product*.

The slope of the collector efficiency line is equal to

$$\text{slope} = F_R U_L \qquad 11.88$$

U_L is the overall collector energy loss coefficient.

The useful energy collected by the collector can be calculated from the water temperature rise, or as a function of the incident radiation, E.

$$q_{useful} = \dot{m}_w c_p (T_{w,\,out} - T_{w,\,in})$$
$$= F_R A_{collector}[E\tau\alpha - U_L(T_{w,\,in} - T_{air})] \qquad 11.89$$

Table 11.15
Transfer Fluids

Medium	Specific Gravity	Viscosity* Centipoise	Specific Heat (BTU/lbm-°F)	Freezing Point, °F
water	1.00	0.5 to 0.9	1.00	+32
50% by wt. water-ethylene glycol	1.05	1.2 to 4.4	.83	−33
50% by wt. water-propylene glycol	1.02	1.4 to 7.0	.85	−28
paraffinic oils	.82	12 to 30	.51	+15
aromatic oils	.85	0.6 to 0.8	.45	−100
silicone oils	.94	10 to 20	.38	−120

*Because viscosity is sensitive to temperature, values are given for a temperature range of approximately 80°F to 140°F.

Appendix A
Typical Heat Loss Factors

Usage	Description	density lbm/ft^3	weight lbm/ft^2	Resistance per inch listed ft^2-°F-hr BTU-in	Resistance for thickness listed ft^2-°F-hr BTU
Building board	asbestos-cement 1/8"	120	1.25	.25	.03
	gypsum or plaster board 1/2"	50	5.00		.45
	gypsum or plaster board 5/8"	50	6.25		0.56
	plywood 1/2"	34	1.42	1.25	.63
	wood fiber board	26		2.38	
	fir/pine sheathing .72"	32	2.08		.98
Flooring	asphalt tile 1/8"	120	1.25		.04
	carpet & fiber pad				2.08
	carpet & rubber pad				1.23
	ceramic tile 1"				.08
	cork tile 1/8"	25	.26	2.22	.28
	linoleum 1/8"	80	.83		.08
	plywood subfloor 5/8"	34	1.77		.78
	rubber/plastic tile 1/8"	110	1.15		.02
	hardwood 3/4"	45	2.81		.68
Insulation (blanket/batt)	fibrous mineral wool including fiberglass	0.8–2.0		3.85	
	wood fiber	3.2–3.6		4.00	
Insulation (board/slab)	glass fiber	9.5		4.00	
	accoustical tile 1/2"	22.4	.93		1.19
	accoustical tile 3/4"	22.4	1.4		1.78
	cellular glass	9.0		2.5	
	foamed plastic	1.62		3.45	
Insulation (loose)	paper/pulp products	2.5–3.5		3.57	
	mineral wool (glass, slag, or rock)	2.0–5.0		3.33	
	expanded vermiculite	7.0		2.08	
Roof insulation	all types for use above deck	15.6	1.3 per inch thickness	2.78	
Masonry materials (concrete)	cement mortar	116		.20	
	lightweight aggregates, depending on density:	120		.19	
		80		.40	
		40		.86	
		20		1.43	
	oven-dried sand, gravel, or stone aggregate	140		.11	
	stucco	116		.20	
Plastering materials	cement (sand) 1/2"	116	4.8	.20	.10
	cement (sand) 3/4"	116	7.2	.20	.15
	gypsum plaster, lightweight aggregate	45	1.88		.32
	lightweight aggregate on metal lath	45		.59	
	sand aggregate 1/2"	105	4.4	.18	.09
	sand aggregate on metal lath 3/4"	105	6.6		.13
Roofing	asbestos-cement shingle	120			.21
	asphalt roll roofing	70			.15
	asphalt shingles	70			.44
	built-up roofing 3/8"	70	2.2		.33
	wood shingles	40			.94

Usage	Description	density lbm/ft³	weight lbm/ft²	Resistance per inch listed ft²-°F-hr / BTU-in	Resistance for thickness listed ft²-°F-hr / BTU
Siding material (on flat face)	16" shingles				
	with 7½" exposure				.87
	with 12" exposure				1.19
	siding				
	asbestos-cement 1/4" lap				.21
	asphalt roll siding				.15
	wood, bevel (½" × 8" lap)				.81
	wood, bevel (¾" × 10" lap)				1.05
	structural glass				.10
Woods	hardwoods	45		.91	
	softwoods	32		1.25	
Masonry	brick, common 4"	120	40	.20	.80
	brick, face 4"	130	43	.11	.44
	hollow clay tile, 4"	48			1.11
	stone (lime or sand)	150		.08	
	concrete blocks (3 oval core)				
	with sand/gravel				
	aggregate 3"	76	19		.40
	6"	64	32		.91
	12"	63	63		1.28
	lightweight aggregate 3"	60	15		.56
	6"	46	23		1.50
	12"	43	43		1.89
Poured concrete	8" thick	30	20		10.0*
		80	53		4.00*
		140	93		1.72*
	12" thick	30	30		14.28*
		80	80		5.55*
		140	140		2.17*
Miscellaneous	adobe 8"		26		2.94*
	glass (single sheet 1/4", vertical)				.88*
	glass (double, vertical)				
	1/4" air space				1.63*
	1/2" air space				1.82*
	glass (single, horizontal)			summer:	1.16*
				winter:	.71*
	glass (double, horizontal)			summer:	2.00*
	1/4" air space			winter:	1.43*
	door, 1" nominal wood				1.45*
	1½" nominal wood				1.92*
	2" nominal wood				2.18*
	door, glass (herculite) 3/4" thick				.95*

*Includes appropriate inside and outside film coefficients.

Appendix B
Psychrometric Chart

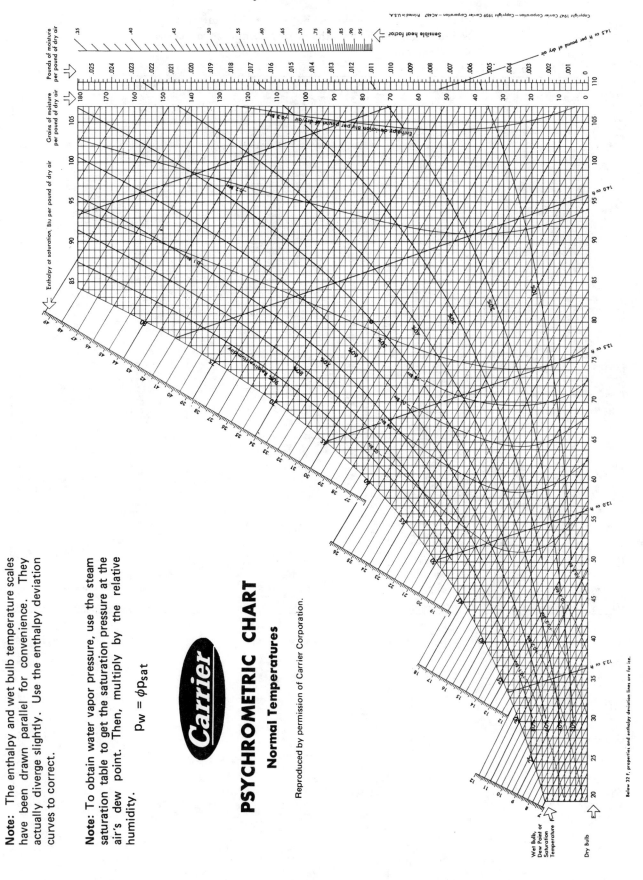

Note: The enthalpy and wet bulb temperature scales have been drawn parallel for convenience. They actually diverge slightly. Use the enthalpy deviation curves to correct.

Note: To obtain water vapor pressure, use the steam saturation table to get the saturation pressure at the air's dew point. Then, multiply by the relative humidity.

$$p_w = \phi p_{sat}$$

PSYCHROMETRIC CHART
Normal Temperatures

Reproduced by permission of Carrier Corporation.

Appendix C
Sample Outside Design Data

(This table contains sample data only. The information is not intended to be exhaustive, complete, or necessarily current.)

STATE	CITY	HEAT	COOL	°DAYS PER YR.
ALA.	Anniston	5	96	2820
	Birmingham	10	97	2780
	Mobile	15	95	1612
	Montgomery	10	98	2137
ARIZ.	Flagstaff	−10	84	7525
	Phoenix	25	108	1698
	Yuma	30	111	951
ARK.	Bentonville	−5	97	4036
	Fort Smith	10	101	3188
	Little Rock	5	99	2982
CAL.	Eureka	30	67	4632
	Fresno	25	101	2532
	Los Angeles	35	94	2015
	Sacramento	30	100	2822
	San Diego	35	86	1574
	San Francisco	25	80	3421
	San Jose	35	90	2410
COLO	Denver	−10	90	6132
	Grand Junction	−15	96	5796
	Pueblo	−20	96	5709
CONN	Hartford	0	90	6139
	New Haven	0	88	6026
D.C.	Washington	0	94	4333
FLA.	Apalachicola	25	92	1307
	Jacksonville	25	96	1243
	Key West	35	90	89
	Miami	35	92	178
	Pensacola	20	92	1435
	Tampa	30	92	674
GA.	Atlanta	10	95	2826
	Augusta	10	98	2138
	Macon	15	98	2049
	Savannah	20	96	1710
IDA.	Boise	−10	96	5890
	Lewiston	−15	98	5483
	Pocatello	−5	94	6976
ILL.	Cairo	0	97	3756
	Chicago	−10	95	6310
	Peoria	−10	94	6087
	Springfield	−10	95	5693
IND.	Evansville	−10	96	4360
	Fort Wayne	−10	93	6287
	Indianapolis	−10	93	5611
IOWA	Charles City	−25	91	7504
	Davenport	−15	94	6091
	Des Moines	−15	95	6446
	Dubuque	−20	92	7271
	Sioux City	−20	96	7012
KAN.	Concordia	−10	101	5323
	Dodge City	−10	99	5058
	Topeka	−10	99	5209
	Wichita	−10	102	4571
KY.	Louisville	0	96	4439
	Lexington	−5	94	4979
LA.	New Orleans	20	93	1317
	Shreveport	20	99	2117
ME.	Eastport	−10	85	8246
	Portland	−5	88	7681
MD.	Baltimore	0	94	4787
MASS.	Boston	0	91	5791
MICH.	Detroit	−10	92	6404
	Escanaba	−20	82	8657
	Grand Rapids	−10	91	7075
	Houghton	−20	—	9030
	Lansing	−10	89	6982
	Ludington	−10	87	7458
	Marquette	−20	88	8529
	Sault Ste. Marie	−20	83	9475
MINN.	Duluth	−25	85	9937
	Minneapolis	−20	92	7853
	Moorehead	−30	92	9327
	Saint Paul	−20	92	7804
MISS.	Corinth	0	98	3087
	Meridian	10	97	2333
	Vicksburgh	10	97	2000
MO.	Columbia	−10	97	5113
	Hannibal	−15	96	5393
	Kansas City	−10	100	4888
	Saint Louis	0	98	4699
	Springfield	−10	97	4693
MONT.	Billings	−35	94	7106
	Havre	−30	91	8213
	Helena	−20	90	8250
	Kalispell	−35	88	8055
	Miles City	−35	97	7850
	Missoula	−20	92	7873
NEBR.	Lincoln	−10	100	6104
	North Platte	−20	97	6546
	Omaha	−20	97	6160
	Valentine	−25	97	7075
NEV.	Reno	−5	92	6036
	Tonopah	−10	92	5813
	Winnemucca	−15	95	6369
N.H.	Concord	−15	88	7612
N.J.	Atlantic City	5	91	4741
	Cape May	—	91	4870
	Newark	0	94	5252
	Sandy Hook	0	—	5369
	Trenton	0	92	5068
N.M.	Albuquerque	0	96	4389
	Roswell	0	99	3424
	Santa Fe	0	88	6123
N.Y.	Albany	−10	88	6962
	Binghamton	−10	80	7537
	Buffalo	−5	86	6838
	Canton	−25	86	8305
	Ithaca	−15	91	6914
	New York	0	94	5050
	Oswego	−10	86	6975
	Rochester	−5	91	6863
	Syracuse	−10	89	6520
N.C.	Asheville	0	91	4072
	Charlotte	10	96	3205
	Hatteras	20	—	2392
	Raleigh	10	95	3369
	Wilmington	15	93	2323
N.D.	Bismark	−30	95	9033
	Devils Lake	−30	93	9940
	Grand Forks	−35	91	9871
	Williston	−35	94	9068
OHIO	Cincinnati	0	94	5195
	Cleveland	0	91	6006
	Columbus	−10	92	5615
	Dayton	0	92	5597
	Sandusky	0	92	5859
	Toledo	−10	92	6394
OKLA.	Oklahoma City	0	100	3519
ORE.	Baker	−5	94	7087
	Medford	5	98	4547
	Portland	10	89	4632
	Roseburg	10	93	4122
PA.	Erie	−5	88	6116
	Harrisburg	0	92	5258
	Philadelphia	0	93	4866
	Pittsburgh	0	90	5905
	Reading	0	92	5060
	Scranton	−5	89	6047
R.I.	Block Island	0	—	5843
	Providence	0	89	6125
S.C.	Charleston	15	95	1973
	Columbia	10	98	2435
	Greenville	10	95	3060
S.D.	Huron	−20	97	7902
	Pierre	−25	98	7283
	Rapid City	−20	96	7535
TENN.	Chattanooga	10	97	3384
	Knoxville	0	95	3590
	Memphis	0	98	3137
	Nashville	0	97	3513
TEX.	Abilene	15	101	2657
	Amarillo	−10	98	4345
	Brownsville	30	94	617
	Corpus Christi	20	95	1011
	Dallas	0	101	2272
	El Paso	20	100	2641
	Ft. Worth	10	102	2361
	Galveston	20	91	1233
	Houston	20	96	1388
	Palestine	15	99	1980
	Port Arthur	20	94	1517
	San Antonio	20	99	1579
	Taylor	10	101	1909
UTAH	Modena	−15	—	6598
	Salt Lake City	−10	97	5866
VT.	Burlington	−10	88	7865
	Northfield	−20	86	8804
VA.	Cape Henry	10	—	3307
	Lynchburg	5	94	4153
	Norfolk	15	94	3454
	Richmond	15	96	3955
	Wytheville	0	92	5103
WASH.	North Head	20	—	5211
	Seattle	15	82	4438
	Seattle Tacoma	10	85	5275
	Spokane	−15	93	6852
	Tacoma	15	85	4866
	Tatoosh Island	15	—	5724
	Walla Walla	−15	98	4848
	Yakima	−5	94	5845
W.VA.	Elkins	−10	87	5773
	Parkersburg	−10	93	4750
WIS.	Green Bay	−20	88	8259
	La Crosse	−25	90	7650
	Madison	−15	92	7417
	Milwaukee	−15	90	7205
WYO.	Cheyenne	−15	89	7562
	Lander	−20	92	8303
	Yellowstone Pk.	−35	90	9605

PRACTICE PROBLEMS: HEATING, VENTILATION, AND AIR CONDITIONING

Warm-ups

1. Find the infiltration around a 4' wide by 5' high double hung wood window in a residence facing south if the due-south wind velocity is 25 mph.

2. An office with 10' high ceilings is 60' × 95'. Using the air-changes method, what should be the ventilation rate? If you also know that 45 people will occupy the room and half of them smoke, what should be the ventilation rate?

3. 150 ppm of methanol and 285 ppm of methylene chloride are found in a plating booth. What should be the ventilation rate if two pints each per hour are evaporated?

4. A room contains 80°F, one atmosphere air at 67°F wet bulb. What are the specific humidity, enthalpy, and specific heat of the mixture?

5. What is the heat gain due to 12,000 watts of fluorescent lighting and 12 90% efficient, 10 horsepower motors operating at 80% of rated capacity? The motors drive various pieces of machinery located in the conditioned space.

6. What is the approximate winter heating cost from October 15 to May 15 for a building located in New York City if the design loss at 0°F is 3.5 EE6 BTUH? The price of Bunker C oil is $.15 per gallon.

7. A flat roof composed of $1\frac{1}{2}''$ insulation, 3" of wood, and a suspended 3/4" acoustical ceiling is exposed to a 95°F outside environment. What is the coefficient of heat transfer if the interior design temperature is 80°F?

8. If one row of cooling coils bypasses $\frac{1}{3}$ of the air passing through it, what is the bypass factor for four rows in series?

9. A 12' × 12' floor is constructed of a concrete slab insulated against radial heat flow on two exposures with two feet of 3" insulation (thermal conductance= 3.75). The two other exposures form part of a heated basement wall. If the outdoor design temperature is −10°F, what is the heat loss?

10. An auditorium is being designed to seat 4500 people. The ventilation rate is to be 60 cfm of outside air. The outside temperature is 0°F, and the pressure is 14.6 psia. Air is exhausted at 70°F dry bulb, and there is no air recirculation. The furnace has been sized assuming a heat loss of 1,250,000 BTUH and the above conditions. (a) At what temperature will the air enter the auditorium? (b) How much heat must be supplied to the incoming air by the heating coils?

Concentrates

1. A 10' high office 100' long by 40' wide is heated to 75°F when the environment is −10°F. One of the 40' walls is shared with an adjacent heated space. The three remaining walls have two 4' × 6' double glass, weatherstripped windows per 20 feet. The basement and the second floor are heated to 75°F. The wall coefficient is .2 BTU/ft²-hr-°F, and the wind velocity is 15 mph. What is the heating load? (Neglect heating and humidification of the ventilation air.)

2. 1000 cfm of 50°F air at 95% relative humidity are mixed with 1500 cfm of recirculated air at 76°F and 45% relative humidity. What are the mixture temperature, the specific humidity, and the dew point?

3. Air at 60°F dry bulb and 45°F wet bulb passes through an air washer with a humidifying efficiency of 70%. What are the bypass factor and the leaving dry bulb temperature?

4. 95°F dry bulb, 75°F wet bulb air passes through a cooling tower and leaves at 85°F dry bulb and 90% relative humidity. What are the heat and moisture additions per cubic foot of air?

5. A room is to be maintained at 75°F dry bulb and 50% relative humidity in an environment of 95°F dry bulb and 75°F wet bulb. There are sensible and latent loads of 200,000 and 50,000 BTUH, respectively. What are the apparatus dew point and the air quantity flowing through the conditioner? The required ventilation is 2000 cfm.

6. Coal with a heating value of 13,000 BTU per pound is used to heat a building in Newark, New Jersey, to 70°F from 8:30 a.m. to 5:30 p.m. The temperature during the rest of the day and the night is allowed to drop to 50°F. A heat loss of 650,000 BTUH has been calculated based on Newark's outside design data. If the burner has an efficiency of 70%, how much coal is required per year? The heating season lasts 245 days, the degree days are 5252, and the outside design temperature is 0°F.

7. An air washer takes 1800 cfm of air at 70°F and 40% relative humidity and discharges it at 75% relative humidity. A constant 50°F water spray is used. (a) How many pounds of water are required per minute? (b) What will be the final condition of the discharged air?

8. Repeat problem 7 using saturated steam at atmospheric pressure.

9. A large theater experiences a sensible heat load of 500,000 BTUH and a moisture load of 175 pounds per hour. Air enters the theater at 65°F and 55% relative humidity and is removed when it reaches 75°F or 60% relative humidity, whichever comes first. (a) What is the ventilation rate in pounds per hour for the theater? (b) What are the leaving air conditions?

10. 500 cfm of air at 80°F and 70% relative humidity are removed from a room. 150 cfm pass through an air conditioner and exit the conditioner saturated at 50°F. The remaining 350 cfm bypass the conditioner and mix with the conditioned air at 14.7 psia. What is the mixture temperature, the humidity ratio, and the relative humidity? What is the tonnage of the air conditioner?

Timed (1 hour allowed for each)

1. A dehumidifier takes 5000 cfm of air at 95°F dry bulb and 70% relative humidity and discharges it at 60°F dry bulb and 95% relative humidity. The dehumidifier uses a refrigeration cycle (Freon-12) operating between 100°F (saturated) and 50°F.

(a) Draw a partial psychrometric chart and locate the entering and leaving points of the air.
(b) Find the quantity of water removed from the air.
(c) Find the quantity of heat removed from the air.
(d) Draw T vs. s and h vs. s diagrams for the refrigeration cycle.
(e) Find values of T, p, h, s, and ν for the two property diagrams.

2. Much has been said recently about the desirability of lowering the thermostat to conserve fuel. At a particular location, the design heat loss of a building is 200,000 BTUH based on a 75°F inside and 0°F outside design temperature. The number of degree days accumulated during an average heating season of 210 days is 4200 at the given location. The building is occupied 24 hours a day. What is the reduction in heating fuel requirements if the thermostat is lowered from 75°F to 68°F?

3. A bypass air conditioning system is shown, which operates with the following characteristics.

supply temperature: 58°F db
sensible load: 200,000 BTUH
latent load: 450,000 grains/hr
air leaving washer is saturated
outside air: 90°F db, 76°F wb
make-up air required: 2000 cfm

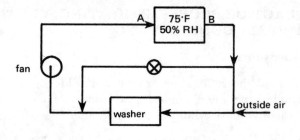

Submit your layout on a psychrometric chart and find the supply rate in cfm, the grains of moisture in the supply air, and the temperature of the air leaving the washer.

4. 1500 cfm of air (25 psia, 100% relative humidity) are heated from 200°F to 400°F in a constant pressure, constant moisture drying process. (a) What is the final relative humidity? (b) What is the final specific humidity? (c) How much heat is required in BTU/lbm of dry air? (d) What is the final dew point?

5. A building has the following construction details.

internal volume: 801,000 ft³
wall area: 11,040 ft² with $U = 0.15$ BTU/hr-ft²-°F
window area: 2,760 ft² with $U = 1.13$
roof area: 26,700 ft² with $U = 0.05$
concrete slab on grade: 690 lft with $U = 1.5$ BTU/hr-lft-°F
gas furnace efficiency: 75%
normal inside temperature: 70°F
ventilation rate: 1 air change per hour when occupied

$\frac{1}{2}$ air change per hour unoccupied

occupied: 8:00 a.m. until 6:00 p.m. Monday through Friday

The building is located where the annual heating season is 21 weeks. The fuel cost is $.25 per therm.

Calculate the annual savings (in dollars) if the temperature is set back 12°F during the unoccupied time.

6. 410 lbm/hr of dry 800°F air must pass through a scrubber to reduce particulate emissions. Since the scrubber is constructed of several elastomer parts, the air temperature must first be reduced to 350°F by passing through a water spray. The spray uses water at 80°F. The pressure in the spray chamber is 100 psia. (a) How much water is required per hour? (b) What will be the relative humidity of the leaving air?

7. An evaporative counter-flow air cooling tower removes 1 EE6 BTU/hr from an air conditioner water flow. The temperature of the water entering the tower is 120°F. The water temperature is reduced to 110°F. Air enters the cooling tower at 91°F and 60% relative humidity. The air leaves at 100°F and 82% relative humidity.

Assume any criteria necessary for calculations and calculate the flow rate (in lbm/hr) for the water-cooled air and for the make-up water.

8. A building is located at 32°N latitude. The inside design temperature is 78°F. The outside design temperature is 95°F. The daily temperature range is 22°F. The building's construction details are

walls: 1600 ft² facing north
         1400 ft² facing south
         1500 ft² facing east
         1400 ft² facing west
         4" brick facing
         3" concrete block
         1" mineral wool
         2" furring

$\frac{3}{8}$" drywall gypsum
$\frac{1}{2}$" plaster

roof: 6000 ft²
        4" concrete
        2" insulation ($k = .29$ BTU/hr-ft²-°F)
        felt
        1" air gap
        ceiling tile

windows: 100 ft² facing east only
           $\frac{1}{4}$" thick, single glazing
           cream-colored Venetian shades
           no exterior shades

What is the sensible transmission load for mid-July at 4:00 p.m. sun time? Why or why not is this the peak cooling load? Ignore the floor loss and the cooling loads from lights and occupants.

RESERVED FOR FUTURE USE

12 STATICS

PART 1: Determinate Structures

1 CONCENTRATED FORCES AND MOMENTS

Forces are vector quantities having magnitude, direction, and location in 3-dimensional space. The direction of a force **F** is given by its *direction cosines*, which are cosines of the true angles made by the force vector with the x, y, and z axes. The components of the force are given by equations 12.1, 12.2, and 12.3.

$$\mathbf{F}_x = \mathbf{F}(\cos\theta_x) \qquad 12.1$$
$$\mathbf{F}_y = \mathbf{F}(\cos\theta_y) \qquad 12.2$$
$$\mathbf{F}_z = \mathbf{F}(\cos\theta_z) \qquad 12.3$$

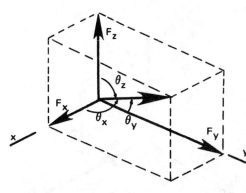

Figure 12.1 Components of a Force **F**

A force which would cause an object to rotate is said to contribute a *moment* to the object. The magnitude of a moment can be found by multiplying the magnitude of the force times the appropriate moment arm. That is, **M=F·d**.

The *moment arm* is a perpendicular distance from the force's line of application to some arbitrary reference point. This reference point should be chosen to eliminate one or more unknowns. This can be done by choosing the reference as a point at which unknown reactions are applied.

Moments also can be treated as vector quantities, and they are shown as double-headed arrows. Using the *right-hand rule* as shown, the direction cosines again are used to give the x, y, and z components of a moment vector.[1]

$$\mathbf{M}_x = \mathbf{M}(\cos\theta_x) \qquad 12.4$$
$$\mathbf{M}_y = \mathbf{M}(\cos\theta_y) \qquad 12.5$$
$$\mathbf{M}_z = \mathbf{M}(\cos\theta_z) \qquad 12.6$$
$$|\mathbf{M}| = \sqrt{\mathbf{M}_x^2 + \mathbf{M}_y^2 + \mathbf{M}_z^2} \qquad 12.7$$

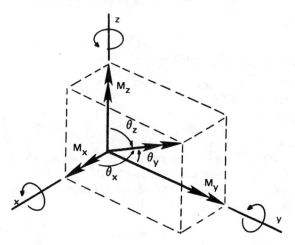

Figure 12.2 Components of a Moment **M**

Moment vectors have the properties of magnitude and direction, but not of location (point of application). Moment vectors can be moved from one location to another without affecting the equilibrium of solid bodies.

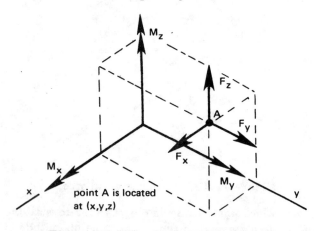

Figure 12.3 Coordinates of a Point A

If a force is not parallel to an axis, it produces a moment around that axis. The moment is evaluated by finding the components of the force and their respective distances to the axis. In figure 12.3, a force acts through point A located at (x, y, z) and produces moments given by equations 12.8, 12.9, and 12.10.

---

[1] The right-hand rule: Close your right hand in such a way that your fingers curl in the direction of the force. Your thumb will point in the direction of the moment.

$$\mathbf{M}_x = y\mathbf{F}_z - z\mathbf{F}_y \qquad 12.8$$
$$\mathbf{M}_y = z\mathbf{F}_x - x\mathbf{F}_z \qquad 12.9$$
$$\mathbf{M}_z = x\mathbf{F}_y - y\mathbf{F}_x \qquad 12.10$$

Any two equal, opposite, and parallel forces constitute a *couple*. A couple is statically equivalent to a single moment vector. In figure 12.4, the two forces, $\mathbf{F}_1$ and $\mathbf{F}_2$, of equal magnitude produce a moment vector $\mathbf{M}_z$ of magnitude Fy. The two forces can be replaced by this moment vector which then can be moved to any location on the object.

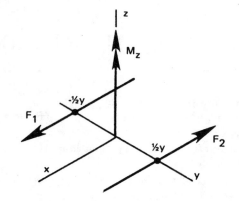

Figure 12.4 A Couple

2 DISTRIBUTED LOADS

If an object is loaded by its own weight or by another type of continuous loading, it is said to be subjected to a *distributed load*. Provided that the load per unit length, w, is acting in the same direction everywhere, the statically equivalent concentrated load can be found from equation 12.11 by integrating over the line of application.

$$\mathbf{F}_R = \int w\,dx \qquad 12.11$$

The location of the resultant is given by equation 12.12.

$$\bar{x} = \frac{\int (wx)\,dx}{\mathbf{F}_R} \qquad 12.12$$

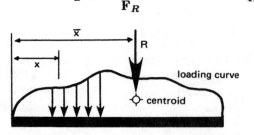

Figure 12.5 A Distributed Load and Resultant

In the case of a straight beam under *transverse loading*, the magnitude of **F** equals the area under the loading curve.[2] The location of the resultant force coincides

---

[2] Loading is said to be *transverse* if its line of action is perpendicular to the length of the beam.

with the centroid of that area. If the distributed load is uniform so that w is constant along the beam,

$$\mathbf{F} = wL \qquad\qquad 12.13$$
$$\bar{x} = \tfrac{1}{2}L \qquad\qquad 12.14$$

If the distribution is triangular and increases to a maximum of w pounds per unit length as x increases,

$$\mathbf{F} = \tfrac{1}{2}wL \qquad\qquad 12.15$$
$$\bar{x} = \tfrac{2}{3}L \qquad\qquad 12.16$$

Example 12.1

Find the magnitude and the location of the resultant of the distributed loads on each span of the beam shown.

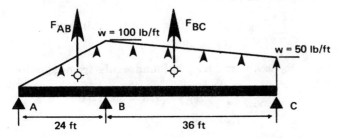

For span A-B: The area under the loading curve is $\tfrac{1}{2}(100)(24) = 1200$ pounds. The centroid of the loading triangle is $(\tfrac{2}{3})(24) = 16$ feet from point A. Therefore, the triangular load on the span A-B can be replaced (for the purposes of statics) with a concentrated load of 1200 pounds located 16 feet from the left end.

For span B-C: The area under the loading curve is

$$(50)(36) + \tfrac{1}{2}(50)(36) = 2700 \text{ lbs}$$

The centroid of the trapezoid is

$$\frac{36\,[(2)(100) + 50]}{3\,(100 + 50)} = 20 \text{ ft from pt } C$$

Therefore, the distributed load on span B-C can be replaced (for the purposes of statics) with a concentrated load of 2700 pounds located 20 feet to the left of point C.

3 PRESSURE LOADS

Hydrostatic pressure is an example of a pressure load that is distributed over an area. The pressure is denoted as p pounds per unit area of surface. It is normal to the surface at every point. If the surface is plane, the statically equivalent concentrated load can be found by integrating over the area. The resultant is numerically equal to the average pressure times the area. The point of application will be the centroid of the area over which the integration was performed.

4 RESOLUTION OF FORCES AND MOMENTS

Any system (collection) of forces and moments is statically equivalent to a single resultant force vector plus a single resultant moment vector in 3-dimensional space. Either or both of these resultants may be zero.

The x-component of the resultant force is the sum of all the x-components of the individual forces, and similarly for the y- and the z-components of the resultant force.

$$\mathbf{F}_{Rx} = \sum \mathbf{F}_i(\cos\theta_{x,i}) \qquad 12.17$$
$$\mathbf{F}_{Ry} = \sum \mathbf{F}_i(\cos\theta_{y,i}) \qquad 12.18$$
$$\mathbf{F}_{Rz} = \sum \mathbf{F}_i(\cos\theta_{z,i}) \qquad 12.19$$

The determination of the resultant moment vector is more complex. The resultant moment vector includes the moments of all system forces around the reference axes plus the components of all system moments.

$$\mathbf{M}_{Rx} = \sum (y_i \mathbf{F}_{z,i} - z_i \mathbf{F}_{y,i}) + \sum \mathbf{M}_i(\cos\theta_{x,i}) \quad 12.20$$
$$\mathbf{M}_{Ry} = \sum (z_i \mathbf{F}_{x,i} - x_i \mathbf{F}_{z,i}) + \sum \mathbf{M}_i(\cos\theta_{y,i}) \quad 12.21$$
$$\mathbf{M}_{Rz} = \sum (x_i \mathbf{F}_{y,i} - y_i \mathbf{F}_{x,i}) + \sum \mathbf{M}_i(\cos\theta_{z,i}) \quad 12.22$$

5 CONDITIONS OF EQUILIBRIUM

An object which is not moving is said to be static. All forces on a static object are in equilibrium. For an object to be in equilibrium, it is necessary that the resultant force vector and the resultant moment vectors be equal to zero.

$$\mathbf{F}_R = \sqrt{\mathbf{F}_{Rx}^2 + \mathbf{F}_{Ry}^2 + \mathbf{F}_{Rz}^2} = 0 \qquad 12.23$$
$$\mathbf{M}_R = \sqrt{\mathbf{M}_{Rx}^2 + \mathbf{M}_{Ry}^2 + \mathbf{M}_{Rz}^2} = 0 \qquad 12.24$$

Since the square of any quantity cannot be negative, equations 12.25 through 12.30 follow directly from equations 12.23 and 12.24.

$$\mathbf{F}_{Rx} = \sum \mathbf{F}_x = 0 \qquad\qquad 12.25$$
$$\mathbf{F}_{Ry} = \sum \mathbf{F}_y = 0 \qquad\qquad 12.26$$
$$\mathbf{F}_{Rz} = \sum \mathbf{F}_z = 0 \qquad\qquad 12.27$$
$$\mathbf{M}_{Rx} = \sum \mathbf{M}_x = 0 \qquad\qquad 12.28$$
$$\mathbf{M}_{Ry} = \sum \mathbf{M}_y = 0 \qquad\qquad 12.29$$
$$\mathbf{M}_{Rz} = \sum \mathbf{M}_z = 0 \qquad\qquad 12.30$$

6 FREE-BODY DIAGRAMS

A *free-body diagram* is a representation of an object in equilibrium, showing all external forces, moments, and support reactions. Since the object is in equilibrium, the resultant of all forces and moments on the free-body is zero.

If any part of the object is removed and replaced by the forces and moments which are exerted on the cut surface, a free-body of the remaining structure is obtained, and the conditions of equilibrium will be satisfied by the new free-body.

By dividing the object into a sufficient number of free-bodies, the internal forces and moments can be found at all points of interest, providing that the conditions of equilibrium are sufficient to give a static solution.

7 REACTIONS

A typical first step in solving statics problems is to determine the supporting reaction forces. The manner in which the structure is supported will determine the type, the location, and the direction of the reactions. Conventional symbols can be used to define the type of reactions which occur at each point of support. Some examples are shown in table 12.1.

Example 12.2

Find the reactions R_1 and R_2.

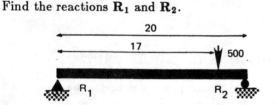

Since the left support is a simple support, its reaction can have any direction. R_1 can, therefore, be written in terms of its x and y components, $R_{1,x}$ and $R_{1,y}$, respectively. R_2 is a roller support which cannot sustain an x component.

From equation 12.25, choosing forces to the right as positive, $R_{1,x}$ is found to be zero.

$$\sum F_x = R_{1,x} = 0$$

Equation 12.26 can be used to obtain a relationship between the y components of force. Forces acting upward are considered positive.

$$\sum F_y = R_{1,y} + R_2 - 500 = 0$$

Since both $R_{1,y}$ and R_2 are unknown, a second equation is needed. Equation 12.30 is used. The reference point is chosen as the left end to make the moment arm for $R_{1,y}$ equal to zero. This eliminates $R_{1,y}$ as an unknown, allowing R_2 to be found directly. Clockwise moments are considered positive.

$$\sum M_{\text{left end}} = (500)(17) - (R_2)(20) = 0$$
$$R_2 = 425$$

Once R_2 is known, $R_{1,y}$ is found easily from equation 12.26.

$$\sum F_y = R_{1,y} + 425 - 500 = 0$$
$$R_{1,y} = 75$$

8 INFLUENCE LINES FOR REACTIONS

An *influence line (influence graph)* is an x-y plot of the magnitude of a reaction (any reaction on the object) as it would vary as the load is placed at different points on the object. The x-axis corresponds to the location along the object (as along the length of a beam); the y-axis corresponds to the magnitude of the reaction. For uniformity, the load is taken as 1 unit. Therefore, for an actual load of P units, the actual reaction would be given by equation 12.31.

$$\frac{\text{actual}}{\text{reaction}} = P \left[\frac{\text{influence graph}}{\text{ordinate}} \right] \qquad 12.31$$

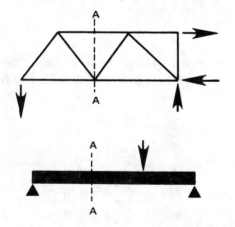
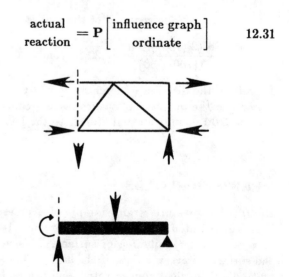

Figure 12.6 Original and Cut Free-bodies

Table 12.1
Common Support Symbols

Type of Support	Symbol	Characteristics
Built-in		Moments and forces in any direction
Simple		Load in any direction; no moment
Roller		Load normal to surface only; no moment
Cable		Load in cable direction; no moment
Guide		No load or moment in guide direction
Hinge		Load in any direction; no moment

Example 12.3

Draw the influence graphs of the left and right reactions for the beam shown.

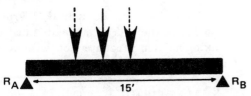

If a unit load is at the left end, reaction R_A will be equal to 1. If the unit load is at the right end, it will be supported entirely by R_B, so R_A will be zero. The influence line for R_A is

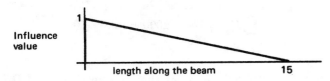

The influence line for R_B is found similarly.

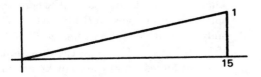

9 AXIAL MEMBERS

A member which is in equilibrium when acted upon by forces at each end and by no other forces or moments is an *axial member*. For equilibrium to exist, the resultant forces at the ends must be equal, opposite, and collinear. In an actual truss, this type of loading can be approached through the use of frictionless bearings or pins at the ends of the axial members. In simple truss analysis, the members are assumed to be axial members, regardless of the end conditions.

A typical inclined axial member is illustrated in figure 12.7. For that member to be in equilibrium, the following equations must hold.

$$\mathbf{F}_{Rx} = \mathbf{F}_{Bx} - \mathbf{F}_{Ax} = 0 \qquad 12.32$$

$$\mathbf{F}_{Ry} = \mathbf{F}_{By} - \mathbf{F}_{Ay} = 0 \qquad 12.33$$

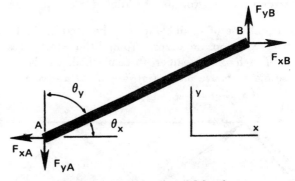

Figure 12.7 An Axial Member

The resultant force, $\mathbf{F}_R$, can be derived from the components by trigonometry and direction cosines.

$$\mathbf{F}_{Rx} = \mathbf{F}_R \cos \theta_x \qquad 12.34$$

$$\mathbf{F}_{Ry} = \mathbf{F}_R \cos \theta_y = \mathbf{F}_R \sin \theta_x \qquad 12.35$$

If, however, the geometry of the axial member is known, similar triangles can be used to find the resultant and/or the components. This is illustrated in example 12.4.

Example 12.4

A 12′ long axial member carrying an internal load of 180 pounds is inclined as shown. What are the x- and y-components of the load?

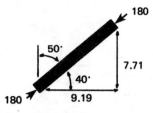

method 1: Direction Cosines

$$F_x = 180 (\cos 40°) = 137.9$$
$$F_y = 180 (\cos 50°) = 115.7.$$

method 2: Similar Triangles

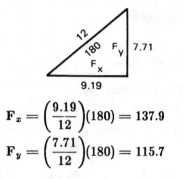

$$F_x = \left(\frac{9.19}{12}\right)(180) = 137.9$$
$$F_y = \left(\frac{7.71}{12}\right)(180) = 115.7$$

10 TRUSSES

This discussion is directed toward 2-dimensional trusses. The loads in truss members are represented by arrows pulling away from the joints for tension, and by arrows pushing toward the joints for compression.[3]

The equations of equilibrium can be used to find the external reactions on a truss. To find the internal resultants in each axial member, three methods can be used. These methods are *method of joints*, *cut-and-sum*, and *method of sections*.

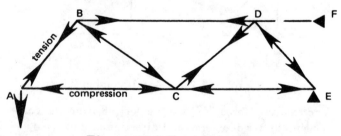

Figure 12.8 Truss Notation

All joints in a truss will be *determinate*, and all member loads can be found if equation 12.36 holds.

$$\text{\# truss members} = 2(\text{\# joints}) - 3 \qquad 12.36$$

If the left-hand side is greater than the right-hand side, indeterminate methods must be used to solve the truss. If the left-hand side is less than the right-hand side, the

[3]The method of showing tension and compression on a truss drawing appears incorrect. This is because the arrows show the forces on the joints, not the forces in the axial members.

truss is not rigid and will collapse under certain types of loading.

A. METHOD OF JOINTS

The *method of joints* is a direct application of equations 12.25 and 12.26. The sums of forces in the x- and y-directions are taken at consecutive joints in the truss. At each joint, there may be up to two unknown axial forces, each of which may have two components. Since there are two equations of equilibrium, a joint with two unknown forces will be determinate.

Example 12.5

Find the force in member **BD** in the truss shown.

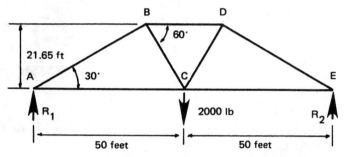

step 1: Find the reactions R_1 and R_2. From equation 12.29, the sum of moments must be zero. Taking moments (counterclockwise as positive) about point A gives R_2.

$$\sum M_A = 100(R_2) - 2000(50) = 0$$
$$R_2 = 1000$$

From equation 12.26, the sum of the forces (vertical positive) in the y direction must be zero.

$$\sum F_y = R_1 - 2000 + 1000 = 0$$
$$R_1 = 1000$$

step 2: Although we want the force in member **BD**, there are three unknowns at joints B and D. Therefore, start with joint A where there are only two unknowns (**AB** and **AC**). The free-body of joint A is shown below. The direction of R_1 is known. However, the directions of the member forces usually are not known and need to be found by inspection or assumption. If an incorrect direction is assumed, the force will show up with a negative sign in later calculations.

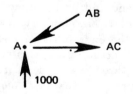

step 3: Resolve all inclined forces on joint A into horizontal and vertical components using trigonometry or similar triangles. R_1 and AC already are parallel to the y and x axes, respectively. Only AB needs to be resolved into components. By observation, it is clear that $AB_y = 1000$. If this were not true, equation 12.26 would not hold.

$$AB_y = AB(\sin 30°)$$
$$1000 = AB(.5) \text{ or } AB = 2000$$
$$AB_x = AB(\cos 30°) = 1732$$

step 4: Draw the free-body diagram of joint B. Notice that the direction of force AB is toward the joint, just as it was for joint A. The direction of load BC is chosen to counteract the vertical component of load AB. The direction of load BD is chosen to counteract the horizontal components of loads AB and BC.

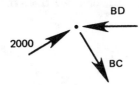

step 5: Resolve all inclined forces into horizontal and vertical components.

$$AB_x = 1732$$
$$AB_y = 1000$$
$$BC_x = BC(\sin 30°) = .5BC$$
$$BC_y = BC(\cos 30°) = .866BC$$

step 6: Write the equations of equilibrium for joint B.

$$\sum F_x = 1732 + .5BC - BD = 0$$
$$\sum F_y = 1000 - .866BC = 0$$

BC from the second equation is found to be 1155. Substituting 1155 into the first equilibrium condition equation gives

$$1732 + .5(1155) - BD = 0$$
$$BD = 2310$$

Since BD turned out to be positive, its direction was chosen correctly. The direction of the arrow indicates that the member is compressing the pin joint. Consequently, the pin is compressing the member, and member BD is in compression.

B. CUT-AND-SUM METHOD

The *cut-and-sum method* can be used if a load in an inclined member in the middle of a truss is wanted. The

method is strictly an application of the equilibrium condition requiring the sum of forces in the vertical direction to be zero. The method is illustrated in example 12.6.

Example 12.6

Find the force in member BC for the truss shown in example 12.5.

step 1: Find the external reactions. This is the same step as in example 12.5. $R_1 = R_2 = 1000$.

step 2: Cut the truss through, making sure that the cut goes through only one member with a vertical component. In this case, that member is BC.

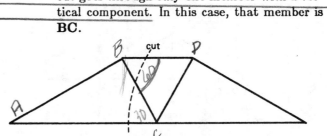

step 3: Draw the free-body of either part of the remaining truss.

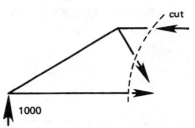

step 4: Resolve the unknown inclined force into vertical and horizontal components.

$$BC_x = .5(BC)$$
$$BC_y = .866(BC)$$

step 5: Sum forces in the y direction for the entire free-body.

$$\sum F_y = 1000 - .866(BC) = 0$$
$$BC = 1155$$

C. METHOD OF SECTIONS

The cut-and-sum method will work only if it is possible to cut the truss without going through two members with vertical components.

The *method of sections* is a direct approach for finding member loads at any point in a truss. In this method, the truss is cut at an appropriate section, and the conditions of equilibrium are applied to the resulting free-body. This is illustrated in example 12.7.

Example 12.7

For the truss shown, find the load in members **CE** and **CD**.

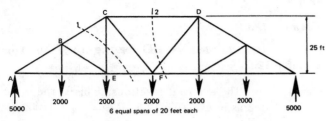

For member force **CE**, the truss is cut at section 1.

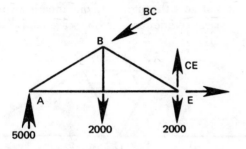

Taking moments about A will eliminate all unknowns except force **CE**.

$$\sum M_A = CE\,(40) - 2000\,(20) - 2000\,(40) = 0$$
$$CE = 3000$$

For member **CD**, the truss is cut at section 2. Taking moments about point F will eliminate all unknowns except **CD**.

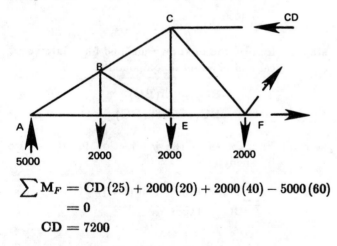

$$\sum M_F = CD\,(25) + 2000\,(20) + 2000\,(40) - 5000\,(60)$$
$$= 0$$
$$CD = 7200$$

11 SUPERPOSITION OF LOADINGS

For any group of forces and moments which satisfies the conditions of equilibrium, the resultant force and moment vectors are zero. The resultant of two zero vectors is another zero vector. Therefore, any number of such equilibrium systems can be combined without disturbing the equilibrium.

Superposition methods must be used with discretion in working with actual structures since some structures change shape significantly under load. If the actual structure were to deflect so that the points of application of loads were quite different from those in the undeflected structure, superposition would not be applicable.

In simple truss analysis, the change of shape under load is neglected when finding the member loads. Superposition, therefore, can be assumed to apply.

12 CABLES

A. CABLES UNDER CONCENTRATED LOADS

An ideal cable is assumed to be completely flexible. It acts as an axial member in tension between any two points of concentrated load application.

The method of joints and sections used in truss analysis applies equally well to cables under concentrated loads. However, no compression members will be found. As in truss analysis, if the cable loads are unknown, some information concerning the geometry of the cable must be known in order to solve for the axial tension in the segments.

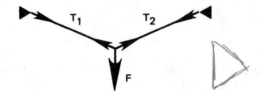

Figure 12.9 Cable Under Transverse Loading

Example 12.8

Find the tension T_2 between points B and C.

step 1:   Take moments about point A to find T_3.

$$\sum M_A = aF_1 + bF_2 + dT_3 \cos\theta_3$$
$$-cT_3 \sin\theta_3 = 0$$

step 2:   Sum forces in the x direction to find T_1.

$$\sum F_x = T_1 \cos\theta_1 - T_3 \cos\theta_3 = 0$$

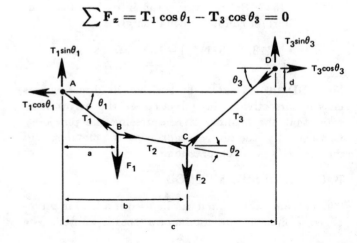

step 3:   Sum forces in the x direction at point B to find T_2.

$$\sum F_x = T_2 \cos\theta_2 - T_1 \cos\theta_1 = 0$$

B. CABLES UNDER DISTRIBUTED LOADS

An idealized tension cable under a distributed load is similar to a linkage made up of a very large number of axial members. The cable is an axial member in the sense that the internal tension acts in a direction which is along the centerline of the cable everywhere.

Figure 12.10 illustrates a cable under a unidirectionally distributed load. A free-body diagram of segment B-C of the cable also is known. **F** is the vertical resultant of the distributed load on the segment.

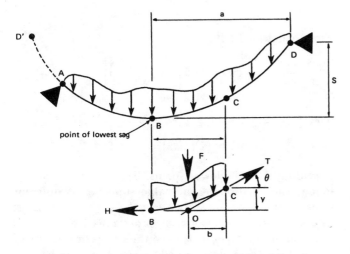

Figure 12.10 Cable Under Distributed Load

T is the cable tension at point C, and **H** is the cable tension at the point of lowest sag. From the conditions of equilibrium for free-body B-C, it is apparent that the three forces, **H**, **F**, and **T**, must be concurrent at point O. Taking moments about point C, the following equations are obtained.

$$\sum M_C = Fb - Hy = 0 \qquad 12.37$$

$$H = \frac{Fb}{y} \qquad 12.38$$

But $\tan\theta = \left(\frac{y}{b}\right)$. So,

$$H = \frac{F}{\tan\theta} \qquad 12.39$$

From the summation of forces in the vertical and horizontal directions,

$$T \cos\theta = H \qquad 12.40$$
$$T \sin\theta = F \qquad 12.41$$
$$T = \sqrt{H^2 + F^2} \qquad 12.42$$

The shape of the cable is a function of the relative amount of sag at point B and the relative distribution (not the absolute magnitude) of the applied running load.

C. PARABOLIC CABLES

If the distribution load per unit length, w, is constant with respect to a horizontal line (as is the load from a bridge floor), the cable will be parabolic in shape. This is illustrated in figure 12.11.

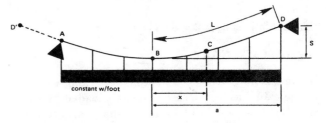

Figure 12.11 Parabolic Cable

The horizontal component of tension can be found from equation 12.38 using $F = wa$, $b = \frac{1}{2}a$, and $y = S$.

$$H = \frac{Fb}{y} = \frac{wa^2}{2S} \qquad 12.43$$

$$T = \sqrt{H^2 + F^2} = \sqrt{\left(\frac{wa^2}{2S}\right)^2 + (wx)^2} \qquad 12.44$$

$$= w\sqrt{x^2 + \left(\frac{a^2}{2S}\right)^2} \qquad 12.45$$

The shape of the cable is given by equation 12.46.

$$y = \frac{wx^2}{2H} \qquad 12.46$$

The approximate length of the cable from the lowest point to the support is given by equation 12.47.

$$L \approx (a)\left[1 + \frac{2}{3}\left(\frac{S}{a}\right)^2 - \frac{2}{5}\left(\frac{S}{a}\right)^4\right] \qquad 12.47$$

Example 12.9

A pedestrian bridge has two suspension cables and a flexible floor. The floor weighs 28 pounds per foot. The span of the bridge is 100 feet between the two end supports. When the bridge is empty, the tension at point A is 1500 pounds. What is the cable sag, S, at the center? What is the approximate cable length?

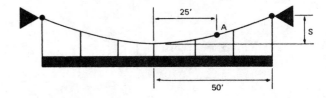

The floor weight per cable is $28/2 = 14$ lb/ft. From equation 12.45,

$$1500 = 14\sqrt{[25]^2 + \left[\frac{(50)^2}{2S}\right]^2}$$

$$S = 12 \text{ feet}$$

From equation 12.47,

$$L = 50\left[1 + \left(\frac{2}{3}\right)\left(\frac{12}{50}\right)^2 - \left(\frac{2}{5}\right)\left(\frac{12}{50}\right)^4\right]$$

$$= 51.9 \text{ feet}$$

The cable length is $2 \times 51.9 = 103.8$ ft.

D. THE CATENARY

If the distributed load, w, is constant along the length of the cable (as in the case of a cable loaded by its own weight), the cable will have the shape of a *catenary*. This is illustrated in figure 12.12.

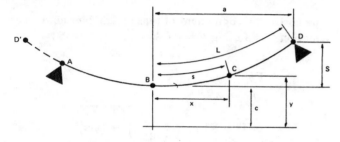

Figure 12.12 A Catenary

As shown in figure 12.12, y is measured from a reference plane located a distance c below the lowest point of the cable, point B. The location of this reference plane is a parameter of the cable which must be determined before equations 12.48 through 12.53 are used. The value of c does not correspond to any physical distance, nor does the reference plane correspond to the ground.

The equations of the catenary are presented below. Some judgment usually is necessary to determine which equations should be used and in which order. To define cable shape, it is necessary to have some initial information which can be entered into the equations. For example, if a and S are given, equation 12.51 can be solved by trial and error to obtain c. Once c is known, the cable geometry and forces are defined by the remaining equations.

$$y = c\left[\cosh\left(\frac{x}{c}\right)\right] \qquad 12.48$$

$$s = c\left[\sinh\left(\frac{x}{c}\right)\right] \qquad 12.49$$

$$y = \sqrt{s^2 + c^2} \qquad 12.50$$

$$S = c\left[\cosh\left(\frac{a}{c}\right) - 1\right] \qquad 12.51$$

$$\tan\theta = \frac{s}{c} \qquad 12.52$$

$$H = wc \qquad 12.53$$

$$F = ws \qquad 12.54$$

$$T = wy \qquad 12.55$$

$$\tan\theta = \frac{ws}{H} \qquad 12.56$$

$$\cos\theta = \frac{H}{T} \qquad 12.57$$

Example 12.10

A cable 100' long is loaded by its own weight. The sag is 25', and the supports are on the same level. What is the distance between the supports?

From equation 12.50 at point D, with $S = 25$,

$$c + S = \sqrt{s^2 + c^2}$$

$$c + 25 = \sqrt{(50)^2 + c^2}$$

$$c = 37.5$$

From equation 12.49,

$$50 = 37.5\left[\sinh\left(\frac{a}{37.5}\right)\right]$$

$$a = 41.2 \text{ feet}$$

The distance between supports is

$$(2a) = (2)(41.2) = 82.4 \text{ feet}$$

Providing that the lowest point, B, is known or can be found, the location of the cable supports at different levels does not significantly affect the analysis of cables. The same procedure is used in proceeding from point B to either support. In fact, once the theoretical shape of the cable has been determined, the supports can be relocated anywhere along the cable line without affecting the equilibrium of the supporting segment.

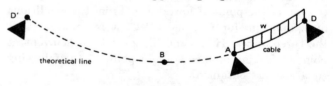

Figure 12.13 Non-Symmetrical Segment of Symmetrical Cable

13 3-DIMENSIONAL STRUCTURES

The static analysis of 3-dimensional structures usually requires the following steps.

step 1: Determine the components of all loads and reactions. This usually is accomplished by finding the x, y, and z coordinates of all points and then using direction cosines.

step 2: Draw three free-bodies of the structure—one each for the x, y, and z components of loads and reactions.

step 3: Solve for unknowns using $\sum \mathbf{F} = 0$ and $\sum \mathbf{M} = 0$.

Example 12.11

Beam ABC is supported by the two cables as shown. The connection at A is pinned (hinged). Find the cable tensions $\mathbf{T}_1$ and $\mathbf{T}_2$.

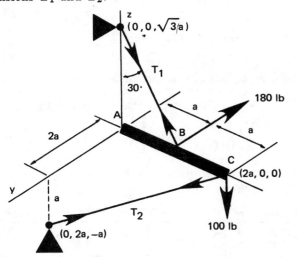

step 1:   For the 180 pound load,

$$\mathbf{F}_x = 0$$
$$\mathbf{F}_y = -180$$
$$\mathbf{F}_z = 0$$

For the 100 pound load,

$$\mathbf{F}_x = 0$$
$$\mathbf{F}_y = 0$$
$$\mathbf{F}_z = -100$$

For cable 1,

$$\cos \theta_x = \cos 120° = -.5$$
$$\cos \theta_y = 0$$
$$\cos \theta_z = \cos 30° = .866$$
$$\mathbf{T}_{1x} = -.5\mathbf{T}_1$$
$$\mathbf{T}_{1y} = 0$$
$$\mathbf{T}_{1z} = .866\mathbf{T}_1$$

For cable 2,

The length of the cable is

$$L = \sqrt{(2a)^2 + (-2a)^2 + (-a)^2} = 3a$$
$$\cos \theta_x = \frac{-2a}{3a} = -.667$$
$$\cos \theta_y = \frac{2a}{3a} = .667$$
$$\cos \theta_z = \frac{-a}{3a} = -.333$$
$$\mathbf{T}_{2x} = -.667\mathbf{T}_2$$

$$\mathbf{T}_{2y} = .667\mathbf{T}_2$$
$$\mathbf{T}_{2z} = -.333\mathbf{T}_2$$

step 2:

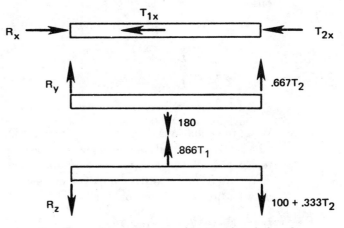

step 3: Summing moments about point A for the z case gives $\mathbf{T}_2$.

$$\sum \mathbf{M}_{Az} = 0.667\mathbf{T}_2(2a) - 180(a) = 0$$
$$\mathbf{T}_2 = 135$$

Summing moments about point A for the y case gives $\mathbf{T}_1$.

$$\sum \mathbf{M}_{Ay} = 0.866\mathbf{T}_1(a) - 0.333(135)(2a) - 100(2a) = 0$$
$$\mathbf{T}_1 = 335$$

14 GENERAL TRIPOD SOLUTION

The procedure given in the preceding section will work with a tripod consisting of three axial pin-ended members with a load in any direction applied at the apex. However, the tripod problem occurs frequently enough to develop a specialized procedure for solution.

step 1: Use the direction cosines of the force, $\mathbf{F}$, to find its components.

$$\mathbf{F}_x = \mathbf{F}(\cos \theta_x) \qquad 12.58$$
$$\mathbf{F}_y = \mathbf{F}(\cos \theta_y) \qquad 12.59$$
$$\mathbf{F}_z = \mathbf{F}(\cos \theta_z) \qquad 12.60$$

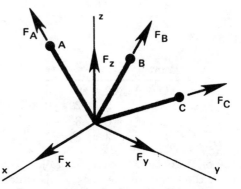

Figure 12.14 A General Tripod

step 2: Using the x, y, and z coordinates of points A, B, and C (taking the origin at the apex), find the direction cosines for the legs. Repeat the following four equations for each member, observing algebraic signs of x, y, and z.

$$L^2 = x^2 + y^2 + z^2 \qquad 12.61$$

$$\cos\theta_x = \frac{x}{L} \qquad 12.62$$

$$\cos\theta_y = \frac{y}{L} \qquad 12.63$$

$$\cos\theta_z = \frac{z}{L} \qquad 12.64$$

step 3: Write the equations of equilibrium for joint O. The following simultaneous equations assume tension in all three members. A minus sign in the solution for any member indicates compression instead of tension.

$$\mathbf{F}_A \cos\theta_{xA} + \mathbf{F}_B \cos\theta_{xB} + \mathbf{F}_C \cos\theta_{xC} + \mathbf{F}_x = 0$$
$$12.65$$

$$\mathbf{F}_A \cos\theta_{yA} + \mathbf{F}_B \cos\theta_{yB} + \mathbf{F}_C \cos\theta_{yC} + \mathbf{F}_y = 0$$
$$12.66$$

$$\mathbf{F}_A \cos\theta_{zA} + \mathbf{F}_B \cos\theta_{zB} + \mathbf{F}_C \cos\theta_{zC} + \mathbf{F}_z = 0$$
$$12.67$$

Example 12.12

Find the load on each leg of the tripod shown.

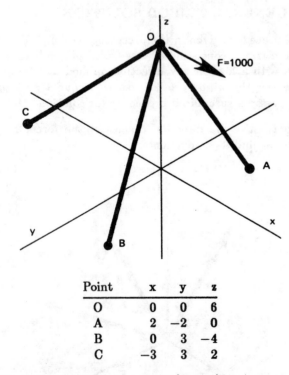

Point	x	y	z
O	0	0	6
A	2	−2	0
B	0	3	−4
C	−3	3	2

Since the origin is not at point (0, 0, 0), it is necessary to transfer the origin to the apex. This is done by the following equations. Only the z values are affected.

$$x' = x - x_0$$
$$y' = y - y_0$$
$$z' = z - z_0$$

The new coordinates with the origin at the apex are

Point	x	y	z
O	0	0	0
A	2	−2	−6
B	0	3	−10
C	−3	3	−4

The components of the applied force are

$$\mathbf{F}_x = \mathbf{F}(\cos\theta_x) = \mathbf{F}[\cos(0°)] = 1000$$
$$\mathbf{F}_y = 0$$
$$\mathbf{F}_z = 0$$

The direction cosines of the legs are found from the following table.

Member	x²	y²	z²	L²	L	cosθ_x	cosθ_y	cosθ_z
O-A	4	4	36	44	6.63	.3015	−.3015	−.9046
O-B	0	9	100	109	10.44	0	.2874	−.9579
O-C	9	9	16	34	5.83	−.5146	.5146	−.6861

From equations 12.65, 12.66, and 12.67, the equilibrium equations are

$$.3015\mathbf{F}_A + \qquad\qquad -.5146\mathbf{F}_C +1000 = 0$$
$$-.3015\mathbf{F}_A +.2874\mathbf{F}_B +.5146\mathbf{F}_C + \qquad 0 = 0$$
$$-.9046\mathbf{F}_A -.9579\mathbf{F}_B -.6861\mathbf{F}_C + \qquad 0 = 0$$

The solution to this set of simultaneous equations is

$$\mathbf{F}_A = +1531 \text{ (tension)}$$
$$\mathbf{F}_B = -3480 \text{ (compression)}$$
$$\mathbf{F}_C = +2841 \text{ (tension)}$$

15 PROPERTIES OF AREAS

A. CENTROIDS

The location of the *centroid* of a 2-dimensional area which is defined mathematically as $y = f(x)$ can be found from equations 12.68 and 12.69. This is illustrated in example 12.13.

$$\bar{x} = \frac{\int x \, dA}{A} \qquad 12.68$$

$$\bar{y} = \frac{\int y \, dA}{A} \qquad 12.69$$

$$A = \int f(x) \, dx \qquad 12.70$$

$$dA = f(x) \, dx = f(y) \, dy \qquad 12.71$$

Example 12.13

Find the x component of the centroid of the area bounded by the x and y axes, $x = 2$, and $y = e^{2x}$.

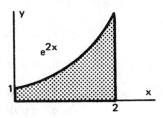

step 1: Find the area.

$$A = \int_0^2 e^{2x} dx = \left[\tfrac{1}{2}e^{2x}\right]_0^2$$
$$= 27.3 - .5 = 26.8$$

step 2: Put dA in terms of dx.

$$dA = f(x)\, dx = e^{2x} dx$$

step 3: Use equation 12.68 to find $\bar{x}$.

$$\bar{x} = \frac{1}{26.8} \int_0^2 xe^{2x} dx = \frac{1}{26.8}\left[\tfrac{1}{2}xe^{2x} - \tfrac{1}{4}e^{2x}\right]_0^2$$
$$= 1.54$$

With few exceptions, most areas for which the centroidal location is needed will be either rectangular or triangular. The locations of the centroids for these and other common shapes are given as an appendix of this chapter.

The centroid of a complex 2-dimensional area which can be divided into the simple shapes in appendix A can be found from equations 12.72 and 12.73.

$$\bar{x}_{composite} = \frac{\sum (A_i \bar{x}_i)}{\sum A_i} \qquad 12.72$$

$$\bar{y}_{composite} = \frac{\sum (A_i \bar{y}_i)}{\sum A_i} \qquad 12.73$$

Example 12.14

Find the y-coordinate of the centroid for the object shown.

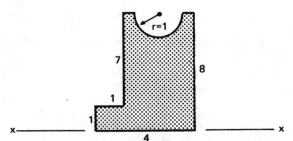

The object is divided into three parts: a 1×4 rectangle, a 3×7 rectangle, and a half-circle of radius 1.

Then, the areas and distances from the x-x axis to the individual centroids are found.

$$A_1 = (1)(4) = 4$$
$$A_2 = (3)(7) = 21$$
$$A_3 = \left(-\tfrac{1}{2}\right)\pi(1)^2 = -1.57$$
$$\bar{y}_1 = \tfrac{1}{2}$$
$$\bar{y}_2 = 4\tfrac{1}{2}$$
$$\bar{y}_3 = 8 - .424 = 7.576$$

Using equation 12.73,

$$\bar{y} = \frac{(4)(\tfrac{1}{2}) + (21)(4\tfrac{1}{2}) - (1.57)(7.576)}{4 + 21 - 1.57} = 3.61$$

B. MOMENT OF INERTIA

The moment of inertia, I, of a 2-dimensional area is a parameter which often is needed in mechanics of materials problems. It has no simple geometric interpretation, and its units (length to the fourth power) add to the mystery of this quantity. However, it is convenient to think of the moment of inertia as a resistance to bending.

If the moment of inertia is a resistance to bending, it is apparent that this quantity always must be positive. Since bending of an object (e.g., a beam) can be in any direction, the resistance to bending must depend on the direction of bending. Therefore, a reference axis or direction must be included when specifying the moment of inertia.

In this chapter, I_x is used to represent a moment of inertia with respect to the x axis. Similarly, I_y is with respect to the y axis. I_x and I_y are not components of the "resultant" moment of inertia. The moment of inertia taken with respect to a line passing through the area's centroid is known as the *centroidal moment of inertia*, I_c. The centroidal moment of inertia is the smallest possible moment of inertia for the shape.

The moments of inertia of a function which can be expressed mathematically as $y = f(x)$ are given by equations 12.74 and 12.75.

$$I_x = \int y^2\, dA \qquad 12.74$$
$$I_y = \int x^2\, dA \qquad 12.75$$

In general, however, moments of inertia will be found from appendix A.

Example 12.15

Find I_y for the area bounded by the y axis, $y = 8$, and $y^2 = 8x$.

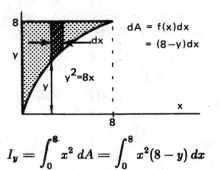

$$I_y = \int_0^8 x^2\, dA = \int_0^8 x^2(8-y)\, dx$$

But $y = \sqrt{8x}$.

$$I_y = \int_0^8 (8x^2 - \sqrt{8}x^{\frac{5}{2}})\, dx = 195.04 \text{ inches}^4$$

Example 12.16

What is the centroidal moment of inertia of the area shown?

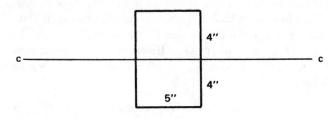

From appendix A,

$$I_c = \frac{(5)(8)^3}{12} = 213.3 \text{ inches}^4$$

The *polar moment of inertia* of a 2-dimensional area can be thought of as a measure of the area's resistance to torsion (twisting). Although the polar moment of inertia can be evaluated mathematically by equation 12.76, it is more expedient to use equation 12.77 if I_x and I_y are known.

$$J_{xy} = \int (x^2 + y^2)\, dA \qquad\qquad 12.76$$

$$J_{xy} = I_x + I_y \qquad\qquad 12.77$$

The *radius of gyration*, k, is a distance at which the entire area can be assumed to exist. The distance is measured from the axis about which the moment of inertia was taken.

$$I = k^2 A \qquad\qquad 12.78$$

$$k = \sqrt{\frac{I}{A}} \qquad\qquad 12.79$$

Example 12.17

What is the radius of gyration for the section shown in example 12.16? What is the significance of this value?

$$A = (5)(4+4) = 40$$

From equation 12.79,

$$k = \sqrt{\frac{213.3}{40}} = 2.31 \text{ inches}$$

$2.31''$ is the distance from the axis c-c that an infinitely long strip (with area of 40 square inches) would have to be located to have a moment of inertia of 213.3 inches[4].

The *parallel axis theorem* usually is needed to evaluate the moment of inertia of a composite object made up of several simple 2-dimensional shapes.[4] The parallel axis theorem relates the moment of inertia of an area taken with respect to any axis to the centroidal moment of inertia. In equation 12.80, A is the 2-dimensional object's area, and d is the distance between the centroidal and new axes.

$$I_{\text{any parallel axis}} = I_c + Ad^2 \qquad\qquad 12.80$$

Example 12.18

Find the moment of inertia about the x axis for the 2-dimensional object shown.

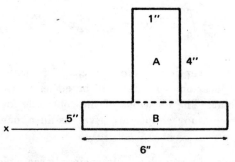

The T-section is divided into two parts: A and B. The moment of inertia of section B can be evaluated readily by using appendix A.

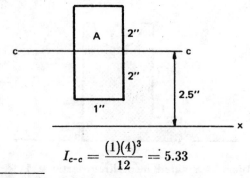

$$I_{x-x} = \frac{(6)(.5)^3}{3} = .25$$

The moment of inertia of the stem about its own centroidal axis is

$$I_{c-c} = \frac{(1)(4)^3}{12} = 5.33$$

---
[4] This theorem also is known as the *transfer axis theorem*.

Using equation 12.80, the moment of inertia of the stem about the x-x axis is

$$I_{x\text{-}x} = 5.33 + (4)(2.5)^2 = 30.33$$

The total moment of inertia of the T-section is

$$.25 + 30.33 = 30.58 \text{ inches}^4$$

C. PRODUCT OF INERTIA

The *product of inertia* of a 2-dimensional object is found by multiplying each differential element of area times its x and y coordinates, and then integrating over the entire area.

$$P_{xy} = \int xy \, dA \qquad 12.81$$

The product of inertia is zero when either axis is an axis of symmetry. The product of inertia may be negative.

16 ROTATION OF AXES

Suppose the various properties of an area are known for one set of axes, x and y. If the axes are rotated through an angle without rotating the area itself, the new properties can be found from the old properties.

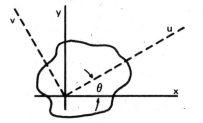

Figure 12.15 Rotation of Axes

$$I_u = I_x \cos^2\theta - 2P_{xy}\sin\theta\cos\theta + I_y\sin^2\theta \qquad 12.82$$
$$= \tfrac{1}{2}(I_x + I_y) + \tfrac{1}{2}(I_x - I_y)\cos 2\theta - P_{xy}\sin 2\theta \qquad 12.83$$
$$I_v = I_x \sin^2\theta + 2P_{xy}\sin\theta\cos\theta + I_y\cos^2\theta \qquad 12.84$$
$$= \tfrac{1}{2}(I_x + I_y) - \tfrac{1}{2}(I_x - I_y)\cos 2\theta + P_{xy}\sin 2\theta \qquad 12.85$$
$$P_{uv} = I_x \sin\theta\cos\theta + P_{xy}(\cos^2\theta - \sin^2\theta)$$
$$\quad - I_y \sin\theta\cos\theta \qquad 12.86$$
$$= \tfrac{1}{2}(I_x - I_y)\sin 2\theta + P_{xy}\cos 2\theta \qquad 12.87$$

Since the polar moment of inertia about a fixed axis is constant, the sum of the two area moments of inertia is also constant.

$$I_x + I_y = I_u + I_v \qquad 12.88$$

There is one angle that will maximize the moment of inertia, I_u. This angle can be found from calculus by setting $dI_u/d\theta = 0$. The resulting equation defines two angles, one of which maximizes I_u, the other of which minimizes I_u.

$$\tan 2\theta = -\frac{2P_{xy}}{I_x - I_y} \qquad 12.89$$

The two angles which satisfy equation 12.89 are 90° apart. These are known as the *principal axes*. The moments of inertia about these two axes are known as the *principal moments of inertia*. These principal moments are given by equation 12.90.

$$I_{\text{max, min}} = \tfrac{1}{2}(I_x + I_y) \pm \sqrt{\tfrac{1}{4}(I_x - I_y)^2 + P_{xy}^2} \qquad 12.90$$

17 PROPERTIES OF MASSES

A. CENTER OF GRAVITY

The *center of gravity* in 3-dimensional objects is analogous to centroids in 2-dimensional areas. The center of gravity can be located mathematically if the object can be described by a mathematical function.

$$\bar{x} = \frac{\int x \, dm}{m} \qquad 12.91$$
$$\bar{y} = \frac{\int y \, dm}{m} \qquad 12.92$$
$$\bar{z} = \frac{\int z \, dm}{m} \qquad 12.93$$

The location of the center of gravity often is obvious for simple objects. It always is located on an axis of symmetry. If the object is complex or composite, the overall center of gravity can be found from the individual centers of gravity of the constituent objects.

$$\bar{x}_{\text{composite}} = \frac{\sum(m_i \bar{x}_i)}{\sum m_i} \qquad 12.94$$
$$\bar{y}_{\text{composite}} = \frac{\sum(m_i \bar{y}_i)}{\sum m_i} \qquad 12.95$$
$$\bar{z}_{\text{composite}} = \frac{\sum(m_i \bar{z}_i)}{\sum m_i} \qquad 12.96$$

B. MASS MOMENT OF INERTIA

The mass moment of inertia can be thought of as a measure of resistance to rotational motion. Although it can be found mathematically from equations 12.97, 12.98, and 12.99, it is more expedient to use appendix B to evaluate simple objects.

$$I_x = \int (y^2 + z^2) \, dm \qquad 12.97$$
$$I_y = \int (x^2 + z^2) \, dm \qquad 12.98$$
$$I_z = \int (x^2 + y^2) \, dm \qquad 12.99$$

The *centroidal mass moment of inertia* is found by evaluating the moment of inertia about an axis passing through the object's center of gravity. Once this centroidal mass moment of inertia is known, the parallel axis theorem can be used to find the moment of inertia about any parallel axis.

$$I_{\text{any parallel axis}} = I_c + md^2 \qquad 12.100$$

The radius of gyration of a 3-dimensional object is defined by equation 12.101.

$$k = \sqrt{\frac{I}{m}} \qquad 12.101$$

$$I = k^2 m \qquad 12.102$$

18 FRICTION

Friction is a force which resists motion or attempted motion. It always acts parallel to the contacting surfaces. The frictional force exerted on a stationary object is known as *static friction* or *coulomb friction*. If the object is moving, the friction is known as *dynamic friction*. Dynamic friction is less than static friction in most situations.

The actual magnitude of the frictional force depends on the *normal force* and the *coefficient of friction*, f, between the object and the surface. For an object resting on a horizontal surface, the normal force is the weight, w.

$$\mathbf{F}_f = f\mathbf{N} = fw \qquad 12.103$$

If the object is resting on an inclined surface, the normal force will be

$$\mathbf{N} = w \cos\theta \qquad 12.104$$

The frictional force again is equal to the product of the normal force and the coefficient of friction.

$$\mathbf{F}_f = f\mathbf{N} = fw \cos\theta \qquad 12.105$$

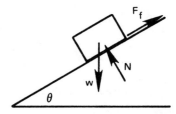

Figure 12.16 Frictional Force

The object shown in figure 12.16 will not slip down the plane until the angle reaches a critical angle known as the *angle of repose*. This angle is given by equation 12.106.

$$\tan\theta = f \qquad 12.106$$

Typical values of the coefficient of friction are given in table 12.2.

Friction also exists between a belt, rope, or band wrapped around a drum, pulley, or sheave. If $\mathbf{T}_1$ is the tight side tension, if $\mathbf{T}_2$ is the slack side tension, and if ϕ is the contact angle in *radians*, the relationship governing *belt friction* is given by equation 12.107. m is the belt mass per unit length. Centrifugal force can be neglected for slow speed operation.

$$\frac{\mathbf{T}_1 - \mathbf{F}_c}{\mathbf{T}_2 - \mathbf{F}_c} = \frac{\mathbf{T}_1 - mv^2}{\mathbf{T}_2 - mv^2} = e^{f\phi} \qquad 12.107$$

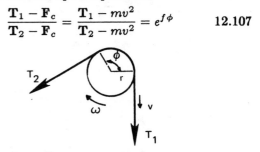

Figure 12.17 Belt Friction

The transmitted torque in ft-lbf is

$$\text{torque} = (\mathbf{T}_1 - \mathbf{T}_2)r \qquad 12.108$$

The power in ft-lbf/sec transmitted by the belt running at speed v in ft/sec is

$$\text{power} = (\mathbf{T}_1 - \mathbf{T}_2)v \qquad 12.109$$

Table 12.2
Approximate Coefficients of Friction
(Also, see table 15.13.)

Material	Condition	Dynamic	Static
cast iron on cast iron	dry	.15	1.10
plastic on steel	dry	.35	
grooved rubber on pavement	dry	.40	.55
bronze on steel	oiled		.09
steel on graphite	dry		.21
steel on steel	dry	.42	.78
steel on steel	oiled	.08	.10
steel on asbestos-faced steel	dry		.15
steel on asbestos-faced steel	oiled		.12
press fits (shaft in hole)	oiled		.10–.15
rubber belt on steel	dry		.32

PART 2: Indeterminate Structures

A structure that is *statically indeterminate (redundant)* is one for which the equations of statics are not sufficient to determine all reactions, moments, and internal stress distributions. Additional formulas involving deflection relationships are required to determine these unknowns completely.

The *degree of redundancy* is equal to the number of reactions of members that would have to be removed in order to make the structure statically determinate. For example, a two-span beam on three simple supports is redundant to the first degree. The degree of redundancy of a truss can be calculated from equation 12.110.

$$\text{degree of redundancy} = \text{\# members} - 2(\text{\# joints}) + 3 \qquad 12.110$$

The *method of consistent deformation* can be used to evaluate simple structures consisting of two or three members in tension or compression. Although this method cannot be proceduralized, it is very simple to learn and to apply. The method makes use of geometry to develop relationships between the deflections (deformations) between different members or locations on the structure.

Example 12.19

A pile is constructed of concrete with a steel jacket. What is the stress in the steel and the concrete if a load, **P**, is applied? Assume the end caps are rigid and the steel-concrete bond is perfect.

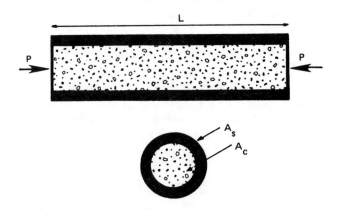

Let P_c and P_s be the loads carried by the concrete and steel, respectively. Then,

$$P_c + P_s = P \qquad 12.111$$

The deformation of the steel is

$$\delta_s = \frac{P_s L}{A_s E_s} \qquad 12.112$$

Similarly, the deflection of the concrete is

$$\delta_c = \frac{P_c L}{A_c E_c} \qquad 12.113$$

But $\delta_c = \delta_s$ since the bonding is perfect. Therefore,

$$\frac{P_c}{A_c E_c} - \frac{P_s}{A_s E_s} = 0 \qquad 12.114$$

Equations 12.111 and 12.114 are solved simultaneously to determine P_c and P_s. The respective stresses are

$$\sigma_s = \frac{P_s}{A_s} \qquad 12.115$$

$$\sigma_c = \frac{P_c}{A_c} \qquad 12.116$$

Example 12.20

A uniform bar is clamped at both ends, and the axial load is applied near one of the supports. What are the reactions?

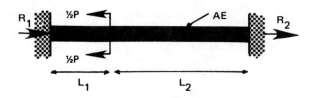

The first required equation is

$$R_1 + R_2 = P \qquad 12.117$$

The shortening of section 1 due to the reaction, R_1, is

$$\delta_1 = \frac{-R_1 L_1}{AE} \qquad 12.118$$

The elongation of section 2 due to the reaction, R_2, is

$$\delta_2 = \frac{R_2 L_2}{AE} \qquad 12.119$$

However, the bar is continuous, so $\delta_1 = -\delta_2$. Therefore,

$$R_1 L_1 = R_2 L_2 \qquad 12.120$$

Equations 12.117 and 12.120 are solved simultaneously to find R_1 and R_2.

Example 12.21

The non-uniform bar shown is clamped at both ends. What are the reactions of both ends if a temperature change of ΔT is experienced?

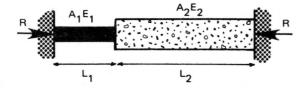

The thermal deformations of sections 1 and 2 can be calculated directly.

$$\delta_1 = \alpha_1 L_1 \Delta T \qquad\qquad 12.121$$

$$\delta_2 = \alpha_2 L_2 \Delta T \qquad\qquad 12.122$$

The total deformation is $\delta = \delta_1 + \delta_2$. However, the deformation also can be calculated from the principles of mechanics of materials.

$$\delta = \frac{RL_1}{A_1 E_1} + \frac{RL_2}{A_2 E_2} \qquad\qquad 12.123$$

Combining equations 12.121, 12.122, and 12.123 gives

$$(\alpha_1 L_1 + \alpha_2 L_2)\Delta T = \left(\frac{L_1}{A_1 E_1} + \frac{L_2}{A_2 E_2}\right)R \qquad 12.124$$

Equation 12.124 can be solved directly for R.

Example 12.22

The beam shown is supported by dissimilar tension members. What are the reactions in the tension members? Assume that the horizontal bar is rigid and remains horizontal.

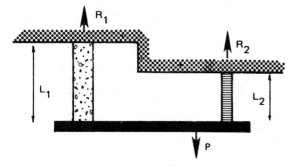

The required equilibrium condition is

$$R_1 + R_2 = P \qquad\qquad 12.125$$

The elongations of the two tension members are

$$\delta_1 = \frac{R_1 L_1}{A_1 E_1} \qquad\qquad 12.126$$

$$\delta_2 = \frac{R_2 L_2}{A_2 E_2} \qquad\qquad 12.127$$

If the horizontal bar remains horizontal, $\delta_1 = \delta_2$. Therefore,

$$\frac{R_1 L_1}{A_1 E_1} = \frac{R_2 L_2}{A_2 E_2} \qquad\qquad 12.128$$

Equations 12.125 and 12.128 are solved simultaneously to find R_1 and R_2.

Example 12.23

Find the forces in the three tension members.

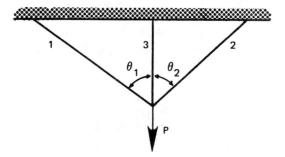

The equilibrium requirement is

$$P_{1y} + P_3 + P_{2y} = P \qquad\qquad 12.129$$

$$P_1 \cos\theta_1 + P_3 + P_2 \cos\theta_2 = P \qquad 12.130$$

The vertical elongations of all three tension members are the same at the junction.

$$\frac{P_1 L_1}{A_1 E_1}(\cos\theta_1) = \frac{P_3 L_3}{A_3 E_3} = \frac{P_2 L_2}{A_2 E_2}(\cos\theta_2) \qquad 12.131$$

Equations 12.130 and 12.131 can be solved simultaneously to find P_1, P_2, and P_3. It may be necessary to work with the x-components of the deflections in order to find a third equation.

Appendix A
Area Moments of Inertia

SHAPE	DIMENSIONS	CENTROID (x_c, y_c)	AREA MOMENT OF INERTIA
Rectangle		$(\tfrac{1}{2}b, \tfrac{1}{2}h)$	$I_{x'} = (1/12)bh^3$ $I_{y'} = (1/12)hb^3$ $I_x = (1/3)bh^3$ $I_y = (1/3)hb^3$ $J_C = (1/12)bh(b^2 + h^2)$
Triangle		$y_c = (h/3)$	$I_{x'} = (1/36)bh^3$ $I_x = (1/12)bh^3$
Trapezoid		$y'_c = \dfrac{h(2B + b)}{3(B + b)}$ Note that this is measured from the top surface.	$I_{x'} = \dfrac{h^3(B^2+4Bb+b^2)}{36(B+b)}$ $I_x = \dfrac{h^3(B+3b)}{12}$
Quarter-Circle, of radius r		$((4r/3\pi), (4r/3\pi))$	$I_{x'} = 0.055r^4$ $I_x = I_y = (1/16)\pi r^4$ $J_O = (1/8)\pi r^4$
Half Circle, of radius r		$(0, (4r/3\pi))$	$I_x = I_y = (1/8)\pi r^4$ $J_O = \tfrac{1}{4}\pi r^4 \quad I_{x'} = .11r^4$
Circle, of radius r		$(0,0)$	$I_x = I_y = \tfrac{1}{4}\pi r^4$ $J_O = \tfrac{1}{2}\pi r^4$
Parabolic Area		$(0, (3h/5))$	$I_x = 4h^3a/7$ $I_y = 4ha^3/15$
Parabolic Spandrel		$((3a/4),(3h/10))$	$I_x = ah^3/21$ $I_y = 3ha^3/15$

Appendix B
Mass Moments of Inertia
(m is in slugs; lengths are in feet)

Slender rod		$I_y = I_z = (1/12)mL^2$ $I_{y'} = I_{z'} = (1/3)mL^2$
Thin rectangular plate		$I_x = (1/12)(b^2+c^2)\,m$ $I_y = (1/12)mc^2$ $I_z = (1/12)mb^2$
Rectangular Parallelepiped		$I_x = (1/12)m(b^2+c^2)$ $I_y = (1/12)m(c^2+a^2)$ $I_z = (1/12)m(a^2+b^2)$ $I_{x'} = (1/12)m(4b^2+c^2)$
Thin disk, radius r		$I_x = \frac{1}{2}mr^2$ $I_y = I_z = \frac{1}{4}mr^2$
Circular cylinder, radius r		$I_x = \frac{1}{2}mr^2$ $I_y = I_z$ $= (1/12)m(3r^2+L^2)$
Circular cone, base radius r		$I_x = (3/10)mr^2$ $I_y = I_z$ $= (3/5)m(\frac{1}{4}r^2+h^2)$
Sphere, radius r		$I_x = I_y = I_z$ $= (2/5)mr^2$
Hollow circular cylinder		$I_x = \frac{1}{2}m(r_o^2 + r_i^2)$ $= \frac{\pi \rho L}{2}(r_o^4 - r_i^4)$

PRACTICE PROBLEMS: STATICS

Warm-ups

1. What torque is necessary to counterbalance a 3 pound force acting at a radius of 3"?

2. What is the equivalent, all-aluminum, centroidal, cross sectional moment of inertia of a bi-metallic spring which is composed of a $1/8'' \times 1\frac{1}{2}''$ aluminum strip bonded on top of a $1/16'' \times 1\frac{1}{2}''$ steel strip?

3. A 60" total diameter cast iron flywheel has a rim which is 6" deep and 12" wide. The hub is 12" wide, 12" outside diameter, and 6" inside diameter. Six 4.25" diameter spokes are used. What is the moment of inertia assuming the entire flywheel (including spokes) is cast iron?

4. Two supports on level ground are 100 feet apart and support a transmission line weighing 2 pounds per foot and which has a mid-point sag of 10'. What are the mid-point and maximum tensions in the line?

Concentrates

1. Two legs of a tripod form a horizontal plane when mounted on a wall. The apex is 12 feet from the wall, and the right leg is 13.4' long. The supports for these two legs are 10 feet apart. A third leg is mounted on the wall 9 feet below the horizontal plane and 6 feet to the left of the 13.4' long leg. A vertical downward load of 200 pounds is supported. What are the reactions?

2. What is the force in member DE?

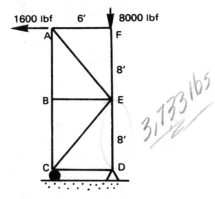

3. Find the centroidal moment of inertia about an axis parallel to the x axis.

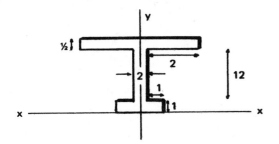

4. Find the forces in each of the legs.

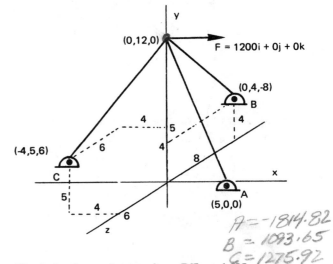

5. Find the forces in members DE and HJ.

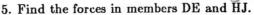

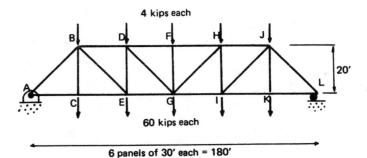

6. What are the x, y, and z components of the forces at A, B, and C? Points A, B, and C are all in the same plane, parallel to the yz plane.

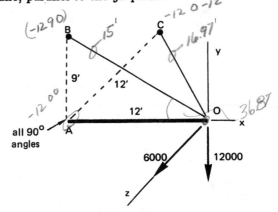

7. A power line weighs 2 pounds per foot of length. It is supported by two equal height towers over a level forest. The tower spacing is 100', and the mid-point sag is 10'. What are the maximum and minimum tensions?

8. What is the sag for the cable described in problem #7 if the maximum tension is 500 pounds?

9. Locate the centroid of the object shown.

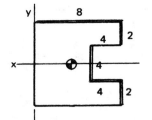

10. Replace the distributed load with three concentrated loads, and indicate the points of application.

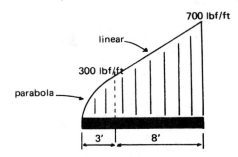

Timed (1 hour allowed for each)

1. A steel railroad rail is welded together to form a continuous horizontal rail one mile long at 70°F. The sun uniformly warms the rail to 99.14°F. Since only one end of the rail is fixed, the rail pops up in the middle to a parabolic shape as the opposite end is forced back to its original position. How high off the ground is the mid-point of the hot rail when the ends are again one mile apart? Assume frictionless lateral support.

RESERVED FOR FUTURE USE

RESERVED FOR FUTURE USE

13 MATERIALS SCIENCE

PART 1: Engineering Materials

Nomenclature

BHN	Brinell hardness number	–
g	local gravitational acceleration	ft/sec^2
g_c	gravitational constant (32.2)	lbm-ft/lbf-sec^2
h	height	ft
m	mass	lbm
S_{ut}	ultimate tensile strength	ksi

1 METALS

A. IRON AND STEEL

The process used to reduce iron ore to *iron* requires coke and limestone. The limestone lowers the melting temperature of the gangue so it can be drawn off as molten slag. The process takes place in a *blast furnace*, as shown in figure 13.1.

The top of the furnace is provided with a pair of conical bells for loading the charge while limiting the escape of gases. High temperature air is injected through openings called *tuyeres* in the lower portion of the furnace. The air oxidizes the coke to produce heat and carbon monoxide. The carbon monoxide rises to the top of the furnace and, at a temperature of approximately 600°F, reduces the iron oxides to FeO.

As the process continues, the FeO drops down to a region where the temperature ranges from 1300°F to 1500°F. The FeO is reduced to iron by the surrounding carbon. This iron drops into an area where the temperature ranges from 1500°F to 2500°F. The iron becomes saturated with carbon in the form of both carbides and free carbon. The absorbed carbon lowers the melting point of the iron so that it runs as a liquid to the bottom of the furnace.

The *slag* melts at the same temperature as the iron and floats on the liquid iron. This allows the slag and the iron to be drawn off into separate receptacles—the iron to molds where it is cast into *pigs*, and the slag to a slag heap (disposal).

Hot air is provided by two or more stoves which adjoin the blast furnace. Each stove alternately is heated by the hot furnace gases and then heats outside air to 1400°F before being injected into the furnace.

Pig iron, which contains approximately 4% carbon, is brittle. Carbon must be removed before the iron can be used. The pigs also contain other undesirable elements such as sulfur, phosphorus, silicon, and manganese.

Four processes are available for the further refinement of the iron: Bessemer, open hearth, electric furnace, and wrought iron processes.

The *Bessemer process* uses a pear-shaped steel crucible lined with refractory materials. The crucible, which is provided with air inlets in its base, is filled with molten pig iron at about 2200°F. High pressure air is blown through the iron, oxidizing the carbon and other impurities and causing a temperature rise to 3500°F.

The burning impurities are consumed in about 20 minutes after which measured amounts of carbon, manganese and other alloying agents are added to obtain the desired grade of steel. The Bessemer process is a crude process since its speed leads to poor control of constituents. The product is used to make cheaper grades of sheet, wire, pipe, and screw stock.

The *open hearth process* is illustrated in figure 13.2. The *charge* is contained in a shallow receptacle and is composed of a mixture of pig iron, scrap iron, iron oxide, and limestone. Heat is provided by burning CO above the iron. Since the combustion products contain very little oxygen, oxidation of the impurities is accomplished by the iron oxide in the charge. The function of the limestone is to provide a protective slag over the melt. Since this is a slow process (8 to 12 hours), continuous monitoring of the steel composition is possible.

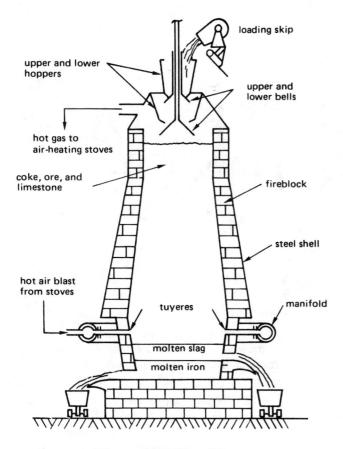

Figure 13.1 Blast Furnace

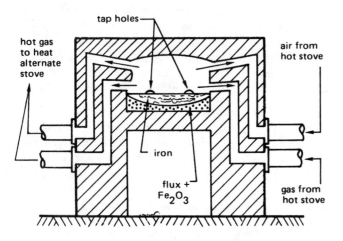

Figure 13.2 Open Hearth Furnace

Electric furnaces employing electric arc or induction heating are used for the production of tool and alloy steels. Since no air or gaseous fuels are required, the impurities introduced by them are eliminated. The furnace walls are lined with either silica (acidic) or magnesite (basic) brick, the selection being based on the iron being refined. This process also depends on the use of iron oxide as the reducing agent.

Wrought iron is steel with small amounts of slag in the form of fibrous inclusions. Wrought iron is produced in a *reverberatory furnace* similar to the open hearth furnace. The molten iron floats on a layer of iron oxide which provides the oxygen for removal of carbon and sulfur. The limestone flux combines with silicon and phosphorus to form slag. Spongy masses of iron and slag are collected on the ends of steel rods inserted into the pool of molten metal. These masses then are removed and hammered or pressed to squeeze out most of the slag. The remaining product consists of slag coated iron particles welded together by the forging process.

Cast iron is available in several forms. *White cast iron* has been cooled quickly from a molten state so that graphite is not produced from the cementite. *Malleable cast iron* is produced by reheating white cast iron to 1600°F, followed by slow or intermediate cooling. The most common type of cast iron is *gray cast iron*, available in both pearlitic and ferritic forms. Since gray cast iron has low strength, magnesium or cerium can be added to the molten alloy. The resulting *nodular cast iron* has superior strength and ductility.

Stainless steel is known for its ability to withstand tarnishing and corrosion. Figure 13.3 can be used to determine some of the properties and characteristics of different stainless compositions.

B. ALUMINUM

Most of the aluminum in use today is recovered from *bauxite ore.* This ore is a mixture of hydroxides of aluminum mixed with iron, silicon, and titanium oxides. The ore is crushed and ground to fine powder. Then it is treated with a hot solution of sodium hydroxide producing sodium aluminate and water. This solution is drawn off into a separate tank, leaving the remaining ore constituents as a solid deposit.

As the solution cools, aluminum hydroxide precipitates out leaving a sodium hydroxide solution. The aluminum hydroxide then is baked to form aluminum oxide, Al_2O_3. Final reduction is accomplished through an electrolytic process which uses molten *cryolite* (Na_3AlF_6) as the electrolyte and large carbon blocks as anodes. The steel process tank is lined with carbon and acts as the cathode. The aluminum oxide dissolves in the cryolite and is separated into molten aluminum and oxygen by the applied electric current. The aluminum collects in the bottom of the tank. Carbon dioxide is released at the anodes. The cryolite does not enter into the reaction and can be reused.

C. COPPER

Copper occurs in the free state as well as in ores containing its oxides, sulfides, and carbonates. *Native cop-*

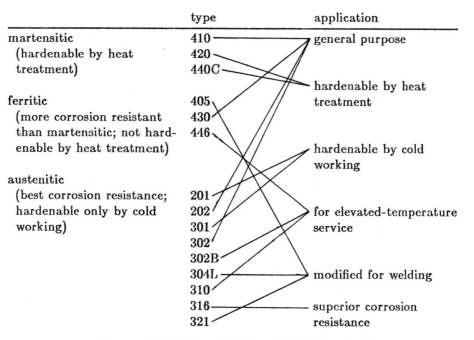

Figure 13.3 Characteristics of Stainless Steel

per is recovered by the simple process of heating finely crushed ore. Molten copper flows to the bottom of the furnace. Oxides and carbonates of copper are reduced in a reverberatory furnace by combining with the hot unburned gases used in the process.

Sulfides are heated in air, and the sulfur is replaced by oxygen. The product, which contains both copper and iron oxides, is reduced in a reverberatory furnace. This removes the oxygen but leaves some sulfur and iron. The final treatment takes place in a furnace similar to the Bessemer converter in which air is injected into the molten copper, removing the sulfur. The iron oxide combines with a silica furnace liner.

Electrolysis is required to remove remaining impurities. Thick sheets of impure copper are immersed, together with thin sheets of pure copper, in an electrolyte of copper sulfate. The pure copper acts as the cathode. A direct current causes copper from the impure material to migrate to the pure sheets. Impurities drop to the bottom of the tank.

2 NATURAL POLYMERS

Many of the natural organic materials (e.g., rubber and asphalt) are polymers. Natural rubber is a polymer of *isoprene*. *Vulcanization* is accomplished by heating raw rubber with small amounts of sulfur. The process raises the tensile strength of the material from approximately 300 psi to 3,000 psi. The addition of carbon black as a reinforcing filler raises this value to 4,500 psi and provides tear resistance and toughness.

3 SYNTHETIC POLYMERS

Polymers are extremely large molecules built up of long chains of repeating units. The basic repeating unit is called the *monomer* or *mer*. The *degree* of the polymer is the average number of units in the compound, typically 75 to 750 mers per molecule.

A compound having only two, three, or four monomers can be called a *di-*, *tri-*, or *tetramer*, respectively. In general, compounds with degrees of less than ten are called *telenomers* or *oligomers*.

The stiffness and hardness of polymers vary with their degree. Polymers with low degrees of polymerization are liquids or oils. With increasing degree, they go through waxy to hard resin stages. High degree polymers have hardness or elastic qualities which make them useful for engineering applications.

Example 13.1

A polyvinyl chloride molecule is found to contain 860 carbon atoms, 1290 hydrogen atoms, and 430 chlorine atoms. What is the degree of polymerization? (The vinyl chloride mer is C_2H_3Cl with a molecular weight of 62.5.)

The molecular weight of the polymer is

$$860(12) + 1290(1) + 430(35.5) = 26,875$$

The degree of polymerization is

$$\frac{26,875}{62.5} = 430$$

Table 13.1
Names of Some Mers

name	repeating unit	combined formula
ethylene	CH_2CH_2	C_2H_4
propylene	$CH_2(HCCH_3)$	C_3H_6
styrene	$CH_2CH(C_6H_5)$	C_8H_8
vinyl acetate	$CH_2CH(C_2H_3O_2)$	$C_4H_6O_2$
vinyl chloride	CH_2CHCl	C_2H_3Cl
isobutylene	$CH_2C(CH_3)_2$	C_4H_8
methyl methacrylate	$CH_2C(CH_3)(COOCH_3)$	$C_5H_8O_2$
acrylonitrile	CH_2CHCN	C_3H_3N
epoxide (ethoxylene)	CH_2CH_2O	C_2H_4O
amide (nylon)	$CONH_2$ or $CONH$	$CONH_2$ or $CONH$

A polymer will contain molecules with different chain lengths. Therefore, the degree of polymerization will vary from molecule to molecule. A determination of the polymer's degree requires statistical analysis.

Example 13.2

What is the degree of polymerization of a polyvinyl acetate sample having the following molecular weight analysis?

range of molecular weights	fraction
5,000–15,000	.30
15,000–25,000	.47
25,000–35,000	.23

Using the midpoint of each range, the average molecular weight is

$$(.30)(10,000) + (.47)(20,000) + (.23)(30,000) = 19,300$$

The vinyl acetate mer is $C_4H_6O_2$ with a molecular weight of 86. The degree of polymerization is

$$\frac{19,300}{86} = 224.4$$

Polymers are named by adding the prefix *poly* to the name of the basic mer. For example, C_2H_4 is the chemical formula for ethylene. Chains of C_2H_4 are called polyethylene. Some important repeating units are listed in table 13.1.

Two processes are used to form polymers: addition polymerization and condensation polymerization. With *addition polymerization*, mers combine sequentially using single covalent bonds to form chains. For example, the formation of polyethylene is

$$2(CH_2 = CH_2) \mapsto -CH_2-CH_2-CH_2-CH_2-$$

Substances called *initiators* are used in addition polymerization. Initiators break down under heat or light and provide free radicals. These free radicals act as chain carriers as well as openers of double bonds. The reaction produces unpaired electrons, and the process continues until the initiator is used up or until another free radical reacts to form a termination molecule. The latter occurrence is called *saturation*.

A typical initiator is benzoil peroxide which, upon decomposition, produces a free radical.

$$(C_6H_5COO_2)_2 \mapsto 2C_6H_5- + 2CO_2$$

Propagation starts with the free radical reacting with the monomer. The radical opens the monomer's double bond and forms a new free radical. The new free radical reacts with an additional monomer, continuing the reaction.

$$C_6H_5- + CH_2 = CHCl \mapsto C_6H_5-CH_2-CHCl-$$
$$C_6H_5-CH_2-CHCl- + CH_2 = CHCl \mapsto$$
$$C_6H_5-CH_2-CHCl-CH_2-CHCl-$$

Condensation polymerization opens bonds in two molecules to form a larger molecule. This formation often is accompanied by the release of small molecules such as H_2O, CO_2, and N_2. The repeating units derived from the condensation process are not the same as the monomers from which they are formed since portions of the original monomer form small molecules and are eliminated.

Phenolic plastics are produced from the condensation of formaldehyde and ammonia into a crystalline solid.

$$6HCHO + 4NH_3 \mapsto (CH_2)_6N_4 + 6H_2O$$

Some types of polymers form multiple chains linked by cross branches. These are called *framework-bonded polymers*. Some framework structures are resistant to heat and will not melt or flow. These are called *thermosetting resins*. Conversely, most chain polymers can be softened and formed under the application of heat and pressure. These are called *thermoplastic resins*.

Table 13.2
Typical Properties of E- and S-Glasses
(room temperature)

property	E-glass	S-glass
specific gravity	2.54	2.48
density (lbm/in^3)	0.092	0.090
ultimate tensile strength (psi)		
· monofilament	5.0 EE5	6.6 EE5
· 12-end roving	3.7 EE5	5.5 EE5
modulus of elasticity (psi)	10.5 EE6	12.5 EE6
coefficient of thermal		
expansion (1/°F)	2.8 EE−6	1.6–2.2 EE−6
specific heat (BTU/lbm-°F)	0.192	0.176

Thermosetting resins are represented by the phenolics (bakelites), epoxies, silicones, and polyesters. The thermoplastic resins include cellulose derivatives, polystyrene, vinyl polymers, polyethylene, nylon, acrylics, and the methacrylics.

4 FIBER/RESIN MIXTURES

The initial process leading to the widespread use of fiber-reinforced polymers was the impregnation of natural fibers (cotton, wool, etc.) with bakelite and phenolic resins. The introduction of various types of glass (rovings,[1] windings, and woven cloth) for reinforcement of polyester resins was the next development.

More recently, the higher-strength epoxy resins have been used as matrices for use with glass, graphite, and boron fibers. The resultant composite materials are highly *anisotropic*. Properties vary with the orientation of monofilament layups as well as with weave orientation in the cloth.

In directions transverse to fiber orientation, tensile and compressive strength is a function of the matrix material. Loads parallel to the fibers are carried by the reinforcement, while flexural strength is limited by the shear bond between the filaments and the matrix.

Typical reinforcing materials for fiberglass are *E*- and *S*-glass.[2] *S*-glass is a silica-alumina-magnesia com-

pound developed for improved tensile properties and is used mainly in nonwoven, monodirectional and wound configurations. *E*-glass is a lime-alumina-borosilicate compound used primarily in woven fabrics.

Graphite fibers are used where high stiffness and low coefficients of thermal expansion are needed. These properties must be balanced against the disadvantages of brittleness and high cost. Ultimate tensile strengths for graphite vary inversely with modulus of elasticity. Graphite fibers range in tensile strength from 180 *ksi* for yarn configurations to 350 *ksi* for *tow* (loose, untwisted fibers), while modulus of elasticity varies from 60 EE6 to 20 EE6 psi.

Boron is being produced in fibrous form for use as a reinforcing material. Its modulus of elasticity is approximately 60 EE6 psi, while tensile strengths can reach 500,000 psi.

Properties of composites (i.e., fibers in their matrix binders) are, of course, different from properties of the fibers themselves. Typical properties of composites are given in table 13.3 for the axial and transverse directions. Data for other fiber orientations (i.e., crisscrossing at 45° off the axial axis) also is available, and such data should be used for composites with those fiber orientations.

---

[1] A *roving* consists of a number of parallel strands of glass fiber. The strands are side by side, forming a flat ribbon. However, the strands are not interwoven or twisted together.

[2] Other types are: A-glass (common soda-lime glass used for windows, bottles, and jars), C-glass (glass developed for greater chemical and corrosion resistance), D-glass (glass possessing a low dielectric constant), and M-glass (glass containing BeO to increase the elastic modulus).

Table 13.3
Typical Composite Properties (Linear Layup)

Material	Fiber content (%)	Density (lbm/in³)	Modulus (EE6 psi)			Ultimate Strength (EE3 psi)		
			Axial	Transverse	Shear	Axial	Transverse	Shear
Graphite-Epoxy								
High strength	65	0.057	20	1.0	0.65	220	6	14
High modulus	65	0.058	29	1.0	0.70	175	5	10
Ultrahigh modulus	65	0.061	44	1.0	0.95	110	4	7
Kevlar 49-Epoxy	65	0.050	12.5	0.8	0.3	220	4	6
E Glass-Epoxy	65	0.072	6	1.5	0.3	180	6	10
Chopped Glass-polyester sheet								
molding compound (SMC)	30	0.068	2.5	2.5	1.0	30	30	20
	65	0.072	3.5	3.5	1.5	50	50	40

PART 2: Materials Testing

The most important test used to predict material properties is the tensile test. This test will be covered in chapter 14.

1 HARDNESS TESTS

Hardness tests measure the capacity of a surface to resist deformation. Through correlation, it is possible to predict the ultimate tensile strength of a metallic material through hardness tests. Hardness tests also are used to verify heat treatments.

The *Brinell hardness number* (BHN) is determined by pressing a hardened steel ball into the surface of the specimen. The diameter of the resulting depression is a measure of the material hardness. The standard ball is 10 mm in diameter, and the loads are 3000 *kg* and 500 *kg* for hard and soft materials, respectively.

The approximate ultimate tensile strength of steel can be calculated from the Brinell hardness number.

$$S_{ut} \approx (500)(BHN) \qquad 13.1$$

The *Rockwell hardness test* is similar to the Brinell test. The depth of penetration of a steel ball or a diamond spheroconical penetrator is measured. The machine applies an initial load (10 *kg*) which sets the penetrator below surface anomalies. Then a load is applied, and the hardness is shown on a dial. Although a number of Rockwell scales exist, the most familiar scales are the *B* and *C* scales. The *B* scale is used for mild steel and high-strength aluminum, while the *C* scale is used for hard steels having ultimate tensile strengths up to 300 ksi.

2 TOUGHNESS TESTING

Toughness is the capability of a metal to absorb (through yielding) highly localized and rapidly applied stresses. In the *Charpy test*, a beam specimen is provided with a 45° *V* notch. This beam is centered on simple supports with the notch down.

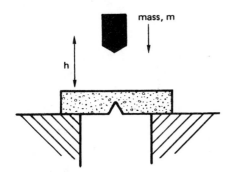

Figure 13.4 The Charpy Test

A falling-weight striker hits the center of the beam. By measuring the height at which the striker is released, the kinetic energy expended at impact can be determined. The objective is to determine the energy required for failure.

$$\text{Energy} = \frac{mgh}{g_c} \qquad 13.2$$

At 70°F failure impact energy ranges from 45 ft-lbs for the carbon steels to 110 ft-lbs for the chromium-manganese steels. As temperature is reduced, the toughness decreases. The *transition temperature* is taken as the point at which an impact of 15 ft-lbs will cause failure.

Table 13.4
Typical Transition Temperatures for Steel

type of steel	transition temperature, °F
carbon steel	30°
high-strength, low-alloy steel	0° to 30°
heat treated, high-strength, carbon steel	−25°
heat treated, construction alloy steel	−40° to −80°

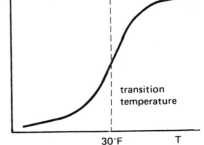

Figure 13.5 Failure Energy versus Temperature for Low-Carbon Steel

Not all materials exhibit the energy-temperature curve shape shown in figure 13.5. Figure 13.6 illustrates the curves for high-strength steels and FCC (face-centered cubic) metals such as aluminum.

Another toughness test is the *Izod test*. This is illustrated in figure 13.7. Equation 13.2 also is applicable to the Izod test.

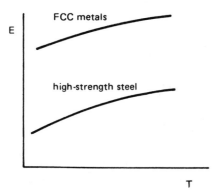

Figure 13.6 Failure Energy Versus Temperature for Other Metals

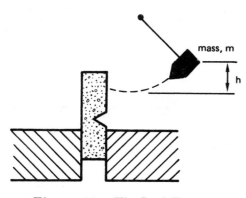

Figure 13.7 The Izod Test

3 CREEP TESTING

During a creep test, a low tensile load of constant magnitude is applied to a sample. The strain is measured as a function of time.

The *creep strength* is the stress which results in a given creep rate, usually .001% or .0001% per hour. The *rupture strength* is the stress which results in failure after a given amount of time, usually 100, 1000, or 10,000 hours.

If the creep is plotted as a function of time (figure 13.8), three different sections on the curve will be apparent. During the first stage, the creep rate ($d\epsilon/dt$) increases since strain hardening occurs at a greater rate than annealing. During the second stage, the creep rate is constant. Strain hardening and annealing occur at the same rate during this second stage. The sample begins to neck down during the third stage, and rupture occurs.

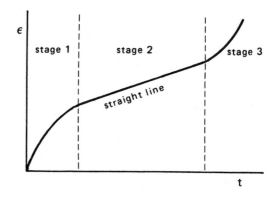

Figure 13.8 Creep versus Time

The *creep rate* is very temperature dependent. The slope of the line during the second stage increases with temperature.

PART 3: Thermal Treatments of Metals

1 EQUILIBRIUM CONDITIONS

Figure 13.9 illustrates the graph of temperature versus time as a pure metal cools from liquid to solid state. At a particular point, the temperature remains constant. This temperature is known as the *freezing point* of the liquid. (The metal continues to lose heat energy—its *heat of fusion*. However, the temperature remains constant during the freezing process.)

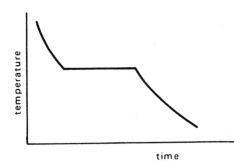

Figure 13.9 Temperature versus Time for a
Pure Metal

If the same experiment is performed with an alloy of two metals, the transition temperature will vary with the proportions of the constituent. The locus of constant temperature points is a line above which the mixture is entirely liquid and below which it is entirely solid.

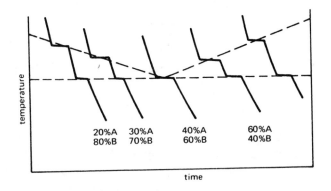

Figure 13.10 Temperature versus Time for a
Binary Alloy

Figure 13.11 illustrates the variations of liquid and solid equilibrium temperatures as a function of the relative concentration of the constituents in a completely miscible alloy. The *liquidus line* represents the limit above which no solid can exist, while the *solidus line* is the limit below which no liquid occurs. The area between these curves denotes a mixture of solid and liquid phase materials.

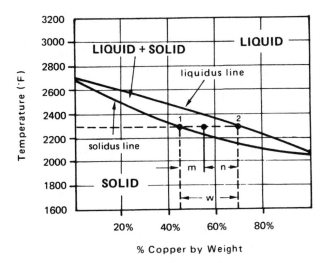

Figure 13.11 Copper-Nickel Phase Diagram

Example 13.3

Consider a mixture of 55% copper and 45% nickel at 2300°F. What are the percentages of solid and liquid materials?

The solid portion of the mixture will have the composition at point 1 (48% copper) while the liquid will be of composition 2 (67.8% copper). To find the relative amounts of solid and liquid in the mixture, the *lever rule* is used. The amounts of each phase are

$$\% \text{ solid} = \frac{n}{w} = \frac{67.8 - 55}{67.8 - 48} = .647 \,(64.7\%)$$
$$\% \text{ liquid} = 100 - 64.7 = 35.3\%$$

The elements of most alloys are not completely miscible in all states and phases. For instance, the elements of a two-component alloy may be perfectly soluble in the liquid state but only partially soluble in the solid state. Moreover, a number of different semi-solid and solid phases can occur, depending on the composition and the temperature of the alloy.

In figure 13.12, the constituents are only perfectly miscible at point C, called the *eutectic point*.[3] The material in area ABC consists of a mixture of crystals of **A** and a liquid solution of **A** and **B**. The eutectic materials will not solidify until the line B-D, the lowest point at which liquid eutectic can exist, is reached. Below the eutectic line, the material will be in the form of a mixture of solid **A** and eutectic crystals. To the right of the eutectic composition line, the inverse relationship will occur with a mixture of **B** crystals and liquid or solid **B** and eutectic crystals, depending on the temperature.

---

[3] The *eutectic point* should not be confused with the *eutectoid point*. A eutectic composition is associated with a liquid-to-solid reaction. A eutectoid composition is associated with a solid-to-solid reaction.

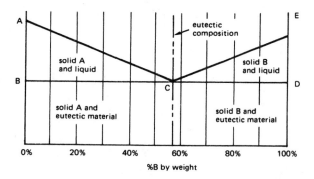

Figure 13.12 Binary Equilibrium Diagram

The *iron-carbon equilibrium* diagram is more complicated. Figure 13.13 is a simplified version.

The following alphabetical list of definitions can be helpful in understanding figure 13.13.

Allotropic changes: Reversible changes which occur in an iron-carbon mixture at the critical points. Compositions remain constant, but properties such as atomic structure, magnetism, and electrical resistance change.

Alpha iron: A BCC (body-centered cubic) structure which exists beneath the A_3 line only. Maximum carbon solubility is 0.03%, the lowest of all forms. Alpha iron is stable from $-460°$ to $1670°F$, quite soft, and strongly magnetic up to $1418°F$.

Austenite: A solid solution of carbon in gamma iron. It is non-magnetic, decomposes on slow cooling, and does not normally exist below $1333°F$. It can be partially preserved through extreme cooling rates.

Bainite: A structure induced through an interrupted quenching process called *austempering*.

Beta iron: A non-magnetic form of alpha iron which exists between $1418°$ and $1670°F$.

Cementite: The chemical compound Fe_3C. It is the hardest of all forms of iron and is quite brittle. Cementite undergoes a magnetic change at the A_0 line.

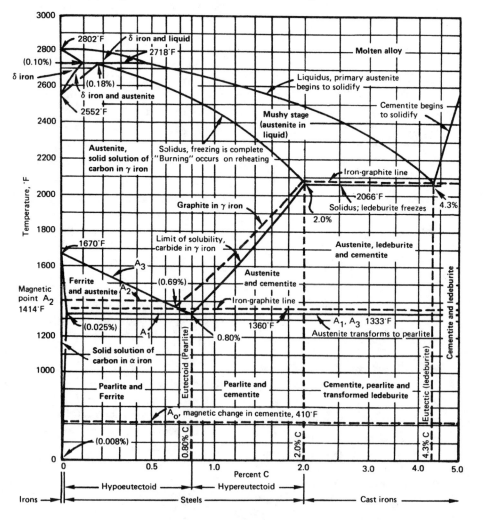

Figure 13.13 Simplified Iron-Carbon Diagram

Critical points: The temperatures at which allotropic changes take place. *Pure iron* has three critical points.

2802°F—liquid iron to delta (BCC) iron
2552°F—delta iron to gamma (FCC) iron
1670°F—gamma iron to alpha (BCC) iron

The critical lines for an iron-carbon mixture are shown in figure 13.13.

- A_0: About 410°F, where cementite becomes non-magnetic.

- A_1: The eutectoid temperature, 1333°F.

- A_2: About 1410°F, below which the proper compositions become magnetic. Also known as the *Curie point*.

- A_3: A temperature dependent on composition. Above A_3, steel is austenite. Below, it is ferrite.

Delta iron: A BCC arrangement existing above 2552°F.

Eutectoid point: 1333°F and .8% carbon, when the iron and carbon are mutually saturated. No free iron (hypoeutectoid) is rejected on cooling austenite from above A_1.

Eutectoid steel: An equilibrium-cooled steel containing .8% carbon, consisting entirely of pearlite.

Ferrite: A solid BCC solution of carbon in alpha iron, generally having very little carbon. Ferrite is magnetic.

Gamma iron: A FCC arrangement of iron atoms which is stable between 1670°F and 2550°F. The maximum carbon solubility is 1.7%, and it is non-magnetic.

Hypereutectoid steel: Steel containing more than .8% carbon, consisting of cementite and pearlite.

Hypoeutectoid steel: Steel containing less than .8% carbon, consisting of ferrite and pearlite.

Iron carbide: See "Cementite."

Martensite: A supersaturated solution of carbon in alpha iron, not formed in an equilibrium process. Results from an operation known as *martempering*.

Pearlite: A result of eutectoid decomposition of austenite into layers of ferrite and cementite.

Table 13.5
Alloying Ingredients in Steel
(XX is the carbon content, 0.XX%)

alloy number	major alloying elements
10XX	plain carbon steel
11XX	resulfurized plain carbon
13XX	manganese
23XX, 25XX	nickel
31XX, 33XX	nickel, chromium
40XX	molybdenum
41XX	chromium, molybdenum
43XX	nickel, chromium, molybdenum
46XX, 48XX	nickel, molybdenum
51XX	chromium
61XX	chromium, vandium
81XX, 86XX, 87XX	nickel, chromium, molybdenum
92XX	silicon

2 HEAT TREATMENTS

Properties of metals can be changed by appropriate heat treatments. Such treatments include heating, rapid cooling, and reheating.

A. STEEL

As low-carbon steel is heated, the grain size remains the same until the A_1 line is reached. Between line A_1 and line A_3, the grain size of austenite in solution decreases. This characteristic is used in heat treatments wherein steel is heated and then cooled rapidly (quenched). Slow heating prevents warping and cracking and allows completion of all phase changes.

The rate of cooling determines the hardness obtained. Rapid quenching in water or brine is necessary to quench low- and medium-carbon steels since low-carbon steels with small amounts of pearlite are difficult to harden. Oil is used to quench high-carbon and alloy steel or parts with non-uniform cross sections because the quenching is less severe. Maximum hardness depends on the carbon content. An upper limit of R_C 66–67 is reached at 0.5% carbon. No increase in hardness can be realized for greater carbon content.

Hardening processes, which are accompanied by a decrease in toughness, consist of heating above line A_3, allowing austenite to form, and then quenching. Hardened steels consist of primarily martensite or bainite. Hardening often is followed by tempering.

Some of the more important heat treatments are listed here in alphabetical order.

Annealing: Heating to just above the critical point and then cooling slowly through the critical range. Refines grain size, softens, and relieves internal stresses.

Austempering: An interrupted quenching process with an austenite-to-bainite transition. Steel is quenched to below 800°F and allowed to reach equilibrium. No martensite is formed, and no tempering is required.

Austenitizing: Quenching after heating above the A_3 line (steel with .8% carbon) or above the A_1 line (high-carbon steel).

Carburizing: Heating for up to 24 hours at about 1650°F in contact with a carbonaceous material followed by rapid cooling. Also known as *cementation*.

Cyaniding: Heating at 1700°F in a cyanide-rich atmosphere to produce a hardened surface.

Flame hardening: Supplying flame heat at the surface in quantities greater than can be conducted into the interior, followed by drastic quenching.

Induction hardening: Using high-frequency electric currents to heat the metal surface, followed by drastic quenching.

Nitriding: Heating for up to 100 hours in an ammonia atmosphere at about 1000°F with slow cooling.

Normalizing: Similar to annealing, but more rapid due to air cooling. Heating is 200°F above critical point.

Tempering: Hypoeutectoid steels are tempered by reheating below the critical temperature after hardening. Tempering changes martensite into pearlite. The higher the temperature, the softer and tougher the steel becomes. Tempering also is known as *drawing* or *toughening*.

Time-temperature transformation (TTT) diagrams have been developed as an aid in designing heat treatment processes. Figure 13.14 is a TTT diagram for high-carbon (.95%) steel.

Curve 1 represents extremely rapid cooling. The transformation begins at 420°F and continues for 8 to 10 seconds, changing all of the austenite to martensite. Curve 2 is a slow quench which converts all of the austenite to fine pearlite. An intermediate curve passing through transition twice would produce combined material (e.g., sorbite and martensite).

If the temperature decreases rapidly to 520°F along curve 1 and then is held constant along curve 3, the structure would be bainite. This is the principle of *austempering*. Performing the same procedure at 350–400°F is *martempering*. Martempering produces a tough and soft steel.

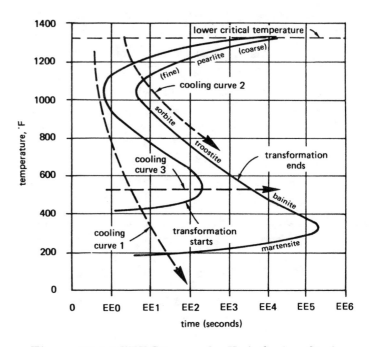

Figure 13.14 TTT Diagram for High-Carbon Steel

B. ALUMINUM

The primary method of heat treating aluminum alloys is *precipitation* or *age hardening*. This involves the formation of a new crystalline structure through the application of controlled quenching and tempering procedures. Precipitation hardening disperses hard particles throughout a ductile matrix. These particles serve to disrupt the long dislocation planes of the matrix, increasing the strength and the stiffness of the alloy. The strength limit is raised to the rupture strength of either the particles or the surrounding matrix.

Solution heat treatment culminates in rapid quenching. Quenching speeds must be consistent with the size of the specimen. Massive specimens may require slower processes using oil or boiling water. Post-treatment cooling for precipitation hardening is relatively unimportant.

Aluminum alloys are identified by a number and a letter (e.g., 2014-T4). The number indicates the chemical composition of the alloy, as determined from table 13.6. The letter indicates the condition of the alloy, as determined from table 13.7.

Table 13.6
Alloying Ingredients in Aluminum

alloy number	major alloying element
1XXX	commercially pure aluminum (99+%)
2XXX	copper
3XXX	manganese
4XXX	silicon
5XXX	magnesium
6XXX	magnesium and silicon
7XXX	zinc
8XXX	other

Table 13.7
Aluminum Alloy Conditions

letter	alloy condition
F	as fabricated
O	soft (after annealing)
H	strain hardened (cold worked)
T	heat treated

The letters *H* and *T* are followed by numbers. These numbers provide additional detail about the type of hardening process used to achieve the material properties. Table 13.8 lists the types of tempers associated with the *T* condition letter.

Table 13.8
Aluminum Tempers

temper	description
T2	annealed (castings only)
T3	solution heat-treated, followed by cold working
T4	solution heat-treated, followed by natural aging
T5	artificial aging
T6	solution heat-treated, followed by artificial aging
T7	solution heat-treated, followed by stabilizing by overaging heat treating
T8	solution heat-treated, followed by cold working and subsequent artificial aging

3 RECRYSTALLIZATION

Recrystallization involves heating a metal specimen to a specific temperature (the *recrystallization temperature*) and holding it there for a long time.[4] This results in the formation and growth of strain-free grains within grains already formed. The grain structure that results is essentially the structure that existed prior to any cold-working (e.g., deep drawing or bending). The new structure is softer and more ductile than the original structure.

Recrystallization can be used with all metals to relieve stresses induced during cold-working. Table 13.9 lists approximate recrystallization temperatures for metals and alloys.

Table 13.9
Approximate Recrystallization Temperatures

material	recrystallization temperature, °F
copper (99.999% pure)	250
(5% zinc)	600
(5% aluminum)	550
(2% beryllium)	700
aluminum (99.999% pure)	175
(99.0+% pure)	550
(alloys)	600
nickel (99.99% pure)	700
(99.4% pure)	1100
(monel metal)	1100
iron (pure)	750
(low-carbon steel)	1000
magnesium (99.99% pure)	150
(alloys)	450
zinc	50
tin	25
lead	25

The recrystallization process is more sensitive to temperature than it is to exposure time. Recrystallization will occur naturally over a wide range of temperatures; however, the reaction rates will vary. The temperatures in table 13.9 will produce complete recrystallization in one hour.

[4]Recrystallization is a specific form of annealing.

PART 4: Corrosion

1 THE CAUSES

Corrosion is a chemical or physical reaction with parts of the environment. Conditions within the aggregate or the crystalline structures can amplify or inhibit corrosion. Three categories of corrosion are galvanic action, stress corrosion, and fretting corrosion.

A. GALVANIC ACTION

Galvanic action results from the electrochemical variation in the potential of metallic ions. If two metals of different potentials are placed in an electrolytic medium, the one with the higher potential will act as an anode and will corrode. The metal with the lower potential, being the cathode, will be unchanged.

Metals often are classified according to their position in the *galvanic series*. Several metals are listed in table 13.10 in the order of their anodic-cathodic characteristics. The exact position of a metal in a galvanic series will depend somewhat on the environment and other conditions, but table 13.10 is accurate for most cases. The greater the separation of two metals on the chart, the greater will be the speed and severity of corrosion.

Table 13.10
The Galvanic Series
(Anodic to Cathodic)

Magnesium alloys
Alclad 3S
Aluminum alloys
Low-carbon steel
Cast iron
Stainless—No. 410
Stainless—No. 430
Stainless—No. 404
Stainless—No. 316
Hastelloy A
Lead-tin alloys
Brass
Copper
Bronze
90/10 Copper-nickel
70/30 Copper-nickel
Inconel
Silver
Stainless steels (passive)
Monel
Hastelloy C
Titanium

B. STRESS CORROSION

When subjected to sustained surface tensile stresses in the presence of corrosive environments, certain metals exhibit *stress corrosion* cracking. This cracking can lead to failure at stresses well below normal working stresses.

Stress corrosion occurs because the more highly-stressed fibers or grains are slightly more anodic than neighboring fibers with lower stresses. The cracks are intergranular and propagate through grain boundaries until failure occurs. One extreme type of stress corrosion, in which open end-grains separate into layers, is called *exfoliation*.

C. FRETTING CORROSION

Fretting corrosion occurs when two highly loaded members have a common surface at which differential movement takes place. The phenomenon is a combination of wear and chemical corrosion. Metals which depend on a surface oxide for protection, such as aluminum, are especially susceptible.

2 PREVENTING CORROSION

In many designs, corrosion can be reduced or eliminated entirely by avoiding conditions conducive to corrosion. Several design principles for avoiding corrosion are available.

- Metals should be chosen on the basis of their potential for corrosion. This requires attention to chemical properties and environment. When two different metals must be in contact, they should be as close together in the galvanic series as possible.

- Protective coatings (e.g., sodium silicate, sodium benzoate, and various organic amines) can be used.

- Dampness should be eliminated. If there is no electrolyte, there can be no corrosion.

- Since each metal exhibits maximum corrosion at a particular pH, acidity or alkalinity of the environment can be controlled.

- An electrical current can be applied to the corrosion circuit to counter the corrosion reaction. This is known as *cathodic protection*.

- *Sacrificial anodes* made from active metals (e.g., magnesium or calcium) can be placed in areas where corrosion would otherwise occur. These anodes intercept the corrosive current (i.e., they give off electrons) and are sacrificed in the process of saving the structure.

PRACTICE PROBLEMS: MATERIALS SCIENCE

Warm-ups

1. Refer to the equilibrium diagram.

(a) For a 4%–96% alloy at temperature T_1, what are the composites of solids α and β?

(b) For a 1%–99% alloy at temperature T_2, how much liquid and how much solid are present?

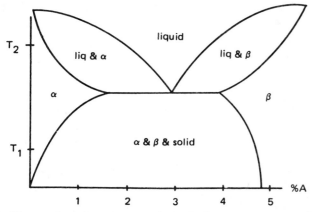

2. The engineering stress and strain for a copper sample (Poisson's ratio = .3) were 20,000 psi and .0200 in/in, respectively. What were the true stress and true strain?

3. An engineering stress-strain curve is shown. Find the .5% parallel offset yield strength, the elastic modulus, the ultimate strength, the fracture strength, and the % elongation at fracture.

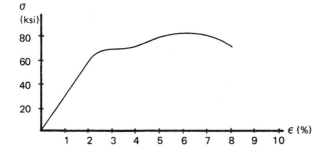

4. What is the shear modulus in problem 3? Assume Poisson's ratio is .3.

5. A 4 in² sample necks down to 3.42 in² before breaking in a tensile test. What is the sample's ductility?

6. What is the toughness of the material in problem 3?

7. The steady state creep rate for a sample is evaluated. A 15,000 psi stress is applied, and the elongation is measured at $t = 5$ hours, 10 hours, 20 hours, ... 70 hours. Measurements were .018, .022, .026, .031, .035, .040, .046, .058. What is the steady state creep rate?

8. How much HCl should be used as an initiator in PVC if the efficiency is 20% and an average molecular weight of 7000 grams/gmole is desired? The final polymer has the following structure.

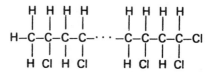

9. 10 ml of a .2% solution (by weight) of hydrogen peroxide is added to 12 grams of ethylene to stabilize the polymer. What is the average degree of polymerization if the hydrogen peroxide is completely utilized? Assume that hydrogen peroxide breaks down according to

$$H_2O_2 \mapsto 2(OH)^- + \cdots$$

Assume that the stabilized polymer has the following structure.

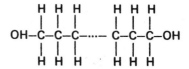

RESERVED FOR FUTURE USE

14 MECHANICS OF MATERIALS

PART 1: Strength of Materials

Nomenclature

A	area	in^2
b	width	in
c	distance from neutral axis to extreme fiber	in
C	end restraint coefficient, or correction	--
D	diameter	in
e	eccentricity	in
E	modulus of elasticity	psi
F	force, or load	lbf
F.S.	factor of safety	--
g	local gravitational acceleration	ft/sec^2
g_c	gravitational constant (32.2)	$\dfrac{\text{lbm-ft}}{\text{lbf-sec}^2}$
G	shear modulus	psi
I	moment of inertia	in^4
J	polar moment of inertia	in^4
k	radius of gyration, or spring constant	in, lbf/in
K	stress concentration factor	--
L	length	in
m	mass	lbm
M	moment	in-lbf
n	ratio, rotational speed, or number	--, rpm, --
N	number of cycles	--
p	pressure	psi
Q	statical moment	in^3
r	radius	in
S	strength, or axial load	psi, lbf
t	thickness	in
T	temperature, or torque	°F, in-lbf
u	virtual truss load	lbf
U	energy	in-lbf
V	shear, or volume	lbf, in^3
w	load per unit length, or width	lbf/in, in
W	work	in-lbf
x	distance, or displacement	in
y	deflection, or distance	in
Z	section modulus	in^3

Symbols

δ	elongation, or displacement	in
θ	angle	degrees
ϕ	angle	radians
σ	normal stress	psi
α	coefficient of linear thermal expansion	1/°F
β	coefficient of volumetric thermal expansion	1/°F
γ	coefficient of area thermal expansion	1/°F
τ	shear stress	psi
ϵ	strain	--
μ	Poisson's ratio	--

Subscripts

a	allowable
b	bending
br	bearing
c	centroidal, or compressive
e	endurance, Euler, or equivalent
ext	external
h	hoop
i	inside
L	long
o	original, or outside
p	pull
s	shear
t	transformed, tension, or temperature
th	thermal
T	torsion
u	ultimate
y	yield

1 PROPERTIES OF STRUCTURAL MATE-RIALS

A. THE TENSILE TEST

Many material properties can be derived from the standard tensile test. In a tensile test, a material sample is loaded axially in tension, and the elongation is measured as the load is increased. A graphical representation of typical test data for steel is shown in figure 14.1, in which the elongation, δ, is plotted against the applied load, F.

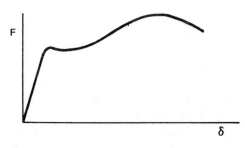

Figure 14.1 Typical Tensile Test Results for Steel

Since this graph is applicable only to an object with the same length and area as the test sample, the data are converted to *stresses* and *strains* by use of equations 14.1 and 14.2. σ is known as the *normal stress*, and ϵ is known as the *strain.* Strain is the percentage elongation of the sample.

$$\sigma = \frac{F}{A} \qquad 14.1$$

$$\epsilon = \frac{\delta}{L} \qquad 14.2$$

The stress-strain data also can be graphed, and the shape of the resulting curve will be the same as figure 14.1 with the scales changed.

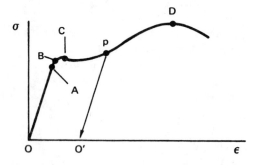

Figure 14.2 A Typical Stress-Strain Curve for Steel

The line O-A in figure 14.2 is a straight line. The relationship between the stress and the strain is given by *Hooke's law*, equation 14.3. E is the *modulus of elasticity (Young's modulus)* and is the slope of the line segment O-A. The stress at point A is known as the *proportionality limit.* The modulus of elasticity for steel is approximately 3 EE7 psi.

$$\sigma = E\epsilon \qquad 14.3$$

Slightly above the proportionality limit is the *elastic limit* (point B). As long as the stress is kept below the elastic limit, there will be no permanent strain when the applied stress is removed. The strain is said to be *elastic*, and the stress is said to be in the *elastic region.*

If the elastic limit stress is exceeded before the load is removed, recovery will be along a line parallel to the straight line portion of the curve, as shown in the line segment p-O'. The strain that results (line O-O') is permanent and is known as *plastic strain* or *permanent set.*

The *yield point* (point C) is very close to the elastic limit. For all practical purposes, the *yield stress*, S_y, can be taken as the stress which accompanies the beginning of plastic strain. Since permanent deformation is to be avoided, the yield stress is used in calculating safe stresses in ductile materials such as steel. A36 structural steel has a minimum yield strength of 36,000 psi.

$$\sigma_a = \frac{S_y}{F.S.} \qquad 14.4$$

Some materials, such as aluminum, do not have a well-defined yield point. This is illustrated in figure 14.3. In such cases, the yield point is taken as the stress which will cause a *.2% parallel offset.*

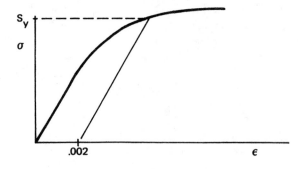

Figure 14.3 A Typical Stress-Strain Curve for Aluminum

The *ultimate tensile strength*, point D in figure 14.2, is the maximum load-carrying ability of the material. However, since stresses near the ultimate strength are accompanied by large plastic strains, this parameter should not be used for the design of ductile materials such as steel and aluminum.

As the sample is elongated during a tensile test, it also will decrease in thickness (width or diameter). The ratio of the lateral strain to the axial strain is known as *Poisson's ratio*, μ. μ typically is taken as .3 for steel and as .33 for aluminum.

$$\mu = \frac{\epsilon_{\text{lateral}}}{\epsilon_{\text{axial}}} = \frac{\frac{\Delta D}{D_o}}{\frac{\Delta L}{L_o}} \qquad 14.5$$

B. FATIGUE TESTS

A part may fail after repeated stress loading even if the stress never exceeds the ultimate fracture strength of the material. This type of failure is known as *fatigue failure.*

The behavior of a material under repeated loadings can be evaluated in a fatigue test. A sample is loaded repeatedly to a known stress, and the number of applications of that stress is counted until the sample fails. This procedure is repeated for different stress levels. The results of many of these tests can be graphed, as is done in figure 14.4.

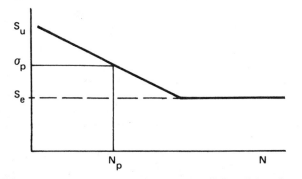

Figure 14.4 Results of Many Fatigue Tests for Steel

For any given stress level, say σ_p in figure 14.4, the corresponding number of applications of the stress which will cause failure is known as the *fatigue life.* That is, the fatigue life is just the number of cycles of stress required to cause failure. If the material is to fail after only one application of stress, the required stress must equal or exceed the ultimate strength of the material.

Below a certain stress level, called the *endurance limit* or the *endurance strength*, the part will be able to withstand an infinite number of stress applications without experiencing failure. Therefore, if a dynamically loaded part is to have an infinite life, the applied stress must be kept below the endurance limit.

Some materials, such as aluminum, do not have a well-defined endurance limit. In such cases, the endurance limit is taken as the stress that will cause failure at EE8 or 5 EE8 applications of the stress.

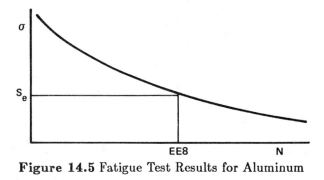

Figure 14.5 Fatigue Test Results for Aluminum

C. ESTIMATES OF MATERIAL PROPERTIES

Although the properties of a material will depend on its classification (ASTM, AISC, etc.), average values are given in table 14.1.

2 DEFORMATION UNDER LOADING

Equation 14.2 can be rearranged to give the elongation of an axially loaded member in compression or tension.

$$\delta = L\epsilon = \frac{L\sigma}{E} = \frac{LF}{AE} \qquad 14.6$$

A tension load is taken as positive, and a compressive load is taken as negative. The actual length of a member under loading is given by equation 14.7 where the algebraic sign of the deformation must be observed.

$$L_{\text{actual}} = L_o + \delta \qquad 14.7$$

The energy stored in a loaded member is equal to the work required to deform it. Below the proportionality limit, this energy is given by equation 14.8.

$$U = \tfrac{1}{2}F\delta = \tfrac{1}{2}\left(\frac{F^2 L}{AE}\right) \qquad 14.8$$

Table 14.1
Typical Material Properties

material	E (psi)	G (psi)	μ	ρ (pcf)	α (1/°F)
steel (hard)	30 EE6	11.5 EE6	.30	489	6.5 EE−6
steel (soft)	29 EE6	11.5 EE6	.30	489	6.5 EE−6
aluminum alloy	10 EE6	3.9 EE6	.33	173	12.8 EE−6
magnesium alloy	6.5 EE6	2.4 EE6	.35	112	14.5 EE−6
titanium alloy	15.4 EE6	6.0 EE6	.34	282	4.9 EE−6
cast iron	20 EE6	8 EE6	.27	442	5.6 EE−6
phosphor bronze	**16 EE6**	**6.0 EE6**	**.33**	**548**	**10.2 EE−6**

3 THERMAL DEFORMATION

If the temperature of an object is changed, the object will experience length, area, and volume changes. These changes can be predicted by equations 14.9, 14.10, and 14.11.

$$\Delta L = \alpha L_o (T_2 - T_1) \qquad\qquad 14.9$$

$$\Delta A = \gamma A_o (T_2 - T_1) \approx 2\alpha A_o (T_2 - T_1) \qquad 14.10$$

$$\Delta V = \beta V_o (T_2 - T_1) \approx 3\alpha V_o (T_2 - T_1) \qquad 14.11$$

If equation 14.9 is rearranged, an expression for the *thermal strain* is obtained. Thermal strain is handled in the same manner as strain due to an applied load.

$$\epsilon_{th} = \frac{\Delta L}{L_o} = \alpha (T_2 - T_1) \qquad\qquad 14.12$$

For example, if a bar is heated but is not allowed to expand, the stress will be given by equation 14.13.

$$\sigma_{th} = E\,\epsilon_{th} \qquad\qquad 14.13$$

4 SHEAR AND MOMENT DIAGRAMS

It was illustrated in chapter 12 that, for an object in equilibrium, the sums of forces and moments are equal to zero everywhere. For example, the sum of moments about point A for the beam shown in figure 14.6 is zero.

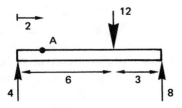

Figure 14.6 A Beam in Equilibrium

Nevertheless, the beam shown in figure 14.6 will bend under the influence of the forces. This bending is evidence of the stress experienced by the beam. Since the sum of moments about any point is zero, the moment used to find stresses and deflection is taken from the point in question to one end of the beam only. This is called the *one-way moment*. The absolute value of the moment will not depend on the end used. This can be illustrated by the beam shown in figure 14.6.

$$\sum M_{A\,(to\ right\ end)} = -8(7) + 4(12) = -8$$
$$\sum M_{A\,(to\ left\ end)} = 4(2) = +8$$

The moment obtained will depend on the location chosen. A graphical representation of the one-way moment at every point along a beam is known as a *moment diagram*. The following guidelines should be observed in constructing moment diagrams.

- Moments should be taken from the left end to the point in question. If the beam is cantilever, place the built-in end at the right.

- Clockwise moments are positive. The *left-hand rule* should be used to determine positive moments.

- Concentrated loads produce linearly increasing lines on the moment diagram.

- Uniformly distributed loads produce parabolic lines on the moment diagram.

- The maximum moment will occur when the shear (V) is zero.

- The moment at any point is equal to the area under the shear diagram up to that point. That is,

$$M = \int V\,dx \qquad\qquad 14.14$$

- The moment is zero at a free end or hinge.

Similarly, the sum of forces in the y direction on a beam in equilibrium is zero. However, the shearing stress at a point along the beam will depend on the sum of forces and reactions from the point in question to one end only.

A *shear diagram* is drawn to represent graphically the shear at any point along a beam. The following guidelines should be observed in constructing a shear diagram.

- Loads and reactions acting up are positive.

- The shear at any point is equal to the sum of the loads and reactions from the left end to the point in question.

- Concentrated loads produce straight (horizontal) lines on the shear diagram.

- Uniformly distributed loads produce straight sloping lines on the shear diagram.

- The magnitude of the shear at any point is equal to the slope of the moment diagram at that point.

$$V = \frac{dM}{dx} \qquad\qquad 14.15$$

Example 14.1

Draw the shear and moment diagrams for the following beam.

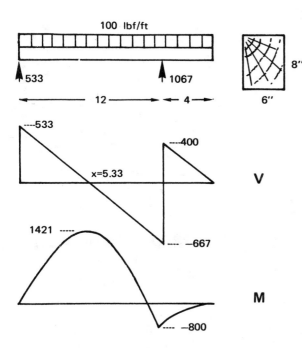

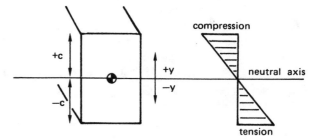

Figure 14.8 Bending Stress Distribution in a Beam

The moment, M, used in equation 14.17 is the *one-way moment* previously discussed. I_c is the centroidal moment of inertia of the beam's cross sectional area. The negative sign in equation 14.17 typically is omitted. However, it is required to be consistent with the convention that compression is negative.

Since the maximum stress will govern the design, y can be set equal to c to obtain the maximum stress. c is the distance from the neutral axis to the *extreme fiber*.

$$\sigma_{b,\max} = \frac{Mc}{I_c} \qquad 14.18$$

For any given structural shape, c and I_c are fixed. Therefore, these two terms can be combined into the *section modulus, Z*.

$$\sigma_{b,\max} = \frac{M}{Z} \qquad 14.19$$

$$Z = \frac{I_c}{c} \qquad 14.20$$

For most beams, the section modulus, Z, is constant along the length of the beam. Equation 14.19 shows that the maximum stress along the length of a beam is proportional to the moment at that point. The location of the maximum bending moment is called the *dangerous section*. The dangerous section can be found directly from a moment or shear diagram of the beam.

If an axial member is loaded eccentrically, it will experience axial stress (equation 14.16) as well as bending stress (equation 14.17). This is illustrated by figure 14.9, in which a load is not applied to the centroid of a column's cross sectional area.

Because the beam bends and supports a compressive load, the stress produced is a sum of bending and normal stress.

$$\sigma_{\max,\ \min} = \frac{F}{A} \pm \frac{Mc}{I_c} = \frac{F}{A} \pm \frac{Fec}{I_c} \qquad 14.21$$

If a cross section is loaded with an eccentric compressive load, part of the section can be in tension. This is illustrated in example 14.3. There will be no stress sign reversal, however, as long as the load is applied within a diamond-shaped area formed from the middle-

5 STRESSES IN BEAMS

A. NORMAL STRESS

Normal stress is the type of stress experienced by a member which is axially loaded. The normal stress is the load divided by the area.

$$\sigma = \frac{F}{A} \qquad 14.16$$

Normal stress also occurs when a beam bends, as shown in figure 14.7. The lower part of the beam experiences normal tensile stress (which causes lengthening). The upper part of the beam experiences a normal compressive stress (which causes shortening). There is no stress along a horizontal plane passing through the centroid of the cross section. This plane is known as the *neutral plane* or the *neutral axis.*

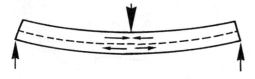

Figure 14.7 Normal Stress Due to Bending

Although it is a normal stress, the stress produced by the bending usually is called *bending stress* or *flexure stress.* Bending stress varies with position within the beam. It is zero at the neutral axis, but it increases linearly with distance from the neutral axis.

$$\sigma_b = \frac{-My}{I_c} \qquad 14.17$$

thirds of the centroidal axis. This area is know as the *kern* or the *kernel*. It is particularly important to keep eccentric compressive loads within the kern on concrete and masonry piers since these materials do not tolerate tension. (The kern of a circular shaft or a round beam is outlined by a circle whose radius is one-quarter of the shaft radius.)

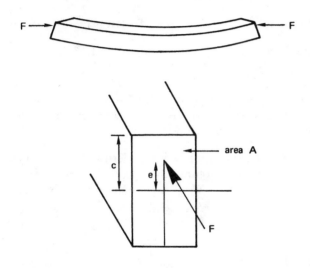

Figure 14.9 Eccentric Loading of an Axial Member

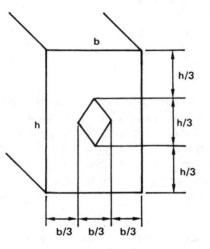

Figure 14.10 The Kern

The *elastic strain energy* stored in a beam experiencing a moment (bending) is

$$U = \tfrac{1}{2} \int \frac{M^2}{EI} dx \qquad 14.22$$

B. SHEAR STRESS

Normal stress is produced when a load is absorbed by an area normal to it. *Shear stress* is produced by a load being carried by an area parallel to the load. This is illustrated in figure 14.11.

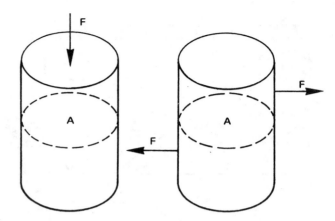

Figure 14.11 Normal and Shear Stresses

The average shear stress experienced by a pin, a bolt, or a rivet in single shear (as illustrated in figure 14.11) is given by equation 14.23. Because it gives an average value over the cross section of the shear member, this equation should be used only when the loading is low or when there is multiple redundancy in the shear group.

$$\tau = \frac{F}{A} \qquad 14.23$$

The actual shear stress in a beam is dependent on the location within the beam, just as was the bending stress. Shear stress is zero at the top and bottom surfaces of a beam and maximum at the neutral axis. This is illustrated in figure 14.12.

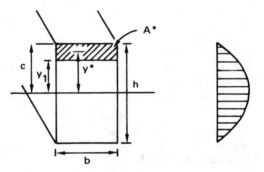

Figure 14.12 Shear Stress Distribution in a Rectangular Beam

The shear stress distribution within a beam is given by equation 14.24.

$$\tau = \frac{QV}{Ib} \qquad 14.24$$

V is the shear (in pounds) at the section where the shear stress is wanted. V can be found from a shear diagram. I is the beam's centroidal moment of inertia. b is the width of the beam at the depth y_1 within the beam where the shear stress is wanted. Q is the *statical moment*[1], as defined by equation 14.25.

[1]The statical moment also is known as the *first moment of the area*.

$$Q = \int_{y_1}^{c} y \, dA \qquad 14.25$$

For rectangular beams, $dA = b \, dy$. Equation 14.25 can be simplified to equation 14.26 for rectangular beams.

$$Q = y^* A^* \qquad 14.26$$

Equation 14.26 says that the statical moment at a location y_1 within a rectangular beam is equal to the product of the area above y_1 and the distance from the centroidal axis to the centroid of A^*.

The maximum shear stress in a rectangular beam is

$$\tau_{max} = \frac{3V}{2A} = \frac{3V}{2bh} \qquad 14.27$$

For a round beam of radius r and area A, the maximum shear stress is

$$\tau_{max} = \frac{4V}{3A} = \frac{4V}{3\pi r^2} \qquad 14.28$$

The shear, V, used in equations 14.27 and 14.28 is the *one-way shear.*

Example 14.2

What are the maximum shear and bending stresses for the beam shown in example 14.1?

From the shear diagram, the maximum shear is -667 pounds. From equation 14.27, the maximum shear stress is

$$\tau_{max} = \frac{(3)(-667)}{(2)(6)(8)} = -20.8 \text{ psi}$$

From the moment diagram, the maximum moment is $+1421$ ft-lbf. The centroidal moment of inertia is

$$I_c = \frac{(6)(8)^3}{12} = 256 \text{ in}^4$$

From equation 14.18, the maximum bending stress is

$$\sigma_{b,max} = \frac{(1421)(12)(4)}{256} = 266.4 \text{ psi}$$

Example 14.3

The chain hook shown carries a load of 500 pounds. What are the minimum and maximum stresses in the vertical portion of the hook?

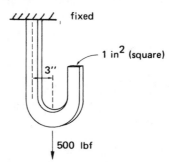

The hook is loaded eccentrically because the load and the supporting force are not in line. The centroidal moment of inertia of the $1'' \times 1''$ section is

$$I_c = \frac{bh^3}{12} = \frac{(1)(1)^3}{12} = .0833 \text{ in}^4$$

From equation 14.21,

$$\begin{aligned} \sigma_{max, \, min} &= \frac{500}{1} \pm \frac{(500)(3)(.5)}{.0833} \\ &= 500 \pm 9{,}000 \\ &= +9{,}500 \text{ and } -8{,}500 \end{aligned}$$

The 500 psi direct stress is tensile. However, the flexural compressive stress of 9,000 psi counteracts this tensile stress, resulting in 8,500 psi compressive stress at the outer face of the hook. The stress is 9,500 psi tension at the inner face.

6 STRESSES IN COMPOSITE STRUCTURES

A *composite structure* is one in which two or more different materials are used, each carrying a part of the load. Unless all the various materials used have the same modulus of elasticity, the stress analysis will be dependent on the assumptions made.

Some simple composite structures can be analyzed using the assumption of *consistent deformations*. This is illustrated in examples 14.4 and 14.5. The technique used to analyze structures for which the strains are consistent is known as the *transformation method*.

step 1: Determine the modulus of elasticity for each of the materials used in the structure.

step 2: For each of the materials used, calculate the ratio

$$n = \frac{E}{E_{weakest}} \qquad 14.29$$

$E_{weakest}$ is the smallest modulus of elasticity of any of the materials used in the composite structure.

step 3: For all of the materials except the weakest, multiply the actual material stress area by n. Consider this expanded (*transformed*) area to have the same composition as the weakest material.

step 4: If the structure is a tension or compression member, the distribution or placement of the transformed area is not important. Just assume that the transformed areas carry the axial load. For beams in bending, the transformed area can add to the width of the beam, but it cannot change the depth of the beam or the thickness of the reinforcing.

step 5: **For compression or tension members, calculate the stresses in the weakest and stronger materials.**

$$\sigma_{weakest} = \frac{F}{A_t} \qquad 14.30$$

$$\sigma_{stronger} = \frac{nF}{A_t} \qquad 14.31$$

step 6: For beams in bending, proceed through step 9. Find the centroid of the transformed beam.

step 7: Find the centroidal moment of inertia of the transformed beam, I_{ct}.

step 8: Find V_{max} and M_{max} by inspection or from the shear and moment diagrams.

step 9: Calculate the stresses in the weakest and stronger materials.

$$\sigma_{weakest} = \frac{M c_{weakest}}{I_{ct}} \qquad 14.32$$

$$\sigma_{stronger} = \frac{n M c_{stronger}}{I_{ct}} \qquad 14.33$$

Example 14.4

Find the stress in the steel inner cylinder and the copper tube which surrounds it if a uniform compressive load of 100 kips is applied axially. The copper and the steel are well bonded. Use $E_{steel} = 3$ EE7 psi and $E_{copper} = 1.75$ EE7 psi.

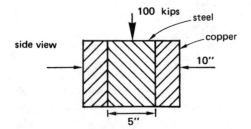

$$n = \frac{3 \text{ EE7}}{1.75 \text{ EE7}} = 1.714$$

The actual steel area is $\frac{1}{4}\pi(5)^2 = 19.63$ in².

The actual copper area is $\frac{1}{4}\pi[(10)^2 - (5)^2] = 58.9$ in².

The transformed area is $A_t = 58.9 + 1.714(19.63) = 92.55$ in².

$$\sigma_{copper} = \frac{100,000}{92.55} = 1080.5 \text{ psi}$$
$$\sigma_{steel} = (1.714)(1080.5) = 1852.0 \text{ psi}$$

Example 14.5

Find the maximum bending stress in the steel-reinforced wood beam shown at a point where the moment is 40,000 ft-lbf. Use $E_{steel} = 3$ EE7 psi and $E_{wood} = 1.5$ EE6 psi.

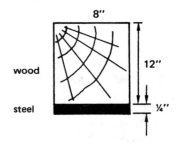

$$n = \frac{3 \text{ EE7}}{1.5 \text{ EE6}} = 20$$

The actual steel area is $(.25)(8) = 2$.

The area of the steel is expanded to $20(2) = 40$. Since the depth of beam and reinforcement cannot be increased, the width must increase. The 160″ dimension is arrived at by dividing the area of 40 square inches by the thickness of $\frac{1}{4}″$.

(Figure on next page.)

The centroid is located at $\bar{y} = 4.45$ inches from the x-x axis. The centroidal moment of inertia of the transformed section is $I_c = 2211.5$ in⁴. Then, from equations 14.32 and 14.33,

$$\sigma_{max, wood} = \frac{(40,000)(12)(7.8)}{(2211.5)} = 1692 \text{ psi}$$

$$\sigma_{max, steel} = \frac{(20)(40,000)(12)(4.45)}{2211.5} = 19,320 \text{ psi}$$

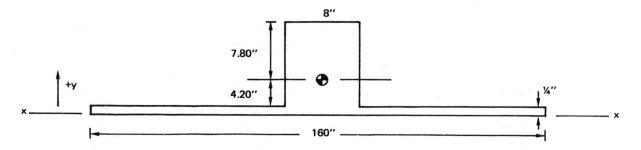

7 ALLOWABLE STRESSES

Once the actual stresses are known, they must be compared to allowable stresses. If the allowable stress is calculated, it should be based on the yield stress and a reasonable factor of safety. This is known as the *allowable stress design method* or the *working stress design method.*

$$\sigma_a = \frac{S_y}{F.S.} \qquad 14.34$$

For steel, the factor of safety ranges from 1.5 to 2.5, depending on the type of steel and application.

The allowable stress method is being replaced in structural work by the *load factor design method,* also known as the *ultimate strength method* and the *plastic design method.* In this method, the applied loads are multiplied by a load factor. The product must be less than the structural member's ultimate strength, usually determined from a table.

8 BEAM DEFLECTIONS

A. DOUBLE INTEGRATION METHOD

The deflection and the slope of a loaded beam are related to the applied moment and shear by equations 14.35 through 14.38.

$$y = \text{deflection} \qquad 14.35$$

$$y' = \frac{dy}{dx} = \text{slope} \qquad 14.36$$

$$y'' = \frac{d^2y}{dx^2} = \frac{M}{EI} \qquad 14.37$$

$$y''' = \frac{d^3y}{dx^3} = \frac{V}{EI} \qquad 14.38$$

If the moment function, M(x), is known for a section of the beam, the deflection at any point can be found from equation 14.39.

$$y = \frac{1}{EI} \int \left[\int M(x)dx \right] dx \qquad 14.39$$

In order to find the deflection, constants must be introduced during the integration process. These constants can be found from table 14.2.

Table 14.2
Beam Boundary Conditions

end condition	y	y'	y''	V	M
simple support	0				
built-in support	0	0			
free end			0	0	0
hinge					0

Example 14.6

Find the tip deflection of the beam shown. *EI* is 5 EE10 lbf-in^2 everywhere on the beam.

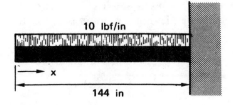

The moment at any point x from the left end of the beam is

$$M(x) = (-10)(x)(\tfrac{1}{2}x) = -5x^2$$

This is negative by the left-hand rule convention. From equation 14.37,

$$y'' = \frac{M}{EI}$$

So,

$$EIy'' = -5x^2$$

$$EIy' = \int -5x^2 \, dx = -\tfrac{5}{3}x^3 + C_1$$

Since $y' = 0$ at a built-in support (table 14.2) and $x = 144$ inches at the built-in support,

$$0 = -\tfrac{5}{3}(144)^3 + C_1$$

$$C_1 = 4.98 \text{ EE6}$$

$$EIy = \int (-\tfrac{5}{3}x^3 + 4.98 \text{ EE6}) \, dx$$

$$= -\tfrac{5}{12}x^4 + (4.98 \text{ EE6})x + C_2$$

Again, $y = 0$ at $x = 144$, so $C_2 = -5.38$ EE8.

Therefore, the deflection as a function of x is

$$y = \left(\frac{1}{EI}\right)\left[(-\tfrac{5}{12})x^4 + (4.98\ EE6)x - 5.38\ EE8\right]$$

At the tip $x = 0$, so the deflection is

$$y_{\text{tip}} = \frac{-5.38\ EE8}{5\ EE10} = -.0108\ \text{inches}$$

B. MOMENT AREA METHOD

The moment area method is a semi-graphical technique which is applicable whenever slopes of deflection beams are not too great. This method is based on the following two theorems.

Theorem I: The angle between tangents at any two points on the *elastic line* of a beam is equal to the area of the moment diagram between the two points divided by EI. That is,

$$\theta = \int \frac{M(x)\,dx}{EI}\qquad\qquad 14.40$$

Theorem II: One point's deflection away from the tangent of another point is equal to the *statical moment* of the bending moment between those two points divided by EI. That is,

$$y = \int \frac{xM(x)\,dx}{EI}\qquad\qquad 14.41$$

The application of these two theorems is aided by the following two comments.

- If EI is constant, the statical moment $\int xM(x)\,dx$ can be calculated as the product of the total moment diagram area times the distance from the point whose deflection is wanted to the centroid of the moment diagram.

- If the moment diagram has positive and negative parts (areas above and below the zero line), the statical moment should be taken as the sum of two products, one for each part of the moment diagram.

Example 14.7

Find the deflection, y, and the angle, θ, at the free end of the cantilever beam shown.

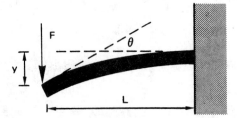

The deflection angle, θ, is the angle between the tangents at the free and built-in ends (Theorem I). The moment diagram is

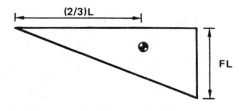

The area of the moment diagram is

$$\tfrac{1}{2}(FL)(L) = \tfrac{1}{2}FL^2$$

From Theorem I,

$$\theta = \frac{FL^2}{2EI}$$

From Theorem II,

$$y = \frac{FL^2}{2EI}\left(\tfrac{2}{3}L\right) = \frac{FL^3}{3EI}$$

Example 14.8

Find the deflection of the free end of the cantilever beam shown.

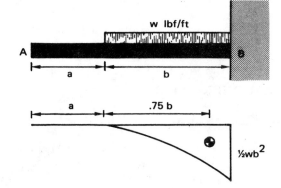

The distance from point A (where the deflection is wanted) to the centroid is $(a + .75b)$. The area of the moment diagram is $(wb^3/6)$. From Theorem II,

$$y = \frac{wb^3}{6EI}(a + .75b)$$

C. STRAIN ENERGY METHOD

The deflection at a point of load application can be found by the strain energy method. This method equates the external work to the total internal strain energy as given by equations 14.8, 14.22, and 14.73. Since work is a force moving through a distance (which in this case is the deflection) we can write equation 14.42.

$$\tfrac{1}{2}Fy = \sum U\qquad\qquad 14.42$$

Example 14.9

Find the deflection at the tip of the stepped beam shown.

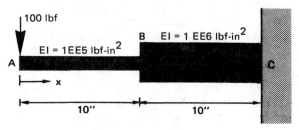

In section A-B: $M = 100x$ in-lbf

From equation 14.22,

$$U = \tfrac{1}{2} \int_0^{10} \frac{(100x)^2}{1\ \text{EE5}}\, dx = 16.67 \text{ in-lbf}$$

In section B-C: $M = 100x$

$$U = \tfrac{1}{2} \int_{10}^{20} \frac{(100x)^2}{1\ \text{EE6}}\, dx = 11.67 \text{ in-lbf}$$

Equating the internal work (U) and the external work,

$$16.67 + 11.67 = \tfrac{1}{2}(100)y$$
$$y = .567 \text{ in}$$

D. CONJUGATE BEAM METHOD

The *conjugate beam method* changes a deflection problem into one of drawing moment diagrams. The method has the advantage of being able to handle beams of varying cross sections and materials. It has the disadvantage of not easily being able to handle beams with two built-in ends. The following steps constitute the conjugate beam method.

step 1:   Draw the moment diagram for the beam as it is actually loaded.

step 2:   Construct the M/EI diagram by dividing the value of M at every point along the beam by the product of EI at that point. If the beam is of constant cross section, EI will be constant, and the M/EI diagram will have the same shape as the moment diagram. However, if the beam cross section varies with x, I will change. In that case, the M/EI diagram will not look the same as the moment diagram.

step 3:   Draw a conjugate beam of the same length as the original beam. The material and the cross sectional area of this conjugate beam are not relevant.

(a) If the actual beam is simply supported at its ends, the conjugate beam will be simply supported at its ends.

(b) If the actual beam is simply supported away from its ends, the conjugate beam has hinges at the support points.

(c) If the actual beam has free ends, the conjugate beam has built-in ends.

(d) If the actual beam has built-in ends, the conjugate beam has free ends.

step 4:   Load the conjugate beam with the M/EI diagram. Find the conjugate reactions by methods of statics. Use the superscript, *, to indicate conjugate parameters.

step 5:   Find the conjugate moment at the point where the deflection is wanted. The deflection is numerically equal to the moment as calculated from the conjugate beam forces.

Example 14.10

Find the deflections at the two load points. EI has a constant value of 2.356 EE7 lbf-in².

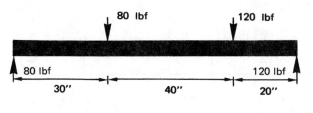

step 1:   The moment diagram for the actual beam is

steps 2, 3, and 4:   Since the cross section is constant, the conjugate load has the same shape as the original moment diagram. The peak load on the conjugate beam is

$$\frac{2400 \text{ in-lbf}}{2.356 \text{ EE7 lbf-in}^2} = 1.019 \text{ EE} - 4\ (1/\text{in})$$

The conjugate reaction, L^*, is found by the following method. The loading diagram is assumed to be made up of a rectangular load and two "negative" triangular loads. The area of the rectangular load (which has a centroid at $x^* = 45$) is $(90)(1.019\ \text{EE}{-}4) = 9.171\ \text{EE}{-}3$.

Similarly, the area of the left triangle (which has a centroid at $x^* = 10$) is $\frac{1}{2}(30)(1.019\ \text{EE}-4) = 1.529\ \text{EE}{-}3$. The area of the right triangle (which has a centroid at $x^* = 83.33$) is $\frac{1}{2}(20)(1.019\ \text{EE}-4) = 1.019\ \text{EE}-3$.

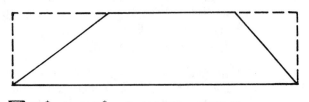

$$\sum M_{L^*}^* = 90R^* + (1.019\ \text{EE}-3)(83.3)$$
$$+ (1.529\ \text{EE}-3)(10) - (9.171\ \text{EE}-3)(45)$$
$$= 0$$
$$R^* = 3.472\ \text{EE}-3\,\frac{1}{in}$$

Then,

$$L^* = (9.171 - 1.019 - 1.529 - 3.472)\text{EE}-3$$
$$= 3.151\ \text{EE}-3\,\frac{1}{in}$$

step 5:   The conjugate moment at $x^* = 30$ is

$$M^* = (3.151\ \text{EE}-3)(30) + (1.529\ \text{EE}-3)(30-10)$$
$$- (9.171\ \text{EE}-3)\left(\frac{30}{90}\right)(15)$$
$$= 7.926\ \text{EE}{-}2\ \text{in}$$

The conjugate moment at the right-most load is

$$M^* = (3.472\ \text{EE}-3)(20) + (1.019\ \text{EE}-3)(13.3)$$
$$- (9.171\ \text{EE}-3)\left(\frac{20}{90}\right)(10)$$
$$= 6.266\ \text{EE}-2\ \text{in}$$

E. TABLE LOOK-UP METHOD

Appendix A is an extensive listing of the most commonly needed beam formulas. The use of these formulas is recommended whenever they can be applied singly or as part of a superposition solution.

F. METHOD OF SUPERPOSITION

If the deflection at a point is due to the combined action of two or more loads, the deflections at that point due to the individual loads can be added to find the total deflection.

9 TRUSS DEFLECTIONS

A. STRAIN-ENERGY METHOD

The deflection of a truss at the point of a single load application can be found by the *strain-energy method* if all member forces are known. This method is illustrated by example 14.11.

Example 14.11

Find the vertical deflection of point A under the external load of 707 pounds. $AE = 10\ \text{EE}5$ pounds for all members. The internal forces have been determined.

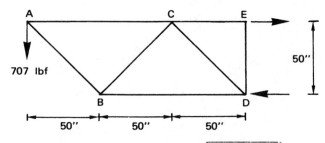

The length of member AB is $\sqrt{(50)^2 + (50)^2} = 70.7$ inches. From equation 14.8, the internal strain energy in member AB is

$$U = \frac{(-1000)^2(70.7)}{2\,(10\ \text{EE}5)} = 35.4\ \text{in-lbf}$$

Similarly, the energy in all members can be determined.

Member	L	F	U
AB	70.7	−1000	+35.4
BC	70.7	+1000	+35.4
AC	100	+707	+25.0
BD	100	−1414	+100.0
CD	70.7	−1000	+35.4
CE	50	+2121	+112.5
DE	50	+707	+12.5
			356.2

The external work is $W_{\text{ext}} = \frac{1}{2}(707)y$, so

$$(\tfrac{1}{2})(707)y = 356.2$$
$$y = 1\ \text{inch}$$

B. VIRTUAL WORK METHOD (HARDY CROSS METHOD)

An extension of the strain-energy method results in an easy procedure for computing the deflection of *any* point on a truss.

step 1:   Draw the truss twice.

step 2:   On the first truss, place all the actual loads.

step 3:   Find the forces, S, due to the actual applied loads in all the members.

step 4: On the second truss, place a dummy one pound load in the direction of the desired displacement.

step 5: Find the forces, u, due to the one pound dummy load in all members.

step 6: Find the desired displacement from equation 14.43.

$$\delta = \sum \frac{SuL}{AE} \qquad 14.43$$

In equation 14.43, the summation is over all truss members which have non-zero forces in *both* trusses.

Example 14.12

What is the horizontal deflection of joint F on the truss shown? Use $E = 3\ EE7$ psi. Joint A is restrained horizontally. (**Member areas have been chosen for convenience.**)

steps 1 and 2: Use the truss as drawn.

step 3: The forces in all the truss members are summarized in step 5.

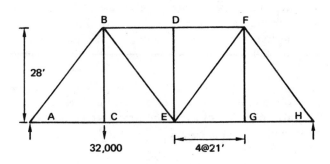

28'

32,000   4@21'

step 4: Draw the truss and load it with a unit horizontal force at point F.

step 5: Find the forces, u, in all members of the second truss. These are summarized in the following table. Notice the sign convention: + for tension and − for compression.

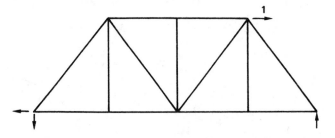

Since 12.01 EE−4 is positive, the deflection is in the direction of the dummy unit load. In this case, the deflection is to the right.

10 COMBINED STRESSES

Most practical cases of combined stresses have normal stresses on two perpendicular planes and a known shear stress acting parallel to these two planes. Based on knowledge of these stresses, the shear and the normal stresses on all other planes can be found from conditions of equilibrium.

Under any condition of stress at a point, a plane can be found where the shear stress is zero. The normal stresses on this plane are known as the *principal stresses*. The principal stresses are the maximum and the minimum stresses at the point in question.

The normal and shear stresses on a plane whose normal is inclined an angle θ from the horizontal are given by equations 14.44 and 14.45.

$$\sigma_\theta = \tfrac{1}{2}(\sigma_x + \sigma_y) + \tfrac{1}{2}(\sigma_x - \sigma_y)\cos 2\theta + \tau \sin 2\theta \qquad 14.44$$
$$\tau_\theta = -\tfrac{1}{2}(\sigma_x - \sigma_y)\sin 2\theta + \tau \cos 2\theta \qquad 14.45$$

member	S(lbf)	u(lbf)	L(ft)	A(in^2)	$\frac{SuL}{AE}$(ft)
AB	−30,000	5/12	35	17.5	−8.33 EE−4
CB	32,000	0	28	14	0
EB	−10,000	−5/12	35	17.5	2.78 EE−4
ED	0	0	28	14	0
EF	10,000	5/12	35	17.5	2.78 EE−4
GF	0	0	28	14	0
HF	−10,000	−5/12	35	17.5	2.78 EE−4
BD	−12,000	1/2	21	10.5	−4.00 EE−4
DF	−12,000	1/2	21	10.5	−4.00 EE−4
AC	18,000	3/4	21	10.5	9.00 EE−4
CE	18,000	3/4	21	10.5	9.00 EE−4
EG	6,000	1/4	21	10.5	1.00 EE−4
GH	6,000	1/4	21	10.5	1.00 EE−4

12.01 EE−4 (ft)

The maximum and minimum values of σ_θ and τ_θ (as θ is varied) are the principal stresses. These are given by equations 14.46 and 14.47.

$$\sigma(\text{max, min}) = \tfrac{1}{2}(\sigma_x + \sigma_y) \pm \tau(\text{max}) \qquad 14.46$$

$$\tau(\text{max, min}) = \pm\tfrac{1}{2}\sqrt{(\sigma_x - \sigma_y)^2 + (2\tau)^2} \qquad 14.47$$

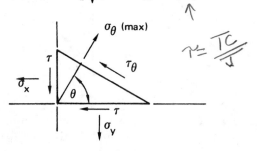

Figure 14.13 Plane of Principal Stresses

The angles of the planes on which the normal stresses are minimum and maximum are given by equation 14.48. θ is measured from the x axis, clockwise if negative and counterclockwise if positive. Equation 14.48 will yield two angles. These angles must be used in equation 14.44 to determine which angle corresponds to the minimum normal stress and which angle corresponds to the maximum normal stress.

$$\theta = \tfrac{1}{2}\arctan\left(\frac{2\tau}{\sigma_x - \sigma_y}\right) \qquad 14.48$$

The angles of the planes on which the shear stress is minimum and maximum are given by equation 14.49. The same angle sign convention used for equation 14.48 applies to equation 14.49.

$$\theta = \tfrac{1}{2}\arctan\left(\frac{\sigma_x - \sigma_y}{-2\tau}\right) \qquad 14.49$$

Proper sign convention must be adhered to when using equations 14.44 through 14.49. Normal tensile stresses are positive; normal compressive stresses are negative. Shear stresses are positive as shown in figure 14.14.

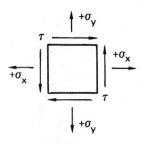

Figure 14.14 Sign Convention

Example 14.13

Find the maximum shear stress and the maximum normal stress on the object shown.

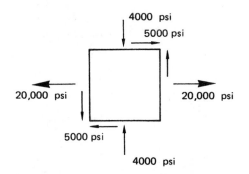

By the sign convention of figure 14.14, the 4000 psi is negative. From equation 14.47, the maximum shear stress is

$$\tau_{\text{max}} = \tfrac{1}{2}\sqrt{[20,000 - (-4000)]^2 + [(2)(5000)]^2}$$
$$= 13,000 \text{ psi}$$

From equation 14.46, the maximum normal stress is

$$\sigma_{\text{max}} = \frac{1}{2}[20,000 + (-4000)] + 13,000$$
$$= 21,000 \text{ psi (tension)}$$

11 DYNAMIC LOADING

If a load is applied suddenly to a structure, the transient response may create stresses greater than would normally be calculated from the concepts of statics and mechanics of materials alone. Although a *dynamic analysis* of the structure is appropriate, the procedure is extremely lengthy and complicated. Therefore, arbitrary dynamic factors are applied to the static stress. For example, if the load is applied quickly compared to the natural period of the structure, a dynamic factor of 2 can be used. This assumes that the load is applied as a ramp function.

12 INFLUENCE DIAGRAMS

Shear, moment, and reaction *influence diagrams* (*influence lines*) can be drawn for any point on a truss. This is a necessary step in the evaluation of stresses induced by moving loads. It is important to realize, however, that the influence line applies to only one point on the truss.

To begin, it is necessary to know if the loads are transmitted to the truss members at the lower chords (a *through truss*) or at the upper chords (a *deck truss*). If the truss is a through truss, the moving load is assumed to move along the lower chords.

Example 14.14

Draw the influence diagram for vertical shear in panel *DF* of the through truss shown.

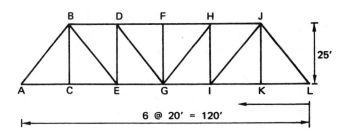

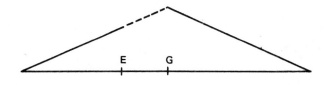

Allow a unit load to move from joint L to joint G along the lower chords. If the unit vertical load is at a distance x from point L, the right reaction will be $+[1-(x/120)]$. The unit load itself has a value of (-1), so the shear at distance x is just $(-x/120)$.

Allow a unit load to move from joint A to joint E along the lower chords. If the unit load is a distance x from point L, the left reaction will be $(x/120)$, and the shear at distance x will be $[(x/120) - 1]$.

These two lines can be graphed.

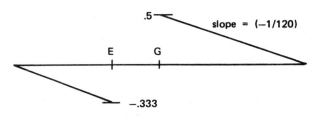

The influence line is completed by connecting the two lines as shown. Therefore, the maximum shear in panel DF will occur when a load is at point G on the truss.

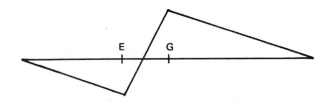

Example 14.15

Draw the moment influence diagram for panel DF on the truss shown in example 14.14.

The left reaction is $(x/120)$ where x is the distance from the unit load to the right end. If the unit load is to the right of point G, the moment can be found by summing moments from point G to the left. The moment is $(x/120)(60) = .5x$.

If the unit load is to the left of point E, the moment will be

$$\left(\frac{x}{120}\right)(60) - (1)(x - 60) = 60 - .5x$$

These two lines can be graphed. The moment for a unit load between points E and G is obtained by connecting the two end points of the lines derived above.

13 MOVING LOADS ON BEAMS

If a beam supports a single moving load, the maximum bending and shearing stresses at any point can be found by drawing the moment and shear influence diagrams for that point. Once the positions of maximum moment and maximum shear are known, the stresses at the point in question can be found from equations 14.18 and 14.24.

If a simply-supported beam carries a set of moving loads (which remain equidistant as they travel across the beam), the following procedure can be used to find the *dominant load*. The dominant load is the one which occurs directly over the point of maximum moment.

step 1: Calculate and locate the resultant of the load group.

step 2: Assume that one of the loads is dominant. Place the group on the beam such that the distance from one support to the assumed dominant load is equal to the distance from the other support to the resultant of the load group.

step 3: Check to see that all loads are on the span and that the shear changes sign under the assumed dominant load. If the shear does not change sign under the assumed dominant load, the maximum moment may occur when only some of the load group is on the beam. If it does change sign, calculate the bending moment under the assumed dominant load.

step 4: Repeat steps 2 and 3, assuming that the other loads are dominant.

step 5: Find the maximum shear by placing the load group such that the resultant is a minimum distance from a support.

14 COLUMNS

The *Euler load* is the theoretical maximum load that an initially straight column can support without buckling. **For columns with pinned ends, this load is given by equation 14.50. (r is the least radius of gyration.)**

$$F_e = \frac{\pi^2 EI}{L^2} = \frac{(k\pi)^2 EA}{L^2} \qquad 14.50$$

The corresponding column stress is

$$\sigma_e = \frac{F_e}{A} = \frac{\pi^2 E}{\left(\frac{L}{k}\right)^2} \qquad 14.51$$

k=radius of gyration

Equations 14.50 and 14.51 assume that the column is long so that the Euler stress is reached before the yield stress is reached. If the column is short, the yield stress of the material may be less than the Euler stress. In that case, short-column curves based on test data are used to predict the allowable column stress.

The value of L/k at the point of intersection of the short column and the Euler curves is known as the critical *slenderness ratio*. The critical slenderness ratio becomes smaller as the compressive yield stress increases. The region in which the short column formulas apply is determined by tests for each particular type of column and material. Typical critical slenderness ratios range from 80 to 120.

In general, the Euler allowable stress formulas can be used if the stress obtained from equation 14.51 does not exceed the compressive yield stress.

Example 14.16

An S-type, 4 × 9.5 A36 steel I-beam 8.5′ long is used as a column. What is the working stress for a safety factor of 3? Use E = 2.9 EE7 psi. The yield stress for A36 steel is 36,000 psi. The required properties of the I beam are $A = 2.79\ \text{in}^2$, $I = .903\ \text{in}^4$, and $k = .569$ in.

From equation 14.51, the Euler stress is

$$\sigma_e = \frac{\pi^2(2.9\text{ EE7})}{\left[\frac{(8.5)(12)}{.569}\right]^2} = 8907\text{ psi}$$

Since 8907 is less than 36,000, the Euler formulas are valid. The allowable working stress is

$$\sigma_a = \frac{8907}{3} = 2969\text{ psi}$$

An ultimate load for any column can be found by using the *secant formula*. The secant formula is particularly suited for use when the column is intermediate in length.

$$\sigma = \frac{F}{A} = \frac{S_y}{1 + \frac{ec}{k^2}\sec\phi} \qquad 14.52$$

$$\phi = \frac{1}{2}\left(\frac{L}{k}\right)\sqrt{\frac{F}{AE}} \qquad 14.53$$

The formula is solved by trial and error for F with the given eccentricity, e. If the value of e is not known, the eccentricity ratio (ec/k^2) is taken as .25. Substituting this value and E = 2.9 EE7 for steel and 1.00 EE7 for aluminum, respectively, the following formulas result which converge quickly to the known L/k ratio when assumed values of F are substituted.

$$\phi = \arccos\left(\frac{0.25F}{S_yA - F}\right) \qquad 14.54$$

$$\frac{L}{k} = 2\phi\sqrt{\frac{EA}{F}} \qquad 14.55$$

$$\left(\frac{L}{k}\right)_{\text{steel}} = \frac{10,770(\phi)}{\sqrt{\frac{F}{A}}} \qquad 14.56$$

$$\left(\frac{L}{k}\right)_{\text{aluminum}} = \frac{6,325(\phi)}{\sqrt{\frac{F}{A}}} \qquad 14.57$$

Example 14.17

A steel member (S_y = 36,000 psi, $A = 17.9\ \text{in}^2$, least $k = 2.45$) is used as a 20 foot column. Use the secant formula and a factor of safety of 2.5 to determine the maximum concentric load.

Even though the loading is intended to be concentric, use $ec/k^2 = 0.25$ to account for uncertainties in construction and loading.

The slenderness ratio is

$$\frac{L}{k} = \frac{(20)(12)}{2.45} = 98$$

Assume a critical load of F = 300,000 lbf. From equation 14.54,

$$\phi = \arccos\left[\frac{(0.25)(300)}{(36)(17.9) - 300}\right] = 1.35\text{ radians}$$

From equation 14.56,

$$\frac{L}{k} = \frac{(10{,}770)(1.35)}{\sqrt{\frac{300{,}000}{17.9}}} = 112.3$$

Since L/k will be smaller when F is larger, try F = 350,000 lbf.

$$\phi = \arccos\left[\frac{(0.25)(350)}{(36)(17.9) - 350}\right] = 1.27\text{ radians}$$

$$\frac{L}{k} = \frac{(10{,}770)(1.27)}{\sqrt{\frac{350{,}000}{17.9}}} = 97.8 \quad \text{(close enough)}$$

$$F_{\text{allowable}} = \frac{350{,}000}{2.5} = 140{,}000\text{ lbf}$$

All the preceding column formulas are for columns with frictionless round or pinned ends. For other end conditions, the *effective length* L' should be used in place of L.

$$L' = CL \qquad 14.58$$

C is the *end restraint coefficient* which varies from .5 to 2. For practical columns, C smaller than .7 should not be used because infinite stiffness of the support structure is not normally achievable.

Table 14.3
End-Restraint Coefficients

illus.	end conditions	ideal C	design C
(a)	both ends pinned	1	1.0*
(b)	both ends built in	.5	.65*–.90
(c)	one end pinned, one end built in	.707	.80*–.90
(d)	one end built in, one end free	2	2.0–2.1*
(e)	one end built in, one end fixed against rotation but free	1	1.2*
(f)	one end pinned, one end fixed against rotation but free	2	2.0*

* AISC values

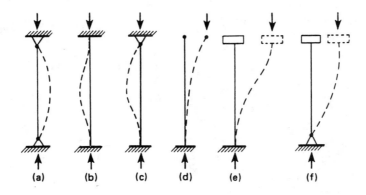

(a)   (b)   (c)   (d)   (e)   (f)

See
Columns
page 14-15

PART 2: Application to Design

1 SPRINGS

Springs are assumed to be perfectly elastic within their working range. *Hooke's law* can be used to predict the amount of compression experienced when a load is placed on a spring.

$$\mathbf{F} = kx \qquad\qquad 14.59$$

k is the *spring constant*. It has units of pounds per unit length.

When a spring is compressed, it stores energy. This energy can be recovered by restoring the spring's original length. It is assumed that no energy is lost through friction or hysteresis when a spring returns to its original length. The energy storage in a spring is given by equation 14.60. This energy is the same as the work required to compress the spring.

$$W = \Delta U = \tfrac{1}{2}kx^2 \qquad\qquad 14.60$$

If a weight is dropped from height h onto a spring, the compression can be found by equating the change in potential energy to the energy storage.

$$m\left(\frac{g}{g_c}\right)(h + x) = \tfrac{1}{2}kx^2 \qquad\qquad 14.61$$

2 THIN-WALLED CYLINDERS

A cylinder can be considered *thin-walled* if its wall thickness-to-diameter ratio is less than .1. The circumferential *hoop stress* for internal pressure can be derived easily from the free-body diagram of a cylinder half.[2] This hoop stress is

$$\sigma_h = \frac{pr}{t} \qquad\qquad 14.62$$

Since the cylinder is assumed to be thin-walled, the radius used in equation 14.62 is taken as the inside radius.

If the cylinder is part of a tank, the axial force on the end plates produces an axial stress. The axial force is equal to the tank pressure times the end plate area. The stress produced is at right angles to the hoop stress. Accordingly, it is called *longitudinal stress* or *long stress.*

$$\sigma_L = \frac{pr}{2t} \qquad\qquad 14.63$$

---

[2]There is no easy method of evaluating stresses in thin-walled cylinders under external pressure, since failure is by collapse, not elongation. However, empirical equations exist for predicting the collapsing pressure.

Equation 14.63 also gives the stress in a spherical tank. In a spherical tank, the hoop and long stresses are the same.

The hoop and long stresses are principal stresses. They do not combine into a larger stress.

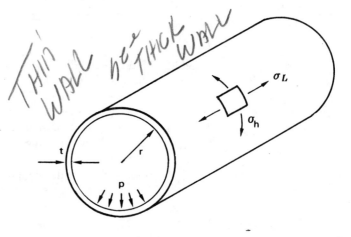

Figure 14.15 Hoop and Long Stresses

3 RIVET AND BOLT CONNECTIONS

A *tension splice* using rivets or bolts can fail in one of three ways: bearing failure, shear failure, or tension failure. All three failure mechanisms must be checked to determine the maximum load the splice can carry.

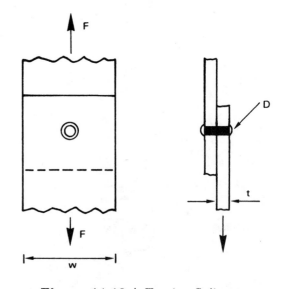

Figure 14.16 A Tension Splice

The plate can fail in bearing. For one connector, the *bearing stress* in the plate is

$$\sigma_{br} = \frac{F}{Dt} \qquad\qquad 14.64$$

The number of rivets required to keep the actual bearing stress below the allowable bearing stress is

$$n_{br} = \frac{\sigma_{br}}{\text{allowable bearing stress}} \qquad 14.65$$

The rivet can fail in shear. The shear stress in the rivet is

$$\tau = \frac{F}{\frac{1}{4}\pi D^2} \qquad 14.66$$

The number of rivets required, as determined by shear, is

$$n_s = \frac{\tau}{\text{allowable shear stress}} \qquad 14.67$$

The plate also can fail in tension. If there are n rivet holes in a line across the width of the plate, the minimum area in tension will be

$$A_t = t(w - nD) \qquad 14.68$$

The tensile stress in the plate at the minimum section is

$$\sigma_t = \frac{F}{A_t} \qquad 14.69$$

The maximum number of rivets across the plate width must be chosen to keep the tensile stress less than the allowable stress.

4 FILLET WELDS

The most common weld type is the *fillet weld*, shown in figure 14.17. Such welds commonly are used to connect one plate to another. The applied load is assumed to be carried by the *effective weld throat* which is related to the weld size, y, by equation 14.70.

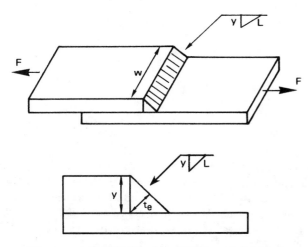

Figure 14.17 Fillet Lap Weld and Symbol

The effective weld throat size is

$$t_e = (.707)y \qquad 14.70$$

Weld sizes (y) of $\frac{3}{16}''$, $\frac{1}{4}''$, and $\frac{5}{16}''$ are desirable because they can be made in a single pass. However, fillet welds

from $\frac{3}{16}''$ to $\frac{1}{2}''$ in $\frac{1}{16}''$ increments are available. The increment is $\frac{1}{8}''$ for larger welds.

Neglecting any effects due to eccentricity, the stress in the fillet lap weld shown in figure 14.17 is

$$\sigma = \frac{F}{wt_e} \qquad 14.71$$

5 SHAFT DESIGN

Shear stress occurs when a shaft is placed in torsion. The shear stress at the outer surface of a bar of radius, r, which is torsionally loaded by a torque, T, is[3]

$$\tau = \frac{Tr}{J} \qquad 14.72$$

The total strain energy due to torsion is

$$U = \frac{T^2 L}{2GJ} \qquad 14.73$$

J is the shaft's polar moment of inertia, as defined in **chapter 12 (page 12-14). For a solid round shaft, J is**

$$J = \frac{\pi r^4}{2} = \frac{\pi D^4}{32} \qquad 14.74$$

For a hollow round shaft, the polar moment of inertia is

$$J = \frac{\pi}{2}[r_o^4 - r_i^4] \qquad 14.75$$

If a shaft of length L carries a torque T, the angle of twist (in radians) will be

$$\phi = \frac{TL}{GJ} \qquad 14.76$$

G is the *shear modulus*, approximately equal to 11.5 EE6 psi for steel. The shear modulus can be calculated from the modulus of elasticity by using equation 14.77.

$$G = \frac{E}{2(1 + \mu)} \qquad 14.77$$

The torque, T, carried by a shaft spinning at n revolutions per minute is related to the transmitted horsepower.[4]

$$T_{\text{in-lbf}} = \frac{(63,025)(\text{horsepower})}{n} \qquad 14.78$$

---

[3]Shear stress in steel shafts commonly is limited to approximately 6000 psi. This represents a factor of safety of approximately 3 based on the torsional yield strength.

[4]The torque is assumed to be steady, as would be supplied by a belt or a pulley. If the load varies, or if the shaft also carries a bending moment, a more complex method is required.

6 ECCENTRIC CONNECTOR ANALYSIS

An eccentric torsion connection is illustrated in figure 14.18. This type of connection gets its name from the load's tendency to rotate the bracket. This rotation must be resisted by the shear stress in the connectors.

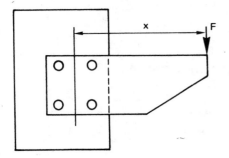

Figure 14.18 Torsion Resistance

An extension of equation 14.72 can be used to evaluate the maximum stresses in the connector group. To use equation 14.72, the following changes in definition must be made.

- The torque, T, is replaced by the moment on the bracket. This moment is the product of the eccentric load, F, and the distance from the load to the centroid of the fastener group, x.

- r is taken as the distance from the centroid of the fastener group to the critical fastener. The critical fastener is the one for which the vector sum of the vertical and torsional shear stresses is the greatest.

- J is based on the parallel-axis theorem. As bolts and rivets have little resistance to twisting in their holes, their polar moments of inertia, J_i, are omitted.

$$J = \sum r_i^2 A_i \qquad 14.79$$

r_i is the distance from the fastener group centroid to the ith fastener, which has an area of A_i.

- The vertical shear stress in the critical fastener must be added in a vector sum to the torsional shear stress. This vertical shear stress is

$$\text{vertical shear stress} = \frac{PA_{\text{critical}}}{\sum A_i} \qquad 14.80$$

Example 14.18

For the bracket shown, find the load on the most critical fastener. All fasteners have a nominal $\frac{1}{2}''$ diameter.

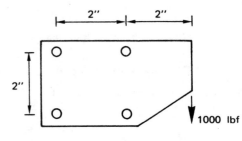

Since the fastener group is symmetrical, the group centroid is centered within the 4 fasteners. This makes the eccentricity of the load equal to 3 inches. Each fastener is located r from the centroid, where

$$r = \sqrt{(1)^2 + (1)^2} = 1.414$$

The area of each fastener is

$$A_i = \tfrac{1}{4}\pi(.5)^2 = .1963$$

Using the parallel axis theorem for polar moments of inertia,

$$J = 4[.1963(1.414)^2] = 1.570 \text{ in}^4$$

The torsional stress on each fastener is

$$\tau_T = \frac{(1000)(3)(1.414)}{(1.570)} = 2702 \text{ psi}$$

This torsional shear stress is directed perpendicularly to a line connecting each fastener with the centroid.

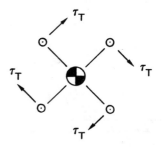

τ_T can be divided into horizontal stresses of $\tau_{T,x}$ and vertical stresses of $\tau_{T,y}$. Both of these components are equal to 1911 psi. In addition, each fastener carries a vertical shear load equal to $(1000/4) = 250$ pounds. The vertical shear stress due to this load is $(250/.1963) = 1274$.

The two right fasteners have vertical downward components of τ_T which add to the vertical downward stress of 1274. Thus, both of the two right fasteners are critical. The total stress in each of these fasteners is

$$\tau = \sqrt{(1911)^2 + (1911 + 1274)^2} = 3714 \text{ psi}$$

7 SURVEYOR'S TAPE CORRECTIONS

The standard surveyor's tape consists of a flat steel ribbon with a length very close to 100 feet. Such tapes are standardized at a particular temperature and with a specific tension and type of support. Since the tape cannot be used in the conditions under which it was standardized, corrections are needed.

A. TEMPERATURE CORRECTION

If the tape is not used at the standardized temperature, the change in length will be

$$C_t = \alpha(T - T_{std})L \qquad 14.81$$

α for steel has an approximate value of 6.5 EE−6 1/°F, although low-coefficient tapes containing nickel can reduce this expansion 75%. The correction given by equation 14.81 can be positive or negative, depending on the values of T and T_{std}. The correction is applied to the distance according to the algebraic operations listed in table 14.4.

Table 14.4
Corrections for Surveyors' Tapes

measuring a distance	setting out points
add C_t	subtract C_t
add C_p	subtract C_p

B. TENSION CORRECTIONS

The correction due to non-standard pull (tension) can be found from equation 14.82. It can be either positive or negative.

$$C_p = \frac{(F - F_{std})L}{AE} \qquad 14.82$$

The correction is applied to the distance according to the algebraic conventions listed in table 14.4.

Example 14.19

A steel surveyor's tape is standardized at 68°F. It is used at 50°F to place two monuments exactly 79 feet apart. What should be the tape reading used to place the monuments?

From equation 14.81,

$$C_t = (6.5\ \text{EE}-6)(50 - 68)(79) = -9.2\ \text{EE}-3$$

From table 14.4,

$$\text{tape reading} = 79.0000 - (-9.2\ \text{EE}-3)$$
$$= 79.0092$$

(The tape cannot be read to the degree of precision indicated by this answer.)

8 STRESS CONCENTRATION FACTORS

Stress concentration factors are correction factors used to account for nonuniform stress distributions within objects.[5] Nonuniform distributions result from nonuniform shapes. Examples of nonuniform shapes requiring stress concentration factors are stepped shafts, plates with holes, shafts with keyways, etc.

The actual stress experienced is the product of the stress concentration factor and the ideal stress. Values of K always are greater than 1.0, and they typically range from 1.2 to 2.5 for most designs. The exact values must be determined graphically from published results of extensive experimentation.

$$\sigma' = K\sigma \qquad 14.83$$

9 CABLES

Cables (*wire ropes*) can be obtained in a wide variety of materials and cross sections to suit the application. Strength and weight properties of steel *standard hoisting rope* (6 strands of 19 wires each) are given in table 14.5.

Table 14.5
6 × 19 (Standard Hoisting) Wire Ropes

Diam. Inches	Approx. Weight per ft., Pounds	Breaking Strength Tons of 2000 Pounds		
		Impr. Plow Steel	Plow Steel	Mild Plow Steel
1/4	0.10	2.74	2.39	2.07
5/16	0.16	4.26	3.71	3.22
3/8	0.23	6.10	5.31	4.62
7/16	0.31	8.27	7.19	6.25
1/2	0.40	10.7	9.35	8.13
9/16	0.51	13.5	11.8	10.2
5/8	0.63	16.7	14.5	12.6
3/4	0.90	23.8	20.7	18.0
7/8	1.23	32.2	28.0	24.3
1	1.60	41.8	36.4	31.6
1 1/8	2.03	52.6	45.7	39.8
1 1/4	2.50	64.6	56.2	48.8
1 3/8	3.03	77.7	67.5	58.8
1 1/2	3.60	92.0	80.0	69.6
1 5/8	4.23	107	93.4	81.2
1 3/4	4.90	124	108	93.6
1 7/8	5.63	141	123	107
2	6.40	160	139	121
2 1/8	7.23	179	156	
2 1/4	8.10	200	174	
2 1/2	10.0	244	212	
2 3/4	12.1	292	254	

For ropes with steel cores, add 7½% to the above strengths.
For galvanized ropes, deduct 10% from the above strengths.

[5]Stress concentration factors also are known as *stress risers*.

In addition to the primary tension load, the design of cables should include the significant effects of bending, friction, and the weight of the cable. Appropriate dynamic factors should be applied to allow for acceleration, deceleration, stops, and starts. In general, the working stress should not exceed 20% of the breaking strength (i.e., a factor of safety of 5).

The stress due to bending a cable, such as bending around a drum, is included as an equivalent added tension load. (For good design, the diameter of the drum on which a cable is wound should be 45 to 90 times the cable diameter.) If d is the cable diameter in inches, R is the bending radius in inches, and N is the number of wires in the cable (114 for a 6×19 cable), the equivalent tensile load from bending is approximately

$$F = \frac{2.8 \ \text{EE}9 d^3}{N^2 R} \qquad 14.84$$

Example 14.20

What is the factor of safety when a $\frac{1}{2}$ inch, mild plow steel, 6×19 standard hoisting cable carrying 10,000 pounds is bent around a 24 inch sheave? Is the factor of safety adequate?

$$F_{\text{bending}} = \frac{(2.8 \ \text{EE}9)(.5)^3}{(114)^2(12)} = 2240 \ \text{lbf}$$

$$F_{\text{total}} = 10,000 + 2240 = 12,240 \ \text{lbf}$$

$$\text{breaking strength} = 8.13(2000) = 16,260 \ \text{lbf}$$

$$\text{factor of safety} = \frac{16,260}{12,240} = 1.33$$

Even without including the cable weight, this is not adequate, since a factor of safety of at least 5 is recommended.

10 THICK-WALLED CYLINDERS UNDER EXTERNAL AND INTERNAL PRESSURE

A. STRESSES

The theory of thick-walled cylinders, *Lamé's solution*, is a continuation of the theory of thin-walled cylinders. The thick-walled cylinder is assumed to be made up of thin laminar rings. The strain variation through the wall is determined such that all the rings are in equilibrium, and the stresses and the deformations are consistent at the boundaries between the rings.

The general equations for stress are given here.[6]

$$\sigma_c = \frac{r_i^2 p_i - r_o^2 p_o + \dfrac{(p_i - p_o) r_i^2 r_o^2}{r^2}}{r_o^2 - r_i^2} \qquad 14.85$$

$$\sigma_r = \frac{r_i^2 p_i - r_o^2 p_o - \dfrac{(p_i - p_o) r_i^2 r_o^2}{r^2}}{r_o^2 - r_i^2} \qquad 14.86$$

[6]It is essential that compressive stresses be given a negative sign in all thick-walled cylinder equations, including those for deflection.

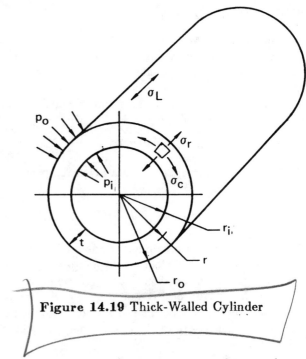

Figure 14.19 Thick-Walled Cylinder

The cases of main interest are those of internal or external pressure only. The stress formulas for these cases are summarized in table 14.6.

The maximum shear and normal stresses occur at the inner surface for both external and internal pressure.

When a longitudinal stress due to internal pressure acting against end plates exists, the longitudinal stress is

$$\sigma_L = \frac{p r_i^2}{(r_o + r_i) t} \qquad 14.87$$

The longitudinal stress is assumed to be uniform across the wall. The magnitude and the location of the maximum shear and normal stress is not changed due to the addition of the longitudinal stress.

At every point in the cylinder, σ_c, σ_r, and σ_L are the principal stresses.

Table 14.6
Stresses in Thick-Walled Cylinders

Stress		External Pressure, p		Internal Pressure, p
σ_{co}		$\dfrac{-(r_o^2 + r_i^2)p}{(r_o + r_i)t}$		$\dfrac{2 r_i^2 p}{(r_o + r_i)t}$
σ_{ro}	m	$-p$		0
σ_{ci}	a	$\dfrac{-2 r_o^2 p}{(r_o + r_i)t}$	m	$\dfrac{(r_o^2 + r_i^2)p}{(r_o + r_i)t}$
	x		a	
σ_{ri}		0	x	$-p$
τ_{max}		$(\frac{1}{2})\sigma_{ci}$		$(\frac{1}{2})(\sigma_{ci} + p)$

B. STRAINS

The *diametral strain*, $\Delta D/D$, and the *circumferential strain*, $\Delta C/C$, are equal in a circular cylinder under pressure loading.

$$\frac{\Delta D}{D} = \frac{\Delta C}{C} = \frac{\Delta r}{r} = \frac{\sigma_c - \mu(\sigma_r + \sigma_L)}{E} \qquad 14.88$$

Example 14.21

A steel cylinder of 1″ I.D. and 2″ O.D. is pressurized internally to 10,000 psi. (a) There are no end caps. What is the change in the inside diameter? (b) What would be the effect on the inside diameter of adding end caps? Use $E = 2.9$ EE7 and $\mu = .3$.

From table 14.6,

$$\sigma_{ci} = \frac{(1^2 + .5^2)(10,000)}{(1 + .5).5} = 16,667 \text{ psi}$$
$$\sigma_{ri} = -10,000 \text{ psi}$$
$$\sigma_L = 0$$

From equation 14.88,

$$\frac{\Delta D}{D} = \frac{[16,667 - .3(-10,000 + 0)]}{2.9 \text{ EE7}} = .000678$$
$$\Delta D = .000678(1) = .000678 \text{ inch}$$

(b)

$$\sigma_L = \frac{(10,000)(.5)^2}{(1 + .5)(.5)} = 3333 \text{ psi}$$
$$\frac{\Delta D}{D} = \frac{[16,667 - .3(-10,000 + 3333)]}{2.9 \text{ EE7}} = .000644$$
$$\Delta D = .000644(1) = .000644 \text{ inch}$$

C. PRESS FITS

If two cylinders are pressed together with an initial interference, I, the pressure, p, acting between them expands the outer cylinder and compresses the inner one. *Interference* usually means diametral interference. Although the total interference can be allocated to inner and outer cylinders in almost any combination, the total interference usually is given to the outer disk in the case of shafts with disks.

$$I = |\Delta D_i|_{\text{outer cylinder}} + |\Delta D_o|_{\text{inner cylinder}} \qquad 14.89$$

If the cylinders are the same length, this method can be used to find the stress conditions after assembly. A stress concentration factor as high as 4 may be needed if the lengths are different.

In the special case where the shaft and the hub have the same modulus of elasticity and the shaft is solid,

the total diametral interference can be found from equation 14.90. The interference pressure, p, can be found if I is known.

$$I = \frac{4r_{\text{shaft}}p}{E}\left[\frac{1}{1 - \left(\frac{r_{\text{shaft}}}{r_{\text{hub}}}\right)^2}\right]. \qquad 14.90$$

When pieces are pressed together, the assembly force can be calculated as a sliding frictional force based on the normal force.

$$F_{\text{max, assembly}} = fN = 2\pi r_{\text{shaft}}Lpf \qquad 14.91$$

The coefficient of friction for press fits is highly variable, having been reported in the range of .03 to .33.

If the hub is acted upon by a torque or a torque-causing force, the maximum resisting torque can be calculated.

$$T_{\text{max}} = 2\pi r_{\text{shaft}}^2 Lpf \qquad 14.92$$

Example 14.22

A brass inner cylinder and an aluminum alloy outer cylinder (both 2.0″ long) have been pressed together with an interference of .004 in. What is the maximum shear stress in the brass? Assuming a coefficient of friction of .25, what is the force required to press them apart?

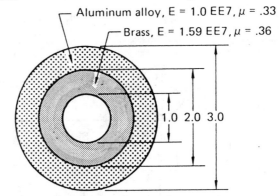

Aluminum alloy, E = 1.0 EE7, μ = .33
Brass, E = 1.59 EE7, μ = .36
1.0 2.0 3.0

Aluminum alloy, internal pressure:

$$\sigma_{ci} = \frac{(1.5^2 + 1^2)p}{(1.5 + 1)(.5)} = 2.6p$$
$$\sigma_{ri} = -p$$
$$\left(\frac{\Delta D}{D}\right)_i = \frac{[2.6p - .33(-p)]}{1.0 \text{ EE7}} = (2.93 \text{ EE} - 7)p$$
$$\Delta D_i = (2.93 \text{ EE} - 7)p(2) = (5.86 \text{ EE} - 7)p$$

Brass, external pressure:

$$\sigma_{co} = \frac{-(1^2 + .5^2)p}{(1 + .5)(.5)} = -1.667p$$
$$\sigma_{ro} = -p$$
$$\left(\frac{\Delta D}{D}\right)_o = \frac{[-1.667p - .36(-p)]}{1.59 \text{ EE7}} = (-.822 \text{ EE} - 7)p$$
$$\Delta D_o = (-.822 \text{ EE} - 7)p(2) = (-1.644 \text{ EE} - 7)p$$

From equation 14.89,

$$|(5.86 \text{ EE} - 7)|p + |(-1.644 \text{ EE} - 7)|p = .004$$

$$p = 5330 \text{ psi}$$

From table 14.6,

$$\sigma_{\text{ci, brass}} = \frac{-2(1^2)5330}{(1 + .5)(.5)} = 14,213 \text{ psi}$$

$$\tau_{\max} = \tfrac{1}{2}(14,213) = 7106.5 \text{ psi}$$

The force to separate the cylinders is

$$F = (5330)[2\pi(1)(2)](.25) = 16,745 \text{ lbf}$$

11 FLAT PLATES UNDER UNIFORM PRESSURE LOADING

Formulas for stresses and deflections of flat plates are given in table 14.7. Their application is subject to the following constraints.

- The plates are of medium thickness, meaning that the thickness is equal to or less than 1/4 the minimum width dimension, and the maximum deflection is equal to or less than 1/2 the thickness.

- The plates are constructed of isentropic, elastic material, and the elastic limit is not exceeded under the applied loading.

Example 14.23

The end of a 10″ inside diameter pipe is capped by welding on an end plate made from mild steel. The safe stress in the end cap is 11,100 psi. The internal pressure in the pipe is 500 psig. What plate thickness is required?

The welding produces a round plate with fixed edges. From table 14.7, the stress is

$$\sigma = \frac{3pr^2}{4t^2} = \frac{(3)(500)(5)^2}{4t^2} = 11,100 \text{ psi}$$

Solving for the thickness results in $t = 0.92''$.

Table 14.7
Flat Plates Under Uniform Pressure of p psi

Shape	Edge Condition	Maximum Stress	Deflection at Center
Circular	Simply Supported	$(3/8)pr^2(3 + \mu)/t^2$ at center	$(3/16)pr^4(1 - \mu)(5 + \mu)/(Et^3)$
	Built-in	$(3/4)pr^2/t^2$ at edge	$(3/16)pr^4(1 - \mu^2)/(Et^3)$
Rectangular	Simply Supported	$C_1 pb^2/t^2$ at center	$C_2 pb^4/(Et^3)$
	Built-in	$C_3 pb^2/t^2$ at centers of long edges	$C_4 pb^4/(Et^3)$

a/b	1.0	1.2	1.4	1.6	1.8	2	3	4	5	∞
C_1	.287	.376	.453	.517	.569	.610	.713	.741	.748	.750
C_2	.044	.062	.077	.091	.102	.111	.134	.140	.142	.142
C_3	.308	.383	.436	.487	.497	.500	.500	.500	.500	.500
C_4	.0138	.0188	.023	.025	.027	.028	.028	.028	.028	.028

PART 3: Designing with Composites

Composite material sections, such as those constructed from carbon, boron, or glass fibers in epoxy fillers, are not designed conventionally. Conventional design is based on a chosen material with a fixed modulus of elasticity. Stresses and strains are limited by choice of cross section. Composite design, however, requires an iterative analysis. The ply orientations are chosen, which affects the modulus and load patterns, and a new set of orientations are selected.

Composite materials are highly anisotropic. That is, their strengths are dependent on the orientation of the fibers. There is essentially complete freedom in choosing the directions of the plies. If a piece is not too highly stressed in one direction, there may be fewer plies or no plies at all in that direction. Indeed, cost considerations in composite materials require designing the material around the stresses.

Designing with composites requires extensive knowledge of the material behavior versus variations in the construction. Figure 14.20 illustrates a family of curves for carbon-epoxy (C-Ep) with 65% high-strength carbon fibers. The plies are arranged at angles of 0°, ±45°, and 90°, although the percentage of fibers in each direction is variable.[7]

Figure 14.20 is used with the following steps to determine the composite plies percentages.

step 1: Determine the modulus of elasticity needed in the x-direction (E_x).

step 2: From figure 14.20 (a), select a percentage of plies at 0°, ±45°, and 90° to meet the required E_x value.[8]

step 3: From figure 14.20 (b), determine the allowable tensile stress in the x-direction for the chosen ply orientations.

step 4: Select a new set of ply orientations if the strength is too high or too low.

Example 14.24

A composite material for room temperature use is needed. For cost purposes, C-Ep with high-strength fibers has been chosen. A minimum strength of 70,000 psi is needed in the x-direction, and a modulus of 10 EE6 psi is needed in the y-direction. What stacking arrangement can be used with 0°, ±45°, and 90° lay-up?

---

[7] There is considerable use of the SI system in composite design. The ply orientations may be specified in radians. In that case, these angles would be 0, 0.79, and 1.6 radians.

[8] This is known as the *stacking choice*.

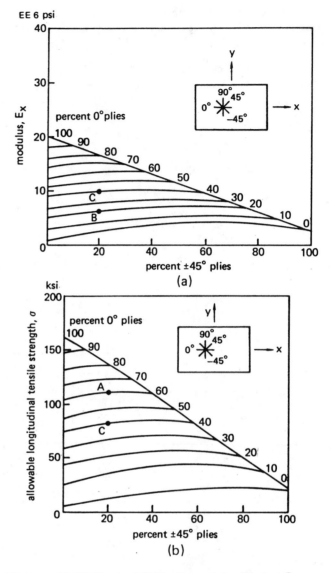

Figure 14.20 Typical C-Ep Composite Design Curves (Room Temperature, 65% fiber content)

This problem has two simultaneous requirements. Therefore, the solution will be iterative.

Start by choosing an arbitrary stacking arrangement of 60% fibers oriented at 0°, 20% at 90°, and 20% at ±45° (i.e., 60-20-20). Then plot this point on figure 14.20 (b).

From figure 14.20 (b), the allowable stress in the x-direction is somewhat greater than 100 ksi, with probably an unnecessary margin. The modulus in the y-direction, however, cannot be checked directly from figure 14.20 (a), since E_x is given.

If the 0° and 90° percentages are reversed (point B), figure 14.20 (a) can be used for the y-direction. In this case, E_y is approximately 6 EE6 psi, too low.

Next move some of the 0° plies to 90°, say a 40-40-20 stacking arrangement. The allowable stress in the x-direction from figure 14.20 (b) (point C) is approximately 85 ksi. Reversing the ply percentages is unnecessary since both are 40%, so E_y is read from figure 14.20 (a) as 10 EE6 psi.

Some minor improvements might be available at this point with additional iterations.

Appendix A
Beam Formulas

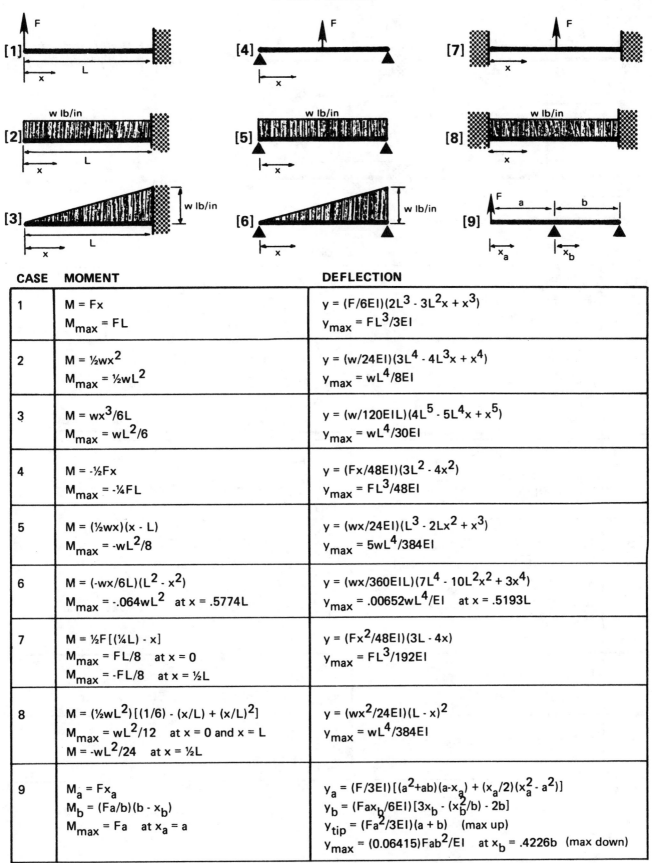

CASE	MOMENT	DEFLECTION
1	$M = Fx$ $M_{max} = FL$	$y = (F/6EI)(2L^3 - 3L^2x + x^3)$ $y_{max} = FL^3/3EI$
2	$M = \frac{1}{2}wx^2$ $M_{max} = \frac{1}{2}wL^2$	$y = (w/24EI)(3L^4 - 4L^3x + x^4)$ $y_{max} = wL^4/8EI$
3	$M = wx^3/6L$ $M_{max} = wL^2/6$	$y = (w/120EIL)(4L^5 - 5L^4x + x^5)$ $y_{max} = wL^4/30EI$
4	$M = -\frac{1}{2}Fx$ $M_{max} = -\frac{1}{4}FL$	$y = (Fx/48EI)(3L^2 - 4x^2)$ $y_{max} = FL^3/48EI$
5	$M = (\frac{1}{2}wx)(x - L)$ $M_{max} = -wL^2/8$	$y = (wx/24EI)(L^3 - 2Lx^2 + x^3)$ $y_{max} = 5wL^4/384EI$
6	$M = (-wx/6L)(L^2 - x^2)$ $M_{max} = -.064wL^2$ at $x = .5774L$	$y = (wx/360EIL)(7L^4 - 10L^2x^2 + 3x^4)$ $y_{max} = .00652wL^4/EI$ at $x = .5193L$
7	$M = \frac{1}{2}F[(\frac{1}{4}L) - x]$ $M_{max} = FL/8$ at $x = 0$ $M_{max} = -FL/8$ at $x = \frac{1}{2}L$	$y = (Fx^2/48EI)(3L - 4x)$ $y_{max} = FL^3/192EI$
8	$M = (\frac{1}{2}wL^2)[(1/6) - (x/L) + (x/L)^2]$ $M_{max} = wL^2/12$ at $x = 0$ and $x = L$ $M = -wL^2/24$ at $x = \frac{1}{2}L$	$y = (wx^2/24EI)(L - x)^2$ $y_{max} = wL^4/384EI$
9	$M_a = Fx_a$ $M_b = (Fa/b)(b - x_b)$ $M_{max} = Fa$ at $x_a = a$	$y_a = (F/3EI)[(a^2+ab)(a-x_a) + (x_a/2)(x_a^2 - a^2)]$ $y_b = (Fax_b/6EI)[3x_b - (x_b^2/b) - 2b]$ $y_{tip} = (Fa^2/3EI)(a + b)$ (max up) $y_{max} = (0.06415)Fab^2/EI$ at $x_b = .4226b$ (max down)

MECHANICS OF MATERIALS

Appendix A, continued

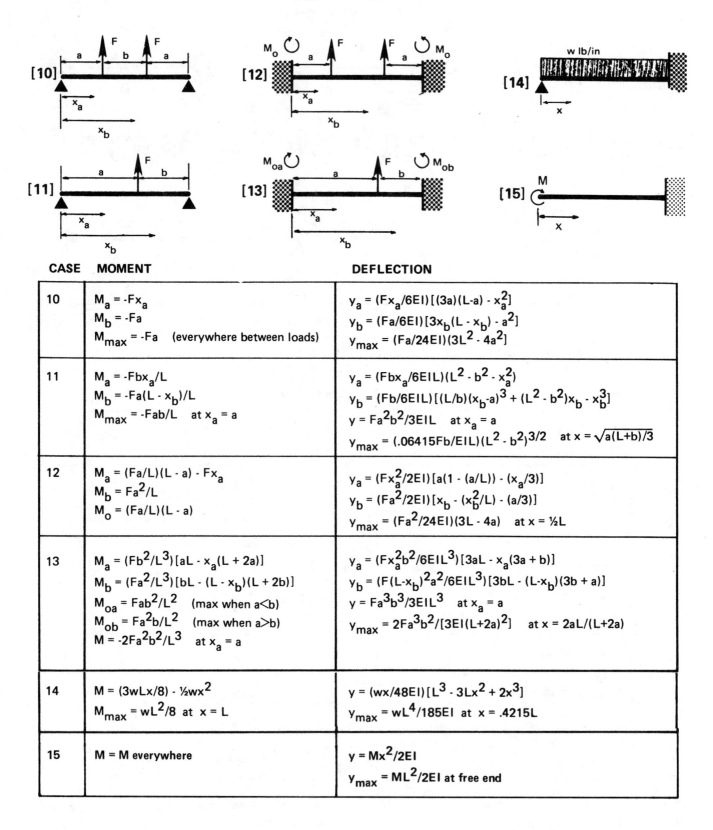

CASE	MOMENT	DEFLECTION
10	$M_a = -Fx_a$ $M_b = -Fa$ $M_{max} = -Fa$ (everywhere between loads)	$y_a = (Fx_a/6EI)[(3a)(L-a) - x_a^2]$ $y_b = (Fa/6EI)[3x_b(L - x_b) - a^2]$ $y_{max} = (Fa/24EI)(3L^2 - 4a^2)$
11	$M_a = -Fbx_a/L$ $M_b = -Fa(L - x_b)/L$ $M_{max} = -Fab/L$ at $x_a = a$	$y_a = (Fbx_a/6EIL)(L^2 - b^2 - x_a^2)$ $y_b = (Fb/6EIL)[(L/b)(x_b-a)^3 + (L^2 - b^2)x_b - x_b^3]$ $y = Fa^2b^2/3EIL$ at $x_a = a$ $y_{max} = (.06415Fb/EIL)(L^2 - b^2)3/2$ at $x = \sqrt{a(L+b)/3}$
12	$M_a = (Fa/L)(L - a) - Fx_a$ $M_b = Fa^2/L$ $M_o = (Fa/L)(L - a)$	$y_a = (Fx_a^2/2EI)[a(1 - (a/L)) - (x_a/3)]$ $y_b = (Fa^2/2EI)[x_b - (x_b^2/L) - (a/3)]$ $y_{max} = (Fa^2/24EI)(3L - 4a)$ at $x = \frac{1}{2}L$
13	$M_a = (Fb^2/L^3)[aL - x_a(L + 2a)]$ $M_b = (Fa^2/L^3)[bL - (L - x_b)(L + 2b)]$ $M_{oa} = Fab^2/L^2$ (max when a<b) $M_{ob} = Fa^2b/L^2$ (max when a>b) $M = -2Fa^2b^2/L^3$ at $x_a = a$	$y_a = (Fx_a^2b^2/6EIL^3)[3aL - x_a(3a + b)]$ $y_b = (F(L-x_b)^2a^2/6EIL^3)[3bL - (L-x_b)(3b + a)]$ $y = Fa^3b^3/3EIL^3$ at $x_a = a$ $y_{max} = 2Fa^3b^2/[3EI(L+2a)^2]$ at $x = 2aL/(L+2a)$
14	$M = (3wLx/8) - \frac{1}{2}wx^2$ $M_{max} = wL^2/8$ at $x = L$	$y = (wx/48EI)[L^3 - 3Lx^2 + 2x^3]$ $y_{max} = wL^4/185EI$ at $x = .4215L$
15	$M = M$ everywhere	$y = Mx^2/2EI$ $y_{max} = ML^2/2EI$ at free end

Appendix B
Mechanical Properties of Representative Metals

The following mechanical properties are not guaranteed since they are averages for various sizes, product forms, and methods of manufacture. Thus, this data is not for design use, but is intended only as a basis for comparing alloys and tempers.

KEY TO HEAT TREATMENT ABBREVIATIONS: **acrt**, air cooled to room temperature; **anat**, annealed at; **ct**, cooled to; **ht**, heated to; **oqf**, oil quench from; **rcf**, rapid cool from; **ta**, tempered at.

Material designation, composition, typical use, and source if applicable	Condition, heat treatment	S_{ut} (ksi)	S_{yt} (ksi)
IRON BASED			
Armco ingot iron, for fresh and salt water piping	normalized (ht 1700·F, acrt)	44	24
AISI 1020, plain carbon steel, for general machine parts and screws, and carburized parts	hot rolled	65	43
	cold worked	78	66
AISI 1030, plain carbon steel, for gears, shafts, levers, seamless tubing, and carburized parts	cold drawn	87	73.9
AISI 1040, plain carbon steel, for high-strength parts, shafts, gears, studs, connecting rods, axles, and crane hooks	hot rolled	91	58
	cold worked	100	88
	hardened (wqf 1525·F, ta 1000·F)	113	86
AISI 1095, plain carbon steel, for hand-tools, music wire springs, leaf springs, knives, saws, agricultural tools such as plows and disks	annealed (ht 1450·F, ct 1200·F, acrt)	100	53
	hot rolled	142	84
	hardened (oqf 1475·F, ta 700·F)	180	118
AISI 1330, manganese steel, for axles, drive shafts	annealed	97	83
	cold drawn	113	93
	hardened (wqf 1525·F, ta 1000·F)	122	100
AISI 4130, chromium molybdenum steel, for high-strength aircraft structures	annealed (ht 1500·F, ct 1230·F, acrt)	81	52
	hardened (wqf 1575·F, ta 900·F)	161	137
AISI 4340, nickel-chromium-molybdenum steel, for large-scale, heavy duty high-strength structures	annealed	119	99
	as-rolled	192	147
	hardened (wqf/oqf 1500·F, ta 800·F)	220	200
AISI 2315, nickel steel, for carburized parts	as rolled	85	56
	cold drawn	95	75
AISI 2330, nickel steel	as rolled	98	65
	cold drawn	110	90
	annealed at 1450·F	80	50
	normalized at 1675·F	95	61
AISI 3115, nickel chromium steel for carburized parts	cold drawn	95	70
	as rolled	75	60
	annealed at 1500·F	71	62
STAINLESS STEELS			
AISI 302 stainless steel, most widely used, same as 18:8.	annealed	90	35
	cold drawn	105	60
AISI 303 austenitic stainless steel, good machineability	annealed, rqf 1950·F	90	35
	cold worked	110	75
AISI 304 austenitic stainless steel, good machineability and weldability	annealed, rqf 1950·F	85	30
	cold worked	110	75
AISI 309 stainless steel, good weldability, high strength at high temperatures, used in furnaces and ovens	annealed	90	35
	cold drawn	110	65
AISI 316 stainlesss steel, excellent corrosion resistance	annealed, rqf 2000·F	85	35
	cold drawn	105	60
AISI 410, magnetic, martensitic, can be quenched and tempered to give varying strength	annealed	60	32
	cold drawn	180	150
	oil quenched and drawn at 1100·F	110	91
AISI 430, magnetic, ferritic, used for auto and architectural trim, and for equipment in food and chemical industries.	annealed	60	35
	cold drawn	100	
AISI 502, magnetic, ferritic, low cost, widely used in oil refineries	annealed	60	25

σ yield

ultimate tensile strength

For steel
$E = 3 \times 10^7$ PSI

$G = \dfrac{E}{2(1+\mu)}$

Appendix B, continued

Material, designation, composition, typical use, and source if applicable	Condition, heat treatment	S_{ut}(ksi)	S_{yt}(ksi)	S_{us}(ksi)
ALUMINUM BASED				
2011, for screw machine parts, excellent machineability, but not weldable and corrosion sensitive	T3	55	43	32
	T8	59	45	35
2014, for aircraft structures; weldable	T3	63	40	37
	T4, T451	61	37	37
	T6, T651	68	60	41
2017, for screw machine parts	T4, T451	62	40	38
2018, for engine cylinders, heads, pistons	T61	61	46	39
2024, for truck wheels, screw machine parts, aircraft structures	T3	65	45	40
	T4, T351	64,	42	40
	T361	72	57	42
2025, for forgings	T6	58	37	35
2117, for rivets	T4	43	24	28
2219, high temperatures applications (up to 600·F), excellent weldability and machineability.	T31, T351	52	36	
	T37	57	46	
	T42	52	27	
3003, for pressure vessels and storage tanks, poor machineability but good weldability. Excellent corrosion resistance.	0	16	6	11
	H12	19	18	12
	H14	22	21	14
	H16	26	25	15
3004, same characteristics as 3003	0	26	10	16
	H32	31	25	17
	H34	35	29	18
	H36	38	33	20
4032 pistons	T6	55	46	38
5083, unfired pressure vessels, cryogenics, towers, drilling rigs	0	42	21	25
	H116, H117, H321	46	33	
5154, salt water services, welded structures, storage tanks	0	35	17	22
	H32	39	30	22
	H34	42	33	24
5454, same characteristics as 5154	0	36	17	23
	H32	40	30	24
	H34	44	35	26
5456, same characteristics as 5154	0	45	23	
	H111	47	33	
	H321, H116, H117	51	37	30
6061 corrosion resistant and good weldability. Used in railroad cars.	T4	33	19	
	T6	42	37	
7178	0	33	15	
	T6	88	78	
COPPER BASED				
Copper, commercial purity	annealed (furnace cool from 400·C)	32	10	
	cold drawn	45	40	
Cartridge brass: 70% Cu, 30% Zn	cold rolled (annealed 400·C, furnace cool)	76	63	
Copper-Beryllium (1.9% Be, .25% Co)	annealed, wqf 1450·F	70		
	cold rolled	200		
	hardened after annealing	200	150	
Phosphor-Bronze, for springs	wire, .025'' and under	145		
	.025'' to .0625''	135		
	.125'' to .250''	125		
Monel metal	cold drawn bars, annealed	70	30	
Red brass	sheet and strip half-hard	51		
	hard	63		
	spring	78		
Yellow brass	sheet and strip half-hard	55		
	hard	68		
	spring	86		
NICKEL BASED				
pure nickel, magnetic, high corrosion resistance	annealed (ht 1400·F, acrt)	46	8.5	
Inconel X, type 550, excellent high temperature properties	annealed at 2050·F	125	75	
	annealed and age hardened	175	110	
K-monel, excellent high temperature properties and corrosion resistance	annealed (wqf 1600·F)	100	45	
	age hardened spring stock	185	160	
Invar, 36% Ni, 64% Fe, low coef of expansion (.9EE6 %/·C)	annealed (wqf 800·C)	71	40	

Appendix B, continued

Material, designation, composition, typical use, and source if applicable	Condition, heat treatment	S_{ut}(ksi)	S_{yt}(ksi)		
REFRACTORY METALS (Properties at room temperature)					
Molybdenum	as rolled	100	75		
Tantalum	annealed at 1050·C in vacuum	60	45		
	as rolled	110	100		
Titanium, commerical purity	annealed at 1200·F	95	80		
Titanium, 6%Al, 4%V	annealed at 1400·F, acrt	135	130		
	heat treated (wqf 1750·F, ht 1000·F, acrt)	170	150		
Titanium, 4%Al, 4%Mn OR 5%Al, 2.75%Cr, 1.25%Fe, OR 5%Al, 1.5%Fe, 1.4%Cr, 1.2%Mo	wqf 1450·F, ht 900·F, acrt	185	170		
Tungsten, commercial purity	hard wire	600	540		
CAST IRON (Note redefinition of columns)		S_{ut}(ksi)	S_{yt}(ksi)	S_{us}(ksi)	S_{uc}(ksi)
Gray cast iron	class 20	20		32.5	80
	class 25	25		34	100
	class 30	30		41	110
	class 35	35		49	125
	class 40	40		52	135
	class 50	50		64	160
	class 60	60		60	150
MAGNESIUM		S_{ut}(ksi)	S_{yt}(ksi)	S_{us}(ksi)	
AZ92, for sand and permanent-mold casting	as cast	24	14		
	solution treated	39	14		
	aged	39	21		
AZ91, for die casting	as cast	33	21		
AZ31X (sheet)	annealed	35	20		
	hard	40	31		
AZ80X, for structural shapes	extruded	48	32		
	extruded and aged	52	37		
ZK60A, for structural shapes	extruded	49	38		
	extruded and aged	51	42		
AZ31B (sheet and plate), for structural shapes in use below 300·F	temper O	32	15	17	
	temper 1124	34	18	18	
	temper 1126	35	21	18	
	temper F	32	16	17	

Appendix C
Centroids

SHAPE	DIMENSIONS	CENTROID (x_c, y_c)	AREA MOMENT OF INERTIA
Rectangle		$(\frac{1}{2}b, \frac{1}{2}h)$	$I_{x'} = (1/12)bh^3$ $I_{y'} = (1/12)hb^3$ $I_x = (1/3)bh^3$ $I_y = (1/3)hb^3$ $J_C = (1/12)bh(b^2 + h^2)$
Triangle		$y_c = (h/3)$	$I_{x'} = (1/36)bh^3$ $I_x = (1/12)bh^3$
Trapezoid		$y'_c = \dfrac{h(2B + b)}{3(B + b)}$ Note that this is measured from the top surface.	$I_{x'} = \dfrac{h^3(B^2 + 4Bb + b^2)}{36(B+b)}$ $I_x = \dfrac{h^3(B + 3b)}{12}$
Quarter-Circle, of radius r		$((4r/3\pi), (4r/3\pi))$	$I_{x'} = 0.055r^4$ $I_x = I_y = (1/16)\pi r^4$ $J_O = (1/8)\pi r^4$
Half Circle, of radius r		$(0, (4r/3\pi))$	$I_x = I_y = (1/8)\pi r^4$ $J_O = \frac{1}{4}\pi r^4$ $I_{x'} = .11r^4$
Circle, of radius r		$(0,0)$	$I_x = I_y = \frac{1}{4}\pi r^4$ $J_O = \frac{1}{2}\pi r^4$
Parabolic Area		$(0, (3h/5))$	$I_x = 4h^3a/7$ $I_y = 4ha^3/15$
Parabolic Spandrel		$((3a/4), (3h/10))$	$I_x = ah^3/21$ $I_y = 3ha^3/15$

Appendix D
Stress Concentration Factors

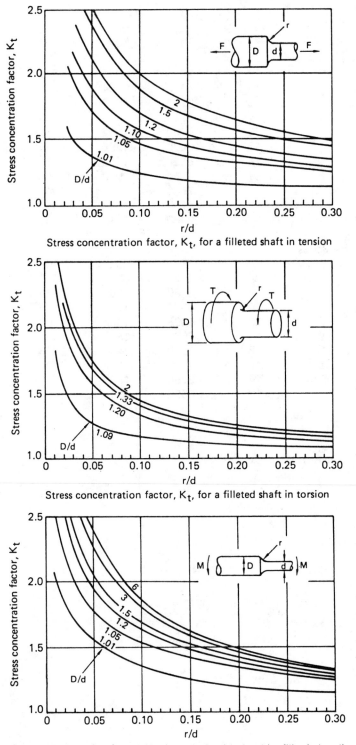

Stress concentration factor, K_t, for a filleted shaft in tension

Stress concentration factor, K_t, for a filleted shaft in torsion

Stress concentration factor, K_t, for a shaft with shoulder fillet in bending

Appendix E
Bolt and External Screw Thread Dimensions
(inches and square inches)

Nominal Size	Threads per inch	Major Diameter	Minor Area	Stress Area *
Coarse Series				
1/4	20	.2500	.0269	.0317
5/16	18	.3125	.0454	.0522
3/8	16	.3750	.0678	.0773
7/16	14	.4375	.0933	.1060
1/2	13	.5000	.1257	.1416
1/2	12	.5000	.1205	.1374
9/16	12	.5625	.1620	.1816
5/8	11	.6250	.2018	.2256
3/4	10	.7500	.3020	.3340
7/8	9	.8750	.4193	.4612
1	8	1.0000	.5510	.6051
Fine Series				
1/4	28	.2500	.0326	.0362
5/16	24	.3125	.0524	.0579
3/8	24	.3750	.0809	.0876
7/16	20	.4375	.1090	.1185
1/2	20	.5000	.1486	.1597
9/16	18	.5625	.1888	.2026
5/8	18	.6250	.2400	.2555
3/4	16	.7500	.3513	.3724
7/8	14	.8750	.4805	.5088
1	12	1.0000	.6245	.6624

*for tension and compression

PRACTICE PROBLEMS: MECHANICS OF MATERIALS

Warm-ups

1. A 12' beam weighs 20 pounds per foot and is supported at the left end and two feet from the right end. A 100 pound load is applied two feet from the left end, and an 80 pound load is applied at the right end. What are the maximum moment and shear?

2. A 6" wide by 4" high cantilever beam is 6' long. It carries a concentrated 200 pound load one foot from the free end and 120 pounds two feet from the free end. What is the tip deflection if the modulus of elasticity is 1.5 EE6 psi?

3. A straight steel beam is 200' long and is supported in such a manner as to allow only $\frac{1}{2}''$ expansion. If the temperature changes 70°F, what will be the stress?

4. A structural steel wide-flange beam will be used as a column with both ends fixed. A vertical load of 75,000 pounds will be supported. The beam is 50' long, and a safety factor of 2.5 is to be used. What is the required moment of inertia?

5. What is the total length of a 1" diameter steel rod carrying a tensile load of 15,000 lbf with an elongation of .158"?

6. A structural steel shape is subjected to a maximum tensile stress of 8240 psi. What is the factor of safety?

Concentrates

1. A 3" diameter horizontal shaft carries a 32" diameter, 600 pound pulley on its 8" overhung end. The belt approaches horizontally and has upper and lower tensions of 1500 and 350 pounds, respectively. What is the maximum stress in the shaft? *NOT To be Confused*
WITH Point In A PLANE

2. A 25' long column ($I = 350$ in^4, $A = 25.6$ in^2, $c = 7''$, yield strength $= 36,000$ psi) has pinned ends. It carries a central load of 100 kips and an eccentric load of 50 kips 10" from the neutral axis. What is the factor of safety?

    W D H
3. A 2' × 3' × 1' high tank is constructed of .25" steel. What is the maximum stress if the tank is pressurized to 2 psig? Neglect stress concentration factors at the corners.

4. A short section of 1.750" inside diameter pipe is to be constructed from turned steel bar stock. The pipe will carry an internal load of 2000 psig. If the allowable stress is 20,000 psi, what outside diameter is needed? **Disregard the longitudinal stress. Do not assume a thin-walled cylinder.**

5. A .742" inside diameter, 1.486" outside diameter aluminum cylinder is subjected to an external pressure of 400 psig. What is the maximum stress developed?

6. A 1" diameter rod is held firmly in a chuck. A 12" wrench applies 60 pounds of twist 8" from the chuck. What are the maximum shear and normal stresses?

7. A horizontal bar with a 1.5 square inch cross section is acted upon by an 18,000 pound compressive load at each end. What are the normal and shearing stresses on a plane inclined +30° from the horizontal if the shear stress at that point is 4000 psi?

8. A 16" outside diameter shell with wall thickness of 0.10" is subjected to a 40,000 pound tensile load and a 400,000 in-lbf torque. What are the principal stresses and the maximum shear stress?

Timed (1 hour allowed for each)

1. The offset wrench handle shown is constructed from a round bar $\frac{5}{8}''$ in diameter with a modulus of elasticity of 29.6 EE6 psi. (a) What are the principal stresses at section A-A? (b) What is the maximum shear at section A-A?

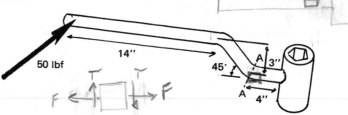

2. A gun barrel is made by shrinking a jacket over a tube.

· tube inside diameter is 4.7"
· tube outside diameter is 7.75"
· jacket outside diameter is 12"

The jacket and the tube are both steel with a modulus of elasticity of 29.6 EE6 psi, and Poisson ratio of .3. (a) What is the diametral interference that will keep the jacket stress at 18,000 psi? (b) What is the maximum circumferential stress in the jacket? (c) What is the minimum circumferential stress in the tube?

3. A brass tube (2.0" O.D., 1" I.D., 6' long) is loaded as a fixed-end cantilever beam. The concentrated load at the free end is 50 pounds. The tip deflection in this condition is excessive, and it is suggested that a 1" O.D. steel rod be inserted into the entire length of the brass tube. (a) Is such a suggestion with merit? (b) What is the change in tip deflection, if any? Express your answer in percent. Neglect the weight of the brass and the steel.

4. A spool valve is constructed of 6061 T4 aluminum as shown below. A 500 psig pressure differential exists across the spool shaft. The shaft is 1.0″ O.D. with a .050″ wall. The end disks are 2″ O.D. and $\frac{3}{8}$″ thick. The end disks are rigidly attached to the tube. Use the distortion energy theory to calculate the factor of safety for this valve.

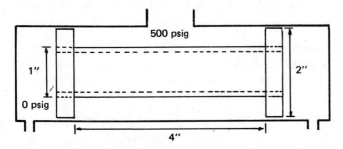

5. Compute the allowable load, P, on the column shown, using a factor of safety of 2. The column, which is supported by a system of cables, is $1\frac{1}{2}$″ schedule 40 mild steel pipe with a yield strength of 30,000 psi.

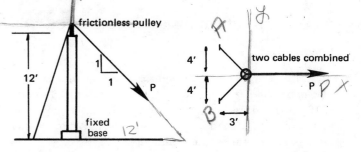

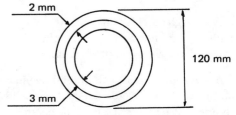

6. Two hollow cylinders are press-fitted together, as shown. $E = 207$ EE9 N/m^2 for both parts. $\mu = .3$ for both parts. The diametral interference is .3 mm. What is the circumferential stress at the interface in both cylinders?

RESERVED FOR FUTURE USE

RESERVED FOR FUTURE USE

15

MACHINE DESIGN

PART 1: Failure Theories

1 STATIC FAILURE THEORIES

A. BRITTLE MATERIALS

Small machine parts, such as gears, springs, and bolts, can be analyzed using formulas derived from general failure theories. The general procedure is to calculate the stresses in a part, and then to compare these stresses with the strength of the material.

Failure theories are used to determine if the part is suitable for an application. Each of the failure theories presented here is based on knowing the highest stresses

in an object. The art of machine design is knowing where the highest stresses occur.

Stresses in a part may be higher than is immediately apparent because of the combining effect of forces in different directions and because of stress concentration factors.

In many cases, the loading is two-dimensional.[1] Knowing all the plane stresses, the principal stresses are found from equation 15.1. (Positive stresses are tensile, and negative stresses are compressive.)

$$\sigma_1, \sigma_2 = \frac{\sigma_x + \sigma_y}{2} \pm \sqrt{\left(\frac{\sigma_x - \sigma_y}{2}\right)^2 + \tau_{xy}^2} \qquad 15.1$$

Example 15.1

A $\frac{1}{2}''$ diameter bolt is loaded as shown. Find the principal stresses. Neglect effects of eccentricity.

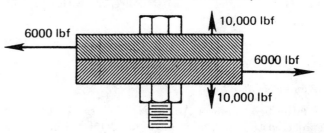

An element located in the bolt at the interface of the two plates will be subjected to stresses caused by the tensile and shear loadings.

$$\sigma_y = \frac{10,000}{\pi \left(\frac{.5}{2}\right)^2} = 50,930 \text{ psi}$$

$$\tau_{xy} = \frac{6,000}{\pi \left(\frac{.5}{2}\right)^2} = 30,560 \text{ psi}$$

---

[1] All of these theories can be extended to three-dimensional loading.

Equation 15.1 is used to find the principal stresses.

$$\sigma_1, \sigma_2 = \frac{0 + 50,930}{2} \pm \sqrt{\left(\frac{0 - 50,930}{2}\right)^2 + (30,560)^2}$$

$$\sigma_1 = 65,244 \text{ psi}$$

$$\sigma_2 = -14,314 \text{ psi}$$

If a material yields less than .5% before fracturing, it is classified as a *brittle material*. Such materials (e.g., cast iron, ceramics, and concrete) usually have a much greater compressive strength than tensile strength. They fail by fracture, not by yielding.

The *maximum normal stress theory* gives reasonably good predictions of the failure stress for brittle materials with static loading. Failure will occur if the largest tensile principal stress is greater than the ultimate tensile strength, or if the largest compressive principal stress is greater than the ultimate compressive strength.

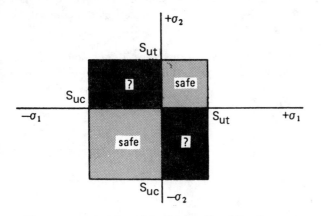

Figure 15.1 Allowable Stress Combinations

The *factor of safety* for brittle materials with constant loading is[2]

$$FS = \left|\frac{S_u}{\sigma}\right| \qquad\qquad 15.2$$

Example 15.2

A lathe bed made of ASTM No. 30 gray cast iron is subjected to a maximum stress of 17,150 psi in tension and 42,800 psi in compression. What is the factor of safety?

The approximate strengths of No. 30 gray cast iron are 30,000 psi in tension and 110,000 psi in compression.

$$FS_{\text{tension}} = \left|\frac{30,000}{17,150}\right| = 1.75$$

$$FS_{\text{compression}} = \left|\frac{110,000}{-42,800}\right| = 2.57$$

---

[2] The factor of safety should not be confused with the *margin of safety*, which is always one less than the factor of safety.

The limiting case is tensile failure with a safety factor of 1.75.

The maximum normal stress theory is somewhat in conflict with experimental evidence. In particular, failures occur in the second and fourth quadrants of figure 15.1, even though the stresses are less than the ultimate strengths. To deal with these failures, two other failure theories are used with brittle materials.

The *Coulomb-Mohr theory* for brittle materials, also known as the *internal friction theory*, is a conservative theory which better defines the failure line in the second and fourth quadrants. The region of safe operation, as specified by this theory, is illustrated in figure 15.2, along with typical points representing failures.

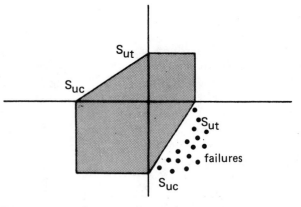

Figure 15.2 Coulomb-Mohr Theory for Brittle Materials

The factor of safety with the Coulomb-Mohr theory is calculated from equation 15.2, just as it was with the maximum normal stress theory.

Since the failure line in figure 15.2 misses the failures by a considerable amount, the theory has been modified to predict more closely the acceptable stress region. This so-called *modified Mohr theory* for brittle materials is illustrated in figure 15.3.

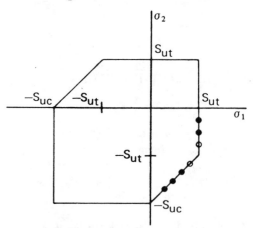

Figure 15.3 Modified Mohr Theory for Brittle Materials

Example 15.3

A .5″ diameter dowel is made from No. 40 cast iron. It carries a compressive load of 15,000 pounds and is subjected to torsion. What shear stress will cause failure? Use the modified Mohr theory.

The approximate properties of No. 40 cast iron are

$$S_{uc} = 135,000 \text{ psi}$$
$$S_{ut} = 40,000 \text{ psi}$$

The negative compressive stress in the dowel is

$$\sigma_c = \frac{15,000}{\frac{\pi}{4}(.5)^2} = -76,400 \text{ psi}$$

The principal stresses are predicted by equation 15.1, but the shear stress, τ_{xy}, is unknown, so a trial and error solution is required.

Try $\tau_{xy} = 10,000$ psi with $\sigma_x = -76,400$ and $\sigma_y = 0$.

$$\sigma_1, \sigma_2 = \tfrac{1}{2}(-76,400) \pm \sqrt{\left(\frac{-76,400}{2}\right)^2 + (10,000)^2}$$
$$= 1300, -77,700 \text{ psi}$$

The next step is to draw the failure envelope and to plot the values of σ_1 and τ_{xy}. If the point is close to or on the failure line, the chosen value of τ_{xy} will be the limiting value. This is done, and the first trial value of τ_{xy} is not sufficiently close to the failure line.

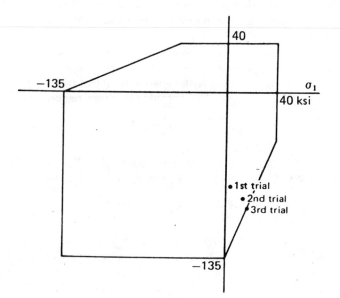

Next try $\tau_{xy} = 30,000$ psi.

$$\sigma_1, \sigma_2 = 10,400, -86,800$$

This point also is not sufficiently close to the failure line.

Try $\tau_{xy} = 40,000$.

$$\sigma_1, \sigma_2 = 17,100, -93,500$$

This point is essentially on the failure line, so it is concluded that the maximum allowable shear stress is 40,000 psi. (It is a coincidence that this also is the value of S_{ut}.)

B. DUCTILE MATERIALS

Materials which yield more than 5% before fracture, such as steel and aluminum, are classified as *ductile materials*. These materials fail by yielding (plastic deformation), not by fracture.

According to the *maximum shear stress theory*, yielding begins when the maximum shear stress equals the maximum shear strength.[3]

$$\tau_{max} = \frac{\sigma_1 - \sigma_2}{2} \qquad 15.3$$

Implicit in this theory is that yield strength in shear is half the tensile yield strength.

$$S_{ys} = \frac{S_{yt}}{2} \qquad 15.4$$

Therefore, the design criterion is

$$\tau_{max} \leq \frac{S_{yt}}{2} \qquad 15.5$$

The allowable stress combinations are shown graphically in figure 15.4. The shape of this failure envelope is similar to that of the Coulomb-Mohr theory for brittle materials, but the limits are based on yield strengths, not on ultimate strengths. Also, since S_{yt} and S_{yc} are assumed equal for ductile materials, the failure envelope is symmetrical.[4]

The factor of safety with the maximum shear stress theory is

$$FS = \frac{S_{yt}}{2\tau_{max}} \qquad 15.6$$

If the symmetrical shape of figure 15.4 is accepted, it is easy to justify equation 15.4, which predicts the shear strength. If the loading is pure shear, the two principal stresses will each be equal, one being the negative of the other and each having the magnitude of τ_{xy}. Plotting the locus of points with $\sigma_1 = -\sigma_2$, the failure envelope is encountered at $\tfrac{1}{2}S_y$.

[3]The maximum shear stress theory is not limited to pure torsional loads. The shear stress is used to indicate failure, but that failure can be due to the combined effects of both shear and normal stresses.

[4]Notice that the limits in figure 15.4 are not divided by 2. The failure envelope is used with the principal stresses. The shear stress is not plotted.

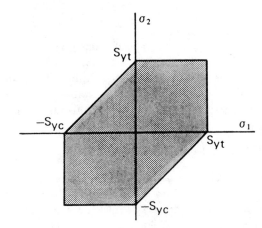

Figure 15.4 Maximum Shear Stress Theory

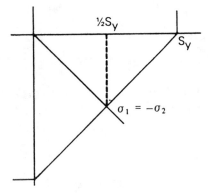

Figure 15.5 Predicting Failure in Pure Shear

Example 15.4

Strain gages on a bearing support show the three principal stresses to be 10,600 psi, 2,400 psi, and $-9,200$ psi. The bearing support is cast aluminum, 356-T6, $S_{yt} = 24,000$ psi. Using the maximum shear stress theory, what is the factor of safety?

$$\tau_{12} = \frac{10,600 - 2,400}{2} = 4,100 \text{ psi}$$

$$\tau_{23} = \frac{2,400 - (-9,200)}{2} = 5,800 \text{ psi}$$

$$\tau_{13} = \frac{10,600 - (-9,200)}{2} = 9,900 \text{ psi}$$

$$\tau_{\max} = 9,900 \text{ psi}$$

From equation 15.6,

$$FS = \frac{24,000}{2(9900)} = 1.21$$

The strain energy theory seldom is used to predict failures. However, it is similar in development to the von Mises theory. Therefore, this older theory is covered here.

The *strain energy theory* makes use of the straight line nature of the stress-strain curve in the elastic region. The area under this curve represents the energy that

is put into one cubic inch of material. This strain energy per unit volume also is known as the *modulus of resilience*.

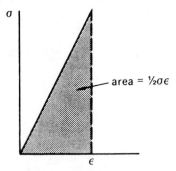

Figure 15.6 Strain Energy as Area Under the Stress-Strain Curve

For two-dimensional loading on the principal planes, the strain energy is calculated from superposition.

$$U = \tfrac{1}{2}(\sigma_1\epsilon_1 + \sigma_2\epsilon_2) \qquad 15.7$$

Unfortunately, the strains usually are not known. They have to be calculated from the stresses, the modulus of elasticity, and Poisson's ratio.

$$\epsilon_1 = \frac{1}{E}(\sigma_1 - \mu\sigma_2) \qquad 15.8$$

$$\epsilon_2 = \frac{1}{E}(\sigma_2 - \mu\sigma_1) \qquad 15.9$$

$$U = \frac{1}{2E}(\sigma_1^2 + \sigma_2^2 - 2\mu\sigma_1\sigma_2) \qquad 15.10$$

One possible design criterion is to keep the actual strain energy less than the strain energy at failure. The strain energy at failure (the modulus of resilience) is determined from a simple tensile test. At failure in that test, $\sigma_1 = S_{yt}$, and $\sigma_2 = 0$. Therefore, the modulus of resilience is determined from equation 15.11 as

$$MR = \frac{1}{2E}\big[S_{yt}^2 + (0)^2 - 2(0)\big] = \frac{S_{yt}^2}{2E} \qquad 15.11$$

Eliminating the $\frac{1}{2E}$ terms, the design criterion with the strain energy theory is given by equation 15.12.

$$\sigma_1^2 + \sigma_2^2 - 2\mu\sigma_1\sigma_2 \le S_{yt}^2 \qquad 15.12$$

The *distortion energy theory* (also known as the *von Mises theory*[5]) is similar in development to the strain energy method, but stricter. It is quite accurate in predicting failure of steel parts in tension and shear. The *von Mises stress* is defined as[6]

$$\sigma' = \sqrt{\sigma_1^2 + \sigma_2^2 - \sigma_1\sigma_2} \qquad 15.13$$

---

[5]Pronounced "von Mee'-sehs"

[6]Remember that σ_1 and σ_2 are the principal stresses, not necessarily the same as σ_x and σ_y.

Failure is assumed to occur when

$$\sigma' > S_{yt} \qquad 15.14$$

The design criterion (similar to equation 15.12) is

$$\sigma_1^2 + \sigma_2^2 - \sigma_1\sigma_2 \leq S_{yt}^2 \qquad 15.15$$

The factor of safety according to the von Mises theory is

$$FS = \frac{S_{yt}}{\sigma'} \qquad 15.16$$

If loading is pure torsion, $\sigma_1 = -\sigma_2 = \pm\tau_{max}$. From equation 15.15, substituting τ_{max} for σ_1 and σ_2,

$$S_{yt}^2 = 3\tau_{max}^2 \qquad 15.17$$

$$\tau_{max} = \frac{S_{yt}}{\sqrt{3}} = .577 S_{yt} \qquad 15.18$$

This implies

$$S_{ys} = .577 S_{yt} \qquad 15.19$$

Equation 15.19 predicts a larger yield point in shear than did the maximum shear stress theory ($.5 S_{yt}$).

Example 15.5

Find the factor of safety of the bearing support in example 15.4 using the distortion energy theory.

$$\sigma' = \sqrt{\tfrac{1}{2}[(10{,}600 - 2400)^2 + (2400 - (-9200))^2 + (-9200 - 10{,}600)^2]}$$
$$= 17{,}231 \text{ psi}$$
$$F.S. = \frac{24{,}000}{17{,}231} = 1.39$$

2 DYNAMIC FAILURE THEORIES

Machine parts subjected to cyclic loading can fail even if stressed below their yield points.[7] This is known as *fatigue failure*. The *fatigue strength* is the peak stress that a material subjected to completed reversed loading can survive without yielding.

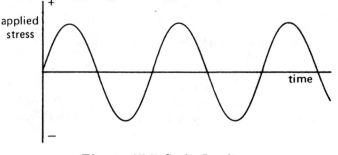

Figure 15.7 Cyclic Loading

---

[7]Cyclic loading usually is shown as a sinusoid. In fact, most endurance tests use sinusoidal loading of one form or another. However, the fatigue and endurance strengths do not seem to depend on the shape of the loading curve. Therefore, these concepts can be used with other types of loading (e.g., sawtooth, square wave, etc.).

When the fatigue strength is plotted logarithmically against the number of cycles (i.e., an *S-N curve*), a linear relationship is found in the range of EE3 to EE6 cycles for steel. Below EE3 cycles, the fatigue strength equals the ultimate strength, and above EE6 there is no further decrease in strength. This strength for an infinite life is called the *endurance strength, S_e'*.

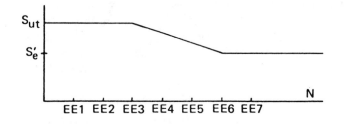

Figure 15.8 A Typical S-N Curve for Steel

For steel, the endurance strength is approximately[8]

$$S_{e,\text{ steel}}' = 0.5 S_{ut} \qquad [S_{ut} < 200{,}000 \text{ psi}]$$
$$= 100{,}000 \text{ psi} \quad [S_{ut} > 200{,}000 \text{ psi}] \qquad 15.20$$

If necessary, the ultimate strength of steel can be calculated from its *Brinell hardness*.

$$S_{ut} \approx (500)(BHN) \qquad 15.21$$

For cast iron, the endurance strength is lower.

$$S_{e,\text{ cast iron}}' = 0.4 S_{ut} \qquad 15.22$$

The strength of aluminum continues to decrease with cyclic loading and never levels off. For this reason, the endurance strength of aluminum usually is taken as the fatigue strength after EE8 or 5 EE8 cycles. If no experimental values are known, the endurance strength can be approximated by equation 15.23.

$$S_{e,\text{ aluminum}}' = \begin{cases} .3 S_{ut} & \text{(cast)} \\ .4 S_{ut} & \text{(wrought)} \end{cases} \qquad 15.23$$

A part experiencing less than 1000 cycles is analyzed as a static part.[9] A steel part subjected to more than EE6 cycles (or aluminum subjected to more than EE8) can be analyzed using the endurance strength. If a part is designed to last some number of cycles in between, an S-N curve should be used to determine the fatigue strength at that number of cycles.

[8]The .5 coefficient in equation 15.20 is actually an average. The coefficient varies between .4 and .6. However, .5 commonly is quoted.

[9]In some instances of ductile construction, such as steel beam building construction, the static case can be assumed with as many as 20,000 cycles.

In actual design problems, the S-N curve cannot be used directly. The curve is actually a composite of many experimental failures, and the line is a locus of average failure stresses. It is possible, with sufficient experimental data, to draw *confidence limits* above and below the S-N curve. A machine part then can be designed with some chosen probability (e.g., 95% or 99%) of not failing in fatigue.

Further complicating the direct use of fatigue test data is the condition of the part. Fatigue strength is a function of the material. However, test data is collected on specimens that are much more carefully controlled than actual production parts. Therefore, the endurance limit must be reduced.

$$S_e = kS_e' \qquad 15.24$$

The k factor in equation 15.24 actually is a combination of many terms affected by surface finish, size, desired reliability, operating temperature, stress concentrations, notch sensitivity, and other factors. The actual values should be determined experimentally for each design, but general reduction factors have been quoted in machine design textbooks.

Example 15.6

A 1″ diameter AISI 4130 steel shaft, machined as shown, is subjected to completely reversed bending. The yield strength of cold drawn 4130 steel is 87,000 psi. The ultimate tensile strength is 98,000 psi. For a 99% reliability, what is the maximum allowable bending stress in the shaft?

Use the following endurance strength reduction factors.

$$k_{\text{surface finish}} = .73 \ (\text{machined})$$
$$k_{\text{size}} = .85 \ (0.3'' \text{ to } 2.0'')$$
$$k_{\text{reliability}} = .814 \quad (99\%)$$
$$k_{\text{temperature}} = 1.0 \ (\text{room temperature})$$
$$k_{\text{stress concentration}} = .70$$
$$k_{\text{plating}} = 1.0 \ (\text{none})$$

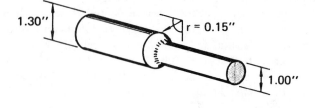

$$S_e' = (0.5)(S_{ut}) = (0.5)(98,000) = 49,000 \text{ psi}$$
$$S_e = (.73)(.85)(.814)(1.0)(.70)(1.0)(49,000)$$
$$= 17,324 \text{ psi}$$

3 FLUCTUATING STRESSES

Many parts are subjected to a combination of static and reversed loading, as illustrated in figure 15.9.

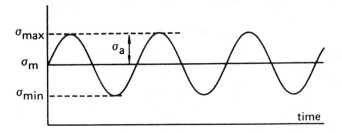

Figure 15.9 Sinusoidal Fluctuating Stress

The *mean stress* is

$$\sigma_m = \frac{\sigma_{\max} + \sigma_{\min}}{2} \qquad 15.25$$

The *alternating stress* is half the *range stress*.

$$\sigma_a = \frac{\sigma_{\max} - \sigma_{\min}}{2} \qquad 15.26$$

For a part subjected to a fluctuating load, failure cannot be determined solely by yield strength or endurance limit. The combined effects of loading must be considered. The stress usually is described graphically on a diagram which plots the mean stress versus the alternating stress. Both of these stresses may be either normal stresses or shear stresses. A criterion for failure is established by relating the yield strength, the ultimate strength, and the endurance limit. One method of relating this information is a *Soderberg line*.[10]

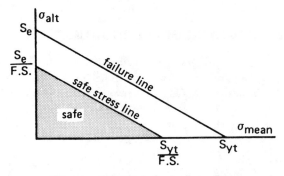

Figure 15.10 A Soderberg Line

Figure 15.10 illustrates how an area of safe operation is enveloped by drawing a straight line from the endurance limit to the yield stress. Both of these values should be divided by a suitable factor of safety to define the *safe operating area*.

Stress concentration factors (with the notable exception of the Wahl correction factor for springs) are applied to the alternating stress only. This is justified since duc-

---

[10] The Soderberg method is particularly suitable for normal stresses in ductile materials for which S_e and S_{yt} are known. However, it is the most conservative of the fluctuating stress failure theories.

tile materials, such as steel, will yield around discontinuities, reducing a constant stress.

Once the operating point is plotted on a Soderberg diagram, the actual factor of safety can be calculated from the equivalent stress, defined in figure 15.11.

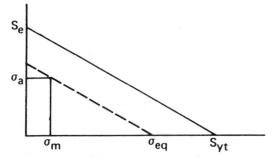

Figure 15.11 The Equivalent Stress

$$FS = \frac{S_{yt}}{\sigma_{eq}} \qquad 15.27$$

The envelope of actual failures is more closely related to a parabolic line extending above the Soderberg line from the endurance limit to the ultimate strength. This so-called *Gerber line* is difficult to work with simply because it is not easily constructed.

The *Goodman diagram* (or *modified Goodman diagram*) is less conservative than the Soderberg digram. It can be used for steel, aluminum, titanium, and some magnesium alloys. The factor of safety is determined from equation 15.28.

$$FS = \frac{S_e}{\sigma_{eq}} \qquad 15.28$$

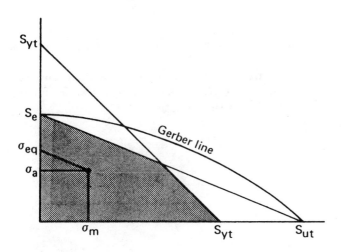

Figure 15.12 Goodman Diagram and Gerber Line

The Goodman diagram should not be used with cast iron and some types of magnesium. The envelope of failure is bounded by the *Smith line* for cast iron. This line is similar to the Gerber line, except that it runs under the $S_e - S_{ut}$ line, not over it.

When a part is stressed repeatedly in shear, the endurance limit in shear can be calculated from equation 15.30.[11] Then figure 15.13 can be used to determine if the operation is in the safe region.

$$S_{ys} = (.577)S_{yt} \qquad 15.29$$
$$S_{es} = (.577)S_e \qquad 15.30$$

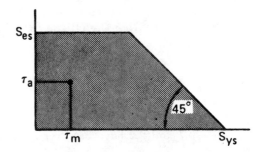

Figure 15.13 Fluctuating Shear Stresses

Example 15.7

An aircraft bellcrank is made from aluminum 6061-T6 alloy. Yield strength is 40,000 psi, ultimate strength is 45,000 psi, and the fatigue strength is 13,500 psi reduced to 8500 psi by various derating factors. The part is to be subjected to a maximum stress of 9500 psi and a minimum stress of 6500 psi, both tensile. If the part is designed for an infinite life with a safety factor of 3, will the material be suitable?

$$\sigma_m = \tfrac{1}{2}(9500 + 6500) = 8000 \text{ psi}$$
$$\sigma_a = \tfrac{1}{2}(9500 - 6500) = 1500 \text{ psi}$$

The material properties are divided by the safety factor prior to being plotted on the modified Goodman diagram. The actual stress values, of course, are not reduced by the safety factor.

$$\frac{S_{yt}}{3} = \frac{40,000}{3} = 13,333 \text{ psi}$$
$$\frac{S_{ut}}{3} = \frac{45,000}{3} = 15,000 \text{ psi}$$
$$\frac{S_e}{3} = \frac{8500}{3} = 2833 \text{ psi}$$

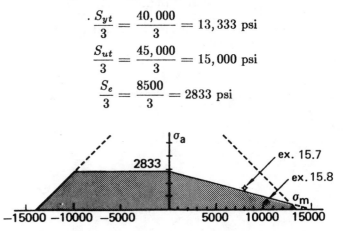

Since the operating point lies above the failure line, the material is not satisfactory.

---

[11]Whether the coefficients are 0.577 or 0.5 depends on whether the maximum shear stress theory or the distortion energy theory is used.

4 CUMULATIVE FATIGUE

The subject of cumulative fatigue is not yet completely researched. However, the *Palmgren-Miner cycle ratio summation theory* (usually called just *Miner's rule*) often is used to evaluate cumulative damage.[12]

If a machine part is subjected to $\sigma_{max, 1}$ for n_1 cycles, $\sigma_{max, 2}$ for n_2 cycles, etc., the part should not fail if equation 15.31 holds. (N_i are the fatigue lives for the various stress levels.)

$$\sum \frac{n_i}{N_i} < C \qquad 15.31$$

A value of 1.0 commonly is used for C. However, it should be determined exactly from experimentation. Values between 0.7 and 2.2 have been reported in the literature.

Miner's rule does not take into account the increase in endurance limit that results when virgin material is understressed.

5 FLUCTUATING COMBINED STRESSES

If there are fluctuating stresses simultaneously present in different directions, a combination of von Mises stresses and the Goodman line is necessary. The following steps can be used as a conservative analysis of such a case.

step 1:   Calculate the principal mean stresses. (These are not x- and y-stresses, but rather principal stresses calculated from equation 15.1.)

step 2:   Calculate the principal alternating stresses.

---

[12]The *cumulative usage factor rule* is another name for this rule.

step 3:   Calculate the mean and alternating von Mises stresses.

$$\sigma'_m = \sqrt{\sigma^2_{m,1} + \sigma^2_{m,2} - \sigma_{m,1}\sigma_{m,2}} \qquad 15.32$$

$$\sigma'_a = \sqrt{\sigma^2_{a,1} + \sigma^2_{a,2} - \sigma_{a,1}\sigma_{a,2}} \qquad 15.33$$

step 4:   Plot the mean and alternating von Mises stresses on a regular Goodman diagram.

Example 15.8

A change in design of the bellcrank from example 15.7 will cause the bracket to be subjected to complicated loading. Strain gages on the part have been used to collect data on the stress variations. Is the design adequate?

Stress	Maximum	Minimum	Mean	Alternating
σ_x	9000	7000	8000	1000
σ_y	−3000	−2000	−2500	500
τ_{xy}	2500	1500	2000	500

From equation 15.1,

$$\sigma_{1m}, \sigma_{2m} = \frac{8000 + (-2500)}{2}$$

$$\pm \sqrt{\left(\frac{8000 - (-2500)}{2}\right)^2 + (2000)^2}$$

$$= 8370 \text{ psi}, -2870 \text{ psi}$$

$$\sigma_{1a}, \sigma_{2a} = \frac{1000 + (500)}{2} \pm \sqrt{\left(\frac{1000 - (500)}{2}\right)^2 + (500)^2}$$

$$= 1020 \text{ psi}, 480 \text{ psi}$$

$$\sigma'_m = \sqrt{(8370)^2 - (8370)(-2870) + (-2870)^2}$$

$$= 10,110 \text{ psi}$$

$$\sigma'_a = \sqrt{(1020)^2 - (1020)(480) + (480)^2} = 880 \text{ psi}$$

σ'_m and σ'_a are plotted on the same Goodman diagram used in example 15.7. This point is below the failure line, so the design is adequate.

PART 2: Specific Design Applications

6 BUILT-UP FLYWHEELS

Special Nomenclature

A	rim cross sectional area	ft^2
b	rim width	ft
C_f	coefficient of fluctuation	–
E	energy	ft-lbf
F	force	lbf
FS	factor of safety	–
g	acceleration due to gravity	ft/sec^2
g_c	gravitational constant (32.2)	$\dfrac{\text{ft-lbm}}{\text{sec}^2\text{-lbf}}$
h	rim depth	ft
J	mass moment of inertia	lbm-ft^2
k	radius of gyration	ft
m	mass	lbm
n	number of holes punched per cycle	–
r	radius	ft
S_{us}	ultimate strength in shear	psi
t	time	sec
v	velocity	ft/sec or in/sec
w	weight	lbf
W	work	ft-lbf

Symbols

α	angular acceleration	rad/sec^2
σ	stress	psf
ρ	density	lbm/ft^3
ω	angular velocity	rad/sec
μ	Poisson's ratio	

Subscripts

ave	average
c	centrifugal
h	hoop
i	inner
k	kinetic
m	mean
o	outer
s	shear

A. LINEAR VELOCITY TERMS

Flywheels store energy as kinetic energy. Ignoring the effects of spokes, hub, and arms, the kinetic energy of rotation can be written in terms of the rim mass and the rim velocity at the mean radius.

$$E_k = \frac{mv^2}{2g_c} \qquad 15.34$$

The flywheel's rim mass is

$$m = 2\pi r_m b h \rho \qquad 15.35$$

$$r_m = \tfrac{1}{2}(r_o + r_i) \qquad 15.36$$

The mean rim velocity is

$$\dot{v} = 2\pi r_m \left(\frac{rpm}{60}\right) \qquad 15.37$$

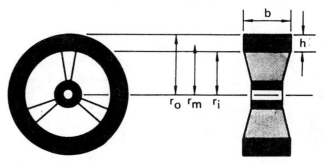

Figure 15.14 Built-up Flywheel

If the velocity fluctuates between velocities v_1 and v_2, the change in the kinetic energy is

$$\Delta E_k = \frac{m(v_1^2 - v_2^2)}{2g_c}$$

$$= \frac{m(v_1 - v_2)(v_1 + v_2)}{2g_c} = \frac{mC_f v_m^2}{g_c} \qquad 15.38$$

The mean velocity is

$$v_m = \tfrac{1}{2}(v_1 + v_2) \qquad 15.39$$

C_f is the *coefficient of fluctuation*.

$$C_f = \frac{v_1 - v_2}{v_m} = \frac{(\text{rpm})_1 - (\text{rpm})_2}{(\text{rpm})_m} \qquad 15.40$$

Speed fluctuations often are the result of work being performed, with the energy required for that work coming from the stored kinetic energy of the flywheel. The change in the kinetic energy is equal to the work done. For example, the work performed by a punch press is

$$W = F_{\text{ave}}(\text{thickness}) \qquad 15.41$$

$$F_{\text{ave}} = \tfrac{1}{2}F_{\text{max}} = \frac{n}{2}\left(\begin{array}{c}\text{hole}\\\text{perimeter}\end{array}\right)(\text{thickness})S_{us} \quad 15.42$$

If the hub and the arms cannot be ignored, it is customary to increase the mass by 5% to 10%. If this accuracy is insufficient, the flywheel will have to be analyzed in terms of angular motion.

B. ANGULAR VELOCITY TERMS

Flywheel kinetic energy can be calculated from the mean angular velocity.

$$E_k = \frac{J\omega^2}{2g_c} \qquad 15.43$$

The *angular velocity* is independent of the radius.

$$\omega = \frac{2\pi(\text{rpm})}{60} \qquad 15.44$$

The mass moment of inertia can be calculated for the entire flywheel. If only the rim is considered,

$$J = \frac{mk^2}{g_c} = \left(\frac{\pi\rho b}{2}\right)(r_o^4 - r_i^4) \qquad 15.45$$

k is the radius of gyration, which is essentially the mean radius, r_m, of the rim. The quantity mk^2 is known as the *flywheel effect*. The time to accelerate a flywheel between two velocities, ω_1 and ω_2, can be found from the applied torque.

$$T = \frac{J\alpha}{g_c} = J\frac{(\omega_1 - \omega_2)}{g_c t} \qquad 15.46$$

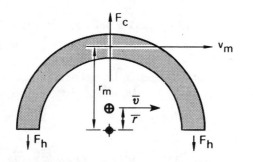

Figure 15.15 Forces on a Flywheel Half

C. FLYWHEEL STRESSES

Assuming that the arms do not restrain the rim from expanding and that the rim thickness is small compared to the mean radius, the *centroidal radius*, $\bar{r}$, is located at

$$\bar{r} = \frac{2r_m}{\pi} \qquad 15.47$$

The centrifugal force can be written in terms of either the mean velocity, v_m, or the velocity, $\bar{v}$, at the centroidal radius.

$$F_c = \frac{m\bar{r}\omega^2}{g_c} = \frac{m\bar{v}^2}{\bar{r}g_c} = \frac{\pi m\bar{v}^2}{2r_m g_c}$$

$$= \frac{mv_m^2}{r_m g_c} \qquad 15.48$$

The hoop force is half the centrifugal force, and the *hoop stress* is

$$\sigma_h = \frac{F_c}{2A} = \frac{F_c}{2bh} = \frac{\rho v_m^2}{g_c} \qquad 15.49$$

If the hoop stress is set equal to the ultimate strength, solving for v will yield the *burst velocity*. The allowable stress is obtained by dividing the ultimate stress by a safety factor between 10 and 13. This also defines the *safe speed*. A rule of thumb is that v should not exceed 6000 fpm.

Example 15.9

A grade 30 built-up cast iron flywheel rotates at a constant speed. The distance from the rotational axis and the rim centroid is 36″. The rim width and depth are 12″ and 4″, respectively. If the density of cast iron is .256 lbm/in³, what is the maximum safe speed?

Choose a safety factor of 10. From equation 15.49, using $g = 386$ in-lbm/sec²-lbf,

$$v_{\text{safe}} = \sqrt{\frac{(386)(30,000)}{(.256)(10)}} = 2127 \text{ in/sec}$$

$$\text{rpm} = \frac{2127(60)}{2\pi(36)} = 564$$

7 DISK FLYWHEELS

A disk flywheel is shown in figure 15.16. Assuming constant thickness, density, and Poisson's ratio, the tangential stress can be determined. The tangential stress will be maximum at the inner boundary.

$$\sigma_{t,\,\text{max}} = \frac{\rho\omega^2}{4g_c}\left[(3+\mu)r_o^2 + (1-\mu)r_i^2\right] \qquad 15.50$$

$$\omega = \frac{2\pi(\text{rpm})}{60} \qquad 15.51$$

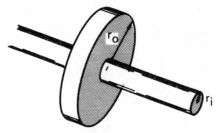

Figure 15.16 A Disk Flywheel

Usually, the tangential stress will control. However, the maximum radial stress also should be checked.

$$\sigma_{r,\,\text{max}} = \left[\frac{3+\mu}{8}\right]\left[\frac{\rho\omega^2(r_o - r_i)^2}{g_c}\right] \qquad 15.52$$

The maximum radial stress occurs at the *geometric mean radius*.

$$r = \sqrt{r_o r_i} \qquad 15.53$$

8 SPRING DESIGN

Nomenclature

A_c	clash allowance	–
C	spring index	–
d	wire diameter	in
D	mean coil diameter	in
E	modulus of elasticity	psi
f	frequency	Hz
F	force exerted by spring (load)	lbf
g_c	gravitational constant (386)	in-lbm/sec²-lbf
G	shear modulus	psi
h	height	in
k	spring rate	lbf/in
L	unwound wire length	in
n_a	number of active coils	–
p	pitch (coil to coil spacing)	in
r	bend radius	in
R	mean coil radius	in
S	material strength	psi
W	Wahl correction factor	in

Symbols

δ	deflection	in
τ	shear stress	psi
ρ	density	lbm/in³
σ	bending stress	psi

Subscripts

a	active
f	free
i	initial
n	the nth
s	solid
t	total
w	working

A. HELICAL COIL COMPRESSION SPRINGS (ROUND WIRE)

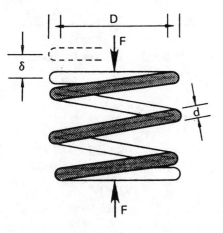

Figure 15.17 Helical Compression Spring

1 Designing for Static Loading

It is assumed that springs designed for static loading follow *Hooke's law*.

$$F = k\delta \qquad 15.54$$

The basic spring design equation is

$$\delta = \frac{F}{k} = \frac{8FD^3 n_a}{Gd^4} \qquad 15.55$$

Deflection also can be expressed in terms of the *spring index, C*.

$$\delta = \frac{8FC^3 n_a}{Gd} = \frac{F}{k} \qquad 15.56$$

$$C = \frac{D}{d} \qquad 15.57$$

C is optimum around 9, although the useful range is approximately 4 to 12.[13] d should be one of the commercially available wire diameters (listed in table 15.2) unless the application warrants a custom wire diameter.

The deflection can be written in terms of the *active wire length, L_a*.

$$\delta = \frac{8FD^2 L_a}{\pi Gd^4} = \frac{8FC^3 L_a}{\pi GdD} \qquad 15.58$$

$$L_a = \pi D n_a \qquad 15.59$$

The number of *active coils* is $n_a = n_t - n^*$. n^* is .5 for *plain ends*, 1.0 for *plain and ground ends*, 1.0 or 1.5 for *squared ends*, and 2.0 for *squared and ground ends*. These end conditions are illustrated in figure 15.18.

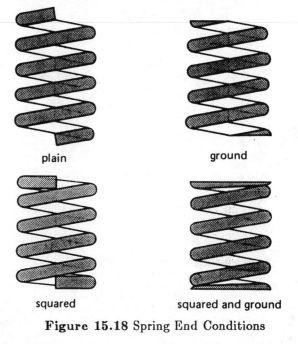

Figure 15.18 Spring End Conditions

[13]It is difficult to wind springs with small spring indexes. The wire cannot be bent to the desired small radius easily. On the other hand, springs with large indexes are flimsy and tend to buckle.

If the *spring rate, k,* is unknown, it can be found from equation 15.60.[14]

$$k = \frac{F}{\delta} = \frac{Gd^4}{8D^3 n_a} = \frac{Gd}{8C^3 n_a} \qquad 15.60$$

The maximum shear stress occurring at the inner face is

$$\tau = \frac{8CFW}{\pi d^2} = \frac{8FDW}{\pi d^3} \qquad 15.61$$

W is the *Wahl correction factor.*[15]

$$W = \frac{4C - 1}{4C - 4} + \frac{0.615}{C} \qquad 15.62$$

The *spring pitch* is the mean coil separation.

$$p = \frac{h_f}{n_a} \qquad 15.63$$

The *solid deflection* is calculated easily.[16]

$$\delta_s = h_f - h_s = h_f - n_a d \qquad 15.64$$

The *solid height, h_s,* should be about .9 times the working height. Alternatively, the *clash allowance* can be specified. A_c usually is about .20.

$$A_c = \frac{\delta_s - \delta_w}{\delta_w} \qquad 15.65$$

Once a spring has been designed, it should be checked for *buckling.* A simple method uses table 15.1. The ratios δ_w/h_f and h_f/D are calculated. The spring will not buckle as long as the maximum value of δ_w/h_f is not exceeded.

Table 15.1
Spring Buckling Limits

h_f/D	maximum δ_w/h_f	
	one end pivoting ball, other squared and ground	both ends squared and ground
4	.45	–
5	.25	–
6	.18	.38
7	.13	.25
8	.10	.20
9	.08	.16
10	.07	.14

[14] This also is known as the *spring constant* and the *spring index.*

[15] Most of the spring design equations are derived easily from superposition of torsional and direct shear stresses, neglecting the spring curvature. The Wahl correction factor corrects for the curvature. As such, it is not a true stress concentration factor but rather a factor which corrects an average stress to the maximum stress occurring at the inside face of the spring wire. The equation is approximate, but the error is less than 2%. Since *W* usually is in the 1.1 to 1.2 range, it can be omitted for initial designs.

[16] To avoid possible failure during improper installations, the stress at solid height should not exceed the ultimate shear strength.

2 Designing for Fatigue Loading

The modified Goodman diagram can be used to predict the safe operating stress of a spring exposed to fluctuating loads. In addition, the natural frequency of the spring should be compared to the operating frequency. Assuming a weightless spring, the nth *harmonic frequency* is given by equation 15.66. If both ends are fixed, test for even values of *n*. If the ends are free, test the odd values of *n*. Specifically, $n = 2$ for common compression springs.

$$f_n = \frac{nd}{16\pi R^2 n_a} \sqrt{\frac{g_c G}{2\rho}} \qquad 15.66$$

If the spring is steel ($\rho = .28$ lbm/in^3, and $G = 11.5$ EE6 psi),

$$f_n \approx \frac{(7100)nd}{n_a D^2} \qquad 15.67$$

This will define the *fundamental frequency.*

To avoid resonance with the harmonics, the spring's fundamental frequency (first harmonic) should be 15 to 20 times higher than the system frequency.

B. HELICAL EXTENSION SPRINGS (ROUND WIRES)

Close-wound *extension springs* can be wound with an initial tension so that a load must be applied to separate the coils. This load is known as the *initial tension.* The initial tension makes it possible to produce springs with highly repetitive free lengths. Of course, if a linear spring rate is required, the initial tension must be zero.

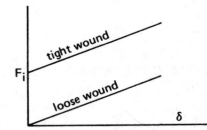

Figure 15.19 Deflection Curves for Extension Springs

The initial stress and tension are related by

$$\tau_i = \frac{8WDF_i}{\pi d^3} \qquad 15.68$$

After the applied load exceeds F_i, the spring should be analyzed as if it were a compression spring.

$$\tau = \tau_i + \frac{8DFW}{\pi d^3} \qquad 15.69$$

$$\delta = \frac{8n_a D^3 (F - F_i)}{Gd^4} \qquad 15.70$$

Extension springs often fail at their hooks from the stress concentration factor caused by the hook curvature. A simplified formula for the maximum bending stress at section A in figure 15.20 is given by equation 15.71.

$$\sigma_A = \frac{16FD}{\pi d^3}\left(\frac{r_1}{r_3}\right) \qquad 15.71$$

A simplified formula for the maximum torsional shear stress at section B is

$$\tau_B = \frac{8FD}{\pi d^3}\left(\frac{r_2}{r_4}\right) \qquad 15.72$$

The allowable working stress in extension springs commonly is limited to 75% of the maximum allowable stress for compression springs of the same material.

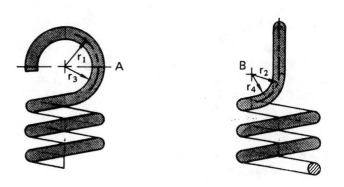

Figure 15.20 Extension Spring Stress Concentrations

Table 15.2
Spring Wire Diameters and Sheet Metal Sizes

Gage	American or B&S (nonferrous sheet and wire) inches	Washburn and Moen (ferrous wire) inches	Piano wire (ferrous wire) inches	Gage	American or B&S (nonferrous sheet and wire) inches	Washburn and Moen (ferrous wire) inches	Piano wire (ferrous wire) inches
0000000	.6513	.4900		18	.0403	.0475	.041
000000	.5800	.4615	.004	19	.0359	.0410	.043
00000	.5165	.4305	.005	20	.0320	.0348	.045
0000	.4600	.3938	.006	21	.0285	.0317	.047
000	.4096	.3625	.007	22	.0253	.0286	.049
00	.3648	.3310	.008	23	.0226	.0258	.051
0	.3249	.3065	.009	24	.0201	.0230	.055
1	.2893	.2830	.010	25	.0179	.0204	.059
2	.2576	.2625	.011	26	.0159	.0181	.063
3	.2294	.2437	.012	27	.0142	.0173	.067
4	.2043	.2253	.013	28	.0126	.0162	.071
5	.1819	.2070	.014	29	.0113	.0150	.075
6	.1620	.1920	.016	30	.0100	.0140	.080
7	.1443	.1770	.018	31	.0089	.0132	.085
8	.1285	.1620	.020	32	.0080	.0128	.090
9	.1144	.1483	.022	33	.0071	.0118	.095
10	.1019	.1350	.024	34	.0063	.0104	.100
11	.0907	.1205	.026	35	.0056	.0095	.106
12	.0808	.1055	.029	36	.0050	.0090	.112
13	.0720	.0915	.031	37	.0045	.0085	.118
14	.0641	.0800	.033	38	.0040	.0080	.124
15	.0571	.0720	.035	39	.0035	.0075	.130
16	.0508	.0625	.037	40	.0031	.0070	.138
17	.0453	.0540	.039				

C. ESTABLISHING ALLOWABLE SPRING STRESSES

1 Static Loading

When loading is static, the maximum allowable stress (before the application of any safety factors) is the yield strength in shear. This can be calculated from the distortion energy theory.[17] A factor of safety of 1.5 is recommended for static and infrequent use.

$$\tau_{max} = S_{ys} = (.577)S_{yt} \qquad 15.73$$

For non-ferrous and stainless steels, the coefficient .577 should be replaced with .35.

2 Fatigue Loading

There are two methods that can be used when springs are repeatedly stressed. One method is to use the modified Goodman diagram for shear stresses. The other method is to design for static loading using a reduced maximum stress.[18] That is,

$$\tau_{max} = (factor)S_{ut} \qquad 15.74$$

The factor applied to the ultimate tensile strength depends on the desired life and condition of the spring wire. Table 15.3 gives factors recommended by the Barnes Group (Associated Spring). These factors can be used with all carbon steel spring materials.

Table 15.3
Strength Reduction Factors for Fatigue Loading

life	unpeened	peened
EE4	.45	.45
EE5	.35	.42
EE6	.33	.40
EE7	.30	.36

[17] In the absence of other information, calculate the tensile yield point as 75% of the ultimate tensile strength.

[18] In effect, equation 15.74 approximates the fatigue strength. Of course, a factor of safety still is needed.

Example 15.10

A steel spring is to be manufactured of W&M No. 4 wire with a spring index of 6. The load fluctuates between 140.2 pounds and 219.8 pounds. The spring material has the following properties: torsional yield point–120,000 psi, and one-way shear endurance limit–100,000 psi. Determine if the design meets a specified factor of safety of 1.5.

From table 15.2, the wire diameter is .2253". The Wahl factor is

$$W = \frac{(4)(6) - 1}{(4)(6) - 4} + \frac{0.615}{6} = 1.2525$$

Both the endurance limit and the torsional yield point are given. Although equation 15.74 could be used, this would ignore the given endurance limit. Therefore, the modified Goodman diagram would be a better approach.

Calculate the mean and alternating stresses. Notice that since the Wahl factor corrects for curvature and is not a true stress concentration factor, it is automatically included in the calculation of the mean and alternating stresses. From equation 15.61,

$$\tau_{max} = \frac{(8)(6)(219.8)(1.2525)}{\pi(.2253)^2} = 82,865 \text{ psi}$$

$$\tau_{min} = \frac{(8)(6)(140.2)(1.2525)}{\pi(.2253)^2} = 52,856 \text{ psi}$$

$$\tau_m = \tfrac{1}{2}(82,865 + 52,856) = 67,861$$

$$\tau_a = \tfrac{1}{2}(82,865 - 52,856) = 15,005 \text{ psi}$$

The allowable region is defined by the material properties reduced by 1.5.

$$\frac{S_{es}}{FS} = \frac{100,000}{1.5} = 66,667 \text{ psi}$$

$$\frac{S_{ys}}{FS} = \frac{120,000}{1.5} = 80,000 \text{ psi}$$

Table 15.4
Typical Spring Material Properties
For 70°F use

name, specification	E EE6 psi	G EE6 psi	ρ lbm/ in³	ultimate tensile strength, EE3, psi wire diameter, inches								
				.02	.04	.06	.08	.10	.15	.20	.30	.40
high-carbon steel												
music, ASTM A228	30	11.5	.284	350	315	295	280	270	255	240		
hard drawn, ASTM A227	30	11.5	.284	285	255	235	225	215	205	190	175	160
oil-tempered, ASTM A229	30	11.5	.284	295	265	250	235	230	215	195	180	170
valve-spring, ASTM A230	30	11.5	.284					210	205	195		
alloy steel												
ASTM 232	30	11.5	.284		280	265	260	245	230	220	205	
stainless steel												
ASTM 313				300	275	260	245	235	215	190	160	

Plotting the actual stress values indicates that the design is just outside the allowable region.

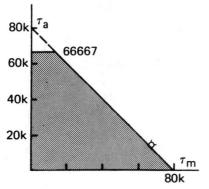

9 GEAR TRAINS

Nomenclature

C	center-to-center distance	in
d_p	pitch circle diameter	in
n	number of teeth	–
P	diametral pitch	1/in
R	gear ratio	–
rpm	revolutions per minute	rpm
s	number	–
TV	train value	–
v_r	pitch circle velocity	fpm

Symbols

ω	angular velocity	rad/sec

Subscripts

c	planet carrier
d	driving
f	driven
p	planet
r	ring
s	sun
t	total

A. BASIC DEFINITIONS

Compound gear: Two gears on a single (usually short) shaft, as in gear B in figure 15.21.

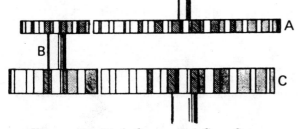

Figure 15.21 A Compound Gear Set

Epicyclic gear train: A gear train consisting of a sun gear and one or more planet gears. An epicyclic gear train is characterized by planet gears which do not have fixed axes of rotation.

External gear: Any gear whose teeth point away from the axis of rotation.

Idler gear: A gear which provides a change in shaft rotation direction without a change in the train ratio, as in gear B in figure 15.22. Also known as *intermediate gear*.

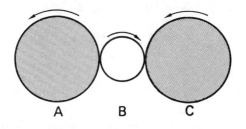

Figure 15.22 Idler Gear

Internal gear: A gear in which the teeth point toward the axis of rotation, as in gear A in figure 15.23.

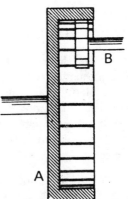

Figure 15.23 Internal Gear

Pitch circle velocity: The linear velocity of the gear teeth at the pitch circle. The pitch circle velocity is the same for all gears which mesh.

$$v_r = \frac{(\text{rpm})(\pi)d_p}{12} \qquad 15.75$$

Planetary gear train: See "Epicyclic gear train."

Reverted gear train: A gear train whose input and output shafts are in-line, usually consisting of one or more compound gears. Only the gears actually in contact need to have the same diametral pitches. A reverted train has an even number of stages.

Simple gear train: A gear train which consists of two external gears in contact. The gears turn in opposite directions. (If one of the gears is an internal gear, the shafts rotate in the same direction.)

Stage: Each pair of contacting gears is known as a stage. The train ratio is the product of the stage ratios.

Train ratio: The train ratio is the ratio of the input to output speeds.

$$R = \frac{\text{input rpm}}{\text{output rpm}} = \frac{n_f}{n_d} = \frac{d_{p,f}}{d_{p,d}} \qquad 15.76$$

If all of the gear centers are fixed, the train ratio can be found from equation 15.77.

$$R = \frac{\text{product of driven gear teeth numbers}}{\text{product of driving gear teeth numbers}} \qquad 15.77$$

Train value: See "velocity ratio."

Velocity ratio: The reciprocal of the train ratio. To avoid teeth interference, the velocity ratio should be kept below 10 for each pair of gears.

$$TV = \frac{1}{R} = \frac{\text{output rpm}}{\text{input rpm}} = \frac{n_d}{n_f} = \frac{d_{p,d}}{d_{p,f}} \qquad 15.78$$

B. DESIGN OF MULTI-STAGE GEAR TRAINS

Finding the number of teeth that each gear should have in order to achieve an arbitrarily chosen train ratio is considerably more difficult than the inverse problem. In general, a given ratio may not be achievable since each gear must contain an integral number of teeth. However, the following procedure can be used to reduce the effort to a minimum.

step 1: Find any numbers, a and b, such that a/b is a common fraction with value close to or equal to that of TV.

step 2: Arbitrarily choose a number, s, as any integer greater than or equal to 1.

step 3: Calculate h. If TV is greater than one, use $1/TV$.

$$h = \frac{s}{a - b(TV)} \qquad 15.79$$

step 4: If h is not an integer, truncate or round up h, giving h'.

step 5: Solve for TV'.

$$TV' = \frac{ah' - s}{bh'} \qquad 15.80$$

step 6: Factor the numerator and the denominator of equation 15.80 individually to give the number of teeth in the driving and the driven gears, respectively. Because of availability, no gear should have more than 120 teeth. If the numerator and the denominator are not factorable (are prime), or if the number of stages is incorrect, return to step 2 or step 4.

Example 15.11

Design a gear train with a train value of 2.54 such that no gear has more than 75 teeth.

$$\frac{1}{TV} = \frac{1}{2.54} = .3937008$$

By trial and error, it is found that (24/61) is close to .3937, so $a = 24$, and $b = 61$. Choose $s = 1$. Then

$$h = \frac{1}{24 - 61(.3937008)} = -63.497$$

Let $h' = -63$. Then

$$TV' = \frac{24(-63) - 1}{61(-63)} = \frac{1513}{3843} = \frac{(17)(89)}{(61)(63)}$$

Since all gears must have fewer than 75 teeth, this solution is unacceptable. Because h is close to -63.5, it could be rounded up, making $h' = -64$. This will result in a satisfactory solution. It also is logical to modify s.

If s is chosen as 2, $h = -126.99$ and $h' = -127$. Then

$$TV' = \frac{24(-127) - 2}{61(-127)} = \frac{3050}{7747} = \frac{3050}{(127)(61)}$$

However, 127 also exceeds 75, and 127 is prime. Choosing $s = 3$, $h = -190.49$, so $h' = -190$.

$$TV' = \frac{24(-190) - 3}{61(-190)} = \frac{4563}{11,590} = \frac{(13)(13)(27)}{(61)(19)(10)}$$

If a 3-stage gear set is allowable, the above numbers represent the number of teeth for each gear.

C. DESIGN OF REVERTED GEAR TRAINS

If each stage of a reverted gear train shares the total reduction equally, the ideal ratio per stage is

$$R = (R_t)^{\frac{1}{\text{stages}}} \qquad 15.81$$

Since this may not be a feasible ratio, factor R_t into numbers which are not too far apart numerically.

$$R_t = R_1 R_2 R_3 \ldots \qquad 15.82$$

An added constraint with reverted gear trains is that the center-to-center distances must be the same for each stage. For a 2-stage gear set, as shown in figure 15.24, the governing equations are 15.83 through 15.85.

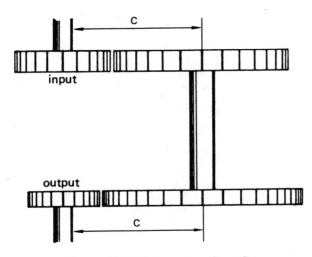

Figure 15.24 Reverted Gear Set

$$R_t = R_1 R_2 \qquad 15.83$$
$$R_1 = \frac{n_B}{n_A} \qquad 15.84$$
$$R_2 = \frac{n_D}{n_C} \qquad 15.85$$

The center-to-center distances must be the same.

$$C = \frac{n_A + n_B}{2P_{AB}} = \frac{n_C + n_D}{2P_{CD}} \qquad 15.86$$

These equations can be written as a system of four simultaneous linear equations.

$$R_1 n_A - n_B \qquad\qquad = 0 \qquad 15.87$$
$$\qquad R_2 n_C - n_D = 0 \qquad 15.88$$
$$n_A + n_B \qquad\qquad = 2P_{AB}C \qquad 15.89$$
$$\qquad n_C + n_D = 2P_{CD}C \qquad 15.90$$

D. ANALYSIS OF EPICYCLIC GEAR TRAINS

The simplest *epicyclic gear train* (*planetary gear set*) consists of a sun, a ring gear, and one or more planet gears, as shown in figure 15.25.

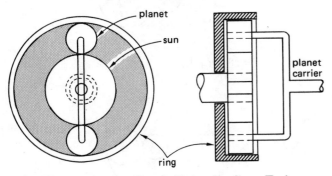

Figure 15.25 Simple Epicyclic Gear Train

The planets rotate about their own axes and also revolve around the sun gear. The number of planets does not affect the output speed, although the load-carrying capacity does increase with the number of planets. If the planet carrier is fixed, the ring gear will rotate. If the ring gear is fixed, the planet carrier will rotate.

A simple epicyclic gear train can be analyzed with the following steps.

step 1: Calculate the *train value*. This value is to be used regardless of whether the gear set is reducing or augmenting. TV will be negative if the sun and ring gears rotate in different directions when the arm is locked. (TV is always negative if the ring gear is an internal gear.)

$$TV = \frac{n_{\text{ring}}}{n_{\text{sun}}} \qquad 15.91$$

step 2: Calculate the unknown velocities from equation 15.92. Be sure to assign correct signs: positive for clockwise rotation, negative for counterclockwise rotation.

$$\omega_s = (TV)\omega_r + \omega_c(1 - TV) \qquad 15.92$$

step 3: If necessary, the angular velocity of the planets can be found from equation 15.93. The quantity is negative because the planet and sun gears turn in opposite directions.

$$\frac{n_p}{n_s} = -\frac{\omega_s - \omega_c}{\omega_p - \omega_c} \qquad 15.93$$

Example 15.12

The ring gear in figure 15.25 is fixed. The sun gear moves 100 rpm clockwise. $n_s = 32$, $n_p = 16$, and $n_r = 64$. Find the direction and speed of the carrier.

$$TV = -\frac{64}{32} = -2$$

$\omega_r = 0$ since the ring gear is fixed. Then from equation 15.92,

$$100 = (0)(-2) + (1 - (-2))\omega_c$$
$$\omega_c = 33.3 \text{ rpm}$$

Since this is positive, the carrier rotates in a clockwise direction.

If the epicyclic gear train is more complex than the simple gear train, the following procedure can be used.

step 1: Identify all the gears which have the same center of rotation as the arm. Form a 3-row table with a column for each of these gears. Include stationary gears but do not include the arm or any gear rigidly attached to it.

step 2: Put "ω_{arm}" into the first row for each gear in the table. (Actually put the variable name in the first row. Do not fill in values at this time.)

step 3: For gear z (whose unknown rotational speed is desired), put "ω_z" into the z column in the third row. If all gear speeds except the arm are known, choose the gear z arbitrarily.

step 4: Put "$\omega_z - \omega_{arm}$" into the second row for column z.

step 5: Determine the velocity relationships between gear z and all other gears in the table (i.e., find ω_i/ω_z for each gear). Use figure 15.26 and be sure to consider directions if other configurations are used in the gear train. Multiply the ratios by "$(\omega_z - \omega_{arm})$" and put the product into row 2.

step 6: Insert all known values of ω into the third row for all empty columns and into ω_{arm} wherever it appears. If the velocity is unknown, insert "ω_i."

step 7: For each column except the pivot column (the column with the unknown speed), write the equation

$$\text{row } 1 + \text{row } 2 = \text{row } 3 \qquad 15.94$$

Example 15.13

In the gear train shown, $n_A = 16$, $n_B = 32$, $n_C = 24$, and $n_D = 72$. If the output (arm) is 20 rpm, what is the input (A)?

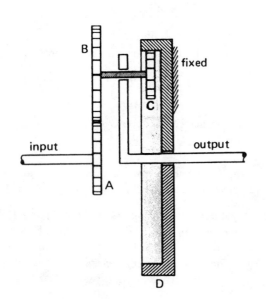

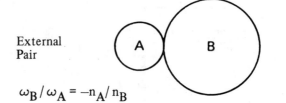

External Pair

$$\omega_B/\omega_A = -n_A/n_B$$

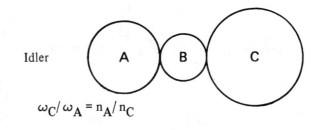

Idler

$$\omega_C/\omega_A = n_A/n_C$$

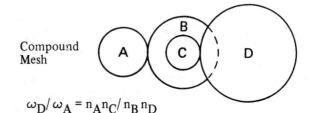

Compound Mesh

$$\omega_D/\omega_A = n_A n_C/n_B n_D$$

Internal Pair

$$\omega_B/\omega_A = n_A/n_B$$

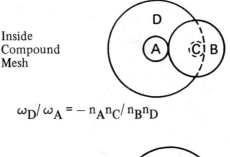

Inside Compound Mesh

$$\omega_D/\omega_A = -n_A n_C/n_B n_D$$

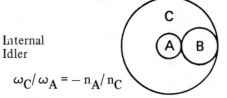

Internal Idler

$$\omega_C/\omega_A = -n_A/n_C$$

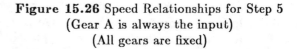

Figure 15.26 Speed Relationships for Step 5
(Gear A is always the input)
(All gears are fixed)

steps 1, 2, 3, and 4: Gears A and D are concentric with the arm. ω_A is the unknown velocity.

	A	D
row 1	ω_{arm}	ω_{arm}
row 2	$\omega_A - \omega_{\text{arm}}$	
row 3	ω_A	

step 5: The path A-to-D is an inside compound mesh with ratio

$$\frac{n_A n_C}{n_B n_D} = -\frac{(16)(24)}{(32)(72)} = -\frac{1}{6}$$

So row 2 for gear D is

$$-\tfrac{1}{6}(\omega_A - \omega_{\text{arm}})$$

step 6: $\omega_{\text{arm}} = 20$, and $\omega_D = 0$. The final table is

	A	D
row 1	20	20
row 2	$\omega_A - 20$	$-(1/6)(\omega_A - 20)$
row 3	ω_A	0

step 7: For gear D, $20 - (1/6)(\omega_A - 20) = 0$, or $\omega_A = 140$. Since this is positive, it rotates in the same direction as the arm.

10 GEAR FORCE ANALYSIS

Nomenclature

a	a number	–
d_p	pitch circle diameter	in
F	force	lbf
HP	transmitted horsepower	hp
k_d	Barth speed factor	–
P	diametral pitch (no. teeth/d_p)	1/in
T	torque	in-lbf
v_r	pitch circle velocity	fpm
w	face width	in
Y	form factor to be used with diametral pitch	–

Symbols

σ	stress	–
ϕ	pressure angle	°

Subscripts

d	dynamic
r	radial
t	tangential

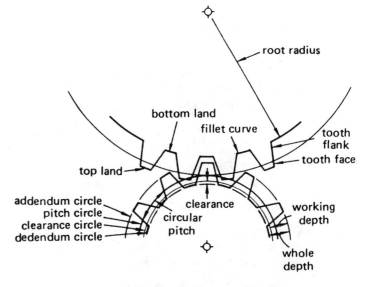

Figure 15.27 Gear Tooth Definitions

A. DEFINITIONS

Addendum: The radial distance from the pitch circle to the tooth top.

Base circle: A circle that is tangent to the line of action.

Center distance: The distance between centers of meshing gears.

Circular pitch: The distance along the pitch circle between corresponding points on two adjacent teeth. It is equal to the tooth thickness plus the width of space along the pitch circle. It is numerically calculated as (π/P).

Clearance: The difference between the dedendum and the addendum.

Clearance circle: The circle that is tangent to the addendum of the meshing gear.

Contact ratio: The average number of teeth in contact. The contact ratio usually is between 1.2 and 1.6. (1.2 means that one pair of teeth is in contact all of the time and a second pair is in contact 20% of the time.)

Dedendum: The radial distance from the pitch circle to the tooth root.

Diametral pitch: The number of teeth per inch of pitch circle diameter. It must be the same in all meshing gears.[19]

Face: The tooth area between the pitch circle and the addendum circle.

----
[19]Not every diametral pitch is available. Design should make use of standard diametral pitches: 1, $1\frac{1}{4}$, $1\frac{1}{2}$, $1\frac{3}{4}$, 2, $2\frac{1}{2}$, 3, 4, 5, 6, 7, 8, 9, 10, 11, 12, 14, 16, 18, 20, 22, 24, 26, 28, 30, and 32.

Face width: The axial width of the tooth.[20]

Flank: The tooth area between the pitch circle and the dedendum.

Line of action: A line passing through the pitch point and tangent to both base circles. Also known as the *pressure line* and *generating line*.

Pitch circle: An imaginary circle on which the gear lever arm is based.

Pitch point: The imaginary point of tangency between the pitch circles of two meshing gears.

Pressure line: See "Line of action."

Whole depth: The distance from the addendum circle to the dedendum circle. It is equal to the working depth plus clearance.

Working depth: The distance that a tooth from a meshing gear extends into the tooth space.

B. FORCE AND STRESS ANALYSIS[21]

The horsepower transmitted by a revolving shaft-carrying torque is

$$HP = \frac{(T)(\text{rpm})}{63,025} \qquad 15.95$$

The *tangential velocity* at the *pitch circle* (where power transmission is assumed to occur) is

$$v_r = \frac{\pi(\text{rpm})d_p}{12} \qquad 15.96$$

Combining equations 15.95 and 15.96 gives

$$HP = \frac{12 T v_r}{(63,025)\pi d_p} \qquad 15.97$$

The *tangential force* at the pitch circle can be found from the transmitted torque, as in equation 15.98.

$$F_t = \frac{2T}{d_p} = \frac{HP(33,000)}{v_r} \qquad 15.98$$

In addition, there is a radial force on the tooth.

$$F_r = F_t \tan\phi \qquad 15.99$$

The total force on the tooth is

$$F = \sqrt{F_t^2 + F_r^2} = \frac{F_t}{\cos\phi} \qquad 15.100$$

At times, one tooth pair will carry the force. At other times, two or more pairs may be in contact. The average number of tooth pairs in contact is the *contact ratio*.

$$\text{contact ratio} = \frac{\text{length of line of action}}{\left(\frac{\pi}{P}\right)\cos\phi} \qquad 15.101$$

If one tooth is assumed to carry the entire tangential load as a cantilever beam, the Lewis beam strength equation gives the maximum tangential load that the tooth can carry. This maximum loading is based on the maximum allowable bending stress. Equation 15.102 assumes that the load is applied parallel to the top land. The radial component of the load customarily is omitted.

$$F_{t,\,max} = \frac{\sigma w Y}{P} \qquad 15.102$$

σ is an allowable stress dependent on the gear material and hardness. Typical values for steel and cast iron are given in table 15.5. Y is a *form factor* for use with the diametral pitch.[22] Y is found in table 15.6.

The allowable stress can be calculated in two different ways. The simplest is to divide the tensile yield strength by a sizeable factor (e.g., 3 to 5). The larger safety factors would be used for designs with heavy shock. An alternate method for determining the allowable stress is to use some fraction (e.g., 75%) of the endurance strength in reversed bending.

Gear stresses vary repeatedly from zero to a maximum value. The use of the endurance limit for completely reversed stresses (as suggested by AGMA) as the maximum allowable stress includes a considerable factor of safety.

Table 15.5
Typical Allowable Gear Stresses

material	BHN	endurance limit
steel (cast steel)	150	36,000 psi
	200	50,000
	240	60,000
	280	70,000
	320	80,000
	360	90,000
	400	100,000
grey cast iron	146	10,000
	163	11,400
	179	12,300
	192	16,500
	196	17,400
	215	19,600
	266	25,200

---

[20] As a general rule, the face width w should be three to five times the circular pitch (π/P).

[21] This method is strictly for spur gear forces. Helical gear tooth forces are analyzed in the following section.

[22] There actually are two factors with the name *form factor*. The *Lewis form factor* typically is written with a lower case y. The *form factor* is written as Y. One can be derived from the other, since $Y = \pi y$.

Table 15.6
Form Factor (Y) for Use in Lewis Equation

Number of teeth	$14\frac{1}{2}°$ full depth	20° full depth	20° stub	Number of teeth	$14\frac{1}{2}°$ full depth	20° full depth	20° stub
12	.210	.245	.311	28	.314	.352	.430
13	.223	.261	.324	30	.317	.359	.437
14	.236	.277	.339	34	.327	.371	.447
15	.245	.290	.346	38	.333	.384	.455
16	.254	.296	.361	43	.339	.397	.462
17	.264	.302	.368	50	.346	.410	.474
18	.270	.308	.377	60	.355	.421	.484
19	.276	.314	.386	75	.361	.434	.496
20	.283	.321	.393	100	.368	.447	.505
21	.289	.327	.399	150	.374	.460	.518
22	.292	.330	.405	300	.383	.472	.534
24	.298	.337	.415	rack	.390	.484	.550
26	.308	.346	.424				

If equation 15.102 is used to predict the tangential force or the transmitted horsepower, both the pinion and the gear must be checked for stress. This is essentially a process of successive iterations, which starts out by assuming one of the gears to be weaker. The design proceeds until the desired quantity (P, horsepower, width, etc.) is determined. Then the problem is worked backwards to get the root stress for the other gear. This stress is compared to the allowable value.

C. DYNAMIC LOADING

The *Barth speed factor* partially accounts for the kinetic loading effects. The speed factor depends on the gear material. For metallic gears,

$$k_d = \frac{a}{a + v_r} \qquad 15.103$$

$a = 600$ for ordinary industrial gears and for gears with cast teeth. Ordinary gears usually run with v_r less than 3000 fpm. $a = 1200$ for accurately cut gears, which **may run as high as 6000 fpm. The bending stress at the tooth root becomes**

$$\sigma = \frac{F_t P}{k_d w Y} \qquad 15.104$$

Although the Barth speed factor has been recommended for many years, its use has been largely superseded by the Buckingham equation.

A considerable safety margin is designed into the gear set if each gear is designed so that the beam strength is greater than the dynamic momentary load carried by the tooth. Such a load will be due to the transmitted power, in addition to errors in gear cutting, inaccuracies in alignment, and inertial effects. The dynamic momentary load is predicted by the *Buckingham equation*.

$$F_{dynamic} = F_t + \left[\frac{0.05v_r(wC^* + F_t)}{0.05v_r + \sqrt{wC^* + F_t}} \right] \qquad 15.105$$

F_t is the transmitted pitch line load, as calculated from equation 15.98.

C^* is a deformation constant which depends on the *error in action* of the gear teeth. If the actual error is not given, likely values can be assumed from the *gear class* and table 15.7. Well-cut commercial gears (typical of shaved, ground, or hobbed spur gears) can be considered as class 1. Class 2 gears are cut with great care. Class 3 gears are precision manufactured.

Table 15.7
Maximum Error in Action Between Gears

Diametral pitch	Class 1	Class 2	Class 3
1	.0048 in.	.0024 in.	.0012
2	.0040	.0020	.0010
3	.0032	.0016	.0008
4	.0026	.0013	.0007
5	.0022	.0011	.0006
6 and finer	.0020	.0010	.0005

Values of C^* depend on the materials in the gear set and the error in action. Table 15.8 can be used to find C^*. The values given for cast iron in table 15.8 also should be used for ordinary semi-steels and bronze since the modulus of elasticity of these materials is essentially the same.

Table 15.8
Values of C* *

Materials	Tooth Form	Error in Action of Gears, in.					
		.0005	.001	.002	.003	.004	.005
cast iron and cast iron	$14\frac{1}{2}°$	400	800	1600	2400	3200	4000
cast iron and steel	$14\frac{1}{2}°$	550	1100	2200	3300	4400	5500
steel and steel	$14\frac{1}{2}°$	800	1600	3200	4800	6400	8000
cast iron and cast iron	20° full-depth	415	830	1660	2490	3320	4150
cast iron and steel	20° full-depth	570	1140	2280	3420	4560	5700
steel and steel	20° full-depth	830	1660	3320	4980	6640	8300
cast iron and cast iron	20° stub	430	860	1720	2580	3440	4300
cast iron and steel	20° stub	590	1180	2360	3540	4720	5900
steel and steel	20° stub	860	1720	3440	5160	6880	8600

Example 15.14

A gear/pinion set is cut with great care in a 20° full-depth design. The 18-tooth steel pinion and the 75-tooth cast iron gear transmit 50 hp when the pinion turns at 1400 rpm. The face width is $3\frac{1}{2}''$, and the diametral pitch is 3. Find the required gear hardness. Neglect stress concentrations at the root.

The pitch diameter of the pinion is $(18/3) = 6$. The pitch line velocity is the same for either gear. It is

$$v_r = \frac{6(\pi)1400}{12} = 2199 \text{ fpm}$$

From table 15.7, the maximum error in action will be .0016″, so the value of C^* is interpolated to be 1824. **The pitch line force from equation 15.98 is**

$$F_t = \frac{33,000(50)}{2199} = 750 \text{ lbf}$$

The dynamic force is found from equation 15.105.

$$F_d = 750 + \frac{(.05)(2199)[(3.5)(1824)+(750)]}{(.05)(2199) + \sqrt{(3.5)(1824)+(750)}}$$
$$= 4785 \text{ lbf}$$

From table 15.6, for a gear with 75 teeth, $Y = .434$. The gear stress is found from equation 15.102.

$$\sigma = \frac{(4785)(3)}{(3.5)(.434)} = 9450 \text{ psi}$$

From table 15.5, a minimum BHN of approximately 140 is required.

11 ANALYSIS OF FORCES ON HELICAL GEARS

There are few standards for helical gears, although the 20° pressure angle commonly is agreed upon. *Helix angles* generally range from 0° to 30°, and angles approaching 45° are not recommended.

The design of helical gears (figure 15.28) is similar to the design of spur gears. The *Lewis beam strength* equation can be used for rough estimates. For more accurate designs, AGMA procedures can be followed.

Figure 15.28 A Helical Gear

Although the Lewis equation can be used to determine the stress in a tooth, it is necessary to calculate the force on the tooth differently. There are three components of force and three angles associated with helical gears. In addition to the *normal pressure angle*, ϕ_n, there is the *tangential pressure angle*, ϕ_t, and the *helix angle*, ψ.

$$\cos\psi = \frac{\tan\phi_n}{\tan\phi_t} \qquad 15.106$$

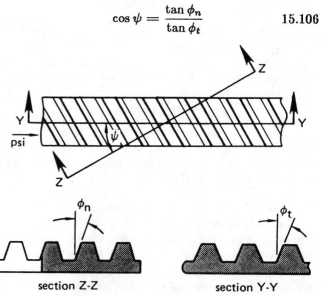

Figure 15.29 Helical Gear Angles

The three force components are the radial, the axial, and the tangential components. The tangential force is calculated from the transmitted horsepower and the rotational velocity, as it was with spur gears.

$$F_t = \frac{(33,000)(HP)}{v_r} \qquad 15.107$$

The radial and tangential components can be calculated from the tangential forces and the various tooth angles.

$$F_r = F_t \tan \phi_t \qquad 15.108$$
$$F_a = F_t \tan \psi \qquad 15.109$$

The three components combine into the total force needed to design the tooth.

$$F = \sqrt{F_r^2 + F_a^2 + F_t^2} \qquad 15.110$$

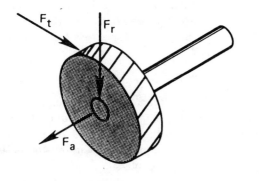

Figure 15.30 Forces on a Helical Gear

12 TRANSMISSION SHAFTING

Nomenclature

c	shaft radius	in
d	shaft diameter	in
E	modulus of elasticity	psi
f	critical speed	rps
FS	factor of safety	–
g	acceleration due to gravity (386)	in/sec²
G	shear modulus	psi
I	bending moment of inertia	in⁴
J	torsional moment of inertia	in⁴
K	derating factor	–
L	shaft length	in
m	mass	lbm
M	moment	in-lbf
r	distance from shaft axis to any point within	in
T	torque	in-lbf
w	weight per unit length	lbf/ft
W	total weight	lbf

Symbols

θ	shaft twist	radians
σ	bending stress	psi
δ_i	deflection at the ith body	in
τ	shear stress	psi
ω	angular velocity	rad/sec
ρ	shaft density	lbm/in³

Subscripts

*	conjugate
ave	average
e	endurance
m	bending
r	range
t	twisting
yp	yield point

A. SHAFT DESIGN—STATIC LOADING

A circular shaft of length L acted upon by a total torque T will experience a torsional deformation in radians given by equation 15.111.

$$\theta = \frac{TL}{GJ} \qquad 15.111$$

The torsional shear stress at radius r is

$$\tau = \frac{Tr}{J} \qquad 15.112$$

$$J = \frac{\pi d^4}{32} \text{ (solid shafts)} \qquad 15.113$$

Combining equations 15.112 and 15.113, the maximum torsional shear stress is

$$\tau = \frac{16T}{\pi d^3} \qquad 15.114$$

If the torque is expressed in in-lbf, the transmitted horsepower is

$$HP = \frac{T\omega}{(12)(550)} \qquad 15.115$$

And, since $\omega = 2\pi \,(\text{rpm})/60$,

$$HP = \frac{T\,(\text{rpm})}{63,025} \qquad 15.116$$

Since many shafts carry both bending and torsional loads, the stress at a point will be a combined stress consisting of torsional shear and flexural stress. If the loading is static, the maximum shearing stress is

$$\tau_{\text{max}} = \sqrt{\tfrac{1}{4}\sigma^2 + \tau^2} \qquad 15.117$$

The torsional stress is given by equation 15.114. The flexural stress is

$$\sigma = \frac{Mc}{I} = \frac{32M}{\pi d^3} \qquad 15.118$$

Then equation 15.117 becomes

$$\tau_{max} = \frac{16}{\pi d^3}\sqrt{M^2 + T^2} \qquad 15.119$$

B. SHAFT DESIGN—DYNAMIC LOADING

Inasmuch as loading often is not static, the various design codes specify that factors be used to account for different loading conditions.[23]

$$\tau_{max} = \frac{16}{\pi d^3}\sqrt{(K_m M)^2 + (K_t T)^2} \qquad 15.120$$

Table 15.9
Loading Factors

Nature of loading	Values for	
	K_m	K_t
Stationary shafts:		
gradually applied load	1.0	1.0
suddenly applied load	1.5–2.0	1.5–2.0
Rotating shafts:		
gradually applied or steady load	1.5	1.0
suddenly applied loads, minor shocks only	1.5–2.0	1.0–1.5
suddenly applied loads, heavy shocks	2.0–3.0	1.5–3.0

The allowable stress based on the distortion energy theory is

$$\tau_{allowable} = \frac{(.577)\sigma_{yield}}{FS} \qquad 15.121$$

It is not uncommon to limit τ to 8000 psi for ordinary steel shafts without keyways. If keyways are present, use 6000 psi. If the shaft properties are known, the shear stress should be limited to the minimum of 30% of S_{yt} and 18% of S_{ut}. A further 25% reduction is called for if a keyway is present.

If sufficient information on the load fluctuations is available to use fatigue analysis, it will be possible to calculate the values of K_t and K_m based on the shaft properties and local stress concentration factors. If k_1

[23]The approach presented by equation 15.120 and the factors in table 15.9 are by no means intended to imply universal acceptance. There are several other theoretical methods, as well as numerous codes developed by companies and government agencies. Without actually specifying the shaft design method to be used, it is impossible to ensure compliance with any particular code.

is the stress concentration factor for bending and k_2 is the stress concentration factor for torsion.

$$M = M_{ave} + k_1 M_r \qquad 15.122$$
$$T = T_{ave} + k_2 T_r \qquad 15.123$$

Assuming that

$$\left(\frac{S_{yt}}{S_e}\right) \approx \left(\frac{S_{ys}}{S_{es}}\right),$$

$$K_m = \frac{M_{ave} + \left(\dfrac{S_{yt}}{S_e}\right)k_1 M_r}{M_{ave} + k_1 M_r} \qquad 15.124$$

$$K_t = \frac{T_{ave} + \left(\dfrac{S_{yt}}{S_e}\right)k_2 T_r}{T_{ave} + k_2 T_r} \qquad 15.125$$

Equations 15.122 through 15.125 can be substituted into equation 15.120 (equations 15.122 and 15.123 will cancel the denominators in equations 15.124 and 15.125) to obtain the maximum shear stress based on fluctuating load theory.

Example 15.15

A keyed shaft carries a steady bending moment of 28,000 in-lbf and a maximum torque of 16,000 in-lbf which may suddenly be applied. What should be the shaft diameter if the yield point in tension is 70,000 psi? Use a factor of safety of 2.0.

From table 15.9, $K_m = 1.5$. Assume $K_t = 1.5$. The allowable stress is

$$\tau_{allowable} = \frac{(.5)(70,000)}{2} = 17,500 \text{ psi}$$

Solving for d from equation 15.120,

$$d^3 = \frac{16}{\pi(17,500)}\sqrt{[(1.5)(28,000)]^2 + [(1.5)(16,000)]^2}$$
$$d = 2.415'' \text{ (say a } 2\tfrac{7}{16}'' \text{ shaft)}$$

C. CRITICAL SPEEDS OF ROTATING SHAFTS

The critical speed of a shaft can be found from its free lateral vibration frequency. The general equation for the fundamental frequency (critical speed in revolutions per second) is

$$f = \frac{1}{2\pi}\sqrt{\frac{g\sum(m_i\delta_i)}{\sum(m_i\delta_i^2)}} \qquad 15.126$$

m_i is the mass of the ith rotating body, and δ_i is the static deflection of the shaft at the ith body.

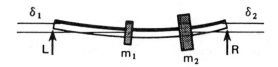

Figure 15.31 Shaft Deflections

Since equation 15.126 is difficult to use, it is common to substitute the *Dunkerly approximation*.[24]

$$\left(\frac{1}{f}\right)^2_{\text{composite}} = \left(\frac{1}{f_1}\right)^2 + \left(\frac{1}{f_2}\right)^2 + \left(\frac{1}{f_3}\right)^2 + \cdots \quad 15.127$$

Deflections for simple configurations with shafts of constant cross sectional areas can be found from the beam formulas. Of course, the deflections will depend on the types of supports. Shafts with single anti-friction (ball and roller) bearings at each end usually are considered to be simply supported. Shafts with sleeve (oil film) bearings or two anti-friction bearings at each end usually are considered to have fixed (built-in) supports.

A round shaft carrying no load other than its own total distributed weight of $W = wL$ has a mid-span deflection given by

$$\delta = \frac{5WL^3}{384EI} \text{ (simple supports)} \quad 15.128$$

$$\delta = \frac{WL^3}{384EI} \text{ (fixed supports)} \quad 15.129$$

The critical frequency in this simple case is

$$f = \frac{1}{2\pi}\sqrt{\frac{g}{\delta}} \quad 15.130$$

The following data and relationships should be used for hard steel shafts.

$$W = \tfrac{1}{4}\pi d^2 L\rho \quad 15.131$$
$$\rho = .28 \text{ lbm/in}^3$$
$$E = 3 \text{ EE7 psi}$$
$$I = \frac{\pi d^4}{64}$$
$$g = 386 \text{ in/sec}^2$$

For an unloaded, simply supported shaft, the critical speed in cycles per second is

$$f = 7.09 \text{ EE4}\left(\frac{d}{L^2}\right) \text{ (simple supports)} \quad 15.132$$

$$f = 1.59 \text{ EE5}\left(\frac{d}{L^2}\right) \text{ (fixed supports)} \quad 15.133$$

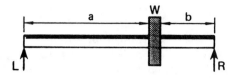

Figure 15.32 Simple Shaft Carrying One Mass

The classical solution to the problem of a shaft carrying a single object of weight W is to assume a weightless shaft.[25] Making the classical assumption of a weightless shaft, the deflection at the load is

$$\delta = \frac{Wa^2b^2}{3EIL} \text{ (simple supports)} \quad 15.134$$

$$\delta = \frac{Wa^3b^3}{3EIL^3} \text{ (fixed supports)} \quad 15.135$$

Making the same substitutions as were made for the unloaded shaft, the critical speeds are

$$f = (6.57 \text{ EE3})\frac{d^2}{ab}\sqrt{\frac{L}{W}} \text{ (simple supports)} \quad 15.136$$

$$f = (6.57 \text{ EE3})d^2\sqrt{\frac{L^3}{Wa^3b^3}} \text{ (fixed supports)} \quad 15.137$$

Example 15.16

Find the lowest speed for which vibration will occur. Neglect the shaft weight.

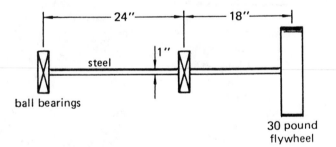

From the beam deflection formulas, the static deflection of the cantilever portion of the shaft will be

$$I = \frac{\pi}{4}(.5)^4 = .049$$

$$\delta = \frac{(30)(18)^2(24 + 18)}{(3)(3 \text{ EE7})(.049)} = .093''$$

From equation 15.130, the critical frequency is

$$f_{\text{nat}} = \frac{1}{2\pi}\sqrt{\frac{386}{.093}} = 10.3 \text{ Hz}$$

---

[24]In order to calculate the deflections by superposition for use with equation 15.126, the number of deflection calculations is the square of the number of rotating masses.

[25]This need not be assumed, as equation 15.127 is capable of handling a shaft with a distributed load as long as the deflections used are the result of both loads.

D. STANDARD SHAFT DIAMETERS

Although equation 15.120 can be solved for d, it is always more economical to select the next highest standard shaft size. Table 15.10 can be used for this purpose.

Table 15.10
Standard Shaft Diameters (inches)

Transmission shafting sizes

$\frac{15}{16}$, $1\frac{3}{16}$, $1\frac{7}{16}$, $1\frac{11}{16}$, $1\frac{15}{16}$, $2\frac{3}{16}$, $2\frac{7}{16}$,

$2\frac{15}{16}$, $3\frac{7}{16}$, $3\frac{15}{16}$, $4\frac{7}{16}$, $4\frac{15}{16}$, $5\frac{7}{16}$, $5\frac{15}{16}$

Machinery shafting sizes

$\frac{1}{2}''$ to $2\frac{1}{2}''$ by $\frac{1}{16}''$ increments

$2\frac{5}{8}''$ to $4''$ by $\frac{1}{8}''$ increments

$4\frac{1}{4}''$ to $6''$ by $\frac{1}{4}''$ increments

13 RADIAL CAM FOLLOWER DYNAMICS

Nomenclature

L	total follower lift	inches
r	radius for displacement x	inches
r_b	base circle (minimum lift) radius	inches
t	time	sec
x	displacement	inches
y_o	vertical separation at minimum lift	inches
z	horizontal follower offset	inches

Symbols

θ	rotational displacement	radians
ω	rotational speed	radians/sec

Subscripts

b	base (minimum lift)
o	full rise

Figure 15.33 illustrates a generalized cam with a roller follower. The cam lobe can be shaped in a variety of ways, depending on what type of displacement and velocity curves are desired. Common lobe shapes are parabolic (constant acceleration), harmonic, and cycloidal.[26]

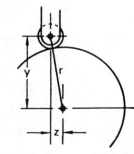

Figure 15.33 A Cam and Its Follower

In figure 15.33, z is the follower offset. It must be taken into consideration when determining the total geometry of the follower rise. r_b is the separation of the cam and the roller centers when the follower is at its lowest point. In equations 15.138 and 15.139, x is the follower displacement (*rise*). From the Pythagorean theorem,

$$y_b^2 = r_b^2 - z^2 \qquad\qquad 15.138$$

$$r^2 = (y_b + x)^2 + z^2 = r_b^2 + 2y_bx + x^2 \qquad 15.139$$

[26]Though easy to manufacture, cam lobes constructed from circular arcs are not covered here.

Table 15.11 Cam Dynamics Equations
(θ in radians, full rise in θ_o)

Table 15.11 May be used for non-symmetrical lobes if the problem is done in two parts, with two values of θ_o. In each part, θ_o is taken as twice the angular range for that phase.

Type of cam	Displacement maximum at $\theta/\theta_o = 1$	Velocity maximum at $\theta/\theta_o = .5$	Acceleration
Parabolic	accelerating, $x = 2L\frac{\theta^2}{\theta_o^2}$	$\frac{dx}{dt} = \frac{4L\omega\theta}{\theta_o^2}$	$\frac{d^2x}{dt^2} = \frac{4L\omega^2}{\theta_o^2}$ (constant)
	decelerating, $x = L\left[1 - 2\left(1 - \frac{\theta}{\theta_o}\right)^2\right]$	$\frac{dx}{dt} = \frac{4L\omega}{\theta_o}\left(1 - \frac{\theta}{\theta_o}\right)$	$\frac{d^2x}{dt^2} = -\frac{4L\omega^2}{\theta_o^2}$ (constant)
Harmonic	$x = \frac{L}{2}\left(1 - \cos\frac{\pi\theta}{\theta_o}\right)$	$\frac{dx}{dt} = \frac{\pi L\omega}{2\theta_o}\sin\frac{\pi\theta}{\theta_o}$	$\frac{d^2x}{dt^2} = \frac{\pi^2 L\omega^2}{2\theta_o^2}\cos\frac{\pi\theta}{\theta_o}$ (maximum at $\frac{\theta}{\theta_o} = 0$)
Cycloidal	$x = \frac{L}{\pi}\left(\frac{\pi\theta}{\theta_o} - \frac{1}{2}\sin\frac{2\pi\theta}{\theta_o}\right)$	$\frac{dx}{dt} = \frac{L\omega}{\theta_o}\left(1 - \cos\frac{2\pi\theta}{\theta_o}\right)$	$\frac{d^2x}{dt^2} = \frac{2\pi L\omega^2}{\theta_o^2}\sin\frac{2\pi\theta}{\theta_o}$ (maximum at $\frac{\theta}{\theta_o} = .25$)

Table 15.11 can be used to calculate the displacement, the velocity, and the acceleration of the follower as a function of the cam rotation. Figure 15.34 illustrates typical performance of the three cam types.

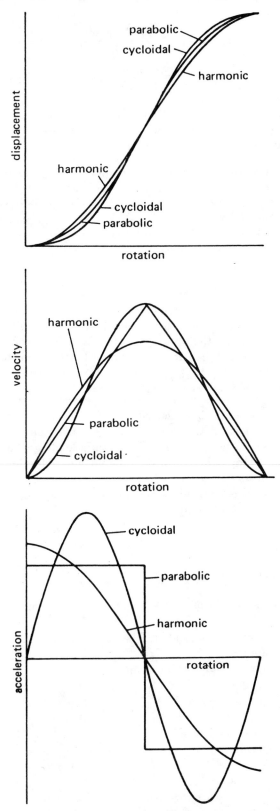

Figure 15.34 Cam Performance Graphs

Jerk is the term used to describe the third derivative of motion. (This sometimes is called *second acceleration*.) The parabolic cam has three infinite jerks per rise. It is the worst of the three cam types for high-speed use.

Similarly, the harmonic cam also has an infinite jerk. The cycloidal cam has a higher acceleration than the other two, but its jerk is finite. Therefore, the cycloidal cam is the best choice for high-speed operation.

The force on a cam or a follower generally will not be in line with the vertical axis of the follower. Rather, it will be offset by the *pressure angle*, as shown in figure 15.35. The torque on the camshaft due to the follower force is

$$T = \frac{F_y \frac{dx}{dt}}{\omega} \qquad 15.140$$

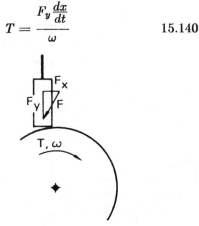

Figure 15.35 Follower Pressure Angle

14 BOLT PRELOADING IN TENSION CONNECTIONS

Preloading is an effective method of reducing the alternating stress in bolted tension connections.[27] The initial tension produces a larger mean stress, but the overall result may be to produce a satisfactory design.

Consider the bolted connection shown in figure 15.36. The load varies from F_{min} to F_{max}. If the bolt is initially snug but without initial tension, the force in the bolt also will vary from F_{min} to F_{max}.

The stress in the bolt depends on the load carrying area of the bolt. It is convenient to define the *spring constant of the bolt*. The effects of the threads usually are ignored, so the area is based on the major (nominal) diameter. The *grip*, L, is the thickness of the parts being connected by the bolt. It is not the bolt length.

$$\sigma_{bolt} = \frac{F}{A} \qquad 15.141$$

$$k_{bolt} = \frac{F}{\Delta L} = \frac{A_{bolt} E_{bolt}}{L_{bolt}} \qquad 15.142$$

---

[27]Grade 0, 1, and 2 bolts without cap markings should not be preloaded.

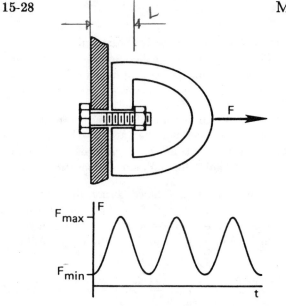

Figure 15.36 A Bolted Tension Joint

If the bolt is tightened so that there is an initial force, F_i, in addition to the applied load, the members being held together will be in compression. The amount of compression will vary since the applied load varies.

The clamped members will carry some of the applied load, since this varying load has to "uncompress" the members as well as lengthen the bolt. The net result is to reduce the variation of the bolt force.[28]

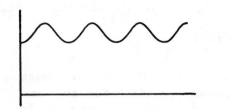

Figure 15.37 A Bolted Joint with Preloading

The spring constant for each of the bolted parts is somewhat difficult to determine if the clamped area is not well defined. If the clamped parts are simply plates, it can be assumed that the bolt force spreads out to three times the bolt diameter. Of course, the hole diameter needs to be considered in calculating the effective force area. If $E_{part} = E_{bolt}$, this larger area results in the parts being 8 times more stiff than the bolts.

$$k_{parts} = \frac{A_{parts} E_{parts}}{L_{parts}} \qquad 15.143$$

If the clamped parts have different elasticities (E values), the composite spring constant can be found from equation 15.144. (If a "soft" washer or gasket is used, its spring constant may control equation 15.144.)

---

[28]It is assumed that the initial tension, F_i, is greater than F_{max}. If the clamped members separate, the bolt once again carries the entire load.

$$\frac{1}{k_{composite}} = \frac{1}{k_1} + \frac{1}{k_2} + \frac{1}{k_3} + \cdots \qquad 15.144$$

Both the bolt and the clamped parts share the applied load.

$$F_{bolt} = F_i + \frac{k_{bolt} F_{applied}}{k_{bolt} + k_{parts}} \qquad 15.145$$

$$F_{parts} = \frac{k_{parts} F_{applied}}{k_{bolt} + k_{parts}} - F_i \qquad 15.146$$

Of course, if the applied load varies, the forces in the bolt and the parts also will vary. In that case, analysis by the Goodman diagram is called for.

For static loading, recommended amounts of preloading often are specified in terms of a percentage (e.g., 90%) of the tensile yield strength. The term *proof strength* (i.e., *proof load* divided by bolt area) can be used in place of tensile yield strength. For fatigue loading, the preload must be determined from an analysis of the Goodman diagram.

Tightening of a tension bolt will induce a torsional stress in the bolt. Where the bolt is to be locked in place, the torsional stress can be removed without greatly affecting the preload by slightly backing off the bolt. If the bolt is subject to cyclic loading, the bolt probably will slip back by itself, and it is reasonable to neglect the effects of torsion in the bolt.

More important than the effects of torsion are stress concentrations at the root. Although stress concentrations frequently are neglected for static loading of ductile connectors, there will be a significant reduction in the endurance limit for cyclic loading. Therefore, the alternating stress should be multiplied by an appropriate factor (e.g., 2.0 to 4.0).

15 BRAKES

Nomenclature

c_p	specific heat	ft-lbf/lbm-°F
E	energy	ft-lbf
f	coefficient of sliding friction	–
F	force	lbf
g_c	gravitational constant (32.2)	$\frac{ft\text{-}lbm}{sec^2 lbf}$
h	height	ft
HP	horsepower	hp
J	mass moment of inertia	lbm-ft²
n	rotational speed	rpm
N	normal force	lbf
p	pressure	psi
r	radius	ft
T	temperature, or torque	°F, ft-lbf
V	brake material volume	ft³
w	drum width	ft

Symbols

θ	contact angle	radians
ω	rotational velocity	rad/sec
ρ	density	lbm/ft^3

Subscripts

f	frictional

A. BLOCK BRAKES

A *block brake* is illustrated in figure 15.38. The pressure will be distributed approximately uniformly if the contact angle θ is less than 60°. If the angle is greater than 60°, the *equivalent coefficient of friction, f'*, from equation 15.150 should be used in place of f.

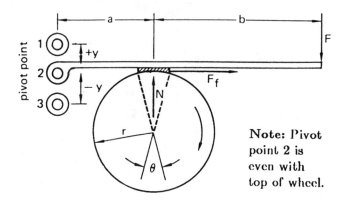

Note: Pivot point 2 is even with top of wheel.

Figure 15.38 A Block Brake

The *frictional clamping force* is calculated easily from the normal force. The normal force may need to be calculated from the torque capacity of the brake.

$$F_f = fN \qquad 15.147$$

$$N = \frac{T}{rf} \qquad 15.148$$

$$T = \frac{(5252)(HP)}{n} \qquad 15.149$$

$$f' = f\left[\frac{4\sin\tfrac{1}{2}\theta}{\theta + \sin\theta}\right] \qquad 15.150$$

The required application force at the end of the band brake can be determined readily by taking moments about the pivot point. If the wheel turns clockwise, pivot point 3 results in a *self-energizing brake*. Pivot point 1 is self-energizing for counterclockwise rotation.

The actuating force, F, can be zero if $a = fy$ for either self-energizing case. If $a < fy$, the brake will be *self-locking*. Since a negative force, F, will be required to disengage the brake, self-locking is not desirable.

B. BAND BRAKE

The band brake shown in figure 15.39 is evaluated from the belt friction equations. (F_1 is assumed to be the tight side force. θ is in radians.)

$$T = (F_1 - F_2)r = rF_2(e^{f\theta} - 1) \qquad 15.151$$

$$\frac{F_1}{F_2} = e^{f\theta} \qquad 15.152$$

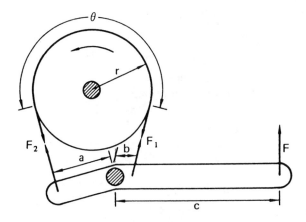

Figure 15.39 A Band Brake

The applied force, F, can be found by taking moments about the pivot.

$$F = \frac{F_2 a - F_1 b}{c} = \left(\frac{T}{cr}\right)\left(\frac{a - be^{f\theta}}{e^{f\theta} - 1}\right) \qquad 15.153$$

If $be^{f\theta} > a$, the applied force F will be negative and the brake will be self-locking.

The contact pressure is maximum at the toe of the brake.

$$p = \frac{F_1}{A} = \frac{F_1}{rw} \qquad 15.154$$

The required width w can be determined from the band material and the allowable contact pressure.

C. ENERGY DISSIPATION

The energy dissipated is equal to the change in rotational kinetic energy of the rotating mass. The maximum temperature increase will occur if all of the kinetic energy is absorbed by the brake.

$$\Delta E = \frac{J(\omega_1^2 - \omega_2^2)}{2g_c} \qquad 15.155$$

$$\omega = \frac{2\pi n}{60} \qquad 15.156$$

$$\Delta T = \frac{\Delta E}{\rho V c_p} \quad (°F) \qquad 15.157$$

Limitations are imposed on maximum energy dissipation rates. Values in table 15.12 should be compared to the product of the average pressure, in psi, and the rubbing velocity, in ft/min.

Table 15.12
Typical Maximum Energy Dissipation Rates
for Brakes

Conditions	$\dfrac{\text{ft-lbf}}{\text{min-in}^2}$
continuous load, poor dissipation	28,000
intermittent load, poor dissipation, long recoveries	55,000
continuous load, good dissipation (as in oil bath)	83,000
vehicle brakes	166,000

D. CONTACT PRESSURES

In order to maintain the integrity of the brake material, it is necessary to keep the contact pressure less than the values in table 15.13.

16 PLATE CLUTCHES

Nomenclature

f	coefficient of static friction	–
F	frictional force	lbf
HP	horsepower	hp
n	rotational speed	rpm
N	normal force	lbf
p	pressure	psf
r	radius	ft
T	torque	ft-lbf

Subscripts

f	frictional
o	outer

If the clutch assembly is rigid, the initial wear will be in the outer areas, since the frictional work is greater in those areas. After a certain amount of wear, the pressure distribution will change, and the wear will become uniform. *Uniform wear* is equivalent to assuming that all work occurs at the average friction radius. If the clutch is semi-flexible and is spring loaded, a *uniform pressure* over the contact area should be assumed.

If in doubt, assume uniform wear. That assumption is more conservative in terms of allowable torque and transmitted horsepower.

The torque capacity is proportional to the number of friction planes. In the case of an automobile clutch disc, there are two friction planes. The maximum transmitted horsepower will be twice the torque values calculated in equations 15.160 and 15.164.

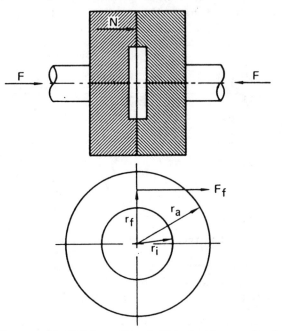

Figure 15.40 Plate Clutch (Motion Impending)

Regardless of whether uniform wear or uniform pressure is assumed, the transmitted horsepower can be found from the applied torque.

$$HP = \frac{Tn}{5252} \qquad\qquad 15.158$$

Table 15.13
Clutch and Brake Design Parameters
(Also, see table 12.2.)

use	wearing	opposing	f oily	f dry	T_{max} (°F)	p_{max} (psi)
clutches:	cast iron	cast iron	.05	.15–.2	600	150–250
		steel	.06	.23	500	120–200
	hard steel	hard steel	.05	.40	500	100
	powdered metal	cast iron/steel	.05–.1	.1–.4	1000	150
	leather	cast iron/steel	.12–.15	.3–.5	200	10–40
	molded asbestos	cast iron/steel	.08–.12	.2–.5	500	50–150
	impregnated asbestos	cast iron/steel	.12	.32	500–750	150
brakes:	molded asbestos lining	cast iron/steel		.47	500	100
	woven asbestos lining	cast iron/steel		.45	400–500	50–100

A. UNIFORM PRESSURE

The axial *application force*, F, is also the normal force causing the friction. Since the pressure is uniform, the maximum pressure occurs everywhere.

$$F = N = \frac{F_f}{f} = \pi p_{max}(r_o^2 - r_i^2) \qquad 15.159$$

The *frictional radius* (the mean radius for frictional purposes) can be used to determine the limiting torque per contact plane.

$$T = F_f r_f = (\tfrac{2}{3})\pi f p_{max}(r_o^3 - r_i^3) \qquad 15.160$$

$$r_f = \frac{2(r_o^3 - r_i^3)}{3(r_o^2 - r_i^2)} \qquad 15.161$$

B. UNIFORM WEAR

The maximum pressure will occur at the inner radius. The pressure at any other radius, r, is

$$p_r = \frac{p_{max}r_i}{r} \qquad 15.162$$

The axial application force is

$$F = 2\pi p_{max}r_i(r_o - r_i) \qquad 15.163$$

The torque can be assumed to be carried at the mean radius.

$$T = \tfrac{1}{2}Ff(r_o + r_i) = \pi f p_{max}r_i(r_o^2 - r_i^2) \qquad 15.164$$

17 JOURNAL BEARINGS

Nomenclature

c_d	diametral clearance	in
c_p	specific heat	BTU/lbm-°F
d	journal diameter	in
D	bearing diameter	in
e	eccentricity	
f	coefficient of friction	–
F	total bearing load	lbf
h_o	minimum film thickness	
L	length	in
n	journal speed	rpm
N	journal speed	rps
N_L	load number	–
p	projected area pressure	psi
q	heat generation rate	BTU/min
r	radius	in
S	Sommerfeld number	–
SSU	Saybolt absolute viscosity	sec
T	temperature, or torque	°F, in-lbf

Symbols

ϵ	eccentricity ratio	–
μ	viscosity	reyns

Subscripts

f	frictional
o	unloaded, or minimum friction

A. LOAD CAPACITY

The load capacity of a short bearing with $L/D < 1$ can be approximated by *Ocvirk's analysis* method. This method requires knowing the *eccentricity ratio*, also known as the *attitude*, of the shaft/bearing combination during operation. The terms used to calculate the eccentricity ratio are shown in figure 15.41.[29]

$$\epsilon = \frac{2e}{c_d} \qquad 15.165$$

$$c_d = D - d \qquad 15.166$$

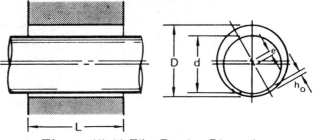

Figure 15.41 Film Bearing Dimensions

The eccentricity ratio also can be determined from the *load number*. The load number is based on the assumption that the pressure on the projected area is

$$p = \frac{\text{shaft load}}{\text{projected area}} = \frac{\text{shaft load}}{Ld} \qquad 15.167$$

The load number can be found graphically from figure 15.42, or it can be calculated from equation 15.168.[30]

$$N_L = \frac{(60)p}{\mu n}\left(\frac{c_d}{d}\right)^2\left(\frac{d}{L}\right)^2 \qquad 15.168$$

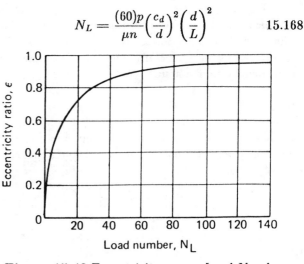

Figure 15.42 Eccentricity versus Load Number

[29] Do not mix diametral and radial clearances.

[30] The ratio, c_d/d, also can be given as c/r. In the latter case, it is understood that the clearance is a radial clearance. Of course, the two ratios are the same.

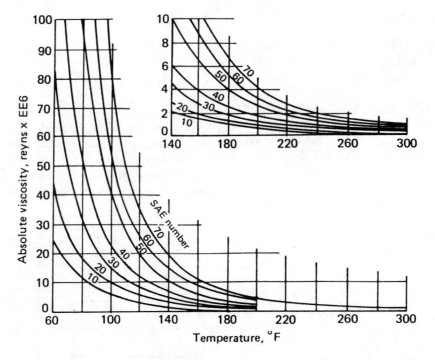

Figure 15.43 Viscosities of Common Lubricating Oils

In order to use equation 15.168, it is necessary to know the oil viscosity. It is not uncommon (and equation 15.168 requires) that the viscosity be expressed in units of *reyns*. This unit is the same as lbf-sec/in². If the viscosity is given in centipoise, it can be converted.

$$\text{reyns} = \frac{\text{centipoise}}{6.9 \text{ EE6}} \qquad 15.169$$

$$\text{centipoise} = \left[(0.22)SSU - \frac{180}{SSU}\right] \qquad 15.170$$

The viscosity should be evaluated at the average temperature of the lubricating oil. Figure 15.43 can be used to find the viscosity. (Be sure to divide the values from figure 15.43 by EE6, since the plotted values have been multiplied by that amount.)

If the eccentricity ratio is known, the minimum film thickness can be determined from equation 15.171. A minimum practical film thickness is 0.00015 inches per inch of bearing diameter.

$$h_o = \tfrac{1}{2}c_d(1 - \epsilon) \qquad 15.171$$

A small, conservative error is introduced if this method is used for bearings with L/d up to 2. In that case, use $L/d = 1$ when calculating the load number. Do not use the actual ratio of L/d. Suggested values of L/d and allowable bearing pressures are listed in table 15.14. These values are averages, not maximums, which may run as high as twice the average.

Table 15.14
Typical Film Bearing Parameters

application	suggested L/d	Pressure, psi
electric motors	2	100–200
pumps	2	80–100
machine tools	2–4	80–100
automotive mains	.5–.8	500–600
automotive crankpins	.5–.8	1500–1800
automotive wristpins	—	2000–3000
steam turbines	—	2–10

B. FRICTION LOSSES

The viscosity of the lubricating oil will cause frictional torque which opposes rotation. If the eccentricity is zero (i.e., the bearing is unloaded), or if the eccentricity is low and the rotational speed is high, the *Petroff equation* can be used to find the *frictional torque*.

$$T_f = \tfrac{1}{2}df(\text{lateral load}) = \frac{\pi^2 \mu d^3 L n}{60 c_d} \qquad 15.172$$

The coefficient of friction can be determined from equation 15.173 if the projected area pressure is known.

$$f = \frac{\pi^2 \mu n d}{30 p c_d} \qquad 15.173$$

The average frictional horsepower (with torque in in-lbf) is

$$HP_f = \frac{T_f n}{63,000} \qquad 15.174$$

If the frictional torque is calculated from equation 15.172 and the load number is known, figure 15.44 can be used to determine the loaded frictional torque.

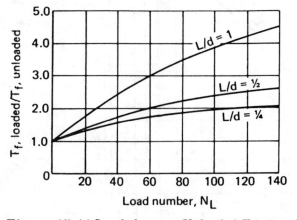

Figure 15.44 Loaded versus Unloaded Frictional Torque

Example 15.17

SAE 20 oil at 150°F average temperature is used in a lightly loaded 3″ diameter bearing. The bearing length is 5″. The shaft turns at 1800 rpm. The diametral clearance is 0.005″. What horsepower is dissipated through friction?

From figure 15.43, the approximate viscosity is 2.5 EE−6 reyns. From equation 15.172, the frictional torque is

$$T_f = \frac{\pi^2 (2.5 \text{ EE} - 6)(3)^3(5)(1800)}{(60)(0.005)} = 20.0 \text{ in-lbf}$$

The frictional horsepower from equation 15.174 is

$$HP_f = \frac{(20.0)(1800)}{63,000} = .57$$

C. RAIMONDI AND BOYD METHOD

In the 1950's, A. A Raimondi and John Boyd (two Westinghouse researchers) investigated film bearing performance as a function of the *bearing characteristic number*, also known as the *Sommerfeld number*.

The results of these investigations have been presented graphically. Graphs useful in determining frictional losses, required lubrication flow, temperature rise, and maximum pressure are available.

The dimensionless bearing characteristic number is

$$S = \left(\frac{d}{c_d}\right)^2 \left(\frac{\mu n}{60p}\right) \qquad 15.175$$

The coefficient of friction can be determined after the *coefficient of friction variable* is read from figure 15.45. Once f is known, the frictional torque and the horsepower can be found from equations 15.172 and 15.174.

$$f = \left(\begin{array}{c} \text{coefficient of} \\ \text{friction variable} \end{array}\right)\left(\frac{c_d}{d}\right) \qquad 15.176$$

The film thickness also has been correlated with bearing characteristic number. The left-most part of the shaded region represents minimum friction operation; the right-most part represents maximum load carrying capacity. (Notice that figure 15.46 has two vertical scales.)

$$\text{minimum film thickness} = r\left(\begin{array}{c} \text{minimum film} \\ \text{thickness variable} \end{array}\right)\left(\frac{c_d}{d}\right)$$
$$15.177$$

D. BEARING HEAT AND OIL TEMPERATURE RISE

The heat generated by bearing friction is determined easily from the generated friction horsepower. This heat varies with the square of the rotational speed. The usual design temperature for oil is 140°F to 160°F. The viscosity changes rapidly above 150°F, and oil deteriorates above 200°F.

$$q_{\text{generated}} = \frac{(HP)_f(33,000)}{778} \qquad 15.178$$

This heat must be dissipated in some manner. As the oil is cooled in a heat exchanger or by contact with a cooler surface, the temperature decreases. The dissipated heat is given by equation 15.179. c_p is approximately .42 − −.49 BTU/lbm-°F for petroleum oils.

$$q_{\text{dissipated}} = c_p(T_{\text{in}} - T_{\text{out}}) \qquad 15.179$$

Self-contained bearings, bearings in housings, and air-cooled bearings should be handled by other methods.

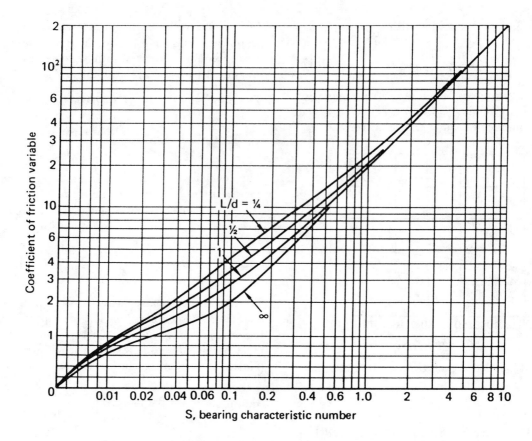

Figure 15.45 Coefficient of Friction Variable

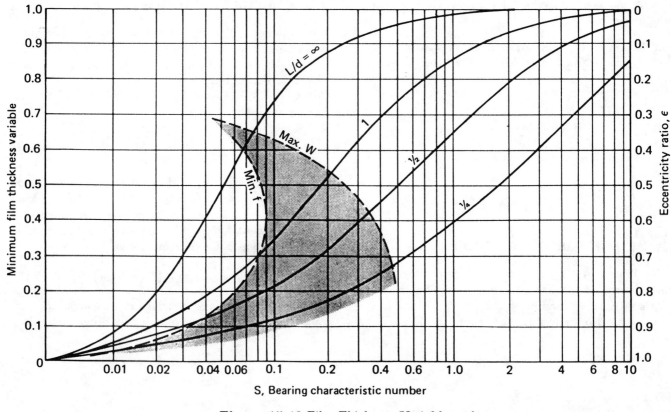

Figure 15.46 Film Thickness Variable and
Eccentricity Ratio

APPENDIX A
Machine Design Criteria

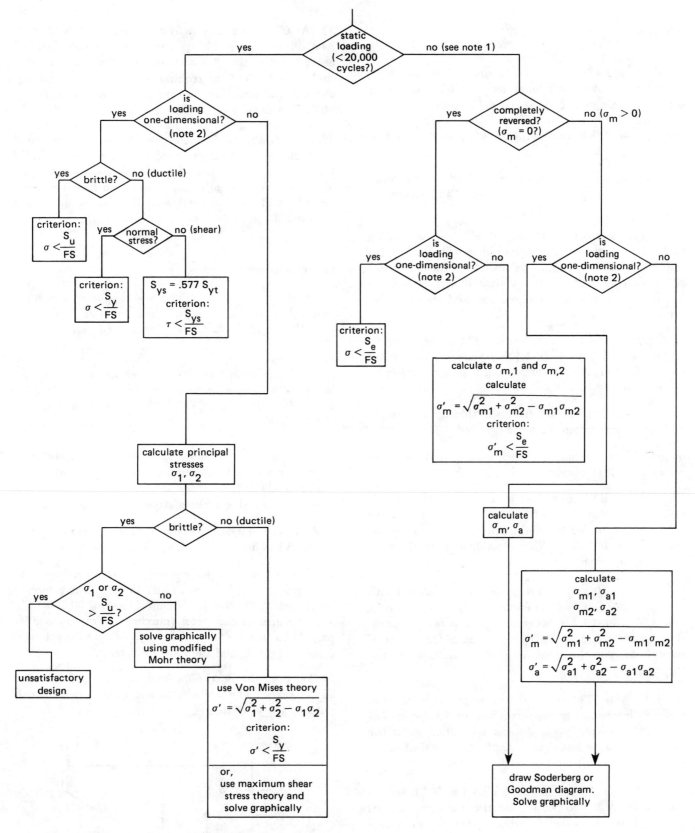

notes: 1. If life is not infinite, substitute the fatigue strength for S_e.

2. One-dimensional loading includes pure shear.

PRACTICE PROBLEMS: MACHINE DESIGN

Warm-ups

1. A device is being designed to run at 270 rpm while being driven by a 550 horsepower, 1200 rpm motor through a pair of $14\frac{1}{2}°$ gears on a 15″ center-to-center distance. The pinion is SAE 1045 steel, and the gear is cast steel. With a safety factor of 3, what should be the face width?

2. Choose gears for a train value of 600: 1 using minimum and maximum tooth numbers of 12 and 96, respectively.

3. A 25 horsepower direct current motor shaft turning at 2000 rpm carries an 8″ fiber pulley. The center distance is 20′ with the tight side down, and the center line is 20° from the vertical. What width of $\frac{5}{16}$″, double-ply oak-tanned belt should be used for continuous operation?

4. A 1.25″ diameter shaft carries 4200 in-lbf of torque by means of a 1030 $\frac{1}{4}$″ × 1.125″ Woodruff key. What are the factors of safety in bearing and shear based on ultimate strengths?

5. A 1.125″ shaft receives 400 in-lbf through a 1030 pin and sleeve. What pin diameter is called for?

6. Prescribe annealing and precipitation hardening procedures for 2011 aluminum.

7. A $\frac{3}{4}$″-16 UNF steel bolt is used to clamp two 2″ rigid steel plates. .75″ of the threaded section of the bolt remains under the nut when tightened to 40,000 psi in the bolt body. What stretching occurs? State your assumptions.

8. A $\frac{5}{8}$″ NC bolt and nut are made from a material with an endurance limit of 30,000 psi. They carry a tensile load which fluctuates between 1000 and 8000 pounds. If the yield stress is 70,000 psi, what is the factor of safety? Include the stress concentration factor in the threads.

9. A 6″ wide, 2′ long cantilever spring is being constructed from steel to support a load of 800 pounds at its tip. What minimum thickness is called for if the deflection is to be less than 2″ at the tip? Neglect bending stress.

10. A 2400′ long mine shaft is inclined at 30° from the horizontal. A 20,000 pound loaded ore car rises at 100 vertical fpm on frictionless wheels. The ore car starts from rest with an acceleration period of 10 seconds. What size wire rope is called for?

11. A 30″ mean diameter flywheel with a 12″ wide cast iron rim rotates at 200 rpm and supplies 1500 ft-lbf of energy each cycle. The rotation drops to 175 rpm after each stroke. Find the required rim thickness assuming the hub and arms increase the mean diameter mass by 10%. The density of the cast iron is .26 lbm/in³.

12. A wet steel-backed asbestos clutch with hardened steel plates is being designed to transmit 300 in-lbf torque. Slip will occur at 300% of the rated torque. The maximum and minimum friction surface diameters are $4\frac{1}{2}$″ and $2\frac{1}{2}$″, respectively. How many plates are required if the contact pressure is 100 psi?

Concentrates

1. A spring with 12 active coils supports 50 pounds wth a deflection of $\frac{1}{2}$″. The shear modulus is 1.2 EE7 psi. What will be the wire size and the mean spring diameter?

2. Design a severe service valve spring with a spring index of 10 to operate between 20 and 30 pounds. The valve lift is .3″. Use ASTM A230 steel wire in standard W&M sizes.

3. A spring with an index of 7 and shear modulus of 1.2 EE7 psi absorbs the impact of a 700 pound object falling freely for 46 inches and deflects 10″. Design a spring to keep the working stress below 50,000 psi.

4. A 12″ class 30 cast iron hub is pressed onto a 6″ steel shaft. What interference is required?

5. Two 20° involute spur gears 15″ apart reduce a 40 horsepower 250 rpm input to 83.33 rpm. The pinion is untreated 1030 steel (maximum stress of 30,000 psi), and the gear is cast steel (maximum stress of 50,000 psi). What are the pitch diameters, the face width, the diametral pitch, and the number of teeth?

6. What is the maximum load that the steel hanger should carry? The rivets in the triangular (symmetrical) pattern are $\frac{3}{4}$″ diameter.

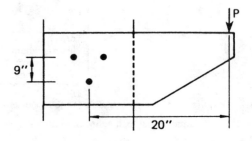

7. What size fillet weld is called for in the steel construction shown?

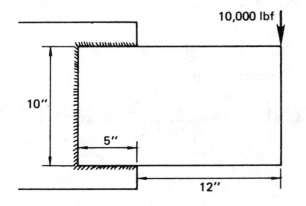

8. A 10-mil adhesive with shear strength of 1500 psi is used to form a lap joint between two .020″ aluminum sheets. The aluminum has a yield strength of 15,000 psi. Assume a stress concentration factor of 2. If the joint is to be as strong as the aluminum, what width overlap is required?

9. A 2500 pound stationary flywheel is mounted on a stepped shaft as shown. If the fillet radius is $\frac{5}{16}″$, what is the maximum stress in the shaft? Ignore local yielding which reduces the stress concentration effects.

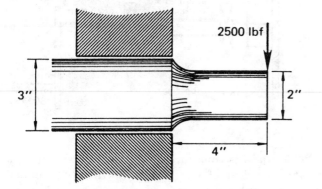

10. A 2″ steel shaft rotates at 3500 rpm and carries a press-fitted steel flywheel of 2″ thickness and 16″ diameter. A contact pressure of 1250 psi is required for torque transmission. What is the required initial interference?

11. A 2″ steel shaft 40″ long is supported on frictionless bearings. The shaft carries a 100 pound disk 15″ from the left bearing and a 75 pound disk 25″ from the left bearing. What is the critical speed of the shaft, neglecting the shaft weight?

12. A standard planetary gearbox with one planet has gears with 24, 40, and 104 teeth on the sun, the planet, and the fixed ring gear, respectively. The sun rotates clockwise at 50 rpm. What is the rotational velocity of the planet carrier?

13. What is the rotational speed of the sun gear if gear A rotates counterclockwise at 100 rpm, and the planet carrier rotates clockwise at 60 rpm?

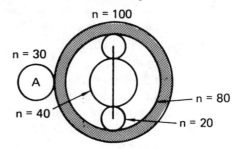

14. An epicyclic gear train consists of a ring gear, 3 planets, and a fixed sun gear. 15 horsepower is transmitted through the input ring gear which turns at 1500 rpm clockwise. The diametral pitch is 10 with a 20° pressure angle. The pitch diameters are 5″, $2\frac{1}{2}″$, and 10″ for the sun, the planets, and the ring gear, respectively. **Except for the planets, determine all speeds, directions, and numbers of teeth. What torque is on the input and output shafts?**

Timed (1 hour allowed for each)

1. A 3″ diameter shaft turns at 1200 rpm in a bearing having a c/r ratio of .001. The load carried by the bearing is 880 pounds. The oil has a viscosity of 1.184 EE−6 lbf-sec/in² (same as 8.16 centipoise). The axial length of the bearing is 3.5″. (a) What is the minimum film thickness? (b) What is the friction loss in horsepower? Assume that the film temperature is 165°F.

2. Pressure in the vessel shown below varies continually between 50 nd 350 psig. The flange is attached to the vessel with six 3/8-24 UNF bolts on a $9\frac{1}{2}″$ circle. The bolts are tightened to an initial preload of 3700 pounds. The bolts have the following properties.
   - 2330 cold rolled steel
   - 90 ksi yield strength
   - 110 ksi ultimate strength

The flange and the vessel are constructed of 1035 steel. Its properties are
   - 30 ksi yield strength
   - 50 ksi ultimate strength

Using a thread concentration factor of 2, what is the factor of safety (for infinite life) for the bolts?

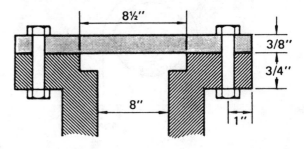

3. Two concentric springs carry a static load of 150 pounds. The spring dimensions and properties are known. (a) What is the deflection of the inner spring? (b) What is the factor of safety for the inner spring based on the maximum shear stress theory? (c) Specify the helix direction for both springs.

     inner spring: .177″ wire
                              1.5″ diameter
                              4.5″ free length
                              squared and ground ends
                              oil hardened steel
                              12.75 coils

     outer spring: .2253″ wire
                              2.0″ diameter
                              3.75″ free length
                              squared and ground ends
                              oil hardened steel
                              10.25 coils

4. A differential gear set is arranged as shown. What is the direction and the speed of gear B?

     gear A: 30 teeth, 50 rpm counterclockwise as viewed from the gear looking up the shaft
     gear B: 30 teeth
     gear C: 18 teeth, 600 rpm counterclockwise as viewed from the gear looking up the shaft
     gear D: 54 teeth
     gear E: 15 teeth
     gear F: 15 teeth

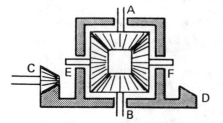

5. The planetary gear set shown has an overall speed reduction of 3:1. The planet carrier is the driven element. The sun gear drives at 1000 rpm. The planet has 20 teeth and a diametral pitch of 10. (a) What is the ratio of the number of teeth on the ring and sun gears? (b) What is the angular velocity (in rpm) of the planet with respect to its bearing?

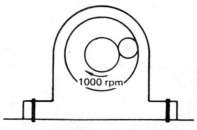

6. A gear train is constructed as shown. The output turns at 250 rpm CCW (counterclockwise). How many teeth does gear C have?

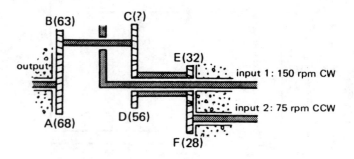

7. The power transmission system shown below is constructed of helical gears (25° helix, 20° normal pressure angle, diametral pitch of 5) and AISI 1045 cold drawn steel shafting. (a) What is the output shaft speed in rpm? (b) What is the output torque if the transmission efficiency of each gear set is 98%? (c) What is the minimum shaft diameter in the section A-A?

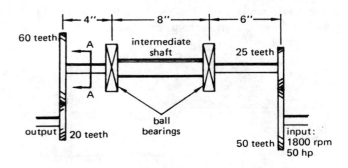

8. A flywheel is constructed of class 30 cast iron. The flywheel consists of a solid disk, 2″ thick, 20″ outside diameter, with a concentric 4″ diameter mounting hole. The density of the cast iron used is .26 lbm/in³. What is the maximum safe speed in rpm for this flywheel?

9. A radial cam follower follows a cam turning at a constant speed of 120 rpm. The follower rises .5″ in constant acceleration as the cam turns 60°. The follower returns to rest at a constant deceleration during the next 90° of cam movement. (a) How far does the follower move during the 150° rotation? (b) What is the magnitude of the acceleration during the first 60°? (c) What is the magnitude of the deceleration during the last 90°?

16 DYNAMICS

1 INTRODUCTION

Dynamics is the study of bodies which are not in equilibrium. The subject is subdivided into kinematics and kinetics. *Kinematics* is the study of motion without consideration of the forces causing motion. Kinematics deals with the relationships between position, velocity, acceleration, and time.

Kinetics is concerned with motion and the forces causing motion. Kinetics deals with the variables of force, mass, and acceleration. The study of kinetics requires the application of Newton's laws.

Bodies in motion are called *particles* if rotation is absent or insignificant. Particles do not possess rotational kinetic energy. All parts of a particle possess the same instantaneous displacement, velocity, and acceleration. *Rigid bodies* are objects whose parts exhibit identical displacements, velocities, and accelerations during translation.

2 NEWTON'S LAWS

Much of this chapter is based on *Newton's laws of motion*. These laws can be stated in many forms. They are presented here, along with *Newton's law of gravitation*, in common wording.

First law: A particle will remain in a state of rest or will continue to move with constant velocity unless an unbalanced external force acts on it.

Second law: The acceleration of a particle is directly proportional to the force acting on it and is inversely proportional to the particle mass.[1]

---

[1] The pound (*lbm*) is used as the unit of mass in this chapter. This choice makes it necessary to include g_c in equation 16.1, 16.2, and others that follow. g_c can be omitted if a consistent set of units is used (e.g., pounds-force, slugs, and ft/sec^2).

The direction of acceleration is the same as the force direction.

$$F = \frac{d\left(\frac{mv}{g_c}\right)}{dt} \qquad 16.1$$

If the mass is constant with respect to time, equation 16.1 can be rewritten as equation 16.2.

$$F = \left(\frac{m}{g_c}\right)\frac{dv}{dt} = \left(\frac{m}{g_c}\right)a \qquad 16.2$$

Third law: There is an equal and opposite reacting force for every acting force.

$$\mathbf{F}_{\text{reacting}} = -\mathbf{F}_{\text{acting}} \qquad 16.3$$

Law of Universal Gravitation: The attractive force between two particles is directly proportional to the product of masses and inversely proportional to the square of distances between their centroids.

$$F = \frac{Gm_1m_2}{(g_c r)^2} \qquad 16.4$$

3 PARTICLE MOTION

A. BASIC VARIABLES

A *linear system* is one in which particles move only in straight lines. The relationships betwee force, position, velocity, and acceleration for a linear system are given by equations 16.5 through 16.8.

$$a = \frac{dv}{dt} = \frac{d^2 s}{dt^2} \qquad 16.5$$

$$v = \frac{ds}{dt} = \int a\,dt \qquad 16.6$$

$$s = \int v\,dt = \int\int a\,dt^2 \qquad 16.7$$

$$F = \left(\frac{m}{g_c}\right)a \qquad 16.8$$

A *rotational system* is one in which the particles move in circular paths. The relationships between torque, angular position, angular velocity, and angular acceleration are given by equations 16.9 through 16.12.

$$\alpha = \frac{d\omega}{dt} = \frac{d^2\theta}{dt^2} \qquad 16.9$$

$$\omega = \frac{d\theta}{dt} = \int \alpha\,dt \qquad 16.10$$

$$\theta = \int \omega\,dt = \int\int \alpha\,dt^2 \qquad 16.11$$

$$T = \left(\frac{I}{g_c}\right)\alpha \qquad 16.12$$

Example 16.1

The velocity of a 40 lbm particle as a function of time is

$$v(t) = 8t - 6t^2 \text{ (ft/sec)}$$

The velocity and position are both zero at $t = 0$. What are the acceleration and position functions?

$$a(t) = \frac{dv(t)}{dt} = 8 - 12t \text{ (ft/sec}^2)$$

$$s(t) = \int v(t)\,dt = 4t^2 - 2t^3 \text{ (ft)}$$

Example 16.2

What is the force acting at $t = 6$ on the particle described in example 16.1?

The acceleration at $t = 6$ is

$$a(6) = 8 - (12)(6) = -64 \text{ ft/sec}^2$$

From equation 16.8, the force is

$$F = \left(\frac{40}{32.2}\right)(-64) = -79.5 \text{ lbf}$$

It is necessary to distinguish between position, displacement, and distance traveled. The *position* of a particle is an actual location. Position is determined from the *position function*, $s(t)$. The *displacement* of a particle is the difference in positions at two different times. Displacement is found by subtracting values of the position function.

$$\Delta s = s(t_2) - s(t_1) \qquad 16.13$$

The *distance traveled* includes distance covered during all direction reversals. It can be found by adding displacements during periods in which the velocity sign does not change.

Example 16.3

What distance is traveled by the particle described in example 16.1 during the period $t = 0$ to $t = 6$?

The velocity changes from positive to negative at $t = \frac{4}{3}$. The initial position is $s = 0$ at $t = 0$. At $t = \frac{4}{3}$, the position is

$$s\left(\frac{4}{3}\right) = 4\left(\frac{4}{3}\right)^2 - 2\left(\frac{4}{3}\right)^3 = 2.37 \text{ ft}$$

The displacement is given by equation 16.13.

$$\Delta s = 2.37 - 0 = 2.37 \text{ ft}$$

The position at $t = 6$ is

$$s(6) = 4(6)^2 - 2(6)^3 = -288 \text{ ft}$$

The displacement between $t = \frac{4}{3}$ and $t = 6$ is

$$s = s(6) - s\left(\frac{4}{3}\right)$$
$$= -288 - 2.37 = -290.37 \text{ ft}$$

The total distance traveled is

$$2.37 + 290.37 = 292.74 \text{ ft}$$

If the acceleration is constant, the a term can be taken out of the integrals in equations 16.6 and 16.7.

$$v(t) = a \int dt = at + v_0 \qquad 16.14$$

$$s(t) = a \iint dt^2 = \tfrac{1}{2}at^2 + v_0 t + s_0 \qquad 16.15$$

Table 16.1 summarizes the equations required to solve uniform acceleration problems. The table can be used for rotational problems by substituting the analogous variables of α, ω, and θ for a, v, and s, respectively.

Table 16.1
Uniform Acceleration Formulas

to find	given these	use this equation
t	$a\ v_0\ v$	$t = \dfrac{v - v_0}{a}$
t	$a\ v_0\ s$	$t = \dfrac{\sqrt{2as + v_0^2} - v_0}{a}$
t	$v_0\ v\ s$	$t = \dfrac{2s}{v_0 + v}$
a	$t\ v_0\ v$	$a = \dfrac{v - v_0}{t}$
a	$t\ v_0\ s$	$a = \dfrac{2s - 2v_0 t}{t^2}$
a	$v_0\ v\ s$	$a = \dfrac{v^2 - v_0^2}{2s}$
v_0	$t\ a\ v$	$v_0 = v - at$
v_0	$t\ a\ s$	$v_0 = \dfrac{s}{t} - \tfrac{1}{2}at$
v_o	$a\ v\ s$	$v_0 = \sqrt{v^2 - 2as}$
v	$t\ a\ v_0$	$v = v_0 + at$
v	$a\ v_0\ s$	$v = \sqrt{v_0^2 + 2as}$
s	$t\ a\ v_0$	$s = v_0 t + \tfrac{1}{2}at^2$
s	$a\ v_0\ v$	$s = \dfrac{v^2 - v_0^2}{2a}$
s	$t\ v_0\ v$	$s = \tfrac{1}{2}t(v_0 + v)$

Example 16.4

A locomotive traveling at 80 mph locks its wheels and skids 580 feet before stopping. If the deceleration is constant, how long (in seconds) will it take for the locomotive to come to a standstill?

Convert the 80 mph to ft/sec.

$$\frac{(80)\,\text{mi/hr}\,(5280)\,\text{ft/mi}}{(3600)\,\text{sec/hr}} = 117.3 \text{ ft/sec}$$

t is unknown. $v_0 = 117.3$ ft/sec. $v = 0$. $s = 580$. From table 16.1,

$$t = \frac{2s}{v_0 + v} = \frac{(2)(580)}{117.3 + 0} = 9.89 \text{ sec}$$

B. RELATIONSHIPS BETWEEN LINEAR AND ROTATIONAL MOTION

If a particle travels in a circular path with instantaneous radius, r, the particle's *tangential velocity* can be calculated from the angular velocity.

$$v_t = \omega r \qquad 16.16$$
$$v_{t,x} = v_t \cos\phi \qquad 16.17$$
$$v_{t,y} = v_t \sin\phi \qquad 16.18$$

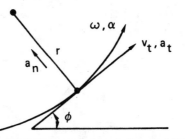

Figure 16.1 Tangential and Normal Variables

If the particle is accelerating as it travels around the curve, its angular velocity will be changing. The relationship between the tangential and angular accelerations is given by equation 16.19.

$$a_t = \alpha r \qquad 16.19$$

A particle traveling in a circular path will tend to continue traveling along its tangent. If the particle is restrained (e.g., a rock being twirled on a string), it will continue to travel in the circular path. This restraint will be due to an applied force (e.g., tension in the string).

From equation 16.2, an acceleration acts whenever a mass experiences a force. The force that keeps the particle in circular motion is directed toward the center of rotation. Therefore, the acceleration also is directed toward the center. This acceleration is known as the *normal acceleration*.

$$a_n = \frac{v_t^2}{r} = r\omega^2 = v_t\omega \qquad 16.20$$

This normal acceleration produces the apparent *centrifugal*[2] *force*.

---

[2] The centrifugal force is an *apparent force* on the object. Centrifugal force is directed outward, away from the center of rotation. It is experienced most commonly by riders in cars. However, the real force on the particle is *toward* the center of rotation. This real force is known as *centripetal force*. The centrifugal and centripetal forces are equal in magnitude but opposite in sign.

$$F_c = \frac{ma_n}{g_c} = \frac{mv_t^2}{g_c r} = \frac{mr\omega^2}{g_c} \qquad 16.21$$

Example 16.5

A 4000 pound car travels at 40 mph around a banked curve with radius of 500 feet. What should the banking angle be such that tire friction is not needed to prevent the car from sliding?

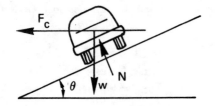

The forces acting on the car are its own weight, the centripetal force, and the normal force.

$$F_c = \frac{mv_t^2}{g_c r} = N \sin \theta \qquad 16.22$$

Solving for N,

$$N = \frac{mv_t^2}{g_c r \sin \theta} \qquad 16.23$$

$$w = \frac{mg}{g_c} = N \cos \theta \qquad 16.24$$

Solving for N,

$$N = \frac{mg}{g_c \cos \theta} \qquad 16.25$$

Equating both expressions for N,

$$\tan \theta = \frac{v_t^2}{gr} \qquad 16.26$$

The car velocity is

$$\frac{(40)\,\text{mi/hr}\,(5280)\,\text{ft/mi}}{(3600)\,\text{sec/hr}} = 58.67\,\text{ft/sec}$$

The required banking angle is

$$\theta = \arctan\left[\frac{(58.67)^2}{(32.2)(500)}\right] = 12.07°$$

C. CORIOLIS ACCELERATION

Consider a particle moving with radial velocity, v_r, on the surface of a disk rotating with velocity, ω. From equation 16.20, the normal acceleration toward the center is $r\omega^2$. The particle's tangential velocity increases as the particle moves outward. This increase in said to be produced by the *coriolis acceleration* acting tangentially.

$$a_c = 2v_r \omega \qquad 16.27$$

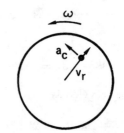

Figure 16.2 Coriolis Acceleration with a Rotating Disk

Motion on the surface of a moving sphere is more complex due to the changing velocity, v_r. Consider an airplane flying with constant airspeed, v, from the equator to the north pole while the earth rotates below it. Three accelerations act on the aircraft. The first is the normal acceleration.

$$a_n = r\omega^2 = R(\cos \phi)\omega^2 \qquad 16.28$$

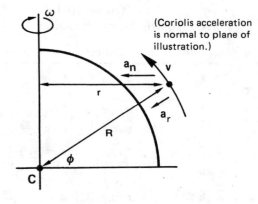

(Coriolis acceleration is normal to plane of illustration.)

Figure 16.3 Coriolis Acceleration with a Rotating Sphere

The second acceleration is the coriolis acceleration, which depends on the latitude, ϕ. It acts perpendicularly to figure 16.3.

$$a_c = 2\omega v_x = 2\omega v(\sin \phi) \qquad 16.29$$

The third acceleration (also a normal acceleration) is directed toward the earth's center. Its magnitude is

$$a = \frac{v^2}{R} \qquad 16.30$$

Example 16.6

A slider moves with a constant velocity of 20 ft/sec along a rod rotating at 5 radians per second. What is the slider's acceleration when it is 4 feet from the center of rotation?

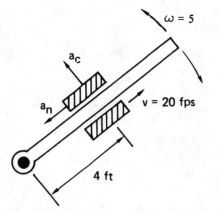

The normal acceleration is

$$a_n = (4)(5)^2 = 100 \text{ ft/sec}^2$$

The coriolis acceleration is

$$a_c = (2)(20)(5) = 200 \text{ ft/sec}^2$$

The resultant acceleration experienced by the slider is

$$a = \sqrt{(100)^2 + (200)^2} = 223.6 \text{ ft/sec}^2$$

D. PROJECTILE MOTION

Consider a projectile launched at an angle of ϕ from the horizontal with initial velocity, v_0. Neglecting air resistance, the trajectory is a parabola with coordinates (x, y) at time t. The impact velocity is equal to the launch velocity. The maximum range is achieved when $\phi = 45°$.

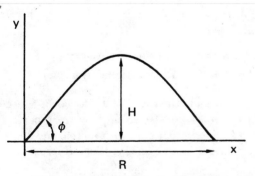

Figure 16.4 Projectile Motion

After a projectile is launched, it is acted upon by a constant gravitational acceleration directed downward. A projectile's altitude above the launch plane varies with time. This altitude is a composite function of the vertical velocity component and gravity.

$$y(t) = (v_0 \sin \phi)t - \tfrac{1}{2}gt^2 \qquad 16.31$$

Other relationships are derived easily from the laws of uniform acceleration. Special nomenclature is used with equations 16.32 through 16.42.

g	acceleration due to gravity
H	maximum altitude
R	total range
t	time after launch
t^*	total flight time
v_0	initial launch velocity
ϕ	launch angle

$$x = (v_0 \cos \phi)t \qquad 16.32$$
$$y = (v_0 \sin \phi)t - \tfrac{1}{2}gt^2 \qquad 16.33$$
$$v = \sqrt{v_0^2 - 2gy} \qquad 16.34$$
$$v_x = v_0 \cos \phi \qquad 16.35$$
$$v_y = v_0 \sin \phi - gt \qquad 16.36$$
$$H = \frac{v_0^2 \sin^2 \phi}{2g} \qquad 16.37$$
$$R = \frac{v_0^2 \sin 2\phi}{g} \qquad 16.38$$
$$t^* = \frac{2v_0 \sin \phi}{g} \qquad 16.39$$

If the projection is horizontal from height H, then $\phi = 0$.

$$x = v_0 t \qquad 16.40$$
$$y = H - \tfrac{1}{2}gt^2 \qquad 16.41$$
$$t^* = \sqrt{\frac{2H}{g}} \qquad 16.42$$

Example 16.7

A bomber flies horizontally at 275 mph at an altitude of 9000 feet. At what angle α should the bombs be dropped? Neglect air friction.

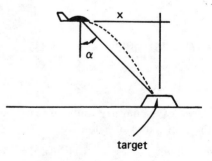

target

This is a case of horizontal projection. The falling time depends on only the height of the bomber.

$$t^* = \sqrt{\frac{(2)(9000)}{32.2}} = 23.64 \text{ sec}$$

The range is

$$x = (23.64)(275)\left(\frac{5280}{3600}\right) = 9535 \text{ ft}$$

Therefore, the angle is

$$\alpha = \arctan\left(\frac{9535}{9000}\right) = 46.7°$$

Example 16.8

A projectile is launched at 600 ft/sec with a 30° inclination from the horizontal. The launch point is 500 feet above the plane of impact. Neglecting friction, find the range and the maximum altitude.

The maximum altitude above the launch elevation can be found from equation 16.37. The total altitude above the impact plane is

$$H_t = 500 + \frac{(600)^2(.5)^2}{(2)(32.2)} = 1897.5 \text{ ft}$$

From equation 16.39, the time to maximum altitude is

$$t_{\max} = \tfrac{1}{2}t^* = \frac{(600)(.5)}{32.2} = 9.32 \text{ sec}$$

The time to fall from the maximum altitude to the impact plane can be found from table 16.1.

$$t_{\text{fall}} = \sqrt{\frac{(2)(1897.5)}{32.2}} = 10.86 \text{ sec}$$

The total flight time is

$$t_{\text{total}} = t_{\max} + t_{\text{fall}} = 9.32 + 10.86 = 20.18 \text{ sec}$$

The x-component of velocity is

$$v_x = 600(\cos 30°) = 519.6 \text{ ft/sec}$$

The maximum range is

$$R = (519.6)(20.18) = 10,485 \text{ ft}$$

E. SATELLITE MOTION

Kepler's laws of planetary motion can be used to describe the movement of satellites.

law 1: Each planet moves in an elliptical path with the sun at one focus.

law 2: The radius vector drawn from the sun to the planet sweeps out equal areas in equal times.

law 3: The squares of the periodic times of the planets are proportional to the cubes of the semi-major axes of the orbits.

For light satellites traveling around the earth, it is possible to make some simplifying assumptions. It is assumed that (1) the satellite's mass is much smaller than the earth's, and (2) the satellite is unaffected by objects other than the earth.

The force exerted on the satellite is given by *Newton's law of universal gravitation.*

$$\mathbf{F} = \frac{Gm_{\text{earth}}m_{\text{satellite}}}{(rg_c)^2} \qquad 16.43$$

G is a gravitational constant with a value of 3.44 EE−8 ft^4/lbf-sec^4. The product (Gm_{earth}) has a value of 4.55 EE17 ft^4-lbm/lbf-sec^4.

4 WORK, ENERGY, AND POWER

There are three general categories of external[3] forces: applied, gravitational, and frictional. Positive work is performed when a force acts in the direction of motion. Negative work is performed when a force opposes motion.

Applied and gravitational forces can do either positive or negative work, depending on the direction of motion. Friction always opposes motion. Therefore, friction can do only negative work.

The work performed by a variable force in the direction of motion is calculated by the scalar products in equations 16.44 and 16.45.

$$W = \int \mathbf{F} \cdot d\mathbf{s} \text{ (linear systems)} \qquad 16.44$$

$$= \int \mathbf{T} \cdot d\theta \text{ (rotational systems)} \qquad 16.45$$

If the force or the torque is constant, the integral can be eliminated.

$$W = \mathbf{F} \cdot \Delta\mathbf{s} \text{ (linear systems)} \qquad 16.46$$

$$= \mathbf{T} \cdot \Delta\theta \text{ (rotational systems)} \qquad 16.47$$

Energy represents the capacity to do work. Since energy cannot be created or destroyed, any work performed on a conservative system goes into increasing the system's energy. This is known as the *Work-Energy Principle.*

$$\Delta E = W \qquad 16.48$$

Potential energy is possessed by a mass, m, due to its relative height, h, in a gravitational field. This energy is equal to the work that would raise the mass a distance, h. Conversely, it is the energy released when the mass falls a distance, h.

$$E_p = \frac{mgh}{g_c} \qquad 16.49$$

---

[3]Internal forces, such as inertia, can do no work on a moving object.

Kinetic energy is the work necessary to bring a moving object to rest. Conversely, it is the work required to accelerate a stationary object to velocity, *v*.

$$E_k = \frac{mv^2}{2g_c} \quad \text{(linear systems)} \qquad 16.50$$

$$= \frac{I\omega^2}{2g_c} \quad \text{(rotational systems)} \qquad 16.51$$

Example 16.9

A 2 lbm projectile is launched straight up with an initial velocity of 700 ft/sec. Neglect air friction and calculate the

(a) kinetic energy immediately after launch
(b) kinetic energy at maximum height
(c) potential energy at maximum height
(d) total energy at an elevation where the velocity is 300 ft/sec
(e) maximum height

(a) From equation 16.50,

$$E_k = \tfrac{1}{2}\left(\frac{2}{32.2}\right)(700)^2 = 15,217 \text{ ft-lbf}$$

(b) At the maximum height, the velocity is zero. Therefore, $E_k = 0$.

(c) At the maximum height, all of the kinetic energy has been converted to potential energy. Therefore, $E_p = 15,217$ ft-lbf.

(d) Although some of the kinetic energy has been transformed into potential energy, the total energy is still 15,217 ft-lbf.

(e) Since all of the kinetic energy has been converted to potential energy, the maximum height can be found from equation 16.49.

$$15,217 = \left(\frac{2}{32.2}\right)(32.2)(h)$$
$$h = 7608.5 \text{ ft}$$

Example 16.10

A lawn mower engine is started by pulling a cord wrapped around a pulley. The pulley radius is 3 inches. The cord is wrapped around the pulley five times. If a constant tension of 20 lbf is maintained in the cord during starting, what work is done?

The torque on the engine is

$$\mathbf{T} = \mathbf{F}r = (20)\left(\frac{3}{12}\right) = 5 \text{ ft-lbf}$$

The cord rotates the engine $(5)(2\pi) = 31.4$ radians. From equation 16.47, the work done is

$$W = (5)(31.4) = 157 \text{ ft-lbf}$$

Power is the amount of work done per unit time.

$$P = \frac{W}{\Delta t} \qquad 16.52$$

Power can be calculated from force and velocity.

$$P = \mathbf{F}v \quad \text{(linear systems)} \qquad 16.53$$

$$= \mathbf{T}\omega \quad \text{(angular systems)} \qquad 16.54$$

Although the general mechanical work unit is the foot-pound, other units are used. Some useful conversions are listed in table 16.2.

Table 16.2
Useful Power Conversions

$$1 \text{ hp} = 550 \,\frac{\text{ft-lbf}}{\text{sec}} = 33,000 \,\frac{\text{ft-lbf}}{\text{min}}$$

$$= .7457 \text{ kW} = .7068 \,\frac{\text{BTU}}{\text{sec}}$$

$$1 \text{ kW} = 737.6 \,\frac{\text{ft-lbf}}{\text{sec}} = 44,250 \,\frac{\text{ft-lbf}}{\text{min}}$$

$$= 1.341 \text{ hp} = .9483 \,\frac{\text{BTU}}{\text{sec}}$$

$$1 \,\frac{\text{BTU}}{\text{sec}} = 778.17 \,\frac{\text{ft-lbf}}{\text{sec}} = 46,680 \,\frac{\text{ft-lbf}}{\text{min}}$$

$$= 1.415 \text{ hp}$$

Example 16.11

A 5-ton ore car, traveling at 4 feet per second, passes point A, rolls down an incline, and is stopped by a spring bumper which compresses 2 feet. A constant frictional force of 50 pounds acts on the ore car. What spring modulus is required?

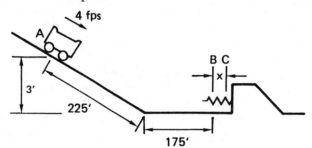

The car's mass is

$$m = (5)(2000) = 10,000 \text{ lbm}$$

The car's total energy at A is

$$E_t = E_k + E_p = \tfrac{1}{2}\left(\frac{10,000}{32.2}\right)(4)^2 + \left(\frac{10,000}{32.2}\right)(32.2)(3)$$
$$= 32,484 \text{ ft-lbf}$$

Since the frictional force does negative work, the total energy at B is

$$E_{t,B} = 32,484 - (50)(225 + 175) = 12,484 \text{ ft-lbf}$$

At point C, the maximum compression point, the energy has gone into compressing the spring and performing a small amount of frictional work.

$$12,484 = \tfrac{1}{2}kx^2 + 50x$$

Since $x = 2$, $k = 6192$ lbf/ft.

Example 16.12

A 200 lbm crate is pushed 25 feet across a warehouse floor. The crate's velocity is constant. What work is done if the coefficient of sliding friction between the crate and the floor is .3?

The frictional force is

$$F_f = \frac{(f)(m)(g)}{g_c} = \frac{(.3)(200)(32.2)}{(32.2)} = 60 \text{ lbf}$$

From equation 16.46, the work done is

$$W = (60)(25) = 1500 \text{ ft-lbf}$$

5 IMPULSE AND MOMENTUM

The vector *impulse* is defined by equations 16.55 and 16.56.

$$\text{impulse} = \mathbf{F}\Delta t \ (\text{linear systems}) \qquad 16.55$$
$$= \mathbf{T}\Delta t \ (\text{angular systems}) \qquad 16.56$$

The vector *momentum* is defined by equations 16.57 and 16.58.

$$p = \frac{mv}{g_c} \ (\text{linear systems}) \qquad 16.57$$
$$= \frac{I\omega}{g_c} \ (\text{angular systems}) \qquad 16.58$$

The *impulse-momentum equations* (equations 16.59 and 16.60) relate an applied impulse to a change in momentum.

$$\mathbf{F}\Delta t = \frac{m\Delta v}{g_c} \ (\text{linear systems}) \qquad 16.59$$
$$\mathbf{T}\Delta t = \frac{I\Delta\omega}{g_c} \ (\text{angular systems}) \qquad 16.60$$

Example 16.13

A 1.62 ounce marble attains a velocity of 170 mph in a hunting slingshot. Contact with the sling is 1/25th of a second. What is the average force on the marble during contact?

Solving equation 16.59 for $\mathbf{F}$,

$$\mathbf{F} = \frac{\frac{1.62}{(16)(32.2)}(170)\left(\frac{5280}{3600}\right)}{\frac{1}{25}} = 19.6 \text{ lbf}$$

6 IMPACT PROBLEMS

According to Newton's second law, momentum is conserved unless the object is acted upon by an external force. In an impact, there are no external forces, only internal forces. Thus, momentum is conserved even though energy may be lost.

Direct impact occurs when the velocities of the two colliding bodies are perpendicular to the contacting surfaces. *Central impact* occurs when the force of impact is along the line of connecting centroids. The disposition of the velocities can be found from the *coefficient of restitution*, e, which corrects for frictional and other losses. It is used even when the particles are smooth and frictionless. However, it should be used only with velocity components along a mutual line.

$$e = \frac{\text{relative separation velocity}}{\text{relative approach velocity}} = \frac{-(v_2' - v_1')}{(v_2 - v_1)} \qquad 16.61$$

A collision is said to be perfectly *elastic* if $e = 1$. A collision is said to be *inelastic* ($e < 1$) if kinetic energy is lost and *perfectly inelastic* ($e = 0$) if the objects stick together. In the case of rebounding objects from a stationary surface,

$$e = \frac{\tan(\phi_{\text{rebound}})}{\tan(\phi_{\text{incident}})} \qquad 16.62$$

Since equation 16.61 generally will have two unknowns, either the *conservation of momentum* equation (equation 16.63) or the *conservation of kinetic energy* equation (equation 16.64) also must be used.[4] (Equation 16.64 is valid only if $e = 1$.)

$$m_1 v_1 + m_2 v_2 = m_1 v_1' + m_2 v_2' \qquad 16.63$$
$$m_1 v_1^2 + m_2 v_2^2 = m_1(v_1')^2 + m_2(v_2')^2 \qquad 16.64$$

If the impact is *oblique*, the coefficient of restitution should be used to find the x-components of the resultant velocities. Then equations 16.65 and 16.66 can be used.

$$v_{1y} = v_{1y}' \qquad 16.65$$
$$v_{2y} = v_{2y}' \qquad 16.66$$

[4]Equations 16.63 and 16.64 contain mass variables in all four terms. The dimensional constant, g_c, should be included if momentum and kinetic energy units are desired. However, the constant will be applied to all four terms in each equation. For that reason, g_c can be omitted without compromising the equations.

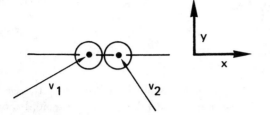

Figure 16.5 Oblique Impact

Example 16.14

A golf ball dropped 8 feet onto a hard surface rebounds $5\frac{1}{2}$ feet. What is the coefficient of restitution?

The impact velocity of the ball is found by equating the initial potential energy to the incident kinetic energy. This yields

$$v_1 = \sqrt{2gh} = \sqrt{(2)(32.2)(8)} = 22.7 \text{ ft/sec}$$

Similarly, the rebound velocity is

$$v_1' = \sqrt{(2)(32.2)(5.5)} = 18.8 \text{ ft/sec}$$

From equation 16.61 with $v_2 = v_2' = 0$,

$$e = \frac{-(0 - 18.8)}{(0 - (-22.7))} = .828$$

Example 16.15

A 1.25 lbm projectile is fired from a cannon at 1200 ft/sec. The mass of the cannon is 350 lbm. What is the cannon's recoil velocity?

Although frictional forces will impede the cannon soon after it begins moving, its initial momentum is due to an impulse internal to the cannon/projectile system. Therefore, the principle of momentum conservation can be used to find the recoil velocity immediately after separation.

From equation 16.63,

$$(350)(0) + (1.25)(0) = (350)(v) + (1.25)(1200)$$
$$v = 4.29 \text{ ft/sec}$$

Since the mass variable appears in all four terms in equation 16.63, it is not necessary to divide each *lbm* by g_c.

7 MOTION OF RIGID BODIES

A. TRANSLATION

When a rigid body experiences pure translation,[5] its position changes without an orientation change. At

---

[5]The word *translation* means movement.

any instant, all points on the object will have the same velocities and accelerations.

The performance of an object experiencing pure translation is governed by equations 16.67, 16.68, and 16.69.

$$\sum \mathbf{F}_x = ma_x \qquad \qquad 16.67$$
$$\sum \mathbf{F}_y = ma_y \qquad \qquad 16.68$$
$$\sum \mathbf{T} = I\alpha \qquad \qquad 16.69$$

A moving object is not in static equilibrium. If internal (inertial) forces are added to the static equilibrium equation, the object will be placed in *dynamic equilibrium*. This procedure, known as *D'Alembert's principle*, is illustrated in example 16.16.

Example 16.16

A 5000 lbm truck skids with a deceleration of -15 ft/sec². What are the horizontal and vertical reactions at the wheels? What is the coefficient of sliding friction?

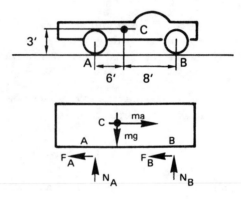

The free-body diagram of the truck in dynamic equilibrium is shown. The equations of dynamic equilibrium are

$$\sum \mathbf{F}_x = 0 : \left(\frac{m}{g_c}\right)a - \mathbf{F}_A - \mathbf{F}_B = 0$$
$$\frac{5000}{32.2}(15) - 5000f = 0$$
$$f = .466$$
$$\sum \mathbf{M}_A = 0 : 14N_B - 6(5000) - 3\left(\frac{5000}{32.2}\right)15 = 0$$
$$N_B = 2642 \text{ lbf}$$
$$\sum \mathbf{F}_y = 0 : N_A + N_B - m\left(\frac{g}{g_c}\right) = 0$$
$$N_A + 2642 - 5000 = 0$$
$$N_A = 2358 \text{ lbf}$$

Since **N** is the total reaction at both wheels,

$$\mathbf{F}_A = (.5)(.466)(2358) = 549 \text{ lbf}$$
$$\mathbf{F}_B = (.5)(.466)(2642) = 616 \text{ lbf}$$

B. ROTATION ABOUT A FIXED AXIS

Rotation is a motion produced from a turning moment or torque. The relationship between turning moment and angular acceleration is

$$\mathbf{T} = \left(\frac{I}{g_c}\right)\alpha \qquad 16.70$$

The rotation usually will be about the centroid axis. If it is not, the correct moment of inertia should be calculated from the *transfer formula*, equation 16.71. d is the distance between the centroidal axis and the axis of rotation

$$I = I_c + md^2 \qquad 16.71$$

The *tangential inertial force* on a rotating object acts normally to the centrifugal force. r in equation 16.72 is the object's distance from the rotational axis.

$$\mathbf{F}_i = \frac{ma_t}{g_c} = \frac{mr\alpha}{g_c} \qquad 16.72$$

Example 16.17

A turntable starts from rest and accelerates uniformly at $\alpha = 1.5$ radian/sec^2. How many revolutions will be made before the rotational speed of $33\frac{1}{3}$ rpm is attained?

The $33\frac{1}{3}$ rpm must be converted to radians per second. Since there are 2π radians per revolution,

$$\omega = \frac{(33\frac{1}{3})(2\pi)}{60} = 3.49 \text{ rad/sec}$$

α, ω_0, and ω are known. θ is the unknown. (This is analogous to knowing a, v_0, and v, and needing s.) From table 16.1,

$$\theta = \frac{\omega^2 - \omega_0^2}{2\alpha} = \frac{(3.49)^2 - (0)^2}{(2)(1.5)} = 4.06 \text{ rad}$$

Since there are 2π radians per revolution, the turntable is up to speed in

$$\frac{4.06}{2\pi} = .65 \text{ revolutions}$$

Example 16.18

A 4 foot diameter pulley with centroidal moment of inertia of 1610 lbm-ft^2 is subjected to tight-side and loose-side tensions of 200 lbf and 100 lbf, respectively. A frictional moment of 15 ft-lbf is acting. What is the angular acceleration?

The net torque is

$$\mathbf{T} = (2)(200 - 100) - 15 = 185 \text{ ft-lbf}$$

From equation 16.70, the angular acceleration is

$$\alpha = \frac{(32.2)(185)}{1610} = 3.7 \text{ radians/sec}^2$$

Example 16.19

A 25 lbm suitcase is placed 6 feet from the center of a revolving table. The table has a centroidal moment of inertia of 9660 lbm-ft^2. A torque of 180 ft-lbf is applied to the table, which initially is at rest. The coefficient of friction between the suitcase and the table is .4. After how many seconds will the suitcase begin to slip off the table?

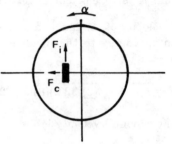

The total moment of inertia is

$$I = 9660 + (25)(6)^2 = 10,560 \text{ lbm-ft}^2$$

From equation 16.69, the angular acceleration is

$$\alpha = \frac{(32.2)(180)}{(10,560)} = .549 \text{ rad/sec}$$

The tangential inertial force on the suitcase is

$$\mathbf{F}_i = \frac{(25)(6)(.549)}{32.2} = 2.56 \text{ lbf}$$

The angular velocity of the table increases with time. At time t, the velocity is $\omega = (.549)t$. From equation 16.21, the centrifugal force at time t is

$$\mathbf{F}_c(t) = \frac{mr\omega^2}{g_c} = \frac{(25 \text{ lbm})(6 \text{ ft})\left(0.549\,t\,\frac{\text{rad}}{\text{sec}}\right)^2}{32.2\,\frac{\text{lbm-ft}}{\text{lbf-sec}^2}}$$

$$= 1.404t^2 \text{ lbf}$$

The total plane reaction on the suitcase is

$$\mathbf{R} = \sqrt{(2.56)^2 + (1.404t^2)^2}$$

The frictional force is

$$\mathbf{F}_f = (25)(.4) = 10 \text{ lbf}$$

Using D'Alembert's principle, the applied frictional and centrifugal forces balance the inertial force.

$$10 = \mathbf{R}$$

Solving for t,

$$t = 2.62 \text{ sec}$$

C. GENERAL PLANE MOTION

Plane motion can be illustrated in two dimensions. Examples are rolling wheels, gear sets, and linkages. Plane motion always can be considered as the sum of a translation and a rotation. This is illustrated in figure 16.6.

Analysis of an object in plane motion sometimes can be simplified by working with the object's *instantaneous center*. The instantaneous center is a point at which the moving object could be pinned without changing the angular velocities of the particles in the body. Thus, as far as the angular velocities are concerned, the body seems to rotate about the instantaneous center.

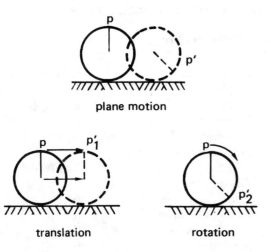

Figure 16.6 Plane Motion Composed of Translation and Rotation

The instantaneous center can be located by finding two particles for which the velocities are different and known in direction. Draw perpendiculars to the velocities through the points. The two perpendiculars will intersect at the instantaneous center. The instantaneous center of a rolling wheel always is the contact point with the supporting surface.

Once the instantaneous center has been located, the velocity of a particle located a distance, l, from the instantaneous center is given by equation 16.73. ω is the actual angular velocity of the object.

$$v = l\omega \qquad\qquad 16.73$$

Use of the instantaneous center reduces problems to simple geometry. However, it is important to use the correct reference point.

The velocity of any point on a wheel (figure 16.7) rolling with the translational velocity, v_0, can be found from its instantaneous center. Assume that the wheel is pinned at point C and rotates with the actual angular velocity. **The absolute velocities and accelerations with respect to point C can be found from geometry.**

$$v = l\omega = \frac{lv_o}{r} \qquad\qquad 16.74$$

$$a_t = l\alpha = \frac{la_o}{r} \qquad\qquad 16.75$$

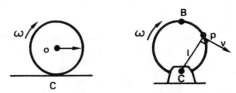

Figure 16.7 Rolling Wheel Pinned at Its Instantaneous Center

Equations 16.74 and 16.75 are valid only for velocities and accelerations referenced to point C. Table 16.3 can be used to find the velocities with respect to other points shown in figure 16.7.

Table 16.3
Relative Velocities on a Rolling Wheel

	reference point		
point	o	C	B
v_o	0	$v_o \rightarrow$	$\leftarrow v_o$
v_C	$\leftarrow v_o$	0	$\leftarrow 2v_o$
v_B	$v_o \rightarrow$	$2v_o \rightarrow$	0

Example 16.20

A car with 35 inch diameter tires is traveling at a constant 35 mph. What is the velocity of point p with respect to the ground?

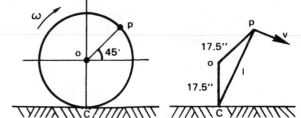

The translational velocity of point o is

$$v_o = 35\left(\frac{5280}{3600}\right) = 51.3 \text{ ft/sec}$$

The angular velocity of the wheel is

$$\omega = \frac{v_o}{r} = \frac{51.3}{\frac{35}{(2)(12)}} = 35.2 \text{ rad/sec}$$

(This angular velocity also could have been found by dividing v_o by the tire circumference and then multiplying by 2π.)

The law of cosines can be used to find the distance l.

$$l^2 = (17.5)^2 + (17.5)^2 - (2)(17.5)(17.5)\cos 135°$$
$$l = 32.3 \text{ inches}$$

The velocity of point p with respect to point C is found from equation 16.74.

$$v = l\omega = \left(\frac{32.3}{12}\right)(35.2) = 94.7 \text{ ft/sec}$$

The instantaneous center, point C, of a *slider rod* assembly can be found from the perpendiculars of the velocity vectors.

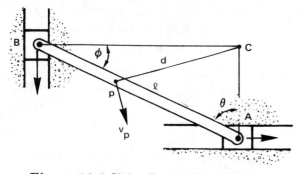

Figure 16.8 Slider Rod in Plane Motion

If the velocity with respect to point C of one end of the slider, say v_A, is known, then v_B can be found from geometry. Since the slider can be assumed to rotate about point C with angular velocity ω,

$$\omega = \frac{v_A}{AC} = \frac{v_A}{l\cos\theta} = \frac{v_B}{BC} = \frac{v_B}{l\cos\phi} \qquad 16.76$$

Since $\cos\phi = \sin\theta$,

$$v_B = v_A(\tan\theta) \qquad 16.77$$

If the velocity with respect to point C of any other point is required, it can be found from

$$v_p = d\omega \qquad 16.78$$

D. RELATIVE PLANE MOTION

The motion of one moving point relative to another moving point should be evaluated using vector subtraction. The following procedure can be used to determine the velocity of one point (B) with respect to another point (A) as illustrated in figure 16.9.

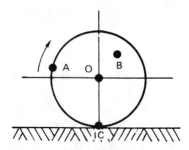

Figure 16.9 Relative Motion of Two Points

step 1: Find the velocity of B with respect to any point. For rolling wheels or rotating pulleys, choose the center as the reference point to simplify the problem. For more complex problems, choose the instantaneous center as the reference point.

step 2: Find the velocity of point A with respect to the same reference point.

step 3: Subtract the v_A vector from v_B. This can be done by reversing the directons of v_A and adding it heads-to-tails to v_B.

8 MECHANICAL VIBRATIONS

Special Nomenclature

a	acceleration	ft/sec^2
A	amplitude	ft
C	coefficient of damping	lbf-sec/ft
f	frequency	1/sec
g	acceleration due to gravity	ft/sec^2
g_c	gravitational constant (32.2)	lbm-ft/lbf-sec^2
h	distance from centroid to suspension point	ft
I	moment of inertia	lbm-ft^2
k	spring constant	lbf/ft
L	pendulum length	ft
m	mass	lbm
P	magnitude of forcing function	lbf
t	time	sec
T	period	sec
v	velocity	ft/sec
w	weight	lbf
x	displacement	ft

Symbols

β	magnification factor	–
ω	frequency	rad/sec
δ	deflection	ft
ς	log decrement	–

Subscripts

d	damped
eq	equivalent
st	static

A mechanical vibration is an oscillatory motion about an equilibrium point. The motion can be the result of a one-time disturbing force (*free vibration*) or a repeated forcing function (*forced vibrations*). If external and internal friction are absent, the vibration is undamped.

A. UNDAMPED, FREE VIBRATIONS

The common example of free vibration (*simple harmonic motion*) is a mass hanging from an ideal spring. After the mass is displaced and released, it will oscillate up and down. Since there is no friction (i.e., the vibration is undamped), the oscillations will continue forever.

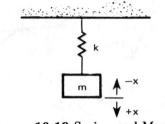

Figure 16.10 Spring and Mass

After the initial disturbing force is removed, the object is acted upon by only the restoring force $(-kx)$ and the inertial force (ma). Acceleration and the restoring force are proportional to the displacement from the equilibrium point, and they are opposite to the displacement. From D'Alembert's principle,

$$-kx = \frac{m}{g_c}\left(\frac{d^2 x}{dt^2}\right) \qquad 16.79$$

The solution to this second-order differential equation is

$$x(t) = x_0 \cos \omega t + \left(\frac{v_0}{\omega}\right)\sin \omega t \qquad 16.80$$

ω is known as the *natural frequency* of the vibrating system. It is not the same as the *linear frequency*, f.

$$\omega = \sqrt{\frac{kg_c}{m}} \qquad 16.81$$

$$f = \frac{\omega}{2\pi} \qquad 16.82$$

$$T = \frac{1}{f} \qquad 16.83$$

An alternate form of the solution is

$$x(t) = A\cos(\omega t - \alpha) \qquad 16.84$$

$$A = \sqrt{(x_0)^2 + \left(\frac{v_o}{\omega}\right)^2} \qquad 16.85$$

$$\alpha = \arctan\left(\frac{v_o}{\omega x_0}\right) \qquad 16.86$$

The maximum values are

$$x_{\max} = A \qquad 16.87$$

$$v_{\max} = A\omega \qquad 16.88$$

$$a_{\max} = A(\omega)^2 \qquad 16.89$$

Equation 16.81 can be written in terms of the weight of the mass suspended from the spring.

$$w = \frac{mg}{g_c} \qquad 16.90$$

$$\omega = \sqrt{\frac{kg}{w}} \qquad 16.91$$

However, w/k is the static deflection[6] experienced by the spring when the mass is attached to it initially.

$$\omega = \sqrt{\frac{g}{\delta_{st}}} \qquad 16.92$$

Springs in parallel share the applied load. The composite spring constant for parallel springs is

$$k_{eq} = k_1 + k_2 + \cdots \qquad 16.93$$

The composite spring constant for springs in series can be found from equation 16.94.

$$\frac{1}{k_{eq}} = \frac{1}{k_1} + \frac{1}{k_2} + \cdots \qquad 16.94$$

Example 16.21

A 120 lbm block is supported by three springs as shown. The initial displacement is 2 inches from the equilibrium position. No external forces act on the block after it is released. What are the maximum velocity and acceleration?

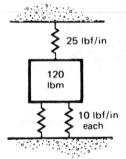

The equivalent spring constant is

$$k_{eq} = 25 + 10 + 10 = 45 \text{ lbf/in}$$

The static deflection is

$$\delta_{st} = \frac{120}{45} = 2.67 \text{ in}$$

---

[6]This initial deflection is known as the *static deflection*. It is not the same as the displacement experienced when the mass is acted upon by a disturbing force.

The natural frequency is given by equation 16.92.

$$\omega = \sqrt{\frac{32.2}{\left(\frac{2.67}{12}\right)}} = 12.0 \text{ rad/sec}$$

The initial conditions are

$$v_0 = 0$$
$$x_0 = \frac{2}{12} = .167 \text{ ft}$$

From equations 16.88 and 16.89,

$$v_{max} = (.167)(12.0) = 2.0 \text{ ft/sec}$$
$$a_{max} = (.167)(12.0)^2 = 24.0 \text{ ft/sec}^2$$

The movement of pendulums which oscillate in small arcs can be described by the equations of simple harmonic motion. A *simple pendulum* (e.g., a mass on a string of length L) oscillates according to equation 16.95.

$$T = 2\pi\sqrt{\frac{L}{g}} \qquad 16.95$$

The period of a *compound pendulum* depends on the moment of inertia taken about the suspension point.

$$T = 2\pi\sqrt{\frac{I}{mgh}} \qquad 16.96$$

B. DAMPED, FREE VIBRATIONS

If damping is present such that the resisting force is proportional to the velocity, the magnitude of the frictional force is

$$\mathbf{F}_d = C\frac{dx}{dt} \qquad 16.97$$

C is the *coefficient of viscous damping*. The differential equation of motion is

$$\frac{m}{g_c}\left(\frac{d^2x}{dt^2}\right) = -kx - C\left(\frac{dx}{dt}\right) \qquad 16.98$$

The general solution is

$$x(t) = e^{-nt}(c_1\cos\omega_d t + c_2\sin\omega_d t) \qquad 16.99$$
$$n = \frac{Cg_c}{2m} \qquad 16.100$$

ω_d is the *damped frequency*. It is not the same as the natural frequency, ω.

$$\omega_d = \sqrt{(\omega)^2 - (n)^2} = \omega\sqrt{1 - \left(\frac{C}{C_{crit}}\right)^2} \qquad 16.101$$

The *damping ratio* is

$$\text{damping ratio} = \frac{n}{\omega} = \frac{C}{C_{crit}} \qquad 16.102$$

case 1: If $n < \omega$, the oscillations will be *underdamped*. Motion will be oscillatory with diminishing amplitude.

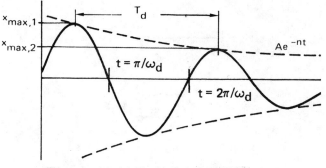

Figure 16.11 Underdamped Vibration

The *logarithmic decrement* can be used to find n.

$$e^{j\varsigma} = \frac{x_{max, \text{ ith cycle}}}{x_{max, \text{ (i+j)th cycle}}} \qquad 16.103$$

$$\varsigma = \frac{2\pi n}{\omega_d} \approx \frac{2\pi n}{\omega} \qquad 16.104$$

case 2: If $n > \omega$, the motion will be *overdamped*. There will be no oscillation, only a gradual return to the equilibrium position.

case 3: *Critical damping*, also known as *dead-beat motion*, occurs if $n = \omega$. There is no overshoot, and the return is the fastest of the three cases. The *critical damping coefficient* is

$$C_{crit} = \frac{2m\omega}{g_c} \qquad 16.105$$

C. DAMPED, FORCED VIBRATIONS[7]

If a system is subjected to a periodic force, as in figure 16.12, vibrations will occur at regular intervals.

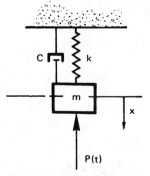

Figure 16.12 Forced Response

[7]These results can be used with undamped, forced vibrations by setting $C = 0$.

If the forcing function is sinusoidal (equation 16.106), the differential equation of motion is given by equation 16.107.

$$P(t) = \mathbf{P} \cos \omega_f t \qquad 16.106$$

$$m\left(\frac{d^2 x}{dt^2}\right) = -kx - C\left(\frac{dx}{dt}\right) + \mathbf{P} \cos \omega_f t \qquad 16.107$$

The solution to equation 16.107 has several terms. Some of the terms incorporate decaying exponentials. These are known as the *transient terms* because their contribution to displacement decreases rapidly. The transient terms do contribute to the initial displacement, however. For this reason, initial cycles may experience displacements greater than the steady state amplitudes.

If a static force $\mathbf{P}$ is applied to a spring with stiffness k, the static deflection will be

$$\delta_{st} = \frac{\mathbf{P}}{k} \qquad 16.108$$

The amplitude of forced vibrations can be found from the pseudo-static deflection and the *magnification factor*, β.

$$A = \beta\left(\frac{\mathbf{P}}{k}\right) \qquad 16.109$$

$$\beta = \frac{1}{\sqrt{\left[1 - \left(\frac{\omega_f}{\omega}\right)^2\right]^2 + \left[\frac{g_c C \omega_f}{m(\omega)^2}\right]^2}}$$

$$= \frac{1}{\sqrt{\left[1 - \left(\frac{\omega_f}{\omega}\right)^2\right]^2 + \left[\frac{2C\omega_f}{C_{\text{crit}}\omega}\right]^2}} \qquad 16.110$$

If the forcing frequency, ω_f, is equal to the natural frequency, ω, the magnification factor will be very large. This condition is known as *resonance*.

9 VIBRATION ISOLATION AND CONTROL

Isolation of vibrating equipment can be accomplished by mounting on resilient pads or springs. The isolating system must have a natural frequency less than $1/\sqrt{2} = .707$ times the disturbing frequency. The equipment forcing frequency, f_f, usually is known. The natural frequency of the isolating system can be found from equation 16.111. (Equations 16.111 and 16.112 require consistent units for g, δ_{st}, and k. Inches commonly are used.)

$$f = \frac{1}{2\pi}\sqrt{\frac{g}{\delta_{st}}} \qquad 16.111$$

$$\delta_{st} = \frac{\text{equipment weight}}{k} \qquad 16.112$$

The amount of isolation is known as the *isolation efficiency*, which depends on the type of equipment.

$$\eta = 1 - \frac{1}{\left(\frac{f_f}{f}\right)^2 - 1} \qquad 16.113$$

Table 16.4
Suggested Isolation Efficiencies

Equipment	Efficiency, %
centrifugal compressors	98
reciprocating compressors	
up to 15 horsepower	85
15 to 75 horsepower	90
75 to 150 horsepower	90
centrifugal fans,	
800 rpm or higher	90–95
centrifugal pumps	95
pipe mounts	95

Once the equipment type is known, the required efficiency will determine the allowable static deflection. Table 16.5 can be used to choose the isolation material.

Table 16.5
Isolation Materials

Deflection	Materials
up to $\frac{1}{16}''$	cork, rubber, felt, lead/asbestos, fiberglass
$\frac{1}{16}''$ to $\frac{1}{4}''$	neoprene pads, neoprene mounts, multiple layers of felt or cork[8]
$\frac{1}{4}''$ to 1.5$''$	steel coil springs, multiple layers of rubber, multiple neoprene pads
1.5$''$ to 15$''$	steel coil or leaf springs

Damping is beneficial only in the case of $(f_f/f) > \sqrt{2}$. Otherwise, damping is detrimental because the transmissibility actually increases above 1.0. The *transmissibility* is the ratio of transmitted force to unbalanced force. For undamped systems $(C = 0)$, the transmissibility is equal to the magnification factor given in equation 16.110. For damped systems, the transmissibility is

$$TR = \frac{P_{\text{transmitted}}}{P_{\text{applied}}}$$

$$= \beta\sqrt{1 + \left(\frac{2C\omega_f}{C_{\text{crit}}\omega}\right)^2} \qquad 16.114$$

---

[8]Neoprene also is known as *synthetic rubber*.

Example 16.22

A 250 lbm motor turns at 1000 rpm and is supported on a resilient pad having a stiffness of 3000 pounds per inch. Due to an unbalanced condition, a periodic force of 20 pounds is applied in the vertical direction once each revolution. If the motor is constrained to move vertically, what is the amplitude of vibration?

The natural frequency of the system is

$$\omega = \sqrt{\frac{k}{m}} = \sqrt{\frac{(3000)(12)}{\left(\frac{250}{32.2}\right)}} = 68.1 \text{ rad/sec}$$

The forcing frequency is

$$\omega_f = \frac{(1000)(2\pi)}{60} = 104.7 \text{ rad/sec}$$

The pseudo-static deflection that would be caused by the unbalanced load is

$$\delta_{st} = \frac{20}{3000} = .00667 \text{ in}$$

The magnification factor (assuming $C = 0$) is found from equation 16.110.

$$\beta = \frac{1}{1 - \left(\frac{104.7}{68.1}\right)^2} = -.733$$

Therefore, the total amplitude is

$$A = (-.733)(.00667) = -.0049 \text{ in}$$

The minus sign indicates that the vibrations are out of phase.

Example 16.23

A 4000 lbm machine is supported by isolation mounts having a combined stiffness of 30,000 lbf/in. In order to reduce the amplitude of vibration, a viscous damper is connected between the machine and its support. The damping ratio is 0.2. A mass of 100 lbm acting with a 2 inch eccentricity rotates at 1000 rpm. What are the amplitude of oscillation and the transmitted force?

The natural frequency is

$$f_n = \frac{1}{2\pi}\sqrt{\frac{(386)(30,000)}{(4000)}} = 8.56 \text{ Hz}$$

The damped frequency is

$$f = 8.56\sqrt{1 - (.2)^2} = 8.38 \text{ Hz}$$

The forcing frequency is

$$f_f = \frac{1000}{60} = 16.67 \text{ Hz}$$

The out-of-balance force caused by the rotating mass is given by the centrifugal force equation.

$$F_f = \frac{m\omega^2}{g_c} \times \text{eccentricity}$$
$$= \frac{100}{386}\left[(2)(\pi)\left(\frac{1000}{60}\right)\right]^2 \times 2 = 5682 \text{ lbf}$$

From equation 16.110, the magnification factor is

$$\beta = \frac{1}{\sqrt{\left[1 - \left(\frac{16.67}{8.56}\right)^2\right]^2 + \left[\frac{(2)(.2)(16.67)}{8.56}\right]^2}} = .345$$

The amplitude of oscillation is

$$x = \frac{(.345)(5682)}{30,000} = .065''$$

The transmitted force is

$$F = (.345)\sqrt{1 + \left[\frac{(2)(.2)(16.67)}{8.56}\right]^2}(5682) = 2485 \text{ lbf}$$

Appendix A
Mass Moments of Inertia

Slender rod		$I_y = I_z = (1/12)mL^2$ $I_{y'} = I_{z'} = (1/3)mL^2$
Thin rectangular plate		$I_x = (1/12)(b^2+c^2)m$ $I_y = (1/12)mc^2$ $I_z = (1/12)mb^2$
Rectangular Parallelepiped		$I_x = (1/12)m(b^2+c^2)$ $I_y = (1/12)m(c^2+a^2)$ $I_z = (1/12)m(a^2+b^2)$ $I_{x'} = (1/12)m(4b^2+c^2)$
Thin disk, radius r		$I_x = \tfrac{1}{2}mr^2$ $I_y = I_z = \tfrac{1}{4}mr^2$
Circular cylinder, radius r		$I_x = \tfrac{1}{2}mr^2$ $I_y = I_z = (1/12)m(3r^2+L^2)$
Circular cone, base radius r		$I_x = (3/10)mr^2$ $I_y = I_z = (3/5)m(\tfrac{1}{4}r^2+h^2)$
Sphere, radius r		$I_x = I_y = I_z = (2/5)mr^2$
Hollow circular cylinder		$I_x = \tfrac{1}{2}m(r_o^2 + r_i^2)$ $= \tfrac{\pi \rho L}{2}(r_o^4 - r_i^4)$

PRACTICE PROBLEMS: DYNAMICS

Warm-ups

1. A 24″ diameter wheel is moving at 28 mph. What is the velocity and the direction of a valve stem mounted 6″ from the center when the valve stem is 45° from the horizontal?

2. Neglecting air friction, at what angle should a cannon be placed to reach a target with a range of 12,000 feet and an elevation of 2000 feet? The initial projectile velocity is 900 fps.

Concentrates

1. A 300 pound electromagnet holds 200 pounds of scrap metal. The current suddenly is removed. If the total equivalent stiffness of the cable and the crane is 1000 pounds/inch, what will be the minimum cable tension? What is the frequency of oscillation?

2. An 800 pound, 1-cylinder vertical compressor is floor mounted and operates at 1200 rpm. Its maximum unbalanced force is 25 pounds. It is desired to reduce this to 3 pounds. What will be the maximum oscillation if four corner springs are used?

3. Determine the natural frequency of the system shown if the static deflection at the spring is .55″. The uniform bar weighs 5 pounds, and the concentrated weight at the end weighs 3 pounds.

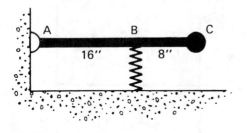

4. A 50 pound, 800 rpm motor is supported by four springs, each with a spring constant of 1000 lbf/in. The rotor unbalance is equivalent to 1 ounce located 5″ from the shaft centerline. The damping factor is $\frac{1}{8}$ of that required for critical damping. What is the maximum vertical amplitude?

Timed (1 hour allowed for each)

1. A 175 pound, single-cylinder air compressor is mounted on four corner springs. Each spring has the same length and stiffness. The motor turns at 1200 rpm. In each cycle, a disturbing force is generated by a 3.6 pound imbalance, which acts at a radius of 3″. Neglect the effect of damping. (a) What individual spring stiffness is required so that only 5% of the dynamic force is transmitted to the base? (b) What is the amplitude of vibration?

2. Eight horizontal 1030 steel plates ($40″ \times 30″ \times \frac{1}{2}″$) are used with rigid rollers to support a 20,000 pound load. Neglect the weight of the rollers and the plates. (a) What is the total vertical deflection of the load? (b) What is the maximum stress in the plates? (c) What is the natural frequency in the vertical direction?

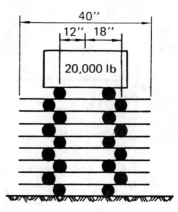

RESERVED FOR FUTURE USE

RESERVED FOR FUTURE USE

17 NOISE CONTROL

Nomenclature

A	surface area	ft^2
B	opening transmission coefficient	–
C	noise exposure time	hr
D	daily noise dose	–
L_{IL}	insertion loss	dB
L_{NR}	noise reduction	dB
L_p	sound pressure level	dB
L_{TL}	transmission loss	dB
L_W	sound power level	dB
NRC	noise reduction coefficient	–
p	sound pressure	N/m^2
PR	power ratio	–
PSIL	preferred speech interference level	dB
Q	directivity factor	–
r	separation of source and observer	ft
R	room constant	–
S	surface absorption	sabins
SR	sabin ratio	–
t	time	hr
t_{60}	time for sound to decay 60 dB	sec
V	volume	ft^3
W	sound power	watts

Symbols

α	absorption coefficient	–
τ	transmission coefficient	–

Subscripts

m	after modification of source or room
o	original condition—before noise treatment
t	total
x	after placing wall, inside newly formed room containing source
z	after placing wall, outside newly formed room containing source

1 PHYSICAL QUANTITIES AND DEFINITIONS

Noise is unwanted sound. Sound propagates as pressure fluctuations in air or fluids. Sound is measured by converting these pressure fluctuations into electrical signals with a sensitive pressure transducer (microphone). In its simplest form, the *sound level meter* is a microphone attached to a voltmeter.

Because the human ear responds to a pressure range of more than a million to one, directly quantifying sound in pressure units is difficult. It also is unnecessary because people perceive large changes in pressure as small changes in sound. Therefore, the logarithmic decibel power ratio scale has been adopted for sound measurements.

A. SOUND PRESSURE LEVEL

The minimum perceptible pressure amplitude has been standardized as 20 $\mu N/m^2$ (formerly expressed as 0.0002 μbar).[1] Sound pressure level is

$$L_p = 10 \log_{10}\left[\frac{p}{20 \text{ EE} - 6}\right]^2 = 20 \log_{10}\left[\frac{p}{20 \text{ EE} - 6}\right]$$
17.1

B. SOUND POWER LEVEL

Sound power level, L_W, while not directly measurable, is used to describe the characteristics of a given noise source. Sound power level is a logarithmic power ratio with a standard reference level of EE−12 watts.

$$L_W = 10 \log_{10}\left[\frac{W}{\text{EE} - 12}\right]$$
17.2

Sound power levels of noise sources are calculated from sound pressure levels measured in a reverberation room to eliminate directionality effects or in an anechoic room to determine directionality. Sound power level is proportional to sound pressure level on a predictable basis

---

[1] N/m^2 is numerically equal to dynes/cm^2.

only for a spherical source in an anechoic room (free field). In actual rooms, sound power levels can be used to estimate sound pressure levels based on knowledge of room reverberation characteristics.

C. NOISE REDUCTION

Noise reduction (L_{NR}) is the difference in sound pressure due to modification of noise sources or the room.

$$L_{NR} = L_{p_o} - L_{p_m} \qquad 17.3$$

D. INSERTION LOSS

Insertion loss, L_{IL}, is used to describe the noise reduction of a noise control device installed between the source and the observer. For example, the insertion loss of an enclosure is the difference between the sound pressure level measured before the enclosure is built and the sound pressure level measured at the same location after installation. Insertion loss can be used to describe sound power attenuating components.

$$L_{IL} = L_{p_o} - L_{p_z} = 10 \log \left(\frac{p_z}{p_o}\right)^2 \text{ for pressure}$$
$$= L_{W_o} - L_{W_z} = 10 \log \left(\frac{W_z}{W_o}\right) \text{ for power} \qquad 17.4$$

E. TRANSMISSION LOSS

Transmission loss, L_{TL}, is related to the portion of sound energy transmitted through a barrier. The fraction of incident energy transmitted through a barrier is the *transmission coefficient*, τ.

$$L_{TL} = 10 \log \frac{1}{\tau} \qquad 17.5$$

2 ADDITION AND SUBTRACTION OF DECIBELS

Summation of decibels cannot be accomplished by simple addition because logarithmic ratios are involved. For example, two 90 dB machines produce a total noise of 93 dB, not 180 dB. Graphical summation of two sound pressure levels utilizing figure 17.1 is fast and is accurate enough for most purposes.

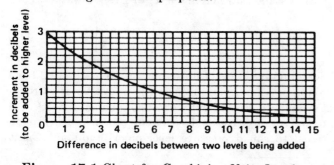

Figure 17.1 Chart for Combining Noise Levels

Example 17.1

Determine the combined sound pressure level produced by two machines with individual sound pressure levels of 78 and 82 dB.

1. Determine the difference in dB: (82 − 78 = 4 dB)
2. Determine increment from figure 17.1: (1.5 dB)
3. Add increment to higher level: (82 + 1.5 = 83.5 dB)

The graphical method can be extended to more than two sources by repeatedly adding the combined sound pressure level to the increment from each additional individual sound pressure level.

A more accurate method requires conversion of decibels back to power (pressure squared) ratios, simple addition, and conversion back to decibels. Equation 17.6 works for any parameter expressed in decibels.

$$L = 10 \log_{10} \sum \left(10^{\frac{dB_i}{10}}\right) \qquad 17.6$$

Example 17.2

What is the total sound pressure level produced by three machines with individual sound pressure levels of 89, 98, and 94 dB?

Divide each sound pressure level by 10 and convert to power ratios.

89 dB machine:	$10^{8.9} =$	794 EE6
98 dB machine:	$10^{9.8} =$	6309 EE6
94 dB machine:	$10^{9.4} =$	2511 EE6
	Total	9614 EE6

$$L = 10 \log_{10}(9614 \text{ EE6}) = 99.8 \text{ dB}$$

Note that total silencing of both the 89 dB and the 94 dB machines will reduce the noise from 99.8 dB to 98 dB, a negligible 1.8 dB noise reduction. By a similar calculation, total silencing of the single 98 dB machine results in a significant 4.6 dB overall reduction. This illustrates an important principle of noise control engineering: the noisiest source should be identified and treated before significant overall noise reduction can be achieved.

3 BACKGROUND NOISE, REFLECTIONS, AND SOUND FIELDS

The environment in which sound is measured can have a large influence on the measured value. Background noise and reflections from walls and obstacles are the most frequent sources of sound measurement errors.

A. BACKGROUND NOISE

Background noise consists of environmental noises which are not components of the sound to be measured. Background noise and total sound are measured separately, if possible; then background noise is subtracted from the total, using decibel addition methods. Background noise that is more than 10 dB below the total measured sound pressure level results in less than 0.5 dB error and generally is ignored.

B. SOUND FIELDS

Measurements made where no significant reflections are present (including those from the ground) are called *free field* measurements. Measurements made near walls or other objects where reflected sound is dominant are called *reverberant field* measurements.

Anechoic rooms (free field throughout) and *reverberation rooms* (reverberant field throughout) represent extreme environments and are used for measuring directionality effects and absorption material performance, respectively.

Measurements made close to a noise source are called *near field* measurements. Near field measurements can be in error due to directionality effects and reflections from the source itself.

Measurements made at large distances from a noise source are called *far field* measurements. Far field measurements can be in error due to background noise and atmospheric effects.

4 ROOM ACOUSTICS, EQUIPMENT ENCLO-SURES, AND ROOM DIVIDERS

A. ROOM CONSTANT

Acoustic absorption characteristics of a room can be described by a *room constant*, R. A room with perfect absorbing surfaces (anechoic) has $R = \infty$, while a room with perfect reflecting surfaces (reverberant) has $R = 0$. The room constant is related to the average *absorption coefficient*, $(\overline{\alpha})$, and the total surface area, A.

$$R = \frac{\overline{\alpha}A}{1 - \overline{\alpha}} \qquad 17.7$$

Since rooms consist of many reflecting and absorbing surfaces, the calculation of the absorption characteristics may not be practical. The *Sabin equation* relates the absorption coefficient to the volume, the surface area, and the time required for a sound to decay to 60 dB below the original level.

$$\overline{\alpha} = \frac{0.049V}{t_{60}A} \qquad 17.8$$

If a room contains absorbing materials with known absorption coefficients, a good estimate of average absorption coefficient can be obtained.

$$\overline{\alpha} = \frac{\sum A_i \alpha_i}{\sum A_i} = \frac{\sum S_i}{\sum A_i} = \frac{\mathbf{S}}{A_{\text{total}}} \qquad 17.9$$

The product, $A\alpha$, is known as the *surface absorption*, **S**, in sabins, where one sabin is one square foot of perfectly absorbing surface. Equation 17.9 assumes that absorption is well distributed so that average absorption applies to a majority of locations in the room.

Example 17.3

Find the 1000 Hz room constant for a room with dimensions of 30' by 60' by 12' high. (Use information from table at bottom of the page.)

From equation 17.9,

$$\overline{\alpha} = \frac{\sum \alpha_i S_i}{\sum A_i} = \frac{356}{5760} = 0.062$$

From equation 17.7,

$$R = \frac{\mathbf{S}}{1 - \overline{\alpha}} = \frac{356}{1 - .062} = 379.5 \text{ ft}^2$$

B. EFFECTS OF INCREASING ROOM CONSTANT

Equation 17.10 can be used to estimate the noise level from a machine with known sound power level at a location in a room where the room constant is known. r is the separation distance in feet. Q is the *directivity*

Table from example 17.3

Material	A	α	S
floor, asphalt tile	1800	.03	54
long wall, plywood paneling	720	.09	65
long wall, 50% plate glass window,	360	.03	11
50% gypsum board	360	.04	14
end walls, 10 ounce undraped velour	720	.17	122
ceiling, rough plaster	1800	.05	90
Totals:	5760	—	356 ft^2

factor. $Q = 1$ for an isotropic source; $Q = 2$ for a source in the center of a surface; $Q = 4$ for a source at an edge halfway between corners; and $Q = 8$ for a source in a corner of three surfaces.

$$L_p = L_W + 10 \log\left[\left(\frac{Q}{4\pi r^2}\right) + \left(\frac{4}{R}\right)\right] + 10.5 \quad 17.10$$

In a room with well-distributed absorption surfaces at large distances from a single noise source (or for many small noise sources distributed in the room), the *noise reduction* can be calculated from equation 17.11.

$$L_{NR} = 10 \log\left(\frac{S_2}{S_1}\right) = 10 \log\left[\frac{R_1(1 - \overline{\alpha}_1)}{R_2(1 - \overline{\alpha}_2)}\right] \quad 17.11$$

5 SUBJECTIVE EFFECTS OF NOISE

Human hearing is not equally sensitive to pressure at different frequencies. In addition, some types of noise are more annoying than others with similar frequency content and level. Subjective noise level is called *loudness.*

The most popular noise scale is the *sone scale.* Sones are calculated from sound pressure level measurements made through frequency filters. Sone values are easier to use than decibel values because a doubling of sones is perceived by most people to be "twice as loud."

The primary engineering disadvantage of utilizing subjective terms is that they are not directly measurable. Subjective terms are used primarily in non-technical communications with architects and building owners, where noise control work is not anticipated.

6 SOUND LEVEL METER RESPONSE CURVES

Subjective terms are so difficult to relate to measurable values that simpler methods for rating noise are used most of the time. The simplest method is to read a sound level meter whose response to noise is a crude approximation of human hearing response. Meter response curves have been standardized worldwide. Most industrial-quality meters respond accurately to at least one of the standard curves. Most respond to three widely used curves called **A**, **B**, and **C** weighted scales. The **A** scale approximates human hearing most accurately, so this scale is used frequently.

The frequency content of noise can be estimated by taking three readings with a different weighting curve selected for each reading. In figure 17.2, the **A** curve discriminates against lower frequencies, the **C** curve is nearly flat, and the **B** curve falls in between. If

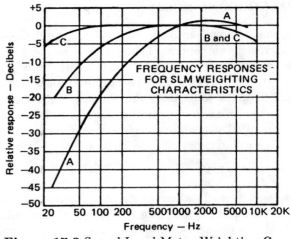

Figure 17.2 Sound Level Meter Weighting Curves

the three readings are the same, the predominant noise frequencies probably are above approximately 500 Hz. If the **C** reading is several dB higher than the **A** and **B** readings, predominant noise frequencies probably are below 500 Hz.

The usual method used to describe weighted measurements is the unit followed by the weighting used. For example, three measurements of a machine taken with **A**, **B**, and **C** weighting curves would be written dBA, dBB, and dBC. **A**-weighting is by far the most common. If the weighting is not specified, **A**-weighting is assumed. However, it is good practice to note the weighting used for all measurements.

Many meters also provide a fourth, unweighted response which is the basic frequency response of the individual microphone and the meter circuitry. It is different for different meters. It is used for making tape recordings and in applications where external filters or special weighting circuits are used. Some names that have been used for *unweighted response* are *flat, 20 kHz,* and *linear response.*

The abbreviation for *sound level* is "L" with subscripts indicating modification of the basic physical measurement. Table 17.1 lists several commonly used sound level terms.

Table 17.1
Common Sound Level Terms

Symbol	
L_A, L_B, L_C	sound level meter weighting circuits
L_{dn}	time weighted average of day and night levels
L_{eq}	equivalent level; time average of energy levels
L_{ob}	filtering through an octave band filter
L_{NR}	subtracting two noise levels

7 HAZARDS OF NOISE EXPOSURE

Research on noise-induced hearing loss has yielded some general principles.

- Hearing loss is proportional to exposure time and noise intensity.
- Long-term noise-induced injuries are not correctable because nerve cells in the inner ear are damaged.
- Low-frequency noise does not damage hearing as much as mid- to high-frequency noise.
- Susceptibility to hearing damage varies among individuals.

The levels at which significant hazard exists are well documented and form the basis for the Occupational Safety and Health Act (OSHA) criteria. Detailed requirements are being argued continually and may be supplemented by state regulations.

Table 17.2
Typical Permissible Noise Exposures

exposure, hours per day	sound level, dBA
8	90
6	92
4	95
3	97
2	100
$1\frac{1}{2}$	102
1	105
$\frac{1}{2}$	110
$\frac{1}{4}$ or less	115

For *impulse noise* (such as chipping hammers and pneumatic drills) the peak level must not exceed 140 dB. *Peak levels* are measured with a special sound level meter circuit that captures the maximum instantaneous noise level and holds it for later reading.

8 NOISE DOSE

Daily *noise dose* is the actual exposure duration divided by the permissible exposure duration listed in table 17.2. If an employee is exposed to one or more different levels during his shift, his daily noise dose can be calculated from equation 17.12.

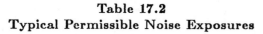

$$D = \left(\frac{C_1}{t_1}\right) + \left(\frac{C_2}{t_2}\right) + \left(\frac{C_3}{t_3}\right) + \cdots \qquad 17.12$$

If daily noise dose is more than unity, one or more of the following actions are required.

- If feasible, reduce the noise source, modify the noise path, or reduce employee exposure time.

- If the above engineering or administrative controls are not sufficient to control exposure, the employer must provide hearing protection and enforce its use.

A good hearing conservation program probably will contain several features. State or revised federal regulations may include some or all of these features as requirements.

- Annually survey the plant and identify all employees exposed to noise doses above 0.5.
- Periodically test the hearing of all employees with noise doses above 0.5.
- Test the hearing of all employees at the time of employment for later comparison.
- Establish detailed procedures for medical surveillance and record keeping.

9 FREQUENCY CONTENT OF NOISE

Detailed knowledge of the frequency content of noise is required for most noise control problems. *Frequency content* is described by dividing noise into a series of frequency ranges called *bands*. The sound level in each frequency band is plotted against the frequency at the center of the band, resulting in a graph of the noise spectrum.

Frequency bands are established by dividing the frequency range of interest into either bandwidth or equal percentage sections. For example, equal bandwidth sections might consist of frequency bands each 100 Hz wide, while equal percentage sections are proportionally wider at higher frequencies.

A. OCTAVE AND ONE-THIRD OCTAVE BANDS

Percentage *bandwidths* and center frequencies have been standardized so that frequency content information can be exchanged accurately. The basic (and widest) standard bandwidth is 50%, meaning that the ratio between the lowest and the highest frequency in the band is 0.5. Musical pitch frequency ratios of 1:2 are called *octaves*, so the name *octave band* for the 50% bandwidth has been adopted.

Center frequencies for sound octave bands have been standardized at 31.5, 63, 125, 250, 500, 1000, 2000, 4000, 8000, and 16,000 Hz. The actual frequency span of any octave band covers a 2:1 ratio. For example, the 1000 Hz (center frequency) octave band includes all frequencies between 707 Hz and 1414 Hz.

A now obsolete series of octave bands was used until the late 1950's, and a great deal of useful data is still available. The older band frequencies are listed in table 17.3.

Table 17.3
Corrections for A-Weighting from Octave Bands

Preferred Series of Octave Bands		Older Series of Octave Bands	
Band Center Frequency (Hz)	Correction to be Added	Octave Band (Hz)	Correction to be Added
31.5	−39.4	18.75– 37.5	−43.4
63	−26.2	37.5– 75	−29.2
125	−16.1	75– 150	−18.3
250	−8.6	150– 300	−10.3
500	−3.2	300– 600	−4.4
1000	0	600– 1200	−0.5
2000	+1.2	1200– 2400	+1.0
4000	+1.0	2400– 4800	+1.1
8000	−1.1	4800– 9600	−0.4

Octave bands are subdivided for better frequency detail. The most common subdivision is the *one-third octave band.* One-third octave bands are centered at 100, 125, 160, 200, 250, 315, 400, 500, 630, and 800 Hz along with multiples and sub-multiples of 10. The usual frequency range for acoustics covers the one-third octave bands between 25 and 16,000 Hz.

B. CONSTANT BANDWIDTH SYSTEMS

Most very narrow bandwidth systems divide the frequency range of interest into a large number of bands that have a constant bandwidth. No standardization of bandwidths exists, partly because narrow band analysis is used more for detailed investigation of noise sources than for reporting absolute levels.

C. CONVERSION OF OCTAVE BAND LEVELS INTO A-WEIGHTED LEVELS

It is not uncommon to see octave band data graphically reported without the measured A-weighted level, even though the A-weighted level is used widely for rating noise. Octave band measurements contain enough information about levels and frequency content to permit calculation of the equivalent A-weighted level. A correction can be made for each octave band center frequency to approximate the A-weighting curve of figure 17.3. Table 17.3 lists the corrections to be added to preferred and older series of octave band data to approximate the proper response curve for A-weighting.

The corrected octave band levels are added to obtain the single A-weighted value.

Example 17.4

Use the octave band data to calculate the equivalent A-weighted level.

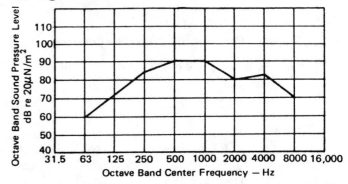

Extract the octave band levels from the graph and add the corrections listed in table 17.3.

Add corrected octave band levels, using equation 17.6.

$$dBA = 10 \log_{10} \left[10^{3.38} + 10^{5.59} + 10^{7.64} + 10^{8.68} \right.$$
$$\left. + 10^{9.0} + 10^{8.12} + 10^{8.4} + 10^{6.89} \right]$$
$$= 10 \log_{10}(1.913\ EE9) = 92.8\ dBA$$

10 MASKING OF SPEECH

If noise interferes with normal communication, it becomes a serious problem. Estimates of the effects of noise on communication are made by arithmetically averaging the octave band levels centered at frequencies of 500, 1000, and 2000 Hz. The quantity is called *speech interference level* and is abbreviated PSIL to denote the use of the preferred series of octave bands.

Table from example 17.4

octave band (Hz)	63	125	250	500	1000	2000	4000	8000
L_{ob} (dB)	60	72	85	90	90	80	83	70
corrections	−26.2	−16.1	−8.6	−3.2	0	+1.2	+1.0	−1.1
corrected L_{ob} (dB)	33.8	55.9	76.4	86.8	90	81.2	84	68.9

Table 17.4
Speech Interference Level Criteria

by room type	maximum recommended PSIL (measured when room is not in use)
small private office	45
conference room for 20	35
conference room for 50	30
movie theater	35
theaters for drama (500 seats, no amplification)	30
concert halls (no amplification)	25
secretarial offices (typing)	60
homes (sleeping areas)	30
assembly halls (no amplification)	30
school rooms	30
by communications type	
satisfactory telephone usage	60
difficult telephone usage	75
normal voice at 9 feet	50
raised voice at 3 feet	60

11 NOISE CONTROL MATERIALS

Special materials are available for noise control purposes. However, common building materials often approach the performance of more expensive acoustic materials if they are applied properly. The basic performance categories for materials used in noise control are summarized here.

Acoustic absorbers prevent reflection of incident sound. Fibrous glass, draperies, and open cell foams are acoustic absorbers. Absorbers generally are used in combination with barriers and dampers since they do not block noise or reduce vibration when used alone. Absorbers work best at higher frequencies unless they are very thick or are tuned to a lower frequency of interest.

The *Noise Reduction Coefficient* rating, NRC, simplifies comparisons of materials. NRC is the arithmetic average of the four absorption coefficients measured at test frequencies of 250, 500, 1000, and 2000 Hz. Coefficients of common building materials and furnishings are listed in appendix A.

Acoustic barriers block sound. The best performing barrier materials have high mass. Gypsum board, plywood, and masonry are classified as *barrier materials*.

A single-number rating for a barrier material is based on the match of the barrier's transmission loss curve to a set of standard ASTM contours and is called *Sound Transmission Class*, STC. Appendix B lists the transmission loss of several common building materials.

12 NOISE REDUCTION TECHNIQUES

Noise in an environment can be reduced in one of two primary ways. Either the generated acoustical energy must be reduced at the source, or shields must be used to reduce acoustical energy.

A. QUIETING THE SOURCE

If the source cannot be replaced with a quieter duplicate, some modification of the source, its mounting, or its noise by-products (exhaust, piping, etc.) will be required.

- Maintain horsepower requirements as low as is possible.
- Use low equipment speed.
- Improve dynamic balance and reduce the ratio of rotating mass to fixed mass. Use center-of-gravity mounting.
- Eliminate large shaft mechanical run-outs.
- Redesign or eliminate two impacting parts.
- Increase or improve effectiveness of lubrication.
- Improve alignment and reduce tolerances.
- Keep air velocities low by maintaining large passageways. Reduce jet velocities.
- Reduce large radiating surfaces by breaking into smaller surface areas.
- Keep peak speed and acceleration of impeller/fan tips low.
- Use vibration isolators during installation.
- Use inlet and discharge baffles.

B. MODIFYING THE PATH

If the source or the observer cannot be separated or repositioned, room treatments are required. Generally, the sabin area (S) should be 20% to 50% of the room area. Otherwise, the room will be reverberatory. In adding noise reduction material to a room, the new sabin area should be 3 to 10 times the old value.

13 CALCULATING INSERTION LOSS

A. PARTIAL ENCLOSURES AROUND EQUIPMENT

This method can be used when equipment with little or no existing enclosure is completely or partially enclosed.

step 1: Calculate SR_1 (sabin ratio for existing partial enclosure). If no partial enclosure exists, use $SR_1 = .433$, which is derived from an imaginary cube enclosure of any size.

$$SR = \frac{\sum A_i B_i}{A_t + \alpha A} \qquad 17.13$$

$A_i =$ area of ith opening in the enclosure

$\alpha =$ absorption coefficient of treated wall

$B_i =$ opening transmission coefficient

$\quad = 1$ if opening faces observer

$\quad = \frac{1}{3}$ for left, right, and top openings

$\quad = \frac{1}{6}$ for rear openings

$A_t = \sum A_i$ (sum over openings only, not walls)

$A =$ acoustically treated wall area

step 2: Repeat step 1, calculating SR_2 for the proposed enclosure.

step 3: Calculate the insertion loss.

$$L_{IL} = 10 \log\left(\frac{SR_1}{SR_2}\right) \qquad 17.14$$

B. FULL-HEIGHT ROOM DIVISION

If a room is divided into two smaller rooms by a new wall (area A, transmission coefficient τ), this procedure can be used to calculate insertion loss. The subscript x refers to the room containing the source after the dividing wall has been installed. The subscript z refers to the room without the source.

step 1: Calculate S, S_x, and S_z.

step 2: Calculate the power ratio.

$$PR = \left(\frac{S_z}{A}\right)\left[\frac{\left(\frac{\tau A}{S_z}\right) + \left(\frac{S_x}{S_z}\right)}{1 + \left(\frac{S_x}{S_z}\right)}\right] \qquad 17.15$$

step 3: Calculate the insertion loss.

$$L_{IL} = 10 \log(PR) \qquad 17.16$$

C. PARTIAL-HEIGHT ROOM DIVISION

This method can be used if a room is divided by a wall which does not reach the ceiling. In that case, some of the sound will go over the top of the partition. This method assumes that no sound is transmitted through the barrier. It considers only the sound passing over the top of the partition.

step 1: Calculate S, S_x, and S_z.

step 2: Calculate the power ratio. $A =$ wall area, and $A_o =$ open area above partial wall.

$$PR = \left(\frac{S_z}{A_o}\right)\left[\frac{\left(\frac{A_o}{S_z}\right) + \left(\frac{S_x}{S_z}\right)}{1 + \left(\frac{S_x}{S_z}\right)}\right] \qquad 17.17$$

step 3: Calculate the insertion loss.

$$L_{IL} = 10 \log(PR) \qquad 17.18$$

Appendix A
NRC and α Coefficients

Materials	125 Hz	250 Hz	α Coefficients 500 Hz	1000 Hz	2000 Hz	4000 Hz	NRC
Brick, unglazed	.03	.03	.03	.04	.05	.07	.04
Brick, unglazed, painted	.01	.01	.02	.02	.02	.03	.02
Carpet							
1/8″ Pile Height	.05	.05	.10	.20	.30	.40	.16
1/4″ Pile Height	.05	.10	.15	.30	.50	.55	.26
3/16″ combined Pile & Foam	.05	.10	.10	.30	.40	.50	.23
5/16″ combined Pile & Foam	.05	.15	.30	.40	.50	.60	.34
Concrete Block, painted	.10	.05	.06	.07	.09	.08	.07
Fabrics							
Light velour, 10 oz. per sq. yd., hung straight, in contact with wall	.03	.04	.11	.17	.24	.35	.14
Medium velour, 14 oz. per sq. yd., draped to half area	.07	.31	.49	.75	.70	.60	.56
Heavy velour, 18 oz. per sq. yd., draped to half area	.14	.35	.55	.72	.70	.65	.62
Floors							
Concrete or Terrazzo	.01	.01	.01	.02	.02	.02	.02
Linoleum, asphalt, rubber or cork tile on concrete	.02	.03	.03	.03	.03	.02	.03
Wood	.15	.11	.10	.07	.06	.07	.09
Wood parquet in asphalt on concrete	.04	.04	.07	.06	.06	.07	.06
Glass							
1/4″, sealed, large panes	.05	.03	.02	.02	.03	.02	.03
24 oz., operable windows (in closed condition)	.10	.05	.04	.03	.03	.03	.04
Gypsum Board, 1/2″ nailed to 2 × 4's 16″ o.c., painted	.10	.08	.05	.03	.03	.03	.05
Marble or Glazed Tile	.01	.01	.01	.01	.02	.02	.01
Plaster, gypsum or lime, rough finish or lath	.02	.03	.04	.05	.04	.03	.04
Same, with smooth finish	.02	.02	.03	.04	.04	03	.03
Hardwood Plywood paneling 1/4″ thick, Wood Frame	.58	.22	.07	.04	.03	.07	.09
Water Surface, as in a swimming pool	.01	.01	.01	.01	.02	.03	.01
Wood Roof Decking, tongue-and-groove cedar	.24	.19	.14	.08	.13	.10	.14
Air, Sabins per 1000 cubic feet @ 50% RH				.9	2.3	7.2	
Audience, seated, depending on spacing and upholstery of seats*	2.5–4.0	3.5–5.0	4.0–5.5	4.5–6.5	5.0–7.0	4.5–7.0	
Seats, heavily upholstered with fabric*	1.5–3.5	3.5–4.5	4.0–5.0	4.0–5.5	3.5–5.5	3.5–4.5	
Seats, heavily upholstered with leather, plastic, etc.*	2.5–3.5	3.0–4.5	3.0–4.0	2.0–4.0	1.5–3.5	1.0–3.0	
Seats, lightly upholstered with leather, plastic, etc.*			1.5–2.0				
Seats, wood veneer, no upholstery*	.15	.20	.25	.30	.50	.50	

*Values given are in sabins per person or unit of seating at the indicated frequency

Appendix B
Transmission Loss of Common Materials
(decibels)

Material	lbm/ft^2	125	250	500	1000	2000	4000	8000
Lead								
1/32-inch thick	2	22	24	29	33	40	43	49
1/64-inch thick	1	19	20	24	27	33	39	43
Plywood								
3/4-inch thick	2	24	22	27	28	25	27	35
1/4-inch thick	0.7	17	15	20	24	28	27	25
Lead vinyl	0.5	11	12	15	20	26	32	37
Lead vinyl	1.0	15	17	21	28	33	37	43
Steel								
18-guage	2.0	15	19	31	32	35	48	53
16-guage	2.5	21	30	34	37	40	47	52
Sheet metal (viscoelastic laminate-core)	2	15	25	28	32	39	42	47
Plexiglass								
1/4-inch thick	1.45	16	17	22	28	33	35	35
1/2-inch thick	2.9	21	23	26	32	32	37	37
1-inch thick	5.8	25	28	32	32	34	46	46
Glass								
1/8-inch thick	1.5	11	17	23	25	26	27	28
1/4-inch thick	3	17	23	25	27	28	29	30
Double glass								
1/4 × 1/2 × 1/4-inch		23	24	24	27	28	30	36
1/4 × 6 × 1/4-inch		25	28	31	37	40	43	47
5/8-inch Gypsum								
On 2 × 2-inch stud		23	28	33	43	50	49	50
On staggered stud		26	35	42	52	57	55	57
Concrete, 4-inch thick	48	29	35	37	43	44	50	55
Concrete block, 6-inch	36	33	34	35	38	46	52	65
Panels of 16-guage steel, 4-inch absorbent, 20-gauge steel		25	35	43	48	52	55	56

PRACTICE PROBLEMS: NOISE CONTROL

Warm-ups

1. Add a 40 dB level to a 35 dB level.

2. If the total level is 45 dBA and the background sound level is 43 dBA, what is the sound level due to the operation of a machine?

3. An unenclosed source produces 100 dBA sound which increases to 110 dBA when enclosed by a 30 dBA transmission loss material. What is the reduction in sound experienced outside the enclosure?

4. How long can an employee be exposed to 95 dBA sound?

5. An isotropic source has a sound level of 92 dB at a distance of 4 feet. What is the sound level at 12 feet?

6. Octave band measurements were made of a sound pressure level. The measurements were 85, 90, 92, 87, 82, 78, 65, and 54 dB. What was the overall sound pressure?

7. What is the speech interference level for the room in question 6?

8. What is the maximum possible reduction in the sound level if the number of sabins is 50% of the total room area?

Concentrates

1. A room has dimensions of $100' \times 400' \times 20'$ with all surfaces consisting of precast concrete. If 40% of the walls are treated accoustically with a material having an absorption coefficient of .8, what will be the reduction in noise?

2. A noise source is located $1\frac{1}{2}$ feet from the floor and $1\frac{1}{2}$ feet from the nearest wall in a very long room. An enclosure 6' long, 3' high, and 3' wide will be built with a $1' \times 1'$ opening in the center of each $3' \times 3'$ end. The noise source will be in the center of this enclosure. What will be the insertion loss?

3. What will be the insertion loss when a 3' cube is constructed around a source, but the back is left open?

4. An office is $20' \times 50' \times 10'$ high, consisting of linoleum flooring, plaster ceiling, and plaster walls with glass windows over 20% of the wall area. There are 15 desks, 15 occupants, and 5 miscellaneous sabins. The ceiling is treated with sound absorbing material with a coefficient of .7. What is the reduction in the sound pressure level due to adding the material?

5. A $20' \times 30' \times 10'$ high room is divided in half by an 8' high partition. The ceiling is covered with an absorption material with a coefficient of .5. What is the insertion loss due to the added partition?

6. A machine with sound *power* of 65 dB is located on the floor in the center of a wall of a room which is $15' \times 20' \times 10'$ high. Ambient sound pressure is 50 dB. If the absorption coefficients are .03, .5, and .06 for the floor, ceiling, and walls, respectively, what is the room constant and the sound *pressure* level 5 feet from the machine?

7. A 600 rpm centrifugal fan has 8 driving blades and 64 fan blades. It is driven by a 1725 rpm, 60 cycle, 4-pole motor through a 72" long belt. The fan and motor pulleys are $11\frac{1}{2}''$ and 4" in diameter, respectively. What frequencies of vibration and sound are produced?

8. A 1725 rpm pump is mounted on a cork pad which is compressed .02". What is the transmissibility?

RESERVED FOR FUTURE USE

18 NUCLEAR ENGINEERING

Nomenclature

A	ratio of moderator mass to neutron mass, activity, or area	–, disintegrations/sec, cm^2
A.W.	atomic weight	amu
B	build-up factor	–
B^2	buckling	$1/cm^2$
c	velocity of light	m/sec
d	extrapolation distance	cm
D	diffusion coefficient, or dose	cm, rad
E	energy	eV, or J
f	thermal utilization	–
g	generation rate	$1/cm^3$-sec
g'	non-(1/v) factor	–
H	equivalent dose	rem
I	ray or particle intensity	$1/cm^2$-sec
J	neutron current density	neutrons/cm^2-sec
k	multiplication (reproduction) constant	–
k_∞	infinite-k	–
l_e	average energy loss fraction per collision	–
l_f	quantity loss fraction	–
L	thermal diffusion length, or length	cm
L_s	epithermal diffusion length	cm
m	mass	kg
M	migration length	cm
MR	moderating ratio	–
n	number of neutrons produced per fission	–
N	number of atoms per unit volume	atoms/cm^3
N_c	number of collisions to thermalize	–
N_o	Avogadro's number (6.023 EE23)	atoms/gmole
p	resonance escape probability	–
P	power	watts
Q	quality factor	–
r^2	mean squared thermal life distance	cm^2

R	range	g/cm^2
S	source activity	1/sec
SDP	slowing down power	1/cm
t	time	sec
T	temperature	°K
v	velocity	cm/sec
V	volume	cm^3
x	distance, or thickness	cm
X	exposure	roentgen

Symbols

ϵ	fast fission factor	–
η	number of neutrons per absorption in fuel	–
η_{th}	thermal efficiency	–
θ	scattering angle	°
λ	mean free path length, or decay constant	cm, 1/time
ξ	lethargy	–
ρ	density	g/cm^3
ρ_N	thermal neutron density	neutrons/cm^3
σ	microscopic cross section	barns
Σ	macroscopic cross section	1/cm
τ	Fermi age (or just age)	cm^2
ϕ	particle flux	particle/cm^2-sec
μ_l	linear absorption coefficient	1/cm
μ_m	mass attenuation coefficient	cm^2/g

Subscripts

a	absorption
b	binding, or build-up
d	doubling
e	enrichment
es	elastic scattering
f	fission
ie	inelastic
m	moderator
n	neutron
o	original, or fast
p	most probable

r   removal
s   scattering
t   total, at time t
th   thermal
tr   transport
1   after one collision

1 NUCLEAR TRANSFORMATIONS

Nuclear reaction equations are similar to chemical reaction equations. Each particle is assigned a symbol, a superscript equal to its mass in *atomic mass units* (*amu*), and a subscript equal to its atomic number or charge. In nuclear reactions, both the superscripts and the subscripts must balance. Shorthand methods can be used to indicate the incident and product particles.

Longhand: $_7N^{14} +_1 H^1 \mapsto_6 C^{11} +_2 He^4$

Shorthand: $_7N^{14} +_1 H^1 \overset{\alpha}{\mapsto} C^{11}$

Shorthand: $N^{14}(p, \alpha)C^{11}$

Extremely accurate measurements of the product masses may show a decrease from the sum of the reacting particles. The decrease in mass (*mass deflect*) will be due to the release of *binding energy*. This energy release is predicted by *Einstein's equation*. The energy is in joules if m is in kilograms and c is in m/sec. If the products have more mass than the reactants, the difference must be made up by the kinetic energy of the reactants.

$$E = mc^2 \qquad 18.1$$

The right-hand side of equation 18.1 is known as the *Q-value* and commonly is expressed in MeV.

Example 18.1

What is the binding energy of the deuterium nucleus?

The deuterium nucleus consists of a proton and a neutron. The mass defect from table 18.1 is $1.007277 + 1.008665 - 2.01345 = .002492$ amu.

The binding energy can be found from equation 18.1.

$$E_b = (.002492) \text{ amu } (1.66053 \text{ EE} - 27) \text{ kg/amu}$$
$$(3 \text{ EE8})^2 (\text{m/sec})^2$$
$$= 3.724 \text{ EE} - 13 \text{ joules} = 2.32 \text{ MeV}$$

2 RADIOACTIVITY

The spontaneous disintegration of nuclei produces alpha particles, beta particles, and gamma rays. Such neutral disintegration occurs because the element has excess or insufficient neutrons. These neutrons have no electrostatic effect, but their presence provides nuclear attraction to balance proton repulsion in the nucleus.

If the nucleus has too many neutrons, one or more neutrons may transform into protons spontaneously with emission of electrons to retain charge neutrality. In addition to the beta particle, gamma radiation may be given off to change the binding energy. Decay with the emission of electrons is known as −*beta decay*.

$$_0n^1 \mapsto_{+1} p^1 +_{-1} e^0$$

If the nucleus has too few neutrons, a proton becomes a neutron with a positron emission. This is known as +*beta decay*.

$$_{+1}p^1 \mapsto_0 n^1 +_{+1} e^0$$

Alpha decay decreases both the number of protons and neutrons by 2. Such an equilateral drop in nucleus mass also may stabilize the atom.

$$2_0n^1 + 2_{+1}p^1 \mapsto_2 He^4$$

Radioactive decay can be described mathematically through the use of a *decay constant*, λ, which is independent of the environment. The number of disintegrations per second is known as the *activity*, A, typically measured in curies. One curie is equal to 3.7 EE10 disintegrations per second (dps).

Table 18.1
Basic Particles
(Physical Carbon Scale)

name	common name	symbol	mass (amu)
proton	nucleus of hydrogen atom	$_1H^1$ or p	1.007277
neutron		$_0n^1$ or n	1.008665
electron	β particle	$_{-1}e^0$ or β	.000548593
positron		$_{+1}e^0$	.000548593
alpha particle	helium nucleus	$_2He^4$ or α	4.00260
gamma	short wavelength photons	γ	0.0000000
deuteron	deuterium nucleus	$_1H^2$ or D	2.01345
carbon		C	12.00000

The decay constant can be determined from the *half-life*, $t_{\frac{1}{2}}$, or from the *mean life expectancy* of the atom, $\bar{t}$.

$$\lambda = \frac{.693}{t_{\frac{1}{2}}} = \frac{1}{\bar{t}} \qquad 18.2$$

The mass, activity, and number of atoms at time t can be calculated from the values at $t = 0$ once λ is known. (N_t is the total number of atoms at time t.)

$$m_t = m_0 e^{-\lambda t} \qquad 18.3$$
$$A_t = A_0 e^{-\lambda t} = \lambda N_t \qquad 18.4$$
$$N_t = N_0 e^{-\lambda t} \qquad 18.5$$

Table 18.2
Half-Lives of Selected Radioactive Isotopes Found in Spent Reactor Fuels

Type	Isotope	Half-Life (yr)
gaseous fission products	H-3	12.3
	Kr-85	10.7
volatile and semi-volatile fission products	Ru-103	0.11
	Ru-106	1.01
	I-129	17×10^6
	I-131	0.02
	Cs-134	2.05
	Cs-135	3×10^6
	Cs-137	30.2
solid fission products	Sr-89	0.14
	Sr-90	28.9
	Y-91	0.16
	Zr-93	0.95×10^6
	Zr-95	0.18
	Nb-95	0.10
	Ce-141	0.09
	Ce-144	0.78
	Pm-147	2.62
	Eu-155	5.01
actinides in U fuel	Np-237	2.14×10^6
	Pu-238	86
	Pu-239	24,400
	Pu-240	6580
	Pu-241	13
	Pu-242	379,000
	Am-241	458
	Am-243	7800
	Cm-242	0.45
	Cm-244	17.6

Example 18.2

60 kg of Pu-239 (half-life = 2.4 EE4 years) are to be sealed in a unit deep within a mountain. If the plutonium is not retrieved until after 2000 years have passed, what will be the activity in disintegrations per second?

$$\lambda = \frac{.693}{t_{\frac{1}{2}}} = \frac{.693}{2.4 \text{ EE4 years}} = 2.888 \text{ EE} - 5 \ (1/\text{yr})$$

The mass of Pu-239 after 2000 years is

$$m_{2000} = 60 e^{(-2.888 \text{ EE}-5)(2000)} = 56.63 \text{ kg}$$

The number of moles of plutonium is

$$\frac{56.63}{239}(1000) \approx 237$$

The number of atoms at $t = 2000$ is

$$N_{2000} = (237 \text{ moles}) \left(6.023 \text{ EE23} \frac{\text{atoms}}{\text{mole}} \right)$$
$$= 1.427 \text{ EE26}$$

The activity is

$$A_{2000} = \frac{(2.888 \text{ EE} - 5) \ 1/\text{yr} \ (1.427 \text{ EE26}) \text{ atoms}}{(3.15 \text{ EE7}) \text{ sec/year}}$$
$$= 1.308 \text{ EE14 dps}$$

3 INTERACTIONS OF PARTICLES WITH MATTER

Nuclear fission reactions require a constant production, flow and capture of neutrons. Calculations for reactor design require an understanding of how these neutrons react with the surrounding materials.

Whether or not a moving neutron interacts with a nearby atom depends on the effective size of the atom. The effective size of atoms is measured in square centimeters, analogous to the size of a target at which the neutrons are beamed. Since the effective area is small, *cross sections* (short for "cross sectional areas") usually are expressed in *barns*, where one barn equals EE–24 cm^2. The symbol for the cross section is σ.

Neutrons can interact with atoms in two basic ways. They can be absorbed or scattered. There are various forms of absorption and scattering interactions.

- Scattering (subscript *s*)

  Inelastic scattering (subscript *is*) occurs when the nucleus is struck by a neutron, gains energy, and then decays by emitting a gamma ray. This reaction is written as (n, n′) since the reflected neutron has a different energy from the incident neutron, the difference having been absorbed by the nucleus. (Inelastic scattering is absent at thermal velocities.)

  Elastic scattering (subscript *es*) occurs when kinetic energy is conserved in the collision. The struck nucleus remains at its original energy level. This is written as an (n, n) reaction.

- Absorption (subscript a)
  · *Radiative capture* (subscript c or γ) occurs when the neutron is captured by the nucleus, which then emits gamma radiation. This is written as (n, γ).
  · *Fission* (subscript f) occurs when the neutron splits the atom into two smaller fragments.
  · *Proton decay* (subscript p) is written as (n, p).
  · *Alpha decay* (subscript α) is written (n, α).

$$\sigma_t = \sigma_{is} + \sigma_{es} + \sigma_c + \sigma_f + \sigma_p + \sigma_\alpha \qquad 18.6$$

Often, it is unnecessary to distinguish between the various forms of absorption and scattering. Equation 18.6 then becomes

$$\sigma_t = \sigma_a + \sigma_s \qquad 18.7$$

It is useful to think of cross sections as being *interaction probabilities* per unit length of path. Thus, the probability of a neutron being absorbed by an atom is

$$p\{absorption\} = \frac{\sigma_a}{\sigma_t} \qquad 18.8$$

The probability that a neutron will interact after traveling distance x through a substance is

$$p\{interacting\} = \sigma_t N x \qquad 18.9$$

N is the number of atoms per unit volume.

$$N = \frac{\rho N_o}{A.W.} \qquad 18.10$$

If a neutron beam with incident density, J, hits a target of area, A, the average number of interactions per second will be

$$interaction\ rate = J A \sigma_t N x \qquad 18.11$$

The product $N\sigma$ occurs so often in nuclear calculations that it is given the special symbol Σ and the special name, *macroscopic cross section*. Thus, equation 18.11 can be written as

$$interaction\ rate = J A \Sigma x \qquad 18.12$$

Example 18.3

A carbon-12 graphite target .05 cm thick and .5 cm^2 in area is struck by a 0.1 cm^2 neutron beam with intensity of 5 EE8 neutrons/cm^2-sec. If the density is 1.6 g/cm^3 and $\sigma_t = 2.6b$ for carbon in this experiment, what is the interaction rate and the probability that there will be an interaction?

From equation 18.10, the atomic density is

$$N = \frac{(1.6)(6.023\ EE23)}{12} = 8.031\ EE22\ atoms/cm^3$$

The probability of interaction is predicted by equation 18.9.

$$p\{interaction\} = (2.6\ EE-24)\ cm^2 \times (8.031\ EE22)\ atoms/cm^3 (.05)\ cm$$
$$= 1.04\ EE-2\ (about\ 1\%)$$

The interaction rate is

$$(5\ EE8)n/cm^2\text{-}sec\ (.1)\ cm^2\ (1.04\ EE-2)$$
$$= 5.2\ EE5\ neutrons/sec$$

4 NEUTRON ENERGIES

It is customary to classify neutrons according to the energies listed in table 18.3.

Table 18.3
Neutron Classifications

energy (eV)	name
.001	cold
.025	thermal
1.	slow (resonant)
100.	slow
EE4	intermediate
EE6	fast
EE8	ultrafast
EE10	relativistic

Most reactors use *thermal neutrons*. The (n, p), (n, α), and (n, n′) reactions are rare at the thermal level. Equation 18.6 can be written as

$$\sigma_t = \sigma_{es} + \sigma_c + \sigma_f \qquad 18.13$$

The term *thermal* implies that the neutrons have experienced sufficient collisions to bring them down to the same energy level as that of the surrounding atoms, usually assumed to be at 20°C.[1] Of course, not all neutrons travel at the same velocity. In fact, a Maxwellian velocity distribution similar to that assumed in the kinetic theory of gases occurs. Such a distribution predicts the *most probable particle velocity* to be

$$v_p = 1.28\ EE4\sqrt{T}\quad (cm/sec) \qquad 18.14$$

Since the velocity is very low, the kinetic energy is

$$E_k = \tfrac{1}{2}mv^2 \qquad 18.15$$

[1] Usually, the terms "thermal", "20°C", ".0253 eV", and "2200 m/sec" are considered synonymous.

Example 18.4

What are the velocity and the kinetic energy of a thermal neutron?

$$T = 20°C = 293°K$$

$$v_p = 1.28 \text{ EE}4\sqrt{293} = 2.19 \text{ EE}5 \text{ (cm/sec)}$$

$$E_k = .5(1.008665)\text{amu}(1.66053 \text{ EE}-27)\frac{\text{kg}}{\text{amu}} \times$$

$$(2.19 \text{ EE}3)^2 \text{ (m/sec)}^2$$

$$= 4.017 \text{ EE}-21 \ J$$

Converting this to *eV*

$$E_k = \frac{4.017 \text{ EE}-18 \ J}{1.60219 \text{ EE}-19 \ J/eV} = .02507 \text{ eV}$$

It is important to know the neutron energy level since cross sections vary with neutron energy. In the low-energy (less than resonant) region, called the *linear* or *(1/v) region*, the actual cross section can be found from the thermal data, which is widely tabulated.

$$\sigma' = \frac{\sigma_{th}v_{th}}{v} = \sigma_{th}\sqrt{\frac{293}{T}} = \sigma_{th}\sqrt{\frac{.0253}{E}} \qquad 18.16$$

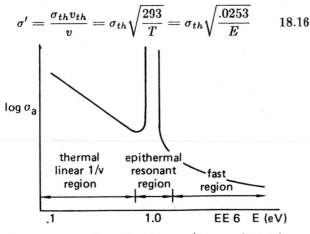

log σ_a

thermal linear 1/v region | epithermal resonant region | fast region

.1   1.0   EE 6   E (eV)

Figure 18.1 Cross Sections vs Energy (U-235)

The cross section obtained in equation 18.16 is based on the most probable speed, which is not the average velocity of the neutron beam. The average value to be used in reactor calculations is obtained by assuming a Maxwellian distribution.

$$\bar{\sigma} = \frac{\sigma_{th}}{1.128} \qquad 18.17$$

If the (1/v) assumption is invalid, as it is for some moderators, uranium, and plutonium, a *non-1/v correction factor*, g', must be used.

$$\bar{\sigma} = \frac{g'\sigma_{th}}{1.128} \qquad 18.18$$

Values of non-1/v factors are temperature dependent.

Table 18.4
Typical Non-1/v Factors (g')

T, °C	Cd	U²³³	
	g_a	g_a	g_f
20	1.3203	0.9983	1.0003
100	1.5990	0.9972	1.0011
200	1.9631	0.9973	1.0025
400	2.5589	1.0010	1.0068
600	2.9031	1.0072	1.0128
800	3.0455	1.0146	1.0201
1000	3.0599	1.0226	1.0284

T, °C	U²³⁵		U²³⁸	Pu²³⁹	
	g_a	g_f	g_a	g_a	g_f
20	0.9780	0.9759	1.0017	1.0723	1.0487
100	0.9610	0.9581	1.0031	1.1611	1.1150
200	0.9457	0.9411	1.0049	1.3388	1.2528
400	0.9294	0.9208	1.0085	1.8905	1.6904
600	0.9229	0.9108	1.0122	2.5321	2.2037
800	0.9182	0.9036	1.0159	3.1006	2.6595
1000	0.9118	0.8956	1.0198	3.5353	3.0079

Finally, if the reaction is not at 20°C, the average value of σ is

$$\bar{\sigma}' = \frac{g'\sigma_{th}}{1.128}\sqrt{\frac{293}{T}} = \frac{g'\sigma_{th}}{1.128}\sqrt{\frac{.0253}{E}} \qquad 18.19$$

Values of σ are revised frequently. Table 18.5 contains approximate data that is useful for non-critical applications.

Table 18.5
Most Probable Thermal Neutron
Cross Sections: Fuels
(in barns)

	σ_{es}	σ_c	σ_f	σ_a	σ_t
U-natural*	8.3	3.43	4.16	7.59	15.89
U-235	10	101	577	678	688
U-238	8.3	2.8	.0005	2.8	11.1
U-233		48	525	573	
Pu-239	9.6	274	741	1015	1025
Pu-241		425	950	1375	

* Assumes composition is .72% U-235 and 99.28% U-238.

Example 18.5

Find the average macroscopic absorption cross section of 1% boron steel (8.0 g/cm³) for .035 eV neutrons. Use the following cross sections.

$$\sigma_{th,Fe} = 2.62 \text{ barns}$$

$$\sigma_{th,B} = 755 \text{ barns}$$

Both iron and boron are 1/v absorbers. From equation 18.19,

$$\bar{\sigma}'_{Fe} = \frac{2.62}{1.128}\sqrt{\frac{.0253}{.035}} = 1.975$$

$$\bar{\sigma}'_{B} = \frac{755}{1.128}\sqrt{\frac{.0253}{.035}} = 569.1$$

From equation 18.10,

$$N = \frac{\rho N_o}{A.W.}$$

$$N_{\text{iron}} = \frac{(.99)(8)(6.023\ \text{EE}23)}{55.85} = 8.54\ \text{EE}22$$

$$N_{\text{boron}} = \frac{(.01)(8)(6.023\ \text{EE}23)}{10.82} = 4.45\ \text{EE}21$$

$$\Sigma = (8.54\ \text{EE}22)(1.975) + (4.45\ \text{EE}21)(569.1)$$

$$= 2.701\ \text{EE}24\frac{\text{barns}}{\text{cm}^3}$$

Since one barn=EE−24 cm², $\Sigma = 2.7$ 1/cm.

5 INTENSITY OF A SOURCE

A. POINT SOURCE

If a point source emits S rays or particles per second isotropically, the flux[2] at a distance x without attenuation is given by

$$\phi(x) = \frac{S}{4\pi x^2} \qquad 18.20$$

B. LINE SOURCE

If an infinite line source emits S rays or particles isotropically per centimeter of length per second, the flux at distance x without attenuation is

$$\phi(x) = \frac{S}{4x} \qquad 18.21$$

For a line source with finite length ($L = L_1 + L_2$), the flux at a point perpendicular to the source and located at distances L_1 and L_2 from the ends is

$$\phi(x) = \frac{S}{4\pi x}\left[\arctan\left(\frac{L_1}{x}\right) + \arctan\left(\frac{L_2}{x}\right)\right] \qquad 18.22$$

C. DISC SOURCE

If a disc of radius R emits S rays or particles per square centimeter, the flux along the disc axis without attenuation is

$$\phi(x) = \frac{S}{4}\ln\left[1 + \left(\frac{R}{x}\right)^2\right] \qquad 18.23$$

6 PARTICLE ATTENUATION[3]

The thickness required to reduce the intensity of a radiating source to a desired level depends on the particle, the particle energy, the type of beam, and the shielding material. *Alpha radiation* with its double charge is short-range radiation due to residual path ionization and electrostatic interaction. *Beta particles* are only singly charged, and they penetrate to greater, though still short, distances.

Gamma radiation has no charge and produces no ionization. Gamma radiation poses the major health threat.

A. GAMMA RADIATION

Gamma radiation is attenuated by

- *Compton scattering* (elastic collisions)—the major attenuating mechanism
- *Photoelectric effect*—gamma energy causes an electron from a nearby atom to be ejected
- *Pair production*—gamma radiation with energy greater than 1.02 MeV transforms itself into a positron and an electron in the vicinity of a nucleus

For gamma rays, the total cross section is

$$\sigma_t = \sigma_{\text{Compton}} + \sigma_{\text{photoelectric}} + \sigma_{\text{pair}} \approx \sigma_{\text{Compton}} \qquad 18.24$$

Traditionally, the macroscopic cross section has been called the *attenuation coefficient* or the *linear absorption coefficient*.

$$\mu_l = \Sigma_t = \sigma_t N \qquad (1/\text{cm}) \qquad 18.25$$

The *mass attenuation coefficient* (or the *mass absorption coefficient*) can be derived from the linear absorption coefficient.

$$\mu_m = \frac{\mu_l}{\rho} \qquad (\text{cm}^2/\text{g}) \qquad 18.26$$

μ_l and μ_m are used to determine the attenuation of a monoenergetic narrow-beam gamma source passing through a shield of thickness, x. The units of ρx are g/cm², a typical measure of shield thickness.[4]

$$I = I_o e^{-\mu_l x} = I_o e^{-\mu_m \rho x} \qquad 18.27$$

The intensity given by equation 18.27 disregards all scattering and energy build-up, factors that usually must be considered.

The *half-value thickness* and the *half-value mass* are defined by equations 18.28 and 18.29. These values will reduce the radiation beam 50%.

[2]Flux intensity also can be represented by the variable ϕ. I generally is used when flux is constant, rather than a function of distance from the source.

[3]Shielding from neutrons is covered later in this chapter.

[4]The *relaxation length* is the shield thickness necessary to reduce the intensity by (1/e).

Table 18.6
Mass Attenuation Coefficients, u_m, for Gamma Radiation

Gamma-Ray Energy, MeV

Material	0.1	0.15	0.2	0.3	0.4	0.5	0.6	0.8	1.0	1.25	1.5	2	3	4	5	6	8	10
H	.295	.265	.243	.212	.189	.173	.160	.140	.126	.113	.103	.0876	.0691	.0579	.0502	.0446	.0371	.0321
Be	.132	.119	.109	.0945	.0847	.0773	.0715	.0628	.0565	.0504	.0459	.0394	.0313	.0266	.0234	.0211	.0180	.0161
C	.149	.134	.122	.106	.0953	.0870	.0805	.0707	.0636	.0568	.0518	.0444	.0356	.0304	.0270	.0245	.0213	.0194
N	.150	.134	.123	.106	.0955	.0869	.0805	.0707	.0636	.0568	.0517	.0445	.0357	.0306	.0273	.0249	.0218	.0200
O	.151	.134	.123	.107	.0953	.0870	.0806	.0708	.0636	.0568	.0518	.0445	.0359	.0309	.0276	.0254	.0224	.0206
Na	.151	.130	.118	.102	.0912	.0833	.0770	.0676	.0608	.0546	.0496	.0427	.0348	.0303	.0274	.0254	.0229	.0215
Mg	.160	.135	.122	.106	.0944	.0860	.0795	.0699	.0627	.0560	.0512	.0442	.0360	.0315	.0286	.0266	.0242	.0228
Al	.161	.134	.120	.103	.0922	.0840	.0777	.0683	.0614	.0548	.0500	.0432	.0353	.0310	.0282	.0264	.0241	.0229
Si	.172	.139	.125	.107	.0954	.0869	.0802	.0706	.0635	.0567	.0517	.0447	.0367	.0323	.0296	.0277	.0254	.0243
P	.174	.137	.122	.104	.0928	.0846	.0780	.0685	.0617	.0551	.0502	.0436	.0358	.0316	.0290	.0273	.0252	.0242
S	.188	.144	.127	.108	.0958	.0874	.0806	.0707	.0635	.0568	.0519	.0448	.0371	.0328	.0302	.0284	.0266	.0255
Ar	.188	.135	.117	.0977	.0867	.0790	.0730	.0638	.0573	.0512	.0468	.0407	.0338	.0301	.0279	.0266	.0248	.0241
K	.215	.149	.127	.106	.0938	.0852	.0786	.0689	.0618	.0552	.0505	.0438	.0365	.0327	.0305	.0289	.0274	.0267
Ca	.238	.158	.132	.109	.0965	.0876	.0809	.0708	.0634	.0566	.0518	.0451	.0376	.0338	.0316	.0302	.0285	.0280
Fe	.344	.183	.138	.106	.0919	.0828	.0762	.0664	.0595	.0531	.0485	.0424	.0361	.0330	.0313	.0304	.0295	.0294
Cu	.427	.206	.147	.108	.0916	.0820	.0751	.0654	.0585	.0521	.0476	.0418	.0357	.0330	.0316	.0309	.0303	.0305
Mo	1.03	.389	.225	.130	.0998	.0851	.0761	.0648	.0575	.0510	.0467	.0414	.0365	.0349	.0344	.0344	.0349	.0359
Sn	1.58	.563	.303	.153	.109	.0886	.0776	.0647	.0568	.0501	.0459	.0408	.0367	.0355	.0355	.0358	.0368	.0383
I	1.83	.648	.339	.165	.114	.0913	.0792	.0653	.0571	.0502	.0460	.0409	.0370	.0360	.0361	.0365	.0377	.0394
W	4.21	1.44	.708	.293	.174	.125	.101	.0763	.0640	.0544	.0492	.0437	.0405	.0402	.0409	.0418	.0438	.0465
Pt	4.75	1.64	.795	.324	.191	.135	.107	.0800	.0659	.0554	.0501	.0445	.0414	.0411	.0418	.0427	.0448	.0477
Tl	5.16	1.80	.866	.346	.204	.143	.112	.0824	.0675	.0563	.0508	.0452	.0420	.0416	.0423	.0433	.0454	.0484
Pb	5.29	1.84	.896	.356	.208	.145	.114	.0836	.0684	.0569	.0512	.0457	.0421	.0420	.0426	.0536	.0459	.0489
U	10.60	2.42	1.17	.452	.259	.176	.136	.0952	.0757	.0615	.0548	.0484	.0445	.0440	.0446	.0455	.0479	.0511
Air	.151	.134	.123	.106	.0953	.0868	.0804	.0706	.0655	.0567	.0517	.0445	.0357	.0307	.0274	.0250	.0220	.0202
H_2O	.167	.149	.136	.118	.106	.0966	.0896	.0786	.0706	.0630	.0575	.0493	.0396	.0339	.0301	.0275	.0240	.0219
Concrete	.169	.139	.124	.107	.0954	.0870	.0804	.0706	.0635	.0567	.0517	.0445	.0363	.0317	.0287	.0268	.0243	.0229
Tissue	.163	.144	.132	.115	.100	.0936	.0867	.0761	.0683	.0600	.0556	.0478	.0384	.0329	.0292	.0267	.0233	.0212

Table 18.7
Energy Absorption Coefficient, u_a
$$cm^2/g$$

Gamma-Ray Energy, MeV

Material	0.1	0.15	0.2	0.3	0.4	0.5	0.6	0.8	1.0	1.25	1.50	2	3	4	5	6	8	10
H	.0411	.0487	.0531	.0575	.0589	.0591	.0590	.0575	.0557	.0533	.0509	.0467	.0401	.0354	.0348	.0291	.0252	.0255
Be	.0183	.0217	.0237	.0256	.0263	.0264	.0263	.0256	.0248	.0237	.0227	.0210	.0183	.0164	.0151	.0141	.0127	.0118
C	.0215	.0246	.0267	.0288	.0296	.0297	.0296	.0289	.0280	.0268	.0256	.0237	.0209	.0190	.0177	.0166	.0153	.0145
N	.0224	.0249	.0267	.0288	.0296	.0297	.0296	.0289	.0280	.0268	.0256	.0236	.0211	.0193	.0180	.0171	.0158	.0151
O	.0233	.0252	.0271	.0289	.0296	.0297	.0296	.0289	.0280	.0268	.0257	.0238	.0212	.0195	.0183	.0175	.0163	.0157
Na	.0258	.0258	.0266	.0279	.0283	.0284	.0284	.0276	.0268	.0257	.0246	.0229	.0207	.0194	.0185	.0179	.0171	.0168
Mg	.0335	.0276	.0278	.0290	.0294	.0293	.0292	.0285	.0276	.0265	.0254	.0237	.0215	.0203	.0194	.0188	.0182	.0180
Al	.0373	.0283	.0275	.0283	.0287	.0286	.0286	.0278	.0270	.0259	.0248	.0232	.0212	.0200	.0192	.0188	.0183	.0182
Si	.0435	.0300	.0286	.0291	.0293	.0290	.0290	.0282	.0274	.0263	.0252	.0236	.0217	.0206	.0198	.0194	.0190	.0189
P	.0501	.0315	.0292	.0289	.0290	.0290	.0287	.0280	.0271	.0260	.0250	.0234	.0216	.0206	.0200	.0197	.0194	.0195
S	.0601	.0351	.0310	.0301	.0301	.0300	.0298	.0288	.0279	.0268	.0258	.0242	.0224	.0215	.0209	.0206	.0206	.0206
Ar	.0729	.0268	.0302	.0278	.0274	.0272	.0270	.0260	.0252	.0242	.0233	.0220	.0206	.0199	.0195	.0195	.0194	.0197
K	.0909	.0433	.0340	.0304	.0298	.0295	.0291	.0282	.0272	.0261	.0251	.0237	.0222	.0217	.0214	.0212	.0215	.0219
Ca	.111	.0489	.0367	.0318	.0309	.0304	.0300	.0290	.0279	.0268	.0258	.0244	.0230	.0225	.0222	.0223	.0225	.0231
Fe	.225	.0810	.0489	.0340	.0307	.0294	.0287	.0274	.0261	.0250	.0242	.0231	.0224	.0224	.0227	.0231	.0239	.0250
Cu	.310	.107	.0594	.0368	.0316	.0296	.0286	.0271	.0260	.0247	.0237	.0229	.0223	.0227	.0231	.0237	.0248	.0261
Mo	.922	.294	.141	.0617	.0422	.0348	.0315	.0281	.0263	.0248	.0239	.0233	.0237	.0250	.0262	.0274	.0296	.0316
Sn	1.469	.471	.222	.0873	.0534	.0403	.0346	.0294	.0268	.0248	.0239	.0233	.0243	.0259	.0276	.0291	.0316	.0339
I	1.726	.557	.260	.100	.0589	.0433	.0366	.0303	.0274	.0252	.0241	.0236	.0247	.0265	.0283	.0299	.0327	.0353
W	4.112	1.356	.631	.230	.121	.0786	.0599	.0426	.0353	.0302	.0281	.0271	.0287	.0311	.0335	.0355	.0390	.0426
Pt	4.645	1.556	.719	.262	.138	.0892	.0666	.0465	.0375	.0315	.0293	.0280	.0296	.0320	.0343	.0365	.0400	.0438
Tl	5.057	1.717	.791	.285	.152	.0972	.0718	.0491	.0393	.0326	.0301	.0288	.0304	.0326	.0349	.0354	.0406	.0446
Pb	5.193	1.753	.821	.294	.156	.0994	.0738	.0505	.0402	.0332	.0306	.0293	.0305	.0330	.0352	.0373	.0412	.0450
U	9.63	2.337	1.096	.392	.208	.132	.0968	.0628	.0482	.0383	.0346	.0324	.0332	.0352	.0374	.0394	.0443	.0474
Air	.0233	.0251	.0268	.0288	.0296	.0297	.0296	.0289	.0280	.0268	.0256	.0238	.0211	.0194	.0181	.0172	.0160	.0153
H_2O	.0253	.0278	.0300	.0321	.0328	.0330	.0329	.0321	.0311	.0298	.0285	.0264	.0233	.0213	.0198	.0188	.0173	.0165
Concrete	.0416	.0300	.0289	.0294	.0297	.0296	.0295	.0287	.0278	.0272	.0256	.0239	.0216	.0203	.0194	.0188	.0180	.0177
Tissue	.0271	.0282	.0293	.0312	.0317	.0320	.0319	.0311	.0300	.0288	.0276	.0256	.0220	.0206	.0192	.0182	.0168	.0160

$$x_{\frac{1}{2}} = \frac{.693}{\mu_l} \qquad 18.28$$

$$m_{\frac{1}{2}} = \frac{.693}{\mu_m} \qquad 18.29$$

Using equation 18.31, the cross section can be obtained from particle attenuation calculations.

$$I = I_o e^{-N\sigma_t x} \qquad 18.30$$

$$\sigma_t = \ln\frac{\left(\dfrac{\phi_o}{\phi}\right)}{Nx} \qquad 18.31$$

The *energy flux density* passing into a shield is given by equation 18.32. E typically is in eV, and I is in rays/cm²-sec.

$$\text{energy flux} = EI \qquad 18.32$$

The energy deposition due to attenuation by Compton scattering, pair production, and the photoelectric effect is $EI\mu_a$. μ_a is the *energy absorption coefficient* given in table 18.7.

B. BETA RADIATION

Beta particles are attenuated according to

$$I = I_o e^{-\mu_m \rho x} \qquad 18.33$$

μ_m for beta radiation is almost independent of the element. An approximate formula for μ_m if E is in MeV is[5]

$$\mu_m = \frac{17.0}{(E)^{1.14}} \qquad (\text{cm}^2/\text{g}) \qquad 18.34$$

The shield thickness required to absorb the most energetic beta particles is called the *maximum range*.[6] R_{max} is in g/cm², and E is in MeV.

$$R_{max} = \left(\frac{1}{\rho}\right)(.412)(E_{max})^{1.265-.0954\,\ln(E_{max})}$$
$$(E < 2.5 \text{ MeV}) \qquad 18.35$$

$$R_{max} = \left(\frac{1}{\rho}\right)(.530)(E_{max} - .106)$$
$$(E > 2.5 \text{ MeV}) \qquad 18.36$$

[5] E is restricted to the range of 0.5 MeV to 6 MeV. This equation also has been reported as $\mu_m = \frac{22}{E^{1.33}}$.

[6] The shield thickness (in g/cm²) necessary to remove the intensity to the background radiation intensity is known as the *extrapolated range*.

Table 18.8
Value of B_m for a Plane Monodirectional Source

Material	E_o, MeV	1	2	4	$\mu_l x$ 7	10	15
Water	0.5	2.63	4.29	9.05	20.0	35.9	74.9
	1.0	2.26	3.39	6.27	11.5	18.0	30.8
	2.0	1.84	2.63	4.28	6.96	9.87	14.4
	3.0	1.69	2.31	3.57	5.51	7.48	10.8
	4.0	1.58	2.10	3.12	4.63	6.19	8.54
	6.0	1.45	1.86	2.63	3.76	4.86	6.78
	8.0	1.36	1.69	2.30	3.16	4.00	5.47
Iron	0.5	2.07	2.94	4.87	8.31	12.4	20.6
	1.0	1.92	2.74	4.57	7.81	11.6	18.9
	2.0	1.69	2.35	3.76	6.11	8.78	13.7
	3.0	1.58	2.13	3.32	5.26	7.41	11.4
	4.0	1.48	1.90	2.95	4.61	6.46	9.92
	6.0	1.35	1.71	2.48	3.81	5.35	8.39
	8.0	1.27	1.55	2.17	3.27	4.58	7.33
	10.0	1.22	1.44	1.95	2.89	4.07	6.70
Lead	0.5	1.24	1.39	1.63	1.87	2.08	
	1.0	1.38	1.68	2.18	2.80	3.40	4.20
	2.0	1.40	1.76	2.41	3.36	4.35	5.94
	3.0	1.36	1.71	2.42	3.55	4.82	7.18
	4.0	1.28	1.56	2.18	3.29	4.69	7.70
	6.0	1.19	1.40	1.87	2.97	4.69	9.53
	8.0	1.14	1.30	1.69	2.61	4.18	9.08
	10.0	1.11	1.24	1.54	2.27	3.54	7.70

7 GAMMA RAY SHIELDING FOR BIOLO-GICAL PURPOSES

The previous calculations cannot be used to determine the shielding requirements for humans for two reasons. First, the source usually is not a collimated beam, and second, the attenuated beam is not lost, just scattered into a wider beam. A *build-up factor*, B, is used to account for this.

A. MONODIRECTIONAL BEAM

With a slab shield of thickness x in place, the emergent flux is called the *build-up flux*, ϕ_b. The build-up flux gives the same exposure rate on the emergent side of the shield as a beam of monoenergetic gamma rays of energy E.

$$\phi_b = \phi_o B_m e^{-\mu_l x} \qquad 18.37$$

The exposure rate is

$$X' = X'_o B_m e^{-\mu_l x} \qquad 18.38$$

B. ISOTROPIC SOURCE

If an isotropic source emitting S gamma rays per second is surrounded by a shield of radius R (e.g., thickness R), the exposure rate is

$$X' = X'_o B_p e^{-\mu_l R} \qquad 18.39$$

$$\phi_b = \frac{S B_p e^{-\mu_l R}}{4\pi R^2} \qquad 18.40$$

8 NEUTRON BEAM SHIELDING

Thermal neutron intensity within a shield follows the exponential decay rule. x is measured along the path.

$$J = J_o e^{-\Sigma_a x} \qquad 18.41$$

It is not possible to absorb fast neutrons, so equation 18.41 can be used only with thermal neutrons. If the neutrons are fast, shielding with heavy elements can be used to slow down the neutrons inelastically.

Table 18.9
Values of B_p for Isotropic Point Sources

Material	E_o, MeV	1	2	4	$\mu_o r$ 7	10	15	20
Water	0.255	3.09	7.14	23.0	72.9	166	456	982
	0.5	2.52	5.14	14.3	38.8	77.6	178	334
	1.0	2.13	3.71	7.68	16.2	27.1	50.4	82.2
	2.0	1.83	2.77	4.88	8.46	12.4	19.5	27.7
	3.0	1.69	2.42	3.91	6.23	8.63	12.8	17.0
	4.0	1.58	2.17	3.34	5.13	6.94	9.97	12.9
	6.0	1.46	1.91	2.76	3.99	5.18	7.09	8.85
	8.0	1.38	1.74	2.40	3.34	4.25	5.66	6.95
	10.0	1.33	1.63	2.19	2.97	3.72	4.90	5.98
Iron	0.5	1.98	3.09	5.98	11.7	19.2	35.4	55.6
	1.0	1.87	2.89	5.39	10.2	16.2	28.3	42.7
	2.0	1.76	2.43	4.13	7.25	10.9	17.6	25.1
	3.0	1.55	2.15	3.51	5.85	8.51	13.5	19.1
	4.0	1.45	1.94	3.03	4.91	7.11	11.2	16.0
	6.0	1.34	1.72	2.58	4.14	6.02	9.89	14.7
	8.0	1.27	1.56	2.23	3.49	5.07	8.50	13.0
	10.0	1.20	1.42	1.95	2.99	4.35	7.54	12.4
Lead	0.5	1.24	1.42	1.69	2.00	2.27	2.65	(2.73)
	1.0	1.37	1.69	2.26	3.02	3.74	4.81	5.86
	2.0	1.39	1.76	2.51	3.66	4.84	6.87	9.00
	3.0	1.34	1.68	2.43	3.75	5.30	8.44	12.3
	4.0	1.27	1.56	2.25	3.61	5.44	9.80	16.3
	5.1	1.21	1.46	2.08	3.44	5.55	11.7	23.6
	6.0	1.18	1.40	1.97	3.34	5.69	13.8	32.7
	8.0	1.14	1.30	1.74	2.89	5.07	14.1	44.6
	10.0	1.11	1.23	1.58	2.52	4.34	12.5	39.2

The *removal cross section* is a measure of the shield material's ability to remove neutrons from the fast group.

$$J_{\text{fast}} = J_{\text{o, fast}} e^{-\Sigma_r x} \qquad 18.42$$

Values of the removal cross section are given in table 18.10.

Table 18.10 Approximate Removal Cross Sections

substance	Σ_r
Al	.0789
Be	.132
C	.065
concrete	.083 ($\rho = 2.3$ g/cm^3)
concrete	.148 (60% Fe)
Cu	.173
Fe	.168
Ni	.173
Pb	.118
water	.103
heavy water	.092

9 RADIOLOGICAL EFFECTS

The term *exposure* is used as a measure of a gamma ray or an x-ray field at the surface of an exposed object. Since this radiation produces an ionization of the air surrounding the object, the exposure is defined as

$$X = \frac{\text{\# of ions produced}}{\text{mass of air}} \left(\frac{\text{coulombs}}{\text{kilogram}} \right) \qquad 18.43$$

The unit of exposure is the roentgen, equal to 2.58 EE−4 C/kg. This is the exposure which produces 1 statcoulomb of ions in one cubic centimeter of air. Expressed in energy deposition, a roentgen is 5.47 EE7 MeV/g or 87.5 ergs/g.

The *exposure rate* is

$$X' = \frac{dX}{dt} = \frac{\text{\# of ions produced}}{\text{mass of air} - \text{time}} \left(\frac{R}{\text{sec}} \right) \qquad 18.44$$

The change in kinetic energy of radiation particles as they pass through an object is ΔE_k. This energy is imparted to the object. The *absorbed dose* is

$$D = \frac{\Delta E_k}{\text{mass of object}} \qquad 18.45$$

The unit of dose is the *rad (radiation absorbed dose)*, with one rad equal to 100 ergs/g (same as .01 J/kg).

The external exposure rate for monoenergetic gamma radiation is

$$X' = 1.83 \text{ EE} - 8\phi E(\mu_a)_{\text{air}} \qquad 18.46$$

X' is in R/sec if I is in rays/cm^2-sec, E is in MeV, and μ_a is in cm^2/g. If the radiation is polyenergetic, equation 18.46 must be summed over the energies.

Similarly, the dose rate in rad/sec is

$$D' = 1.6 \text{ EE} - 8\phi E(u_a)_{\text{tissue}}$$
$$= \frac{.874 X'(u_a)_{\text{tissue}}}{(u_a)_{\text{air}}} \qquad 18.47$$

Because damage is not proportional to dose alone, some types of radiation are more damaging than others. To account for this, the *relative biological effectiveness* (*RBE*) is used to increase the actual dose. RBE is determined experimentally and is too difficult to use in normal calculations. A rounded up version of RBE is called the *quality factor*, Q. The *equivalent dose* is

$$H = QD \qquad 18.48$$

The unit of equivalent dose is the *rem*, (roentgen equivalent man).

**Table 18.11
Typical Values of Q**

x-rays and γ-rays	1.0
β-rays $< .03$ MeV	1.7
β-rays $> .03$ MeV	1.0
thermal neutrons	2.0
fast (or unspecified) neutrons	10.0
α-particles (or more heavy)	20.0
protons	10.0

Table 18.12 shows the expected early (within 60 days of exposure) effects of whole-body exposure. Fatalities are expected with doses above 200 rem. With medical treatment, the fatality threshold can be extended to 500 rem. The whole-body lethal dose that leads to death for 50% of an exposed group within T days of the exposure is known as LD_{50}/T. $LD_{50}/30$ is approximately 500 rem.

**Table 18.12
Early Effects of Whole-Body Exposure**

exposure, rems	effects
0–50	no observable effects
50–100	some blood changes
100–200	minor fatigue, vomiting, and blood changes; recovery by all
200–600	severe illness; fatal to 20%
600–1000	severe illness; fatal to 80%–100%

Table 18.13
Typical Industrial Radiation Limits
for Adults*

trunk, organs, eyes, and gonads	1.25 rem per quarter, with cumulative dose not to exceed 5 (age 18)
hands, forearms, feet, and ankles	18.75 rem per quarter
skin of body	7.5 rem per quarter

* No exposure allowed under age 18.

The *maximum permissible dose (MPD)* of 1.25 rem per quarter sometimes is quoted as 100 mrem per week, and 2.5 mrem per hour if distributed over 40 hours. For general population individuals not working in a nuclear setting by choice, a suggested MPD is $\frac{1}{2}$ rem per year, not to exceed 5 rems before age 30 (approximately 170 mrem/year).·

Example 18.6

1 MeV gamma radiation with an intensity of 2 EE5 rays/cm^2-sec deposits energy at the rate of EE−2 ergs/g-sec. What is the absorbed tissue dose rate?

$$D' = \text{EE} - 2 \text{ ergs/g} = \frac{\text{EE}{-}2 \text{ ergs/g-sec}}{100 \text{ ergs/g-rad}}$$

$$= \text{EE} - 4 \text{ rad/sec}$$

Since $Q = 1$ for gamma radiation,

$$H' = \text{EE} - 4 \text{ rem/sec} = 360 \text{ mrem/hr}$$

10 MODERATION OF MONOENERGETIC NEUTRONS

Most cross sections increase with a decrease in velocity. A slow neutron has a larger cross section since it is in the vicinity of a nucleus for a longer time, allowing nuclear attraction a greater chance to occur.

Neutrons in a moderator will be slowed by successive glancing collisions. However, the neutrons cannot be slowed below the energy possessed by the molecules in the moderating materials. This is the *thermal equilibrium energy.*

A fast neutron with energy E_o coming into the vicinity of a nucleus will experience a glancing collision, deflecting the neutron and reducing its energy to E'.

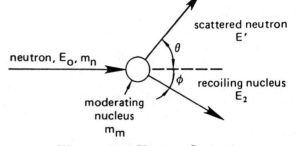

Figure 18.2 Neutron Scattering

The ratio of E' to E_o is

$$\left(\frac{E'}{E_o}\right) = \frac{A^2 + 2A\cos\theta + 1}{(1 + A)^2} \qquad 18.49$$

$$A = \frac{m_m}{m_n} \approx \text{moderator atomic weight} \qquad 18.50$$

$E'/E_o = 1$ when $\theta = 0$. E' is minimum when $\theta = \pi$. (The neutron is returned to its source by a head-on collision.)

$$E'_{\min} = E_o\alpha \qquad 18.51$$

$$\alpha = \left(\frac{E'}{E_o}\right)_{\min} = \left(\frac{A-1}{A+1}\right)^2 \qquad 18.52$$

The maximum *energy loss fraction per collision* occurs when $\theta = \pi$.

$$(l_e)_{\min} = 1 - \alpha \qquad 18.53$$

The average energy and average loss fraction of scattered neutrons is

$$\overline{E'} = \tfrac{1}{2}E_o(1 + \alpha) \qquad 18.54$$

$$\overline{l_e} = \tfrac{1}{2}(1 - \alpha) \qquad 18.55$$

The *average cosine of the scattering angle* is

$$\overline{\cos\theta} = \frac{2}{3A} \qquad 18.56$$

The *transport cross section* is defined as

$$\sigma_{tr} = \sigma_s(1 - \overline{\cos\theta}) \qquad 18.57$$

When A is large, $\overline{\cos\theta}$ is small, and scattering is isotropic. Neutrons are scattered forward and backward almost equally often. If A is small, more neutrons will be scattered forward.

A variable called *lethargy* also is used in reactor studies. This is the *logarithmic energy decrement* per collision.

$$\xi = \ln\left(\frac{1}{l_e}\right) = \frac{\ln\alpha}{1-\alpha} \approx \frac{2}{A + .667} \qquad 18.58$$

The approximation in formula 18.58 is good for values of A greater than 2. Lethargy is defined as 1 for hydrogen.

The average *number of collisions* to bring neutron energy from E_o to E_{th} is

$$N_c = \frac{\ln\left(\dfrac{E_o}{E_{th}}\right)}{\xi} \qquad 18.59$$

If $E_o = 2\ EE6\ eV$ and $E_{th} = .025\ eV$,

$$N_c = \frac{18.2}{\xi} \qquad 18.60$$

The *average energy loss per collision* is

$$\bar{l}_e = \frac{E_1}{E_o} = 1 - \left(\frac{E_{th}}{E_o}\right)^{1/N_c} = 1 - \left(\frac{E_{th}}{E_o}\right)^{\xi/18.2} \qquad 18.61$$

However, N_c is insufficient for predicting good moderators since the cross section of the moderator also must be small enough that neutrons are not captured radiatively. Therefore, two other measures are used to evaluate moderators. The *slowing down power* (SDP) of a moderator is

$$SDP = \xi N \sigma_s = \xi \Sigma_s \qquad 18.62$$

SDP can be interpreted as the logarithmic energy loss per collision per centimeter of travel. It should be high for a good moderator. The *moderating ratio* also should be high.

$$MR = \frac{\xi \sigma_s}{\sigma_a} \qquad 18.63$$

Large absorption cross sections make the use of lithium and boron impractical. Beryllium, carbon, hydrogen, and deuterium water are used as moderators. When cost is a factor, water and graphite usually are chosen over deuterium.

11 IRRADIATION AND ACTIVATION

Substances form radioactive atoms when irradiated by a neutron beam. The *generation rate* per unit volume of activated sample is

$$g = \phi N \sigma_a \quad \left(\frac{\text{atoms}}{\text{cm}^3\text{-sec}}\right) \qquad 18.64$$

The number of activated atoms at time t during the irradiation is

$$\begin{aligned} N_t &= N_o e^{-\lambda t} + \frac{(\text{volume})g}{\lambda}(1 - e^{-\lambda t}) \\ &= N_o e^{-\lambda t} + \frac{N_s \sigma_a \phi}{\lambda}(1 - e^{-\lambda t}) \end{aligned} \qquad 18.65$$

N_o is the original number of unstable atoms (if any). λ is the decay constant. N_s is the total number of atoms in the sample.

The activity at any time is

$$A = \lambda N_t \qquad 18.66$$

After irradiation stops at t_2,

$$N_t = N_{t2} e^{-\lambda t} \qquad 18.67$$

Table 18.14
Approximate Collision Parameters

element	thermal σ_a	σ_s	l_e	N_c	α	ξ	SDP	MR	τ	A
hydrogen	.33	38	.63	18–19	0	1.00	1.425	62	27–30	1
H_2O			—	19	—	≈.927	1.425	62		—
deuterium	.0005	7	.52	25	.111	.725		47,500	125	2
D_2O	.00133	13.6	—	35	—	≈.510	.177	4830		—
helium	0	1	.35	42	.36	.427	≈0	51		4
lithium*	71	1.4	.27	62	.56	.265		.0083		7
beryllium	.01	7	.18	86	.640	.207	.1538	126	98	9
boron**	795	4	.17	105	.690	.174	.092	.00086		11
carbon (graphite)	.0035	4.8	.14	114	.716	.158	.083	216	364	12
oxygen	.0002	4.2	.11	150	.779	.120	.022	316		16
nitrogen	1.8	10	.12	132	.751	.137	.045	.74		14
sodium	.53	4.0	.080	221	.840	.0825	.007	.63		23
iron	2.62	11.0	.035	510	.931	.0357		.59		56
uranium-238	2.8	8.3	.0085	2172	.983	.00838		.025		238

*   A natural mixture of 92.48% Li-7 and 7.52% Li-6

** A natural mixture of 80.2% B-11 and 19.8% B-10

12 DIFFUSION OF THERMAL NEUTRONS

Thermal *neutron flux* is the number of neutrons crossing a one square centimeter area from all directions in one second. It usually is found by multiplying the neutron density by the most probable velocity.

$$\phi = \rho_n v_p = 2.2 \text{ EE5 for } 20° \text{ C} \qquad 18.68$$

Neutron current density, J, with the same units as flux, is the number of unidirectional neutrons crossing a one square centimeter boundary in one second. ρ_n, ϕ, and J are essentially zero outside the reactor. They are constant throughout only in a reactor of infinite size. In most real reactors, they are at a maximum at the center.

According to Fick's general law, the diffusion of neutrons is

$$J = -D\frac{d\phi}{dt} = -Dv\frac{dN}{dx} \qquad 18.69$$

During the diffusion process, the neutrons experience various events, such as scattering, absorption, and fission. The average *mean free path lengths* between these events are

$$\lambda_s = \frac{1}{\Sigma_s} \qquad 18.70$$

$$\lambda_a = \frac{1}{\Sigma_a} \qquad 18.71$$

$$\lambda_f = \frac{1}{\Sigma_f} \qquad 18.72$$

$$\lambda_t = \frac{1}{N\sigma_t} = \frac{1}{\Sigma_t} = \frac{1}{\Sigma_a + \Sigma_s} \qquad 18.73$$

$$\lambda_{tr} = \frac{1}{N\sigma_{tr}} = \frac{1}{\Sigma_{tr}}$$

$$= \lambda_s[1 + \overline{(\cos\theta)} + \overline{(\cos\theta)}^2 + \overline{(\cos\theta)}^3 + \cdots] \qquad 18.74$$

Notice that the Σ's are additive but the λ's are not. The physical significant of the mean free paths is shown in figure 18.3.

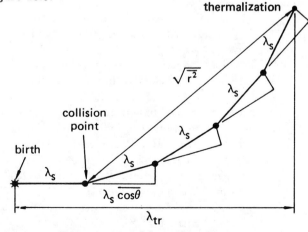

Figure 18.3 Mean Free Paths

The *diffusion coefficient*, D, is

$$D = \frac{\lambda_{tr}}{3} = \frac{1}{3\Sigma_s(1 - \overline{\cos\theta})} = \frac{1}{3\Sigma_{tr}} \approx \frac{d}{2.13} \qquad 18.75$$

d is the *extrapolation distance*, the distance to the point where the neutron flux is assumed to be zero.

$$d = \frac{2}{3\Sigma_{tr}} \approx 2.13D \qquad 18.76$$

The *thermal diffusion length*, also known as the *diffusion length for pure moderation*, is a measure of, but not the actual, average distance traveled by a neutron between thermalization and absorption.

$$L = \frac{1}{\sqrt{3\Sigma_{tr}\Sigma_a}} = \sqrt{\frac{\lambda_{tr}\lambda_a}{3}} = \sqrt{\frac{D}{\Sigma_a}} \qquad 18.77$$

The *diffusion area* is one-sixth of the mean squared distance of *thermal life*.

$$L^2 = (\tfrac{1}{6})\overline{(r_{\text{thermal}})^2} = (\tfrac{1}{3})\lambda_{tr}\lambda_a \qquad 18.78$$

The *slowing down length*, also known as the *epithermal diffusion length*, is a measure of the average distance traveled by the neutron between birth and thermalization. The *epithermal diffusion area* also is known as the *Fermi age* or just *age*.

$$\tau = L_s^2 = (\tfrac{1}{6})\overline{(r_{\text{epithermal}})^2} = (\tfrac{1}{3})(\lambda_s\lambda_{tr})_{\text{epithermal}} \qquad 18.79$$

The *migration length*, M, is a measure of the average distance traveled by a neutron between birth and absorption. The *migration area* is M^2 and is a measure of the opportunity for a neutron to escape the reactor.

$$M^2 = L_s^2 + L^2 = \tau + L^2 \qquad 18.80$$

13 NUCLEAR FISSION

There are three types of isotopes that take part in nuclear reactions. A *fertile isotope* absorbs neutrons and becomes fissionable. A *fissionable and fissile isotope* absorbs thermal neutrons, which increases the energy above the critical level. U-233, U-235, and Pu-239 are fissionable and fissile. A *fissionable and non-fissile isotope* requires high-energy neutrons to cause fission. Isotopes with even atomic weights are in this category.

Fission can occur only in heavy nuclei. The absorption of a neutron with energy above the threshold energy causes the target nucleus to split into smaller fission fragments. These fragments emit one or two 2 MeV neutrons each. The fission fragments still are radioactive and continue to decay several times. Although

Table 18.15
Approximate Constants of Moderators (20°C)[7]

one-group material	$\rho(g/cm^3)$	N	L(cm)	L_s(cm)	$\tau(cm^2)$	σ_{tr}(b)	σ_a(b)	λ_{tr}(cm)	λ_a(cm)	$\overline{D}$(cm)
H_2O	1.00	.0334 EE24	2.85	5.7	27–30	62.9	.660	.48	51.2	.16
D_2O (pure)	1.10	.0331 EE24	170	11.0	125	12.6	.00093	2.4	36,700	.87
Be	1.85	.1235 EE24	20.8	9.9	100	5.65	.0101	1.43	906	.50
C, graphite (pure)	1.60	.0803 EE24	59	18.7	364	4.54	.0032	2.52	4167	.84
BeO	2.95		28						1666	.47
two-group										
H_2O					27				23.9	1.13
D_2O					131				101.5	1.29
Be					102				181.5	.562
C, graphite					368				362.3	1.016

[7]Reactor analyses which consider only thermal neutrons are known as *one-group calculations*. Of course, neutrons are born fast, taking N_c collisions before being slowed down. If an analysis looks at fast and slow neutrons as two distinct sets, it is known as a *two-group analysis*. Multi-group calculations give greatest accuracy, but not for hand solutions. Generally, one-group calculations are adequate for graphite-moderated reactors, but are not for water-moderated reactors.

the exact fission fragments vary considerably,[8] a typical reaction is

$$_{92}U^{235} +_0 n^1 \mapsto _{92} U^{236} \mapsto_{54} Xe^{140} +_{38} Sr^{94} + 2_0n^1$$
$$+ \gamma + 200 \text{ MeV}$$

The energy release in all fission reactions, uranium, plutonium, and thorium, is the same—around 200 MeV per fission. This energy is shared by various fragments, rays, and particles. All but neutrino energy is useful.

—kinetic energy of the fragments	84%
—kinetic energy of the neutrons	2.5
—prompt gamma rays	2.5
—decay of fission fragments (delayed gamma, beta decay, and neutrinos)	11

Fission is a chain reaction because only one neutron is required to produce two or three other neutrons. The fission fragments also decay and emit delayed neutrons, so the average number of neutrons obtained in uranium fission is 2.43. In plutonium fission, the value is 2.89.

Thermal neutrons will fission fuel with odd atomic masses, such as U-235 and Pu-239. However, fast neutrons are required to fission the stable isotopes with even atomic masses, such as U-238 and Pu-240. Fission also can occur when fuel is exposed to high-energy gamma or x-rays. Such fission is known as *photofission*.

[8]It is required only that mass and charge be conserved.

Table 18.16
Fission Threshold Energies (MeV)

fuel	neutrons	deuterons	alpha	x, γ rays
U-235	.025 eV	8	21	5.31±.25
U-238	1.0 ± .1	8	21	5.8±.15
Pu-239	.025 eV			5.31±.27
Th-232	1.1 ± .05	8	21	5.4±.22
Pa-231	.45			
Np-237	.25			
U-234	.3			
U-233	.025 eV			
Pu-241	.025 eV			

14 THE NEUTRON FISSION CYCLE

In this explanation, a mixture of U-235 and U-238 is used. If pure U-235 is used, both ϵ and p are 1.0. All other processes are unchanged.

1. A thermal neutron fissions the nucleus of U-235, producing fission fragments and n fast neutrons.
2. The fast neutrons have sufficient energy to fission U-238, resulting in more fast neutrons. However, only a small number of the original neutrons cause fission in U-238. The sum of the remaining original neutrons and the neutrons from the U-238 is $n\epsilon$, where ϵ is the *fast fission factor*. ϵ is equal to 1 for homogeneous reactors. It normally is 1.03 to 1.05 for heterogeneous reactors.
3. $n\epsilon l_f$ neutrons are lost outside the reactor, where l_f is the *loss fraction*. $L_f = (1 - l_f) = exp(-B^2\tau)$ is the *fast non-leakage probability*.

L_f is approximately .623 for water-moderated U-235 reactors.

4. The remaining $n\epsilon(1 - l_f)$ neutrons are slowed down to thermal speeds.

5. A minority fraction $(1 - p)$ of the moderated neutrons undergoes *radiative capture*, transforming U-238 into U-239 and eventually into Pu-239. p is known as the *resonance escape probability*. The remaining thermal neutrons number $n\epsilon(1 - l_f)p$. p typically is .7 to .8 and is the fraction of neutrons becoming thermal without being captured radiatively. $p = 1$ for simple reactors fueled only with U-235 and containing no resonant absorbers.

6. A fraction, l_t, is lost outside the reactor. The number lost is $n\epsilon(1 - l_f)pl_t$. L_t is the *thermal non-leakage probability*, equal to $(1 - l_t)$. L_t can be found from $1/(1 + B^2L^2)$.

7. Of the remaining $n\epsilon(1 - l_f)p(1 - l_t)$ neutrons, $(1 - f)$ are absorbed by the moderator and the structure. f is the *thermal utilization* and the neutron fraction entering the fuel. Then, $n\epsilon(1 - l_f)p(1 - l_t)f$ neutrons are available for carrying on the reaction. f can be found from

$$f = \frac{\Sigma_{a,f}}{\Sigma_{a,f} + \Sigma_{a,m}} \qquad 18.81$$

8. Not all available thermal neutrons cause fission. The *absorption ratio*, σ_f/σ_a, is the fraction that causes fission. The *multiplication constant* is the number of neutrons left over to carry on the reaction.

$$k = n\epsilon\left(\frac{\sigma_f}{\sigma_a}\right)(1 - l_f)p(1 - l_t)f = \eta\epsilon pfL_fL_t \qquad 18.82$$

$$\eta = n\left(\frac{\sigma_f}{\sigma_a}\right) \qquad 18.83$$

If the reactor is infinite in size, L_f and L_t each will be one. Then equation 18.83 can be written for the ideal case, giving the value of *infinite-k*.

$$k_\infty = n\epsilon pf\left(\frac{\sigma_f}{\sigma_a}\right) = \eta\epsilon pf \qquad 18.84$$

Equation 18.84 is known as the *four-factor formula*.

15 SIMPLIFIED CORE CALCULATIONS

If k_∞ is less than one, any amount of fission material will be subcritical, and the reaction will die out quickly. Only when k_∞ is greater than one will there be the possibility of sustained fission. Thus, the *critical equation* is

$$k_\infty > 1 \qquad 18.85$$

If k_∞ is greater than one, the multiplication constant, k, determines what occurs in the reactor. If k is less than one, the reaction rate will be an exponential decay; if k is greater than one, the rate will be an exponential growth. If $k = 1$, the reaction rate will be constant.

$(k_\infty - k)$ is the percentage of neutrons which leak out of the reactor. This percentage is proportional to the migration area.

$$k_\infty - k \propto M^2 \qquad 18.86$$

Nuclear fuels with k_∞ less than one can be enriched to increase k_∞.

$$(k_\infty)_{\text{mixture}} = \frac{n_1N_1\sigma_{f1} + n_2N_2\sigma_{f2}}{N_1\sigma_{a1} + N_2\sigma_{a2}} = \frac{n_1\Sigma_{f1} + n_2\Sigma_{f2}}{\Sigma_{a1} + \Sigma_{a2}} \qquad 18.87$$

N_i is the number of nuclei per unit volume of fuel i or the percentage of component i. The relative concentrations of N_1 and N_2 can be found by setting $k_\infty = 1$.

Usually, a reactor is built so that $k = 1 + \delta$, where δ is the *excess reactivity*, usually .01 to .02. The control system is adjusted to keep $k = 1$.

Material buckling is the name given to the quantity in formula 18.88. B_m^2 appears in the critical core size formulas and is a measure of the neutron flux.

$$B_m^2 = \frac{k_\infty - 1}{M^2} = \frac{k_\infty - 1}{L^2 + \tau} \qquad 18.88$$

Geometric buckling (usually just called *buckling*), B_g^2, is a measure of the core size.

Table 18.17
Fast Neutrons Emitted per Fission (n)
Fast Neutrons Emitted per Neutron Absorbed (η)

incident neutron energy	U-233 n η	U-235 n η	U-238 n η	Pu-239 n η	Pu-241 n η	natural uranium n η
thermal	2.50/2.30	2.43/2.07		2.89/2.11	3.00/2.21	2.50/1.31
1 MeV	2.62/2.54	2.58/2.38		3.00/2.92		
2 MeV	2.73/2.57	2.70/2.54	2.69/2.46	3.11/2.99		

At criticality, $B_m^2 = B_g^2$. This defines the critical dimensions and the one-group *critical equation*.

$$k = \frac{k_\infty}{(1 + L^2 B^2)} = 1 \qquad 18.89$$

Table 18.18 defines the geometric buckling for regular core shapes, as well as specifying flux at various locations in the core. Table 18.18 uses the following special terms.

E_R—recoverable energy per fission, expressed in joules. If the energy per fission is 200 MeV, E_R is 3.2 EE−11 J.

P—power density in the reactor core, expressed in watts per cm^2 (slab), watts/cm^3 (sphere), or watts/cm length (cylinder).

J_o—ordinary Bessel function of the first kind, with an argument of $(B_m r)$, no units.

A—a flux constant with units of neutrons/cm^2-sec.

The dimensions and the volume of a critical sphere can be determined from table 18.18 and equation 18.88.

$$B^2 = \left(\frac{\pi}{R}\right)^2 \qquad 18.90$$

$$R = \frac{\pi}{B} \qquad 18.91$$

$$V = \tfrac{4}{3}\pi R^3 = \frac{130}{B} \qquad 18.92$$

$$B = \sqrt{\frac{k_\infty - 1}{M^2}} \qquad 18.93$$

The volume and the side of a *critical cube* are

$$V = \frac{161}{B^{1.5}} \qquad 18.94$$

$$t = \pi\sqrt{\frac{3}{B}} \qquad 18.95$$

Minimum *leakage* in a parallelepiped occurs when $a = b = c = t$.

The volume of a *critical cylinder* is

$$V = \frac{148}{B^{1.5}} \qquad 18.96$$

Minimum *leakage* occurs in a finite cylinder when $h = 1.82r$. The minimum leakage radius is

$$r = \sqrt{\frac{8.763}{B}} \qquad 18.97$$

16 OTHER FORMS OF THE CRITICAL EQUATION

Equation 18.89 is the *one-group critical equation*. It assumes that $\tau = 0$ or $\tau \ll L_t$. It is useful for rough estimates only, since it assumes that neutrons are generated at one energy level only. Values of ϕ and σ must be known or specified since the distribution of neutron velocities is not Maxwellian. No reflector (blanket) can be used.

Equation 18.98 is the *modified one-group critical equation*. It can be used for large bare reactors with $\sqrt{\tau} \ll$ reactor dimensions. It should be used for water-moderated reactors since $\tau > L$.

$$k = \frac{k_\infty}{1 + B^2 M^2} = 1 \qquad 18.98$$

Equation 18.99 is a *two-group critical equation* for bare reactors. It is needed for fast reactors.

$$k = \frac{k_\infty e^{-B^2 \tau}}{1 + B^2 L^2} = 1 \qquad 18.99$$

Equation 18.98 can be derived from equation 18.99 by expanding the exponential term.

Table 18.18
One-Group Buckling for Critical Bare Reactors

Geometry	Dimensions	Buckling, B^2	Flux, ϕ_r	A	ϕ_{max}/ϕ_{ave}
Infinite slab	Thickness a	$\left(\frac{\pi}{a}\right)^2$	$A\cos\left(\frac{\pi x}{a}\right)$	$1.57 P/a E_R \Sigma_f$	1.57
Rectangular parallelepiped	$a \times b \times c$	$\left(\frac{\pi}{a}\right)^2 + \left(\frac{\pi}{b}\right)^2 + \left(\frac{\pi}{c}\right)^2$	$A\cos\left(\frac{\pi x}{a}\right)\cos\left(\frac{\pi y}{b}\right)\cos\left(\frac{\pi z}{c}\right)$	$3.87 P/V E_R \Sigma_f$	3.88
Infinite cylinder	Radius R	$\left(\frac{2.405}{R}\right)^2$	$AJ_0\left(\frac{2.405r}{R}\right)$	$0.738 P/R^2 E_R \Sigma_f$	2.32
Finite cylinder	Radius R Height H	$\left(\frac{2.405}{R}\right)^2 + \left(\frac{\pi}{H}\right)^2$	$AJ_0\left(\frac{2.405r}{R}\right)\cos\left(\frac{\pi z}{H}\right)$	$3.63 P/V E_R \Sigma_f$	3.64
Sphere	Radius R	$\left(\frac{\pi}{R}\right)^2$	$A\frac{1}{r}\sin\left(\frac{\pi r}{R}\right)$	$P/4R^2 E_R \Sigma_f$	3.29

$$e^{-B^2\tau} = \frac{1}{1 + B^2\tau + \frac{1}{2}(B^2\tau)^2 + \frac{1}{6}(B^2\tau)^3 + \cdots} \qquad 18.100$$

If the reactor is large, B^2 is small, and B^4 is insignificant.

$$e^{-B^2\tau} \approx \frac{1}{1 + B^2\tau} \qquad 18.101$$

Combining with equation 18.99 and disregarding the B^4 term,

$$k = \frac{k_\infty}{(1 + B^2 L^2)(1 + B^2\tau)} \approx \frac{k_\infty}{1 + B^2(L^2 + \tau)}$$
$$= \frac{k_\infty}{1 + B^2 M^2} \qquad 18.102$$

17 REACTOR POWER

The *reactor rate* depends on the average flux and is defined as $\Sigma_a \overline{\phi}$. The number of fissions per second per cubic centimeter of core is $\Sigma_f \overline{\phi}$. The *power density*, or *watt density*, is $\Sigma_f \overline{\phi}/3.1 \text{ EE}10$.[9]

The *thermal power* from a reactor is

$$P_{th} = \frac{\Sigma_f \overline{\phi} V}{3.1 \text{ EE}10} = \frac{\overline{\phi} \sigma_f N_f V}{3.1 \text{ EE}10}$$
$$\approx 3.92 \text{ EE} - 8\overline{\phi} m_f \qquad 18.103$$

P_{th} is in watts if the mass is in kilograms and the volume is in cm^3.

The number of fuel molecules per unit volume is

$$N_f = \frac{f_e m_f N_o}{V(A.W.)_f} = \frac{f_e \rho N_o}{(A.W.)_f} \qquad 18.104$$

f_e is the *enrichment fraction*, the percent of fissionable molecules by weight.

In power calculations, the reactor power (*watts-thermal*) must be distinguished from the transmitted power (*watts-electric*). The transmitted power is

$$P_e(\text{watts}) = \eta P_{th}(\% \text{ of reactor capacity}) \qquad 18.105$$

η typically is in the order of 40%.

18 DESIGN OF HETEROGENEOUS REACTORS

The following procedure is valid for bare reactors where the core is a square lattice of square fuel elements with fuel rods of radius r_o and moderator width s. It also can

be used as an approximation when the fuel rod element is round with radius r_1. In this case, s is solved from

$$\pi r_1^2 = s^2 \qquad 18.106$$

The procedure can be used as a rough approximation when there is a coolant void between the fuel and the moderator. Then the assumption is made that the coolant void is filled with moderator.

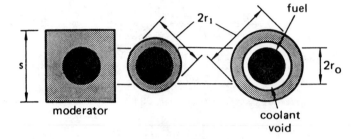

Figure 18.4 Valid Fuel Elements

step 1: Calculate the moderator macroscopic cross section. Assume that the moderator cross section is not dependent on the energy of the neutrons.

$$N_m = \frac{\rho_m N_o}{(A.W.)_m} \qquad 18.107$$
$$\Sigma_m = N_m \sigma_{a,m} \qquad 18.108$$

step 2: Convert thermal cross section data to effective cross section data by using the non-(1/v) factors for the temperature at which the reactor operates.

$$\sigma_{\text{eff}} = \frac{g' \sigma_{th}}{1.128} \qquad 18.109$$

step 3: Calculate fission and absorption cross sections for the fuel mixture. N_i is the atom fraction for component i. i is the fuel component index and is equal to 1 if only U-235 is used. It is equal to 2 if both U-235 and U-238 are used. j is the index over fissionable components only.

$$\sigma_{a,f} = \frac{\Sigma N_i \sigma_{a,i}}{\Sigma N_i} \qquad 18.110$$
$$\sigma_{f,f} = \frac{\Sigma N_j \sigma_{f,j}}{\Sigma N_j} \qquad 18.111$$

step 4: Calculate the number of neutrons causing fission per initial reaction.

$$\eta = n\left(\frac{\sigma_{f,f}}{\sigma_{a,f}}\right) \qquad 18.112$$

step 5: Calculate N_f, $\Sigma_{a,f}$, and $\Sigma_{tr,f}$. ($\Sigma_{tr,f}$ is approximately .18 for all U, Pu, and Th oxides.)

$$N_f = \frac{\rho_f N_o}{(A.W.)} \qquad \text{18.113}$$

$$\Sigma_{a,f} = N_f \sigma_{a,f} \qquad \text{18.114}$$

$$\Sigma_{tr,f} = \Sigma_{a,f} + \Sigma_{s,f} \qquad \text{18.115}$$

step 6: Calculate the thermal utilization. V is the volume per unit length. (Note: All fluxes are average values.)

$$f = \frac{\overline{\phi}_f \Sigma_{a,f} V_f}{\overline{\phi}_f \Sigma_{a,f} V_f + \overline{\phi}_m \Sigma_{a,m} V_m} \qquad \text{18.116}$$

$$\left(\frac{1}{f}\right) = 1 + \left(\frac{\overline{\phi}_m}{\overline{\phi}_f}\right)\left(\frac{\Sigma_{a,m}}{\Sigma_{a,f}}\right)\left(\frac{V_m}{V_f}\right) \qquad \text{18.117}$$

The *thermal disadvantage factor* is

$$\left(\frac{\overline{\phi}_m}{\overline{\phi}_f}\right) = F + \left(\frac{V_f \Sigma_{a,f}}{V_m \Sigma_{a,m}}\right)(E-1) \approx 1 \qquad \text{18.118}$$

The *lattice functions* are

$$F \approx 1 + \frac{1}{2}\left(\frac{r_o}{2L_f}\right)^2 - \left(\frac{1}{12}\right)\left(\frac{r_o}{2L_f}\right)^4 + \left(\frac{1}{48}\right)\left(\frac{r_o}{.2L_f}\right)^6 \quad \text{18.119}$$

$$E \approx 1 + \frac{r_1^2}{2L_m^2}\left[\frac{\ln\left(\frac{r_1}{r_o}\right)}{1 - \left(\frac{r_o}{r_1}\right)^2} - \frac{3}{4} + \left(\frac{r_o}{2r_1}\right)^2\right] \qquad \text{18.120}$$

step 7: Calculate the resonance escape probability.

$$p = \exp\left[-\left(\frac{V_f}{V_m}\right)\left(\frac{1}{\xi_m}\right)\left(\frac{N_f}{N_m}\right)\left(\frac{I_{re}}{\sigma_{s,m}}\right)\left(\frac{\overline{\phi}_f}{\overline{\phi}_m}\right)\right]$$

$$\approx \exp\left[-\left(\frac{V_f}{V_m}\right)\left(\frac{1}{\xi_m}\right)\left(\frac{\Sigma_{re}}{\Sigma_{s,m}}\right)\right]$$

$$= \exp\left[-\left(\frac{V_f}{V_m}\right)\left(\frac{1}{\xi_m}\right)\frac{(N_f I_{re})}{\Sigma_{s,m}}\right] \qquad \text{18.121}$$

I_{re} is the *effective resonance integral* with units of barns.

$$I_{re} = A + B\left(\frac{S_f}{m_f}\right) = C + \left(\frac{D}{\sqrt{r_o \rho_f}}\right) \qquad \text{18.122}$$

S_f is the fuel rod cross sectional area, $2\pi r_o$, and m_f is the fuel rod mass, $\pi r_o^2 \rho_f$. Values of A, B, C, and D are taken from table 18.19. (See table 18.15 also.)

step 8: Calculate the fast fission factor from the *Spinrad formula.*

$$\epsilon = 1 + .4853R\left(\frac{1 - 3.726R}{1 - 2.431R}\right) \qquad \text{18.123}$$

R is the ratio of fast to thermal fissions. p_c is the *collision probability.*

$$R = \frac{.561 p_c N_c\left(\frac{\sigma_f}{\sigma_t}\right)}{1 - p_c\left(\frac{.561 \sigma_f N_c + I_{re}}{\sigma_t}\right)} \qquad \text{18.124}$$

step 9: Calculate infinite-k.

$$k_\infty = \epsilon \eta p f \qquad \text{18.125}$$

step 10: Calculate the diffusion length.

$$L = \sqrt{\frac{1}{(3\Sigma_a \Sigma_{tr})}} \qquad \text{18.126}$$

The cross sections in equation 18.126 are flux and volume weighted averages.

$$\Sigma = \frac{\phi_f V_f \Sigma_f + \phi_m V_m \Sigma_m}{\phi_f V_f + \phi_m V_m} \qquad \text{18.127}$$

step 11: Calculate the *corrected age*. This correction ignores the effects of inelastic scattering.

$$\tau_{\text{corrected}} = \tau_{\text{reference}}\left(\frac{\rho_{m,\text{ref}}}{\rho_{m,\text{act}}}\right)^2 \qquad \text{18.128}$$

Table 18.19
Moderator/Fuel Properties

Fuel	A	B	C	D	$(\xi\Sigma_{s,m})$	$\Sigma_{s,f}$	ρ	Σ_{tr}	L
U-238 metal			2.8	38.3			19.0		
U-natural									1.55
238 UO_2	9.25	24.7	3.0	39.6		.4	.18		
U_3O_8									3.7
Th-232			3.9	20.9			11.5		
232 ThO_2			3.4	24.5					
water					1.46		1.0		
D_2O					.178		1.1		
Be					.155		1.82		
C (graphite)					.0608		1.6–2.7		

step 12: Calculate the buckling. A good approximation for k_∞ near one is

$$B^2 \approx \frac{\ln(k_\infty)}{L^2 + \tau} \qquad 18.129$$

This can be used to find B^2 by trial and error with

$$k = 1 = \frac{k_\infty \exp(-B^2 \tau)}{1 + B^2 L^2} \qquad 18.130$$

step 13: Calculate the reactor size using the critical size equations and the value of buckling.

19 BREEDER REACTORS AND SYNTHETIC FUELS

Fissionable material is produced in a breeder reactor by placing a blanket of fertile material (U-238 or Th-232) around a core of concentrated fissionable material, such as Pu-239 or U-233. Core neutrons are captured by the blanket to breed more fissionable fuel.

$$_{92}U^{238} + _0n^1 \xrightarrow{\text{fast}} {}_{92}U^{239} \xrightarrow[\beta-]{(23 \text{ min})} {}_{93}Np^{239} \xrightarrow[\beta-]{(2.3 \text{ days})} {}_{94}Pu^{239}$$

$$_{90}Th^{232} + _0n^1 \xrightarrow{\text{fast}} {}_{90}Th^{233} \xrightarrow[\beta-]{(23 \text{ min})} {}_{91}Pa^{233} \xrightarrow[\beta-]{(27 \text{ days})} {}_{92}U^{233}$$

An approximate formula for the *doubling time* (the time to double the number of plutonium atoms) given the original plutonium mass in grams, the reactor power in megawatts, the fuel consumption per unit power, m' (typically 1.23 to 1.3 grams/MW-day), and the conversion ratio (CR), is

$$\text{linear } t_d = \frac{m}{(CR - 1)(m')(P)} \quad \text{(days)} \quad 18.131$$

$$\text{exponential } t_d = (.693)(\text{linear } t_d) \quad \text{(days)} \quad 18.132$$

The *linear doubling time* assumes that all excess plutonium remains in the reactor. The *exponential doubling time* assumes that excess plutonium is removed regularly and is placed in a second breeder reactor.

The *conversion ratio*, CR, is the ratio of the number of plutonium atoms created to the number used. CR-1 is the *breeding gain*. CR will be less than 1 for conversion operation and greater than 1 for breeding.

$$CR = (\eta\epsilon - l_f - l_t - 1)$$

$$\approx (\eta\epsilon - 1) \text{ for larger reactors} \qquad 18.133$$

If the core load is a composite of two components (e.g., U-238 and Pu-239), the value of η is weighted.

$$\eta = (\%U)n_U\left(\frac{\sigma_{f,U}}{\sigma_{a,U}}\right) + (\%Pu)n_{Pu}\left(\frac{\sigma_{f,Pu}}{\sigma_{a,Pu}}\right)$$

$$= (\%U)\eta_U + (\%Pu)\eta_{Pu} \qquad 18.134$$

20 FUSION

If two or more light atoms have sufficient energy (available only at high temperatures), they can fuse together to form one or more heavier atoms. Because of the high temperatures required, the reaction is called a *thermonuclear reaction*. During fusion, mass is lost and converted to energy, according to equation 18.1.

Cold stars (2 EE6 °K) produce energy by the proton-proton cycle. The *proton-proton* cycle is

$$_1H^1 + _1H^1 \mapsto {}_1H^2 + .4 \text{ MeV}$$
$$_1H^1 + _1H^2 \mapsto {}_2He^3 + 5.5 \text{ MeV}$$
$$_2He^3 + _2He^3 \mapsto {}_2He^4 + 2(_1H^1) + 12.9 \text{ MeV}$$

As each of the first two reactions must occur twice for each $_2He^4$ produced, the total energy is more than apparent. The energy released per helium atom production after neutrino energy is subtracted is approximately 26.2 MeV. Hotter stars fuse from the *carbon cycle*, which produces about 24.7 MeV per cycle.

The most promising commercial cycles are

$$_1H^2 + _1H^2 \mapsto {}_1H^3 + _1H^1 + 4.0 \text{ MeV}$$
$$_1H^2 + _1H^2 \mapsto {}_2He^3 + _0n^1 + 3.3 \text{ MeV}$$

Both reactions have approximately equal probability of occurring. However, the high temperatures (EE6 °K) required and the confinement problems make commercial fusion difficult. Recent advances have replaced the thermal energy requirements with kinetic energy from high-velocity particles. Very short-duration fusion reactions have been obtained by this method.

Appendix A
Nuclear Conversion Factors

1 amu	1.66053×10^{-24} gm
	931.481 MeV
1 curie	3.7×10^{10} disintegrations/sec
1 coulomb	2.998×10^{9} esu
1 erg	1 gm cm^2/sec^2
	0.6242×10^{6} MeV
1 eV	1.60210×10^{-12} erg
	1.517×10^{-22} BTU
	4.44×10^{-26} kW-hr
	1.60219 EE$-$19 joule
1 fission	$\sim$ 200 MeV (total)
	8.9×10^{-18} kW-hr (total)
	$\sim$ 180 MeV (in fuel)
1 joule	10^7 ergs
1 hp	2545 BTU/hr
	0.7457 kW
	550 ft-lbf/sec
1 unit electronic charge	4.77×10^{-10} esu
1 watt	1 joule/sec
	3.413 BTU/hr

PRACTICE PROBLEMS: NUCLEAR ENGINEERING

Warm-ups

1. Radioactive sodium emits 2.75 MeV gammas to initiate the photodisintegration of deuterium.

$$\gamma +_1 D^2 \mapsto _1 H^1 +_0 N^1$$

What will be the kinetic energy of the neutron?

2. The half-life of cesium-132 is approximately 6.47 days. How long will it take to reduce its activity to 5% of its original activity?

3. A 2-curie source emits 2 MeV gammas. How thick must a lead shield be constructed to reduce the activity level to 1%?

4. If a 10 cm thick lead plate shields a 2 MeV gamma source (EE6 rays/cm^2-s), calculate the exit flux and the exposure rate.

5. A neutron flux of $J = $ EE8 neutrons/cm^2-s irradiates a 50°C gold foil for 24 hours. If the foil diameter and volume are 2.5 cm and .4909 cm^3, respectively, what is the removal activity?

6. If a point source (EE7 neutrons/s) is located in 20°C water, what is the neutron flux 20 cm away?

7. What is the individual neutron probability of causing fission in natural uranium by 20°C neutrons?

8. What thickness of spherical iron shield must be used if an isotropic 1 MeV gamma source (EE8 gammas/s) must be reduced to 1 mr/hr?

Concentrates

1. What is the exponential doubling time of a 1000 MW breeder reactor containing 2000 kg of 20% Pu-239 and 80% U-238?

2. Given that $\Sigma_f = .005$ cm^{-1}, what is the power generated in a bare spherical reactor of 40 cm radius if the maximum flux is 4.5 EE15 neutrons/cm^2-s?

3. What is the thermal utilization of a graphite moderated square lattice of 2.04 cm diameter natural uranium rods with a lattice pitch of 25.4 cm?

Timed (1 hour allowed for each)

1. A nuclear reactor produces 500,000 kw of thermal energy. The size of the core is 10' diameter by 10' height. It contains 100,000 lbs of uranium enriched to 2% in U^{235}. If the fission cross section of U^{235} is 547 barns, what is the average thermal neutron flux (neutrons/cm^2-sec) in the reactor required to sustain the rate of energy production?

RESERVED FOR FUTURE USE

19

MODELING OF ENGINEERING SYSTEMS

PART 1: Mechanical Systems

Nomenclature

a	acceleration
B	coefficient of damping
F	force
k	spring constant (stiffness)
m	mass
t	time
x	position
v	velocity

1 MECHANICAL ELEMENTS

A *mechanical system* consists of interconnected masses, springs, dampers,[1] and energy sources. System modeling is used to predict the performance of a mechanical system without actually observing the system in operation.

A *lumped mass* is a rigid body which behaves like a particle. All parts of the mass experience identical velocities, accelerations, and displacements. The lumped mass is *ideal* if Newtonian physics applies (i.e., there are no relativistic changes in mass). The performance of ideal lumped masses is predicted by *Newton's second law*, equation 19.1.

$$F(t) = m\,a(t) \qquad 19.1$$

Various symbols are used to diagram masses in system models. Typical symbols are illustrated in figure 19.1.

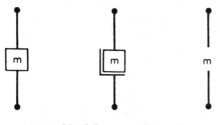

Figure 19.1 Ideal Lumped Mass Symbols

---

[1]Dampers also are known as *dashpots*. An automobile shock absorber is an example of a damping device.

An ideal *spring* is massless, has a constant *stiffness*, and is immune to set, creep, and fatigue. The performance of a spring is predicted by *Hooke's law*, equation 19.2.

$$F(t) = k[x_1(t) - x_2(t)] \qquad 19.2$$

x_1 and x_2 in equation 19.2 are displacements from the initial equilibrium positions of the two spring ends. If both ends are displaced the same distance, there will be no net extension or compression. Consequently, the spring will not experience a force.

Typical symbols used to diagram springs are illustrated in figure 19.2.

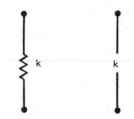

Figure 19.2 Ideal Spring Symbols

An ideal *damper* is massless, has no stiffness, and behaves linearly according to equation 19.3. B is the damping coefficient. The damping force is proportional to the difference in velocities of the two damper ends.

$$F(t) = B[v_1(t) - v_2(t)] \qquad 19.3$$

Typical symbols used to diagram dampers are illustrated in figure 19.3.

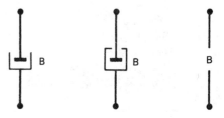

Figure 19.3 Ideal Damper Symbols

2 RESPONSE VARIABLES

A dependent variable which predicts the performance of a mechanical system is known as a *response variable*.[2] Position, velocity, and acceleration vary with time and are the dependent response variables. Time is usually the independent variable.

Springs and dashpots are *two-port devices*. They have two ends, each of which can have a different value of the response variable. For example, the velocity of both ends of a damper need not be the same. Even though masses do not have ends in the traditional sense of the word, masses also are considered to be two-port devices. Rules for assigning values of response variables to these ends are covered later in this chapter.

The goal of system modeling is to determine the *response* of a system. This means finding the position, the velocity, and the acceleration of each part of the system as a function of time.

It is not necessary to work separate problems to find $x(t)$, $v(t)$, and $a(t)$ functions. If $x(t)$ is known, it can be differentiated consecutively to find $v(t)$ and $a(t)$. If any one of the three response functions is known, the other two can be derived.

If a system can be defined completely by a single response variable (e.g., x), it is known as a *single-degree-of-freedom (SDOF) system*. Examples of SDOF systems are a mass on a spring, a swinging pendulum, and a rotating pulley. In each case, only one variable (x or θ) is needed to define the position of the primary system object.

Systems in which multiple components can take on independent values (i.e., positions) are known as *multiple-degree-of-freedom (MDOF) systems*. The number of response variables needed is equal to the number of independent system objects. This number is known as the *degree of freedom* of the system.

3 ENERGY SOURCES

Masses, springs, and dampers are known as *passive devices* because they dissipate and absorb energy. Energy is required to start a mechanical system oscillating. Energy sources are known as *active components*.

An energy source need not be an actual component such as a battery, a fuel cell, or a wound spring. Anything that produces motion in the system is an energy source. Mechanical system models usually consider the energy source to be a pure force, without regard to the origin of that force.

---

[2]Response variables also are known as *state variables*.

Force is known as a *through variable* since it can be passed through objects. Consider a mass on a string. The gravitational force on the mass is passed through the string to the support. Through variables have the same magnitude at both ends of an element.

The velocity attained by a mass acted upon by a force will depend on the magnitude of the force. Another type of energy source, a *velocity source*, produces a specific velocity regardless of the system mass. Velocity is known as an *across variable* since it must be measured with respect to another reference frame. Across variables have different magnitudes at the two ends of an element.

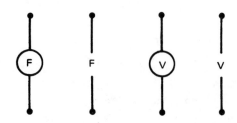

Figure 19.4 Symbols for Energy Sources

4 SYSTEM DIAGRAMS

The system elements are interconnected to produce the system diagram. This diagram then is used to write the differential equations which describe the response of the system.

step 1:   Choose a response variable. Although v is a common choice, x or a can be used.

step 2:   Identify all parts of the mechanical system which have unique values of the chosen response variable. It is not necessary to know the actual values of the response variable. It is necessary to know only which parts have different values.

step 3:   **On paper, draw horizontal lines and label** these lines with the different values of the response variable.

step 4:   Insert the passive elements (m, B, or k) between the appropriate horizontal lines. Connect the elements to the horizontal lines with vertical lines.

rule 19.1:   Masses always connect to the lowest (ground) level.

step 5:   Insert the active sources (F or v) between the appropriate horizontal lines and connect with vertical lines.

rule 19.2:   Energy sources always connect to the lowest (ground) level.

Example 19.1

A mass is connected through a damper to a solid wall as shown. The mass slides without friction on its support. A force is applied to the mass. What is the system diagram?

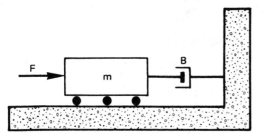

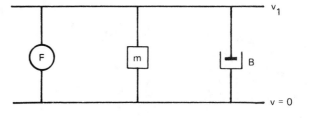

step 1: Choose velocity as the response variable.

step 2: All parts of the mass move with the same velocity. Call this velocity v_1. The plunger also moves with velocity v_1 since it is attached to the mass. The body of the damper is attached rigidly to the wall, which remains stationary. Thus, there are two unique velocities in this system: v_1 and $v = 0$.

step 3: Two horizontal lines are drawn. The top line is associated with v_1. The lower line is associated with $v = 0$.

_____ v_1

_____ $v = 0$

step 4: One end of the damper travels at v_1; the other is stationary. Therefore, insert the damper symbol so that its two ends connect to the v_1 and $v = 0$ lines. The mass moves at v_1, so one of its vertical lines should connect to v_1. By rule 19.1, the other line connects to $v = 0$.

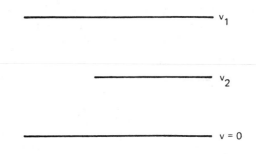

step 5: The force contacts the mass which moves at v_1. Therefore, one end of the force symbol contacts to v_1. By rule 19.2, the other line connects to $v = 0$.

Example 19.2

A force is applied through a damper to a mass. The mass rolls on frictionless bearings. What is the system diagram?

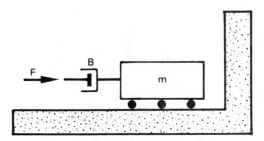

step 1: Choose velocity as the response variable.

step 2: The damper plunger moves. Call its velocity v_1. The mass moves. Call its velocity v_2. The support does not move $(v = 0)$.

step 3: Draw three lines corresponding to v_1, v_2, and $v = 0$.

_____ v_1

_____ v_2

_____ $v = 0$

step 4: One end of the damper moves at v_1; the other moves at v_2. Connect the damper between lines v_1 and v_2. The mass moves at v_2 and connects to $v = 0$ by rule 19.1.

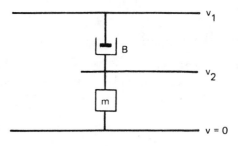

step 5: The force contacts the damper plunger moving at v_1 and connects to $v = 0$ by rule 19.2.

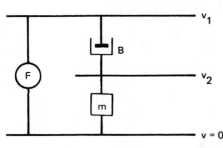

Example 19.3

A high-performance shock absorber is constructed with an integral coil spring as shown. What is the system diagram if a force is applied to one end?

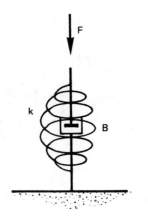

step 1:   Choose velocity as the response variable.

step 2:   The plunger and the top end of the spring move with the same velocity, v_1. The damper body and the lower end of the spring do not move ($v = 0$).

steps 3, 4, and 5:

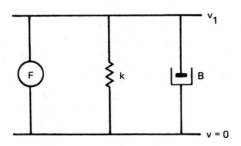

5 LINE DIAGRAMS

Line diagrams are similar to system diagrams. However, the horizontal lines are replaced by nodes, and the element symbols are removed, leaving only their variables.

Example 19.4

Draw the system and line diagrams for the spring and the dashpot shown.

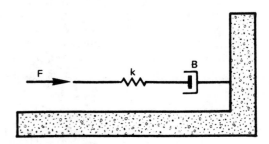

The system diagram is

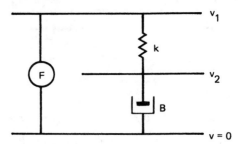

The line diagram is

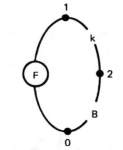

6 SYSTEM EQUATIONS

Once the system diagram is drawn, the *system equations* can be derived. The system equations describe the response variables as functions of the independent variable.

The system equations derived from the system diagram do not give the response variables explicitly. Rather, the system equations are differential equations which must be solved before the response functions can be used. Solution procedures for differential equations (e.g., traditional methods and Laplace transforms) are covered in other chapters of this book.

The principle used to derive the system equations is analogous to Kirchoff's current law and other conservation laws. These conservation laws say "...what goes in must come out..." In the case of system modeling, the quantity being conserved is force. The total force being supplied by the energy source must equal the sum of the forces leaving through all parallel legs.

rule 19.3: The force passing through a leg consisting of series elements can be determined from the conditions across the first element in that leg.

rule 19.4: When writing a difference in a response variable (e.g., $v_2 - v_1$, $x_2 - x_1$), the first subscript is the same as the node number for which the equation is being written.

Example 19.5

What is the system equation for the coil and the shock absorber described in example 19.3?

The force is considered to be a quantity flowing up through its own leg. This force splits, with part of it (F_B) going down the first leg and the rest (F_k) going down the second leg.

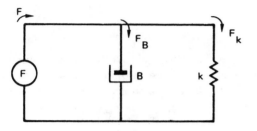

Since force is conserved, the system equation is

$$F(t) = F_B(t) + F_k(t) = Bv(t) + kx(t)$$

However, $v(t)$ is the first derivative of $x(t)$. Therefore, this is a first-order differential equation.

$$F(t) = Bx' + kx$$

Example 19.6

The coupling of a railroad car is modeled as the mechanical system shown. Assume all elements are linear. What are the system equations which describe the positions x_1 and x_2 as functions of time?

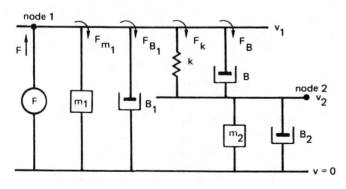

The system diagram is

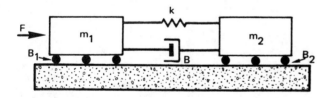

The force that enters node 1 is equal to the sum of forces leaving node 1. The system equation is

$$F = F_{m_1} + F_{B_1} + F_k + F_B$$

It is correct to write $F_{B_1} = B_1 v_1$ since the lower end of the damper has a zero velocity. However, it would be incorrect to write F_B in terms of v_1 only, since the two ends of damper B move with velocities v_1 and v_2.

Using rule 19.3, the force F_{B_1} can be written in terms of Δv. It may not be clear if Δv means $(v_1 - v_2)$ or $(v_2 - v_1)$. Since this system equation is being written for node 1, rule 19.4 requires that the difference be $(v_1 - v_2)$.

The system equation in terms of the response variables is

$$F = m_1 a_1 + B_1 v_1 + B(v_1 - v_2) + k(x_1 - x_2)$$

This can be written as a differential equation.

$$F = m_1 x_1'' + B_1 x_1' + B(x_1' - x_2') + k(x_1 - x_2)$$

This differential equation contains two variables—x_1 and x_2. A second differential equation is required. Once it is found, the two system differential equations can be solved simultaneously.

The second system equation can be found from node 2. The same principle used to write the node 1 system equation applies to node 2—the forces into and out of node 2 are equal. There is no force going into node 2 since there is no force source across the terminals on the right side of the system diagram.

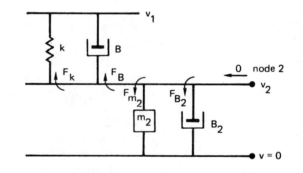

The node 2 system equation is

$$0 = F_{B_2} + F_{m_2} + F_k + F_B$$
$$= B_2 v_2 + m_2 a_2 + B\Delta v + k\Delta x$$

Using rule 19.4, Δv and Δx become $(v_2 - v_1)$ and $(x_2 - x_1)$.

$$0 = B_2 v_2 + m_2 a_2 + B(v_2 - v_1) + k(x_2 - x_1)$$
$$= B_2 x_2' + m_2 x_2'' + B(x_2' - x_1') + k(x_2 - x_1)$$

7 ENERGY TRANSFORMATION

Levers can be used to transform one force into another. This transformation is represented in system diagrams by the symbol for an electrical transformer.

Example 19.7

What are the system equations for the mechanical system shown?

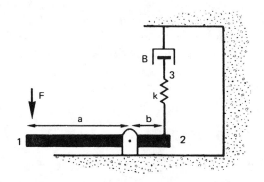

The lever transforms the force, displacement, velocity, and acceleration at point 1 into force, displacement, velocity, and acceleration at point 2. The system diagram is

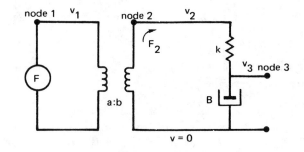

The system equations are

$$node\ 2: \quad F_2 = k(x_2 - x_3)$$
$$node\ 3: \quad 0 = k(x_3 - x_2) + B(x_3')$$

Since F_2, x_2, and x_3 are unknown, additional equations are needed. These additional equations are based on the lever's ratio of transformation.

$$x_2 = \left(\frac{b}{a}\right)x_1$$
$$F_2 = \left(\frac{a}{b}\right)F_1$$

PART 2: Rotational Systems

Nomenclature

a   ratio of transformation
B   coefficient of damping
J   rotational moment of inertia
k   rotational spring stiffness
N   number of teeth

Symbols

α angular acceleration
θ angular rotation
τ torque
ω angular velocity

1 ROTATIONAL ELEMENTS

Rotational elements are the rotational mass (flywheel), the rotational spring (torsional spring), and the rotational damper (fluid coupling).

When a *rotational mass* is acted upon by a torque, it behaves according to equation 19.4.

$$\tau = J\alpha \qquad\qquad 19.4$$

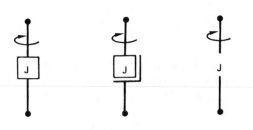

Figure 19.5 Symbols for Rotational Mass

The torque needed to twist a *rotational spring* is given by equation 19.5. k is the spring's *stiffness*.

$$\tau = k\Delta\theta \qquad\qquad 19.5$$

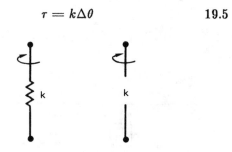

Figure 19.6 Symbols for Rotational Spring

The performance of a *rotational damper* is governed by equation 19.6.

$$\tau = B(\omega_2 - \omega_1) \qquad\qquad 19.6$$

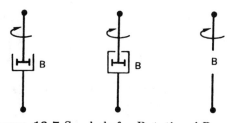

Figure 19.7 Symbols for Rotational Damper

Gearsets can be used to transform one torque into another. The ratio of speeds and torques depends on the number of teeth possessed by each gear.

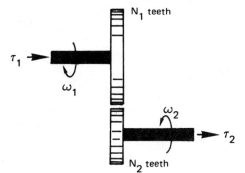

Figure 19.8 Rotational Transformer

The transformation ratio is

$$a = \frac{N_1}{N_2} \qquad\qquad 19.7$$

The transformed torques and speeds can be found from equation 19.8. The minus signs imply a direction change.

$$a = -\frac{\tau_1}{\tau_2} = -\frac{\omega_2}{\omega_1} \qquad\qquad 19.8$$

2 RESPONSE VARIABLES

The rotational response variables are angular displacement, θ, angular velocity, ω, and angular acceleration, α. These response variables usually are expressed in terms of *radians*, although *revolutions* also can be used.

As with translational mechanical systems, any of these three response variables can be used to write the system equations. The three variables are related. If one is known, the others can be found by integration or differentiation.

3 ENERGY SOURCES

Energy can be provided to rotational systems by constant-torque sources or constant-velocity sources. Symbols for these energy sources are shown in figure 19.9.

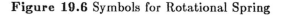

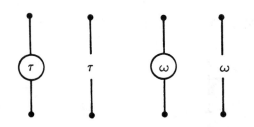

Figure 19.9 Symbols for Rotational Energy Sources

Example 19.8

Two rotating flywheels are connected by a flexible shaft with known stiffness. The second flywheel is acted upon by a viscous force proportional to the velocity. Draw the system diagram and write the differential equations of motion.

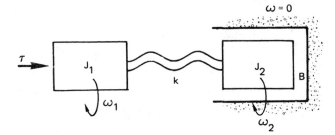

step 1:   Choose ω as the response variable.

step 2:   There are three different rotational speeds: ω_1, ω_2, and $\omega = 0$.

steps 3, 4, and 5:

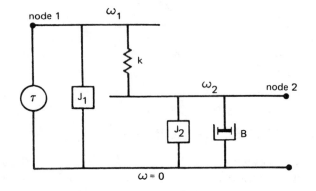

The system equations in terms of the response variables are

$$node\ 1: \quad \tau = J_1\alpha_1 + k(\theta_1 - \theta_2)$$
$$node\ 2: \quad 0 = B\omega_2 + J_2\alpha_2 + k(\theta_2 - \theta_1)$$

These simultaneous equations can be written as differential equations.

$$node\ 1: \quad \tau = J_1\theta_1'' + k(\theta_1 - \theta_2)$$
$$node\ 2: \quad 0 = B\theta_2' + J_2\theta_2'' + k(\theta_2 - \theta_1)$$

Example 19.9

A motor drives a flywheel through a set of reduction gears. The flywheel is connected to the driven gear by a flexible shaft. What are the system equations?

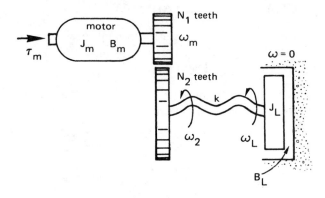

The system diagram uses the symbol for an electrical transformer to represent the gear set. The transformation ratio is $a = (N_1/N_2)$, a number less than *one*.

$$node\ m: \quad \tau_m = J_m\alpha_m + B_m\omega_m + a\tau_2$$
$$node\ 2: \quad \tau_2 = k(\theta_2 - \theta_L)$$
$$node\ L: \quad 0 = B_L\omega_L + J_L\alpha_L + k(\theta_L - \theta_2)$$

Since ω_m, ω_2, ω_L, and τ_2 all are unknown, a fourth equation, $a\omega_m = \omega_2$ is needed.

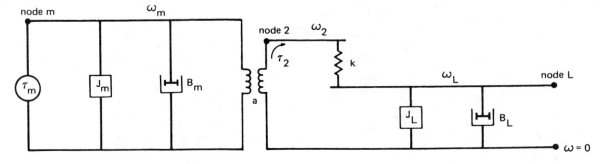

System Diagram for *Example 19.9*

PART 3: Electrical Systems

Nomenclature

C   capacitance
I   current
L   inductance
R   resistance
t   time
V   voltage

1 ELECTRICAL ELEMENTS

The passive elements in an electrical circuit are the resistor, the capacitor, and the inductor. Their governing equations are presented here.

$$\text{resistor}: \quad I = \frac{1}{R}(V_1 - V_2) \qquad 19.9$$

$$\text{capacitor}: \quad I = C\frac{d(V_1 - V_2)}{dt} \qquad 19.10$$

$$\text{inductance}: \quad I = \frac{1}{L}\int (V_1 - V_2)dt \qquad 19.11$$

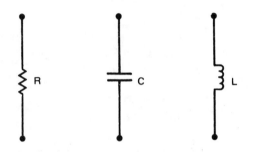

Figure 19.10 Passive Electrical Symbols

2 RESPONSE VARIABLES

Voltage is the response variable chosen when working with electrical systems. The integral and the derivative of voltage commonly are not encountered.

3 ENERGY SOURCES

Voltage and current sources both can be used with electrical systems. Current is a *through variable* since it passes through the circuit elements. Voltage is an *across variable* since it must be measured across two terminals.

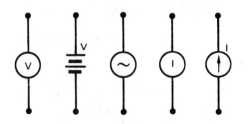

Figure 19.11 Electrical Energy Source Symbols

4 SYSTEM DIAGRAMS

The system diagram and the electrical current are identical. Therefore, it is not necessary to draw a different system diagram if the electrical circuit is shown.

Example 19.10

What are the system equations for the electrical circuit shown?

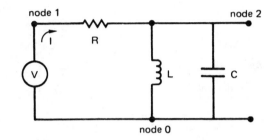

Nodes 1 and 2 have different potentials (voltages).

$$\text{node 1}: \quad I = \frac{1}{R}(V_1 - V_2)$$

$$\text{node 2}: \quad 0 = C\frac{dV_2}{dt} + \frac{1}{L}\int V_2 dt + \frac{1}{R}(V_2 - V_1)$$

Example 19.11

What are the system equations for the electrical circuit shown?

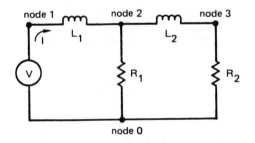

Nodes 0, 1, 2, and 3 all have different potentials.

$$\text{node 1}: \quad I = \frac{1}{L_1}\int (V_1 - V_2)dt$$

$$\text{node 2}: \quad 0 = \frac{V_2}{R_1} + \frac{1}{L_2}\int (V_2 - V_3)dt$$

$$+ \frac{1}{L_1}\int (V_2 - V_1)dt$$

$$\text{node 3}: \quad 0 = \frac{1}{L_2}\int (V_3 - V_2)dt + \frac{V_3}{R_2}$$

PART 4: Fluid Systems

Nomenclature

A   area
C_f fluid capacitance
F   force
h   height
I   fluid inertance
L   length
m   mass
p   pressure
Q   volumetric flow rate
R_f fluid resistance
t   time
v   velocity

Symbols

ρ fluid density

Just as electrical capacitors store charge, *reservoirs* (*fluid capacitors*) store fluid. Figure 19.12 shows a reservoir with constant cross sectional area.

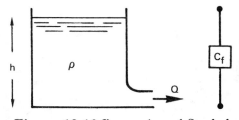

Figure 19.12 Reservoir and Symbol

The flow quantity can be calculated from the reservoir cross sectional area and the rate of change in height.

$$Q = A\frac{dh}{dt} \qquad 19.12$$

Since $h = p/\rho$, equation 19.12 can be rewritten as equation 19.13.

$$Q = \frac{A}{\rho}\frac{dp}{dt} = C_f\frac{dp}{dt} \qquad 19.13$$

C_f is known as the *fluid capacitance*. Open reservoirs always connect to the lowest (ground) level.

Flow resistance (friction loss) is a fluid element which must be considered with long pipes, porous plugs, and flow through orifices. The flow quantity is proportional to the pressure drop across the pipe length.

$$Q = \frac{1}{R_f}(p_2 - p_1) \qquad 19.14$$

R_f is the *fluid resistance coefficient*. Its form depends on the type of flow, the quantity of flow, and the physical configuration. In most cases, the pressure drop is proportional to the *square* of the flow quantity.

Therefore, equation 19.14 is an approximation which is valid over limited ranges of the flow quantity. For that reason, R_f usually is found by direct experimentation.

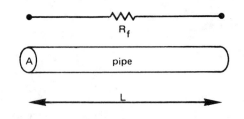

Figure 19.13 Fluid Resistance Symbol

Fluid inertance accounts for the inertia of the fluid flow. The impulse-momentum principle (equation 19.15) is the basis for fluid inertance.

$$F\,dt = m\,dv \qquad 19.15$$

Equation 19.15 can be reorganized by dividing both sides by dt.

$$F = m\frac{dv}{dt} \qquad 19.16$$

F, m, and v in equation 19.16 can be replaced by equations 19.17, 19.18, and 19.19.

$$F = A(p_2 - p_1) \qquad 19.17$$
$$m = \rho AL \qquad 19.18$$
$$v = \frac{Q}{A} \qquad 19.19$$

Combining equations 19.16 through 19.19 produces equation 19.20.

$$A(p_2 - p_1) = \rho AL\left[\frac{1}{A}\left(\frac{dQ}{dt}\right)\right] \qquad 19.20$$

$$p_2 - p_1 = \frac{\rho L}{A}\left(\frac{dQ}{dt}\right) = I\left(\frac{dQ}{dt}\right) \qquad 19.21$$

I is known as the *fluid inertance*. It also is known as the *fluid inductance*.

Integrating both sides of equation 19.21 results in an expression for the flow quantity.

$$\int (p_2 - p_1)dt = I\,dQ \qquad 19.22$$

$$Q = \frac{1}{I}\int (p_2 - p_1)dt \qquad 19.23$$

Figure 19.14 Fluid Inertance Symbol

Example 19.12

A pump is used to keep a liquid flowing through a filter pack as shown. What is the system equation? (Neglect the pipe friction.)

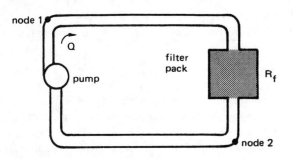

step 1: Pressure is the response variable.

step 2: Neglecting pressure drop in the pipe, there are two different pressures in the system: p_1 before the filter pack and p_2 after the filter pack.

steps 3, 4, and 5:

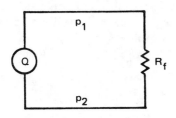

The system equation is

$$Q = \frac{1}{R_f}(p_2 - p_1)$$

Example 19.13

The reservoir shown discharges through a long pipe. The reservoir is not refilled. What is the system equation?

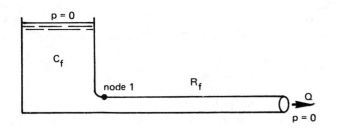

Although there is flow, there is no energy source (e.g., pump) in the system. The system diagram is

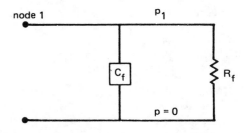

The system equation is

$$0 = \frac{p_1}{R_f} + C_f\left(\frac{dp_1}{dt}\right)$$

Example 19.14

A pump is used to transfer liquid from one reservoir to another as shown. What are the system equations? (Neglect pipe friction.)

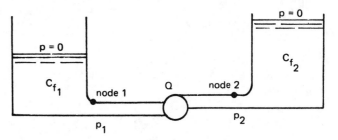

Since open reservoirs always connect to the lowest level, the system diagram is

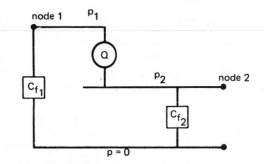

The system equations are

$$node\ 1: \quad Q = C_{f1}\left(\frac{dp_1}{dt}\right)$$

$$node\ 2: \quad Q = C_{f2}\left(\frac{dp_2}{dt}\right)$$

PART 5: Analysis of Engineering Systems

Nomenclature

a	imaginary part
B	bandwidth
E	error
F	forcing function
G	forward transfer function
h	height of pulse
H	reverse transfer function
j	square root of -1
K	scalar
Q	quality factor
r	real part
R	response function
s	Laplace variable
S	sensitivity
t	time

Symbols

ω	angular frequency

Subscripts

f	feedback
n	natural
o	output
s	signal

1 NATURAL, FORCED, AND TOTAL RESPONSE

If motion or change is induced in an engineering system, the system equation can be solved to determine the response function. This function gives the condition (e.g., position, voltage, pressure) as a function of time. The response function can have constant terms, sine and cosine terms, exponential terms, or multiplicative combinations of sinusoids and exponentials.

Natural response is induced when energy is applied to an engineering system and subsequently is removed. The system is left alone and is allowed to do what it would do naturally, without the application of further disturbing forces.

Natural response is characterized by pure sine and cosine terms in the system equation if friction is absent. Frictionless systems oscillate continuously without decay. If friction is present, the natural terms will be products of sinusoids and exponentials.

If a system is acted upon by a force which repeats at regular intervals, the system will move in accordance with that force. This is known as *forced response*. It is characterized by sinusoidal terms having the same frequency as the forcing function.

The natural and forced responses are present simultaneously in forced systems. The sum of the two responses is known as the *total response*. Since the natural effects usually disappear after a few cycles, they also are known as *transient effects* or *transient response*.

2 FORCING FUNCTIONS

An equation which describes the introduction of energy into the system as a function of time is known as a *forcing function*. Although a wide variety of forcing functions are possible, engineering system problems are too complicated for manual calculations if not limited to the simpler types.

A *homogeneous forcing function* is a constant zero force. A zero forcing function does not preclude initial disturbance. For example, a spring/mass system which is displaced, released, and allowed to oscillate freely is an example of a homogeneous forcing function. The forcing function is homogeneous if it ceases to act after the system begins to move.

$$F(t) = 0 \qquad\qquad 19.24$$

A *unit step* is a forcing function which has zero magnitude up to a particular instant (say t_1) and a magnitude of *one* thereafter.

$$F(t) = \begin{cases} 0 & t < t_1 \\ 1 & t \geq t_1 \end{cases} \qquad 19.25$$

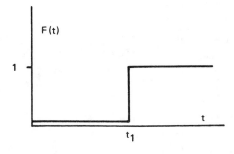

Figure 19.15 Unit Step

The unit step can be multiplied by a scalar if the actual force appears at t_1 with a magnitude other than *one*.

The *unit pulse* is a limited duration force whose total area under the curve is *one*. A necessary condition of a unit pulse is that the product of its magnitude and duration is *one*.

$$F(t) = \begin{cases} 0 & t < t_1 \\ \frac{1}{\Delta t} & t_1 \leq t < t_1 + \Delta t \\ 0 & t \geq t_1 + \Delta t \end{cases} \qquad 19.26$$

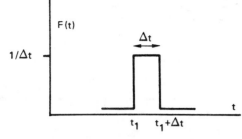

Figure 19.16 Unit Pulse

The unit pulse can be multiplied by a scalar if the actual pulse has a magnitude greater than $1/\Delta t$.

The terms *impulse* and *pulse* often are used interchangeably. However, *impulse* is more appropriate for very short impacts with large magnitudes. Hitting a bell with a hammer is illustrative of an impulse.

The most common forcing functions used in the analysis of engineering systems are the sine and cosine functions. Sinusoids, when combined with Fourier analysis, can be used to approximate all other forcing functions.

$$F(t) = \sin \omega t \qquad 19.27$$

3 TRANSFER FUNCTIONS

A system model can be thought of as a *black box*. The input to the black box is the forcing function. The output is the performance of the system, known as the *system response function*, $R(t)$. This analogy is particularly valid with electrical systems. The input voltage is applied across the input terminals. The output voltage is measured across the output terminals.

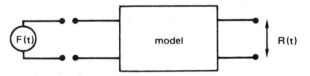

Figure 19.17 A Two-Port Black Box

The ratio of the system response (output) to the forcing function (input) is known as the *transfer function*[3], $T(t)$. The transfer function can be a simple scalar. However, it also can be a phasor (vector) and can possess both magnitude and phase change capability.

$$T(t) = \frac{R(t)}{F(t)} \qquad 19.28$$

Transfer functions generally are written in terms of the s variable.[4] This is accomplished, if $T(t)$ is known, by taking the Laplace transform of the transfer function.

---

[3]The transfer function also is known as the *rational function, system function, closed-loop function,* and *control ratio.*

[4]A review of the material on Laplace transforms in chapter 1 is recommended.

The result is the *transform of the transfer function.*[5]

$$T(s) = \mathcal{L}[T(t)] \qquad 19.29$$

Transforming $T(t)$ into $T(s)$ is more than a simple change of variables. The variable s can be thought of as an operator which performs differentiation. Specifically, x' could be written in terms of s as sx. Similarly, x'' could be written as $s^2 x$. The operation of integration can be represented by the reciprocal of s.

Example 19.15

Write the following system differential equation in terms of the s operator.

$$I(t) = \frac{1}{L} \int (V_1 - V_2)dt + C \frac{d(V_2 - \dot{V}_1)}{dt}$$

If all derivatives are replaced by s and the integral is replaced by $1/s$, the transformed system equation is

$$I(s) = \frac{V_1}{sL} - \frac{V_2}{sL} + CsV_2 - CsV_1$$

Example 19.16

Determine the transformed transfer function for the electrical system shown.

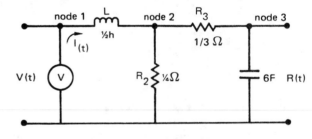

node 1:

$$I(t) = \frac{1}{L} \int (V_1 - V_2)dt$$

Substituting $\frac{1}{2}$ for L and using the s operator,

$$I(s) = \frac{2(V_1 - V_2)}{s} \qquad \text{Equation I}$$

node 2:

$$0 = \frac{1}{L} \int (V_2 - V_1)dt + \frac{1}{R_2}(V_2) + \frac{1}{R_3}(V_2 - V_3)$$

$$0 = \frac{2(V_2 - V_1)}{s} + 4V_2 + 3(V_2 - V_3) \qquad \text{Equation II}$$

node 3:

$$0 = C \frac{dV_3}{dt} + \frac{1}{R_3}(V_3 - V_2)$$

$$0 = 6sV_3 + 3(V_3 - V_2) \qquad \text{Equation III}$$

---

[5]$T(s)$ frequently is called just the *transfer function*. In actual practice, $T(t)$ almost never is encountered. It is acceptable, therefore, for $T(s)$ to share the name of $T(t)$.

The transfer function is the ratio of the output to the input. It does not depend on the intermediate voltage, V_2.

$$T(t) = \frac{V_3(t)}{V_1(t)}$$

Since $V_2(t)$ does not appear in $T(t)$, V_2 must be eliminated from equations **II** and **III**. (Equation **I** cannot be used unless the current is known. It generally is unknown.)

From equation **II**,

$$V_2 = \frac{2V_1 + 3sV_3}{2 + 7s}$$

From equation **III**,

$$V_2 = V_3(1 + 2s)$$

Setting these two expressions for V_2 equal, the ratio V_3/V_1 can be determined.

$$T(s) = \frac{V_3}{V_1} = \frac{1}{7s^2 + 4s + 1}$$

The black box which transforms a signal into the same output as produced by the original circuit is

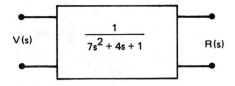

4 BLOCK DIAGRAM ALGEBRA

Example 19.16 illustrated how an engineering system can be modeled by a single operational black box. Several such boxes can be grouped together to obtain the desired response.

Complex systems of several box diagrams can be simplified by using the equivalent structures shown in table 19.1.

Example 19.17

A complex block system is constructed from five blocks and two summing points as shown.

Simplify the system and determine its overall effect on an input.

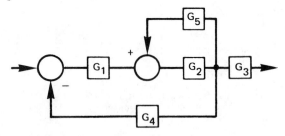

Use case 5 from table 19.1 to move the second summing point back to the first summing point.

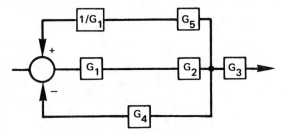

Use case 1 to combine boxes in series.

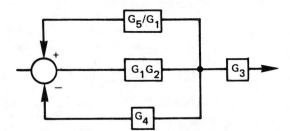

Use case 2 to combine the two feedback loops.

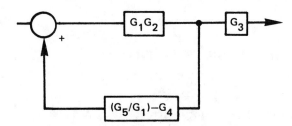

Use case 8 to move the pick-off point outside of the G_3 box.

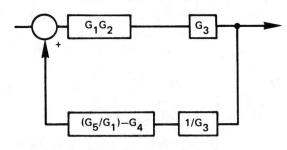

Use case 1 to combine the boxes in series.

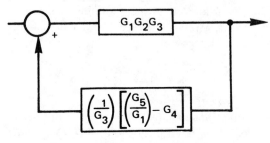

Use case 3 to determine the system gain.

$$G = \frac{G_1 G_2 G_3}{1 - (G_1 G_2 G_3)\left(\dfrac{1}{G_3}\right)\left[\left(\dfrac{G_5}{G_1}\right) - G_4\right]}$$

Table 19.1
Equivalent Block Diagrams

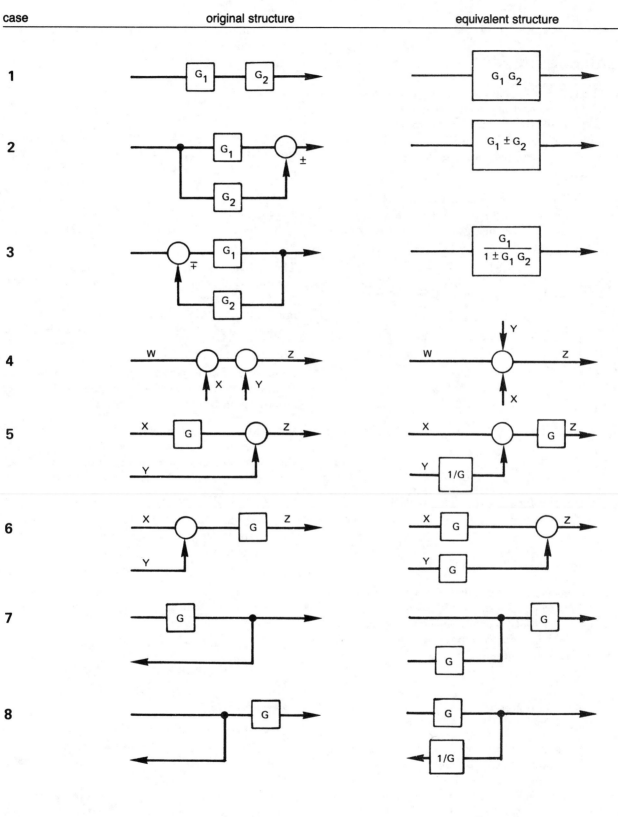

The system gain equation can be simplified.

$$G = \frac{G_1 G_2 G_3}{1 - G_2 G_5 + G_1 G_2 G_4}$$

5 FEEDBACK THEORY

If a small part of a device's output is returned to the input, the device is part of a *feedback system*. Feedback can be used to control a system or to improve the system's stability, efficiency, or sensitivity. Feedback also can decrease such negative effects as distortion and frequency dependence.

The basic feedback system consists of a *dynamic unit*, a *feedback element*, a *pick-off point*, and a *summing point*. The summing point is assumed to perform positive addition unless a minus sign is present.

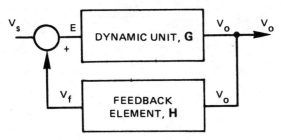

Figure 19.18 Positive Feedback System

The dynamic unit transforms E into V_o according to the *forward transfer function*, G.

$$V_o = GE \qquad\qquad 19.30$$

For amplifiers, the forward transfer function is known as the *direct* or *forward gain*. G can be a simple scalar if the dynamic unit merely scales its input. However, E, V_o, and G also can be phasors if the dynamic unit shifts the signal phase.

The difference between the signal and the feedback is known as the *error*. Whether addition or subtraction occurs in equation 19.31 depends on the summing point. Addition is used with additive summing points, subtraction with subtractive summing points.

$$E = V_s \pm V_f \qquad\qquad 19.31$$

E/V_s is known as the *error ratio*. V_f/V_s is known as the *primary feedback ratio*.

The pick-off point transmits V_o back to the feedback element. The output of the dynamic unit is not reduced by the pick-off point. As the picked-off signal travels through the feedback loop, it is acted upon by the *feedback* or *reverse transfer function*, H. H can be a simple scalar or a phasor. For typical control systems, $H = 1$ (i.e., $HV_o = V_o$), and there is no transformation at all.

The output of a positive feedback system is

$$V_o = GV_s + GHV_o \qquad\qquad 19.32$$

The *closed-loop transfer function* (also known as the *control ratio* or the *system function*) is the ratio of the output to the signal.[6]

$$G_{\text{loop}} = \frac{V_o}{V_s} = \frac{G}{1 - GH} \qquad\qquad 19.33$$

The *sensitivity*, S, of a feedback system is the percent change in the loop transfer function divided by the percent change in the forward transfer function.

$$S = \frac{\frac{dG_{\text{loop}}}{G_{\text{loop}}}}{\frac{dG}{G}} = \frac{1}{1 - GH} \qquad\qquad 19.34$$

The network is said to have *positive feedback* if the product, GH, is positive and less than *one*. In that case, the denominator in equation 19.33 will be less than *one*, making G_{loop} larger than G. Thus, a characteristic of a positive feedback system is an increase in the gain.

If the product, GH, approaches one, G_{loop} approaches infinity. Such performance may be undesirable. However, oscillators make use of this ability to produce an output in the absence of an input.

If the product, GH, is negative, the network is said to have *negative feedback*. Although the G_{loop} will be less than G, there may be other desirable effects. Some of these effects are listed here.

- reduction in sensitivity to temperature
- reduction in sensitivity to changes in the operating characteristics of the circuit components
- reduction in sensitivity to frequency of the signal
- reduction in sensitivity to noise and other variations in the signal
- improvement in the circuit's input and output impedences.[7]

Example 19.18

Three inverting amplifier stages with gains of -100 each are cascaded as shown. Feedback is provided by a resistor network. What is the overall gain with feedback?

---

[6]The *open-loop transfer function* is V_f/E.

[7]For circuits to be connected directly in series without affecting the performance of each, all input impedances must be infinite, and all output impedances must be zero.

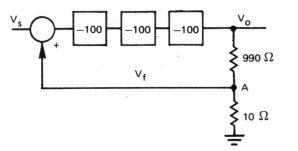

The two resistors form a voltage divider. The fraction of V_o fed back to the summing point is

$$H = \frac{10}{10 + 990} = .01$$

The forward transfer function is

$$G = (-100)^3 = -1,000,000$$

Without feedback, the output would be $(-1,000,000)V_s$. The minus sign indicates that the output is 180° out of phase with the input.

Since the summing point adds the feedback to the signal, equation 19.33 should be used to calculate the overall gain.

$$G_{\text{loop}} = \frac{-1,000,000}{1 - (-1,000,000)(.01)} \approx -100$$

Example 19.19

A closed-loop gain of -100 and stability of 99% is required from an inverting amplifier. The forward gain of the amplifier varies by 20%. What values of G and H satisfy the design requirements?

From equation 19.33,

$$G_{\text{loop}} = \frac{G}{1 - GH} = -100$$

The sensitivity requirement limits $dG_{\text{loop}}/G_{\text{loop}}$ to .01 when dG/G is .2. From equation 19.34,

$$.01 = \left[\frac{1}{1 - GH}\right](.2)$$

Solving for GH as as unknown results in $GH = -19$. Substituting GH into the equation for G_{loop} gives the value $G = +2000$. H is found to be $-.009$.

6 PREDICTING SYSTEM RESPONSE FROM TRANSFER FUNCTIONS

The transfer function alone is not sufficient to predict the response of a system. It also is necessary to know the input to the system. (The transfer function was derived without prior knowledge of the input.) The response of the system will depend on whether the input is homogeneous, a step function, sinusoid, etc.

The *transformed transfer function* is the ratio of the transformed response function to the transformed forcing function.

$$T(s) = \frac{\mathcal{L}[R(T)]}{\mathcal{L}[F(t)]} = \frac{R(s)}{F(s)} \qquad 19.35$$

Assuming that $T(s)$ and $\mathcal{L}[F(t)]$ are known, the response function can be found by performing an inverse transformation.

$$R(t) = \mathcal{L}^{-1}[R(s)] = \mathcal{L}^{-1}[\mathcal{L}[F(t)]T(s)] \qquad 19.36$$

If the extreme (initial and final) values of $R(t)$ are wanted, they can be found from the *initial value theorem* and the *final value theorem* for Laplace transforms.

$$R(0^+) = \lim_{s \to \infty} [sR(s)] \qquad 19.37$$
$$R(\infty) = \lim_{s \to 0} [sR(s)] \qquad 19.38$$

Example 19.20

What is the transformed forcing function for a step of height 8 occurring at $t = 0$?

The Laplace transform of a unit step is $(1/s)$. The Laplace transform of a step with height 8 is

$$F(s) = \mathcal{L}[F(t)] = \frac{8}{s}$$

Example 19.21

A mechanical system is acted upon by a constant force of 8 pounds starting at $t = 0$. What is the response if the transfer function is $T(s)$?

$$T(s) = \frac{6}{(s + 2)(s + 4)}$$

In example 19.20 $F(s)$ was found to be $(8/s)$. From equation 19.36,

$$R(t) = \mathcal{L}^{-1}\left[\left(\frac{8}{s}\right)\left(\frac{6}{(s + 2)(s + 4)}\right)\right]$$
$$= \mathcal{L}^{-1}\left[\frac{48}{s(s + 2)(s + 4)}\right]$$

$R(t)$ is found by taking the inverse transform. Using a table of Laplace transforms and recognizing that the product of linear terms in the denominator translates into a sum of exponential terms gives $R(t)$.

$$R(t) = 6 - 12e^{-2t} + 6e^{-4t}$$

The last two terms in $R(t)$ are decaying exponentials. They represent the *transient natural response*. The **6** in $R(t)$ does not vary with time. It is the *steady state response*. The sum of the transient and steady state responses is the *total response*.

Example 19.22

What is the final value of $R(t)$?

$$R(s) = \frac{1}{s(s+1)}$$

The limit is taken to zero for final values.

$$R(\infty) = \lim_{s \to 0}[sR(s)] = \lim_{s \to 0}\left[\frac{s}{s(s+1)}\right]$$
$$= \lim_{s \to 0}\left[\frac{1}{s+1}\right] = 1$$

Abbreviated methods are available if only the steady state response is wanted.

 unit step: The steady state response for a unit step can be obtained by substituting 0 for s everywhere in $T(s)$. If the step has a magnitude h, the steady state response must be multiplied by h.

 unit pulse: The steady state response for a unit pulse is the Laplace inverse of the transfer function. That is, a pulse has no long-term effect on an engineering system.

 sinusoids: The steady state response can be found by the following procedure.

  step 1: Substitute $j\omega$ for s in $T(s)$.
  step 2: Convert $T(j\omega)$ into phasor form.
  step 3: Convert the input sinusoid to phasor form.
  step 4: Multiply $T(j\omega)$ by the input phasor.

Example 19.23

Determine the steady state response when the system described in example 19.21 is acted upon by a step of height 8 at $t = 0$.

Substitute 0 for s in $T(s)$ and multiply by 8.

$$R(t)_{\text{steady state}} = 8\left[\frac{6}{(0+2)(0+4)}\right] = 6$$

Example 19.24

$4[\sin(2t + \frac{\pi}{4})]$ is applied as a sinusoidal input to a system with transfer function $T(s)$. What is the steady state response?

$$T(s) = \frac{-1}{7s^2 + 7s + 1}$$

step 1: The angular frequency is $\omega = 2$. Substituting $j2$ for s in $T(s)$,

$$T(j2) = \frac{-1}{7(j2)^2 + 7(j2) + 1}$$

Simplify this expression by recognizing that $j^2 = -1$.

$$T(j2) = \frac{-1}{-28 + 14j + 1} = \frac{-1}{14j - 27}$$

step 2: $(14j - 27)$ in phasor form is $30.4\angle152.6°$. The negative inverse of this is

$$T(j2) = \frac{-1}{30.4\angle152.6°} = -.033\angle - 152.6°$$

This system has a magnitude gain of $-.033$ and a phase shift of $-152.6°$.

step 3: The input phasor is $4\angle45°$.

step 4: The steady state phasor is

$$V(t) = (-.033\angle - 152.6°)(4\angle45°)$$
$$= -.132\angle - 107.6°$$

7 POLES AND ZEROS

A *pole* of the transfer function is a value of s which makes $T(s)$ infinite. Specifically, a pole is a value of s which makes the denominator of $T(s)$ zero. A *zero* of the transfer function makes the numerator of $T(s)$ zero. Poles and zeros can be real or complex quantities. Poles and zeros can be repeated within a given transfer function—they need not be unique.

A rectangular coordinate system based on the real-imaginary axes is known as an *s-plane*. If poles and zeros are plotted on the s-plane, the result is a *pole-zero diagram*.

Poles are represented on the pole-zero diagram as $\times$'s. Zeros are represented as $\bigcirc$'s.

Example 19.25

Draw the pole-zero diagram for the transfer function, $T(s)$.

$$T(s) = \frac{(5)(s+3)}{(s+2)(s^2 + 2s + 2)}$$

The numerator can be zero only if $s = -3$. This is the only zero of the transfer function.

The denominator can be zero if $s = -2$ or if $s = -1 \pm j$. These three values are the poles of the transfer function.

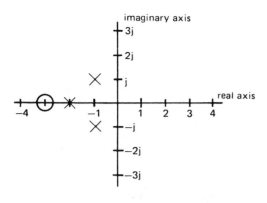

The pole-zero diagram can be used partially to find the transfer function, $T(s)$. The pole-zero diagram effectively gives the factors of the numerator and the denominator. It does not give any scalars (*scale factors*) in the numerator.

Example 19.26

A pole-zero diagram has a pole at $s = -2$ and a zero at $s = -7$. What is the transfer function?

$$T(s) = \frac{K(s+7)}{s+2}$$

The scalar K must be determined separately. This analysis assumes that the forcing function $F(s)$ has not been included in the pole-zero diagram.

8 FREQUENCY RESPONSE TO SINUSOIDAL FORCING FUNCTIONS

The gain and phase angle response of a system will change as ω is varied. The *gain characteristic* is the relationship between gain and frequency. The *phase characteristic* is the relationship between phase and frequency.

A. GAIN CHARACTERISTIC

A plot of gain versus frequency can be obtained from the following procedure.

step 1: Factor the denominator of the transfer function, $T(s)$, to determine its roots. Poles with imaginary parts correspond to values of $j\omega$ (in radians/sec) where the output peaks in magnitude.

step 2: Calculate the system gain for all values of the roots.

step 3: Set the input frequency to zero and determine the system gain.

step 4: Assume an infinite input frequency and calculate the system gain.

step 5: Choose several points around and close to each peak. Determine the system gain for these points.

step 6: Determine the *half-power points*. These are the frequencies for which the gain is .707 of the maximum values determined in step 2.

step 7: Determine the *bandwidth*. This is the difference between the upper and the lower half-power points.

step 8: Calculate the *quality factor*.

$$Q = \frac{\text{frequency at peak}}{\text{bandwidth}} \qquad 19.39$$

Figure 19.19 illustrates the gain characteristics of transfer functions with the form given by equation 19.40. (The coefficient of the s^2 term must be 1.)

$$T(s) = \frac{as + b}{s^2 + Bs + \omega_n^2} \qquad 19.40$$

The zero defined by a and b is not significant. ω_n is much larger than B, so that the pole is close to the imaginary axis. B will correspond to the bandwidth (in radians/sec). ω_n will be the *natural* or *resonant frequency* of the system, the frequency at which the gain peaks.

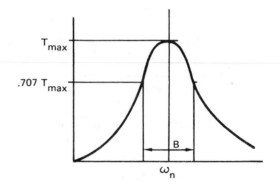

Figure 19.19 Gain Characteristic

Example 19.27

Predict the resonant frequency and the bandwidth for the transfer function, $T(s)$.

$$T(s) = \frac{s + 19}{s^2 + 7s + 1000}$$

The coefficient of the s^2 term is 1, and the form is the same as equation 19.40. Therefore, the bandwidth is 7 radians/sec, and the resonant frequency is $\sqrt{1000} = 31.6$ radians/sec.

B. PHASE CHARACTERISTIC

Figure 19.20 illustrates the phase characteristic for a transfer function with the form given by equation 19.40. Values of the angle versus the frequency are determined simultaneously with the derivation of the gain characteristic.

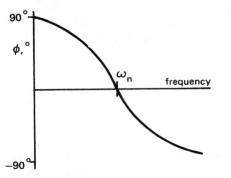

Figure 19.20 Phase Characteristic

9 PREDICTING SYSTEM RESPONSE FROM POLE-ZERO DIAGRAMS

Poles on the pole-zero diagram can be used to predict the usual response of engineering systems. Zeros are not used.

- *pure oscillation*: Sinusoidal oscillation will occur if a pole-pair[8] falls on the imaginary axis as in figure 19.21(a). A pole with a value of $\pm ja$ will produce oscillation with a natural frequency of $\omega = a$ radians/sec.

- *exponential decay*: Pure exponential decay is indicated when a pole falls on the real axis as in figure 19.21(b). A pole with a value of $-r$ will produce an exponential with time constant $(1/r)$.

- *damped oscillation*: Decaying sinusoids result from pole-pairs in the second and third quadrants of the s-plane. A pole-pair having the

[8]Poles off the real axis always occur in conjugate pairs.

value $r \pm ja$ will produce oscillation with natural frequency of

$$\omega_n = \sqrt{r^2 + a^2} \qquad 19.41$$

The closer the poles are to the real axis, the greater will be the damping effect. The closer the poles are to the imaginary axis, the greater will be the oscillatory effect.

Example 19.28

What is the natural response of a system with the pole-zero diagram shown?

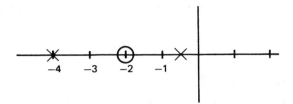

The poles are at $r = -\frac{1}{2}$ and $r = -4$. The response is

$$R(t) = C_1 e^{-0.5t} + C_2 e^{-4t}$$

C_1 and C_2 can be found by methods not covered in this book.

Since the response of an engineering system depends on the type of input (step, pulse, etc.), a pole-zero diagram based on $T(s)$ alone cannot be used to predict the system response. The system response must be determined from the product of the transfer function and the forcing function. This is equivalent to plotting $T(s)$ and $F(s)$ simultaneously on the pole-zero diagram.

Example 19.29

What is the system response if $T(s) = \dfrac{s+2}{s+3}$ and the input is a unit step?

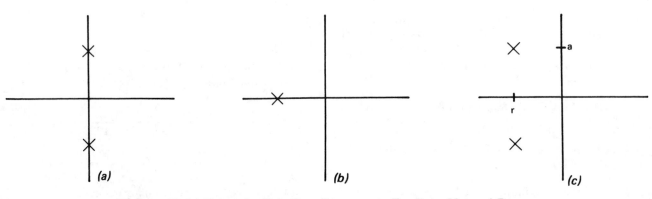

Figure 19.21 Using the Pole-Zero Diagram to Predict Natural Response

The response is found from equation 19.35.

$$R(s) = F(s)T(s) = \frac{s+2}{s(s+3)}$$

The pole-zero diagram is

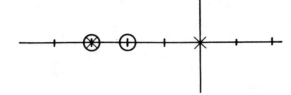

The pole at $r = 0$ contributes the exponential $C_1 e^{-0t}$ to the total response. Since this is equal to C_1, the total response is

$$R(t) = C_1 + C_2 e^{-3t}$$

10 ROOT-LOCUS DIAGRAMS

A root-locus diagram is a pole-zero diagram in which one system parameter is varied. The locus of points defined by the poles can be used to predict critical operating points (e.g., instability). The gain factor (scalar multiplier) of a system frequently is the varied parameter.

The root-locus diagram gets its name from the necessity of finding roots for the denominator of the transfer function.

Example 19.30

Draw the root-locus diagram for the feedback system shown. K is a scalar constant which can be varied.

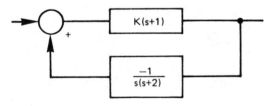

The transfer function is

$$T(s) = \frac{K(s+1)}{1 - [K(s+1)]\left[\dfrac{-1}{s(s+2)}\right]}$$

$$= \frac{K(s+1)[s(s+2)]}{s(s+2) + K(s+1)}$$

The poles are found by solving for the roots of the denominator.

$$s_1, s_2 = -\tfrac{1}{2}(2+K) \pm \sqrt{1 + \tfrac{1}{4}K^2}$$

Since the second term can be either added or subtracted, there are two roots for each chosen value of K. Allowing K to vary from zero to infinity produces the following root-locus diagram.

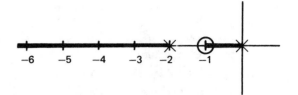

The root-locus diagram shows two distinct parts known as *branches*. The first branch extends from the pole at the origin to the zero corresponding to $s = -1$. The second branch extends from the pole corresponding to $s = -2$ to infinity.

Root-locus diagrams do not always result in loci confined to the real axis. They frequently leave the real axis at *break-away points* and continue on with constant or varying slopes.

11 STABILITY

A pole with a value of $-r$ on the real axis corresponds to an exponential response of e^{-rt}. Similarly, a pole with a value of $+r$ on the real axis corresponds to an exponential response of e^{rt}. However, e^{rt} increases without limit. For that reason, such a pole is said to be unstable.

Since any pole in the first and fourth quadrants of the s-plane will correspond to a positive exponential, a stable system must have poles limited to the left half of the s-plane (i.e., quadrants two and three).

Passive systems always are stable. A passive system does not contain an energy source. The system may experience an initial disturbance, but once movement begins the energy source is removed. In the absence of an energy source, exponential growth cannot occur.

Active systems contain energy sources. Not all active sources, however, are unstable.

- If any coefficient in the denominator of the transfer function for an active system is negative, the system is unstable.

- If all coefficients in the denominator of the transfer function for an active system are positive, the system can be stable or unstable.

If a single pole exists on the imaginary axis (between the left and right parts of the s-plane), the response is stable. However, a double pole on the imaginary axis (resulting from an s^2 term in the denominator of the transfer function) usually corresponds to a linearly increasing term in the response. Such a system is unstable.

A number of tabular and graphical methods exist for evaluating the stability of engineering systems.

- *root-locus diagrams*: If a parameter is varied until the locus-line crosses the imaginary axis, the root-locus diagram can be used to predict the parameter values critical to stability.

- *Routh stability criterion*: This is a tabular method for checking the stability of an nth order characteristic equation.

- *Hurwitz stability criterion*: This method calculates determinants from the coefficients of the characteristic equation to determine stability.

- *Nyquist analysis*: This is a graphical method for checking stability. It is particularly useful when time delays are present in a system or when frequency response data is available on a system.

- *Bode plots*: This is another graphical approach for checking stability against variations in frequency. It uses two graphs plotted on logarithmic scales. One graph is phase angle versus frequency. The other graph is gain (in decibels) versus frequency.

PART 6: Instrumentation

1 POTENTIOMETERS

A potentiometer converts linear or rotary motion into a variable voltage. It consists of a resistance with a variable-position tap. A voltage is applied across the resistance ends. The voltage at the tap will vary with its position. This device also is known as a *potentiometer transducer*.

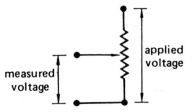

Figure 19.22 A Potentiometer

2 LINEAR VARIABLE DIFFERENTIAL TRANSFORMER

The linear variable differential transformer (LVDT) converts linear motion into a change in voltage. The transformer is supplied with low A.C. voltage. Movement of the core changes the flux linkage between the primary and the two secondary windings. The output voltage is proportional to the displacement of the core from the center (null) position. The voltage will change phase 180° as the core passes through the null position.

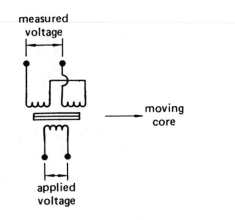

Figure 19.23 Linear Variable Differential Transformer

3 STRAIN GAGES

A strain gage is a folded wire which exhibits a resistance change as the wire length changes. The resistance change will be small, and temperature effects should be compensated by using a second unstrained gage as part of the bridge measurement system. Nichrome wire with a total resistance under 1000 ohms commonly is used.

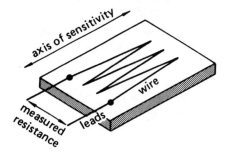

Figure 19.24 A Bonded Strain Gage

The *strain sensitivity (gage factor)* is defined as

$$K = \left(\frac{\Delta R}{R_o}\right)\left(\frac{L_o}{\Delta L}\right) \qquad 19.42$$

The strain (generally in microinches per inch) is related to the resistance change by equation 19.43.

$$\epsilon = \frac{\Delta R}{K R_o} \qquad 19.43$$

4 PHOTOSENSITIVE CONDUCTORS

Cadmium sulfide and cadmium selenide are two compounds that decrease in resistance as light is applied. Cadmium sulfide is most sensitive to light in the 5000 to 6000 angstrom range, while cadmium selenide shows peak sensitivities in the 7000 to 8000 angstrom range. Due to a *hysteresis effect*, photosensitive conductors do not react instantaneously to changes in intensity. High speed operation requires high light intensities and careful design.

5 PHOTOVOLTAIC CELLS

Photovoltaic cells convert light to voltage. The magnitude of the voltage (or the current) depends on the illumination. If the cell is reverse-biased by an external battery, its operation is similar to a constant-current source.

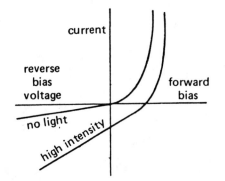

Figure 19.25 Photovoltaic Characteristic Curve

6 TEMPERATURE-SENSITIVE RESISTORS

Resistance in conductors will vary with temperature. The change can be positive or negative, and the variation is nonlinear. The variation can be calculated if the coefficients of thermal resistance are known. (β generally is small and is not considered in most analyses.)

$$R_T = R_o(1 + \alpha\Delta T + \beta\Delta T^2) \qquad 19.44$$
$$\Delta T = T - T_o \qquad 19.45$$

Table 19.2
Coefficients of Thermal Resistance (1/°C)

conductors

aluminum	0.0043	manganin	0.00002
brass	0.0020	nichrome	0.0004
constantan	0.000002	nickel	0.0068
copper	0.0067	platinum	0.0039
gold	0.0040	silver	0.0041
iron	0.0061	tin	0.0046
lead	0.0039		

semiconductors

manganese oxide @	25°C	−.044
manganese oxide @	100°C	−.030
manganese oxide @	-25°C	−.061
silicon		−.040

A *thermistor* is a temperature-sensitive semiconductor. Thermistors typically are less precise and more unpredictable than metallic resistors.

7 THERMOCOUPLE

A thermocouple consists of two dissimilar metals joined at both ends. (One set of ends may be connected through a measurement circuit.) A voltage will be generated when the temperatures of the two joined ends are different.

The thermally generated voltage is small, and thermocouples are calibrated in microvolts per degree Celsius. An amplifier is required to provide usable signal levels.

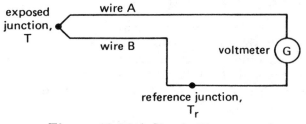

Figure 19.26 A Simple Thermocouple

The common thermocouple materials and their useful ranges are

copper/constantan	(−300° to +100°F)
iron/constantan	(0° to 1300°F)
chromel/alumel	(600° to 1800°F)
platinum/platinum-rhodium	(1300° to 2900°F)

The output voltage of a thermocouple is given by equation 19.46. The constant k is taken from table 19.3. If one of the wires is not platinum, the constant is the difference between the two values in table 19.3.[9] (Equation 19.46 ignores higher order terms.)

$$\text{voltage} \approx k(T - T_r) \qquad 19.46$$

Table 19.3
Thermoelectric Constants[10] at 0°C
(microvolts/°C, platinum reference)

bismuth	−72	lead	4
constantan	−35	silver	6.5
nickel	−15	copper	6.5
platinum	0.0	gold	6.5
mercury	0.6	iron	18.5
aluminum	3.5	nichrome	25

[9]Because k varies with temperature, most thermocouple problems are solved from published tables of generated voltage versus temperature.

[10] Multiply microvolts/°C by 0.555 to get microvolts/°F.

SPECIAL TOPIC
TACHOMETER CONTROL

The d.c. tachometer commonly is used to measure rotational speeds. One common form is a simple d.c. generator or magneto with field produced from permanent magnets. The generated voltage is directly proportional to the speed of armature rotation. For speeds much higher than 2000 rpm, the tachometer may need to be geared down.

Another form is the *pulse tachometer*. Each rotation of the shaft produces one or more pulses. Inductive and photoelectric processes can be used to generate the pulses, which vary in profile according to the method of generation. By counting and scaling the number of pulses per period, the rotational speed can be determined. If the tachometer is to be used to control a shafted machine, its output also must be converted to a d.c. voltage. A *low-pass filter* circuit will accomplish this conversion.

A simple *constant-speed control system* is shown in figure 19.27. This system consists of a tachometer, a difference amplifier, and the motor. The tachometer output voltage is compared to a reference voltage. The difference is amplified by the *difference amplifier*. The output of the amplifier is known as the *error, e*. This error is used to modify the motor speed by passing it through a second *booster amplifier*. The output of the booster amplifier carries sufficient power to drive the motor's field winding.[11]

---

[11] The method of controlling the motor speed is not important in these problems. The output of the booster voltage may feed the armature instead of the field. Alternately, the booster voltage may control a generator, and the output of the generator output would feed the motor. However, the transfer function of this control mechanism must be known.

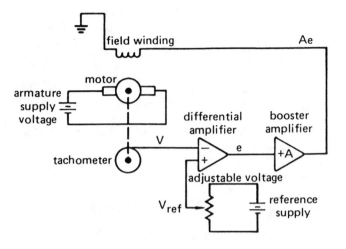

Figure 19.27 A Constant-Motor Speed System Schematic

Applying a load to the motor, which is running, will tend to decrease the motor speed. There will be a proportional decrease in tachometer output. The output of the differential amplifier becomes more positive due to the use of inverting terminals. The increased booster voltage increases the motor speed.

Speed is set by varying the reference voltage. There always should be a differential voltage. The booster voltage maintains the quiescent field current.

Figure 19.28 illustrates the constant-motor speed system in traditional block diagram form. Notice that the battery voltages are not part of the control loop. The *control mechanism* will depend on the type of controller. The summing point replaces the differential amplifier.

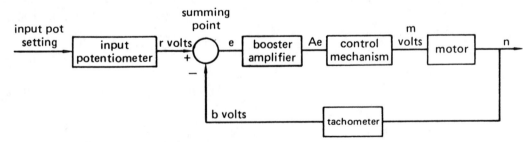

Figure 19.28 Constant-Motor Speed Block Diagram

SPECIAL TOPIC
BOOLEAN ALGEBRA

Boolean algebra is a numbering system in which the variables are limited to two values, usually **0** and **1**. The basic operations and corresponding gate symbols are shown in figure 19.29.

Gates can have more than two inputs, except for the **not** gate, which can have only one input.

The basic algebraic laws governing Boolean variables are listed here.

Commutative: $A + B = B + A$
$A \cdot B = B \cdot A$

Associative: $A + (B + C) = (A + B) + C$
$A \cdot (B \cdot C) = (A \cdot B) \cdot C$

Distributive: $A \cdot (B + C) = (A \cdot B) + (A \cdot C)$
$A + (B \cdot C) = (A + B) \cdot (A + C)$

Absorptive: $A + (A \cdot B) = A$
$A \cdot (A + B) = A$

$$0 + 0 = 0$$
$$0 + 1 = 1$$
$$1 + 0 = 1$$
$$1 + 1 = 1$$
$$0 \cdot 0 = 0$$
$$0 \cdot 1 = 0$$
$$1 \cdot 0 = 0$$
$$1 \cdot 1 = 1$$
$$A + 0 = A$$
$$A + 1 = 1$$
$$A + A = A$$
$$A + (-A) = 1$$
$$A \cdot 0 = 0$$
$$A \cdot 1 = A$$
$$A \cdot A = A$$
$$A \cdot (-A) = 0$$
$$-0 = 1$$
$$-1 = 0$$
$$-(-A) = A$$

De Morgan's theorems are

$$\overline{(A + B)} = \overline{A} \cdot \overline{B}$$
$$\overline{A \cdot B} = \overline{A} + \overline{B}$$

Truth tables are used to document the results of a Boolean operation. Table 19.4 gives the output of each gate shown in figure 19.29.)

Table 19.4
Truth Table for Logic Gates

A	B	$A \cdot B$	$A + B$	$-A$	$\overline{A \cdot B}$	$\overline{A + B}$	$A \oplus B$
0	0	0	0	1	1	1	0
0	1	0	1	1	1	0	1
1	0	0	1	0	1	0	1
1	1	1	1	0	0	0	0

Example 19.31

Simplify and write the truth table for the following system.

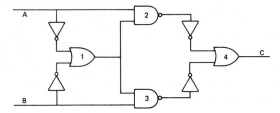

step 1:　The inputs and the outputs of each gate are listed below.

gate	inputs	output
1	$\overline{A}, \overline{B}$	$(\overline{A} + \overline{B})$
2	$A, (\overline{A} + \overline{B})$	$\overline{A \cdot (\overline{A} + \overline{B})}$
3	$B, (\overline{A} + \overline{B})$	$\overline{B \cdot (\overline{A} + \overline{B})}$
4	$A \cdot (\overline{A} + \overline{B}), B \cdot (\overline{A} + \overline{B})$	$A \cdot (\overline{A} + \overline{B}) + B \cdot (\overline{A} + \overline{B})$

step 2:　The output of gate 4 can be simplified.

$A \cdot (\overline{A} + \overline{B}) + B \cdot (\overline{A} + \overline{B})$	(original)
$A \cdot \overline{A} + A \cdot \overline{B} + B \cdot \overline{A} + B \cdot \overline{B}$	(distributive)
$A \cdot \overline{B} + B \cdot \overline{A}$	(since $A \cdot \overline{A} = 0$)
$A \oplus B$	(definition)

step 3:　The truth table is

A	B	C
0	0	0
0	1	1
1	0	1
1	1	0

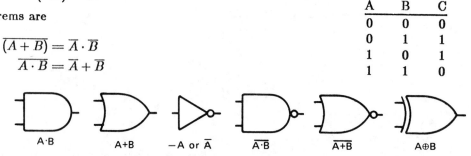

A·B	A+B	−A or $\overline{A}$	$\overline{A \cdot B}$	$\overline{A + B}$	A⊕B

Figure 19.29 Logic Gate Symbols

PRACTICE PROBLEMS: MODELING OF ENGINEERING SYSTEMS

Warm-ups

1. Simplify the following block diagram and determine the system gain.

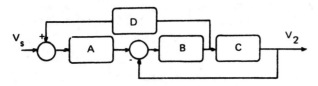

2. Simplify the following block diagram and determine the system gain.

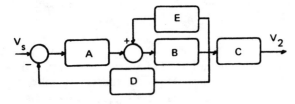

3. For each of the systems shown,
   (a) draw the circuit diagrams using idealized elements
   (b) write the differential equation or the system of differential equations which describes the motion of the system

3.1

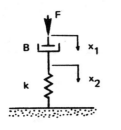

3.2

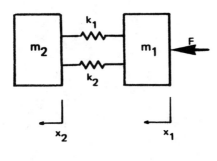

3.3

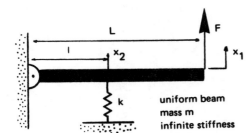

uniform beam
mass m
infinite stiffness

3.4

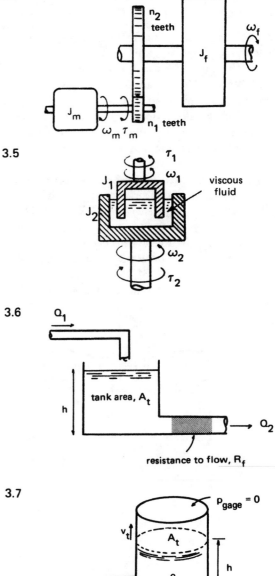

3.5

3.6

3.7

Timed (1 hour allowed for each)

1. A constant speed motor/magnetic clutch drive train is monitored and controlled by a speed-sensing tachometer. The entire system is modeled as a feedback system. The lower case variables represent small signal increments from the equilibrium (reference) values.

By way of operation, the desired motor speed (n, in rpm) is set with a speed-setting potentiometer. From there the input signal is compared to the tachometer output. The comparator output error (e, in volts) controls the clutch. A current (i, in amps) passes through the clutch coil. The external load torque (t_L, in in-lbf) is seen by the clutch and is countered by the clutch output torque (t, in in-lbf).

(a) Plot the open-loop frequency response. What is the open-loop steady-state gain?
(b) Plot the unity feedback closed-loop frequency response. What is the closed-loop steady-state gain?
(c) Plot the system sensitivity.
(d) Describe the closed-loop response to a step change in the desired output angular velocity. Is it damped or oscillatory? Is there a steady-state error? Why or why not?
(e) Describe the closed-loop response to a step change in the load torque. Is the response damped or oscillatory? Is there a steady-state error? Why or why not?
(f) Assume that you have to select the gain in the comparator and that it doesn't necessarily have to be .1. Using the root locus or Routh's or Nyquist stability criteria, find the limits of the comparator gain that cause the closed-loop system to be unstable.

(g) How can you easily improve the steady-state response of the closed-loop system to constant disturbances in the load torque, t_L?

2. A 100 lbm mass is supported by a spring (1200 lbf/ft). A dashpot ($B = 60$ lbf-sec/ft) has been installed to damp vibrations.

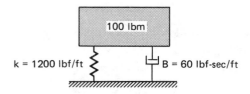

(a) What is the undamped natural frequency? (b) What is the damping ratio? (c) Sketch the magnitude and phase characteristics of the frequency response. (d) Sketch the response to a unit step input, $F(t)$.

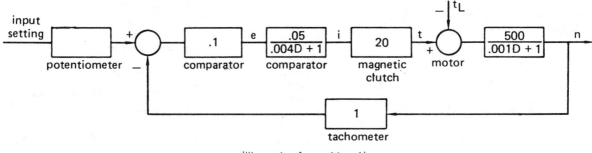

(illustration for problem 1)

RESERVED FOR FUTURE USE

RESERVED FOR FUTURE USE

20 MANAGEMENT

1 INTRODUCTION

Effective management techniques are based on behavioral science studies. Behavioral science is an outgrowth of the *human relations theories* of the 1930's, in which the happiness of employees was the goal (i.e., a happy employee is a productive employee...). Current behavioral science theories emphasize minimizing tensions that inhibit productivity.

There is no evidence yet that employees want a social aspect to their jobs. Nor is there evidence that employees desire job enlargement or autonomy. Behavioral science makes these assumptions anyway.[1]

2 BEHAVIORAL SCIENCE KEY WORDS

Cognitive system: method we use to interpret our environment.

Collaboration: influence through a mutual agreement, relationship, respect, or understanding, without a formal or contractual authority relationship.

Equilibrium: maintaining the status quo of a group or an individual.

Job enrichment: letting employees have more control over their activities and working conditions.

Manipulation: influencing others by recognizing and building upon their needs.

MBO: management by objectives—setting job responsibilities and standards for each group and employee.

Normative judgment: judging others according to our own values.

Paternalism: corporate subsidy—showering employees with benefits and expecting submission in return.

Personal map: a person's expectations of his environment.

Selective perception: seeing what we want to see—a form of defense mechanism, since things first must be perceived to be ignored.

Superordinate goals: goals which are outside of the individual, such as corporate goals.

3 HISTORY OF BEHAVIORAL SCIENCE STUDIES AND THEORIES

A. HAWTHORNE EXPERIMENTS

During 1927–1932, the Hawthorne (Chicago) Works of the Western Electric Company experimented with working conditions in an attempt to determine what factors affected output.[2] Six average employees were chosen to assemble and inspect phone relays.

Many factors were investigated in this exhaustive test. After weeks of observation without making any changes (to establish a baseline), Western Electric varied the number and length of breaks, the length of the work day, the length of the work week, and the illumination level of the lighting. Group incentive plans were tried, and in several tests, the company even provided food during breaks and lunch periods.

The employees reacted in the ways they thought they should. Output (as measured in relay production) increased after every change was implemented. In effect, the employees reacted to the attention they received, regardless of the working conditions.

Western Electric concluded that there was no relationship between illumination and other conditions to productivity. The increase in productivity during the testing procedure was attributed to the sense of value each employee felt in being part of an important test. The employees also became a social group, even after

---

[1] In an exhaustive literature survey up through 1955, researchers found no conclusive relationships between satisfaction and productivity. There was, however, a relationship between lack of satisfaction and absenteeism and turnover.

[2] Experiments were conducted by Elton Mayo from the Harvard Business School.

hours. Leadership and common purpose developed, and even though the employees were watched more than ever, they felt no supervision anxiety since they were, in effect, free to react in any way they wanted.

One employee summed up the test when she said, "It was fun."

B. BANK WIRING OBSERVATION ROOM EXPERIMENTS

In an attempt to devise an experiment which would not suffer from the problems associated with the Hawthorne studies, Western Electric conducted experiments in 1931 and 1932 on the effects of wage incentives.

The group of nine wiremen, three soldermen, and two inspectors was interdependent. This was supposed to prevent any individual from slacking off. However, wage incentives failed to improve productivity. In fact, fast employees slowed down to protect their slower friends. Illicit activities, such as job trading and helping, also occurred.

The group was reacting to the notion of a *proper day's work*. When the day's work (or what the group considered to be a day's work) was assured, the whole group slacked off. The group also varied what it reported as having been accomplished and claimed more unavoidable delays than actually occurred. The output was essentially constant.

Western Electric concluded that social groups form as protection against arbitrary management decisions, even when such decisions have never been experienced. The effort to form the social groups, to protect slow workers, and to develop the notion of a proper day's work is not conscious. It develops automatically when the company fails to communicate to the contrary.

C. NEED HIERARCHY THEORY

During World War II, Dr. Abraham Maslow's *need hierarchy theory* was implemented into leadership training for the U.S. Air Force. This theory claims that certain needs become dominant when lesser needs are satisfied. Although some needs can be sublimated and others overlap, the need hierarchy theory generally requires the lower-level needs to be satisfied before the higher-level needs are realized. (The ego and self-fulfillment needs rarely are satisfied.)

The need hierarchy theory explains why money is a poor motivator of an affluent individual. The theory does not explain how management should apply the need hierarchy to improve productivity.

Table 20.1
The Need Hierarchy
(In order of lower to higher needs)

1. Physiological needs: air, food, water.

2. Safety needs: protection against danger, threat, deprivation, arbitrary management decisions. Need for security in a dependent relationship.

3. Social needs: belonging, association, acceptance, giving and receiving of love and friendship.

4. Ego needs: self-respect and confidence, achievement, self-image. Group image and reputation, status, recognition, appreciation.

5. Self-fulfillment needs: realizing self potential, self development, creativity.

D. THEORY OF INFLUENCE

In 1948, the Human Relations Program (under the direction of Donald C. Pelz) at Detroit Edison studied the effectiveness of its supervisors. The most effective supervisors were those who helped their employees benefit. Supervisors who were close to their employees (and sided with them in disputes) were effective only if they were influential enough to help the employees. The study results were formulated into the *theory of influence*.

- Employees think well of supervisors who help them reach their goals and meet their needs.

- An influential supervisor will be able to help employees.

- An influential supervisor who is also a disciplinarian will breed dissatisfaction.

- A supervisor with no influence will not be able to affect worker satisfaction in any way.

The implication of the theory of influence is that whether or not a supervisor is effective depends on his influence. Training of supervisors is useless unless they have the power to implement what they have learned. Also, increases in supervisor influence are necessary to increase employee satisfaction.

E. HERZBERG MOTIVATION STUDIES

Frederick W. Herzberg interviewed 200 technical personnel in 11 firms during the late 1950's. Herzberg was especially interested in exceptional occurrences resulting in increases in job satisfaction and performance. From those interviews, Herzberg formulated his *motivation-maintenance theory*.

According to this theory, there are satisfiers and dissatisfiers which influence employee behavior. The *dissatisfiers* (also called *maintenance/motivation fac-*

tors) do not motivate employees; they can only dissatisfy them. However, the dissatisfiers must be eliminated before the satisfiers work. Dissatisfiers include company policy, administration, supervision, salary, working conditions (environment), and interpersonal relations.

Satisfiers (also known as *motivators*) determine job satisfaction. Common satisfiers are achievement, recognition, the type of work itself, responsibility, and advancement.

An interesting conclusion based on the motivation/maintenance theory is that fringe benefits and company paternalism do not motivate employees since they are related to dissatisfiers only.

F. THEORY X AND THEORY Y

During the 1950's Douglas McGregor (Sloan School of Industrial Management at MIT) introduced the concept that management had two ways of thinking about its employees. One way of thinking, which was largely pessimistic, was theory X. The other theory, theory Y, was largely optimistic.

Theory X is based on the assumption that the average employee inherently dislikes and avoids work. Therefore, employees must be coerced into working by threats of punishment. Rewards are not sufficient. The average employee wants to be directed, avoids responsibility, and seeks the security of an employer-employee relationship.

This assumption is supported by much evidence. Employees exist in a continuum of wants, needs, and desires. Many of the need satisfiers (salary, fringe benefits, etc.) are effective only off the job. Therefore, work is considered a punishment or a price paid for off-the-job satisfaction.

Theory X is pessimistic about the effectiveness of employers to satisfy or motivate their employees. By satisfying the physiological and safety (lower level) needs, employers have shifted the emphasis to higher level needs which they cannot satisfy. Employees, unable to derive satisfaction from their work, behave according to theory X.

Theory Y, on the other hand, assumes that the expenditure of effort is natural and is not inherently disliked. It assumes that the average employee can learn to accept and enjoy responsibility. Creativity is widely distributed among employees, and the potentials of average employees are only partially realized.

Theory Y places the blame for worker laziness, indifference, and lack of cooperation in the lap of management, since the integration of individual and organization needs is required. This theory is not fully validated, nor is its full use ever likely to be implemented.[3]

4 JOB ENRICHMENT

In an effort to make their employees happier, companies have tried to enrich the jobs performed by employees. Enrichment is a subjective result felt by employees when their jobs are made more flexible or are enlarged. Adding flexibility to a job allows an employee to move from one task to another, rather than doing the same thing continually. Horizontal job enlargement adds new production activities to a job. Vertical job enlargement adds planning, inspection, and other non-production tasks to the job.

There are advantages to keeping a job small in scope. Learning time is low, employee mental effort is reduced, and the pay rate can be lower for untrained labor. Supervision is reduced. Such simple jobs, however, also result in high turnover, absenteeism, and lower pride in job (and subsequent low quality rates).

Job enlargement generally results in better quality products, reduces inspection and material handling, and counteracts the disadvantages previously mentioned. However, training time is greater, tooling costs are higher, and inventory records are more complex.

5 QUALITY IMPROVEMENT PROGRAMS

A. ZERO DEFECTS PROGRAM

Employees have been conditioned to believe that they are not perfect and that errors are natural. However, we demand zero defects from some professions (e.g., doctors, lawyers, engineers). The philosophy of a zero defect program is to expect zero defects from everybody.

Zero defects programs develop a constant, conscious desire to do the job right the first time. This is accomplished by giving employees constant awareness that their jobs are important, that the product is important, and that management thinks their efforts are important.

Zero defects programs try to correct the faults of other types of employee programs.[4] Programs are based on

---

[3]Theory Y is not synonymous with soft management. Rather than emphasize tough management (as does theory X), theory Y **depends on commitment of employees to achieve mutual goals.**

[4]Motivational programs are not honest, according to the zero-defects theory, since management tries to convince employees to do what management wants. Wage incentive programs encourage employee dishonesty and errors by emphasizing quantity, not quality. Theory X management, with its implied punitive action if goals are not achieved, never has been effective.

what the employee has for his own: pride and desire. The programs present the challenge of perfection and explain the importance of that perfection. Management sets an example by expecting zero defects of itself. Standards of performance are set and are related to each employee. Employees are checked against these performance requirements periodically, and recognition is given when goals are met.

B. QUALITY CIRCLES/TEAM PROGRAMS

Quality circle programs are voluntary or required programs in which employees within a department actively participate in measuring and improving quality and performance. It involves periodic meetings on a weekly or a monthly basis. Workers are encouraged to participate in volunteering ideas for improvement.

RESERVED FOR FUTURE USE

RESERVED FOR FUTURE USE

21 MISCELLANEOUS TOPICS

PART 1: Accuracy and Precision in Experiments

1 ACCURACY

An experiment is said to be *accurate* if it is unaffected by experimental error. In this case, *error* is not synonymous with *mistake*, but rather includes all variations not within the experimenter's control.

For example, suppose a gun is aimed at a point on a target and five shots are fired. The mean distance from the point of impact to the sight-in point is a measure of the alignment accuracy between the barrel and sights. The difference between the actual value and the experimental value is known as *bias*.

2 PRECISION

Precision is not synonymous with accuracy. Precision is concerned with the repeatability of the experimental results. If an experiment is repeated with identical results, the experiment is said to be precise.

In the previous example, the average distance of each impact from the centroid of the impact group is a meas-

ure of the precision of the experiment. Thus, it is possible to have a highly precise experiment with a large bias.

Most techniques applied to experiments to improve the accuracy of the experimental results (e.g., repeating the experiment, refining the experimental methods, or reducing variability) actually increase the precision.

Sometimes the word *reliability* is used with regards to the precision of an experiment. A reliable estimate is used in the same sense as a precise estimate.

3 STABILITY

Stability and *insensitivity* are synonymous terms. A stable experiment is insensitive to minor changes in the experiment parameters. Suppose the centroid of a bullet group is 2.1 inches away from the sight-in point at 65°F and 2.3 inches away at 80°F. The experiment's sensitivity to temperature changes would be $(2.3 - 2.1)/(80 - 65) = .0133$ inches/°F.

PART 2: Dimensional Analysis

Nomenclature

c_p	specific heat	BTU/lbm-°F
C_i	a constant	–
D	diameter	ft
F	force	lbf
g_c	gravitational constant (32.2)	lbm-ft/sec²-lbf
$\bar{h}$	average film coefficient	BTU/hr-ft²-°F
J	Joule's constant (778)	ft-lbf/BTU
k	number of pi-groups $(m - n)$	–
L	length	ft
m	number of relevant independent variables	–
M	mass	lbm
n	number of independent dimensional quantities	–
N_{Nu}	Nusselt number	–
N_{Pe}	Peclet number	–
N_{Re}	Reynolds number	–
v	velocity	ft/sec
x_i	the ith independent variable	various
y	dependent variable	various

Symbols

ρ	density	lbm/ft³
θ	time	sec
π_i	ith dimensionless group	–
μ	viscosity	lbm/ft-sec

Dimensional analysis is a means of obtaining an equation for some phenomenon without understanding the inner mechanism of the phenomenon. The most serious limitation to this method is the need to know beforehand which variables influence the phenomenon. Once these variables are known or are assumed, dimensional analysis can be applied by a routine procedure.

The first step is to select a system of primary dimensions. Usually the MLθT system (mass, length, time, and temperature) is used, although this choice may require the use of g_c and J in the final results. The dimensional formulas and symbols for variables most frequently encountered are given in table 21.1.

The second step is to write a functional relationship between the dependent variable and the independent variables, x_i.

$$y = \mathbf{f}(x_1, x_2, \cdots, x_m) \qquad 21.1$$

This function can be expressed as an exponentiated series.

$$y = C_1 x_1^{a_1} x_2^{b_1} x_3^{c_1} \cdots x_m^{z_1} + C_2 x_1^{a_2} x_2^{b_2} x_3^{c_2} \cdots x_m^{z_2} + \cdots \quad 21.2$$

The C_i, a_i, b_i, $\cdots z_i$ in equation 21.2 are unknown constants.

The key to solving the above equation is that each term on the right-hand side must have the same dimensions as y. Simultaneous equations are used to determine some of the a_i, b_i, c_i, and z_i. Experimental data is required to determine the C_i and the remaining exponents. In most analyses, it is assumed that the $C_i = 0$ for $i = 2$ and up.

Example 21.1

A sphere submerged in a fluid rolls down an incline. Find an equation for the velocity, v.

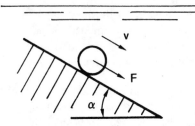

It is assumed that the velocity depends on the force, F, due to the inclination, the diameter of the sphere, D, the density of the fluid, ρ, and the viscosity of the fluid, μ.

$$v = \mathbf{f}(F, D, \rho, \mu)$$

This equation can be written in terms of the dimensions of the variables.

$$\frac{L}{\theta} = C\left(\frac{ML}{\theta^2}\right)^a (L)^b \left(\frac{M}{L^3}\right)^c \left(\frac{M}{L\theta}\right)^d$$

Since L on the left-hand side has an implied exponent of one, the necessary equation is

$$1 = a + b - 3c - d \qquad (L)$$

Similarly, the other necessary equations are

$$-1 = -2a - d \qquad (\theta)$$
$$0 = a + c + d \qquad (M)$$

Solving simultaneously yields

$$b = -1$$
$$c = a - 1$$
$$d = 1 - 2a$$

or

$$v = C\left(\frac{\mu}{D\rho}\right)\left(\frac{F\rho}{\mu^2}\right)^a$$

C and a would have to be determined experimentally.

Table 21.1
Units and Dimensions of Typical Variables

Quantity	Symbol	$ML\theta T$ System	$ML\theta TFQ$ System	Units in Engineering System
length	L or x	L	L	ft
time	θ	θ	θ	sec or hour
mass	M	M	M	lbm
force	F	ML/θ^2	F	lbf
temperature	T	T	T	°F
heat	Q	ML^2/θ^2	Q	BTU
velocity	V	L/θ	L/θ	ft/sec
acceleration	a or g	L/θ^2	L/θ^2	ft/sec²
dimensional conversion factor	g_c	none	ML/θ^2F	32.2 lbm-ft/sec²-lbf
energy conversion factor	J	none	FL/Q	778 ft-lbf/BTU
work	W	ML^2/θ^2	FL	ft-lbf
pressure	p	M/θ^2L	F/L^2	lbf/ft²
density	ρ	M/L^3	M/L^3	lbm/ft³
internal energy and enthalpy	u, h	L^2/θ^2	Q/M	BTU/lbm
specific heat	c	L^2/θ^2T	Q/MT	BTU/lbm-°F
dynamic viscosity	μ_f	$M/L\theta$	$F\theta/L^2$	lbf-sec/ft²
absolute viscosity	μ	$M/L\theta$	$M/L\theta$	lbm/ft-sec
kinematic viscosity	$\nu = u/\rho$	L^2/θ	L^2/θ	ft²/sec
thermal conductivity	k	ML/θ^3T	$Q/LT\theta$	BTU/hr-ft-°F
coefficient of expansion	β	$1/T$	$1/T$	1/°F
surface tension	σ	M/θ^2	F/L	lbf/ft
stress	σ or τ	$M/L\theta^2$	F/L^2	lbf/ft²
film coefficient	h	M/θ^3T	$Q/\theta L^2T$	BTU/hr-ft²-°F
mass flow rate	m	M/θ	M/θ	lbm/sec

Since the above method requires working with m different variables and n different independent dimensional quantities (such as M, L, T, and θ), an easier method is desirable. One simplification is to combine the m variables into dimensionless groups, called *pi-groups*.

If these dimensionless groups are represented by π_1, π_2, π_3, $\cdots\pi_k$, the equation expressing the relationship between the variables is given by the *Buckingham π-theorem*.

$$\mathbf{f}(\pi_1, \pi_2, \pi_3, \cdots\pi_k) = 0 \qquad 21.3$$

$$k = m - n \qquad 21.4$$

The dimensionless pi-groups usually are found from the m variables according to an intuitive process. A formalized method is possible as long as the following conditions are met.

- The dependent variable and independent variables chosen contain all of the variables affecting the phenomenon. Extraneous variables can be included at the expense of obtaining extra pi-groups.
- The pi-groups must include all of the original x_i at least once.
- The dimensions all must be independent.

The formal procedure is to select n variables (x_i) out of the total m as repeating variables to appear in all k pi-groups. These variables are used in turn with the remaining variables in each successive pi-group. Each of the repeating variables must have different dimensions, and the repeating variables collectively must contain all of the dimensions. This procedure is illustrated in example 21.2.

Example 21.2

It is desired to determine a relationship giving the heat transfer to air flowing across a heated tube. The following variables affect the heat flow.

Variable	Symbol	Dimensional Equation
tube diameter	D	L
fluid conductivity	k	ML/θ^3T
fluid velocity	v	L/θ
fluid density	ρ	M/L^3
fluid viscosity	μ	$M/L\theta$
fluid specific heat	c_p	L^2/θ^2T
film coefficient	h	M/θ^3T

There are $m = 7$ variables and $n = 4$ primary dimensions (L, M, θ, and T). Accordingly, there are $k = 7 - 4 = 3$ dimensionless groups that are required to correlate the data. The four repeating variables are chosen such that all dimensions are represented. Then the π_i are written as functions of these repeating variables in turn with the remaining variables.

The repeating variables should not include any of the unknown quantities. For example, $\overline{h}$ should not be chosen as a repeating variable since it is directly related to the unknown heat flow. In addition, important material properties, such as c_p and k, often are omitted. Trial and error is required to include all four primary dimensions.

Using trial and error, omitting $\overline{h}$ as a repeating variable, and representing all four primary dimensions, arbitrarily choose the variables as D, k, v, and ρ.

The pi-groups are

$$\pi_1 = D^{a_1} k^{a_2} v^{a_3} \rho^{a_4} \mu$$
$$\pi_2 = D^{a_5} k^{a_6} v^{a_7} \rho^{a_8} c_p$$
$$\pi_3 = D^{a_9} k^{a_{10}} v^{a_{11}} \rho^{a_{12}} \overline{h}$$

Since the π_i are dimensionless, we write for π_1

$$0 = a_1 + a_2 + a_3 - 3a_4 - 1 \quad \text{(L)}$$
$$0 = a_2 + a_4 + 1 \quad \text{(M)}$$
$$0 = -3a_2 - a_3 - 1 \quad (\theta)$$
$$0 = -a_2 \quad \text{(T)}$$

Therefore,

$$a_2 = 0 \quad a_3 = -1 \quad a_4 = -1 \quad a_1 = -1$$
$$\pi_1 = \frac{\mu}{Dv\rho}$$

π_1 is the reciprocal of the Reynolds number. Proceeding similarly with π_2,

$$0 = a_5 + a_6 + a_7 = 3a_8 + 2 \quad \text{(L)}$$
$$0 = a_6 + a_8 \quad \text{(M)}$$
$$0 = -3a_6 - a_7 - 2 \quad (\theta)$$
$$0 = -a_6 - 1 \quad \text{(T)}$$

Therefore,

$$a_6 = -1 \quad a_7 = 1 \quad a_8 = 1 \quad a_5 = 1$$
$$\pi_2 = \frac{Dv\rho c_p}{k}$$

π_2 is the *Peclet number* (product of the Reynolds number and the Prandtl number).

π_3 is found to be $\frac{D\overline{h}}{k}$, which is the *Nusselt number*.

The seven original variables have been combined into three dimensionless groups, making data correlation much easier. The implicit equation for heat transfer is

$$\mathbf{f}_1(\pi_1, \pi_2, \pi_3) = \mathbf{f}_1(N_{Re}, N_{Nu}, N_{Pe}) = 0$$

Rearrangement of the pi-groups is needed to isolate the dependent variable (in this case, $\overline{h}$).

$$N_{Nu} = \mathbf{f}_2(N_{Re}, N_{Pe}) = C(N_{Re})^{e_1}(N_{Pe})^{e_2}$$

C, e_1, and e_2 are found experimentally.

The selection of the repeating and non-repeating variables is the key step. The choice of repeating variables determines which dimensionless groups are obtained. The theoretical maximum number of valid dimensionless groups is

$$\frac{m!}{(n+1)!(m-n-1)!} \qquad \text{21.5}$$

Not all dimensionless groups obtained are equally useful to researchers. For example, the Peclet number was obtained in the above example. However, researchers would have chosen D, k, ρ, and μ as repeating variables in order to obtain the Prandtl number as a dimensionless group. This choice of repeating variables is a matter of intuition.

PART 3: Reliability

Nomenclature

$f(t)$	probability density function	–
$F(t)$	cumulative density function	–
k	minimum number for operation	–
MTBF	mean time before failure	time
n	number of items in the system	–
R^*	system reliability	–
$R_i(t)$	ith item reliability	–
t	time	time
x	number of failures	–
X	binary ith item performance variable	–
Y	arbitrary event	–
$z(t)$	hazard function	1/time

Symbols

λ	constant failure or hazard rate	1/time
ϕ	binary system performance variable	–

1 ITEM RELIABILITY

Reliability as a function of time, $R(t)$, is the probability that an item will continue to operate satisfactorily up to time t. Although other distributions are possible, reliability often is described by the *negative exponential distribution*. Specifically, it is assumed that an item's reliability is

$$R(t) = 1 - F(t) = e^{-\lambda t} = e^{-t/MTBF} \qquad 21.6$$

This infers that the probability of x failures in a period of time is given by the Poisson distribution.

$$p\{x\} = \frac{e^{-\lambda}\lambda^x}{x!} \qquad 21.7$$

The negative exponential distribution is appropriate whenever an item fails only by random causes but never experiences deterioration during its life. This implies that the *expected future life* of an item is independent of the previous duration of operation.

Example 21.3

An item exhibits an exponential time to failure distribution with MTBF of 1000 hours. What is the maximum operating time such that the reliability does not drop below .99?

$$.99 = e^{-t/1000}$$

$$t = 10.05 \text{ hours}$$

The *hazard function* is defined as the conditional probability of failure in the next time interval given that no failure has occurred thus far. For the exponential distribution, the hazard function is

$$z(t) = \lambda \qquad 21.8$$

Since this is not a function of t, exponential failure rates are not dependent on the length of time previously in operation.

In general,

$$z(t) = \frac{f(t)}{R(t)} = \frac{\dfrac{dF(t)}{dt}}{1 - F(t)} \qquad 21.9$$

The exponential distribution is summarized by equations 21.10 through 21.13.

$$f(t) = \lambda e^{-\lambda t} \qquad 21.10$$

$$F(t) = 1 - e^{-\lambda t} \qquad 21.11$$

$$R(t) = 1 - F(t) = e^{-\lambda t} \qquad 21.12$$

$$z(t) = \frac{\lambda e^{-\lambda t}}{e^{-\lambda t}} = \lambda \qquad 21.13$$

2 SYSTEM RELIABILITY

The binary variable, X_i, is defined as 1 if item i operates satisfactorily and 0 otherwise. Similarly, the binary variable, ϕ, is 1 only if the system operates satisfactorily. ϕ will be a function of the X_i.

A. SERIAL SYSTEMS

The *performance function* for a system of n serial items is

$$\phi = X_1 X_2 X_3 \cdots X_n = \min\{X_i\} \qquad 21.14$$

Equation 21.14 implies that the system will fail if any of the individual items fail. The system reliability is

$$R^* = R_1 R_2 R_3 \cdots R_n \qquad 21.15$$

Example 21.4

A block diagram of a system with item reliabilities is shown. What is the performance function and the system reliability?

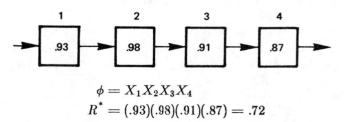

$$\phi = X_1 X_2 X_3 X_4$$
$$R^* = (.93)(.98)(.91)(.87) = .72$$

B. PARALLEL SYSTEMS

A parallel system with n items will fail only if all n items fail. This property is called *redundancy*, and such a system is said to be redundant. Using redundancy, a highly reliable system can be produced from components with relatively low individual reliabilities.

The performance function of a redundant system is

$$\phi = 1 - (1 - X_1)(1 - X_2)(1 - X_3) \cdots (1 - X_n)$$
$$= \max\{X_i\} \qquad\qquad 21.16$$

The reliablity is

$$R^* = 1 - (1 - R_1)(1 - R_2)(1 - R_3) \cdots (1 - R_n) \quad 21.17$$

Example 21.5

What is the reliability of the system shown?

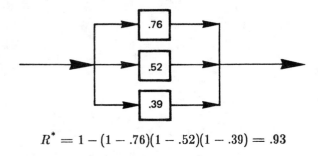

$$R^* = 1 - (1 - .76)(1 - .52)(1 - .39) = .93$$

C. k-out-of-n SYSTEMS

If the system operates with an k of its elements operational, it is said to be a k-out-of-n system. The performance function is

$$\phi = \begin{cases} 1 \text{ if } \Sigma X_i \geq k \\ 0 \text{ if } \Sigma X_i < k \end{cases} \qquad\qquad 21.18$$

The evaluation of the system reliability is quite difficult unless all elements are identical and have identical reliabilities, **R**. In that case, the system reliability follows the binomial distribution.

$$R^* = \sum_{j=k}^{n} \binom{n}{j} R^j (1 - R)^{n-j} \qquad 21.19$$

D. GENERAL SYSTEM RELIABILITY

A general system can be represented by a graphical network. Each path through the network from the starting node to the finishing node represents a possible operating path. For the 5-path network below, even if BD and AC are cut, the system will operate by way of path ABCD.

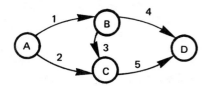

The reliability of the system will be the sum of the serial reliabilities, summed over all possible paths in the system. However, the concepts of minimal paths and minimal cuts are required to facilitate the evaluation of the system reliability.

A *minimal path* is a set of components that, if operational, will ensure the system's functioning. In the previous example, components [1 with 4] are a minimal path, as are [2 with 5] and [1 with 3 with 5]. A *minimal cut* is a set of components that, if non-functional, inhibits the system from functioning. Minimal cuts in the previous example are [1 with 2], [4 with 5], [1 with 5], and [2 with 3 with 4].

Since it usually is easier to determine all minimal paths, a method of finding the exact system reliability from the set of minimal paths is needed. In general, the probability of a union of n events contains $(2^n - 1)$ terms and is given by

$$\begin{aligned}
p\{Y_1 \text{ or } Y_2 \text{ or } \cdots Y_n\} &= p\{Y_1\} + p\{Y_2\} + p\{Y_3\} \\
&+ \cdots + p\{Y_n\} - p\{Y_1 \text{ and } Y_2\} - p\{Y_1 \text{ and } Y_3\} \\
&- \cdots - p\{Y_1 \text{ and } Y_n\} - p\{Y_1 \text{ and } Y_2 \text{ and } Y_3\} \\
&- p\{Y_1 \text{ and } Y_2 \text{ and } Y_4\} - \cdots - p\{Y_i \text{ and } Y_j \text{ and } Y_k\} \\
&\text{all } i \neq j \neq k \\
&+ \{-1\}^{n-1} p\{Y_1 \text{ and } Y_2 \text{ and } Y_3 \text{ and } \cdots \text{ and } Y_n\}
\end{aligned}$$
$$21.20$$

Returning to the 5-path example,

$$Y_1 = [1 \text{ with } 4]$$
$$Y_2 = [2 \text{ with } 5]$$
$$Y_3 = [1 \text{ with } 3 \text{ with } 5]$$

Then,

$$\begin{aligned}
p\{\phi = 1\} &= p\{Y_1 \text{ or } Y_2 \text{ or } Y_3\} \\
&= p\{X_1 X_4 = 1\} + p\{X_2 X_5 = 1\} \\
&+ p\{X_1 X_3 X_5 = 1\} - p\{X_1 X_2 X_4 X_5 = 1\} \\
&- p\{X_1 X_3 X_4 X_5 = 1\} \\
&- p\{X_1 X_2 X_3 X_5 = 1\} \\
&+ p\{X_1 X_2 X_3 X_4 X_5 = 1\}
\end{aligned}$$

In terms of the individual item reliabilities, this is

$$\begin{aligned}
R^* &= R_1 R_4 + R_2 R_5 + R_1 R_3 R_5 \\
&- R_1 R_2 R_4 R_5 - R_1 R_3 R_4 R_5 - R_1 R_2 R_3 R_5 \\
&+ R_1 R_2 R_3 R_4 R_5
\end{aligned}$$

In the 5-path example given,

$$R^* \leq p\{X_1 X_4 = 1\} + p\{X_2 X_5 = 1\} + p\{X_1 X_3 X_5 = 1\}$$

This method requires considerable computation, and an upper bound on R^* would be sufficient. Such an upper bound is close to R^* since the product of individual reliabilities is small. The upper bound is given by

$$p\{\phi = 1\} \leq p\{Y_1\} + p\{Y_2\} + \cdots + p\{Y_n\} \qquad 21.21$$

PART 4: Replacement

Nomenclature

C_1	item replacement cost with group replacement
C_2	item replacement cost after individual failure
$F(t)$	number of units failing in the interval ending at t
$K(t)$	total cost of operating from $t = 0$ to $t = T$
MTBF	mean time before failure
n	number of units in original system
$p\{t\}$	probability of failing in the interval ending at t
$S(t)$	number of survivors at the end of time t
t	time
$v\{t\}$	conditional probability of failure in the interval $(t-1)$ to t given non-failure before $(t-1)$

1 INTRODUCTION

Replacement and renewal models determine the most economical time to replace existing equipment. Replacement processes fall into two categories, depending on the life pattern of the equipment, which either deteriorates gradually (becomes obsolete or less efficient) or fails suddenly.

In the case of gradual deterioration, the solution consists of balancing the cost of new equipment against the cost of maintenance or decreased efficiency of the old equipment. Several models are available for cases with specialized assumptions, but no general solution methods exist.

In the case of sudden failure, of which light bulbs are examples, the solution consists of finding a replacement frequency which minimizes the costs of the required new items, the labor for replacement, and the expected cost of failure. The solution is made difficult by the probabilistic nature of the life spans.

2 DETERIORATION MODELS

The replacement criterion with deterioration models is the present worth of all future costs associated with each policy. Solution is by trial and error, calculating the present worth of each policy and incrementing the replacement period by one time period for each iteration.

Example 21.6

Item A currently is in use. Its maintenance cost is $400 this year, increasing each year by $30. Item A can be replaced by item B at a current cost of $3500. However, the cost of B is increasing by $50 each year. Item B has no maintenance costs. Disregarding income taxes, find the optimum replacement year. Use 10% as the interest rate.

Calculate the present worth of the various policies.

policy 1: Replacement at $t = 5$ (starting the 6th year)

$$PW(A) = -400\left(\frac{P}{A}, 10\%, 5\right) - 30\left(\frac{P}{G}, 10\%, 5\right)$$
$$= -1722$$
$$PW(B) = -[3500 + 5(50)]\left(\frac{P}{F}, 10\%, 5\right) = -2328$$

policy 2: Replacement at $t = 6$

$$PW(A) = -400\left(\frac{P}{A}, 10\%, 6\right) - 30\left(\frac{P}{G}, 10\%, 6\right)$$
$$= -2033$$
$$PW(B) = -[3500 + 6(50)]\left(\frac{P}{F}, 10\%, 6\right) = -2145$$

policy 3: Replacement at $t = 7$

$$PW(A) = -400\left(\frac{P}{A}, 10\%, 7\right) - 30\left(\frac{P}{G}, 10\%, 7\right)$$
$$= -2330$$
$$PW(B) = -[3500 + 7(50)]\left(\frac{P}{F}, 10\%, 7\right) = -1976$$

The present worth of A drops below the present worth of B at $t = 7$. Replacement should take place at that time.

3 FAILURE MODELS

The time between installation and failure is not constant for members in the general equipment population. Therefore, in order to solve a failure model, it is necessary to have the distribution of individual item lives (*mortality curve*). The *conditional probability of failure* in a small time interval, say from t to $(t + \delta t)$, is calculated from the mortality curve. This probability is *conditional* since it is conditioned on non-failure up to time t.

The conditional probability can decrease with time (e.g., *infant mortality*), remain constant (as with an exponential reliability distribution and failure from random causes), or increase with time (as with items that deteriorate with use). If the conditional probability decreases or remains constant over time, operating items should never be replaced prior to failure.

It usually is assumed that all failures occur at the end of a period. The problem is to find the period which minimizes the total cost.

Example 21.7

100 items are tested to failure. Two failed at $t = 1$, five at $t = 2$, seven at $t = 3$, 20 at $t = 4$, 35 at $t = 5$, and 31 at $t = 6$. Find the probability of failure in any period, the conditional probability of failure, and the mean time before failure.

The MTBF is

$$\frac{(2)(1) + (5)(2) + (7)(3) + (20)(4) + (35)(5) + (31)(6)}{100}$$

$$= 4.74$$

elapsed time t	failures F(t)	survivors S(t)	probability of failure p{t} = .01F(t)	conditional probability of failure v{t} = F(t)/S(t−1)
0	0	100	---	---
1	2	98	.02	.02
2	5	93	.05	.051
3	7	86	.07	.075
4	20	66	.20	.233
5	35	31	.35	.530
6	31	0	.31	1.00

4 REPLACEMENT POLICY

The expression for the number of units failing in time t is

$$F(t) = n\left[p\{t\} + \sum_{i=1}^{t-1} p\{i\}p\{t-i\} \right.$$
$$\left. + \sum_{j=2}^{t-1} \left[\sum_{i=1}^{j-1} p\{i\}p\{j-i\} \right] p\{t-j\} + \cdots \right] \quad 21.22$$

The term $np\{t\}$ gives the number of failures in time t from the original group.

The term $n\sum p\{i\}p\{t-i\}$ gives the number of failures in time t from the set of items which replaced the original items.

The third probability term times n gives the number of failures in time t from the set of items which replaced the first replacement set.

It can be shown that $F(t)$ with replacement will converge to a steady state limiting rate of

$$\overline{F(t)} = \frac{n}{MTBF} \quad 21.23$$

The optimum policy is to replace all items in the group, including items just installed, when the total cost per period is minimized. That is, we want to find T such that $K(T)/T$ is minimized.

$$K(T) = nC_1 + C_2 \sum_{t=0}^{T-1} F(t) \quad 21.24$$

Discounting usually is not included in the total cost formula since the time periods are considered short. If the equipment has an unusually long life, discounting is required.

There are some cases where group replacement always is more expensive than replacing just the failures as they occur. Group replacement will be the most economical policy if equation 21.25 holds.

$$C_2[\overline{F(t)}] > \frac{K(T)}{T}\bigg|_{\text{minimum}} \quad 21.25$$

If the opposite inequality holds, group replacement still may be the optimum policy. Further analysis is required.

PART 5: FORTRAN Programming

The FORTRAN language currently exists in several versions. Although differences exist between compilers, these are relatively minor. However, some of the instructions listed in this chapter may not be compatible with all compilers.

This section is not intended as instruction in FORTRAN programming, but rather serves as a documentation of the language.

1 STRUCTURAL ELEMENTS

Symbols are limited to the upper-case alphabet, digits 0 through 9, the blank, and the following special characters.

$$+ = -\ ^{*}/\ (\ )\ ,\ .\ \$$$

Statements written with these characters generally are prepared in an 80-column format. Statements are executed sequentially regardless of the statement numbers.

position	use
1	The letter C is used for a *comment*. Comments are not executed.
2–5	The statement number, if used, is placed in positions 2 through 5. Statement numbers can be any integers from 1 through 9999.
6	Any character except *zero* can be placed in position 6 to indicate a continuation from the previous statement.
7–72	The FORTRAN statement is placed here.
73–80	These positions are available for any use and are ignored by the compiler. Usually, the final debugged program is numbered sequentially in these positions.

FORTRAN compilers pack the characters. Therefore, blanks can be inserted at any place in most statements. For example, the following statements are compiled the same way.

    IF (AGE.LT.YEARS) GO TO 10
    IF(AGE.LT.YEARS)GOTO10

2 DATA

Numerical data can be either real or integer. *Integers* usually are limited to nine digits. Unsigned integers and integers preceded by a plus sign are the same. Commas are not allowed in integer constants. For example, ninety thousand would be written as 90000, not 90,000.

Real numbers are distinguished from integers by a decimal point and may contain a fractional part. Scientific notation is indicated by the single letter E. Real numbers are limited to one decimal point and usually seven digits.

value	FORTRAN Notation
2 million	2. E6
.00074	7.4 E−4
2.	2.

3 VARIABLES

Variable names can be formed from up to six alphanumeric characters. The first character must be a letter. Variable names starting with the letters I, J, K, L, M, or N are assumed by the compiler to be integers unless defined otherwise by an *explicit typing statement*. All other variable names represent real variables, unless explicitly typed.

The type convention can be overridden in an explicit typing statement. This is done by defining the desired variable type in the first part of the program with an INTEGER or REAL statement. For example, the statements

    INTEGER TIME,CLOCK
    REAL INSTANT

would establish TIME and CLOCK as integer variables and INSTANT as a real variable. The order of such declarations is unimportant. Variables following the standard type convention (implicit typing) do not have to be declared.

Subscripted variables with up to seven dimensions are allowed. They always must be defined in size by the DIMENSION statement. For example, the statements

    DIMENSION SAMPLE(5)
    REAL DIMENSION INCOME(2,7)

would establish a 1×5 real *array* called SAMPLE and a 2×7 real array called INCOME. INCOME would have been an integer array without the REAL declaration.

Elements of arrays are addressed by placing the subscripts in parentheses.

    SAMPLE(2)
    INCOME(1,6)

The subscripts also can be variables. SAMPLE(K) would be permitted as long as K was defined, was between 1 and 5, and was an integer.

Variables and arrays once defined and declared are not initialized automatically. If it is necessary to initialize a storage location prior to use, the DATA statement can be used. Consider the following statements.

```
REAL X,Y,Z
DIMENSION ONEDIM(5)
DIMENSION TWODIM(2,3)
DATA X,Y,Z/3*0.0/(ONEDIM(I),I=1,5)
1/5*0.0/ ((TWODIM(I,J),J=1,3),I=1,2)
2/1.,2.,3.,4.,5.,6./
```

Variables X, Y, and Z will be initialized to 0.0. The entries in ONEDIM will have the values (0,0,0,0,0). The TWODIM array will be initialized to

$$\begin{pmatrix} 1.0 & 2.0 & 3.0 \\ 4.0 & 5.0 & 6.0 \end{pmatrix}$$

After being initialized with a DATA statement, variables can have their values changed by arithmetic operations.

4 ARITHMETIC OPERATIONS

FORTRAN provides for the usual arithmetic operations. These are listed in table 21.2.

Table 21.2
FORTRAN OPERATORS

symbol	meaning
=	replacement
+	addition
−	subtraction
*	multiplication
/	division
**	exponentiation
. ( )	preferred operation

The *equals* symbol is used to replace one quantity with another. For example, the following statement is algebraically incorrect. However, it is a valid FORTRAN statement.

$$Z = Z + 1$$

Each statement is scanned from left-to-right (except that a right-to-left scan is made for exponentiation). Operations are performed in the following order.

exponentiation first
multiplication and division second
addition and subtraction last

Parentheses can be used to modify this hierarchy.

Each operation must be stated explicitly and unambiguously. Thus, AB is not a substitute for A^*B. Two

operations in a row, as in $(A + -B)$ also are unacceptable. Some FORTRAN compilers allow mixed-mode arithmetic. Others, ANSI FORTRAN among them, require all variables in an expression to be either integer or real. Where mixed-mode arithmetic is permitted, care must be taken in the conversion of real data to the integer mode.

Integer variables used to hold the results of a mixed-mode calculation will have their values truncated. This is illustrated in the following example.

Example 21.8

Evaluate J in the following expression.

$$J = (6.0 + 3.0)^*3.0/6.0 + 5.0 - 6.0^{**}2.0$$

The expressions within parentheses are evaluated in the first pass.

$$J = 9.0^*3.0/6.0 + 5.0 - 6.0^{**}2.0$$

The exponentiation is performed in the second pass.

$$J = 9.0^*3.0/6.0 + 5.0 - 36.0$$

The multiplication and division are performed in the third pass.

$$J = 4.5 + 5.0 - 36.0$$

The addition and subtraction are performed in the fourth pass.

$$J = -26.5$$

However, J is an integer variable, so the real number -26.5 is truncated and converted to integer. The final result is -26.

5 PROGRAM LOOPS

Loops can be constructed from IF and GO TO statements. However, the DO statement is a convenient method of creating loops. The general form of the DO statement is

$$\text{DO } s\, i = j, k, l$$

where s is a statement number.

i is the integer loop variable.
j is the initial value assigned to i.
k, which must exceed j, is an inclusive upper bound on i.
l is the increment for i, with a default value of 1 if omitted.

The DO statement causes the execution of the statements immediately following it through statement s until i equals k or greater.[1] A loop can be *nested* by plac-

---

[1] A peculiarity of FORTRAN DO statements is that they are executed at least once, regardless of the values of j and k.

ing it within another loop. The loop variable may be used to index arrays.

When i equals or exceeds k, the statement following s is executed. However, the loop may be exited at any time before i reaches k if the logic of the loop provides for it.

6 INPUT/OUTPUT STATEMENTS

The READ, WRITE, and FORMAT statements are FORTRAN's main I/O statements. Forms of the READ and WRITE statements are

READ (u_1, s) [list]
WRITE (u_2, s) [list]

where u_1 is the unit number designation for the desired input device, usually 5 for the card reader.
u_2 is the unit designation for the desired output device, usually 6 for the line printer.
s is the statement number of an associated FORMAT statement.
[list] is a list of variables separated by commas whose values are being read or written.

The [list] also can include an implicit DO loop. The following example reads six values, the first five into the array PLACE and the last into SHOW.

READ (5,85) (PLACE(J),J=1,5) SHOW

The purpose of the FORMAT statement is to define the location, size, and type of the data being read. The form of the FORMAT statement is

s FORMAT [field list]

As before, s is the statement number. [field list] consists of specifications, set apart by commas, defining the I/O fields. [list] can be shorter than [field list].

The format code for integer values is nIw.

n is an optional repeat counter which indicates the number of consecutive variables with the same format. w is the number of character positions.

The format codes for real values are

$$nFw.d \quad \text{or} \quad nEw.d$$

Again, w is the number of character positions allocated, including the space required for the decimal point. d is the implied number of spaces to the right of the decimal point. In the case of input data, decimal points in any position take precedence over the value of d.

The F format will print a total of $(w-1)$ digits or blanks representing the number. The E format will print a total of $(w-2)$ digits or blanks and give the data in a standard scientific notation with an exponent.

Other formats which can be used in the FORMAT statement are

X	horizontal blanks
/	skipping lines
H	alphanumeric data
D	double precision real
T	position (column) indicator
Z	hexadecimal
P	decimal point modification
L	logical data
' '	literal data

The usual output device is a line printer with 133 print positions. The first print position is used for *carriage control*. The data (control character) in the first output position will control the printer advance according to the rules in table 21.3.

Table 21.3
FORTRAN Printer Control Characters

control character	meaning
blank	advance one line
0	advance two lines
1	skip to line one on next page
+	do not advance (overprint)

Carriage control usually is accomplished by the use of literal data. Consider the following statements.

```
INTEGER K
K=193
WRITE (6,100) K
WRITE (6,101) K
100 FORMAT (' ',I3)
101 FORMAT (I3)
```

The above program would print the number 193 on the next line of the current page and the number 93 on the first line of the next page.

Data can be written to or read from an array by including the array subscripts in the I/O statement.

```
DIMENSION CLASS (2,5)
READ (5,15) ((CLASS(I,J),I=1,2),J=1,5)
15    FORMAT (10F3.0)
```

7 CONTROL STATEMENTS

The STOP statement is used to indicate the logical end of the program. The format is s STOP.

STOP should not be the last statement. When it is reached, program execution is terminated. The value of s is printed out or made available to the next program step. The use of STOP rarely is recommended.

The END statement is required as the last statement. It tells the compiler that there are no more lines in the program to be compiled. A program cannot be compiled or executed without an END statement.

The PAUSE statement will cause execution to stop temporarily. Its format is *s* PAUSE.

When the PAUSE statement is reached, the number *s* is transmitted to the computer operator. This gives the operator a chance to set various control switches on the console (the choice of switches being dependent on the value of *s* and the program logic), prior to pushing the START button. The PAUSE statement is used only if the programmer is operating the computer.

The CALL statement is used to transfer execution to a *subroutine*. CALL EXIT will terminate execution and turn control over to the operating system. The CALL EXIT and STOP statements have similar effects. The RETURN statement ends execution of a program called subroutine and passes execution to the main program. The CONTINUE statement does nothing. It can be used with a statement number as the last line of a DO loop.

The GO TO [s] statement transfers control to statement *s*.

The arithmetic IF statement is written

$$IF[e]s_1, s_2, s_3$$

[e] is any numerical variable or arithmetic expression, and the s_i are statement numbers. The transfer occurs according to the following table.

[e]	statement executed
[e] < 0	s_1
[e] = 0	s_2
[e] > 0	s_3

The logical IF statement has the form

$$IF[le][statement]$$

[le] is a logical expression, and [statement] is any executable statement except DO and IF. Only if [le] is true will [statement] be executed. Otherwise, the next instruction will be executed.

The logical expression [le] is a relational expression using one of several operators.

Logical expressions also can incorporate the connectors .AND., .OR., and .NOT..

Table 21.4
FORTRAN Logical Operations

operator	meaning
.LT.	less than
.LE.	less than or equal to
.EQ.	equal to
.NE.	not equal to
.GT.	greater than
.GE.	greater than or equal to

Example 21.9

$$IF\ (A.GT.25.6)\ A = 27.0$$

Meaning: If A is greater than 25.6, set A equal to 27.0.

$$IF\ (Z.EQ.(T-4.0).OR.Z.EQ.0.)\ GO\ TO\ 17$$

Meaning: If Z is equal to (T−4.0) or if Z is equal to *zero*, go to statement 17.

8 LIBRARY FUNCTIONS

The following single-precision library functions are available. Most are accessed by placing the argument in parentheses after the function name. Placing the letter D before the function name will cause the calculation to be performed in double precision. Arguments for trigonometric functions are expressed in radians.

Table 21.5
Some FORTRAN Library Functions

function	use
EXP	e^x
ALOG	natural logarithm
ALOG10	common logarithm
SIN	sine
COS	cosine
TAN	tangent
SINH	hyperbolic sine
SQRT	square root
ASIN	arcsine
MOD	remaindering modulus (integer)
AMOD	remaindering modulus (real)
ABS	absolute value (real)
IABS	absolute value (integer)
FLOAT	convert integer to real
FIX	convert real to integer

9 USER FUNCTIONS

A user-defined function can be created with the FUNCTION statement. Such functions are governed by the following rules.

- The function is defined as a variable in the main program even though it is a function.

- When used in the main program, the function is followed by its arguments in parentheses.

- In the function itself, the function name is type-declared and defined by the word FUNCTION.

- The arguments (parameters) need not have the same names in the main program and function.

- Only the function has a RETURN statement.

- Both the main program and the function have END statements.

- The arguments (parameters) must agree in number, order, type, and length.

These construction rules are illustrated by example 21.10.

Example 21.10

```
REAL HEIGHT, WIDTH, AREA, MULT
HEIGHT= 2.5
WIDTH= 7.5
AREA=MULT(HEIGHT,WIDTH)
END

REAL FUNCTION MULT(HEIGHT,WIDTH)
REAL HEIGHT, WIDTH
MULT=HEIGHT*WIDTH
RETURN
END
```

10 SUBROUTINES

A subroutine is a user-defined subprogram. It is more versatile than a user-defined function as it is not limited to mathematical calculations. Subroutines are governed by the following rules.

- The subroutine is activated by the CALL statement.

- The subroutine has no type.

- The subroutine does not take on a value. It performs operations on the arguments (parameters) which are passed back to the main program.

- A subroutine has a RETURN statement.

- Both the main program and the subroutine have END statements.

- The arguments (parameters) need not have the same names in the main program and the sub-routine.

- The arguments (parameters) must agree in number, order, type, and length.

- It is possible to return to any part of the main program. It is not necessary to return to the statement immediately below the CALL statement.

These rules are illustrated by example 21.11.

Example 21.11

```
    REAL HEIGHT, WIDTH, AREA
    CALL GET(HEIGHT, WIDTH)
    AREA=HEIGHT*WIDTH
    END

    SUBROUTINE GET(A,B)
    REAL A,B
    READ (5,100) A,B
100 FORMAT(2F3.1)
    RETURN
    END
```

Variables in functions and subroutines are completely independent of the main program. Subroutine and main program variables which have the same names will not have the same values. A link between the main program and the subroutine can be established, however, with the COMMON statement.

The COMMON statement assigns storage locations in memory to be shared by the main program and all of its subroutines. Even the COMMON statement, however, allows different names. It is the order of the common variables which fixes their position in upper memory.

Example 21.12

What are the values of X and Y in the subroutine?

```
COMMON X, Y      main program
    X=2.0
    Y=10.0

COMMON Y, X      subroutine
```

Since Y is the first common subroutine variable which corresponds to X in the main program, Y=2.0. Similarly, X=10.0.

If variables are to be shared with only some of the subroutines, the *named* COMMON statement is required. Whereas there can be only one regular COMMON statement, there can be multiple-named COMMON statements.

```
COMMON/PLACE/CAT, COW, DOG main program
COMMON/PLACE/HORSE, PIG, EXPENSE
                                subroutine
```

PART 6: Fire Safety Systems

1 INTRODUCTION

In many cases, the design of fire detection, fire alarm, and sprinkler systems is governed by state or local codes. It is necessary to review all applicable codes and to meet the most stringent of them. Generally, insurance carrier requirements are more stringent than code minimums. Although codes mandate minimum standards, common sense and professional prudence should be used in specifying the level of fire protection in a building.

In buildings that are partially or wholly sprinklered, provisions for alarm and evacuation as well as provisions for sprinkler supervision must be made.

The type of occupancy greatly affects the degree of protection. Nursing homes, schools, hospitals, and office buildings all have greatly different needs. Furthermore, multi-story buildings with limited escape routes affect the degree of protection.

2 DETECTION DEVICES

- **Manual Fire Alarm Stations**
  Mandatory in any system, large or small. Locate in the natural path of exit with the maximum traveling distance to any manual station of 200 feet. Identification of an activated manual station which when opened should be readily visible from the side, down a corridor for at least 200 feet.

- **Heat Detectors—Fixed Temperature**
  135°F in open spaces. 190°–200°F generally used in enclosed or confined spaces such as boiler rooms, closets, etc., where the heat build-up will be fast and confined.

- **Heat Detectors—Rate of Rise**
  Combination rate of rise and fixed temperature of 135° or 200°F. Rate of rise portion operates when the temperature rises in excess of 15° per minute. More sensitive than the fixed temperature detector.

- **Heat Detector—Rate Compensated**
  Considered to be the most responsive of all thermal detectors. Operates at 135° or 200°F. Detects both slow and fast developing fires by anticipating the temperature increase and moving towards the alarm point as the temperature gradually increases.

- **Photoelectric Smoke Detectors**
  Operates on a photo beam or light scattering principle. The photoelectric detector responds best to products of combustion or smoke with a particle size from approximately 10.0 microns down to 0.1 micron and of the proper concentration. Proper concentration is defined by Underwriters Laboratories as the ability to sense smoke in the 0.2 to 4.0 percent obscuration per foot range. Photoelectric detectors generally are considered to be the best for cold smoke fires.

- **Ionization Detectors**
  Ionization detectors detect products of combustion by sensing the disruption of conductivity in an ionized chamber due to the presence of smoke. The ionization detector responds best to fast burning fires where particle sizes range from approximately 1.0 micron down to 0.01 micron and of the proper concentration. Proper concentration is defined by Underwriters Laboratories as the ability to sense smoke from 0.2 to 4.0 percent obscuration per foot range. **Note:** Ionization and photoelectric detectors can be intermixed within a system to provide the best form of detection suitable to the environment.

- **Infrared Flame Detectors**
  Generally, infrared detectors respond to radiation in the 6,500 to 8,500 angstrom range. Good detectors will filter out solar interference and respond to radiation in the 4,000 to 5,500 angstrom range. It is preferrable to use a detector with a dual sensing circuit in order to discriminate against unwanted or false alarms.

- **Ultraviolet Flame Detectors**
  Ultraviolet flame detectors respond to radiation in the spectral range of 1,700 to 2,900 angstroms. It is not sensitive to solar interference. Built-in time delays prevent false alarms.

- **Waterflow Detectors**
  Used in wet sprinkler systems to indicate a flow of water. Use on the main sprinkler risers and throughout the building to indicate sub-sections or floors of the building to locate sprinkler discharges quickly. These detectors employ a retard mechanism to prevent false alarms from water surges.

- **Pressure Switches**
  Used in dry or pre-action sprinkler systems to provide an alarm when water is discharged.

- **Valve Monitor Switches**
  Closed water supply valves are a major weakness of sprinkler systems. This is the most frequently neglected and forgotten item in the sprinkler system. Closed valves also have accounted for countless millions of dollars worth of damage because the system would not operate.

- **Low Temperature Monitor Switches**
  Low air temperature (under 40°F) or low water temperature is detrimental to a sprinkler system and creates a great nuisance by freezing and bursting the sprinkler system pipes.

3 ALARM DEVICES

- **Bells**

  This is the most commonly accepted form of alarm. However, bells should not be used in schools where bells are used for other signaling purposes. Bells should not be used in any area where the same sound is used for any other function.

- **Chimes**

  Single stroke devices in coded systems and used in certain types of applications such as quiet areas of hospitals or nursing homes.

- **Horns**

  Horns generally are capable of producing a higher sound level than either a bell or a chime.

- **Alarm Lights**

  Used individually as a fire alarm visual indicator or used in conjunction with a horn. The light can be either a flashing incandescent bulb or a flashing strobe. It sometimes is required by certain codes or types of occupancy in order to provide an alarm for handicapped persons.

- **Remote Annunciators**

  Generally, these duplicate the main control panel and have a light for every fire alarm zone. They are used at the second entrance or at the main entrance to assist the fire department in locating the fire zone. Also used in nursing homes, at nursing stations in hospitals or in the engineering room of a factory or a building.

- **Speakers**

  Used in emergency voice evacuation systems, generally in high-rise buildings. Specific quality, construction, and performance have been established by the NFPA code for speakers used in voice evacuation systems.

4 SIGNAL TRANSMISSION

- **Reverse Polarity**

  Uses a dedicated leased telephone directly between the protected premises and the municipal fire department or a commercial central station.

- **Central Station**

  These are central monitoring facilities which monitor fire and security alarms from protected premises for a monthly fee.

- **Telephone Dialers**

  Tape dialers have had a very bad reputation due to high failure rate and susceptibility to false alarms. Solid state digital dialers with compatible solid state digital receivers have increased the reliability and dependability of this product very dramatically and

are becoming more acceptable as an alternate form of signal transmission.

- **Radio Transmitters**

  Radio transmitter boxes are used in certain applications to transmit alarms between the protected premises and some monitoring point. Before using, it is advisable to check that the line of transmission is clear from obstruction and interference.

5 AUXILIARY CONTROL

- **Smoke Doors**

  Generally held open with floor or wall mounted electromagnets. Generally all doors close in all parts of the building on the first alarm.

- **Fire Doors**

  Generally treated the same way as smoke doors.

- **Stairwell Exit Doors**

  Sometimes for security reasons, these doors are held latched with a door strike. The door also must employ a panic exit bar to override and manually open the door.

- **Elevator Capture**

  Generally accepted by all codes that the elevator immediately return to the first floor or to some designated alternate floor. The specification should designate that the elevator manufacturer is responsible for accepting low voltage signal or a dry contact from the fire alarm panel and programming the elevator to return to the designated floor.

- **Fire Dampers**

  Either motorized or fusible link type are closed to prevent the spread of smoke to other areas. Care should be exercised in connecting these to a fire alarm system because a fire alarm system generally operates with small amounts of D.C. power. The responsibility should be stated for coordinating the voltages and contact ratings necessary to do the job.

- **Fans—Supply and Exhaust**

  Sometimes all fans are shut down on the first alarm. However, some buildings and other considerations make it desirable to exhaust the smoke from the building and shut down the input supply fans.

- **Pressurization**

  Pressurization creates positive air pressure to inhibit the influx of smoke from adjacent areas or floors above or below. Exit stairwell escape routes out of the high-rise building should be pressurized to provide a smoke-free exit path.

6 SPECIAL CONSIDERATIONS

- **Handicapped Persons**

  There may be requirements from Federal health officials (HEW) or OSHA to provide consideration for handicapped persons.

- **Weather Protection of Devices**
  Check to make sure that none of the devices will be exposed to adverse conditions such as excessive moisture.

- **Open Plenums**
  Drop ceilings that use the space above as an open air plenum present special problems and are treated as a potential hazard in most codes. Smoke detectors located in open air plenums need to be located properly, and it is advisable to locate them so that they are accessible for maintenance. A remote alarm lamp should be brought down and mounted below the ceiling level to indicate which detector is an alarm.

- **Duct Detectors**
  Duct heat detectors are considered to be of negligible value. Duct smoke detectors can provide warning of smoke in an air duct and can prevent costly damage to expensive HVAC equipment. Duct detectors are not a substitute for open area smoke detectors. Problem areas can develop with duct detectors from excessive humidity, poor or no maintenance, and improper location.

- **Sound Pressure Level of Alarm Devices**
  The sound pressure level required to provide an adequate alarm in the environment in which it is intended to operate generally is considered to have been accomplished if the alarm sound is 12 decibels above the normal ambient level.

- **Emergency Generators**
  If the fire alarm system does not use its own standby battery pack, it is advisable to coordinate the details of how the emergency generator feeds power back into the building distribution system and to insure that the fire alarm system will be provided with emergency power. In some cases, it is advisable to provide the fire alarm system with a small amount of standby battery power in order to keep the fire alarm system on line until the generator gets to full power.

- **Fire Pump Supervision**
  Insure that fire pumps, if used, are supervised adequately for all critical functions and that the building fire alarm system is provided with one or more zones to interface with fire pump signals.

7 SPECIAL SUPPRESSION SYSTEMS

- **Deluge Systems**
  The control panel for the water deluge system is specialized and generally is mounted in a hostile environment. The control equipment is mounted inside an enclosure which is watertight and dust tight. Deluge systems are actuated manually and also by thermal detectors or flame detectors. The entire system, including the electronic control panel, is provided by the sprinkler contractor.

- **Foam Systems**
  High expansion foam or water suppression systems are specialized systems requiring special handling. They generally are activated by infrared or ultraviolet flame detectors.

- **Halon Systems**
  Halon systems generally are used in clean environments such as computer rooms or any room that contains high value equipment. Examples of this are tapes, microfilms, computer records, laboratories for research and design, and medical laboratories. Generally, the criterion is the high value of the equipment or the data stored, and it is desirable that water not touch the equipment. Halon systems, because of their great expense, usually are engineered specially for the particular room in which they are to be located. The control panel uses cross-zoning techniques to prevent unnecessary discharges.

- **High/Low Pressure CO_2 Systems**
  CO_2 systems are used in industrial applications to suppress fires where the use of water, extinguishing powders, or Halon would be unsuitable because of expense, hazard, or the size of the equipment to be protected. Because CO_2 is hazardous to life, it has to be an engineered system designed for the particular application.

PART 7: Nondestructive Testing

1 MAGNETIC PARTICLE TESTING

This procedure is based on the attraction of magnetic particles to leakage flux at surface flaws. The particles accumulate and become visible at the flaw. This method works for magnetic materials in locating cracks, laps, seams, and in some cases, subsurface flaws. The test is fast and simple and is easy to interpret. However, parts must be relatively clean and demagnetized. A high current (power) source must be available.

2 EDDY CURRENT TESTING

Alternating currents from a source coil induce eddy currents in metallic objects. Flaws and other material properties affect the current flow. The change in current flow is observed on a meter or a screen. This method can be used to locate defects of many types, including changes in composition, structure, and hardness, as well as locating cracks, voids, inclusions, weld defects, and changes in porosity.

Intimate contact between the material and the test coil is not required. Operation can be continuous, automatic, and monitored electronically. Therefore, this method is ideal for unattended continuous processing. Sensitivity is easily adjustable. Many variables, however, can affect the current flow.

3 LIQUID PENETRANT TESTING

Liquid penetrant (dye) is drawn into surface defects by capillary action. A developer substance then is used to develop the penetrant to aid in visual inspection. This method can be used with any nonporous material, including metals, plastics, and glazed ceramics. It locates cracks, porosities, pits, seams, and laps.

Liquid penetrant tests are simple to perform, can be used with complex shapes, and can be performed on site. Parts must be clean, and only small surface defects are detectable.

4 ULTRASONIC TESTING

Mechanical vibrations in the 0.1–25 MHz range are induced in an object. The transmitted energy is reflected and scattered by interior defects. The results are in-terpreted from a screen or a meter. The method can be used for metals, plastics, glass, rubber, graphite, and concrete. It detects inclusions, cracks, porosity, laminations, changes in structure, and other interior defects.

This test is extremely flexible. It can be automated and is very fast. Results can be recorded or interpreted electronically. Penetration of up to 60 feet of steel is possible. Only one surface needs to be accessed. However, rough surfaces or complex shapes may cause difficulties.

5 INFRARED TESTING

Infrared radiation emitted from objects can be detected and correlated with quality. The detection can be recorded electronically. Any discontinuity that interrupts heat flow, such as flaws, voids, and inclusions, can be detected.

Infrared testing requires access to only one side, and it is highly sensitive. It is applicable to complex shapes and assemblies of dissimilar components, but it is relatively slow. Results are affected by material variations, coatings, and colors, and hot spots can be hidden by cool surface layers.

6 RADIOGRAPHY

X-ray and gamma-ray sources can be used to penetrate objects. The intensity is reduced in passing through, and the intensity changes are recorded on film or screen. This method can be used with most materials to detect internal defects, material structure, and thickness. It also can be used to detect the absence of internal parts.

Up to 30 inches of steel can be penetrated by x-ray sources. Gamma sources, which are more portable and lower in cost than x-ray sources, are applicable to 10″ thickness of steel.

There are health and government standards associated with these tests. Electric power and cooling water may be required in large installations. Shielding and film processing also is required, making this the most expensive nondestructive test.

PART 8: Nomograph Construction

A nomograph for an equation relating two variable quantities consists of two scales and a pivot point. The solution of the equation is achieved by drawing a straight *tie line* to connect the scales.

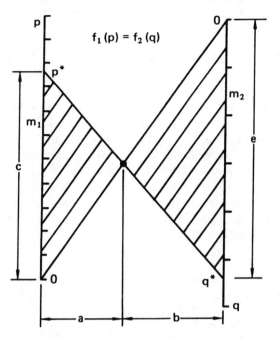

Figure 21.1 A Nomograph

The construction of a conversion chart is shown in figure 21.1. Assume that the relationship between two variable quantities, p and q, is expressed by a conversion equation of the form

$$\mathbf{f}_1(p) = \mathbf{f}_2(q)$$

$\mathbf{f}_1(p)$ and $\mathbf{f}_2(q)$ are algebraic functions of a single variable. The vertical p and q lines are located a con-

venient distance apart. The two scales run in opposite directions from a base line joining their origins and are graduated according to the scale equations.

$$d_p = m_1 \mathbf{f}_1(p)$$
$$d_q = m_2 \mathbf{f}_2(q)$$

m_1 and m_2 are the moduli of the p and the q scales, respectively. A modulus is a constant introduced into an equation to control the length of scale for a desired range of values of the variable.

A straight line can be drawn to connect any two values, p^* and q^*, on the two scales. The distance c on the p scale is

$$c = m_1 \mathbf{f}_1(p^*)$$

Similarly,

$$e = m_2 \mathbf{f}_2(q^*)$$

The tie line connecting p^* and q^* intersects the base line at point p to form two similar triangles in which corresponding sides and altitudes are in the same ratio, $c/e = a/b$. Substituting equivalent values for c and e,

$$\frac{m_1 \mathbf{f}_1(p^*)}{m_2 \mathbf{f}_2(q^*)} = \frac{a}{b}$$

Hence, $bm_1 \mathbf{f}_1(p^*) = am_2 \mathbf{f}_2(q^*)$. In order to reduce this equation to the original equation, $\mathbf{f}_1(p) = \mathbf{f}_2(q)$, the coefficients of the two terms must be equal.

$$bm_1 = am_2 \text{ or } \frac{a}{b} = \frac{m_1}{m_2}$$

Thus, the pivot point location must be such that the distances, a and b, are in the same ratio as the moduli of the two scales.

PART 9: Pressure Vessels and Piping

Nomenclature

E	minimum joint efficiency	–
H	hydrostatic test pressure	psi
P	design pressure	psi
R	inside radius, before corrosion allowance	inches
S	maximum allowable stress	psi
t	minimum thickness, exclusive of corrosion allowance	in

1 INTRODUCTION

In 1911, ASME set up the Boiler and Pressure Vessel Committee. Its function is to establish rules governing the design, fabrication, and inspection and repair during construction of boilers and pressure vessels; and to interpret these rules when questions arise. The rules are embodied in the ASME Boiler and Pressure Vessel Code.[2]

2 PIPING DRAWINGS

To simplify the preparation of working drawings of piping systems, the set of symbols shown in table 21.6 has been developed to represent the various pipes, fittings, and valves in common use.

3 DIVISION I REQUIREMENTS FOR CONSTRUCTION OF PRESSURE VESSELS

When the wall thickness does not exceed half the inside radius or when P does not exceed $(0.385)SE$, equation 21.26 should be used for cylindrical shells.

---

[2]This section is a brief summary of division 1 of Section VIII of the ASME Boiler and Pressure Vessel Code, "Pressure Vessels." Division 2 allows thinner walls (or higher pressures) but requires more complex analysis and extensive testing/inspection.

$$t = \frac{PR}{SE - 0.6P} \qquad \text{21.26}$$

When the thickness of a wholly-spherical shell or a hemispherical shell does not exceed $(0.356)R$ or when P does not exceed $(0.665)SE$, the following formula should be used.

$$t = \frac{PR}{2SE - 0.2P} \qquad \text{21.27}$$

The *design pressure*, P, is the maximum difference in pressure between the inside and the outside of a vessel or between any two chambers of a combination unit. For existing vessels, the maximum allowable working pressure replaces the design pressure in the preceding formulas. The maximum allowable working pressure for a vessel is the maximum pressure permissible at the top of the vessel in its normal operating position at the operating temperature.

For seamless shells, $E = 1.00$. For arc and gas welded vessels, the joint efficiency depends on the type of joint and the degree of examination. $E = 1.00$ for butt welds with fully radiographed joints, $E = 0.85$ for spot radiographed joints, and $E = 0.70$ for joints not radiographed.

Vessels designed for internal pressure shall be subjected to a hydrostatic test pressure of at least $1\frac{1}{2}$ times the design pressure multiplied by the lowest ratio (for the materials of which the vessel is constructed) of the stress value, S, for the test temperature to the stress value, S, for the design temperature.

$$H = \frac{(1.5)PS_{\text{test temperature}}}{S_{\text{design temperature}}} \qquad \text{21.28}$$

Table 21.6
American Standard Piping Symbols

	Flanged	Screwed	Bell & Spigot	Welded	Soldered
Joint					
Elbow—90°					
Elbow—45°					
Elbow—Turned Up					
Elbow—Turned Down					
Elbow—Long Radius					
Reducing Elbow					
Tee					
Tee—Outlet Up					
Tee—Outlet Down					
Side Outlet Tee—Outlet Up					
Cross					
Reducer—Concentric					
Reducer—Eccentric					
Lateral					
Gate Valve					
Globe Valve					
Check Valve					
Stop Cock					
Safety Valve					
Expansion Joint					
Union					
Sleeve					
Bushing					

PART 10: Methods of Machine Safeguarding

Safeguards must meet the following minimum requirements.

- Prevent contact: The safeguard must prevent hands, arms, or any other part of a worker's body from making contact with dangerous moving parts. A good safeguarding system eliminates the possibility of the operator or another worker placing his hands near hazardous moving parts.

- Be secure: Workers should not be able to remove easily or to tamper with the safeguard. A safeguard that can be easily made ineffective is no safeguard at all. Guards and safety devices should be made of durable material that will withstand the conditions of normal use. They must be secured firmly to the machine.

- Protect from falling objects: The safeguard should ensure that no objects can fall into moving parts. A small tool which is dropped into a cycling machine could become a projectile that could strike and injure someone.

- Create no new hazards: A safeguard defeats its purpose if it creates a hazard of its own, such as a shear point, a jagged edge, or an unfinished surface which can cause a laceration. The edges of guards, for instance, should be rolled or bolted in such a way that they eliminate sharp edges.

- Create no interference: Any safeguard which impedes a worker from performing the job quickly and comfortably might soon be overridden or disregarded. Proper safeguarding actually can enhance efficiency since it can relieve the worker's apprehensions about injury.

- Allow safe lubrication: If possible, one should be able to lubricate the machine without removing the safeguards. Locating oil reservoirs outside the guard, with a line leading to the lubrication point, will reduce the need for the operator or the maintenance worker to enter the hazardous area.

There are many ways to safeguard machinery. The type of operation, the size and shape of stock, the method of handling, the physical layout of the work area, the type of material, and production requirements or limitations will help determine the appropriate safeguarding method for the individual machine.

As a general rule, power transmission apparatus is best protected by fixed guards that enclose the danger area. For hazards at the point of operation, where moving parts actually perform work on stock, several kinds of safeguarding are possible.

- Guards
  —fixed
  —interlocked
  —adjustable
  —self-adjusting

- Devices
  —presence sensing
    1. photoelectric (optical)
    2. radiofrequency (capacitance)
    3. electromechanical
  —pullback
  —restraint
  —safety controls
    1. safety trip control
       a) pressure-sensitive body bar
       b) safety tripod
       c) safety tripwire cable
    2. two-hand control
    3. two-hand trip
  —gates
    1. interlocked
    2. other

- Remote or Distant Operation

- Feeding and Ejection Methods for the Operator
  —automatic feed
  —semi-automatic feed
  —automatic ejection
  —semi-automatic ejection
  —robot

- Miscellaneous Aids
  —awareness barriers
  —miscellaneous protective shields
  —hand-feeding tools and holding fixtures

RESERVED FOR FUTURE USE

RESERVED FOR FUTURE USE

22

SYSTEMS OF UNITS

1 CONSISTENT SYSTEMS OF UNITS

A set of units used in a problem is said to be *consistent*[1] if no conversion factors are needed. For example, a moment with units of foot-pounds cannot be obtained directly from a moment arm with units of inches. In this illustration, a conversion factor of $\frac{1}{12}$ feet/inch is needed, and the set of units used is said to be *inconsistent*.

On a larger scale, a system of units is said to be consistent if Newton's second law of motion can be written without conversion factors. Newton's law states that the force required to accelerate an object is proportional to the amount of matter in the item.

$$F = ma \qquad 22.1$$

The definitions of the symbols, F, m, and a, are familiar to every engineer. However, the use of Newton's second law is complicated by the multiplicity of available unit systems. For example, m may be in kilograms, pounds, or slugs. All three of these are units of mass. However, as figure 22.1 illustrates, these three units do not represent the same amount of mass.

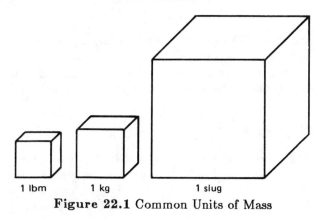

1 lbm   1 kg   1 slug

Figure 22.1 Common Units of Mass

It should be mentioned that the decision to work with a consistent set of units is arbitrary and unnecessary. Problems in fluid flow and thermodynamics commonly

are solved with inconsistent units. This causes no more of a problem than working with inches and feet in the calculation of a moment.

2 THE ABSOLUTE ENGLISH SYSTEM

Engineers are accustomed to using pounds as a unit of mass. For example, density typically is given in pounds per cubic foot. The abbreviation *pcf* tends to obscure the fact that the true units are pounds of *mass* per cubic foot.

If pounds are the units for mass, and feet per second squared are the units of acceleration, the units of force for a consistent system can be found from Newton's second law.

$$\text{units of } F = (\text{units of } m)(\text{units of } a)$$
$$= (\text{lbm})(\text{ft/sec}^2) = \frac{\text{lbm-ft}}{\text{sec}^2} \qquad 22.2$$

The units for F cannot be simplified any more than they are in equation 22.2. This particular combination of units is known as a *poundal*.[2]

The absolute English system, which requires the poundal as a unit of force, seldom is used, but it does exist. This existence is a direct outgrowth of the requirement to have a consistent system of units.

3 THE ENGLISH GRAVITATIONAL SYSTEM

Force frequently is measured in pounds. When the thrust on an accelerating rocket is given as so many pounds, it is understood that the pound is being used as a unit of force.

If acceleration is given in feet per second squared, the units of mass for a consistent system of units can be determined from Newton's second law.

---

[1] The terms *homogeneous* and *coherent* also are used to describe a consistent set of units.

[2] A poundal is equal to .03108 pounds force.

$$\text{units of } m = \frac{\text{units of } F}{\text{units of } a}$$

$$= \frac{\text{lbf}}{\text{ft/sec}^2} = \frac{\text{lbf-sec}^2}{\text{ft}} \qquad 22.3$$

The combination of units in equation 22.3 is known as a *slug*.[3] Slugs and pounds-mass are not the same, as illustrated in figure 22.1. However, units of mass can be converted, using equation 22.4.

$$\# \text{ slugs} = \frac{\# \text{ lbm}}{g_c} \qquad 22.4$$

g_c[4] is a dimensional conversion factor having the following value.

$$g_c = 32.1740 \frac{\text{lbm-ft}}{\text{lbf-sec}^2} \qquad 22.5$$

32.1740 commonly is rounded to 32.2 when six significant digits are unjustified. That practice is followed in this book.

Notice that the number of slugs cannot be determined from the number of pounds-mass by dividing by the local gravity. g_c is used regardless of the local gravity. However, the local gravity can be used to find the weight of an object. *Weight* is defined as the force exerted on a mass by the local gravitational field.

$$\text{weight in lbf} = (m \text{ in slugs})(g \text{ in ft/sec}^2) \qquad 22.6$$

If the effects of large land and water masses are neglected, the following formula can be used to estimate the local acceleration of gravity in ft/sec^2 at the earth's surface. ϕ is the latitude in degrees.

$$g_{\text{surface}} = 32.088[1 + (5.305 \text{ EE} - 3)\sin^2 \phi$$
$$- (5.9 \text{ EE} - 6)\sin^2 2\phi] \qquad 22.7$$

If the effects of the earth's rotation are neglected, the gravitational acceleration at an altitude, h, in miles is given by equation 22.8. R is the earth's radius—approximately 3960 miles.

$$g_h = g_{\text{surface}}\left[\frac{R}{R+h}\right]^2 \qquad 22.8$$

4 THE ENGLISH ENGINEERING SYSTEM

Many thermodynamics and fluid flow problems freely combine variables containing pound-mass and pound-force terms. For example, the steady-flow energy equa-

---

[3] A slug is equal to 32.1740 lbm.

[4] Three different meanings of the symbol g commonly are used. g_c is the dimensional conversion factor given in equation 22.5. g_o is the standard acceleration due to gravity with a value of 32.1740 ft/sec^2. g is the local acceleration due to gravity in ft/sec^2.

tion (SFEE) used in chapter 6 mixes enthalpy terms in BTU/lbm with pressure terms in lbf/ft^2. This requires the use of g_c as a mass conversion factor.

Newton's second law becomes

$$F \text{ in lbf} = \frac{(m \text{ in lbm})(a \text{ in ft/sec}^2)}{\left(g_c \text{ in } \frac{\text{lbm-ft}}{\text{lbf-sec}^2}\right)} \qquad 22.9$$

Since g_c is required, the English Engineering System is inconsistent. However, that is not particularly troublesome, and the use of g_c does not overly complicate the solution procedure.

Example 22.1

Calculate the weight of a 1.0 lbm object in a gravitational field of 27.5 ft/sec^2.

Since weight commonly is given in pounds-force, the mass of the object must be converted from pounds-mass to slugs.

$$F = \frac{ma}{g_c} = \frac{(1) \text{ lbm } (27.5) \text{ ft/sec}^2}{(32.2)\frac{\text{lbm-ft}}{\text{lbf-sec}^2}} = .854 \text{ lbf}$$

Example 22.2

A rocket with a mass of 4000 lbm is traveling at 27,000 ft/sec. What is its kinetic energy in ft-lbf?

The usual kinetic energy equation is $E_k = \frac{1}{2}mv^2$. However, this assumes consistent units. Since energy is wanted in foot-pounds-force, g_c is needed to convert m to units of slugs.

$$E_k = \frac{mv^2}{2g_c} = \frac{(4000) \text{ lbm } (27,000)^2 \text{ ft}^2/\text{sec}^2}{(2)(32.2)\frac{\text{lbm-ft}}{\text{lbf-sec}^2}}$$
$$= 4.53 \text{ EE10 ft-lbf}$$

In the English Engineering System, work and energy typically are measured in ft-lbf (mechanical systems) or in British Thermal Units, BTU (thermal and fluid systems). One BTU equals 778.26 ft-lbf.

5 THE cgs SYSTEM

The cgs system has been used widely by chemists and physicists. It is named for the three primary units used to construct its derived variables. The *c*entimeter, the *g*ram, and the *s*econd form the basis of this system.

The cgs system avoids the lbm versus lbf type of ambiguity in two ways. First, the concept of weight is not used at all. All quantities of matter are specified in grams, a mass unit. Second, force and mass units do not share a common name.

When Newton's second law is written in the cgs system, the following combination of units results.

$$\text{units of force} = (m \text{ in } g)\left(a \text{ in } \frac{cm}{sec^2}\right)$$

$$= \frac{g - cm}{sec^2} \qquad\qquad 22.10$$

This combination of units for force is known as a *dyne*.

Energy variables in the cgs system have units of dyne-cm or, equivalently, of

$$\frac{g\text{-}cm^2}{sec^2}$$

These combinations are known as an *erg*. There is no uniformly accepted unit of power in the cgs system, although calories per second frequently is used. Ergs can be converted to calories by multiplying by 2.389 EE−8.

The fundamental volume unit in the cgs system is the cubic centimeter (cc). Since this is the same volume as one thousandth of a liter, units of millimeters (ml) are used freely in this system.

6 THE mks SYSTEM

The mks system is appropriate when variables take on values larger than can be accomodated by the cgs system. This system uses the *meter*, the *kilogram*, and the *second* as its primary units. The mks system avoids the lbm versus lbf ambiguity in the same ways as does the cgs system.

The units of force can be derived from Newton's second law.

$$\text{units of force} = (m \text{ in } kg)\left(a \text{ in } \frac{m}{sec^2}\right)$$

$$= \frac{kg - m}{sec^2} \qquad\qquad 22.11$$

This combination of units for force is known as a *newton*.

Energy variables in the mks system have units of N-m or, equivalently, $\frac{kg-m^2}{sec^2}$. Both of these combinations are known as a *joule*. The units of power are joules per second, equivalent to a *watt*. The common volume unit is the liter, equivalent to one-thousandth of a cubic meter.

Example 22.3

A 10 kg block hangs from a cable. What is the tension in the cable?

$$F = ma = (10) \text{ kg } (9.8)\frac{m}{sec^2}$$

$$= 98\frac{kg\text{-}m}{sec^2} = 98 \ N$$

Example 22.4

A 10 kg block is raised vertically 3 meters. What is the change in potential energy?

$$\Delta E_p = mg\Delta h = (10) \text{ kg } (9.8)\frac{m}{sec^2} (3) \text{ m}$$

$$= 294\frac{kg - m^2}{sec^2} = 294 \ J$$

7 THE SI SYSTEM

Strictly speaking, both the cgs and the mks systems are *metric* systems. Although the metric units simplify solutions to problems, the multiplicity of possible units for each variable sometimes is confusing.

The SI system (International System of Units) was established in 1960 by the General Conference of Weights and Measures, an international treaty organization. The SI system is derived from the earlier metric systems, but it is intended to supersede them all.

The SI system has the following features.

(a) There is only one recognized unit for each variable.
(b) The system is fully consistent.
(c) Scaling of units is done in multiples of 1000.
(d) Prefixes, abbreviations, and symbol-syntax are defined rigidly.

Table 22.1
SI Prefixes

Prefix	Symbol	Value
exa	E	EE18
peta	P	EE15
tera	T	EE12
giga	G	EE9
mega	M	EE6
kilo	k	EE3
hecto	h	EE2
deca	da	EE1
deci	d	EE−1
centi	c	EE−2
milli	m	EE−3
micro	μ	EE−6
nano	n	EE−9
pico	p	EE−12
femto	f	EE−15
atto	a	EE−18

Three types of units are used: base units, supplementary units, and derived units. The base units (table 22.2) are dependent on only accepted standards or reproducible phenomena. The supplementary units (table 22.3) have not yet been classified as being base units or derived units. The derived units (tables 22.4 and 22.5) are made up of combinations of base and supplementary units.

The expressions for the derived units in symbolic form are obtained by using the mathematical signs of multiplication and division. For example, units of velocity are m/s. Units of torque are $N \cdot m$ (not $N\text{-}m$ or Nm).

Table 22.2
SI Base Units

Quantity	Name	Symbol
length	meter	m
mass	kilogram	kg
time	second	s
electric current	ampere	A
temperature	kelvin	K
amount of substance	mole	mol
luminous intensity	candela	cd

Table 22.3
SI Supplementary Units

Quantity	Name	Symbol
plane angle	radian	rad
solid angle	steradian	sr

In addition, there is a set of non-SI units which can be used. This temporary concession is due primarily to the significance and widespread acceptance of these units. Use of the non-SI units listed in table 22.6 usually will create an inconsistent expression requiring conversion factors.

In addition to having standardized units, the SI system also specifies syntax rules for writing the units and combinations of units. Each unit is abbreviated with a specific *symbol*. The rules for writing these symbols should be followed.

(a) The symbols always are printed in roman type, irrespective of the type used in the rest of the text. The only exception to this is in the use of the symbol for *liter*, where the lower case l (ell) may be confused with the number 1 (one). In this case, *liter* should be written out in full

or the script l used. There is no problem with such symbols as cl (centiliter) or ml (milliliter).

(b) Symbols are never pluralized: 1 kg, 45 kg (not 45 kgs).

(c) A period is not used after a symbol, except when the symbol occurs at the end of a sentence.

(d) When symbols consist of letters, there always is a full space between the quantity and the symbols; e.g. 45 kg (not 45kg). However, when the first character of a symbol is not a letter, no space is left, e.g. 32°C (not 32° C or 32 °C) or 42°12′45″ (not 42 ° 12 ′ 45 ″).

(e) All symbols are written in lower case, except when the unit is derived from a proper name. For example, m for meter, s for second, but A for ampere, Wb for weber, N for newton, W for watt. Prefixes are printed roman type without spacing between the prefix and the unit symbol, e.g., km for kilometer.

(f) In text, symbols should be used when associated with a number. When no number is involved, the unit should be spelled out. For example, the area of a carpet is 16 m^2, not 16 square meters, and carpet is sold by the square meter, not by the m^2.

(g) A practice in some countries is to use a comma as a decimal marker, while the practice in North America, the United Kingdom, and some other countries is to use a period (or a dot) as the decimal marker. Further, in some countries using the decimal comma, a dot frequently is used to divide long numbers into groups of three. Because of these differing practices, spaces must be used instead of commas to separate long lines of digits into easily-readable blocks of three digits with respect to the decimal marker, e.g. 32 453.246 072 5. A space (a half space is preferred) is optional with a four-digit number, e.g; 1 234, 1 234, or 1234.

(h) Where a decimal fraction of a unit is used, a zero should be placed before the decimal marker; e.g., 0.45 kg (not .45 kg). This practice draws attention to the decimal marker and helps avoid errors of scale.

(i) Some confusion may arise with the word *tonne* (1 000 kg). When this word occurs in French text of Canadian origin, the meaning may be a ton or 2 000 pounds.

Table 22.4
Some SI Derived Units with Special Names

Quantity	Name	Symbol	Expressed in Terms of Other Units
frequency	hertz	Hz	
force	newton	N	
pressure, stress	pascal	Pa	N/m^2
energy, work, quantity of heat	joule	J	N·m
power, radiant flux	watt	W	J/s
quantity of electricity, electric charge	coulomb	C	
electric potential, potential difference, electromotive force	volt	V	W/A
electric capacitance	farad	F	C/V
electric resistance	ohm	Ω	V/A
electric conductance	siemen	S	A/V
magnetic flux	weber	Wb	V·s
magnetic flux density	tesla	T	Wb/m^2
inductance	henry	H	Wb/A
luminous flux	lumen	lm	
illuminance	lux	lx	lm/m^2

Table 22.5
Some SI Derived Units

Quantity	Description	Expressed in Terms of Other Units
area	square meter	m^2
volume	cubic meter	m^3
speed—linear	meter per second	m/s
angular	radian per second	rad/s
acceleration—linear	meter per second squared	m/s^2
angular	radian per second squared	rad/s^2
density, mass density	kilogram per cubic meter	kg/m^3
concentration (of amount of substance)	mole per cubic meter	mol/m^3
specific volume	cubic meter per kilogram	m^3/kg
luminance	candela per square meter	cd/m^2
dynamic viscosity	pascal second	Pa·s
moment of force	newton meter	N·m
surface tension	newton per meter	N/m
heat flux density, irradiance	watt per square meter	W/m^2
heat capacity, entropy	joule per kelvin	J/K
specific heat capacity, specific entropy	joule per kilogram kelvin	J/(kg·K)
specific energy	joule per kilogram	J/kg
thermal conductivity	watts per meter kelvin	W/(m·K)
energy density	joule per cubic meter	J/m^3
electric field strength	volt per meter	V/m
electric charge density	coulomb per cubic meter	C/m^3
surface density of charge, flux density	coulomb per square meter	C/m^2
permittivity	farad per meter	F/m
current density	ampere per square meter	A/m^2
magnetic field strength	ampere per meter	A/m
permeability	henry per meter	H/m
molar energy	joule per mole	J/mol
molar entropy, molar heat capacity	joule per mole kelvin	J/(mol·K)
radiant intensity	watt per steradian	W/sr

Table 22.6
Acceptable Non-SI Units

Quantity	Unit Name	Symbol	Relationship to SI Unit
area	hectare	ha	$1 \text{ ha} = 10\,000 \text{ m}^2$
energy	kilowatt-hour	kWh	$1 \text{ kWh} = 3.6 \text{ MJ}$
mass	metric ton[5]	t	$1 \text{ t} = 1000 \text{ kg}$
plane angle	degree (of arc)	°	$1° = 0.017\,453 \text{ rad}$
speed of rotation	revolution per minute	r/min	$1 \text{ r/min} = \frac{2\pi}{60} \text{ rad/s}$
temperature interval	degree Celsius	°C	$1°C = 1 \text{ K}$
time	minute	min	$1 \text{ min} = 60 \text{ s}$
	hour	h	$1 \text{ h} = 3600 \text{ s}$
	day (mean solar)	d	$1 \text{ d} = 86\,400 \text{ s}$
	year (calendar)	a	$1 \text{ a} = 31\,536\,000 \text{ s}$
velocity	kilometer per hour	km/h	$1 \text{ km/h} = 0.278 \text{ m/s}$
volume	liter[6]	l	$1\,l = 0.001 \text{ m}^3$

Numbers in parentheses are the number of ESU or EMU units per single SI unit, except for the permittivity and the permeability of free space, where actual values of ϵ_o and μ_o are given.

---

[5] The international name for metric ton is *tonne*. The metric ton is equal to the *megagram*, Mg.

---

[6] The international symbol for liter is the lowercase "l," which can be confused easily with the numeral "1." Several English speaking countries have adopted the script l as the symbol for liter in order to avoid any misinterpretation.

Appendix A
Selected Conversion Factors to SI Units

	SI Symbol	Multiplier to Convert From Existing Unit to SI Unit	Multiplier to Convert From SI Unit to Existing Unit
Area			
Circular Mil	μm²	506.7	0.001 974
Foot Squared	m²	0.092 9	10.764
Mile Squared	km²	2.590	0.386 1
Yard Squared	m²	0.836 1	1.196
Energy			
Btu (International)	kJ	1.055 1	0.947 8
Erg	μJ	0.1	10.0
Foot Pound-Force	J	1.355 8	0.737 6
Horsepower Hour	MJ	2.684 5	0.372 5
Kilowatt Hour	MJ	3.6	0.277 8
Meter Kilogram-Force	J	9.806 7	0.101 97
Therm	MJ	105.506	0.009 478
Kilogram Calorie (International)	kJ	4.186 8	0.238 8
Force			
Dyne	μN	10.	0.1
Kilogram-Force	N	9.806 7	0.101 97
Ounce-Force	N	0.278 0	3.597
Pound-Force	N	4.448 2	0.224 8
KIP	N	4 448.2	0.000 224 8
Heat & Power			
Btu Per Hour	W	0.293 1	3.412 1
Btu Per (Square Foot Hour)	W/m²	3.154 6	0.317 0
Btu Per (Square Foot Hour °F)	W/(m² · °C)	5.678 3	0.176 1
Btu Inch Per (Square Foot Hour °F)	W/(m·°C)	0.144 2	6.933
Btu Per (Cubic Foot °F)	MJ/(m³·°C)	0.067 1	14.911
Btu Per (Pound °F)	J/(kg·°C)	4 186.8	0.000 238 8
Btu Per Cubic Foot	MJ/m³	0.037 3	26.839
Btu Per Pound	J/kg	2 326.	0.000 430
Length			
Angstrom	nm	0.1	10.0
Foot	m	0.304 8	3.280 8
Inch	mm	25.4	0.039 4
Mil	mm	0.025 4	39.370
Mile	km	1.609 3	0.621 4
Mile (International Nautical)	km	1.852	0.540
Micron	μm	1.0	1.0
Yard	m	0.914 4	1.093 6
Mass (weight)			
Grain	mg	64.799	0.015 4
Ounce (Avoirdupois)	g	28.350	0.035 3
Ounce (Troy)	g	31.103 5	0.032 15
Ton (short 2000 lb.)	kg	907.185	0.001 102
Ton (long 2240 lb.)	kg	1 016.047	0.000 984 2
Slug	kg	14.593 9	0.068 522
Pressure			
Bar	kPa	100.0	0.01
Inch of Water Column (20°C)	kPa	0.248 6	4.021 9
Inch of Mercury (20°C)	kPa	3.374 1	0.296 4
Kilogram-force per Centimeter Squared	kPa	98.067	0.010 2
Millimeters of Mercury (mm·Hg) (20°C)	kPa	0.132 84	7.528
Pounds Per Square Inch (P.S.I.)	kPa	6.894 8	0.145 0
Standard Atmosphere (760 torr)	kPa	101.325	0.009 869
Torr	kPa	0.133 32	7.500 6

Appendix A (continued)
Selected Conversion Factors to SI Units

	SI Symbol	Multiplier to Convert From Existing Unit to SI Unit	Multiplier to Convert From SI Unit to Existing Unit
Power			
Btu (International) Per Hour	W	0.293 1	3.412 2
Foot Pound-Force Per Second	W	1.355 8	0.737 6
Horsepower	kW	0.745 7	1.341
Meter Kilogram-Force Per Second	W	9.806 7	0.101 97
Tons of Refrigeration	kW	3.517	0.284 3
Torque			
Kilogram-Force Meter (kg·m)	N·m	9.806 7	0.101 97
Pound-Force Foot	N·m	1.355 8	0.737 6
Pound-Force Inch	N·m	0.113 0	8.849 5
Gram-Force Centimeter	mN·m	0.098 067	10.197
Temperature			
Fahrenheit	°C	$\frac{5}{9}(°F-32)$	$(\frac{9}{5}°C)+32$
Rankine	K	$(°F+459.67)\frac{5}{9}$	$(°C+273.16)\frac{9}{5}$
Velocity			
Foot Per Second	m/s	0.304 8	3.280 8
Mile Per Hour	m/s	0.447 04	2.236 9
	or	or	or
	km/h	1.609 34	0.621 4
Viscosity			
Centipoise	mPa·s	1.0	1.0
Centistoke	µm²/s	1.0	1.0
Volume (Capacity)			
Cubic Foot	l (dm³)	28.316 8	0.035 31
Cubic Inch	cm³	16.387 1	0.061 02
Cubic Yard	m³	0.764 6	1.308
Gallon (U.S.)	l	3.785	0.264 2
Ounce (U.S. Fluid)	ml	29.574	0.033 8
Pint (U.S. Fluid)	l	0.473 2	2.113
Quart (U.S. Fluid)	l	0.946 4	1.056 7
Volume Flow (Gas-Air)			
Standard Cubic Foot Per Minute	m³/s	0.000 471 9	2119.
	or	or	or
	l/s	0.471 9	2.119
	or	or	or
	ml/s	471.947	0.002 119
Standard Cubic Foot Per Hour	ml/s	7.865 8	0.127 133
	or	or	or
	µl/s	7 866.	0.000 127
Volume Liquid Flow			
Gallons Per Hour (U.S.)	l/s	0.001 052	951.02
Gallons Per Minute (U.S.)	l/s	0.063 09	15.850

Appendix B
Consistent Electric/Magnetic Units

Variable	Symbol	SI System	Gaussian ESU System	Gaussian EMU System
length	l	meter	cm (100)	cm (100)
mass	m	kg	g (1000)	g (1000)
time	t	sec	sec	sec
force	F	newton	dyne (EE5)	dyne (EE5)
work, energy	W, E	joule	erg (EE7)	erg (EE7)
power	P	watt	erg/sec (EE7)	erg/sec (EE7)
charge	q	coulomb	statcoulomb (3 EE9)	abcoulomb (EE−1)
current	i	ampere	statcoulomb/sec (3 EE9)	abcoulomb/sec (EE−1)
electric flux	ϕ	coulomb	statcoulomb (3 EE9)	abcoulomb (EE−1)
electric flux density	D	coulomb/m^2	— (12π EE5)	— (4π EE−5)
electric field intensity	E	volt/meter =newtons/coulomb	statvolt/cm (3.33 EE−5)	abvolt/cm (EE6)
electric potential	V	volt	statvolt (3.33 EE−3)	abvolt (EE8)
resistance	R	ohm	statohm (1.11 EE−12)	abohm (EE9)
capacitance	C	farad	statfarad (9 EE11)	abfarad (EE−9)
magnetic pole strength	m	weber	$-\left(\frac{1}{12\pi\,EE2}\right)$	$-\left(\frac{EE8}{4\pi}\right)$
magnetic flux	ϕ	weber	statweber (3.33 EE−3)	maxwell (EE8)
magnetic flux density	B	tesla=wb/m^2	— (3.33 EE−7)	gauss (EE4)
magnetic field intensity	H	amp-turn/meter =N/wb	— (12π EE7)	oersted (4π EE−3)
magnetomotive force	MMF	amp-turn	— (12π EE9)	gilbert (4π EE−1)
inductance	L	henry=wb/amp	abhenry (1.11 EE−12)	stathenry (EE9)
reluctance	$\mathcal{R}$	amp-turn/weber	— (36π EE11)	gilbert/maxwell (4π EE−9)
permittivity of free space	ϵ_o	$\frac{1}{(36\pi\ EE9)}\frac{farad^*}{meter}$	$1\frac{statfarad}{cm}$	$\frac{1}{(9\ EE20)}\frac{abfarad}{cm}$
permeability of free space	μ_o	$4\pi\ EE{-}7\frac{henry^{**}}{meter}$	$\frac{1}{9\ EE20}\frac{stathenry^{***}}{cm}$	$1\frac{abhenry}{cm}$

* same as coul^2/n-m^2

** same as (unit-poles)^2/n-m^2 and webers/amp-turn-meter

*** same as (sec/cm)^2

RESERVED FOR FUTURE USE

23 POSTSCRIPTS

This chapter collects comments, revisions, and commentary that cannot be incorporated into the body of the text until the next edition. New postscript sections are added as needed when the *Mechanical Engineering Reference Manual* is reprinted. Subjects in this chapter are not necessarily represented by entries in the index.

Introduction

ABOUT THE NEW EXAM SUBJECTS

When NCEES performed its 1982 task analysis, it identified several activities performed by a minority of mechanical engineers. These activities were grouped into the new subject areas of "Management" and "Controls." This edition of the *Mechanical Engineering Reference Manual* covers these subjects in chapters 19 and 20, as well as in related sections of other chapters.

Some of these activities are very general, and although identified as being part of a mechanical engineer's job, are difficult to cover in this book.

- Management
    - tool design
    - material handling
    - plant layout
    - metal fabrication
    - shop work
- Controls
    - fluidics
    - automatic assembly

MISCELLANEOUS DATA

Several problems have required knowledge of drill bit sizes and sheet metal gauges. Such information is included in *Machinery's Handbook* (but not necessarily in other common references), so that book should be part of your examination kit.

Chapter 1: VARIANCE

The statistical term *variance* is used in example 1.42 (page 1-26) but is not defined in the text. The variance is the square of the standard deviation. Since there are two standard deviations (s and σ), there are two variances. The *sample variance* is s^2. The *population variance* is σ^2.

Chapter 2: USING THE MARR

When a MARR is given, it is not always necessary to calculate an alternative's ROR (to compare against) in order to qualify or disqualify the alternative. It is easier to calculate the alternative's present worth using the MARR as an interest rate. If the present worth is positive, then ROR $\geq$ MARR.

Chapter 2: PAYBACK PERIOD

Although it is tempting to incorporate the time value of money (i.e., compounding), the *payback period* of an investment is simplistically defined as the length of time (usually in years) required for the annual net cash flow to equal the initial cost.

Chapter 3: BUTTERFLY VALVES

Loss coefficients for *butterfly valves* are calculated from the friction factor for the pipe with completely turbulent fluid. The sizes given are inside diameter measurements of the pipe.

$$\text{sizes } 2'' \text{ to } 8'': \quad K = 45 f_t$$
$$\text{sizes } 10'' \text{ to } 14'': \quad K = 35 f_t$$
$$\text{sizes } 16'' \text{ to } 24'': \quad K = 25 f_t$$

Chapter 3: PRESSURE DROP ACROSS AN ORIFICE PLATE

The *pressure difference* used in equations 3.130 through 3.140 for orifice plates is not the permanent *pressure loss* due to friction. The permanent pressure loss depends on the ratio D_o/D_i (see figure 3.23) and is 73% of $p_1 - p_2$ for $D_o/D_i = .5$, 56% for $D_o/D_i = .65$, and 38% for $D_o/D_i = .8$.

Chapter 3: THE FROUDE NUMBER

There is considerable confusion regarding the definition of the Froude number. Dimensional analysis determines it to be v^2/gL, a form which is also used in model similitude analysis. However, in open channel flow analysis, the Froude number is taken as the square root of the derived form. Whether the derived form or its square root is used can be determined from the application. If the Froude number is squared (as in $dE/dd = 1 - N_{Fr}^2$), then the square root form is necessary.

Chapter 3: WATER HAMMER IN DUCTILE PIPE

The speed of sound used in water hammer calculations (i.e., c in equations 3.176 and 3.177) must account for the expansion of ductile pipe walls as the water pressure builds up. Equation 3.20 can be used to calculate c, but the modulus of elasticity used should include the elastic contributions of both the water and pipe material. In the following equation, t_{pipe} is the pipe wall thickness, and d_{pipe} is the inside diameter.

$$E = \frac{E_{\text{water}} t_{\text{pipe}} E_{\text{pipe}}}{t_{\text{pipe}} E_{\text{pipe}} + d_{\text{pipe}} E_{\text{water}}}$$

Chapter 4: MOTOR SERVICE FACTOR

General-purpose A-C motors can be overloaded up to 40% provided that the nameplate voltage and frequency are maintained. The actual overload horsepower suitable for continuous operation is obtained by multiplying the nameplate horsepower by the *motor service factor*. The service factor will be approximately 1.4 (up to 1/8 hp motors), 1.35 (1/6 to 1/3 hp motors), 1.25 (1/2 to 1 hp motors), and 1.15 (1 1/2 hp and above). When running over the rated power, the motor speed, temperature, power factor, and efficiency will differ from the nameplate value. However, the locked-rotor and breakdown torques will remain the same, as will the starting current.

Chapter 5: NONSTANDARD REGAIN COEFFICIENTS

For nonstandard regain coefficients (R), the coefficient c in equation 5.4 should be replaced with

$$c = \frac{6.256 \,\text{EE} - 2}{R}$$

Chapter 6: NOZZLE EFFICIENCY

Equation 6.93 shows that a fluid achieves velocity at the expense of energy (enthalpy). The *nozzle efficiency* is defined as

$$\eta_{\text{nozzle}} = \frac{\Delta h_{\text{ideal}}}{\Delta h_{\text{actual}}} = \left(\frac{v_{\text{actual}}}{v_{\text{ideal}}} \right)^2$$

Chapter 7: MBH UNITS

The abbreviation MBH stands for "one thousand BTU's per hour."

Chapter 10: HEAT TRANSFER VISCOSITY UNITS

Viscosity in the heat transfer chapter contains units of *hours*. Since most data tables give viscosity with units of seconds, a conversion factor of 3600 is needed when calculating dimensionless numbers (i.e., Prandtl number, Grashoff number, etc.) from raw data.

Chapter 11: PARTIAL PRESSURE OF WATER VAPOR

Equation 11.33 is the so-called *Carrier formula* for calculating the partial pressure of water vapor, as the formula was originally presented. Modern practice has refined the equation somewhat, although the additional accuracy may be unnecessary.

$$p_w = p_{sat,wb} - \frac{p_t - (p_{sat,wb})(T_{db} - T_{wb})}{2830 - 1.44\,T_{wb}}$$

Chapter 11: HEAT TRANSFER TO AND FROM AIR CONDITIONING DUCTS

The calculation of heat transfer to and from air conditioning and heating ducts, respectively, is not sophisticated. Such heat transfers are generally estimated from tables or figures that are based on some length of duct and standard temperature difference (e.g., heat loss per 10 feet of duct and 10 degrees of temperature difference). For other duct lengths and temperature differences, linear extrapolation is used.

In an instance where the heat transfer is to be calculated from a film coefficient, it makes sense to use the logarithmic mean temperature difference (LMTD), ΔT_m, since the temperature of the air in the duct changes along the duct length. However, this is seldom done by engineers working in the field of HVAC, probably because the accuracy of the other data does not warrant this level of sophisticated analysis.

The heat transfer usually is based on the temperature difference between the environment and the mid-length temperature of the duct. If the mid-length temperature is not known, one or more iterations may be needed to calculate the temperature drop.

Chapter 14: STEEL STRENGTHS

You will want to augment Appendix B in chapter 14 with additional grades of steel. In addition, information on how those properties vary with increased temperatures is needed. Such information is included in Baumeister's (Marks') *Standard Handbook for Mechanical Engineers.*

Chapter 14: CAST IRON CLASSES

The modulus of elasticity and strength of cast iron varies greatly with its class. This variation is not reflected in table 14.1, so the following information is provided. (All information is in psi.)

class	E (tension)	S_{ut} (minimum)
20	10–14 EE6	20 EE3
25	12–15	25
30	13–16.5	30
35	14.5–17	35
40	16–20	40
50	18.8–23	50
60	20.4–23.5	50

Chapter 14: DEFLECTION OF A PIER

Appendix A of chapter 14 omits the equation for calculating the performance of a fixed pier (i.e., a column with one end that can translate laterally, with both ends remaining vertical).

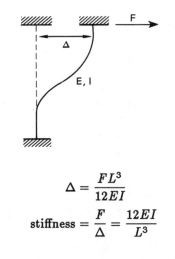

$$\Delta = \frac{FL^3}{12EI}$$

$$\text{stiffness} = \frac{F}{\Delta} = \frac{12EI}{L^3}$$

Chapter 15: THREE-DIMENSIONAL VON MISES STRESS

Example 15.5 is solved using the three-dimensional form of equation 15.13. That equation is provided here, where σ_1, σ_2, and σ_3 are principal stresses.

$$\sigma' = \sqrt{\frac{1}{2}\Big[(\sigma_1 - \sigma_2)^2 + (\sigma_2 - \sigma_3)^2 + (\sigma_3 - \sigma_1)^2\Big]}$$

Chapter 15: SHAFT INTERFERENCE IN DISK FLYWHEELS

The tangential stress at the inside radius of the disk, $\sigma_{t,i}$, can be calculated from equation 15.50 and used to find the radial interference at that point.

$$\sigma_{t,i} = E\epsilon_{t,i} = E\frac{\Delta d_i}{d_i} = E\frac{\Delta r_i}{r_i}$$

Therefore, the radial interference is

$$\Delta r_i = \frac{r_i \sigma_{t,i}}{E}$$

Chapter 15: CHOOSING GEAR TOOTH WIDTHS

While there is considerable latitude in choice of widths of gear teeth, usual practice is to select a width such that

$$\frac{3\pi}{P} \le w \le \frac{5\pi}{P}$$

Chapter 15: GEAR TOOTH CONTACT STRESS

Gear tooth design in chapter 15 neglects the subject of tooth contact stress. Such stress may be the limiting factor in sizing gear widths.

Chapter 15: HIGHER VIBRATIONAL MODES

Equation 15.128 gives the fundamental (first modal) frequency of lateral vibration assuming uniformly distributed mass and simple supports. The nth modal frequency is calculated as n^2 times the fundamental frequency. If both ends are fixed (i.e., as equation 15.129 requires), the higher modal frequencies are very nearly n^2 times the fundamental frequency, but not exactly. Cantilever beams with distributed mass are not handled so easily. And, beams with lumped masses do not have higher modes at all.

Chapter 15: MODULUS OF ELASTICITY FOR CRITICAL SHAFT SPEEDS

In calculating the critical speeds of rotating steel shafts, it is more appropriate to assume a modulus of elasticity of $E = 2.9$ EE7 psi unless the shaft is known to have a different value throughout. The discussion of critical speeds starting on page 15-24 uses $E = 3.0$ EE7 psi. Although the difference in critical speed is not significant, this higher value is relevant only to hardened steel.

Chapter 15: TYPICAL BOLT STRENGTHS

SAE grade	No. of cap marks*	ASTM grade	Nominal diameter in.	Proof strength kpsi	Tensile strength kpsi
1	0	A307	$\frac{1}{4}$ to $1\frac{1}{2}$	33	55
2	0		$\frac{1}{4}$ to $\frac{1}{2}$	55	69
			over $\frac{1}{2}$ to $\frac{3}{4}$	52	64
			over $\frac{3}{4}$ to $1\frac{1}{2}$	28	55
3	2		$\frac{1}{4}$ to $\frac{1}{2}$	85	110
			over $\frac{1}{2}$ to $\frac{5}{8}$	80	100
5	3	A449	$\frac{1}{4}$ to $\frac{3}{4}$	85	120
			over $\frac{3}{4}$ to 1	78	115
			over 1 to $1\frac{1}{2}$	74	105
7	5		$\frac{1}{4}$ to $1\frac{1}{2}$	105	133
8	6	A354	$\frac{1}{4}$ to $1\frac{1}{2}$	120	150

* Grade 0 has no cap marks; grade 6 has four cap marks.

Chapter 16: RAYLEIGH'S METHOD

The mass of a spring element (beam, bar, shaft, etc.) is usually disregarded when calculating the frequency or period of vibration of a simple system. This is done to simplify the solution, although the mass of the spring element actually does affect the frequency. The exact method is complex, but *Rayleigh's method* can be used to derive answers that will usually be less than five percent in error. This method increases the mass of the oscillating object by a fraction of the spring element's mass.

- For spring-mass systems, add 1/3 of the spring mass to the oscillating object's mass.
- For simply supported beams carrying a mass at the midpoint, add 17/35 of the beam mass to the carried mass.
- For cantilever beams loaded at the free end, add 33/140 of the beam mass to the carried mass.
- For circular shafts in torsion, add 1/3 of the shaft mass moment of inertia to the mass moment of inertia of the rotating load.

RESERVED FOR FUTURE USE

INDEX

(Index of Tables and Figures begins on page 21.)

H

I

J

K

L

M

N